WARFARE AND SOCIETY

WARFARE AND SOCIETY

Archaeological and Social Anthropological Perspectives

EDITED BY TON OTTO
HENRIK THRANE AND
HELLE VANDKILDE

AARHUS UNIVERSITY PRESS 2006

WARFARE AND SOCIETY
Archaeological and Social Anthropological Perspectives

© Aarhus University Press, Ton Otto, Henrik Thrane
& Helle Vandkilde 2006

ISBN 87 7934 110 1

Editors: Ton Otto, Henrik Thrane & Helle Vandkilde
English revision: Stacey Cozart, Nick Thorpe,
Mary Waters Lund
Proofreading: Steffen Dalsgaard
Layout & cover: Hanne Kolding
Printed by: Narayana Press
Type face: Stone Serif, Stone Sans
Paper: Arctic Volume

Aarhus University Press
Langelandsgade 177
DK-8200 Aarhus N
Denmark
www.unipress.dk

White Cross Mills
Hightown
Lancaster
LA1 4XS
UK

Box 511
Oakville, CT 06779
USA

Cover: "War Magic", 1975, screenprint by the Papua New Guinean
artist Timothy Akis (deceased 1984). The picture illustrates the connection between warfare and social identities. In some Melanesian
societies war magic is used to transform men into warriors, so that they
can kill people and thereby establish group identities and social
boundaries.

*Published with the financial support of
the Danish Research Council for the Humanities*

Contents

Warfare and Society: Archaeological and Social Anthropological Perspectives

TON OTTO, HENRIK THRANE
AND HELLE VANDKILDE

/1

...you can't understand what the war has done to us. At first sight everything may look normal, but it's not. Nothing is normal. The war has changed everything.[1]

The present book deals with the interrelationship between society and war seen through the analytical eyes of anthropologists and archaeologists. The opening quote – spoken by an informant to Torsten Kolind and published in his thesis about discursive practices in Bosnia just after the war in 1992-95 – captures the problems we face when we study war. Archaeologists and anthropologists alike rarely possess war experiences of their own: we study past and present wars, but remain total outsiders who depend on numerous and complex discursive layers – material, written, and spoken – to bring us insight on this subject, so demanding and so necessary to deal with. War is a ghastly thing, which unfortunately is thriving almost everywhere in the world at present: we need to understand better what war does to people and their societies. We are trained analysts, but to insiders war is mostly chaos and death and hence in a sense beyond analysis. It is a challenge in our studies to both ignore and include the compassion and feeling this subject is also about. Nevertheless, under the chaotic conditions of war and its aftermath people are fully aware of the changes happening to their world even if they cannot describe them sociologically. Doubtless, war always affects society and its agents. War does produce change, and archaeologists and anthropologists are analytically equipped to pinpoint its direction, patterning, scale and content. The perspective – and filter – of time provides one important tool, context and comparison other tools. Looking at the history of war studies, war is quite often perceived of and treated as something set aside from other practices; almost personified. However, the results published in this book allow us to say that it is never autonomous and self-regulating. War always forms part of something else. Numerous questions arise and at least some answers, often tentative and multifaceted, are provided in the collection of studies published below. They certainly add to an ongoing debate, hopefully qualifying it as well.

The book is the end product of the research project 'Archaeological and Social Anthropological Perspectives on War and Society' at the Institute of Anthropology, Archaeology and Linguistics at Aarhus University, Moesgård. This project formed part of the Danish Research Council for the Humanities' special initiative on the subject of 'Civilisation and War'. It began the 1st of January 1999 and was officially concluded by the end of 2002, but continued on a lesser scale throughout 2003 and 2004. This book reports on the results, and in so doing incorporates a series of edited articles originating from seminars and work meetings that took place within the framework of the project. Most of all, the book presents the research conducted by members of the research team from about 1999 to 2004. The publication deals with a series of related research fields, notably war in the context of theory, philosophy, and research history, but also takes up the discussion of the position and role of war in non-state and state societies. In addition, the relationship to rituals, social identification and material and non-material forms of discourse are among the themes discussed, notably on a cooperative basis across institutions and across the two major disciplines of archaeology and anthropology. The curriculum and outcome of the War & Society project are summarised below.

The research team

The research team on the project consisted of an average of five or six members. The project was headed by Professor Ton Otto, Professor Henrik Thrane and Associate Professor Helle Vandkilde, who all contributed with co-financed research, the last-mentioned as coordinator of the project and the day-to-day work. Ton Otto held the primary administrative responsibility for the project. These three researchers have contributed to the project in particular through the working meetings. The project group also comprised two doctoral students, Andreas Hårde and Torsten Kolind, who began their work on the project on 1 August 1999 and 1 November 1999, respectively. The latter recently defended his doctoral dissertation at the University of Aarhus (Kolind 2004). At the beginning of the project, anthropologist Dr. Kristoffer Brix Bertelsen made his mark on the project but left it in favour of a position with the Research Council for the Humanities. Anthropologist Dr. Claus Bossen was employed as a research fellow on the project until 31 January 2001, but fortunately continued his involvement and participation through working meetings and seminars.

In addition, visiting researchers contributed to the project: curator Nick Araho, Dr. Erik Brandt, Professor Polly Wiessner and Professor Jürg Helbling, who have all served as external supervisors for the doctoral students and as resource persons in various fields (cp. chapters 6, 9, and 11). Furthermore, the project has drawn on a number of researchers associated with the project as external resource persons. In particular, Dr. David Warburton (cp. chapter 4) should be mentioned by name for having contributed with his theoretical expertise and knowledge of the Middle East, and Jürg Helbling for his thoroughgoing assistance with the editorial work as peer-reviewer.

Seminars and workshops

The project invested considerable energy into organising seminars and working meetings where war and warriors were discussed thematically and from various angles. Invited guests and the project members presented their thoughts and research results at international seminars that resulted in many fruitful and in-depth discussions as well as substantial contributions. More informal working meetings for the project members were held on a regular basis and created a fruitful basis for developing concepts and interpretations. In this way the project created a common platform for the individual projects under the general umbrella of War and Society. Personal opinions and points of view were typically greatly influenced by the debates that took place at the seminars and working meetings, which also rubbed off on the content of the written production, especially the present book. It is characterised positively by a combination of archaeology and social anthropology. Even though it was not always simple to direct archaeology and anthropology towards each other, it certainly proved to be worth the effort. A close collaboration between the two fields has in reality not occurred in Denmark in recent times, but the War and Society project has allowed for mutual enrichment, which may be considered one of the important outcomes of the project. This will hardly be the last project where both fields are involved on equal footing. Beyond the productive collaboration between anthropology and archaeology, the project has also received considerable input from history, politology and philosophy.

The English-language seminars have included the following activities:

1. *'Civilisation and war'* (focus on source materials and theory), 18.6.1999.
2. *'Warfare and Social Structure'* (warfare, violence and social structure; warfare and warriors in prehistory). 28.-29.4. 2000.
3. *'Warfare and State Formation'*. 5.10. 2000.
4. *'Warrior Identities and Warrior Ideals in Past and Present Societies'*. 26.01. 2001.
5. *'Warfare and Sacrificial Rituals'*. 10.5. 2001.
6. *'Identity and Discourse in Post-War Communities'*. 9.11. 2001.
7. *'The junction between archaeology and anthropology'* was the main heading for four activities that took place in connection with a visit by Professor Polly Wiessner and Professor Chris Gosden, 30.4.-6.5. 2002 at Moesgård.
   A. *'Material Culture, the Individual and the Collective'*. 30.4. Seminar.
   B. *'Anthropology & Archaeology: A Changing Relationship'*. 2.5. Lecture.
   C. *'Warfare in the South Pacific: Strategies, Histories, and Politics'*. 3.5. Seminar.
   D. *'Changes in Economy, Social Networks, Material Culture and Identity among the Bushmen in the 20th Century'*. 6.5. Lecture.

Visiting scholars

Quite a few foreign researchers have contributed to the project. Below is a list of these researchers, five of whom – Erik Brandt, Ivana Macek, Polly Wiessner, Jürg Helbling and Nick Araho – were part of the project for a period of time, ranging from one week to one month. Several of these researchers do both archaeological and anthropological work and have therefore been able to give a high degree of positive input to the project (cp. chapters in this volume).

- *Jan Abbink*, Professor, African Studies Centre, University of Leiden and Department of Anthropology, Free University of Amsterdam.
- *Miranda Aldhouse-Green*, Professor, Department of Archaeology, University of Wales, Newport.
- *Nick Araho*, Curator, the National Museum of Papua New Guinea, Port Moresby.
- *Martijn van Beek*, Associate Professor, Department of Ethnography and Social Anthropology, Moesgård, University of Aarhus.
- *Pia Bennike*, Senior Researcher, Laboratory of Biological Anthropology, University of Copenhagen.
- *Erik Brandt*, Ph.D, Department of Anthropology, University of Nijmegen.
- *Henri Claessen*, Professor, Department of Anthropology, University of Leiden.
- *Raymond Corbey*, Associate Professor, Department of Anthropology, Universities of Leiden and Tilburg.
- *Chris Gosden*, Professor, Pitt Rivers Museum Oxford, Department of Anthropology, Oxford University.
- *Anthony Harding*, Professor, Department of Archaeology, University of Durham.
- *Jürg Helbling*, Professor, Department of Anthropology, University of Zürich.
- *Christian K. Højbjerg*, Senior Researcher, Danish Institute of Advanced Studies in the Humanities, Copenhagen.
- *Stef Jansen*, Assistant Professor, Department of Anthropology, University of Hull.
- *Kristian Kristiansen*, Professor, Department of Archaeology, University of Göteborg.
- *Staffan Löfving*, Assistant Professor, Department of Cultural Anthropology and Ethnology, University of Uppsala.
- *Ivana Macek*, Assistant Professor, Peace and conflict research group, University of Uppsala.
- *Ron May*, Senior Research Fellow, Research School of Pacific and Asian Studies, Australian National University, Canberra.
- *Lena Holmquist Olausson*, Assistant Professor, Department of Archaeological Science, University of Stockholm.
- *Michael Olausson*, Curator, Swedish National Heritage Board (RAÄ), Stockholm.
- *Richard Osgood*, Archaeologist, South Gloucestershire Council.
- *Sanimir Resic*, Associate Professor, Department of History, University of Lund.
- *Henrik Rønsbo*, Associate Professor, Rehabilitation and Research Centre for Torture Victims, Copenhagen.
- *Heiko Steuer*, Professor, Institut für Ur- und Frühgeschichte & Archäologie des Mittelalters, University of Freiburg.
- *Marie Louise Stig Sørensen*, Associate Professor, Department of Archaeology, University of Cambridge.
- *Nick Thorpe*, Associate Professor, Department of Archaeology, King Alfred's College, Winchester.
- *David Warburton*, Research Assistant, Department of the Study of Religion, University of Aarhus.
- *Polly Wiessner*, Professor, Department of Anthropology, University of Salt Lake City, Utah.

The individual projects

The War and Society project served as an umbrella for six individual projects that included the disciplines of archaeology and anthropology and in some cases both.

Claus Bossen's studies concerned the connection between early state formation and war in Hawaii and Fiji (chapter 17) and recent theories of war within social anthropology (chapter 7). On the one hand, according to his studies it appears probable that war plays a role in state formation but, on the other hand, that war and military organisation cannot stand alone. Military power should be combined with ideological, economic and political power in order for a state to form. Subsequently, the question arises of how people come to accept a ruler's sovereignty and power?

Andreas Hårde studied war in the Early Bronze Age cultures of Nitra, Únětice- and Věteřov-Mad'arovce in Eastern Europe, in particular in Moravia and Slovakia (chapter 24). The main issue of concern to him was how to identify acts of violence within a prehistoric material by regarding war as a social phenomenon rather than as military history. It was also important to consider warfare as a phenomenon divided into several phases, of which the preliminaries to war and its effects are just as important to study as the act of war itself. War, just like other means of power, requires a social decision-making which for one thing is expressed in social rituals. The employment of violence can thus create or strengthen a socio-political identity. The work on war in the early Bronze Age consists of two studies, the first of which concerns the relationship between warriors and social change, and the other violence – in the form of human sacrifices – as a means of power.

In sum, Andreas Hårde's studies show that warfare within the Early Bronze Age was closely connected to economic and political power. Evidence of war is most obvious in the periods when socio-political changes occur. The frequency of skeletal trauma, grave plundering and warrior cenotaphs increases along with changes within burial customs among the social elite and with the introduction of new prestige goods and objects of metal. In addition, violence in the form of human sacrifice was used as a means to gain power over life and death.

Torsten Kolind's work in the Warfare and Society project resulted in his recently completed doctoral dissertation about 'Post-war identifications. Counter-discursive practices in a Bosnian town', based on six months of field work among a Muslim population in ethnically mixed Stolac (a town in southwest Bosnia). Kolind examined the connections between war-related violence and identification analysing the informants' experiences of a world in ruins, destroyed by war, and the politically over-heated post-war situation. Focus was on the most central identifications of 'the others' that the Muslims in Stolac employ, the general conclusion being that these can be regarded as part of a counterdiscourse characterised precisely by the rejection of the nationalistic and ethnic categorisations and explanations existing in the public and political sphere (chapter 29). The conclusion here is that the nationalistic as well as religious identifications that were key to the war have lost their relevance. Instead, people identify themselves in respect to a local patriotism, an ideal of tolerance, the discursive construction of the Balkans as part of Europe, and the role of the victim. Apart from the role of the victim, these identifications can also be seen as part of the Muslims' everyday counterdiscourse.

Ton Otto was especially involved in the theoretical discussions, in particular in developing a conceptual framework for comparatively analysing war as a power factor and as a cultural phenomenon. He was Torsten Kolind's main supervisor. In addition, Ton Otto presented and worked on empirical material from Manus (Papua New Guinea), especially historical data concerning war in a society without central authorities (chapter 12). It is normally assumed that exchange unites while war divides, but this is too simple. War creates not only groups of allies and enemies, but it also leads to networks of connections, inasmuch as there exists a responsibility to either retaliate (otherwise lose prestige and status) or mediate between fighting parties (achieve prestige). In pre-colonial Manus, war was a factor that maintained relatively small political units (through fission) but that at the same time created connections between the units and therefore integrated the region into a larger system of exchange relations. War was a strategic, but risky possibility for local entrepreneurs to increase their status. The colonial power's policy of pacification put a stop to this option. Therefore, the focus for status politics shifted entirely from waging war to organising great exchange ceremonies.

Henrik Thrane's research focused on territorial organisation and armament in the Scandinavian Bronze Age, above all in the study of sword production and sword function on the basis of quantitative methods (chapter 32). It has been quite a few years since active research has been carried out on this subject; in part the material has become more accessible, in part the theoretical apparatus and viewpoints concerning context and social roles have changed decisively in recent years. Henrik Thrane's principal interest was to relate the sources to the theories and understandings of war and warrior roles on which the project has worked, considering it essential to reveal how the sources support or contradict these. He was Andreas Hårde's main supervisor.

Helle Vandkilde examined warrior identities in the European societies of the later Stone Age and Bronze Age. Organised warrior bands often seem to have played a decisive role. It is probable that these warrior groups were recruited according to hierarchical principles, not unlike, for example, the system that can be deduced from Homer's *Iliad* and that is also evident in a large number of ethnographically studied cases (chapters 26 and 34). Vandkilde's analysis of the history of research (chapter 5) furthermore points out that war and violence do not really enter the archaeological interpretations until c. 1995. Two opposing myths have generally characterised archaeology – one of them regarding prehistory as populated with potentially violent warriors that repeatedly changed society, the other presenting prehistory as populated with peaceful peasants in harmonious and static societies. It is finally suggested that both the ideal and real sides of war and warriors in prehistory should be studied, and also that interpretative stereotypes can be avoided through the use of theories that view humans as participating both routinely and strategically in societal frameworks. Consequently, another dimension of her work has concentrated on writing war and warriors into sociological theories of material culture, social practice, power, and social identity such as notably gender.

Project outcome – an outline

The subprojects typically covered more than one subject area. Below is an outline of some of the general considerations and results.

Within social anthropology war has long been an object of study. In archaeology, however, war did not become an established area of study until the past decade, and it must be assumed that the many ethnic wars and genocides of the 1990s as well as the massive media coverage have played a decisive role. The horror and awful chaos of war are now analysed in social anthropological studies, but shall henceforth also be incorporated in archaeological studies, which still do not portray prehistoric war realistically enough. This is especially due to the fact that the discourse is still influenced by some myths of heroic warrior elites.

War should be understood as a collective and violent social practice which is always based on a cultural logic and therefore cannot merely be explained with reference to biology, genetics or evolution. Warriorhood is a social identity closely connected to military actions, but also motivated by stereotypic myths of men and war. Helle Vandkilde's studies focus in particular on this aspect. Warrior organisations are clubs with a military objective that generally have male members. A certain degree of support is found for the hypothesis that warrior organisations themselves carry a potential for social change, but apparently it can only be activated during crises and considerable external pressure. The warrior institutions can be separated into three categories on the basis of whether the access is regulated through the criteria of age, status/prestige or social rank. The first category is found, for instance, among nomadic tribes in Eastern Africa, the other among prairie Indians and the Central European Corded Ware Culture. All three categories integrate elements of 'Gefolgschaft' in the sense of a long-term reciprocal relationship between a leader and his group of warrior-followers, who are bound by economic interests and moral rules. Gender is a relevant aspect to study. War is waged as a rule (but not always) by men. Often women take on the responsibility for the families' and the society's honour and contribute by rousing to war and by assisting before, during and after the acts of war. The border line between soldier and warrior is rather fluid, but the role of the warrior is decidedly more marked by an individualistic mode of thought and organisation.

Material culture and personal appearance organise and maintain all kinds of identities, among these warrior identities as they exist in many prehistoric, historical and ethnographic contexts. Weapons and special dress and body attitudes are strategically used to form and manipulate the image of the warrior as identity and ideal within the warrior group, between warrior groups, and in respect to the outside world, but at the same time have an effect on the individual warrior by influencing his self-understanding and personal appearance. Furthermore, advances in weapon technology can escalate conflicts and in some cases (e.g. horses and swords) actually precipitate social change.

Ritual war is a rather unclear concept that has been misused to postulate peaceful conditions in societies without centralised political power. It must be pointed out that 'ritual war' will always merely be one facet of a military reality, with all its implications of human suffering and death. On the other hand, war is almost always related to different kinds of rituals carried out before, during and after acts of violence. Sacrifices of weapons and people in prehistory can be regarded as part of a series of actions that includes war. In addition to this, there are certain religious aspects by which appeals are made to 'higher powers' for a positive intervention. Through his Bronze Age case study (chapter 24), Andreas Hårde shows that violence in the shape of human sacrifices was used by the political elite as a means to consolidate their control over life and death and to frighten outer and inner enemies. The mass graves that mar the past and the present should on the one hand be associated with military acts, but they also have distinct functions in the way of debasing and deterring defeated enemies as well as demonstrating power.

Power is a key concept in the understanding of war and warriors. Power – i.e., dominance – can be achieved either through persuasion or force; in the case of the latter, through war and violence or threats of violence. War can certainly be part of groups' and individuals' strategic effort to achieve overall dominance. On the other hand, there are a number of examples where war is carried out by warrior groups operating autonomously and in isolation in respect to the more primary authorities of society and here it is not directly related to dominance. In certain decentralised societies war is not directly accessible as a source of dominance, but these societies are nevertheless often extremely marked by war that seems to have the effect of maintaining rather than changing the society.

War is a key ingredient in social change and for this reason alone it is relevant to study. There is no one-sided relationship between input and output, and perhaps more than any other kind of strategic act war tends to create unintended effects. Since war is a violent form of social practice, it can be said to always contribute in some measure to social change even if its aim is maintaining the political status quo. War is thus in a very general sense a processual force. States have, for instance, always attempted to maintain themselves through war. Also other kinds of centralised societies have used war and the military as a source of power, for instance, to strengthen an existing base of power. This was true, for example, in the complex Bronze Age societies in Southern Scandinavia and the so-called chiefdoms on Fiji, Hawaii, and in the Grand Chaco. War is therefore often used for reproductive purposes, but can war also change society more radically?

This question has in particular been discussed in connection with theories of state formation. Claus Bossen (chapters 7 and 17) evaluates the relationship between war and state formation, and concludes that there is a connection, but that many other factors come into play. The same question is, however, relevant to discuss in cases where the social structure in 'egalitarian' societies quite suddenly moves in the direction of institutionalised hierarchy, such as in north and central Europe with the emergence of the Battle-axe or Corded Ware cultures (2800-2500 BC) or in certain hot spots in the Early Bronze Age of Central Europe and the Balkans (2000-1500 BC). War was also, for example, a strategic but risky opportunity for local entrepreneurs on New Guinea to develop their status, but egalitarian institutions pulled hard in the opposite direction. In this area, our studies have not been able to indicate clear regularities or patterns in either the archaeological or the social anthropological material, but it should be emphasised that the topic deserves further illumination. Warfare is part of most state formations and of the formation of the above-mentioned hierarchies, but other factors enter into a complex interaction with war. Furthermore, there are a number of cases, historically and in recent times, in which war has wiped out societies rather than contributed to creating something new. A regularity that can be pointed out, however, is that war tends to create more war.

This particular logic of war has been scrutinised by Jürg Helbling among tribal societies (chapter 9). Contemporary tribal wars always take place in the context of expanding or deteriorating states and in the wider context of the world economy influencing the course and intensity of war, but it is nevertheless imperative to search for the internal logic of these indigenous wars. Two structural conditions may explain the high level of war in these societies. First, the local groups operate autonomously in a political system that can best be described as anarchic. Second, these local groups are relatively immobile being dependent on locally concentrated resources. People do not wage war because they are fond of it. Despite high economic and personal costs and despite the fact that peaceful cooperation will yield the highest gain for all groups, each group is compelled to adopt a bellicose strategy. Game theoretical considerations may explain this apparent paradox: engagement in peaceful strategies is simply too risky because a one-sided bellicose strategy will potentially bring the highest gains while a one-sided peaceful strategy may lead to the highest losses. Only when both parties engage in peaceful strategies, both will gain, but none can be certain of this. Therefore the military superiority of one group inevitably constitutes a threat to the others, forcing them to attempt achieving superiority in turn. Helbling concludes that the two structural conditions of tribal societies create an environment in which war is prevalent. The societies adapt to this social environment and this explains a number of their characteristics which often – but according to Helbling mistakenly – are considered as causes of tribal warfare, such as the centrality of warrior values, political status competition and conflicts over scarce resources.

War is always waged against 'the others', and in this sense it may be said that war often originates from narrowly defined groups, but on the other hand war often appears to strengthen these groups as well as create new groupings. The connection between war and identity is thus quite complex as demonstrated by Torsten Kolind concerning the Bosnian material (chapter 29). His conclusion is that everyday identifications can be regarded as part of a counterdiscourse – against the nationalistic and religious categorisations that on the public and political level were the reasons and aims for the war in Yugoslavia. The direction and kind of the changes can seldom be pinpointed in advance due to the presence of crucial unpredictable elements, in part because identity is formed in various ways at several levels ranging from everyday life to overriding political authorities.

New problems and questions

The Warfare and Society project can, *qua* the perspectives and results described above, point out a number of new problem areas and questions that require profound study through new research. In particular, three complexes of problems should be mentioned:

More research is necessary in the limitations that seem to be in force in societies with egalitarian institutions – as on Papua New Guinea – , especially the potential of war to create political inequalities and structural social change. It is also necessary to further analyse the qualitative changes in war brought about by the use of firearms or other new technology. In Papua New Guinea a destabilisation of the existing exchange systems occurred and as a result an acceptance of the colonial power and its efforts at pacification; in fact, an external state's monopolisation of violence. In general, it must be considered relevant to theorise warfare as a form of transaction unlike, yet in many ways also complementary to, other forms of exchange in societies without centralised power.

Violence and war articulate existing identities and create new identities often in a determining way, but it is also important to analyse the discursive strategies that people use to adapt these general identities to everyday life, which is precisely where a need exists to create new exchange relations and connections.

The relationship between the formation of social identities and material culture needs further illumination. The understanding of war and the role of warriors in prehistoric societies is still not profound enough, and henceforth the focus should be directed more toward using the archaeological material and relevant theoretical tools interactively; along the lines of Vandkilde's proposal in this volume (chapter 26). The creative and preserving role of material culture in respect to a large number of violent and non-violent identities within and across lines of gender, age, family, status, rank, occupation and ethnicity still requires thorough investigation. Concrete investigations with theoretical superstructures can clearly occur through interdisciplinary collaboration, especially between social anthropology and archaeology. The warrior is for instance often particularly visible in European prehistory, especially in the funerary domain, but the question remains of the extent to which these presentations represent contemporary ideals and myths. Other questions that remain unanswered are when the first warrior institutions appeared in Europe and what their social and economic background was. The appearance of institutionalised warriorhood (probably in certain hotspots around 5000 BC and again, more massively, around 2800 BC) seems to coincide with three other phenomena, namely, a clear gender differentiation in funerary etiquette, the formation of an elite, and a drastic expansion in the use and production of copper objects. But for the present this must remain a qualified hypothesis.

About this book

The structure of this book reflects the six areas upon which the project activities and debate were focussed during the four years it ran: war as presented in philosophy, social theory and the discourses of anthropology and archaeology; war in non-state societies; war and the state; war, rituals and mass graves; war, discourse and identity, and war and material culture. The publication gathers in total thirty-four contributions from a selection of seminar participants, among these the project participants. Included in these are the editors' introductory articles, which serve as critically annotating introductions to each of the six subject areas.

Both archaeologists and anthropologists have contributed to the subject areas, which occur quite mixed in this respect. It also appears that many of the authors are inspired by their 'neighbouring discipline' and consequently incorporate other perspectives. Several articles are definitely situated in the intersection between archaeology and anthropology. Through its seminar activities and this book, the War and Society project has demonstrated a potential for new insight to be gained through combining theories, methods and results from different disciplines. The essence of archaeology is by nature far-sighted and material, although when operating in historical periods it is able to add the evidence of written discourse. Social anthropology is more contemporaneous and based especially on spoken discourse. The data patterns of the disciplines should however be interpreted within a social context, and in this way it becomes possible to compare and integrate results.

Considering the scope and quality of the contributions, the three editors also consider the book an important contribution to the international discussions in this field, which are increasing currently due to the escalating situation in the Middle East and disturbing reports from other war-stricken areas in the world.

Acknowledgment

The War and Society Project, including the present publication, was generously supported by a grant from the Danish Research Council for the Humanities, for which we express our gratitude. The editors wish to thank all participants in the project for their contributions in the form of discussions, presentations, working papers and, eventually, the chapters in this volume. We are much indebted to Professor Jürg Helbling from the University of Zürich for having peer-reviewed all articles. In addition we collectively thank all scholars, who have agreed to review and comment on one or more chapters. Furthermore, we express special thanks to those who have helped with the production of this sizeable book: Nick Thorpe, Stacey Cozart, and Mary Waters Lund for language correction; Toke Bjerregaard and Steffen Dalsgaard for formatting the articles and checking references; Steffen also for proofreading and producing the index; Hanne Kolding for layout and cover; and finally, Sanne Lind Hansen, our editor at Aarhus University Press.

NOTE

1. Kolind, T. 2004. Post-war Identifications: Counterdiscursive practices in a Bosnian Town. Ph.D. dissertation. Aarhus University: Institute of Anthropology, Archaeology and linguistics, p. 65

Warfare and Society

Conceptions of Warfare
in Western Thought and Research

/2-7

Conceptions of Warfare
in Western Thought and Research:
An Introduction

TON OTTO

/2

This section addresses the overall conceptual frameworks that have informed Western thinking about warfare and explores how these frameworks have impacted on anthropological and archaeological research into war and violence. One of the central questions, of course, concerns the origin of war: has it always existed, if not in reality then as a potential of human nature, or is it a product of the development of human society? It is clear that in order to ask and answer such a question we must first agree on what to regard as warfare. Even though there is widespread agreement that warfare can and should be distinguished from phenomena such as homicide and feuding, authors, also in this section, disagree about the nature of the political units that can wage wars. David Warburton argues that only states make wars and he thus places himself in a long tradition of Western thought that sees statehood and warfare as intrinsically connected. One should be aware, however, that this tradition of thought arises in a period of Western history when states were the common form of organising polities – and thus wars. The disciplines of archaeology and anthropology, also products of Western history, extend the empirical horizon to societies without centralising authorities and this makes it necessary to consider whether the violent interactions in which these societies engage should also be called war.

In this introduction, Otterbein's definition is used as a guideline: war is a planned and organised armed dispute between political units (Otterbein 1985: 3). In this definition these units do not necessarily have the character of states (cp. also Ferguson 1984: 5), thus extending the phenomenon of warfare to a large range of societies. The idea that warfare has evolved in relation to the transformation of human societies has strongly influenced anthropological and archaeological research, but before I develop this central assumption further I want to highlight another central idea that has impacted on Western thinking (and acting) up to the present day: that of the morally justified war.

The idea of morally justified war

Warburton (chapter 4) sketches two main lines of thought concerning war in the Western history of ideas. One line is exemplified by the Greek historian Thucydides (5th century BC) as well as by the German general Carl von Clausewitz (1780-1831). In this line of thought war is the exercise of power to impose one's will – 'politics by other means' as Clausewitz (1989) formulated it. Thucydides emphasised that we should not have an idealised concept of warfare. According to him the strong do what they can and the weak suffer what they must. Ideas of morality play no role in this perspective: war is the application of violence to achieve one's goals. However, there is another line of thought that has played a dominant role in Western history, namely that of war as a morally justified activity. Thomas Aquinas clearly formulates a concept of the 'just war', which among other things is characterised by the right intention to engage in war. Warburton follows this conceptual thread among other Western thinkers, in particular Rousseau and Hegel. Obviously we cannot find the roots of this concept in ancient Greece, which apart from Thucydides' cynical view of human nature also gave rise to the older Homeric worldview of war as a game of honour and revenge. According to Warburton the roots of this idea have to be found in the Near East. Ancient Egyptian and Assyrian warfare was connected with the idea that their military and political expansion was sanctioned by their gods, who gave victory in war.

The Hebrew bible incorporated this idea of divine justification, but here we have a God who also used war to punish his own people. Being in fact the only God, a concept of absolute and universal justice became connected with warfare. Through the Hebrew bible the idea of a just war became absorbed into Western thought, and adapted to various forms of states: from medieval kingdoms, which had to relate to the overarching church organisation, to states characterised by the Reformation, and later modern democratic states claiming the right to call on their citizens to take up arms for just causes.

Warburton continues his sketch of the history of the idea of just wars by observing an interesting contrast between Europe and the United States of America. European states have become weary of their numerous conflicts which showed that territorial expansion – however morally defended – was in practice unsustainable over time because dominated peoples always fight back. By the end of the 20th century most European states shared a determination to avoid the use of war as a political instrument. For many Europeans warfare no longer was morally justified, rather the opposite. The USA however continued its belief in the justification of war. Its conceptions were structured to a high degree by the context of the Cold War with the communist regimes, which was seen as an ideological conflict. At the same time it was generally conceived that real war was impossible because of the implicit risk of total destruction caused by modern nuclear weapons. This war was no longer territorial but based on ideological principles. America was not seen as defending (only) its own interests, but rather as fighting for universal values such as individual rights, democracy and economic growth.

While the Cold War petered out due to dramatic internal changes in former communist societies, a new ideological war has taken centre stage: that between fundamentalist Islamic terrorist groups on the one hand, and Western states on the other, with the USA and its allies as the central targets and combatants. This war has taken new forms, as one of the warring parties is not organised as a state.

This is clearly causing problems for the Western states, which continue a strategy of attacking 'brigand' states, supposedly harbouring or supporting the terrorists. Perhaps even stronger than during the Cold War, this war is informed by ideas of justified war from both sides. Religious rhetoric abounds and one cannot avoid speculating whether a more pragmatic attitude in line with Clausewitz and Thucydides would perhaps lead to different, less violent, strategies.

The origin and evolution of war

Apart from the morality of warfare, another central question has occupied the minds of Western thinkers as well as archaeologists, anthropologists and primatologists, namely concerning the origin and evolution of war. Is warfare a heritage from mankind's biological origin or is it rather the opposite, a result of human history?

Raymond Corbey (chapter 3) presents some of the key issues and key thinkers to inform this debate, which hinges on the contrast between nature and culture. A central figure is Hobbes, who bases his argument on the idea of an original, natural state, where everyone could potentially be attacked by everyone else: a 'warre' of all against all. This original situation could only be transcended by a social contract, investing the power of violence in the sovereign state. A newer version of this transition from an original state of war to more peaceful interaction is provided by the anthropologist Marcel Mauss. In his view, exchange is the earliest and therefore most fundamental human solution for overcoming warfare; relations of exchange replace and prevent violent interactions: in order to trade one has to be able to lay aside the spear. This idea has engendered a strong line of anthropological research and theorising, exemplified by key theoreticians like Claude Lévi-Strauss, Louis Dumont and Marshall Sahlins. They present a view of a cultural solution to a fundamental biological tendency in humankind: man's original aggression is constrained by social institutions, in particular exchange, exogamous marriage, and collective representations (ideas and values about altruism and collaboration).

Even though highly appreciative of this approach, Corbey also identifies a central problem, namely its assumption of the duality of human nature: emotion versus reason, primordial war versus pacifying gift. He argues convincingly that human nature is the result of the co-evolution of genetic make-up and socio-cultural behaviour. Culture and nature have evolved in relationship to each other and therefore human nature is also the product of culture. He further argues that a comprehensive approach that analyses the integration of nature and culture in human societies would be clearly in line with Mauss' heuristic principle: to study social phenomena and people in their totality (think of his famous concept of the 'fait social total').

I would like to point to another limitation of the Maussian line of thought, namely that its focus on cultural 'solutions' does not offer an explanation for the enormous variability of warlike phenomena that exist in human societies, in particular concerning the frequency and intensity of war. The Maussian focus on exchange implies a concern with more or less egalitarian, non-state societies, but even there great variety exists. In addition, some anthropologists argue that exchange is *not* always a solution to war: exchange may in fact also lead to war, or, to put it otherwise, exchange and warfare can apparently be well integrated in a regional system of interacting, exchanging and warring small social units (see Brandt chapter 6; Wiessner chapter 11; Otto chapter 12; M. Strathern 1985; A. Strathern 1992).

The various forms and appearances of warfare in connection with different types of societies, in particular those characterised by a central authority (state) and those without, is the subject matter of the next two sections in this volume. Here I would like to draw attention to an interesting perspective that focuses on the borderline between nature and culture and thus relates directly to the issues raised above.

In an important article Bruce Knauft (1991) discusses the literature concerning great-ape and simple human societies, defining the latter as lacking recognisable leadership roles and status differentials among adult men. Simple human societies cover by far the largest period of the evolution of Homo Sapiens, but are represented poorly in the ethnographic and archaeological record. They have to be distinguished from more complex pre-state societies, called 'middle-range', where sedentism, property ownership and male status differentiation are more developed, namely complex hunter-gatherer societies, 'tribes', and 'chiefdoms'. Knauft argues that simple human societies differ from both great-ape and middle-range human societies in that they show a relative absence of competitive male hierarchies and of systematic violence between closed social groups. They are more egalitarian among the adult males, sexually, politically and in terms of sharing resources. Thus, he argues, the invention of cultural rules of cooperation and exchange has had a clear impact on the use of violence in these societies, which sets them apart from the high level of violence in middle-range societies and from the competition and violence observed among apes.

There is thus not a lineal development from pre-human to human societies, but rather something that resembles the Maussian model: a (temporary) constraint on competition and violence through cultural institutions that have given simple human societies an evolutionary advantage over their non-human environment. This is an interesting hypothesis that certainly deserves further research, but it is complicated and possibly weakened by the observation that lethal violence may actually be high in these societies, even though the cultural ethos is against it, and even though the violence may relate more to status levelling than to status elevation (cp. Knauft 1991: 391).

Conceptions of warfare and present-day research

The contributions in this section by Helle Vandkilde (chapter 5) and Erik Brandt (chapter 6) are concerned with a third question, namely how different conceptions of warfare have impacted upon the actual research conducted by archaeologists and anthropologists. This question received high visibility through the publication of Lawrence Keeley's book *War before civilisation* (1996), which argues that anthropological and archaeological research into warfare has been hampered by the conception that primitive warfare was much less serious and destructive than modern warfare. Keeley relates this 'myth' of the rather peaceful savage to the horrible experiences of the two World Wars, which made scholars more susceptible to imagining alternatives to the horrors of modern war. Keeley's idea is supported by the anthropologist Keith Otterbein (2000), who has worked on warfare for more than three decades.

However, Otterbein wishes to correct Keeley's historical sketch on two accounts. In the first place the myth of the peaceful savage arose already before

World War II and, secondly, its driving force was a framework of evolutionary theory, later conceptually nurtured by cultural relativism. He further argues that Keeley has produced a 'replacement myth' which depicts pre-state societies as bellicose. Unfortunately this new myth has caused a polarisation among researchers, dividing them into Hawks and Doves

Helle Vandkilde describes the situation for the discipline of archaeology in Europe which, in her view, has been dominated by two different tales of prehistoric society. In the one tale prehistoric society is perceived as changing radically in certain periods, caused by human agents migrating and revolutionising existing societies. Even though this view, classically exemplified by V.G. Childe, implies the existence of warriors, it has not emphasised warfare as an important element of the historical changes. The other tale sees prehistoric society as changing slowly, through gradual evolution instead of revolution. The main characters in this vision are hunters, peasants and traders, while warriors are apparently neglected, and prehistoric society is imagined as basically peaceful. The later view has been dominant since World War II and this appears to accord with Keeley's periodisation of the myth of the peaceful savage.

Vandkilde observes a greater interest in, and a more realistic evaluation of warfare in recent archaeology, but finds that there is still much to do. She suggests that anthropological research, which has had only a relatively modest impact on archaeology, should be used more extensively, while research into warrior identities should be open to conceiving more variation in the status and role of warriors according to context and period.

With regard to anthropological research on warfare, particularly in New Guinea, Erik Brandt's contribution criticises and modifies Keeley's and Otterbein's hypothesis concerning the impact of the myth of the peaceful savage (cp. Brandt 2000; Knauft 1990). He does not deny that the myth has existed, also in relation to New Guinean research, but it has not hampered anthropological research in the way envisaged by Keeley and Otterbein. Brandt shows that Malinowski had already made a sharp distinction between modern war and savage war. Whereas modern war was considered as total, affecting every single cultural activity, savage war was seen rather as a form of physical exercise devoid of political relevancy. This depiction appears to support Otterbein's rendering of the origin of the myth of the peaceful savage, but Brandt shows that another view can also be detected in Malinowski's writings, one that accepts tribal war as a serious and destructive phenomenon for the people concerned, who try to constrain and overcome it by means of exchange.

This Maussian view was later reproduced and refined by the focusing of influential anthropologists, such as Andrew and Marilyn Strathern, on the role of local leaders – 'big men', who engage in exchange as an alternative and preferred way of gaining status in contrast to warfare. Their work has to be seen in the context of earlier work on New Guinea, which had not refrained from making ethnographic descriptions of ubiquitous and pervasive warfare that in all aspects were reminiscent of total war. According to Brandt, it was in opposition to such a view of total war that the work of the Stratherns should be understood, but this did not lead them to assume predominantly peaceful savages. Thus, Brandt concludes, the concept of total war, rather than the alternative notion of the peaceful savage, has burdened the ethnography of New Guinean warfare, but ethnographers have equally been influenced by the realities of war and violence they met in the field.

Empirical studies and theoretical modelling

Brandt's conclusion is a crucial motivation and legitimation for the studies that are included in the following sections of this volume. Research cannot avoid being informed and partly determined by the conceptual frameworks that are available at the time of investigation and that are in dynamic relationship with the wider social experiences of that period. But research is also informed and influenced by the empirical findings carefully produced by anthropologists, archaeologists and other researchers in their various projects.

The first four contributors in the present section reflect on the conceptual and ideological context of research into warfare: the philosophical questions, the history of these questions in Western thought and the treatment of them in the disciplines of archaeology and anthropology. The final chapter by Claus Bossen (chapter 7) is an attempt to integrate existing theoretical perspectives into one framework. The author's central concern is to understand the relationship between warfare and social change, and the chapter provides a useful overview of anthropological, sociological and archaeological theories of warfare. Bossen critically assesses their potential to explain social change in relation to warfare and argues that the analysis of war and social change involves three perspectives or 'levels': praxis, society and process.

At the level of praxis Bossen identifies three aspects of violent acts which comprise meaning, technology and organisation. At the level of society Bossen adopts Michael Mann's four fields of social organisation: economy, politics, ideology and the military. Finally Bossen assesses the ways in which warfare and military organisation can contribute to social change, namely via internal effects within a society, via submission of one society to another, and via the general context of a warlike environment. Bossen argues that these three perspectives are mutually interdependent and therefore should be integrated into one conceptual framework. The explanatory value of such an integrated model obviously needs testing in relation to concrete cases, but as it stands Bossen's model may serve as a welcome heuristic tool to ask relevant questions about possible links and dependencies between warfare, social practice and societal change.

REFERENCES

- Brandt, E. 2000. 'Images of war and savagery: Thinking anthropologically about warfare and civilisaton, 1871-1930'. *History and Anthropology,* 12(1), pp. 1-36.
- Clausewitz, C. von 1989. *On War.* M. Howard and P. Paret (eds. and trans.) Princeton: Princeton University Press.
- Ferguson, R.B. 1984, 'Introduction: Studying war'. In: R.B. Ferguson (ed.): *Warfare, Culture and Environment.* Orlando: Academic Press, pp. 1-81.
- Keeley, L. 1996. *War Before Civilisation.* Oxford: Oxford University Press.
- Knauft, B.M. 1990, 'Melanesian warfare: A theoretical history', *Oceania* 60(4), pp. 250-311.
- Knauft, B.M. 1991, 'Violence and sociality in human evolution'. *Current Anthropology,* 32(4), pp. 391-428.
- Otterbein, K. 1985. *The Evolution of War* (3rd edition). New Haven: HRAF Press.
- Otterbein, K. 2000, 'A history of research on warfare in anthropology'. *American Anthropologist,* 101(4), pp. 794-805.
- Strathern, A. 1992. 'Let the bow go down'. In: R.B. Ferguson and N. Whitehead (eds.): *War in the Tribal Zone: Expanding States and Indigenous Warfare.* Santa Fe: School of American Research Press, pp. 229-250.
- Strathern, M. 1985. 'Discovering social control'. *Journal of Law and Society,* 12, pp. 111-134.

Laying Aside the Spear:
Hobbesian *Warre* and the Maussian Gift

RAYMOND CORBEY

/3

Ethnologists inevitably come to their subjects with a certain philosophical baggage which is part of their own, North Atlantic universe of cosmological and moral meaning, and influences the way they gather and interpret their data. In the following, I will examine one particular, widespread assumption informing Maussian and structuralist theorising on gifts and reciprocity: the idea of violence as a basic tendency of human nature. While the other contributions to this volume focus on detailed archaeological and ethnographic data pertaining to conflict and violence more directly, the present one looks at historical and epistemological backgrounds of one particular, quite influential way of handling such data theoretically and conceptually.

From violence to sociality

The notion of disorder and conflict as a 'natural', 'primordial', or 'original' state of humankind, and at the same time of human nature, is continually present in Marcel Mauss' *Essai sur le don*. It guides his empirically directed work as a conceptual or ontological presupposition, linking his thought to that of the leading social theorists of the Enlightenment. In his analysis, *état naturel* refers to both humankind before history and civilisation – its natural history –

and a state of 'raw nature' that is partly constitutive for human society, as a condition that must continually be transcended to make humanness possible. Becoming human, as Mauss analyses it, happened in (pre)history, but it is also, ontologically speaking, a permanent, structural feature of humans who, according to this view, continually transcend the state of nature, by exchanging. The 'natural state' is seen as primordial, both ontologically and phylogenetically, and social order as discontinuous with nature in both respects.

Mauss holds exchange to be constitutive of social life and social order because it is the chronologically earliest and ontologically most fundamental solution to the Hobbesian *warre* of all against all that, in Hobbes' view, ensues from man's selfish nature. 'Societies have progressed', he writes in the conclusion to the *Essai sur le don*,

in so far as they themselves, their subgroups, and, lastly, the individuals in them, have succeeded in *stabilizing* relationships, giving, receiving, and finally, giving in return. To trade, the first condition was to be able to lay *aside the spear*. From then onwards, they succeeded in exchanging goods and persons, no longer only between clans, but between tribes and nations, and, above all, between individuals. *Only then* did people learn how to create mutual interests, giving mutual satisfaction, and, in the end, to defend them *without having to resort to arms*. (Mauss 1990: 82, my italics)

In the 'natural' state which is overcome through gift exchange the 'fundamental motives for human action: emulation between individuals of the same sex, that "basic imperialism of human being"' (Mauss 1990: 65) still had free reign, but in the end reason overcomes the folly of unbridled primeval *warre*: 'It is by opposing reason to feeling, by pitting the will to peace against sudden outbursts of insanity of this kind that people succeed in substituting alliances, gifts, and trade for war, isolation and stagnation' (Mauss 1990: 65). Social order is conceived of as the constraining, taming, subduing of a primitive, primordial, condition of violence and warfare; the pacifying gift brings about co-operation and sociality.

More recently a number of ethnologists have put the insights of Marcel Mauss and his disciple Louis Dumont to use in their research on a number of mostly tribal societies, focussing on patterns of exchange. Time and again they show by detailed ethnographic analysis how in such societies certain 'ideas/values' – *idées-valeurs* – perpetuate themselves beyond the life or death of particular individuals, imposing themselves in all the various sorts of social relations (e.g., Platenkamp 1988; Geirnaert-Martin 1992; Barraud *et al.* 1994). The plethora of exchanges going on in a village every day form, and constantly renew, the value-orientated matrix of the 'sociocosmos' which is constitutive for social order and, at the same time, for the individuals involved, including the dead and the spirits. 'Subjects and objects intertwine ceaselessly', Barraud *et al.* write, underlining one of the key insights of Mauss' analysis of the gift, 'in a tissue of relations which make of exchanges the permanent locus where these societies reaffirm, again and again, their highest values.' (Barraud *et al.* 1994; 105). Exchange is here not taken in a narrow economic sense, but as *symbolic* exchange, as a *fait social total*, a 'total social phenomenon', with many, non-separated aspects, normative, economic, jural, religious, and so on. It is, with what is probably the best known dictum from Mauss' *Essai*, 'one of the human foundations on which our societies are built' (Mauss 1990: 4).

According to these Durkheimians, in small-scale traditional, non-state societies not only social order but also personal identity, Mauss' *personnage* (Mauss 1995: 331-61), is constituted through gifts and exchange. Persons are not primarily seen as particular biological organisms, but as coming about and being transformed – for instance, from living to dead – by the ritually and intergenerationally bestowing upon each other of souls, names, titles, rights, and duties that are part of the family clan. This happens not only in birth ceremonies, marriages, funerals, and other important rituals that punctuate the life cycle, but also in the context of subsistence activities such as hunting and horticulture, usually conceived of as an exchange with spirits inhabiting the landscape, as well as in the context of such seemingly trivial everyday activities as greeting, gossiping, and sharing food.

The influential structuralist approach of Claude Lévi-Strauss shares the Maussian presupposition of social order as a *human* imposition upon a relatively unstructured, chaotic, brute state of nature:

The social life of monkeys does not lend itself to the formulation of any norm ... [The] monkey's behaviour is surprisingly *changeable*. Not only is the behaviour of a single subset inconsistent, but there is *no regular pattern* to be discerned in collective behaviour (Levi-Strauss 1969: 6-7; my italics; cp. Rodseth *et al.* 1991: 222, 233)

In Lévi-Strauss' opinion, a particular animal became human, and social organisation came into being, only by the prohibition of incest. '[Humankind] has understood very early', he states in *The Elementary Structures of Kinship*,

that, in order to free itself from a wild *struggle for existence*, it was confronted with the very simple choice of "either marrying-out or being killed-out". The alternative was between biological families living in juxtaposition and endeavouring to remain closed, self-perpetuating units, *overridden by their fears, hatreds and ignorances*, and the systematic establishment, through the incest prohibition, of links of intermarriage between them, thus succeeding to build, out of the artificial bounds of affinity, a true human society ... (Levi-Strauss 1956: 277-78; my italics)

Social, political, and economic order, in this view, come about by giving; they are a consequence of the exchange – of giving and receiving, giving and giving-in-return – of women between male-dominated descent groups. Hereby the natural state is transcended and a truly human existence is attained.

This particular way of conceptualising the relation between society and nature is analogous to Thomas Hobbes' social contract theory. 'The finall Cause, End or Designe of men,' Hobbes wrote in the first part of *Leviathan*,

... in the introduction of ... restraint upon themselves, ... is the foresight of their own preservation, and of a more contented life thereby; that is to say, of getting themselves out from *that miserable condition of Warre, which is necessarily consequent to the naturall Passions of men*, when there is no visible Power to keep them in awe. (Hobbes 1972 [1651]: 223, my italics)

Social order, according to Hobbes, is not in human's nature, but is installed by a social contract that constrains and pacifies the natural state of humankind and the solitary individual's brutish natural tendencies. Most Hobbes commentators stress that the natural state is but a hypothesis, an imagined, fictional condition facilitating the analysis of how social order is constituted, but Hobbes regularly alludes to American Indians and the prehistory of humankind.

At a certain point, the analogy stops, for while in Hobbes' thought the state is an instrument of selfish individuals, for the Durkheimians, the social fabric which is constituted by exchange is a moral and religious order. Their approach was critically pitted against the liberal, in their view, too individualistic, voluntaristic, and utilitarianist *homo oeconomicus* approach to the foundations of society of the social contract theorists of the Enlightenment. The Maussian gift can be seen as 'the primitive analogue of the social contract ... the primitive *way of achieving the peace* that in civil society is secured by the State' (Sahlins 1972: 169, my italics).

The Durkheimian view of the 'primordial nature' of humans is clearly quite close to that of Enlightenment authors who postulate a progress from a savage primordial state to the civilised condition. Both positions are heir to a typically European, dualistic perception of humans and reality that issues from Platonic and Christian ideas on the spirit and the flesh, innate sinfulness and redemption (Corbey 1993; Sahlins 1996; Carrithers 1996). More specifically, the view of humans and nature underlying Mauss' analysis of the transition of war of all against all to exchange of all with all, or, at least, of many with many, is the Durkheimian one of *homo duplex*. It was formulated succinctly by Emile Durkheim in an article from 1914. The individual, in his view, has 'a double existence ... the one purely individual and rooted in our organisms, the other social and nothing but an extension of society' (Durkheim 1960: 337; cp. Sahlins 1996: 402; Rapport 1996).

A deep antagonism between the demands of the individual organism and those of social order is postulated, a conflict in which Durkheim and Mauss are firmly on the side of the *morale de la réciprocité*, which triumphs over the primordial *intérêt personnel*. One particular animal species becomes human phylogenetically, ontogenetically, ontologically, and morally, through inculcation in a different order of existence: the spiritually, morally, and intellectually superior world of society, language, and culture, thus rising above its naturally selfish animal individuality which is directly rooted in the organism. Such and similar dualistic views of humans and society, nature and culture, determine how most ethnologists, in the French Durkheimian-cum-Maussian tradition, but also, along slightly different lines, in the American Boasian tradition, conceive of their discipline: as a *human* science.

Society as biology or society as culture?

There are baffling divergences in styles of scientific explanation between, as well as within, disciplines depending on whether natural sciences types of approach are followed or interpretive, typically human sciences, ones. Explanations of human violence and warfare are a case in point. Support has been lent to the Hobbesian perception of human nature in recent decades by a number of researchers working with biological, evolutionary approaches, in the wake of the ethology of Konrad Lorenz, Irenäeus Eibl-Eibesfeld, and Nico Tinbergen; the sociobiology of Edmund Wilson; Richard Alexander's theory of the maximisation of reproductive success; the biosocial anthropology of Robin Fox; as well as, more recently, the inclusive fitness theory, dual inheritance theory, and evolutionary psychology. Johan van der Dennen, for example, in his 1995 analysis of the evolutionary origin of war, takes a rigorously biological approach, analysing warfare as a highly effective, high-risk/high-gain male-coalitional adaptive and reproductive strategy. This is a Hobbesian *bellum omnium contra omnes*; not Malthusian society read into nature, as Karl Marx once wrote to Friedrich Engels upon reading Darwin, but exactly the opposite. For the Maussians, altruism means the suppression of selfish instincts; for inclusive fitness theory, their articulation.

Combining human sociobiological insights with cultural ecological ones, anthropologist Napoleon Chagnon, in his research on the Yanomami of Venezuela, stresses the inclusive fitness of male warriors in the complex interrelationship between individuals, groups, and their natural environment (e.g., Chagnon 1988). The more women they realise access to, the better the proliferation of their genes. Most cultural anthropologists, however, conceive of their discipline as a typically human science, and, unlike Chagnon, conceive of society as a cultural and normative order, not primarily a biological one. Accordingly, they interpret warfare and violence as predominantly cultural phenomena, following collective rules and values, or rationally responding to historical or environmental circumstances such as resource scarcity, rather than issuing from individual basic drives. 'Indeed', Leslie Sponsel (1996: 909) writes,

in recent decades diverse lines of evidence have converged to strongly suggest, if not to demonstrate, to everyone's satisfaction, that human aggression, including warfare, is overwhelmingly determined by culture.

This sharply contrasts with van der Dennen's claim that warfare is overwhelmingly determined by biology, and explainable only by a rigorously neodarwinist approach.

It has been argued (by the authors contributing to Sponsel and Gregor 1994, among others) that peacefulness, not war, sociality, not aggression, is the natural, normal condition. In their view, aggression and warfare do not issue from basic human nature, but are triggered by specific historical and cultural circumstance. The Hobbesian idea of aggression as germane to the human condition has accordingly been criticised as social Darwinist ideology. Biologically orientated authors have retorted to ethnologists of that persuasion that peacefulness is but a romantic, utopian dream – a case of primitivist wishful thinking.

Controversies between, broadly conceived, researchers who stress natural determinants and those who favour a culturalist approach have led to characterisations of certain peoples as explicitly aggressive and fierce or, alternatively, unambiguously gentle and peaceful. Against Chagnon's vengeful Yanomani aggressors, beating up women and warring constantly, the Chewong and Semai Senoi from Malaysia, and the Sakkudei from Indonesia, among others, have been thrown in the balance as decisively peaceful peoples. The Kalahari Desert !Kung too were initially cast as a gentle, harmless people. However, the considerable role of preconceived ideas in research is shown once again by the fact that most of such claims have been contested. Jacques Lizot, for example, has sharply criticised the image of the Yanomami as a 'fierce people' (Lizot 1994). Biologically orientated authors, on the other hand, have highlighted the occurrence of violence among the !Kung as well as the Semai Senoi. Something similar happened to Margaret Mead's fieldwork among Samoan adolescents in the 1920s; her underestimation of the role of jealousy, abuse, rape and violence was criticised as a culturalist bias by, again, a biologically orientated anthropologist (Freeman 1984).

Violence and peacefulness as interpretive concepts have a remarkably analogous role to play in primatology. Traditionally, violence has been one of the main ascribed characteristics not only of non-western peoples, but also of nonhuman primates. Both categories were perceived as primitive, brute and unrestrained, and associated with the savage beginnings of humankind's progress to civilisation (Corbey 1989). In recent primatology, a divergence similar to the one just described for ethnography exists. Primatologist Frans de Waal on the one hand, in publications with such telltale titles as *Peacemaking Among Primates* (1989) and *Good Natured: The Origins of Right and Wrong in Humans and Other Animals* (1996), stresses mechanisms for avoiding, reducing and resolving conflicts in the social life of all primates, including humans. Whereas in the Hobbesian-cum-Durkheimian view morality and reciprocal altruism are to be found in culture as a layer superimposed upon the violent and selfish nature of humans, De Waal sees it as part and parcel of the biological make-up of humans and other primates. Empathy and sympathy, reconciliation and forgiveness in his view are ultimately more adaptive than aggression.

Harvard primatologist Richard Wrangham, on the other hand, sums up his, in this respect at least diametrically opposed, Hobbesian approach in his 1996 book with Dale Peterson on *Demonic Males: Apes and the Origin of Human Violence*. Males are selected by females for exploitive and aggressive behaviours, leading to reproductive success. Like de

Waal's books, this one too was written for a broad audience, but reports on a substantial body of detailed empirical studies. 'We are cursed', Wrangham and Peterson conclude, 'with a demonic male temperament and a Machiavellian capacity to express it – a 5-million stain of our ape past.' (Wrangham and Peterson 1996: 258). In the seventies, the positive image of chimpanzees that had emerged in the 1960s had started to change to a less positive, more ambiguous one – in much the same way as that of !Kung Bushmen – when it was discovered that neither the Gombe reserve nor the Kalahari desert were idyllic Shangri-La's after all: in both cases murder and violence turned out to be present next to gentleness and cooperation. The same goes for the Arnhem Burgers Zoo colony of chimpanzees studied by de Waal.

Summing up, we have seen how the Maussian perception of a violent primordial condition of humankind which is to be transcended for real morality and sociality to be possible is partly supported by biological approaches such as those of Wrangham and van der Dennen. Primatologist de Waal, however, stresses innate morality instead of innate aggression. Most ethnologists, on the other hand, seek to explain war and violence on the level of culture and history, and relativise the role of aggression, explicitly posing peacefulness as the basic human condition.

Researchers in a functionalist, Malinowskian tradition have taken issue with Durkheimian views of social exchange stressing moral altruism. The latter take reciprocity as elementary morality and as a means of maintaining equality within the total moral universe, within, in the terminology of Dumont, De Coppet, Barraud and others, the 'sociocosmos' of 'ideas/values' (Barraud *et al.* 1994). Mauss himself, for example, has criticised Malinowski's work on Melanesia as too individualist and utilitarianist (Mauss 1990: 71ff). However, in spite of his eye for agonistic aspects of ritual exchange, Mauss' own work has in recent decades been criticised for underestimating precisely the – indeed Hobbesian – dimension of utility. Annette Weiner and others have pointed to the neglect, in the Maussian camp, of the calculation of outputs and the maximisation of returns, and analysed ritual exchanges not so much as adhering to basic values, but as strategic action increasing power and inequality. Weiner re-examines Maussian and other

classic anthropological exchange theories and the ethnographies that validated these theories [in order] to demystify the ahistorical essentialism in the norm of reciprocity which has masked the political dynamics and gender-based power constituted through keeping-while-giving. (Weiner 1992: 17)

Here again – in parallel to the aforementioned accusations of ideology between proponents of *warre* and proponents of sociality – the Maussian assumption that modern, western exchange is predominantly Malinowskian (utilitarian), while 'archaic', pre-modern exchange is Maussian (reciprocal), has been criticised as primitivist. The Malinowskian view of pre-modern exchange, on the other hand, has been accused of fallaciously and ethnocentrically reading modernity into non-modern cultures.

Regrettably, as another instance of the traditional cleavage between biological and ethnological approaches, the Malinowskian viewpoint is largely out of touch with biology, even though it converges considerably with such viewpoints as reciprocal altruism and inclusive fitness, as a quote from sociobiologist van der Dennen clearly shows. 'I regard human beings', he writes,

as shrewd social strategists, clever manipulators, and conscious, intelligent decision-makers in the service of their inclusive fitness, operating within the constraints of their cultural semantics: the signification and interpretive frameworks ... provided by the culture they happen to be born in. (van der Dennen 1995: 9)

It is but a small step from here to Malinowski's and Weiner's self-interested actors who constantly calculate their costs and benefits, also on the level of sacral and spiritual esteem. According to Weiner's line of argument with its stress on utility, social life and the Maussian gift do not eclipse, but are the very expression of, Hobbesian selfishness and Machiavellian manoeuvring.

Anthropology held captive by *homo duplex*

We have by now encountered at least three different stances with respect to the idea of Hobbesian *warre* as the quintessential mark of the human beast. The first, Durkheimian one, sees it subdued and transcended by a holistic 'sociocosmos' of ideas/values, reproduced in pacifying gift exchange. A second, biological one, in terms of inclusive fitness, supports it, in an updated and more subtle version. A third major theoretical stance are individualist and functionalist approaches in ethnology, which lend indirect support to the preconception of war as a fundamental condition by regarding utility as all-important in social life.

Of course, nobody would subscribe to such ahistorical essentialisations as either war or peacefulness as the basic nature of humankind, or of certain groups; everyone would agree that fierceness and gentleness do not exclude one another and have both roles to play; nobody would deny that there are biological *and* environmental *and* cultural *and* historical aspects to warfare. Still, and nevertheless, as is clear from the foregoing, such preconceptions are capable of creating considerable theoretical divergence. One of the challenges for 21st-century anthropology lies in combining the efforts of biological and ethnological approaches. This may not be simple, because the repudiation of biology is nearly constitutive of much of anthropology's disciplinary identity, especially in the French, Durkheimian tradition and the American, Boasian one. In both traditions, culture is taken as what transcends human biology, and thus gives anthropology its own identity vis-à-vis the biological sciences.

Mauss himself has provided a starting point for overcoming the unproductive *homo duplex* view – which opposes emotion and reason, primordial war and the pacifying gift – with his programmatic heuristic of *phénomènes de totalité* and *hommes totaux*. We hardly ever find man divided into several faculties ('L'homme divisé en facultés'), he wrote in 1924 (Mauss 1995: 303); we always come across the whole human body and mentality, given totally and at the same time, and basically, body, soul, society, everything is mixed up here ('Au fond, corps, âme, société, tout ici se mèle'). The gift is perhaps the best example of such a total social phenomenon (*fait social total or prestation totale*).

Given what was known in Mauss' day of behavioural genetics, kin selection, reciprocal altruism, gene-culture coevolution, the neurological basis of cultural behaviour, and epigenetic development, it may not be held against him that he did not entirely live up to this valuable methodological adage as far as corporeality and the biology of behaviour were concerned. Durkheimian-orientated authors from recent decades, however, are here confronted with an exciting and important challenge. In fact, as we now

know, and as quite a few theoreticians on exchange fail to realise, human nature results from the co-evolution of genetic make-up and cultural as well as social behaviour. Our hands, for example, were shaped while wielding chopping tools and handaxes; parts of our brains and respiratory tracts when our ancestors started to use arbitrary symbols. Stone tools and spoken language are thus integral parts of our biological existence. Similarly, the acquisition and intergenerational, partly symbolic, transmission of cultural and social abilities in humans is crucially dependent upon a whole gamut of cognitive and motivational capabilities that are part of our specific biological equipment. A complex, subtle, and well-timed interaction of these capacities with social environmental influences is of vital importance – an interaction which can also be described on the level of epigenetic neuronal development.

There is a clear biological dimension to various forms of reciprocity in humans, who, as experiments show, solve abstract logical problems more quickly when framed in terms of compliance or cheating with social rules. This shows the importance of social calculation in humans, more specifically their aptness at tallying mutual benefits as an adaptive feature (Cosmides and Tooby 1992). Analogously, *Desmodus rotundus* vampire bats in Costa Rica exchange blood they have sucked with others in the group according to strictly registered and respected patterns of reciprocity. Thus they enhance their chances of survival considerably, for not all flights are successful, and three unsuccessful nightly flights in order may lead to death by starvation (Denault and McFarlane 1995). According to socioecology, the mutual exchange of gifts, services, or women in hominids benefits all parties in such transactions, and is highly adaptive, for example as an ecological safety network to fall back upon in difficult times. Such practices are part of the sociality of hominid and human kin, evolving through selective pressures on reproductive success.

Our nature thus was, and is, social and cultural from its very beginning. That there is a brutish, impulsive animal nature deep within us, in the need of being restrained and subdued in order to make civilisation and social order possible, is a conviction that, at least in this form, does not hold in the light of recent insights. Much of what is social does not come about through a symbolic exchange or contract that restrains the biological, but is biological, that is, natural, itself. How much the Maussian perspective underestimates the role of that purportedly 'raw' organic nature, both phylogenetically and ontologically, is illustrated forcefully by recent work on sociality, individuality, politics, motivation, and communication in nonhuman primates, for example, chimpanzees (Rodseth *et al.* 1991; Quiatt and Reynolds 1993; Ducros, Ducros and Joulian 1998). The challenge here is to understand better the interaction between biology, sociality and cultural meaning.

Evolutionary perspectives dealing with biological aspects of living together cannot replace cultural interpretive ones which deal with subjective, symbolic dimensions of life, but they can supplement them in approaches which focus on the interaction and reciprocal relations between biological, symbolic, sociodemographic, politicoeconomic, and other dimensions of the wielding and the laying aside of spears, of conflict and contract.

In general, evolutionary-biological approaches have adduced solid evidence that human individuals 'favour genetic over classificatory kin, that patterns of residence, descent and marriage are in part affected by reproductive considerations, and that wealth and status are converted into reproductive advantage' (Borgerhoff Mulder 1987: 8; cp. Betzig, Borgerhoff Mulder and Turke 1988). Napoleon Chagnon's aforementioned work on the Yanomami provides a convincing example of this line of argument, as does the analysis of kinship and marriage in humans and other primates in Quiatt and Reynolds (1993: Ch. 9 and 10; cp. Fox 1989). In addition, and more specifically, Robert Trivers' (1985) reciprocal altruism theory, which claims that individuals may donate resources to nonkin if equivalent aid is returned in the future, provides an interesting complementary perspective on patterns of exchange as analysed in detail by Maussian and other ethnographers.

Conclusion

Mauss' and the Maussians' stimulating views of exchange can be put to good use in confrontation and concurrence with recent biological insights such as the aforementioned. To some extent this goes beyond what Mauss intended, but it remains faithful to his heuristic principle of a 'totalising' approach which must take the natural into account too, against the grain of *homo duplex* approaches. It is worthwhile to try and bring *homo symbolicus*,

constitutive of much of cultural anthropology as a discipline, back down to earth, to nature, by bringing the richness of evolutionary biology to bear upon the idea of man the symbolic and cultural animal. Such approaches as dual inheritance theory, behavioural socioecology, and evolutionary psychology are 'total' in Mauss' sense and well in tune with the traditional holistic intention of anthropology, while approaching cultural symbolism as a biological phenomenon. They can help in taking human culture, sociality and society not too one-sidedly as a predominantly Darwinian instinctual order, nor as an exclusively Durkheimian normative order, but as a complex and subtle interaction of both.

The Maussian paradigm, valuable though it is, is flawed by a too radically dualistic view of man, and stands in the need not so much of being overthrown, but of being rethought and updated. While for Durkheim and Mauss altruism meant the cultural suppression of selfish instincts, for evolutionary biology it is precisely the opposite: the expression of such – altruistic, but on another level of analysis selfish – instincts. More can be learned by asking how the symbolic behaviour, altruism and Maussian gifts that make human societies and identities possible may be *rooted* in nature than by asserting that they constitute the difference that *sets humans apart from* nature (cp. Quiatt and Reynolds 1993: 265).

REFERENCES

- Barraud, C., de Coppet, D., Iteanu, A. and Jamous, R. 1994. Of Relations and the Dead: *Four Societies Viewed from the Angle of Their Exchanges.* Oxford: Berg Publishers.
- Betzig, L., Borgerhoff Mulder, M. and Turke, P. (eds.) 1988. *Evolutionary Biology and Human Social Behaviour: An Anthropological Perspective.* Cambridge: Cambridge University Press.
- Borgerhoff Mulder, M. 1987. 'Progress in sociobiology'. *Anthropology Today,* 3(1), pp. 5-8.
- Carrithers, M. 1996. 'Nature and culture'. In: A. Barnard and J. Spencer (eds.): *Encyclopedia of Social and Cultural Anthropology.* London and New York: Routledge, pp. 393-96.
- Chagnon, N. 1988. 'Male Yanomami manipulations of kinship classifications of female kin for reproductive advantage'. In: L. Betzig, M. Borgerhoff Mulder and P. Turke (eds.): *Evolutionary Biology and Human Social Behaviour: An Anthropological Perspective.* Cambridge: Cambridge University Press, pp. 23-48.
- Corbey, R. 1989. *Wildheid en Beschaving: De Europese verbeelding van Afrika.* Baarn: Ambo.

- Corbey, R. 1993. 'Freud et le sauvage'. In: C. Blanckaert (ed.): *Des Sciences contre L'homme II: Au Nom du Bien*. Paris: Editions Autrement, pp. 83-103.
- Cosmides, L. and Tooby, J. 1992. 'Cognitive adaptations for social exchange'. In: J.H. Barker, L. Cosmides and J. Tooby (eds.): *The Adapted Mind: Evolutionary Psychology and the Generation of Culture*. New York and Oxford: Oxford University Press, pp. 163-228.
- Denault, L. and McFarlane, D.A. 1995. 'Reciprocal altruism between male vampire bats, *Desmodus rotundus*'. *Animal Behaviour*, 49, pp. 855-56.
- Dennen, J.M.G. van der 1995. *The Origin of War: The Evolution of a Male-Coalitional Reproductive Strategy*. Vol. I and II. Ph.D. dissertation, Groningen University, Groningen.
- Ducros, A., Ducros, J. and Joulian, F. 1998. *La Nature est-elle Naturelle? Histoire, Épistémologie et Applications Récentes du Concept de Culture*, Paris: Éditions Errance.
- Durkheim, E. 1960. 'The dualism of human nature and its social conditions'. In: K.H. Wolff (ed.): *Emile Durkheim, 1858-1917*. Columbus: Ohio State University Press, pp. 325-40.
- Fox, R. 1989. *The Search for Society: Quest for a Biosocial Science and Morality*. New Brunswick and London: Rutgers University Press.
- Fox, R. 1991. 'Comment (on Rodseth et al. 1991)'. *Current Anthropology*, 32(3), pp. 242-43.
- Freeman, D. 1984. *Margaret Mead and Samoa: The Making and Unmaking of an Anthropological Myth*. Harmondsworth: Penguin Books.
- Geirnaert-Martin, D. 1992. *The Woven Land of Laboya: Socio-cosmic Ideas and Values in West Sumba, Eastern Indonesia*. Leiden: CNWS Publications.
- Hobbes, T. 1972(1651). *Leviathan*. Harmondsworth: Penguin.
- Lévi-Strauss, C. 1956. 'The family'. In: H.L. Shapiro (ed.): *Man, Culture and Society*. New York: Oxford University Press, pp. 261-85.
- Lévi-Strauss, C. 1969. *The Elementary Structures of Kinship*. J.H. Bell., J.R. von Sturmer and R. Needham (trans. and eds.). London: Eyre and Spottiswoode.
- Lizot, J. 1994. 'Words in the night: The ceremonial dialogue – One expression of peaceful relations among the Yanomami'. In: L.E. Sponsel and T. Gregor (eds.): *The Anthropology of Peace and Nonviolence*. Boulder: Lynne Rienner, pp. 231-240.
- Mauss, M. 1990. *The Gift: The Form and Reason for Exchange in Archaic Societies*. W.D. Halls (trans.). London: Routledge.
- Mauss, M. 1995. *Sociologie et Anthropologie*. Paris: PUF.
- Platenkamp, J.D.M. 1988. *Tobelo: Ideas and Values of a North Moluccan Society*. Ph.D. dissertation, Leiden University, Leiden.
- Quiatt, D. and Reynolds, V. 1993. *Primate Behaviour: Information, Social Knowledge, and the Evolution of Culture*. Cambridge: Cambridge University Press.
- Rapport, N. 1996. 'Individualism'. In: A. Barnard and J. Spencer (eds.): *Encyclopedia of Social and Cultural Anthropology*. London and New York: Routledge, pp. 298-302.
- Rodseth, L., Wrangham, R., Harrigan, A. and Smuts, B. 1991. 'The human community as a primate society'. *Current Anthropology* 32(3), pp. 221-54.
- Sahlins, M. 1972. 'The spirit of the gift'. In: *Stone Age Economics*. New York: Aldine de Gruyter, pp. 149-83.
- Sahlins, M. 1996. 'The sadness of sweetness: The native anthropology of western cosmology'. *Current Anthropology*, 37(3), pp. 395-428.
- Sponsel, L.E. 1996. 'Peace and nonviolence'. In: D. Levinson and M. Ember (eds.): *Encyclopedia of Cultural Anthropology*. New York: Henry Holt, pp. 908-12.
- Sponsel, L.E. and Gregor, T. (eds.) 1994. *The Anthropology of Peace and Nonviolence*. Boulder: Lynne Rienner.
- Trivers, R. 1985. *Social Evolution*. Menlo Park: Benjamin/Cummings.
- Waal, F. de 1989. *Peace-making among Primates*. Cambridge (Mass.): Harvard University Press.
- Waal, F. de 1996. *Good Natured: The Origins of Right and Wrong in Humans and Other Animals*. Cambridge (Mass.): Harvard University Press.
- Weiner, A. 1992. *Inalienable Possessions: The Paradox of Keeping-While-Giving*. Berkeley, Los Angeles and Oxford: University of California Press.
- Wrangham, R. and Peterson, D. 1996. *Demonic Males: Apes and the Origin of Human Violence*. Boston: Houghton Mifflin.

Aspects of War and Warfare
in Western Philosophy and History

Of the gods we believe, and of men we know, that by a necessary law of their nature they rule wherever they can. And it is not as if we were the first to make this law, or to act upon it when made; we found it existing before us, and shall leave it to exist forever after us...
(Thucydides V: 105, 2)

/4

DAVID WARBURTON

Western philosophy of war: 1000 BC-1900 AD

In the Melian dialogue, Thucydides (460?-395? BC), one of the first European observers to comment on the character of war, has the Athenians state the laws of international relations. Obeying this law in the 5th century BC, the Athenians pursued a war of conquest in the Aegean. The Melian dialogue ended when the Athenians massacred the adult men and sold the women and children into slavery; the war ended when the Spartans destroyed the Athenian fleet.

In the 19th century AD the Prussian general Carl von Clausewitz (1780-1831 AD) viewed war as an act of political violence, with the object of imposing one's will on the enemy (Clausewitz 1991; 1989). Political organisation, goals, territorial borders, strategic method, hostility and violence were basic to warfare as understood by Thucydides and Clausewitz. Of equal significance for Clausewitz was the concept of reciprocity: that both parties observed the same reciprocal and symmetrical attitude to war. The limits on violence were the character of the political goals and the activity of the enemy. The political goal of compelling the enemy to obey one's will was fundamental. The method was the utmost use of violence; the clash of armies.

Clausewitz realised that this ideal form of warfare was never achieved. Among the principal reasons given in the published form of the book were the importance of intelligence, logistics, friction and luck. Obviously, these had an impact, diminishing the pure violence in warfare. Clausewitz was also conscious of other forms of warfare, and the 'limited war' in particular, where the goal was not total victory and thus the total application of force was not required. This did not play a major role in the book as published. Before his death, however, Clausewitz had begun to develop a dual approach to the analysis of war, assuming that it would be possible to view two types of war, one with the object of total victory, and the other with the object of limited conquests. He intended to revise the entire work, taking account of this dichotomy, distinguishing 'total' and 'limited' war (Clausewitz 1991: 179).

Thucydides (V: 89) has the Athenians state that 'the strong do what they can and the weak suffer what they must.' As Clausewitz' book stands, it formulates the principles this implied, and Thucydides' categorisation of war as the utmost use of force for the pursuit of political goals was his testament to his heirs, as the European philosophy of war for more than two millennia.

Introduction

Archaeologists and anthropologists are perfectly aware that Clausewitz' assumptions about political goals, territorial frontiers, reciprocal responses, etc. may not characterise all activity classified as 'warfare', whereas Thucydides' observation about the 'strong' and the 'weak' would appear to be so accurate that it is virtually a tautology.

In neither approach is there any implication of 'morality' or 'social responsibility', whereas 'warfare' in the West today is understood as being a 'moral' activity. This remains true whether or not the observer is opposed to the use of war, or in favour of it: it is invariably associated with some 'higher purpose' and not just 'interest' and 'policy'. At the same time, 'warfare' is associated with 'violence', and particularly the use of force for achieving goals without recourse to reason or negotiation.

One of the central assumptions implicit in the current argument is that the origins of warfare can lie in inter-communal violence, but that this violence itself is not necessarily warfare, i.e., the 'origins' and 'character' of warfare are two separate issues. The use of violence in human communities doubtless preceded the appearance of states and the pursuit of power and political interest. In this chapter I suggest that inter-communal violence be distinguished from 'warfare', and argue that the distinction be made by distinguishing the 'political' character of warfare.

'Warfare' as treated in the current chapter is essentially that reflected in the last five millennia of recorded history, as this chapter is devoted to the changing character of warfare in Western philosophy, and not to the origins of warfare. This chapter aims to explore the development of the Western understanding of war, and how 'policy', 'interest', 'morality', 'justice', 'violence' and the 'state' came to play the roles that they do, as well as the impact that this has had on the conduct of war.

Origins and development of European concepts of warfare

Although among the first to analyse warfare, Thucydides was not the first European known to have recorded it. According to the poet Homer, warfare involved honour, revenge and bloody carnage. Fear, anger, courage, pauses, and divine intervention play an important role in his narrative, while goals, policy and absolute force do not. Divine behaviour – in their own world and among humans – could be as arbitrary as the activity of the humans.

The political purpose of Homeric warfare is not evident. The utmost use of force is avoided. Nor would the war appear to have involved a policy of territorial expansion. Although plunder did play a role, it is not clear that the concept of calculated gain underlies either the goal or the execution. Homeric warfare would, therefore, appear to defy the criteria set by Clausewitz. The interruptions in the Trojan War – due to pride etc. – would also appear to be quite foreign to Clausewitz. With his pithy description, however, Thucydides would appear to have grasped the character of war, as applied both in his own time – for imperial expansion – and in Homer's (where it was apparently a form of amusement).

Those European philosophers who approached the subject of warfare from the Middle Ages onwards had access to the works of Thucydides and Homer, as well as Arrian, Polybius, Livy, and Caesar. Like Thucydides and Clausewitz, none viewed war as a moral activity. For them, warfare was a communal activity involving violence with the object of subjecting the weaker to the will of the stronger. Courage and deception were equally valuable in contributing to victory. Neglect of ritual observances was mere pretext – or ruse. Divine guidance was only required to aid in strategy; victory was herself a goddess. This is the character of war as understood by Europeans as the Roman Empire began its gradual decline.

The situation was quite different when Europe began its gradual re-awakening in the Middle Ages. In contrast to Clausewitz, St. Thomas Aquinas insisted on the absolute minimum use of force, even in war. Aquinas advocated a 'just war', declared by a monarch and pursued with a right intention in support of a just cause. Aquinas' concept of the 'just war' included three principles, (a) a declaration, (b) a just cause and (c) a right intention. The declaration should be produced by a monarch and self-defence was among the better 'causes'. In discussing Aquinas' doctrine of the 'just war', Sigmund (1997: 227) suggested that precedents could be found in Augustine and Cicero. It is striking that Aristotle – who could usually be relied upon to turn up as a reference for almost everything in medieval philosophy – signally failed Aquinas here. Aquinas's concept was quite outside European tradition. His concept of war was expressed in terms of monarchy, but this alone did not suffice to guarantee that war was just, even if the monarch was the divinely selected administrator of the commonwealth. Aquinas insisted on a just cause and a right intention as well.

Precedents for details can be found in Augustine's justification for the use of violence by Christians. St. Augustine assumed that a 'just war' was possible; assuming that war carried out in obedience to the Christian God would be a 'just war'. This allowed him to absolve Moses of making war, for Moses could hardly be expected to refuse a command of the Lord (Paolucci 1962: 170). Defining the state as 'a multitude of men bound together by some bond of accord', and assuming that Christianity could provide the spirit of concord, Augustine had a doctrine whereby the Christian state could wage a just war with the support of God. This allowed ample precedent for a theory of just war as Aquinas understood it.

Rousseau to Hegel

As is to be expected of the translator of Thucydides, Thomas Hobbes viewed civil society as the only means of protecting the weak from the strong. Hobbes assumed that 'the life of man was nasty, brutish and short'. Jean-Jacques Rousseau (Hoffmann and Fidler 1991: 1-100) took issue with Hobbes quite strongly, (a) seeing no reason why he should believe that people were as greedy as Hobbes assumed, and (b) assuming that people only came into armed conflict under duress. Rousseau assumed that the state was not confined by the natural boundaries set by nature. Instead, the state was bounded only by other states and a state of war would exist between them as the natural form of frontier separating artificial entities without natural boundaries. He assumed that existence gave the state power to coerce individuals into warfare. Whether he viewed this as a violation of either their natural rights or inclinations is immaterial. Given the various limits imposed on the state by its size and the might of its neighbours, Rousseau assumed it would be impossible for a European monarch to assemble an army large enough to defeat all of the other powers. He therefore concluded that world conquest was impossible.

While denying a natural tendency towards violence and greed, and recognising the concept of a legitimate war which was just, Rousseau also appreciated that avarice and selfishness could lead a democracy to wage an unjust war. Rousseau not only conceded the existence of avarice, but that it could undermine the state, and therefore he appealed to patriotism. While dismissing Hobbes' view of human nature, Rousseau recognised that the behaviour of states corresponded to the behaviour of the individual posited by Hobbes.

Rousseau proposed an alternative role for the state itself, assuming that truth and justice guide the state, assuring freedom to its citizens. Rousseau recognised the right to self-defence and revenge. Ultimately, the idea that service to the state could be construed as morally correct depended upon self-lessness. Based on private property, Rousseau assumed that the state could guarantee the rights of citizens by providing regular income from public funds for its magistrates. He contended that the 'triumph of private over public good' could be marginalised. Arguing that Hobbes' state of nature did not correspond to the reality of mankind's social bonds, Rousseau assumed that deficiencies were due to injustice, rather than flaws in human nature.

For Rousseau, therefore, the state can be transformed from an instrument of despotic oppression into a public corporation and an instrument of good, which can wage just wars. This was not, however, the natural state of the state, merely an ideal. The natural state of the state was to oppress and to wage wars, bounded only by the powers of other states, largely as Thucydides had perceived it.

While likewise arguing that humans need not be completely bestial, Kant also associated violence and unruliness with pre-state and non-state societies. However, Immanuel Kant joined his predecessors in making the heads of states responsible for war, thus linking war to the state. He complemented his ambivalent view of human nature with the observation that despite the allegedly evil nature of the human race, statesmen invariably attempted to accompany their declarations of war with legal and moral arguments. To this realistic assessment – that humans are violent and states wage war – Kant advocated an agreement among states that could lead to 'perpetual peace' (Kant 1917). The agreement of states would guarantee its success, since states alone caused war. In this seemingly Hegelian dialectic, Kant, like Rousseau, was able to transform an instrument of oppression and war into an instrument of peace.

G.W.F. Hegel (e.g. Hegel 1999a; 1999b) viewed Kant's project of 'perpetual peace' with disdain. Hegel assumed that states would resolve conflict through war, and that any possible peace would depend upon the consent of the governments of the states and that this was not a reliable basis. Whereas Hobbes, Rousseau and Kant viewed the state as a convenience, Hegel viewed it as the end of history. Through reason and science, Hegel argued that the states of the Germanic peoples would allow the state itself to take its rightful place as a manifestation of the divine, alongside the natural and ideal worlds.

For Hegel, the basic principles of the world were 'spirit' and 'power'. The state was the substantial form of both 'freedom' and 'will'. Hegel viewed the state as the 'unification of the principle of the family and civil society'. While recognising that 'the state is divine will', Hegel specifically refuted the principle that the state was based on religion. Religion was the sphere of the absolute truth, and 'state, laws and obligations' corresponded to the ideal norms, but not any specific form of religion providing a foundation for these norms. 'State' and 'religion' were thus different forms of divine will. Hegel assumed that a nation without a monarch would be a formless mass and no longer a state, which allowed him to identify the 'person of the monarch' as the 'sovereignty of the state'. Legitimacy came through birth.

Given its absolute worldly moral authority, the state was the source of the rights and obligations of the individual, meaning that the individual was subordinated to the state. The constitution of the state was the codification of the conscious substantial realisation of the will of civil society. The state could thus expect and demand taxes on property and military service. In return for guaranteeing property and protection, the state could abolish the individual's right to property, freedom and life.

While assigning the state absolute power, Hegel assumed that war 'was not an absolute evil' but rather a 'necessity' which should be viewed as 'incidental' and inevitable in the concert of nations. The state was the absolute power on earth and depended upon recognition for its existence, and it was therefore obliged to assure its recognition. Conflict between states was resolved by war, which threatened the state's independence. The two forms of state merged: the internal role meant the state was assured the support of citizens, while the external role demanded that it be recognised by other states.

The obligation to perform military service in the standing army of the state was no different from the necessity of observing civil or commercial laws concerning marriage or business. The state had the right to depend upon people to defend its borders, and successful application of force could transform a defensive war into a war of conquest.

Thus, while Rousseau assumed that humans were peaceful by inclination and that the state was the source of evil, Hegel viewed the state as the ideal combination of power and spirit, assigning it a higher moral power. Rousseau perceived that the power of the state lay in its people; Hegel assumed that the power of the state lay in its divinity. In either case, the state was given the highest moral position and responsibility for warfare: in the one case in violation of alleged human nature, in the other case as part of the state's moral responsibility to protect the citizens of a civil society.

War as policy and the just war

For diametrically opposing reasons, Rousseau and Hegel viewed war and the state as a moral issue related to property and rights, whereas Kant did not. Although Kant's intent was to establish 'perpetual peace', it effectively opened the way for Clausewitz' understanding of war as a political instrument employed in the pursuit of power, quite separate from moral values or material greed. This revived the notion of war familiar from Thucydides.

The moral element has several different sources. Rousseau's and Hegel's concerns for justice were part of a European heritage. Aquinas' use of morals to justify war represents a different tradition, one to which we will turn in an instant, one which can be traced back to the Ancient Near East, through the Hebrew *Bible*, and not Plato's *Republic*.

Although superficially similar to Aquinas's concept of monarchy, Hegel's view of monarchy lay not in the medieval world, but that of the 19th century nation-state. For Augustine, the Christian state was not a concept but a real political possibility, and thus the unity of 'justice' and the 'state' went hand in hand. The feudal order brought about the end of the unitary Christian state, but not the unity of the Christian world. For Aquinas, aristocracy and church were legitimate constraints on the absolute power of the monarch, while the importance of belief and adherence to the Catholic faith was absolute.

For Hegel, the church and the aristocracy were not recognised limits and the importance of any particular religious faith was of no import. Hegel's state and Hegel's monarch were thus powers in themselves, not different from Augustine's.

By combining Hegel's state with the ancient doctrine of power and eschewing morality and justice and any restrictions on the use of force in the interest of state policy, Clausewitz had opened the way to pure military power as pure political power. By maintaining Aquinas's concept of the just war while rejecting the limits he imposed on the use of force and adopting Clausewitz' doctrine of absolute force, by the end of the 20th century AD, warfare had become an absolute, involving the absolute use of force and absolute justice.

It is conceivable that there was a transformation in European warfare between the heroic futility of Homer and the imperial arrogance of Athens, but Thucydides was correct in letting the Athenians state that they found their laws of domination existing when they began their program of conquest (see above).

Aquinas' theory of the 'just war' can be genetically traced back through Augustine's, but not to Thucydides. Although ostensibly citing Cicero and the New Testament, Augustine was elaborating a far older tradition in which religion and state were identified as one, with the 'just war' sanctioned by a supreme national deity. Ultimately, Aquinas' just war can only be understood when traced through Augustine, Jesus, the Old Testament, the Egyptians, and the Assyrians. Wars in this age were divinely inspired and merely executed by monarchs, thereby differing from Aquinas' which were only just when the conditions were met. In antiquity, gods and men sought domination: strong and weak were divided by defeat, whether gods or men; justice was divine, correct and absolute, not based on equity.

The origins of western war: the Near East and the Mediterranean

Virtually all philosophers build their theories on assumptions about the origin of war. Despite an abundance of archaeological evidence, it remains idle to speculate about war among early human communities, and we can leap through early prehistory. The end of the Palaeolithic in the Near East was a long affair, lasting from ca. 20,000 B.P. to ca. 10,000 B.P. during which individual groups settled the mountains of the Levant before the development of agriculture. The sedentary life preceded – but led to – food production and the Neolithic, hesitantly beginning from ca. 12,000 B.P.

Arrowheads are found in a number of human skeletons buried together in Egypt around 10,000 BC, with several projectile points in each of several skeletons (Hendrickx and Vermeersch 2000: 30). At Wadi Hammeh 27, in contemporary Palestine, 'many burnt skull fragments' lay among the settlement debris. A 'skull with atlas and axis' lay on a living floor beside a hearth at Mallaha, likewise in Palestine. A cranial cap 'cut off, apparently deliberately' lay on another floor at the same site (Valla 1998: 176). Inter-communal violence was increasing before the Neolithic.

The abundance of arrowheads in the Neolithic of the Levant is such that they are used to distinguish chronological and social boundaries (Gopher 1994). Examining the defensive settlement structures and the typology of the period, the terms 'raiding', 'conquest' and 'warfare' were used in a discussion of the late Levantine Pre-Pottery Neolithic (Eichmann, Gebel and Simmons 1997). Cauvin (2000: 126) dismissed the evidence of early warfare, remarking that 'One arrow-head retrieved here or there stuck in a human vertebra is perhaps not as adequate proof as has been claimed'.

The logic of such an argument escapes the current author, as does Cauvin's (2000: 126) suggestion that arrowheads had a 'symbolic' significance, while disputing that arrowheads had any specific meaning. If arrowheads were 'symbolic' they must have had a 'meaning' to the farmers of the early Neolithic: elsewhere arrowheads are frequently found in human bones. In the literate societies of the Bronze and Iron Ages, armour-piercing arrowheads played a political role. The wide-spread 'skull cults' of the late Levantine Pre-Pottery Neolithic indicate that death was a primary concern in this age. In the Near East, the evidence is still coming in, and some argue that the evidence supports an interpretation in terms of inter-communal warfare, even if Cauvin rejects the interpretation.

Virtually from their appearance near the end of the Upper Palaeolithic, arrowheads have left their traces in human bones (Bachechi et al. 1997). It is possible to assume that the arrowheads of the Neolithic represent an explosion of improved hunting technology. However, this would imply that substantial effort was going into hunting technology at precisely the moment when the greater part of subsistence concerns were dedicated to transforming economic life from hunting and gathering to farming. Such an argument would reverse the interpretation of the economic transformation in the Near Eastern Neolithic, which is considered to be among the most significant in human history.

Not only is it invalid to claim that arrowheads are exclusively hunting tools, but one can plausibly link their invention and development to inter-communal violence: (a) arrowheads are the typical typological artefact of early sedentary societies and (b) they are found in human bones. Other evidence of conflict is not absent. Although unparalleled for another three millennia, the walls of Jericho were erected at the very start of the Neolithic, possibly suggesting a defensive purpose. Certainly, the tactical location of subsequent Neolithic villages suggests that villagers were conscious of security (Gebel and Bienert 1997).

If warfare is communal violence involving the deliberate death of humans, then this would indicate the beginnings of warfare. The symbolic use of arrowheads can be linked to the social consciousness and recognition of their role. The context would be simple – but fatal – conflict between competing groups. It would be difficult to allege that disputes over land or property could antedate the era of significant food production, and yet this lethal inter-communal violence clearly antedates the era when property and land became significant. The gradual development of projectile points can be linked to earliest sedentary communities in Palestine, which can be dated to the ten thousand years preceding the Neolithic revolution. The Neolithic was the culmination of a development whereby the movement of individual hunting communities was increasingly limited by the presence of neighbouring groups (Goring-Morris 1998; Valla 1998). The development of arrowheads accelerated during the era when these human groups were living in close proximity, in carefully protected settlements.

For the following six millennia, settlement was determined by two main axes. On the one hand, people moved out of the mountains and the fringes of the desert into the plains (Matthews 2000: 12, 30, 42, 56). At the same time, more groups took up the sedentary way of life and farming. This early sedentary activity culminated in a move into the plains of

southern Mesopotamia. Settlement density in the plains increased between ca. 6000 and 3000 BC; the first states had emerged by the end of this period. This new settlement pattern was decisive, for the most successful states emerged in the plains: along the Nile Valley and in Mesopotamia.

Walled settlements appeared in Mesopotamia (e.g. Tell es-Sawwan) and Anatolia (e.g. Hacilar) before the fifth millennium. By 3500 BC, small settlements were scattered from the Upper Nile Valley across the Levant and Anatolia, the plains of Syria, and as far as the mountains of Afghanistan; aside from the growing cities of southern Mesopotamia. Around the middle of the fourth millennium, an expansion from the urban centres of Mesopotamia ended the evolution of the small pre-state settlements in Syria and Anatolia. These settlements participated in trade networks, but eventually the evolution was interrupted as the settlements contracted at the end of the fourth millennium. The disturbance was partly economic and partly political. Regardless of its character, Anatolia was cut off from developments for almost a millennium, and its progress curtailed again as the recovery was again disturbed at the end of the third millennium BC (For references to the Near East one can consult, e.g. Warburton 2001, and the relevant articles in Sasson 1995, and the Oxford Encyclopaedias of Egypt and the Near East).

Throughout this entire period, cities and city walls grew apace. From 3000 BC onwards, arrowheads, sling bullets, maces, daggers, spears, axes, and other tools of war became more abundant and widespread. By 2200 BC sieges were a common activity: siege ladders and wagon-trains with logistical supplies accompanied armies. Egypt was in conflict with peoples in Palestine and Nubia; the Mesopotamian states were in conflict with powers in Syria, Anatolia, Iran and the Gulf. The expansion of the Mesopotamian system of city-states and the Egyptian Empire in the Nile represented two alternative modes of political development, but the two did not come into direct conflict in the third millennium. Most of the political conflicts in the third millennium were between small state entities, or between the major powers and the settlements on their periphery, as these were incorporated into their spheres of power. The unprotected settlements scattered across the Near East were exposed to the power of the centres, just as the centres were exposed to conquest by competing entities. The most significant military change was the wide-spread destruction following the conquests of the kings of the city state of Agade. They managed to subdue or destroy most of the powers in the Near East, from Syria to Iran and the Gulf but were unable to assemble a coherent state from the pieces. The collapse of the empire led to a fragmentation of power throughout the Near East.

The second millennium was characterised by widespread warfare. In southern Mesopotamia at the end of the third millennium, power fell into the hands of several city-states before the city of Ur was able to assert its pre-eminence. Nevertheless, the kings of the third dynasty of Ur were in a virtually constant state of warfare with their neighbours on all sides. This unstable situation meant that weakness would spell the end of their power, and thus the dynasty was short-lived, lasting only slightly longer than a century. Initially, political power fell to the region straddling the trade routes between Iran and Iraq. Eventually, however, the Assyrian king Shamshi-Adad cut out a large empire in northern Mesopotamia based on the fertile plains of the North. This broke up with his death, opening the way for the expansion of another kingdom based on the south Mesopotamian alluvium, under Hammurabi and his successors in Babylon. After Shamshi-Adad's death, Syria broke free of Mesopotamian influence; the cities of Palestine fought each other and achieved a temporary hegemony in the Egyptian Delta; Babylon found itself in constant conflict with Iran. The Indo-Europeans moved into the Aegean, Anatolia and the Indus Region. Local and regional conflicts characterised political activity during the second millennium BC.

Having consolidated their hold on Anatolia, the Indo-European Hittites were able to expel the Assyrians, eventually sweeping across the cities of northern Syria and destroying Babylon. This opened up a new power vacuum; power in southern Mesopotamia fell to a new dynasty without expansionist inclinations; power in the North fell to another Indo-European kingdom, that of Mitanni. This opened the way for a competition between the Assyrians and Mitanni in the North, because the Hittites were unable to consolidate their hold on the North in the aftermath of their conquests. Ultimately, however, the Egyptians placed pressure on the southern reaches of the Mitanni holdings

while Assyria pressed it from the East and the Hittites from the North.

The Hittites were able to destroy Mitanni, but then faced Egypt and Assyria in the South and East. They were also under pressure from the West after the Mycenaean conquest of Crete and the expansion to the Aegean coasts of Anatolia. However, the Mycenaeans and the Hittites were eliminated as political powers. Egypt was forced out of Asia and the Assyrian expansion into Syria was halted.

For an era at the end of the Bronze Age (from around 1200 to perhaps 900 BC), non-state peoples brought about the end of large scale political units. The original epics which ultimately became the Homeric poems were composed during this era, and reflect such violence. It is impossible to state with certainty that those who waged these wars against the Egyptians and Hittites were not organised as states, but it can be confirmed that their victory resulted in the disappearance of imperial power in parts of the Near East. In some cases, small states appeared in the aftermath of the collapse, but it is not certain that the beneficiaries of the power vacuum were the authors of the collapse. In any case, the Near East and the Aegean were reduced to small-scale political units, each vying with the other in small-scale warfare far removed from the major imperial conflicts which characterised the Bronze Age of the second millennium and the Iron Age Empires of the first.

The Assyrians ultimately recovered their power in Syria, but the expansion was preceded by a significant contraction. For an age, Syria and the Aegean were dominated by small states with the Assyrians driven back to the region along the present Syro-Iraqi border. Eventually, the Assyrians subdued the smaller states until their empire reached from the Nile to Anatolia and Iran. The first millennium BC was characterised by the Assyrian expansion, and the inheritance of their empire by the Babylonians, Persians and Macedonians.

The ideological underpinnings of these conflicts are easily grasped. The pretext for the Egyptian expansion into Asia was a desire for revenge, as the Egyptians clearly stated that they held the leaders of Palestine responsible for the earlier conquest of Egypt by the Hyksos. The Assyrian conquest of Babylonia can be traced back to feelings of inferiority, and the expansion eastwards spurred on the Medes, who aided the Babylonians in defeating Assyria. Since the earliest settled communities, eradication of enemies and territorial expansion leading to political control made conflict endemic as each expansionist move triggered a response.

From an ideological standpoint – even when the opponents were political entities ostensibly outside the Egyptian sphere of influence – the Egyptians viewed opposition to their rule as rebellion. The Assyrians simply viewed it as unwise. For both the Egyptians and the Assyrians, their expansion can be linked to divine sanction, since both assumed that the gods gave victory in war and that failure was tantamount to a failure to satisfy the gods. Like the Ancient Romans and the Chinese, the Babylonian and Assyrian kings consulted oracles to determine whether the outcome of war might be favourable. For the Babylonians this was of the utmost importance. It is clear that the decision to wage the war was thus explicitly political, and the god was simply consulted to suggest whether the moment was propitious. Despite some similarities, there is an enormous difference between a war waged in the name of a god – such as Assur or Amun-Re – and a politically motivated war which a god then approves. The concept of partial divine sanctions for war can therefore be found in the Babylonian material. However, Babylonian campaigns were frequently anchored in a clear political program, 'omens' serving merely as guides.

Influences on western philosophy: Homer, the Bible and history

Homer emerged on the periphery of the Bronze Age empires, but most modern Europeans dismiss Homer's romantic view of war. Even when reading Homer, they assume that the war must have had some rational purpose, such as plunder, if no other. Following Thucydides, modern observers tend to assume a practical – state and power oriented – form of warfare. However, basic to both Homer and Thucydides was that justice cannot possibly be included as a motive, purpose or feature of warfare.

This makes it incomprehensible that one of the first post-classical European philosophers – Aquinas – viewed warfare as a moral activity. Two aspects contributed to this: (a) the absence of a state system in Europe after the fall of the Roman Empire meant that 'political goals' in the ordinary sense of Clausewitz and Thucydides were impossible; (b) the

second was due to events some two thousand years earlier.

We noted that the start of the Iron Age in the Near East was marked by small states. The eclipse of Assyrian power between 1100 and 800 BC can be dismissed as an interlude in the cyclical conquests of the Near East. However, Western philosophy of war would be decisively formed by the fact that among those small states were Israel and Greece. Their epic struggles with the Assyrians and Babylonians, and the Persians, respectively, were immortalised for the West, in history and philosophy.

The poets and prophets celebrated contests of wills couched in terms of their own society. Neither made sense in Clausewitz' definition of war, or even Thucydides'. The tale of the Trojan War is senseless violence; divine intervention has no more purpose than the violence itself. By contrast, the tale of the Old Testament is that of national salvation through purgatory; divine intervention is the national god punishing his own people.

Both contrast greatly with Assur: the name of the god, the city, the country and the Empire were all the same; conquest ordained by Assur was absolute. The mirror image of this was the concept that the gods could use foreign enemies to destroy those who failed to comply with their wishes. In these cases, cities feared that their gods would abandon them, and abandoned their gods if they failed them. Jahweh was an exception, for the god of Israel used the gods of other lands to execute his will against his own people.

For the Assyrians and the Egyptians – the two most important neighbours of the Hebrews – war was part of national policy. Both viewed war as a divinely driven instrument of territorial expansion. More important than mere 'territorial expansion' is the role of 'justice', which is firmly tied to the state and the rulers and the gods, rather than the individual, from the beginning of Ancient Near Eastern history. The expansion of the Egyptians and Assyrians is explicitly territorial, expressed in terms of extending boundaries. The role of the state itself was, however, ideological, expressed in terms of justice. The link between the state, warfare and justice was thus woven into the fabric of the Ancient Near East. This legacy was then integrated into Hebrew traditions.

Israel was an ephemeral peripheral player, quashed by the Assyrians, Babylonians, and Romans. Israel could not play a major role, and thus the god of Israel could only act by exploiting the armies of adjoining powers. Thus, whereas the Assyrians owed their victories to Assur, the Prophets explained that Jahweh was using the Assyrians to punish Israel. This would be of no significance, had not the West adopted Christianity.

Christianity meant that Western philosophy adopted ancient Near Eastern notions, including warfare guided by justice and divine will. This was absent from ancient Greece, where 'Justice' was but one god among many. The Hebrew prophets viewed warfare as punishment ordained by god; only god or piety could ward off defeat. Written in the aftermath of the Assyrian and Babylonian conquests, the *Bible* brought ancient Near Eastern concepts of warfare to the West, a West which was forsaking Thucydides and Homer.

The ancient Near Eastern concepts would have been lost to us, but for their preservation in the *Bible*, even if in modified form. Because of the *Bible*, the Assyrian concept of warfare is quite easily recognisable for us. Two opposing concepts provided the fuel for the debate about the purpose and origins of war.

On the one hand was the Christian Church spreading the Biblical version of world history while simultaneously preventing the emergence of a political entity which could be independent of the Church. On the other were those political entities which did emerge, and which were dependent upon feudal arrangements with vassals. Emperors, kings, vassals, knights and guilds represented one set of relations, while popes, bishops, the clergy, the monasteries and the monks represented an alternative set. Vassals vied with kings, kings with popes. Land tenure was an issue for both, church and vassal alike. The peasants were at the mercy of all.

Ultimately, the nation-states of Europe appropriated church lands while individual land-owners increased their holdings. For Europe land tenure was a bone of contention, contributing to the central importance of territorial boundaries in the wars of the 18th century, and to the concept of private property. Territory and land tenure played a central role for Clausewitz and Rousseau.

Before the dissolution of the feudal order, the Church was of central significance. Pre-Christian warfare was based on strength, honour, power and the state. The Hebrew *Bible* introduced war as an instrument of divine retribution, meaning the defeat of the state of Israel by its own god. Although

it was in opposition to the state (for practical and ecclesiastical reasons), Christian doctrine also emphasised weakness. Viewing the Hebrew god as a deity of universal appeal and relevance, and a god of justice, it introduced the concept of absolute justice and morality into warfare, yet tempered this with defeat and lack of purpose. This effectively eclipsed the role of the state. The weakness of the state was a virtue for the Church, and thus the doctrines matched.

Following the Reformation, power fell to the state. Despite disagreement about the legitimacy of the government and the distinctions between the government and the state, there was virtually no opposition to the concept that the state had the capacity to wage war and ensure peace. The transformation of the state from the private possession of a monarch to a popular institution transformed war from a state affair to a national affair. In the West, the concept of morality and justice remained associated with warfare, even as the state severed the links between warfare and religion. This paved the way for the concept of just wars fought by states able to call upon their citizens.

Napoleon appreciated Rousseau's observation that world conquest would depend upon assembling a sufficiently large army. Whereas Rousseau viewed this as impossible, Napoleon unleashed the patriotism of the French Revolution. He was, however, defeated as the other nations of Europe reacted in the same – reciprocal – fashion. Subsequent history has demonstrated that the 'balance of power' exists in precisely this fashion, due merely to the failure of any one actor to assemble sufficient forces to execute a design of world conquest. Rousseau's fundamental observation remains valid: the state has no natural boundaries, and existing boundaries are defined by the powers of other states.

In linking sovereignty and the state to popular will and territorial boundaries, Rousseau conceded that a democratic state could wage an unjust war. Hegel shaved off the moral aspect of war, vesting legitimacy in the king, assigning the institution of the state moral superiority as a divine institution. Hegel's state would be incapable of waging an unjust war.

While conceding the possibility that 'the Deity is the ultimate author of all government', Hume differed from Hegel in assuming that no one person could *a priori* represent divine will more than any other (cf. e.g. Aiken 1964). Authority and legitimacy could be associated with magistrates. Rousseau wanted the magistrates paid by the public purse, making them public servants. The spread of democracy and the concept that the people were the source of legitimacy eventually severed the link between the state and the divine, leaving the state responsible for waging war, for its own purposes. Gradually, moral rectitude became the prerogative of democratic states.

At the same time that the ancient teachings remained valid – that the power of the state could only be constrained by other states – the power of the state was now associated with justice and morality on an international plane. In ancient Israel, justice and morality were the domain of Jahweh. In ancient Greece, justice and morality were the domain of the citizens and the concern of the community; morality did not extend to interstate conflict. In the ancient Near East, interstate conflict was the domain of the gods, and thus – through the ancient Near East and the Old Testament – Jahweh's association with morality placed morality at the centre of inter-state conflict in medieval Christian thought. This contributed to a strain of early modern thought – through Hegel's association of the state's divine legitimacy. However, Hegel did not assume that inter-state conflict would differ from the character Thucydides and Clausewitz assigned to it: a conflict of will with political goals. In the modern West, however, the concept of democratic legitimacy and moral rectitude came to play a similar role in underpinning the state, and thus also, by extension, the behaviour of states in inter-state conflict. Rousseau had, however, also realised that a democratic state could wage an unjust war. The paradox was not clearly resolved.

Warfare would be understood and defined as Clausewitz had defined it: determined as an ordeal of reciprocal violence among states with goals defined in territorial terms. This was the case for most of the period from the 17th century through the 20th. Among others, the Swedes, Russians, French, English, and Germans came to grief in their pursuit of territorial expansion, at least partially due to the reciprocal character of war. Superficially, it was, of course, their logistical systems which were overstretched, but in practice, it was the reciprocal character of the opposition which rendered the invaders incapable of consolidating their gains in distant lands. By the end of the 20th century, most European states had abandoned any hope of territorial expansion and shared a determination to avoid the use of war as a political instrument.

The unification of Europe by conquest had failed. Consciousness of the failure to make territorial gains was enhanced by the human and material costs, and thus warfare became associated with losses. Previously, warfare had inspired hope of gain and glory. Repeated defeat and fruitless victory had left Europe weakened. Lacking hope of gain and an appetite for military renown opened the way to a more balanced view of warfare. For Europeans, warfare ceased to have a purpose, becoming synonymous with senseless destructive violence. Unification became a political project expressly designed to prevent further wars on European territory.

Many political commentators and observers began to assume that 'peace' was the object of foreign policy. Quite aside from the general public, generals (e.g. Fuller 1972) and historians (e.g. Taylor 1996) alike despaired at the growing pointlessness of war, as planned, used, conducted and ended. This was quite a different attitude than that which viewed 'warfare as the continuation of politics by other means'. Previously, the state treasury had existed almost solely to fund warfare; now warfare was viewed as an undesirable expense competing with other priorities. Warfare was a burden on – rather than the purpose of – the treasury. This economical attitude also extended to the concept of sparing lives – both those of the enemy and one's own nation. In a paradoxical reversal of values, for many – particularly in Germany – warfare itself became an 'immoral' activity. Europe had taken leave of 'European war'.

The American way of war from the mid-20th century

'European' war-making was, however, pursued by the major power across the Atlantic. 'Warfare' and 'International Relations' were viewed as one, without recourse to morality. Hoffmann's (1965) riposte to Aron's *Paix et guerre entre les nations* was simply entitled *The State of War*. Whereas Europeans viewed *peace* as *a* natural state of affairs, the USA came to view *war* as *the* natural state of affairs (for the early part, cf. e.g. Weigley 1973; the continuation is still unfolding).

Total war and the balance of power

Superficially, this appeared to represent the concept of war as state policy. It began in the ideological conflict with the Soviet Union; viewed as a struggle for global dominance, it could be analysed in the framework of territorial disputes. This particular conflict had roots in other conflicts. The ideological conflict between (a) the 'East' and (b) the 'West' pitted communism against capitalism, with the USA leading the 'Western' camp and the Soviet Union the 'Eastern' camp. This ideological conflict took place in the same global geographical and temporal context as the process of de-colonisation after the Second World War. The process of territorial expansion had led to the occupation of large parts of the world by France and England. In the course of (c) a liberation struggle, large parts of the world were freed from colonial rule. While recognising self-determination, the USA assumed the mantle of the colonial powers, opposing liberation movements cloaked as communist movements (or vice versa), and (d) respecting the territorial integrity of the newly independent states against internal and external opponents.

The war was not just ideological and conceptually global, but also truly global in geographical terms, due to the existence of American bases around the world. The establishment of foreign bases was a by-product of two separate wars. The first group were the bases created in the aftermath of the Second World War. Occupation of the vanquished followed the pattern of war, except that it merged with the global ideological conflict which followed, for the occupying forces in Germany and Japan became the front lines of the Cold War. The second group were the bases established during the Cold War, never intended as occupation, but merely as 'support'. Reformulating the colonial legacy, the USA aimed to secure the Persian Gulf from the Soviets and to oppose the advance of communism in Asia.

This could have been viewed as 'politics as usual', except for the nuance that the goals were not clear, since the character of war itself was obscured by an innovation transforming attitudes towards war. The atomic bomb, strategic bomber, ballistic missile, and the hydrogen bomb appeared against the background of the global struggle between 'East' and 'West', construed as an irresolvable ideological conflict in which (a) military war was impossible and (b) the war would be won by ideological victory. For both sides, victory was viewed in economic and political terms. The most successful system would produce the most goods and acquire the most allies among the newly independent nations.

Military war was viewed as impossible: the thought of armies battling for victory had been replaced by the image of civilisations reduced to ashes. American thought was dominated by the belief that nuclear war was the only significant war, and that this was 'unthinkable'. For Americans, warfare was 'total', yet in this case 'total' meant defeat for victor and vanquished alike. The concept of 'limited war' was restricted to scenarios of 'escalation' (Kahn 1968), which did not offer any hope. Twice, Americans had gone to war to end war. Now, terror of war appeared to be the way to end war.

Although similar to the anti-war attitude shared by many Europeans, the American one differed. In the USA, repugnance of war was not born from the ashes of war-torn Europe. In some cases rational appreciation rejected violence for political purposes as un-Christian (following Aquinas). In reality, however, it was part of an unconscious mutation: the complement of the idea that another war was inconceivable. For Europeans rejection of war was the outcome of fruitless losses, and the despair of gain.

Total 'limited' war

Americans had neither suffered occupation and defeat, nor fruitless victory. However, Americans were persuaded that 'nuclear war' was 'war', and 'unthinkable' because 'war' meant 'destruction', not victory or defeat. Policy-makers propagated the message that war was pointless and that military preparations were to prevent, not fight, war. They failed to develop the concept of war as an instrument for the defence of national interests. Instead they had developed the concept of ideological war which could not be fought. A war pitting the USA – perceived as a legacy of imperialism – against a nationalist movement was complicated by these doctrines, as well as a lack of understanding of local society. Aside from lack of purpose, the defeat in Vietnam can be linked to (a) the modern Western attitude about the sheer undesirability of war, (b) a failure to prepare for 'limited war', and (c) a failure to appreciate interests in terms of the state rather than the ideology (cf. e.g. McNamara 1995).

The Hegelian concept of protecting civil society was dominant during the Cold War, as political scientists and military strategists developed scenarios of avoiding 'unthinkable' wars. The 'preservation of the peace' became an end in itself; the 'balance of power' the only means. Modern weapons necessitated peace; imposing one's will on the enemy was unrealistic; security lay in a 'balance of terror'. Kissinger (e.g. 1969; 1995) employed diplomacy to avoid war, and even to preserve the potential foe whose existence rendered war unthinkable – until the Soviet Union unexpectedly dematerialised.

The disappearance of the foe did not, however, arrest the system of thought. For decades, ideological deadlock had accustomed Americans (a) to the thought that victory in war was meaningless, yet also simultaneously (b) to the idea that conflict between nations was ideological (rather than territorial). The global character of the conflict – which had retained the territorial element – was forgotten while savouring the ideological victory. Throughout, war had played a prominent role in thinking. The policy element of defence and territorial expansion had disappeared, replaced by the concept of ideological war. 'Limited war' was incompatible with ideological war, and with the notion that victory in war would be defeat.

Easy victory in the Cold War left the USA as the greatest power in world history, after an unexpected and bloodless ideological victory, for which neither political nor military thinking had been prepared. This was accompanied by the simultaneous easy victory in the liberation of Kuwait for which the Americans were not psychologically prepared. The justice of the American cause was not – and never had been – in doubt; America was not perceived as fighting for its interests, but for 'universal values' (economic growth, democracy, individual rights) which it wished to impose on other peoples. This attitude survived the Cold War. America struggled with unexpected victory. Relief at the ease of the victories was combined with unparalleled power and self-confidence. Works on 'just wars' and 'morality' in warfare (e.g. Walzer 1992; Johnson 1999) proliferated.

The end of the Cold War – which had effectively excluded the use of war in conflict – and the character of Saddam Hussein meant that the warless ideological conflict of the Cold War was transformed into a military struggle against 'evil', identified in terms of 'terrorism' and 'rogue states'. Conflict was defined as 'good' and 'evil'; 'with us (and good)' or 'against us (and evil)'.

The danger of war was not reduced by diminishing potential sources of conflict. Instead those opposed to American hegemony would be punished. The purpose was neither territorial expansion, nor the resolution of irreconcilable conflicts of interest. It was simply global self-defence. While 'self-defence' is viewed as legitimate, there is no way in which a nation can legitimately identify its definition of 'security' in a fashion which clearly threatens others. Such a policy can only expect a reciprocal response. Therefore, in military terms this policy is irreconcilable with the concepts developed by Thucydides and Clausewitz, while also being as far from the amoral contest celebrated by Homer as it is from Jahweh's national war of divine retribution.

Clausewitz (1991: 215) stressed that war had to end with crushing the 'will' of the vanquished, as well as his armed forces. Indeed, this was the purpose of war, not a mere detail. This implied defeat of the enemy, and occupation, otherwise the 'spirit' of the foe would revive to resume the 'reciprocal' struggle. However, the USA neither sought to defeat the enemy nor to occupy their territory. The policy opposed governments, not states and peoples. The philosophy of 'firepower in limited war' (Scales 1995) even stressed waging war with limited casualties among friend and foe alike.

The policy of global self-defence pursued narrowly defined interests, with foes and friends defined by convenience. The intervention in Kuwait expelled the army of one artificial postcolonial state from another. However, this arguably Clausewitzian territorial conflict was transformed as the USA waged war against Iraq throughout the 1990s, with the concepts of 'victory' and 'defeat' redefined. Neither Iraq nor the USA accepted the 'limited' results of 1991. The 'exit strategy' of the 'limited war' failed to provide an 'exit'. The war for the restoration of Kuwait's territorial integrity was continued to prevent Iraq from acquiring 'Weapons of Mass Destruction', and subsequently redefined as part of a 'war on terrorism', against an 'axis of evil'.

These ideological and geopolitical changes in the character of warfare (previously defined in terms of 'interests', 'territorial expansion', etc.) must be set beside the unconscious transformation in attitudes towards battlefield victory which had prevailed on either side of the Cold War. Before the Cold War, battlefield victory was viewed as a matter of chance. During the Cold War, battlefield victory was excluded because it would result in annihilation. After the Cold War, the technological superiority of the USA meant that it was an issue of 'logistics', not chance. However, during the Cold War, such a victory would have ended the conflict. After the Cold War, a technologically devastating battlefield victory which spared the lives of the enemy population could not guarantee a similar result. And this result which was neglected was exactly that purpose which Clausewitz assigned to war: crushing the enemy's will.

The USA failed to appreciate that ending a war without complete victory meant that the spirit of the foes was not broken; resentment would smoulder. In a paradoxical reversal, wars fought against governments (e.g. Iraq, Serbia) fed the resentment of individuals. This led to the most bizarre conflict in world history where a single individual struck the American heartland – not an overseas outpost. It was an expression of personal resentment arising from a nationalist feeling that the Saudi state failed to satisfy popular needs by allowing the 'unbelievers' to occupy the land of the 'holy sites'.

This response was conceived in terms of states as defined by the West, and it was an accident of history that the modern state of 'Saudi Arabia' included the 'holy sites' of Islam. The American troops were far from the holy cities, and interested not in them, but the oil on the other side of the peninsula. American intervention had been conceptually defensive, yet defence became 'occupation' when viewed from the opposite perspective, given the modern state boundaries. In the same fashion, in order to maintain the 'war on terror', the USA government viewed their foes as states allegedly harbouring terrorists, and thus eliminated the governments of the states of Iraq and Afghanistan. Therefore, peculiarly, bin Laden's concept of war as a conflict defined by states matched the American conception, although neither was actually pursuing a state policy rational in terms of interest.

Paradoxically, during the 'war on terror', the USA established yet more foreign bases, even while officials conceded that further attacks on American soil could not be excluded. American military policy ceased to follow the logic of interest – pursuing instead the path of the possible, responding to the actions of others, yet without attempting to understand their grievances, let alone to defuse conflicts which routinely aroused animosity towards the USA. It was assumed that foes were acting without motivation, and would merely act where possible, disregarding motives and interest – exactly as the USA was.

This attitude can only be grasped when viewing another aspect to the American understanding of 'Weapons of Mass Destruction'. The invention of the ballistic missile and the hydrogen bomb had an impact on American thinking about warfare, in the sense that war was excluded due to the technological developments rendering it senseless. American military thinking had long been dominated by technology. The use of the Bomb against Japan was an outgrowth of this policy, and American attitudes towards technological development continue to hamper strategic thinking. Technological superiority is itself superiority; technology substitutes policy and strategy. Although masked by ideological pretext, this war is based on violence by the strong against the weak.

Moral warfare

This attitude towards the technological character of weapons assumes that other powers should be deprived of access to such weapons, and that seeking access to such weapons is *eo ipso* 'evil'. Technological and ideological components are united, while will, purpose and interest are neglected, implying that the purpose of warfare is the use of weapons to impose one's will on those with whom one disagrees. Weapons are not viewed as a tool of state interests. A corollary is that warfare does not require absolute force, i.e., deliberate civilian casualties are avoided. Imposing one's will simply for the sake of ideological differences, regardless of state interest, means that war has some other character than that advocated by Clausewitz and Thucydides.

The concept of will and purpose is neglected as much as state interest. Just as Kissinger's policy neglected Clausewitz' understanding of the political role of war, this new approach failed to appreciate Clausewitz on the reciprocal nature of war. The first reciprocal response to this form of war was the equally irrational mirror image of American ideology, bin Laden. It matched the most powerful state in world history against a single individual. Technology had failed; war became a senseless heroic duel again.

American war is, therefore, easily recognisable as Homeric warfare, dominated by honour, revenge, violence, pauses, unstable coalitions, unreliable allies and divine intervention: titanic and heroic violence, 'us' against 'them'. While morally absolute – expressed in Biblical terms of 'good' and 'evil' – policy has stipulated that 'evil' foes were 'weak' and politically convenient. The policy thus includes elements of (a) the Hobbesian power of the state, as the technological expression of (b) teleological Hegelian divine will and (c) Biblical 'right', as well as (d) Rousseau's concept that a democracy could wage unjust war. However, Rousseau understood war as a conflict between states.

The fundamental problem of this rarefied form of warfare is that it does not fit into the neat analytical categories. Major-General Fuller pointed out the fundamental difficulty of reconciling the concept of rational warfare with the catastrophic results of the first and second world wars. The experience of the 20th century AD was that the use of violence or the threat of the use of violence was consistently matched by a reciprocal increase in the use of violence or threat. Such a procedure itself implied that the use of violence led to further violence, without necessarily resolving the underlying political conflicts, which continue to smoulder, promising new outbreaks of violence.

Fuller (1972: 12) cited Clausewitz to the effect that the responsibility of the statesman and general demanded that war be undertaken realistically, and 'not to take it for something ... which by the nature of its relations, it is impossible for it to be.' Clausewitz did not consider the possibility of an 'absolute war' waged without reason in a limited fashion, just as Thucydides did not consider the possibility of a war waged without clearly defined goals and interest. National (state) interests would be paramount, and moral reasoning would justify, not motivate, warfare. The conscious link between purpose and execution would determine the end. During the 20th century, the value of war as a tool of policy was neglected. It remained, however, intimately linked with the state, even if the state was deprived of responsible statesmen.

The nature of war

The cause of territorial expansion has been lost, and the moral pretext has become cause, yet the ill-defined enemy ('terrorism', mere 'possession of weapons of mass destruction') cannot be politically isolated. The result is the use of force in pursuit of goals which cannot be achieved with force.

It is, however, the state which is the precondition for such 'warlike' conflict. Although the means chosen to respond may be fundamentally flawed, it is the state which unleashes the ultimate military response, or accepts defeat. The state is the object as well as the agent in any reciprocal military conflict of opposing political interests, even if one actor is not a state, and even if one state actor is not pursuing its interests.

The concept of warfare is treated here as a political phenomenon. Surveys dedicated to mere violence can demonstrate the presence of violence at all periods of human history. Distinguishing systematic violence from warfare is the issue, since human sacrifice, executions, torture, duelling, and other activities with lethal results cannot be confused with warfare in any of the senses used here.

Here the naiveté of Rousseau's assumptions about human nature is revealed, since the evidence implies that violence was widespread among human communities long before the appearance of the state. Rousseau trusted that Hobbes was wrong about the state of nature, observing that mankind did not invariably seek to injure others.

By identifying the individual with the group and encouraging patriotism, the individual assumes an identity by which the world can be divided: 'us' and 'them'. As units, the individuals composing the larger political units thus defined can cause injury to neighbours, with or without cause. Self-defence is the response to aggression, and both forms can be associated with groups of people. This allowed Hegel to espouse the concept of the power of the state and compel the individual to relinquish his rights in the interest of the community. Hegel was concerned about the state's right to call upon its citizens to protect themselves and the state. Clausewitz then posited that war was the natural form of intercourse in relations between states. Clausewitz assumed that war was the prerogative of states, completing the circle begun with Rousseau's assumption that the boundaries of the state would be defined by the power of other states. This clearly differs from an argument based on ideological and political conflict unrelated to territorial conquest.

'Religion', 'ideology', 'nationalism', 'civilisation', etc., can be subsumed under the concept of 'identity'. Where groups sharing a common identity also share common geographical and ethic borders, these can be viewed as 'states' or 'potential states' with territorial ambitions. Where conflict takes place over territory, the expression will be similar to that described by Thucydides and Clausewitz. Where the destruction of the opposing group is not followed by occupation, and was not motivated by aggression on the part of the vanquished, this analysis fails. The wars at the end of the Bronze Age – at least as portrayed by Homer – are of this type. In most cases, however, the conquerors eventually select a territory to settle, even after wreaking seemingly senseless havoc across wide swathes of territory. The Mongols eventually settled in China after withdrawing from elsewhere, and the Turks settled Anatolia as well as Central Asia. There are few examples of a people spontaneously, without provocation, conquering another people, eradicating or defeating them *and then departing*. State warfare is the form known to us in historical and philosophical contexts, where a war is a response to aggression, or leads to occupation. Regardless of how it is perceived, the American 'war on terrorism' is couched in terms of 'self-defence'. Bin Laden likewise views his war as a defensive war against American aggression, interpreted as the occupation of the land of the Holy Places.

Reciprocity returned, with the limited war in Iraq unable to stay limited. For Clausewitz (1991: 179, 210-11) warfare was politics; violence and force were the ultimate arbiters of political intercourse. Clausewitz (1991: 199) stressed that the result of any given war was 'never absolute', and his concept of 'limited war' may have been matched by 'limited peace', implying that both peace and war were determined by subjective appraisals of the power relations at any one time. Any war which does not end in total defeat must be viewed as ending in a temporary balance, confirming war as being political in character, and therefore limited in terms of violence. If one side views the war as limited, the other may not. A failure to match violence with equal violence means that defeat will follow. We may assume that Clausewitz could hardly have reached any other conclusion than that warfare was the 'utmost use of violence' to achieve 'political goals' in a 'reciprocal contest'. War only remains 'limited' if both sides view it reciprocally as 'limited'; the reciprocal character of war would be the only relevant feature.

Superficially, wars of identity and 'divine will' cannot be understood in Clausewitzian terms, whereas Clausewitzian and Biblical wars can be understood in Homeric terms. It is, however, possible to assume that in some cases, aggression can be understood merely in terms of a hostile encounter of two groups – such as the Mycenaeans and Troy, or bin Laden and the USA – in which territorial expansion is not a motivation. The role of the state is crucial in such conflicts. The American state is as essential to bin Laden's war as the Crusader state was to Saladin. Two states may go to war, with the outcome leaving one state, two states or none. In state warfare, one participant must have a territorial homeland and a political identity, which can be perceived as injured – or serve as the basis for aggression. Victory need not lead to territorial conquest: it can be expressed by imposing the values of the victor on the vanquished, and thus breaking their will.

Ultimately, a conflict originating in a contest of two wills can take the form of Clausewitzian territorial war between states, even where it does not originate in this fashion. By contrast, disparate groups probably cannot resist the force of a state without uniting to form one. It can, therefore, be argued that the use of violence by state entities could not find any other expression than the form identified by Clausewitz, precisely because the unwise use of violence will call forth a reciprocal response: even where there is not a state enemy before the war.

It can even be argued that there are no known cases of state formation which did not involve a war. Even where a state did not exist on either side prior to the conflict, one or two states may exist afterwards (as in Claessen's example from Madagascar, chapter 15). In other cases, a group will band together to inflict defeat on an enemy, and thus form a larger state, as in the first unification of Egypt (Warburton 2001). States will also appear in political vacuums created by state conflict, such as the statelets after the collapse of the major Bronze Age powers (Homeric Greece and Israel, among others). War would therefore be a necessary precondition for state formation, but not a sufficient condition. By conceptually linking social violence with warfare, the role of warfare in 'state formation' becomes virtually incidental, either denied or assumed, but assigned no specific role. The current writer believes that it will be difficult to find a single example of state formation which did not involve armed conflict, either defensive or offensive.

A note on archaeological views of warfare

In my opinion some of the Western philosophical debate on warfare has not been integrated into archaeological theory, while some of the recent thought has had an impact on archaeological thinking which is more implicit than explicit, and finally archaeological thinking has a way of affecting its subject matter which is highly relevant to a discussion. Before proceeding to the conclusions, we will briefly cover some of the points.

The first note concerns this final point, about archaeological thinking. As usual, one comes across the idea that 'Ancient Warfare' (as studied by archaeologists) is somehow different from other types of warfare (in the same way that, e.g. 'palaeopsychology' or 'palaeoeconomics' apparently differ from psychology and economics). It is possible that there is a distinction, but the issue must be investigated in terms of methodology and definitions, rather than positing a difference and then establishing the characteristics of the difference. Particularly important is the fact that – if it is accepted as a definition that 'warfare' is a state activity – then it follows that 'prehistoric' and 'historic' archaeology are not studying the same subject.

This last point is a mere matter of definition, but the treatment of archaeological data on warfare gives the impression that more is involved. While surveys of warfare as seen by prehistoric and historical archaeologists are dominated by violence or evidence of violence, stressing victims, victors, weapons, battles and fortifications (e.g. Carman and Harding 1999; Yadin 1963), studies of ancient political interest in historical periods concentrate on trade and diplomacy, not purposeful warfare (e.g. Liverani 1990; Cohen and Westbrook 2000). In my view the implication is that the violence and wastage visible in the archaeological record seem to imply that the futility of war can be thrown back to antiquity by emphasising the violence while denying that warfare had a conscious purpose.

In this chapter, warfare has been linked to state activity, even where the goals of warfare are not rational or do not involve the goal of territorial conquest. The lethal results of social conflict cannot legitimately be compared with execution or sacrifice. Evidence of violence is insufficient to demonstrate the presence of warfare, and should not be permitted to be mistaken for it. By contrast, purposeful warfare has a political aspect. If a distinction is to be made between 'warfare' and 'violence', 'warfare' should employ violence with a political purpose. The 'social' and 'territorial' purpose is recognisable in state conflict, and most political theory has dealt with war as an attribute of states. This is a matter of identification.

Explanations are another matter, and in fact, one result of modern Western hostility to war (born of the 20th century futility) has been expressed by finding 'explanations' for it. We would argue that this hostility to war has been integrated into archaeological theory. On the most simplistic level, the explanation for the horror of war lies in its reciprocal character which is elementary. The essence of state sponsored military conflict is wasteful carnage. This does not, however, mean that it is pointless. Therefore, this character should not conceal the fact that warfare differs from mere violence in that warfare has a *purpose* and not a mere 'origin' or 'explanation'.

Such organised violence cannot have existed before the emergence of states, yet this does not mean that Rousseau was correct and that violence was unknown before states, because foreign to human nature. Despite the prehistoric evidence of violence, Rousseau's assumption has been incorporated into archaeological theory in several ways, among which is Carneiro's 'Circumscription Theory', which is directly related to states and warfare.

Carneiro's theory assumes that demographic needs outstrip resources and lead to conquest. Applied to the early development of the state in ancient Egypt, Carneiro (Carneiro 1970; Bard and Carneiro 1989) contends that the population density 'reached a point where conquest became a necessity' (Janssen 1992: 315). This appears to provide an 'explanation' or 'justification' for conquest. However, despite the apparent certainty, there is no consensus on the issue. For Egypt, Janssen, Bard and Carneiro assume its validity; Kemp (1989: 31), Eyre (1997) and the current author (Warburton 2001: 244-45, 282) dispute it. For Egypt, an equally useful alternative would combine Mann's (1997) concept of 'caging' with Clausewitz' 'reciprocity'. This is important, since (aside from Egypt) I know of no instance where Carneiro's hypothesis has been tested. Obviously, the court is out on this, but this example should advise against simply 'applying' it as an 'explanatory' model elsewhere, at least until more evidence is in.

More important, however, is the fact that the principle of the use of 'Circumscription Theory' excuses conquest by combining elements of Rousseau and Malthus, opening the way to claims that climatic causes or demographic growth alone triggered early conflicts. These attempts 'justify' the use of violence rationally, defined in terms of economic constraints. This logic is then extended to 'explain' later wars of conquest in terms of 'economic motives'.

Western economic growth has encouraged Western philosophers to associate economic growth and material gain with 'reason'. It is viewed as 'reasonable' that wars are fought over economic goods. It is occasionally argued that ancient wars will have been fought for similar reasons (Mayer 1995: 1), yet it is not even clear that this occurs today (although figuring among pretexts). The origin in constraints, and the later pursuit of conquest for economic reasons are thus assumed to be identical, and the character of warfare thus assigned a rational and consistent character. Although easily recognisable to the Western student, the logic is purely circular. Yet there is no reason to assume that it really explains the wars of the Bronze Age Near East in a more convincing fashion than the one which Thucydides and Clausewitz would offer.

Conversely, there are scholars who will attempt either (a) to deny the importance of violence before it is unequivocally documented, or (b) to deny its purpose. The repeated denial that arrowheads of the Levantine Neolithic were designed and used to kill humans is symptomatic of this approach (cf. Cauvin, above). It is true that evidence is still sparse. However, elsewhere arrowheads are documented in human bones, and Bronze Age armour-piercing arrowheads served no other purpose, as the Egyptian documents confirm. On the other hand, however, conceding the possibility of violence while denying a purpose compounds the difficulties of understanding the use of force in human societies.

This confuses two issues: one is the issue of violence and the other is the use of violence to pursue a 'political goals'. An approach to 'ancient warfare' might better be directed at the issue of violence in inter-communal relations, and not mere killing and violence, with a distinction between 'violence' as 'personal or social lethal interaction' and 'warfare' as 'political lethal interaction'. Among the 'origins' one could also seek not just 'causes' and 'explanations', but also 'purposes'.

We would therefore advise students of archaeological aspects of warfare to pay close attention to the definitions they use, and to appreciate that the 'moral' element of warfare is itself part of the history of the philosophy of warfare, and the experience of war, but not necessarily part of warfare itself.

Conclusions

The capacity of a state to wage war and impose its will depends upon 'power'. Aron (1962) identified territorial space, resources, manpower and capacity for collective action as the essential attributes of 'power'. War is therefore the expression of the spirit of the state. According to Clausewitz, the power of a state served a political purpose. Rousseau and Hegel would have viewed this as a tautology.

At different times, Western philosophy has approached the issue of state-organised violence in terms as different as 'interest' and 'morality', with a view to 'justifying' or 'explaining' warfare. None of the proposed roles has matched the actual performance of warfare. In fact, therefore, it is not the violence, but the state's use of violence which constitutes the essence of warfare, rendering it distinctly different from mere lethal conflict. The origins of inter-communal lethal violence lie in the depths of pre-history, and these contributed to the development of warfare which was transformed into a new activity with the appearance of the first states, some five thousand years ago. This opened the path to 'interest', 'policy' and the 'just war'.

The Hegelian assumption that the state was the spirit of reason in substantial form depended upon the assumption that the state was divine. This 'divinity' can be traced back to the absolute 'divinity' and 'justice' of ancient Near Eastern warfare which fed into Christian doctrine. This differed substantially from the approach of Mediterranean antiquity where state interest dominated, but the political goals stipulated by Clausewitz can be used to frame any kind of warfare: even an epic duel, with a state on one side.

The concept of war as an instrument of policy whereby the weak were subjected to the will of the strong was apparent in Thucydides, and Thucydides clearly stated that the moral reasons for the war were merely a pretext. This view was adopted by Hobbes, who assigned the state a role in protecting the individual. Hegel extended this and ultimately combined it with Aquinas's view to form his own conception of the state. Clausewitz then analysed war as state activity.

Here, Hobbes, Hegel, Clausewitz and Rousseau were all in agreement, differing only in the attitude with which they viewed the state and human nature. Rousseau took leave of the others in assuming that the individual suffered from the impositions of the state, rather than benefiting from the rights granted by the state. Rousseau assumed that human beings were basically peaceful and that they did not deliberately seek to harm others; Kant suggested that peace could be a goal of state policy. This matched modern Western distaste for war, born of defeat and the consciousness of futile destruction.

While Aquinas, Rousseau and Kant described ideals they perceived as practical, Clausewitz and Thucydides perceived themselves to be describing an ideal version of reality. Both types of ideals were incorporated into modern Western warfare. At the same time modern views of warfare – as senseless violence which does not serve the interests of the state – are transformed into assumptions in archaeological theory.

I have tried to argue that war independent of states is virtually impossible, and that 'state interest' and 'warfare' thus move together. Given the contrast between war as a political instrument employed to subject the weak, and the concept of war as a moral instrument, it is clear that the key to understanding the Western philosophies of war is the Biblical interpretation of warfare. Via Jahweh, the Christian concept of a morally just and righteous god guiding history was married to the concept of the national gods of Egypt and Assyria. The concept of the divinely sanctioned state waging divinely sanctioned war was then incorporated into the paradigm of war as an instrument of policy.

We may distinguish between (a) the origins of warfare – probably in inter-communal conflict at the end of the Palaeolithic before property and the first states – and (b) the character of warfare as defined by Clausewitz. The origins of a phenomenon do not necessarily correspond to the character of the same phenomenon at a later stage in its development. History would tend to indicate that 'state warfare' represented an irreversible transformation in the use of violence. Violent conflict continued to exist, but a new form of violent conflict emerged.

Once states came into being, the rules of reciprocal violence in Clausewitzian warfare left little choice in its conduct. I would argue that neglecting the link between divine legitimacy and political power in the earliest Near Eastern states, and the link between policy and violence in modern warfare permits some latitude in understanding and defining 'warfare'. I would, however, argue that this approach cannot aid in understanding the phenomenon of 'warfare' discussed in this chapter, nor in achieving general agreement about the concept, as generally understood.

'Warfare' is an aspect of the state, with its own system of 'values'. 'Violent conflict' is a characteristic of human behaviour. Reciprocity may be present in either, but the social system is pre-eminent and decisive in the former, the individual in the latter. Although organised violence and the state are essential features in my understanding of war, rational goals are not. However, the reciprocal nature of warfare means that irrational goals pursued with the systematic use of violence will meet with opposition, and thus re-introduce rationality.

This is the defining characteristic of warfare as opposed to mere inter-communal violence. The state can employ force, but the state itself cannot guarantee the pursuit of interest or rational goals: only a reciprocal opposing force can compel the state to behave rationally. Contending that 'warfare' is a state activity means deciding whether this is (a) a logical and valid conclusion, (b) a possible assumption, (c) a definition, (d) a criterion or (e) a tautology.

I argue that 'warfare' is a state activity and that Clausewitz would have been unable to revise his book by assuming a form of 'limited warfare' compatible with the 'reciprocity' inherent in state warfare, which is what gives the purposefulness of war its awful form.

BIBLIOGRAPHY

- Aiken, H.D. (ed.) 1964. *Hume's Moral and Political Philosophy*. New York: Hafner.
- Aron, R. 1962. *Paix et guerre entre les nations*. Paris: Calman-Lévy.
- Bachechi, L., Fabri, P.-F. and Mallegni, F. 1997. 'An arrow-caused lesion in a Late Upper Palaeolithic human pelvis'. *Current Anthropology*, 38, pp. 135-40.
- Bard, K.A. and Carneiro, R.L. 1989. 'Patterns of predynastic settlement location, social evolution, and the circumscription theory'. *Cahiers de recherches de l'institut de payrologie et d'Égyptologie de Lille*, 11, pp. 15-23.

- Carman, J. and Harding, A. (eds.) 1999. *Ancient Warfare*. Stroud: Sutton.
- Carneiro, R.L. 1970. 'A theory of the origin of the state'. *Science*, 1169, pp. 733-38.
- Cauvin, J. 2000. *The Birth of the Gods and the Origins of Agriculture*. Cambridge: Cambridge University Press.
- Clausewitz, C. von 1989. *On War*. M. Howard and P. Paret (eds. and trans.) Princeton: Princeton University Press.
- Clausewitz, C. von 1991. *Vom Kriege*. Bonn: Ferd. Dümmlers Verlag.
- Cohen, R. and Westbrook, R. (eds.) 2000. *Amarna Diplomacy*. Baltimore: Johns Hopkins University Press.
- Eichmann, R., Gebel, H.G.K. and Simmons, A. 1997. 'Symposium general discussion'. *Neolithics*, 2, pp. 10-11.
- Eyre, Ch. 1997. 'Peasants and "modern" leasing strategies in ancient Egypt'. *Journal of the Economic and Social History of the Orient*, 40, pp. 367-90.
- Fuller, J.F.C. 1972. *The Conduct of War 1789-1961*. London: Methuen.
- Gebel, H.G.K. and Bienert, H.D. 1997. 'Ba'ja hidden in the Petra Mountains'. In: H.G.K. Gebel, Z. Kafafi and G.O. Rollefson (eds.): *The Prehistory of Jordan, II*. (Studies in Early Near Eastern Production, Subsistence, and Environment, 4). Berlin: Ex Oriente, pp. 221-62.
- Gopher, A. 1994. *Arrowheads of the Neolithic Levant*. (American Schools of Oriental Research Dissertation Series, 10). Winona Lake: Eisenbrauns
- Goring-Morris, N. 1998. 'Complex hunter/gatherers at the end of the Palaeolithic'. In: Th. Levy (ed.): *The Archaeology of Society in the Holy Land*. London: Leicester University Press, pp. 141-68.
- Hegel, G.W.F. 1999a(1830). *Enzyklopädie der philosophischen Wissenschaften im Grundrisse. Hauptwerke in sechs Bänden*. Vol. VI. Darmstadt: Wissenschaftliche Buchgesellschaft.
- Hegel, G.W.F. 1999b(1830). *Grundlinien der Philosophie des Rechts. Hauptwerke in sechs Bänden*. Vol. V. Darmstadt: Wissenschaftliche Buchgesellschaft.
- Hendrickx, S. and Veermeersch, P. 2000. 'Prehistory: From the Palaeolithic to the Badarian Culture'. In: I. Shaw (ed.): *The Oxford History of Ancient Egypt*. Oxford: Oxford University Press, pp. 17-43.
- Hoffmann, S. 1965. *The State of War*. New York: Praeger.
- Hoffmann, S. and Fidler, D.P. (eds.) 1991. *Rousseau on International Relations*. Oxford: Clarendon Press.
- Janssen, J.J. 1992. 'Review of Kemp 1989'. *Journal of Egyptian Archaeology*, 78, pp. 313-17.
- Johnson, J.T. 1999. *Morality and Contemporary Warfare*. New Haven: Yale University Press.
- Kahn, H. 1965. *On Escalation*. Baltimore: Penguin.
- Kant, I. 1917. *Zum Ewigen Frieden*. Leipzig: Insel.
- Kemp, B.J. 1989. *Ancient Egypt*. London: Routledge.
- Kissinger, H. 1969. *Nuclear Weapons and Foreign Policy*. New York: Norton.
- Kissinger, H. 1995. *Diplomacy*. New York: Touchstone.
- Levy, Th. (ed.) 1998. *The Archaeology of Society in the Holy Land*. London: Sheffield Academic Press.
- Liverani, M. 1990. *Prestige and Interest*. (History of the Ancient Near East Studies, 1). Padua: Sargon.
- Mann, M. 1997. *The Sources of Social Power*. Cambridge: Cambridge University Press.
- Matthews, R. 2000. *The Early Prehistory of Mesopotamia*. (Subartu, 5). Turnhout: Brepols Publishers.
- Mayer, W. 1995. *Politik und Kriegskunst der Assyrer*. (Abhandlungen zur Literatur Alt-Syrien-Palästinas und Mesopotamiens, 9). Münster: Ugarit-Verlag.
- McNamara, R.S. 1995. *In Retrospect: The Tragedy and Lessons of Vietnam*. New York: Times Books.
- *Oxford Encyclopedia of Ancient Egypt*. D. Redford (ed.) 2001. *Oxford Encyclopedia of Ancient Egypt*. 3 vols. Oxford: Oxford University Press.
- *Oxford Encyclopedia of Archaeology in the Near East*. E. Meyers (ed.) 1997. *Oxford Encyclopedia of Archaeology in the Near East*. 5 vols. New York: Oxford University Press.
- Paolucci, H. (ed.) 1962. *The Political Writings of St. Augustine*. Washington D.C.: Regnery Gateway.
- Raaflaub, K. and Rosenstein, N. (eds.) 1999. *War and Society in the Ancient and Medieval Worlds*. (Center for Hellenic Studies Colloquia, 3). Cambridge (Mass.): Harvard University Press.
- Sasson, J. (ed.) 1995. *Civilizations of the Ancient Near East*. 4 vols. New York: Scribners.
- Scales, R.H. 1995. *Firepower in Limited War*. Novato (Calif.): Presidio.
- Shaw, I. (ed.) 2000. *The Oxford History of Ancient Egypt*. Oxford: Oxford University Press.
- Sigmund, P.E. 1997. 'Law and politics'. In: N. Kretzmann and E. Stump (eds.): *The Cambridge Companion to Aquinas*. Cambridge: Cambridge University Press, pp. 217-31.
- Strassler, R.B. (ed.) 1996. *The Landmark Thucydides: A Comprehensive Guide to the Peloponnesian War*. New York: Simon and Schuster.
- Taylor, A.J.P. 1996. *The Origins of the Second World War*. New York: Touchstone.
- Thucydides 1953. *Historiae*. 2 vols. Oxford: Oxford University Press.
- Valla, F. 1998. 'The First Settled Societies – Natufian'. In: Th. Levy (ed.): *The Archaeology of Society in the Holy Land*. London: Leicester University Press, pp. 169-87.
- Walzer, M. 1992. *Just and Unjust Wars*. (2nd edn.). New York: HarperCollins.
- Warburton, D. 2001. *Egypt and the Near East: Politics in the Bronze Age*. (Civilisations du Proche-Orient, Série IV, Histoire-Essais, 1). Neuchâtel: Recherches et Publications.
- Weigley, R.F. 1973. *The American Way of War*. Bloomington: Indiana University Press.
- Yadin, Y. 1963. *The Art of Warfare in Biblical Lands*. London: Weidenfield and Nicolson.

Archaeology and War: Presentations of Warriors and Peasants in Archaeological Interpretations

The horse-mounted and axe-wielding pastoral tribes migrated into Jutland from the south, rapidly making themselves rulers of the central and western half of the peninsula. Up through southern Jutland the burial mounds mark the routes of colonization, which can only be loosely followed. The oldest occurrences of axes show how comprehensive this first influx was. The wide and leafy river-valleys of central and western Jutland became the first resort of these foreigners, since here was abundant food for their beasts. The old fishing and hunting folk who lived close to lakes and streams was rapidly subdued in most places, and the same destiny undoubtedly stroke the dispersed farming communities, unless they succeeded in reaching eastern Jutland, where their kinsfolk, the megalithic people, was densely situated. (Glob 1945: 242, author's translation from Danish)

/5

HELLE VANDKILDE

This article critically examines archaeologies of the Stone and Bronze Ages by looking at them through a broad contemporary framework. Although clearly including warriors in explanations of material transitions from one culture to another, the academic discourse of archaeology has strangely ignored warfare and violence as relevant aspects of past human activity, an apparent contradiction that this article will examine and debate.[1]

Power, dominance and coercion are almost inevitably connected to warfare and its principal actors, soldiers and warriors, brutally interfering with human existence almost everywhere in our late modern world. These factors, embedded in a 21st century setting, make it obvious that warfare should be an object of archaeological study. More generally, war seems to be a central ingredient in social reproduction and change, which constitutes another reason for engaging in the study of war, warfare and warriors. However, looking back at the Stone and Bronze Age archaeology of the 20th century, it becomes clear that archaeologists have studied weaponry, and in some measure warriors, but not war. There are notable exceptions, but possible reasons for the general absence of an interest in violence need to be outlined and debated. Warfare and violence began to enter the archaeological discourse only after c.

1995. Compared to the general implementation of anthropological and sociological theories in archaeology (late 1960s and early 1970s), war studies thus arrive on the scene much delayed. Even after this date the theme is quite often embarked upon as something set aside from the rest of social practice.

Vencl (1984) has argued that the absence of warfare studies in prehistoric archaeology is linked to the inadequacy of archaeological sources. It is undoubtedly true that archaeological data do not reflect the ratio of war in prehistory. Trauma is probably underrepresented, and so are weapons of organic materials (Capelle 1982). However, direct and indirect evidence of war-related violence is by no means non-existent (Figs. 1-4). The number of prehistoric weapons, including fortifications, is huge, and iconographic presentations of war and warriors in art and rituals supplement the picture, as do examinations of patterns of wear and damage on swords (cf. Bridgford 1997; Kristiansen 2002). Skeletal traumata are, in fact, relatively frequent in European prehistory when it is taken into account that skeletons are often not well preserved, they are not routinely examined for marks of violence and that much physical violence does not leave visible traces on the skeleton. The evidence is most certainly adequate as a basis for studies of violence and war.

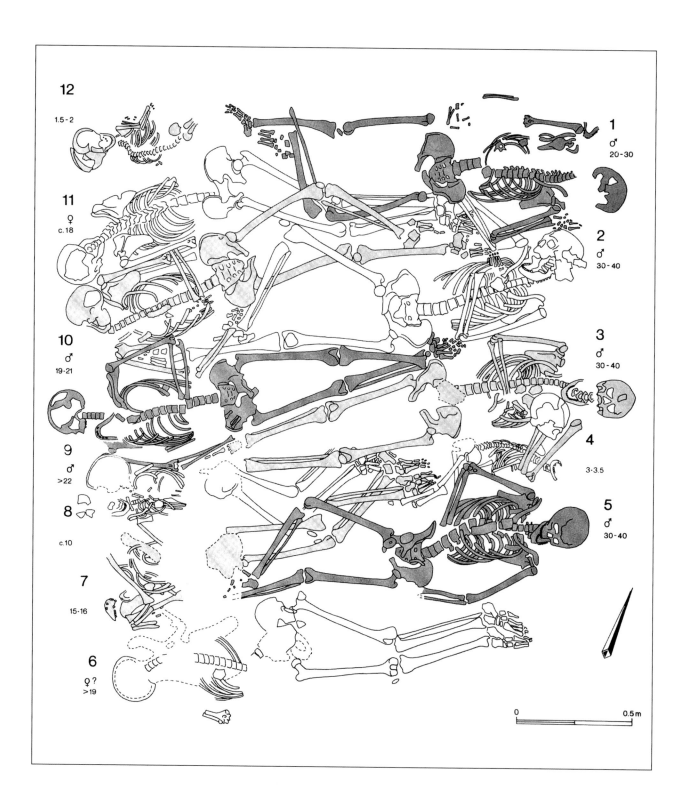

FIG. 1: *The mass grave from Wassenaar in the Netherlands (after Louwe Kooijmans 1993) provides concrete evidence of war-related violence at the end of the Early Bronze Age, c. 1700 BC. It also testifies to the brutality, horror, and finality of prehistoric war hence belying the celebrative tales of brave warriors and peaceful peasants. Twelve individuals – six men of warrior age, four adolescents and children in addition to one or two women – had been interred in a large grave pit following a certain regulated pattern. This suggests that despite the abnormal burial practice the bodies had been at least provisionally cared for by surviving kin or women taken captive by the victors. Three of the males had received cutting blows on their skull and arms, while a fourth male had been wounded, probably fatally, by a flint arrowhead sitting in situ in the rib cage; one child had been decapitated.*

The reason why war has played a small part in the archaeology of the 20th century must be searched for elsewhere. Insufficient evidence is hardly the core of the matter. Rather the existent evidence has been ignored, underestimated, rationalised, or idealised – for example, through the use of soft metaphors for war (unrest, troubled times, etc.). The initially weak presence of war and the belated appearance of war studies are arguably linked to the politics and wars of contemporary society, but they are simultaneously entrenched in two myths about the primitive other, which have persuasively influenced European thought at least since the 17th century. These myths have had a considerable impact on archaeological interpretations, which are divided into two tales positioned at either end of the scale commemorating certain stereotypical identities and societies with contemporary political meanings; these interpretations have circulated within the discipline and have also been communicated to the public.

The two tales of prehistoric society are interpretive traditions, or trends, rather than schools of thought, inasmuch as they contain different theoretical stances which may not consent to the classification undertaken here. They partly co-exist and they even compete with each other. Each trend is joined together by a related understanding of society and of how and to what degree social change is generated. The first trend conjures up warriors – though not always explicitly – and advocates prehistoric society as an organism that changes through human agency, often suddenly and radically. The second trend evokes peaceful hunters, peasants and traders – indeed the antitheses of the warrior – by proposing a view of prehistoric society as a mainly reproductive organism, involving a stepwise or slow long-term social evolution.

I shall insist that it is not sufficient merely to add war to the themes that archaeology can study. The seriousness of the topic demands that we discuss how war should be studied and portrayed. Is it really indispensable to incorporate the vicious face of war in archaeological studies of this phenomenon? This is a relevant question because the archaeology of the late 1990s has analysed war in a strictly rational fashion. In this respect the newest war anthropology can potentially inform the archaeological study of war. Furthermore, the present contribution wishes to promote an understanding of warfare as a social phenomenon which cannot be studied isolated from its social context and which needs an adequate theoretical framework. The position taken is, in short, that warfare is a flow of communally based social action aimed at violent confrontation with the other. Being a warrior is consequently a social identity founded in warfare.

The following account is intended to lay out the history of research in broad brush strokes and in doing so it draws on examples from European, and particularly Scandinavian, Stone and Bronze Age archaeology. No attempt is made at comprehensiveness.

Warriors, but little war

A vigorous tradition in 20th century archaeology has envisioned prehistoric society as an unstable entity which was transformed through radical events. Sudden material changes were explained as a result of migration or revolution, often with fierce warriors as front figures. Economic and social forces of power, tensions and underlying contradictions generated revolutions, whereas invading aristocracies formed the essence of migrations. This warrior tradition took shape in the beginning of the century and in its earlier phase it included first and foremost V. G. Childe and contemporaries of the empiricist school. Childe was indeed the inventor of revolutions in prehistory, though not of migration, which emerged in archaeological thought at the end of the 19th century and continued well into the 20th (Trigger 1980: 24ff, 102ff; Champion 1990). Later, Grahame Clark diagnosed the obsession with migrations as 'invasion neurosis' (Clark 1966: 173). This was no exaggeration: especially in central Europe, depositions of valuables were associated with an unsteady political situation in the wake of ethnic migration (Bradley 1990: 15).

Whilst Childe characteristically drew a complex map of social change, he also composed vivid scenes of migrating people with armed warriors in front, sometimes on horseback. Waves of migrating Beaker folk, for example, put an end to egalitarian clan society, forever changing the European societies of the Neolithic Period (1946: 41).

The last warrior group to appear in the archaeological record from Western and Central Europe played a far more constructive part than hitherto mentioned. For though they travelled fast and far in small well-armed bands, their objectives were not only pastures and arable lands, but also raw materials for trade and industry, and smiths accompanied them. (Childe 1958: 144)

He imposed the same model upon a series of other archaeological cultures exhibiting a sudden and massive geographical expansion, notably the forerunner of Bell Beakers, the Battle-Axe warriors, who became pastoral overlords of the local peasants they subdued during their conquest of new land (op. cit.: 142f). Despite his fondness of the warrior model, Childe was reluctant to incorporate war as a force in history. He nevertheless included warfare in some of his later works, but merely ascribed it to the economic greed of aristocrats (1951[1936]: 134; 1941: 133).

A similar line of thought recurred in P.V. Glob's study of the Single Grave Culture, the Danish version of the Battle-Axe Culture. Presumably influenced by Childe, Glob evoked a lively scenario of invading warrior nomads, axe wielding and on horseback (cf. the opening quotation). In his great survey of Danish prehistory, published first in 1938-40, Johannes Brøndsted conjured up several intrusions of warrior groups during the Neolithic Period. He even employed a Marxist-Childean vocabulary, not least when categorising the Bronze Age as a class society with an aristocratic upper class of warriors and an oppressed peasant class (Brøndsted I, 1957; II, 1958: 10). War and violence were, however, absent.

Structural Marxist approaches of the late 1970s and 1980s continued this tradition even if internal social dynamics and structural contradictions replaced migration as the cause of rapid social change. In studies by Susan Frankenstein and Michael Rowlands (1978; Rowlands 1980), warriors were in particular related to competition for rank. Kristian Kristiansen has similarly made use of warriors in seeking to explain the archaeological record of the Bronze Age (1982; 1984a; 1984b; 1991b; 1998; 1999). An examination of wear traces on Bronze Age swords showed that they were used in real fighting as well as in the display of social rank (Kristiansen 1984a; 2002). The notion of the rise of a warrior class on the background of structural change was also present in my own studies of the earliest Bronze Age (Vandkilde 1996; 1998; 1999). Yet these studies rarely mentioned war and violence.

Hedeager and Kristiansen (1985) reused the structural Marxist model in their pioneering article that made war an object of study. The article highlights the social functions of warfare, notably describing fighting as a route to social success. Jarl Nordbladh (1989) resumed this thread in his study of 'armour and fighting in the south Scandinavian Bronze Age'. Petroglyphs and graves alike celebrated war heroes in a socially unbalanced society in which

fighting was probably very ritualised and often unequal with rules guaranteeing the safety of the most noble and prescribing more spectacular and serious duels for persons of lower rank. (Nordbladh 1989: 331)

What was envisioned was a rather bloodless theatre kind of war.

It is possible to similarly categorise one branch of the post-Processual archaeology of the late 1980s and early 1990s because of the key role attributed to social domination and inherent conflict (Miller and Tilley 1984: 5ff; Shanks and Tilley 1987: 72f). This branch considered power as central to social life, but hardly mentioned violence and bloodshed at all. In her studies of the Danish Middle Neolithic, Charlotte Damm was inspired by post-processual ideas, especially Christopher Tilley's work on the Swedish Battle-Axe culture (Tilley 1984). Widespread social upheaval instituted the Single Grave Culture in Jutland:

I consider it likely that a break with the existing society in one of these groups led to general uprising and the emergence of a new social and material order in large parts of the North European lowland. (Damm 1993: 202)

Damm believed her views of the Single Grave culture to be in opposition to the earlier studies by Glob (1945) and Kristiansen (1991a), who argued for ethnic migration rather than cultural construction. What they had in common, however, was an emphasis on rapid change through human agency, and, furthermore, a lack of definite reference to the waging of war even if their theoretical framework presupposed such activities.

To sum up: despite the fact that the model of migration and revolution does not envision peaceful interaction, warfare and violence are seldom mentioned, and this is regardless of the precise theoretical persuasion. The entire explanatory trend underplays the violence it so clearly implies. Even when the armed individuals – implicitly males – are termed 'warriors', brutality and killing do not form part of their actions. The violent face of massive ethnic migration and social revolution is ignored, understated or simply not realised. The scholars who mention or examine warfare – notably Childe, Rowlands, Hedeager and Kristiansen, and Nordbladh – underplay its deadly and destructive effects on human life and society while emphasising the heroic aspect or its socio-economic functions; hence, they describe fighting as mostly ritualised and related to social presentation and rivalry. Richard Bradley (1990: 139ff) similarly suggests an interpretation of the flamboyant weapon sacrifices of the Late Bronze Age in terms of 'potlatch as surrogate warfare'. A relationship between rituals and warfare is certainly valid, but so-called ritual war is an ambiguous concept. Even if it occurred it would only be one component of a warfare pattern (Otterbein 1999: 796ff). We are in other words presented with an idealised image of revolution, migration and warriors in prehistory.

Various subjective influences may hide a key to the lack of realism. The above portraits of Stone and Bronze Age society, and the warrior trend in general, undoubtedly respond to war-related events in the contemporary world. Several of the scholars behind the warrior tale have sympathised with left-wing politics and ideology up through the 20th century – maybe the most violent and warlike ever and thus encroaching upon most people's lives in one way or

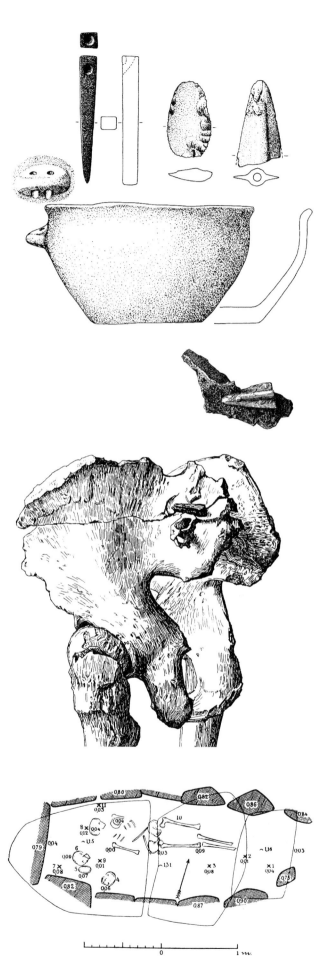

FIG. 2: *The gallery grave at Over Vindinge in Præstø County in southern Zealand (after Vandkilde 2000) with remains of several interments and a small collection of humble grave goods. The pelvis of a mature male preserved the tip of a bronze spearhead, shot into his lower back from behind (cp. close-up). The wound was not immediately fatal since there are signs of regrowth/healing. The spearhead can be tentatively identified as of Valsømagle type, datable to Bronze Age Period IB (1600-1500 BC). Bronze weapons were then clearly employed in acts of violence and war, simultaneously however with other potential uses and meanings in the social field. This period saw the emergence of a new social order with emphasis on new forms of social conduct and material culture among an elite. Evidently, struggles for power did not only pertain to ideology and socio-politics; warfare violently affected people's lives.*

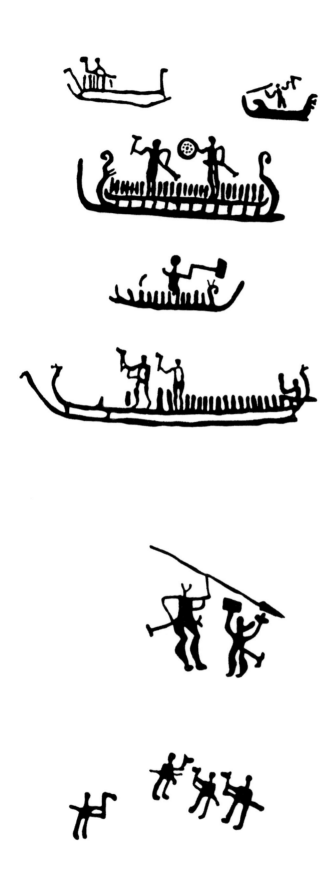

the other. With the exception of extremist versions, the left wing was from the onset tied up with anti-war movements. One possible reason for the missing, or understated, violence could be that pacifist attitudes coloured the archaeology of revolution and migration; another reason could be rooted in idealist attitudes to riots and radical social transformation. Quite possibly, these reasons have interacted. A proportion of revolutionary romanticism does seem to inhabit the work of Childe and contemporaries like Glob and Brøndsted as well as later structural-Marxists and post-processualists.

Women as prehistoric agents entered the discourse rather late, and then rarely mixed up with warriors, power and migration. The later work of Marija Gimbutas from the mid-1970s was a forceful and feminist exception. In her vision, Old Europe was a peaceful place ruled by women and structured by female values. Migrating bands of horse-mounted warriors from the Eurasian steppes destroyed this paradise, and Old Europe never recovered. Social order and cultural values were reversed, and New Europe became a thoroughly male-dominated and violent place (Gimbutas 1982[1974]: 9).

This explanation of the archaeological record has naturally not escaped criticism since it does not really agree with the evidence (Häusler 1994; Chapman 1999). It is nevertheless interesting because of its unique position. Gimbutas treads in the footsteps of Childe with her emphasis upon warriors and migration, but her explanation contrasts with those of her contemporaries in two ways: firstly, it has an inherent binary opposition of warlike maleness and peaceful femaleness, female gender representing the dominated part. This certainly recalls 'radical' feminist ideology of the 1970s. Secondly, it contains an unusually direct reference to a state of violence and warfare, which is considered fatal for human life and values of equality. These attitudes correlate with

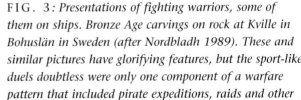

FIG. 3: *Presentations of fighting warriors, some of them on ships. Bronze Age carvings on rock at Kville in Bohuslän in Sweden (after Nordbladh 1989). These and similar pictures have glorifying features, but the sport-like duels doubtless were only one component of a warfare pattern that included pirate expeditions, raids and other kinds of warfare.*

her feminist stance. It is also related to her life outside the protected sphere of Western European academics. John Chapman has convincingly traced the origin of Gimbutas's dichotomous perception of an Old and a New Europe in her idealised Lithuanian childhood as opposed to the horrors of the Russian invasion (Chapman 1998). There is certainly no romanticised attitude to revolutions here.

Peaceful peasants

The reverse tradition had its breakthrough between 1940 and 1950. Here we find harmonious and egalitarian societies described as mostly static and without latent conflict or underlying contradictions. Hardworking and peaceful hunters, peasants and traders replaced the armed warriors of the opposite tradition, and emphasis was placed on the agrarian, and otherwise economic, foundation of prehistoric society, especially in the influential work of Grahame Clark (e.g., 1939; 1952; 1975), whose focus upon ecology and the function of culture, rather than upon explaining cultural change, was a distinct reaction against Childe's Marxist approach (Trigger 1989: 264ff). Clark and Piggott (1978[1965]) characterised the earlier Neolithic as peasant communities and the later Neolithic and the Bronze Age as trading communities based upon an expanding metal industry.

Clark's work inspired studies in settlement and subsistence. A similar understanding of prehistoric society with impact from the natural sciences underlay much empiricist archaeology outside the circles of Cambridge University. Scandinavian settlement archaeology effectively mediated a peaceful picture of prehistoric society based on nature, subsistence and cultural continuity (Fabech *et al.* 1999: 18). This was evident in economist-ecological analyses of Mesolithic settlement sites in Denmark. Søren H. Andersen (e.g., 1972; 1975) described a life preoccupied with the daily necessities of hunting and gathering, a peaceful society in harmony with nature and with other people. In the face of frequent skeletal trauma in Late Mesolithic Ertebølle burials, Andersen, quite possibly due to his basic attitude, underrated the significance of violence and conflict by deeming it non-lethal, marginal and occasional (Andersen 1981: 71f).[2]

A mainly reproductive vision of prehistory was also clear-cut in studies by C. J. Becker and Mats Malmer, who refuted that animosities between Battle-Axe people and Funnel Beaker people ever took place (Becker 1954: 132ff; Malmer 1962; 1989: 8ff). Becker expressed it this way:

When the two ethnic groups had the opportunity to meet – according to our current knowledge of the finds – this probably happened quite peacefully and hardly in a warlike fashion. (Becker 1954: 143, author's translation from German)

They envisioned peaceful interaction between the various Stone Age cultures, hence stressing cultural and social continuity. Battle-axes were status symbols, not tools of war, since prehistoric society was inherently peaceful (Malmer 1989: 8). H.C. Broholm used much the same vocabulary, interpreting the Nordic Bronze Age as a primitive and peaceable peasant culture without marked social differentiation (1943-44). In the early 1980s, Poul Otto Nielsen even claimed that society largely remained static throughout the Neolithic and the Bronze Age (1981: 154ff). These studies clearly underestimated the potential social significance of large-scale material changes and ignored the possibility of warlike encounters. This tradition correlated remarkably with the views of the Danish social anthropologist Kaj Birket-Smith, who in an influential study from the early 1940s characterised the primitive other encountered by ethnographers as essentially peaceful and preoccupied with subsistence (Birket-Smith 1941-42: e.g., 138ff).[3]

Although concerned with processes of social change, and with explaining it, New Archaeology performed the reorientation of the discipline without warriors and war, interpreting the weapons – undeniable there – as symbols of social status (cf. Vandkilde 2000: 6ff). Social evolution became a core point due to substantial influence from the neo-evolutionism that developed in social anthropology from the 1940s onwards. Prehistoric society accordingly progressed towards still greater complexity in evolutionary sequences from band to tribe to chiefdom and eventually the state. Change was either imperceptibly slow or occurred at the transition between these societal categories brought about by population pressure and ecological crises. Thus, in a long-term perspective social reproduction was thought to be much more normal than social transformation. Jørgen Jensen's analyses of Danish prehistory (1979; 1982) represented a Processual – neo-evolutionist view; he emphasised population pressure, ecology, economy and long-term social trends, and even when he recognised war as a characteristic feature of ethnographic tribes (Jensen 1979: 113), this insight was not really used to explain the frequency of weaponry in the Danish past. He mainly regarded weapons as symbols of superior social rank and performing key social functions in gift exchange between high ranking members of society (ibid.: 147ff).

This essentially static view of society in New Archaeology accords with the empiricist view of prehistoric society as largely unchanging. They agree in describing prehistoric man as hunter, peasant and trader rather than warrior – the female half of society still being without a place in history. Prehistoric society was a peaceful place characterised by evolution rather than revolution and migration, socially balanced and economically prosperous due to the efforts of skilled hunters, successful traders and, most particularly, hard-working peasants.

This major chronicle of Stone and Bronze Age society also responded socially to important war-related events in the contemporary world. The described perception of the past can be linked to a very similar ideal of modern – i.e., post Second World War – Western society's concern with technological development, human progress, harmony, welfare and peace; hence confirming it (cf. Trigger 1989: 289). This optimistic attitude to life was ultimately a reaction to years of hardship, wars and genocide, and it had a profound impact on how European prehistory was perceived and mediated. After the Second World War the assessment of the prehistoric being as innately non-violent and preoccupied with subsistence, production and trade became quite dominant. But the opposite stance survived and had a revival, especially in the years after c.1980 with the Marxist recovery – now in Structuralist bedding – and with the power-branch of post-Processual archaeology.

Chiefly warriors, and more war

During the last ten years the warrior tale has prevailed, and social categories like warrior elite, warrior aristocracy, warrior or martial society increasingly inhabit archaeological interpretations. Warfare is usually allowed, but the language often retains celebrative undertones:

The appearance of warrior aristocracies represents the formation of a new chiefly elite culture in Europe. It was embedded in new rituals, in new ideas of social behaviour and life style (body care, clothing, etc), and in a new architecture of housing and landscape. It centred around values and rituals of heroic warfare, power and honour, and it was surrounded by a set of new ceremonies and practices. They included ritual drinking, the employment of trumpets or lurs in warfare and ritual, special dress, special stools, and sometimes chariots. It meant that chiefs were both ritual leaders and leaders of war.
(Kristiansen 1999: 180-81)

Similar views are being widely communicated and also transmitted to the public, not least by scholars studying the Bronze Age (cf. Demakopoulou *et al.* 1998). They mediate an elitist and heroic stereotype of the Bronze Age, stressing cultural similarities on a pan-European scale, and thus, incidentally, hinting at a tie to a modern European project, which also conceals cultural differences (Gröhn 2004; Gramsch 2000). The increasing popularity of warriors in archaeological interpretations at the transition to the 21st century is a follow-up of the celebrative tale of the warrior so distinct during the previous century, in which warriors are viewed as the brave-hearted heads of society. There is likewise an inclination to focus one-sidedly upon the privileged upper classes as if these were the only agents of significance. Sameness is contended in warriorhood in Europe regardless of time and place. The paired institution of the war chieftain and his retinue described in later writings (such as 'Indo-European' sources, Tacitus' Germania and medieval sources to the feudal *Gefolgschaft*) is in the process of finding its way back in time into the Neolithic and the Bronze Age (e.g., Treherne 1995). In recent writings, the warrior thus tends to be an unchangeable identity, embedded in social constancy, despite the often underlying evolutionary idea of prehistory society as tending towards increasing complexity. It remains a rigid construct that is not really negotiated with archaeological data.

The current prevalence of the warrior tale is presumably related to the belated appearance around 1995 of specific studies of violence and war in archaeology. This is an important event, which is marked by Lawrence Keeley's influential publication *War before Civilisation: the Myth of the Peaceful Savage* (1996), followed by a veritable explosion in studies of war. Quite possibly, the dramatic increase in ethnic

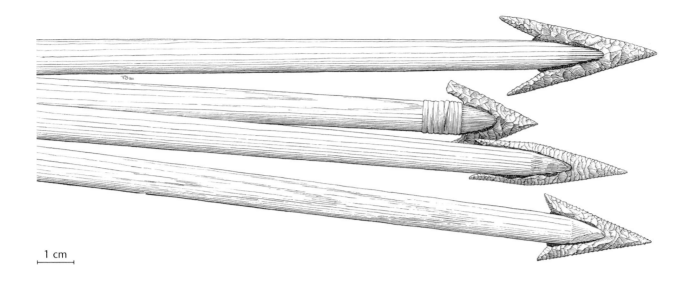

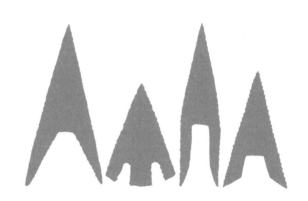

FIG. 4: *Late Neolithic pressure-flaked flint arrowheads – hafting reconstructed – from Denmark (after Nielsen 1981; reproduced with the kind permission of Flemming Bau). Their elaborate shape suggests they were meant for warfare, simultaneously however with other functions and meanings relating to social identification. The number of weapons preserved from Neolithic and Bronze Age Europe is huge, and it is very unlikely that their uses were restricted to the sociable peace of community life and religion.*

wars and genocides of the 1990s among dissolving national states on several continents has contributed to this sudden escalation. Especially, the massive media coverage may have made it difficult to carry on ignoring or idealising archaeological data of war and violence.

Studies of violence and war were undertaken somewhat earlier in Iron Age research (cf. Brun 1988; Jørgensen and Clausen 1997; Ringtved 1999 with references). Research on the Late Bronze Age and the Iron Age has been generally more willing to incorporate war in its interpretations. I believe that the evolutionary thinking that has also influenced social anthropology is an important factor here. If indigenous people without war, or with so-called ritual war, are placed at the bottom of the evolutionary ladder then warfare has to get more common as the complexity of society increases.

Commemorative myths

The archaeological traditions of warriors and peasants mostly pacify the past and populate it with idealised figures of male identity. The scholars involved in the debate, predominantly men, consciously produce their views of prehistory in disagreement or agreement with existent views and this may have contributed to the academic reproduction of two stereotyped perceptions of prehistoric society positioned at each end of the scale: a disruptive and discordant society or a peaceful and harmonious society. As argued above they reflect a politicised impact from contemporary events, but much deeper-lying myths have doubtless been influential in creating and sustaining the two tales of warriors and peasants.

Even if war and violence are not really central to the former of the stereotypes, I believe both of them have their origin in the history of European identity created and recreated in the face of growing colonialism. The warrior tale and the peasant tale can be claimed to incorporate two dominant myths about the primitive other, which have been reproduced probably on a continuous scale at least since the 17th century. The bellicose and brutal savage emerges in Thomas Hobbes' Leviathan (1958[1651]) and reappears in a different form in Karl Marx' conflicting and latently violent society. This being is in contrast to the noble and peaceful savage who originates in Jean-Jacques Rousseau's romantic writings and who in a sense recurs in Max Weber's consensual society.

The two tales are, in effect, a modern commemoration of heroes in the same way as heroes have been celebrated in past societies through material means – a social presentation with limited bearing on the real world of the past, or the present for that matter. The past – insofar as this can be treated as a unit – was no doubt characterised by rivalling myths about heroes and heroines, probably in stark contrast to the realities of most people's lives. This is also valid for our own time. Sanimir Resic (1999; chapter 27) has recently identified celebrations of warriorhood and combative actions before, during and after the Vietnam War (1965-1973). This romanticisation included not only the official propaganda, but also the stories the soldiers told family and friends in letters. Idolising warriors and the idealisation of warlike behaviour have deeply permeated contemporary society as well as its interpretation of history. It may be added here that nothing indicates that war in so-called primitive societies should be less bloody than in so-called civilised societies (Keeley 1996; Helbling 1996; Wiessner and Tumu 1998: 119ff, 152f). Rather, the frequency of war and violence varies with the specific cultural and social setting.

Keeley's book from 1996 was pioneering in that it helped to encourage war studies in archaeology. His key message is that archaeologists and anthropologist who have nourished the Rousseauian notion of the noble savage have utterly pacified the past.[5] While this portrayal of the historiography is not entirely wrong, it is inadequate. The recurrence of the warrior theme throughout the 20th century supplements the picture, as do early studies by Childe, Gimbutas, Rowlands, Hedeager and Kristiansen, Nordbladh and Vencl, who to varying extents incorporate or study war. If anything, two myths have coexisted: the peaceful savage and a fierce and potentially warlike

savage. Otterbein has expressed a similar critique towards the anthropological part of Keeley's book (1999: 794, 800ff).

The history of research can now be summarised. Two opposite tales of prehistoric society have coexisted, and probably still do. The warrior tale with its emphasis on revolution and migration grew strong during the first half of the century while the peasant tale with its emphasis on harmonious, industrious and imperceptibly changing societies gradually came into focus after c. 1945 and then receded again in the 1980s. It is characteristic that the brutal and deadly side of the warrior tale is usually left out or transformed into soft warfare, while the possibility of violence and war is completely ignored by those advocating the peasant tale. During the last decade or so the warrior tale has resumed a predominant position, coinciding in part with a veritable boom in war and violence studies from c. 1995. This whole trajectory reflects differential social responses to contemporary politics and wars, while simultaneously, however, incorporating deeply rooted European myths that celebrate opposite ideals of society and masculinity.

Anthropology, archaeology and war

Social anthropology has likewise had 'hawks' and 'doves' advocating opposite societal stereotypes based on the myths of the peaceful or warlike savage (Otterbein 1999). Anthropology has, however, been much more willing to make war an object of study. An explosion in anthropological warfare studies occurred between 1960 and 1980, including classic ethnographies as well as theoretical analyses of the causes and effects of war (ibid.). Developments in the anthropology of war can potentially improve on the understanding of archaeological views on the subject and ultimately enrich archaeological studies of prehistoric war. A comparison is therefore undertaken below.

Otterbein (1999) operates with a four-phased process of development in the research history of the anthropology of war: during the Foundation Period (c.1850-c.1920) warfare was not a central concern for the prevailing evolutionary approaches to ethnography, but a strong database was produced showing that very few societies were without war. During the Classical Period (c.1920-c.1960) these data were largely ignored. The myth of the peaceful savage emerged from evolutionary thinking inasmuch as indigenous people without war or with so-called ritual war were considered to be at the lower stages of development. During the Golden Age (c.1960-c.1980), two opposite sides were formed: those that believed in the 'peaceful savage' and those that believed in the 'warlike savage'. A veritable explosion in warfare studies occurred, classic ethnographies as well as diverse theoretical analyses concerned with the causes and effects of war. In the Recent Period (c.1980 onward) the handling of warfare has been simplified to comprise fewer theories, and a new interest has formed regarding ethnic wars and genocide. A controversy has developed between 'hawks' arguing that man's nature is to wage war and 'doves' arguing that man's nature is to live peacefully. The latter position considers war as a result of expanding and warring states, the so-called 'tribal zone theory', or the result of states dissolving into warring ethnic groups (ibid.).

The differences between the historiographies of the two disciplines are striking, even if there is also a similarity in the fact that opposite myths have partly coexisted and interacted. Whilst the partial entanglement of archaeology and anthropology is generally apparent (Gosden 1999) the two disciplines have informed each other surprisingly little as regards the issue of warfare and warriors despite the possibility that new insight might have been achieved through a cross-dialogue. The tribal-zone theory proposed by Ferguson and Whitehead (1992), for instance, would not have survived unmodified for long if archaeological sources of prehistory had been consulted. The anthropological database of peace and war might similarly have been used to assess critically the two tales of warriors and peasants in archaeology. Moreover, the dramatic increase in warfare studies during anthropology's Golden Age failed to achieve any profound impact on archaeology. It is therefore unlikely that the breakthrough of archaeological war studies in the mid-1990s is directly linked to anthropology. Rather it is connected to the growth in ethnic-based wars after c. 1990 as suggested above. A related aspect is that neo-evolutionist approaches in social anthropology heavily influenced the first theoretical applications in archaeology in the 1960s and 1970s and even later. According to this view, peace ruled the lower stages of societal development, and this state of affairs was thus automatically transferred to prehistory. A general feature is furthermore that in anthropology the emphasis

has clearly been on warfare as structure, whereas in archaeology the focus has been upon the specialised agent of war, the warrior, in a heavily idolised version. The anthropology of war, by comparison, contains little reference to the agents of war, notably the warriors, who are out of focus and undertheorised.

Stereotyped understandings of society are definitely not absent in anthropology, as the debate between 'hawks' and 'doves' demonstrates, but the realist constituent has nevertheless in general been more substantially present. This difference could be rooted in the material sources of archaeology, which perhaps makes it more innately disposed to produce stereotyped visions of society. The precise handling of war in anthropology has more recently undergone considerable changes, thus bringing forth another dimension of the historiography. There is generally a development from explaining war through cross-cultural comparisons to understanding war in its specific social context. In addition, the realist component has moved more and more into the focus of research. At least four traditions can be outlined,[6] even if they do not occur completely separated from each other.

First, a materialist-functionalist approach locates the causes of war in the competition over scarce resources (females, food, land, etc.). It is sometimes argued that war has a positive effect in that it redistributes agents more fittingly across the landscape. Second, a structural approach claims that the explanation of war lies in the patterns of social structures; the individual actor is without much significance. Warfare either results from a breakdown of social norms, or war is capable of reproducing and changing social norms. A third tradition is the structure-agency approach, which considers action/agent and social structure as mutually dependent and inseparable, and moreover does not see a dichotomy between examining war as 'verstehen' and 'erklären'. War and violent conflict are regarded as strategic action situated within the continuities of social practice. This approach has so far only been used in modern settings, but doubtless also has potential in studies of prehistoric war and violence. Fourth, a fairly new approach strives to understand war by focussing directly upon the violent acts and their meanings in the cultural and social contexts that created them. A related approach is concerned with the subjective sphere of the war victims; that is, their personal feelings of pain and hopelessness when they experience war and post-war. One focal point has been the breakdown of social life, another the re-building of it.

This fourth development in some ways breaks with the classic anthropology of warfare of the first three traditions,[7] which may be summarised thus: despite the experienced disorder of war it can be analysed on its social and historical background and it can be compared cross-culturally; structured and ordered patterns will then reveal themselves in the rear-view mirror. From an archaeological point of view this is a reasonable approach, but it needs to be supplemented by elements from the fourth tradition: the terrible face of war as experienced by the participants should be added to the interpretation. Likewise, the culturally specific meanings of war and violence should be considered – even the possibility that they may deviate substantially from our own values in this respect. A past with deviating cultural values should be allowed, but nevertheless critically assessed. Simon Harrison has made this important point by comparing warfare in the Highlands and coastal areas of Papua New Guinea (1989), illustrating how the attitude to war and violence varied even within this region.

In the Highlands, warfare was conducted and perceived in a manner that Europeans find fairly easy to understand. Warfare was regarded simply as a violent form of sociability, and anger and aggression were conceptualised rationally as a drive to use violence. In a strategy of revenge, killings were reciprocated as a harmful alternative to gift exchange, simultaneously emphasising and confirming male companionship (Harrison 1989: 586). By contrast, in the endogamous villages of the Lowlands all outsiders were regarded as enemies inasmuch as they threatened the internal universe of everyday social life; this demonisation of the other, however, was a mainly male point of view. Warfare was aimed at outsiders, legitimated through the use of hunting metaphors and undertaken as ritual action centred on the village male cult. Through a ritual process each of the men became transformed into another person: the initiated men took on a ritual mask of war making them capable of extreme and indiscriminate violence; their 'spirits' went ahead of them performing the atrocities (ibid.: 586ff). This is not aggression in the Western sense; rather it is impassiveness, withdrawal from emotion and suspension of any feeling (which was probably not shared by the victims). Marshall Sahlins explained warfare

among tribes as a total breakdown of existent normative rules of sociability, but this theory is not valid in Lowland New Guinea, says Harrison (ibid.: 583, 590f). Through secret rituals of magic the initiated men divorced themselves entirely from the social world of their community and outsiders alike, thus setting aside morals and norms (Harrison 1989: 591; 1993; cf. also 1996). Rituals are thus in this context used to create a social space for the enactment of violence.

In summary, prior to 1995, when archaeology finally broached the topic of warfare more consistently, social anthropology experienced a renewal of the subject in the direction of a marked interest in uncovering the multiple cultural meanings of atrocities committed during ethnic wars and in revealing the human pain and disaster involved in all wars. The archaeology of war, as performed hitherto, compares best with the first and second of the above four attitudes to war in social anthropology and sociology. There is, however, no doubt that the remaining positions can provide useful alternatives and supplements. The structure-agency approach may prove especially valuable to archaeology, but obviously the fourth approach contains subjective elements that cannot be ignored.

Towards an archaeology of warfare and warriors

The question is how to avoid the pitfalls associated with dealing with the issues of warfare and warriors in the past? The answer can never be definite, because we all fall victim to subjective influences. A few suggestions should nevertheless be made. One possibility for a better understanding of warfare and warriors clearly lies in theoretical reflection, but of course also in taking into account the empirical database as it stands. In addition, archaeological studies have to mediate the viciousness of war-related violence even if the participants in prehistoric wars can no longer be interviewed.

Archaeology must resist thinking in dichotomies, and thus discard the historically and ideologically rooted, contrasting pre-understandings of the other. We are in the habit of thinking in rigid categories, often in contrasts, as Ian Hodder points out (1997). The possibility of a variety of in-between positions must be considered, the growing database optimistically exercising an increasing constraint on the number of possible interpretations. There is evidently a need to diversify a rather static view of the prehistoric warrior by taking into account the variegated social realms of this being, archaeologically, sociologically and ethnographically. The archaeological database of the European Neolithic and the Bronze Age, for example, possesses the potential to gain insight into the ways of the warrior. The precise nature of prehistoric discourses, being mainly material and thus silent, should not be considered a disadvantage, inasmuch as social and material practice always interact. People at all times have produced and utilised material culture, which therefore embodies visible patterns and underlying structures of human actions and thoughts, hence also the more violent aspects of these (cp. chapters 31-34).

Viewed across time and space, warriorhood is a complex social institution (see Vandkilde, chapter 26) which thrives in eras of prolonged war as well as in periods of more limited warfare and even peace. Warriorhood as social identity feeds on very different human qualities, notably bravery and brutality, and interpretations have to include these. Within the two extreme positions described above queries about aggression or peace too easily become a question of being: humans are by nature aggressive or peaceful (Vandkilde and Bertelsen 1999). An alternative approach is to regard warfare – when it occurs – as part of social practice; one kind of social action amongst many others. This promotes a relational understanding of warriors as a social identity constantly being negotiated with other social identities within society. This furthermore highlights the fact that the meaning of warfare and warriors – apart from being culturally specific – also depends on the perspective of the agents, and on their varied and changing identifications in society, including whether they are victors or victims.

Anthropological literature sometimes describes societies as being in a constant state of war or peace, but these absolutist expressions may well hide variations in scale, purpose, meaning and frequency. This point is evident in Polly Wiessner and Akii Tumu's recent historical-anthropological analysis of the Mae Enga in Highland Papua New Guinea (1998: 119ff, 152f). Throughout their history Enga were concerned with warfare, but seen through historical glasses the scale of warlike activities nevertheless fluctuated, the degree of viciousness being dependent on the objectives of the conflicting parties.

A similar message is conferred in Simon Harrison's *The Mask of War* (1993), about the Lowland New Guinea Avatip for whom war belonged in a ritual dimension separated from the sociable peace of community life. Even among the notoriously warlike Yanomami in southern Venezuela and northern Brazil, the frequency of war varied according to the region and time: Jürg Helbling (1996) argues that hostility and warlike behaviour prevail in the region merely because it is too risky to engage in a strategy of peace. These cases accord with the suggestion made here, namely, to regard warfare as strategy and action rather than a trait rooted in biology. Warfare and related social activities thus constitute an identity-forming frame that may result in identification as warrior and, ultimately, exclusionist institutions comprising warriors.

Archaeology has tended to treat warfare as if divorced from the rest of social practice (cf. Thorpe 2001). Approaches to early warfare ought to become more nuanced with recognition of its ideal and real aspects and examination of its social setting and historical background. The same is of course valid for studies of warriors: their real and ideal features must be approached and related to other identities of age, gender, rank and profession (cf. Robb 1997; Shepherd 1999). Recent anthropology and sociology can be used here as a source of inspiration in that warfare is generally regarded as intentional action situated within the continuities of social practice (cf. Jabri 1996). The perspective of warfare as a violent kind of social action is furthermore useful because archaeological remains are essentially fragments of past social action. War and violence are, moreover, never accidental, since they are always embedded in a cultural logic; therefore, attention should be paid to the violent acts themselves, the cultural meanings they carry and the cultural landscape that forms them.

At the threshold of the 21st century war finally entered the archaeological agenda, hence making interpretations accord better with the archaeological sources. However, the terrible side of war has not yet become part of archaeological histories, which are usually written in a strictly analytical language. The recent approach to war in archaeology recalls the rationalisation of war that has been accomplished in the sciences of the 19th and 20th centuries. From Carl von Clausewitz onwards war has been systematically reduced to rules, procedures, functions, causes and effects as part of a modern political agenda, thereby legitimising the slaughter on and off the battlefield (Pick 1993: 165ff). Recent developments in the anthropology of war suggest that it is possible to enhance the degree of realism.

During the last ten years the anthropological handling of war has changed. A concern with understanding the actions, experiences, motives and feelings of combatants, civilians and victims under the chaotic conditions of war and post-war has supplemented the impersonal political-science analysis (e.g., Nordstrom and Robben 1995; Nordstrom 1998; Macek 2000; Kolind 2004; this volume chapter 29). This particular interest is thus one of expression. Physical pain is notably considered an irreducible bodily experience which can never be wholly communicated and which therefore forms an instrument of power (Scarry 1988). Feelings of meaninglessness and pain under circumstances of war are on the other hand universal in character (cf. Tarlow 1999: 20ff, 138ff). This means that the human suffering involved can at least to some extent be communicated in our reports regardless whether the war took place in Jutland during the Middle Neolithic or in Bosnia merely a few years ago.

Violence, hardship and death are situated at the core of warfare and warriors; if left out the story becomes incomplete and too easily a commemoration of heroes. Skeletal trauma and weaponry in various archaeological contexts can be taken as direct or indirect evidence of the presence of war and violence with everything it implies in terms of cultural meaning, agency and human suffering. It is crucial in our archaeologies to include the vicious face of war and warriors. This is because idealisations of violent identities and one-sided rationalisations of prehistoric war inevitably run the risk of legitimising the use of violence and waging of war in our own time.

NOTES

1 The present article is a much-extended version of Vandkilde 2003, especially as regards the history of the research and the relationship between archaeology and social anthropology.
2 The increase in evidence of trauma during the transition to food production in Scandinavia is currently being re-examined and debated (Thorpe n.d.; 2000 with references; cp. also the present volume chapter 10).

3 Candidates in archaeology attended Kaj Birket-Smith's lectures at the University of Copenhagen. Birket-Smith was influenced by the structural-functionalist tradition in ethnography and cultural anthropology, notably by Alfred Reginald Radcliffe-Brown (1881-1955).
4 War studies occur in the form of conference reports, anthologies, monographs and articles: for instance, Randsborg 1995; Carman 1997; Martin and Frayer 1997; Maschner and Reedy-Maschner 1998; Osgood 1998; Laffineur 1999; Louwe Kooijmans 1993; Fokkens 1999; Earle 1997; Carman and Harding 1999; Osgood et al. 2000; Runciman 1999; Thorpe 2000; 2001; n.d.
5 Keeley (1996) renamed the Rousseauian notion of the noble savage 'myth of the peaceful savage'.
6 Here I am relying on Torsten Kolind (2004.) and a summary by Simon Harrison (1996 with references).
7 For example, Carneiro 1970; Chagnon 1968; Clastres 1994; Fergusson and Whitehead 1992; Haas 1990; Knauft 1990; 1991; Otterbein 1970; 1999 (with references); Vayda 1960; Wiessner and Tumu 1998 and Wolf 1987.

BIBLIOGRAPHY

- Andersen, S.H. 1972. 'Ertebøllekulturens harpuner'. *Kuml* 1971, pp. 73-125.
- Andersen, S.H. 1975. 'Ringkloster, en jysk indlandsboplads med Ertebøllekultur'. *Kuml* 1973-74, pp. 11-108.
- Andersen, S.H. 1981. *Stenalderen: Jægerstenalderen* (Danmarkshistorien Oldtiden). Copenhagen: Sesam.
- Becker, C.J. 1954. 'Die mittel-neolithischen Kulturen in Südskandinavien'. *Acta Archaeologica*, XXV, pp. 49-150.
- Birket-Smith, K. 1941-1942. *Kulturens Veje. Naturfolk og kulturfolk*. Vol. I-III. Copenhagen: Politikens Forlag.
- Bradley, R. 1990. *The Passage of Arms. An Archaeological Analysis of Prehistoric Hoards and Votive Deposits*. Cambridge: Cambridge University Press.
- Bridgford, S.D. 1997. 'Mightier than the pen? An edgewise look at Irish Bronze Age swords'. In: J. Carman (ed.): *Material Harm. Archaeological Studies of War and Violence*. Glasgow: Cruithne Press, pp. 95-115.
- Broholm, H.C. 1943-44. *Danmarks Bronzealder*. Vol. I and II. Copenhagen: Nyt Nordisk Forlag.
- Brøndsted, J. 1957-58. *Danmarks Oldtid*. Vol. I and II (Stone Age and Bronze Age). Copenhagen: Gyldendal.
- Brun, P. 1988. 'L'entité Rhin-Suisse-France orientale: nature et évolution'. In: P. Brun et C. Mordant (eds.): *Le Groupe Rhin-Suisse-France Orientale et la Notion de Civilisation des Champs d'Urnes*. (Actes du Colloque International de Nemours 1986, Mémoires du Musée Préhistoire d'Ile-de-France 1). Nemours, pp. 599-620.
- Capelle, T. 1982. 'Erkenntnismöglichkeiten ur- und frühgeschichtlicher Bewaffnungsformen. Zum Problem von Waffen aus organischem Material'. *Bonner Jahrbücher*, 182, pp. 265-88.

- Carman, J. (ed.) 1997. *Material Harm. Archaeological Studies of War and Violence.* Glasgow: Cruithne Press.
- Carman, J. and Harding A. (eds.) 1999. *Ancient Warfare. Archaeological Perspectives.* Stroud: Sutton Publishing.
- Carneiro, R.L. 1970. 'A theory of the origin of the state'. Science, 169, pp. 733-38.
- Chagnon, N.A. 1968. *Yanomami. The Fierce People.* New York, London and Chicago: Holt, Rinehart and Winston.
- Champion, T. 1990. 'Migration revived'. *Journal of Danish Archaeology*, 9, pp. 214-18.
- Chapman, J. 1998. 'The impact of modern invasions and migrations on archaeological explanation. A biological sketch of Marija Gimbutas'. In: M. Díaz-Andreu and M.L.S. Sørensen (eds.): *Excavating Women: A History of Women in European Archaeology.* London: Routledge, pp. 295-314.
- Chapman, J. 1999. 'The origins of warfare in the prehistory of Central and Eastern Europe'. In: Carman J. and Harding A. (eds.): *Ancient Warfare. Archaeological Perspectives.* Stroud: Sutton Publishing, pp. 39-55.
- Childe, V.G. 1941. 'War in Prehistoric Societies'. *The Sociological Review*, 33, pp. 126-39.
- Childe, V.G. 1946. *Scotland Before the Scots.* London: Methuen.
- Childe, V.G. 1951(1936). *Man Makes Himself.* London: Watts and Co.
- Childe, V.G. 1958. *The Prehistory of European Society.* Harmondsworth: Penguin.
- Clark, J.G.D. 1939. *Archaeology and Society.* London: Methuen.
- Clark, J.G.D. 1952. *Prehistoric Europe: The Economic basis.* London: Methuen.
- Clark, J.G.D. 1966. 'The invasion hypothesis in British archaeology'. *Antiquity*, 40, pp. 172-89.
- Clark, J.G.D. 1975. *The Earlier Stone Age Settlement of Scandinavia.* Cambridge: Cambridge University Press.
- Clark, J.G.D. and Piggott, S. 1978(1965). *Prehistoric Societies.* Harmondsworth: Penguin.
- Clastres, P. 1994. *Archeology of Violence.* New York: Semiotext(e).
- Damm, C. 1993. 'The Danish Single Grave Culture – ethnic migration or social construction?' *Journal of Danish Archaeology*, 10, pp. 199-204.
- Demakopoulou, K., Eluère, C., Jensen J., Jockenhövel, A. and Mohen, J.P. (eds.) 1998. *Gods and Heroes of the Bronze Age. Europe at the Time of Ulysses.* Copenhagen: the National Museum of Denmark; Bonn: Kunst und Ausstellungshalle der Bundesrepublik Deutschland; Paris: Galeries nationales du Grand Palais; Athens: National Archaeological Museum.
- Earle, T.K. 1997. *How Chiefs come to Power: the Political Economy in Prehistory.* Stanford: Stanford University Press.
- Fabech, C., Hvass, S., Näsmann, U. and Ringtved, J. 1999. '"Settlement and Landscape" – a presentation of a research programme and a conference'. In: C. Fabech and J. Ringtved (eds.): *Settlement and Landscape. Proceedings of a conference in Århus, Denmark, May 4-7 1998.* Aarhus: Jutland Archaeological Society, pp. 13-28.
- Ferguson, R.B. and Whitehead N.L. (eds.) 1992. *War in the Tribal Zone. Expanding States and Indigenous Warfare.* Santa Fee: School of American Research.
- Fokkens, H. 1999. 'Cattle and martiality: changing relations between man and landscape in the Late Neolithic and the Bronze Age'. In: C. Fabech, S. Hvass, U. Näsmann and J. Ringtved (eds.): *Settlement and Landscape. Proceedings of a Conference in Århus, Denmark, May 4-7, 1998.* Aarhus: Jutland Archaeological Society, pp. 35-43.
- Frankenstein, S. and Rowlands, M. 1978. 'The internal structure and regional context of Early Iron Age Society in South-West Germany'. *Bulletin of the Institute of Archaeology, London*, 15, pp. 73-112.
- Gimbutas, M. 1982(1974). *The Godesses and Gods of Old Europe 6500-3500 BC. Myths and Cult Images.* London: Thames and Hudson.
- Glob, P.V. 1945. 'Studier over den jyske enkeltgravskultur'. *Aarbøger for Nordisk Oldkyndighed og Historie*, 1944, pp. 5-282.
- Gosden, C. 1999. *Anthropology and Archaeology. A Changing Relationship.* London and New York: Routledge.
- Gramsch, A. 2000. '"Reflexiveness" in archaeology, nationalism, and Europeanism'. *Archaeological Dialogues*, 7(1), pp. 4-19.
- Gröhn, A. 2004. *Positioning the Bronze Age in social theory and research contexts.* Lund University: Department of Archaeology and Ancient History – Almqvist & Wiksell (Acta Archaeologica Lundensia Series).
- Haas, J. (ed.) 1990. *The Anthropology of War.* New York: Cambridge University Press.
- Harrison, S. 1989. 'The Symbolic Construction of Aggression and War in a Sepik River Society'. *Man*, 24, pp. 583-99.
- Harrison, S. 1993. *The Mask of War: Violence, Ritual and the Self in Melanesia.* Manchester: Manchester University Press.
- Harrison, S. 1996. 'War, warfare'. In: A. Barnard and J. Spencer (eds.): *Encyclopaedia of Social and Cultural Anthropology.* London: Routledge, pp.651-52.
- Häusler, A. 1994. 'Grab- und Bestattungssitten des Neolithikums und der frühen Bronzezeit in Mitteleuropa'. *Zeitschrift für Archäologie*, 28, pp. 23-61.
- Hedeager, L. and Kristiansen, K. 1985. 'Krig og samfund i Danmarks Oldtid'. *Den Jyske Historiker*, 31-32, pp. 9-25.
- Helbling, J. 1996. 'Weshalb bekriegen sich die Yanomami? Versuch einer spieltheoreretischen Erklärung'. In: P. Braünlein and A. Lauser (eds.): *Krieg und Frieden.* Bremen: Keo.
- Hobbes, T. 1958(1651). *Leviathan.* Indianapolis: Bobbs-Merrill.
- Hodder, I. 1997. 'Relativising Relativism'. *Archaeological Dialogues*, 4(2), pp. 192-94.
- Jabri, V. 1996. *Discourses on Violence. Conflict analysis reconsidered.* Manchester and New York: Manchester University Press.
- Jensen, J. 1979. *Oldtidens samfund. Tiden indtil år 800.* (Dansk Social Historie, 1). København: Gyldendal.
- Jensen, J. 1982. *The Prehistory of Denmark.* London and New York: Methuen.

- Jørgensen, A. and Clausen, B. (eds.) 1997. *Military Aspects of Scandinavian Society in a European Perspective*, A.D. 1-1300. København: Nationalmuseet.
- Keeley, L.H. 1996. War Before Civilization. *The Myth of the Peaceful Savage*. New York and Oxford: Oxford University Press.
- Knauft, B.M. 1990. 'Melanesian warfare: A theoretical history'. *Oceania*, 60, pp. 250-311.
- Knauft, B. M. 1991. 'Violence and sociality in human evolution'. *Current Anthropology*, 32(4), pp. 391-428.
- Kolind, T.N. 2004. *War, Violence, and Identification. Resisting the Ethnification of Everyday Life in Post-War Bosnia*. Ph.D. thesis. Aarhus: Department of Anthropology and Ethnography, Aarhus University.
- Kristiansen, K. 1982. 'The Formation of Tribal Systems in later European Prehistory: northern Europe 4000-800 BC'. In: C. Renfrew, M. Rowlands and B. Seagraves (eds.): *Theory and Explanation in Archaeology*. (The Southampton Conference). New York: Academic Press, pp. 241-80.
- Kristiansen, K. 1984a. 'Krieger und Häuptling in der Bronzezeit Dänemarks: Ein Beitrag zur Geschichte des bronzezeitliches Schwertes'. *Jahrbuch des römisch-germanisches Zentralmuseums Mainz*, 31, pp. 187-208.
- Kristiansen, K. 1984b. 'Ideology and Material Culture: An Archaeological Perspective'. In: M. Spriggs (ed.): *Marxist Perspectives in Archaeology*. Cambridge: Cambridge University Press, pp. 72-100.
- Kristiansen, K. 1991a. 'Prehistoric Migrations – the case of the Single Grave and Corded Ware Cultures'. *Journal of Danish Archaeology*, 8, pp. 211-25.
- Kristiansen, K. 1991b. 'Chiefdoms, States and Systems of Social Evolution'. In: T.K. Earle (ed.): *Chiefdoms: Economy, Power and Ideology*. Cambridge: Cambridge University Press, pp. 16-43.
- Kristiansen, K. 1998. *Europe Before History*. Cambridge: Cambridge University Press.
- Kristiansen, K. 1999. 'The emergence of warrior aristocracies in later European prehistory and their long-term history'. In: J. Carman and A. Harding (eds.): *Ancient Warfare. Archaeological Perspectives*. Stroud: Sutton Publishing, pp. 175-89.
- Kristiansen, K. 2002. 'The Tale of the Sword – Swords and Swordfighters in Bronze Age Europe'. *Oxford Journal of Archaeology*, 21(4), pp. 319-32.
- Laffineur, R. (ed.) 1999. *Polemos. Le contexte guerrier en Égée à l'age du Bronze. Actes de la 7e Rencontres egéenne internationale Université de Liège, 14-17 avril 1998*. (Aegaeum 19). Université de Liège, Liège and University of Texas, Austin.
- Louwe Kooijmans, L.P. 1993. 'An Early/Middle Bronze Age multiple burial at Wassenaar, the Netherlands'. *Analectica Praehistorica Leidensia*, 26, pp. 1-20.
- Macek, I. 2000. *War Within. Everyday Life in Sarajevo under Siege*. Uppsala: Acta Univeritatis Upsaliensis.
- Malmer, M.P. 1962. *Jungneolithischen Studien*. Lund: Acta Archaeologica Lundensia.
- Malmer, M.P. 1989. 'Etnoarkeologiska synspunkter på stridsyxakulturen. Stridsyxakultur i Sydskandinavien'. In: L. Larsson (ed.): *Rapport från det andra nordiska symposiet om Stridsyxatid i sydskandinavien*. (Institute of Archaeology Report Series, 36). Lund: Institute of Archaeology, Lund University, pp. 7-12.
- Martin, D.L. and Frayer, D.W. (eds.) 1997. *Troubled Times. Violence and Warfare in the Past*. Langhorne: Gordon and Breach Publishers.
- Maschner, H.D.G. and Reedy-Maschner, K.L. 1998. 'Raid, Retreat, Defend (Repeat): The Archaeology and Ethnohistory of Warfare on the North Pacific Rim'. *Journal of Anthropological Archaeology*, 17, pp. 19-51.
- Miller, D. and C. Tilley 1984. 'Ideology, Power and Prehistory: an introduction'. In: D. Miller and C. Tilley (eds.): *Ideology, Power and Prehistory*. Cambridge: Cambridge University Press, pp. 1-15.
- Nielsen, P.O. 1981. *Stenalderen: Bondestenalderen* (Danmarkshistorien Oldtiden). København: Sesam.
- Nordbladh, J. 1989. 'Armour and fighting in the south Scandinavian Bronze Age, especially in view of rock art representations'. In: T.B. Larson and H. Lundmark (eds.): Approaches to Swedish Prehistory: *A Spectrum of Problems and Perspectives in Contemporary Research*. (British Archaeological Reports, International Series, 500). Oxford: British Archaeological Reports, pp. 323-33.
- Nordstrom, C. 1998. *A Different Kind of War Story*. Philadelphia: University of Pennsylvania Press.
- Nordström, C. and Robben, A. (eds.) 1995. *Fieldwork under Fire: Contemporary Studies of Violence and Survival*. Berkeley: University of California Press.
- Osgood, R.H. 1998. *Warfare in the Late Bronze Age of North Europe*. (British Archaeological Reports International Series, 694). Oxford: Archaeopress.
- Osgood, R., Monks, S. and Toms, J., (eds.) 2000. *Bronze Age Warfare*. Stroud: Sutton Publishing.
- Otterbein, K.F. 1970. *The Evolution of War. A Cross Cultural Study*. New Haven: HRAF Press.
- Otterbein, K.F. 1999. 'A History of Research on Warfare in Anthropology'. *American Anthropologist*, 101(4), pp. 794-805.
- Pick, D. 1993. *War Machine. The Rationalisation of Slaughter in the Modern Age*. New Haven and London: Yale University Press.
- Randsborg, K. 1995. *Hjortspring. Warfare and Sacrifice in Early Europe*. Aarhus: Aarhus University Press.
- Resic, S. 1999. *American Warriors in Vietnam. Warrior Values and the Myth of the War Experience During the Vietnam War 1965-1973*. Malmö: Team Offset and Media.
- Ringtved, J. 1999. 'Settlement organisation in times of war'. In: C. Fabech and J. Ringtved (eds.): *Settlement and Landscape. Proceedings of a conference in Århus, Denmark, May 4-7 1998*. Aarhus: Jutland Archaeological Society, pp. 361-82.
- Robb, J. 1997. 'Violence and gender in Early Italy'. In: D.L. Martin and D.W. Frayer (eds.): *Troubled Times: vio-*

lence and Warfare in the Past. Langhorne: Gordon and Breach, pp. 111-44.

- Rowlands, M. 1980. 'Kinship, alliance and exchange in the European Bronze Age'. In: J. Barrett and R.J. Bradley (eds.): *Settlement and Society in the British Later Bronze Age*. (British Archaeological Reports, 83). Oxford: British Archaeological Reports, pp. 15-55.
- Runciman, W.G. 1999. 'Greek Hoplites, Warrior Culture, and Indirect Bias'. *Journal of the Royal Anthropological Institute (N.S.)*, 4, pp. 731-51.
- Scarry, E. 1988. *The Body in Pain. The Making and Unmaking of the World*. Oxford: Oxford Paperbacks.
- Shanks, M. and Tilley, C. 1987. *Social Theory and Archaeology*. Cambridge: Polity Press.
- Shepherd, D. 1999. 'The Elusive Warrior Maiden Tradition: Bearing Weapons in Anglo-Saxon Society'. In: J. Carman and A. Harding (eds.): *Ancient Warfare. Archaeological Perspectives*. Stroud: Sutton Publishing, pp. 219-43.
- Tarlow, S. 1999. *Bereavement and Commemoration. An Archaeology of Mortality*. Oxford and Malden: Blackwell Publishers Ltd.
- Thorpe, I.J.N. 2000. 'Origins of War. Mesolithic Conflict in Europe'. *British Archaeology*, 52, pp. 9-13.
- Thorpe, N. 2001. 'A War of Words. Ancient Warfare: Archaeological perspectives'. J. Carman and A. Harding (eds.). 1999. *Cambridge Archaeological Journal*, 11(1), pp. 1-3.
- Thorpe, I.J.N. n.d. Food, Death and Violence - the later Mesolithic of Southern Scandinavia. King Alfred's College: Winchester UK. Unpublished manuscript.
- Tilley, C. 1984. 'Ideology and the legitimation of power in the Middle Neolithic of Southern Sweden'. In: D. Miller and C. Tilley (eds.): *Ideology, Power and Prehistory*. Cambridge: Cambridge University Press, pp. 111-46.
- Treherne, P. 1995. 'The Warrior's Beauty: The masculine body and self-identity in Bronze-Age Europe'. *Journal of European Archaeology*, 3(1), pp. 105-44.
- Trigger, B.G. 1980. *Gordon Childe. Revolutions in Archæology*. London: Thames and Hudson.
- Trigger, B.G. 1989. *A History of Archaeological Thought*. Cambridge: Cambridge University Press.
- Vandkilde, H. 1996. *From Stone to Bronze. The Metalwork of the Late Neolithic and Earliest Bronze Age in Denmark*. Aarhus: Jutland Archaeological Society and Aarhus University Press.

- Vandkilde, H. 1998. 'Metalwork, Depositional Structure and Social Practice in the Danish Late Neolithic and Earliest Bronze Age'. In : C. Mordant, M. Pernot and V. Rychner (eds.) : *L'atelier du bronze en Europe du XX au VIII siècle avant notre ère. Actes du colloque international Bronze' 96. Tome III (session de Dijon): L'atelier du bronzier: élaboration, transformation et consommation du bronze en Europe du XXe au VIIIe siècle avant notre ére*. Paris: Comite des Travaux Historiques et Scientifiques, pp. 243-57.
- Vandkilde, H. 1999. 'Social distinction and ethnic reconstruction in the earliest Danish Bronze Age'. In: C. Clausing and M. Egg (eds.): *Eliten in der Bronzezeit. Ergebnisse zweier Kolloquien in Mainz und Athen*. (Monographien des Römisch-Germanischen Zentralmuseums 43). Mainz: Römisch-Germanischen Zentralmuseums, pp. 245-76.
- Vandkilde, H. 2000. 'Material Culture and Scandinavian Archaeology: A Review of the Concepts of Form, Function, and Context'. In: D. Olausson and H. Vandkilde (eds.): *Form, Function and Context. Material Culture Studies in Scandinavian Archaeology*. (Acta Archaeologica Lundensia. Series in 8.o. 31). Lund: Almqvist and Wiksell International, pp. 3-49.
- Vandkilde, H. 2003. 'Commemorative tales: archaeological responses to modern myth, politics, and war'. *World Archaeology*, 35(1), pp. 126-44.
- Vandkilde, H. and Bertelsen, K.B. 2000. 'Krig og samfund i et arkæologisk og socialantropologisk perspektiv'. In: O. Høiris, H.J. Madsen, T. Madsen og J. Vellev (eds.): *Menneskelivets Mangfoldighed. Arkæologisk og Antropologisk Forskning på Moesgård. Moesgårds Jubilæumsskrift*. Aarhus: Jysk Arkæologisk Selskabs Skrifter, pp. 115-26.
- Vayda, A.P. 1960. *Maori Warfare*. Auckland and Wellington: Polynesian Society.
- Vencl, S. 1984. 'War and Warfare in Archaeology'. *Journal of Anthropological Archaeology*, 3, pp. 116-132.
- Wiessner, P. and Tumu, A. 1998. *Historical Vines: Enga Networks of Exchange, Ritual and Warfare in Papua New Guinea*. Washington: Smithsonian Institution Press.
- Wolf, E. 1987. 'Cycles of war'. In: K. Moore (ed.): Waymarks: *The Notre Dame Inaugural Lectures in Anthropology*. Notre Dame, Indiana: University of Notre Dame Press, pp. 127-50.

'Total War' and the Ethnography of New Guinea

ERIK BRANDT

/6

'I may add at once that I am a convinced pacifist. I believe that modern warfare cannot in any way be shown to be beneficent', Bronislaw Malinowski wrote in 1922. Having declared so, Malinowski continued that

savage warfare is something quite different, the toll in human life and the suffering which it takes is as a rule relatively small. On the other hand, it provides a wide field for physical exercise, the development of personal courage, cunning and initiative, and the sort of dramatic and romantic interest, the wide vision of possibilities and ideals, which probably nothing else can replace. (1961[1922]: 212)

He illustrated his point by relating something he had learned during fieldwork:

Round the east end of New Guinea, where cannibalism and headhunting flourish, the natives had the unpleasant habit of making nocturnal raids, and of killing without any necessity, and in unsportmanslike manner, women and children as well as combatants. But when investigated more closely and concretely, such raids appear rather as daring and dangerous enterprises, crowned, as a rule, with but small success – half-a-dozen victims or so – rather than as wholesale slaughter, which indeed they never were. For the weaker communities used to live in inaccessible fastnesses, perched high up above

precipitous slopes, and they used to keep good watch over the coast. Now, when European rule has established peace and security, these communities have come down to the sea-coast and to swampy and unhealthy districts, and their numbers have rapidly diminished. (ibid.: 213)

Malinowski concluded this argument by addressing the colonial authorities with the suggestion that their effort to first pacify all their legal subjects was 'by no means an unmixed blessing' (ibid.).

Such a conclusion was no longer on Malinowski's mind when, in 1940, he set himself down to write 'An Anthropological Analysis of War', which was to provide backup for the fight against Nazi Germany. But however much his political concerns had changed, the Malinowski of 1940 still maintained that wars which were known ethnographically were quite different to the wars of the modern era. Comparing instances of 'man-hunting in search for anatomic trophies' with '[w]arfare as the political expression of early nationalism' and 'expeditions of organized pillage, slave-raiding, and collective robbery', the anthropologist declared that '[i]n a competent analysis of warfare as a factor in human evolution, they must be kept apart' (1941: 538-41, 538). At the evolutionary stages of 'savagery' and 'barbarism', Malinowski found that most of the fighting belongs to an interesting, highly complicated, and somewhat exotic type: ... it is devoid of any political relevancy, nor can it be considered as any systematic pursuit of intertribal policy. (ibid.: 538)

Such wars most certainly differed from 'the cultural pathology of today', for Malinowski considered that in his own time '[t]he influence of present warfare on culture is so total that it poses the problem whether the integral organization for effective violence – which we call totalitarianism – is compatible with the survival of culture' (ibid.: 543-44). To explicate his meaning, Malinowski turned to 'the war of 1914-1918':

In its technique, in its influence on national life, and also in its reference to the international situation, it became a total war. Fighting goes on now not merely on all frontiers geographically possible; it is waged on land, on sea, and in the air. Modern war makes it impossible to distinguish between the military personnel of an army and the civilians; between military objectives and the cultural portion of national wealth; and the means of production, the monuments, the churches, and the laboratories. ... The total character of war, however, goes much further. War has to transform every single cultural activity within a belligerent nation. The family and the school, the factory and the courts of law, are affected so profoundly that their work – the exercise of culture through autonomous self-contained institutions – is temporarily paralyzed or distorted. (ibid.: 544-45)

'This development', Malinowski underscored his point, 'is not due to the barbarism of a nation or of a dictator. It is inevitable, for it is dictated by the modern technique of violence' (ibid.: 545).

Sixty years after Malinowski wrote these words, the contradistinction that he made between 'savage war' and 'modern war' is still much debated among anthropologists. These days, however, authors tend to be rather critical of the idea that anthropology should conceptualise 'savage' and 'modern' war as two different phenomena. Most explicit on the issue, both Lawrence Keeley (1996) and Keith Otterbein (2000) agree that a myth was created that has much hampered the anthropology of war when historic anthropologists, such as Malinowski (1941: 543), differentiated 'the civilizations of savages' from 'the savagery of civilization'. Quoting Keeley, Otterbein (2000: 795) asserts that 'the myth includes three aspects: the notion of prehistoric peace or the "paci-

fied past" (prehistoric peoples did not have warfare) (1996: 17-24), the belief that hunter-gatherers or band-level societies did not engage in warfare (disputed by Ember [1978] and Dentan [1988]), and the assumption that when war occurred among tribal level societies it was ritualistic, game-like in nature'. According to Keeley (1996: 9), anthropologists aimed to

save the Rousseauian notion of the Noble Savage, not by making him peaceful (as this was clearly contrary to fact), but by arguing that tribesmen conducted a more stylized, less horrible form of warfare than their civilized counterparts waged. This view was systematized and elaborated into the theory that there existed a special type of 'primitive war' very different from 'real', 'true', or 'civilized' war.

Adding that modern anthropology denied thereby 'a brutal reality that modern Westerners seem very loath to accept' (Keeley 1996: 174), Keeley and Otterbein made a point that well fits the modernist history of anthropology known from authorities like Stocking (1992) and Kuklick, who wrote of 'the horrors of World War I' that made for a 'disenchantment with progress' that ensured that '[p]ost-World War I anthropologists typically portrayed the simplest societies as the realization of a cultural ideal' (1991: 23, 277, 270). However, the Rousseauian anthropology of war that Keeley and Otterbein imagine has little resemblance to the historic anthropology that has left its traces in the archives and libraries of our departments. At least, the archived texts strongly suggest that Malinowski's call to recognise the differences between different sorts of war was lost on most of those, who like him, wrote on New Guinea. Many New Guineasts, including at times Malinowski himself, proceeded as though all warfare is of 'an ugly sameness' (Keeley 1996: 173-74). Discussion of the history of the anthropology of war should take this into account, for New Guinea has long been one of 'the prestige zones of anthropological theory' (Appadurai 1986: 357). Ethnographies relating to this area have had much impact on the anthropology of war, and continue to do so (Simons 1999).

Where, against Keeley and Otterbein, I reconstruct an anthropology that has considered all war as being essentially the same, my work is not without precedent; it recaptures one of the conclusions of a reflective strand in the anthropology of New Guinea of the 1980s-1990s. But this contribution also challenges the historical narrative that has been presented by the authors of this reflective study of New Guinea, Marilyn Strathern (1985), Simon Harrison (1989; 1993) and Bruce Knauft (1999). Just as anthropological understanding of New Guinean warfare was not blinded by a myth about noble savagery and primitive war, it did not fall victim to some stable 'tradition of political thought' either (Harrison 1993: 3). In suggesting that the anthropological analysis of war in New Guinea derived its direction from a traditional set of assumptions known already from Thomas Hobbes's (1651) *Leviathan*, Strathern, Harrison and Knauft were no more accurate than Keeley and Otterbein. As my reading of the relevant historic ethnographies demonstrates, the anthropological debate about war in New Guinea from roughly 1940 to 1980 was in effect a debate over whether or not New Guinea was to be depicted as a site of 'total war'. Aided by Malinowski, anthropologists brought some neo-Hobbesian ideas to bear on this issue, but their work was not therefore a mindless reproduction of the view on war that Hobbes provided.

To suggest either that after World War I anthropology transformed into a primitivist discourse on non-violence, or that New Guineasts were mindless Hobbesians, is to misrepresent the past. And this is not a mere historiographic issue, for where the past is thus misrepresented misguided visions of the anthropology of war that is to be imagined today may well profit. The reconstruction I present in this paper is therefore also meant to be meaningful for the future. Before I can articulate the lesson I derive from the past I must, however, first detail the history of the anthropology of war in New Guinea. Hence, I will first present a further analysis of Malinowski's writing on war in New Guinea, and discuss the interpretive model of New Guinean life that Malinowski helped develop in his work on exchange. Next I turn to the critical representation of this model that some New Guineast came to articulate in the 1980s and '90s. My objections to this representation then brings me back to the first half of the twentieth century when not only Malinowski, but also Margaret Mead and Ruth Benedict, wrote influential works about New Guinea. These authors first suggested an image of New Guinea as a site of total war. Many studies of the next decades elaborated that image, which, in turn, provoked others, in the 1960s, to return to Malinowski's work of the 1920s and 1930s.

On 'savage war', and the war of ethnographies of the gift

As recent historical work demonstrates, the past offers many opportunities to doubt whether indeed warfare in Europe and North America was more 'total' in the twentieth century than it was before.[1] In the early nineteenth century, von Clausewitz (1832) believed he had good reason to think that 'absolute war' had arrived with the Napoleonic Wars of about 1800, and most of the European fighting that Wright (1942) designated as 'general war' occurred between 1700 and 1783. With the Thirty Years' War of the seventeenth century civilians and soldiers were equally vulnerable to violence, and 'the "war aims" of all sides in this conflict addressed fundamental questions of the social and moral order' (Chikering 1999: 23). Concerning the American experience of violence and destruction, it can be argued that at least the first of the world wars was less 'total' than the Civil War of the nineteenth century. Moreover, the imperial wars that were fought in the same period belong to the most violent wars that modern Westerners ever initiated. Nevertheless, a good deal of history has been written that, along with Malinowski's anthropology, brands the twentieth century the 'Century of Total War' (Marwick 1967). Military historians who have subscribed to this view have tended to proceed like the Malinowski I quoted above, branding all warfare before 1914 as less violent, less destructive, and less economically, socially, and culturally consequential than the wars of modernity. In the context of such claims, many military historians have stated that the anthropology of war reveals a 'non-violent' type of warfare. Thus, Michael Howard wrote in *The Laws of War*:

Anthropological studies show that although war in some form was endemic in most primitive societies, it was often highly ritualised and sometimes almost bloodless. It could be a rite de passage for adolescents, a quasi-religious ceremonial substituting for legal process, or a legitimised form of violent competition comparable to team sports in contemporary society (1994: 2, cp. Dawson 1996: 13-24)

Ever since Malinowski's day the anthropological literature on war in New Guinea has been more complex, though. While in some of his writing he explicitly denied the horrors of modern total war to New Guinean warfare, specialists on New Guinea have been more influenced by another strand in his work, in which he hinted at an analysis that pre-supposes that war is a violent horror also to the people of New Guinea.

When Malinowski (1961[1922]) wrote about the functions, and limited violence, of war in New Guinea, he reproduced a claim that the influential W.H.R. Rivers (1920) had presented just before, and Camilla Wedgwood (1930) would soon elaborate. In *Argonauts of the Western Pacific* Malinowski (1961[1922]) hinted, however, at the possibility of an alternative analysis, which proposed that at least to the New Guineans themselves war was not a functional institution of limited violence but a serious problem. Before he had gone to New Guinea, Malinowski had worked at the London School of Economics, where Edward Westermarck had encouraged his students to question the assertion by militarist anthropologists, such as Pearson, that 'struggle and ... suffering have been the stages by which the white man has reached his present stage of development, and they account for the fact that he no longer lives in caves and feeds on roots and nuts' (1901: 27). In that context, Malinowski had come to value arguments to the effect that among savages war was just one more ritual, not something to be singled out as the key to progress, but also arguments suggesting 'that there are among them germs of what is styled "international law"' (Westermarck 1910: vi). While his writing on the functions of 'savage war' recaptured much of that first argument, constructing premodern war as just another ritual, in *Argonauts*, Malinowski returned to this second type of argument. Formulating a conclusion that would underline the value of his description of the ritual exchange of bracelets and necklaces in savage New Guinea, he suggested that his work on exchange could well be read in search for an understanding of 'the evolution of intertribal intercourse and of primitive international law' (1961[1922]: 515). Though in writing so Malinowski did not yet explicate this, this implied that war was a violent horror also in the New Guinean experience. Those who in Malinowski's time concerned themselves with International Law were, after all, interested in law as a means to prevent war. The contemporary discourse on International Law to which the anthropologist appealed centred around the idea that one had to find a way to prevent a second world war and that International Law was that way, since, as a widely read interpretation of the First World War had it, it was 'the condition of international anarchy' that 'has produced war, and always must' (Dickinson 1926: 47, 41, cf. 1916).[2]

Whereas, in presenting exchange as primitive International Law, Malinowski himself did not yet explicitly articulate an interpretation of war as a violent problem, in following years Marcel Mauss' comments on the gift in New Guinea would make it quite clear that Malinowski's suggestion entailed that New Guineans tried to avoid war, and that they did so because their war too was a horror. Just as convinced as Malinowski that all had to be done to establish peace in Europe, the author of *L'Essai sur le don* declared that '[s]ocieties have progressed in the measure in which they, their sub-groups and their members, have been able to stabilise their contacts [and] people can create, can satisfy their interests mutually and define them without recourse to arms' (1954[1923-34]: 80). That savage people had managed to do so, the French ethnologist explained as in part derived from their reverence for their gods who taught them to share their wealth. Nevertheless, in the final pages of his study Mauss pointed out that the exchange practice of savages was a rational act inspired by dread of war and the recognition that exchange was the way to avoid its 'rash follies'. Here, Mauss reported that '[t]he people of Kiriwina said to Malinowski: "the Dobu man is not good as we are. He is fierce, he is a man-eater. When we come to Dobu, we fear him, he might kill us!"' (ibid.: 79-80). Following Richard Thurnwald (1912: Vol.3, Tab. 35, n.2), Mauss related that elsewhere in New Guinea, 'Buleau, a chief, had invited Bobal, another chief, and his people to a feast which was probably to be the first in a long series. Dances were performed all night long. By morning everyone was excited by the sleepless night of song and dance. On a remark made by Buleau one of Bobal's men killed him; and the troop of men massacred and pillaged and ran off with the women of the village' (1954[1923-24]: 80). According to Mauss, it was awareness that such things could happen that made the Savage exchange. 'It is by opposing reason to emotion and setting up the will for peace against follies of this kind that people succeed in substituting alliance, gift and commerce for war, isolation and stagnation', he wrote, adding:

In tribal feasts ... men meet in a curious frame of mind with exaggerated fear and an equally exaggerated generosity which appear stupid in no one's eyes but our own. In these primitive and archaic societies there is no middle path. There is either complete trust or mistrust. One lay down one's arms, renounces magic and gives everything away, from casual hospitality to one's daughter or one's property. It is in such conditions that men, despite themselves, learnt to renounce what was theirs and made contracts to give and repay. (ibid.)

War is war, in this argument. Among savages, as among moderns, war is something to be avoided in order to allow for progress and happiness.

Highly successful at the time, this analysis was soon also regarded as a most important contribution to anthropology by Malinowski, who, in *Coral Gardens and their Magic*, made it clear that reading Mauss had helped him to see that there was need for further inquiries into ceremonial exchange as a 'substitute for head-hunting and war' (1935: vol. 2, 246). When in the mid-1930s, he wrote about the possibility of analysing the substitution of head-hunting by exchange, Malinowski suggested that it would be possible to present a future study of this issue as an ethnographic exploration of the 'heroic' activities of the Trobriand islanders. This suggests that Malinowski still considered it appropriate to describe warfare in New Guinea in the idiom of aristocratic romance he had used in his writings about 'savage war' of the early 1920s, when he had declared Trobriand warfare 'open and chivalrous', 'with a considerable amount of fairness and loyalty', and 'rather a form of social "duel," in which one side earned glory and humiliated the other, than warfare' (1920: 10, 11). At the same time, Malinowski (1933) remained explicitly critical of colleagues who were referring at the time to 'modern war' in this romanticising idiom. Even when he hinted at the substitution of war by exchange, Malinowski thus continued to treat 'savage war' as significantly different from 'modern war'. Precisely how he thought he could integrate the image of war as a problem with an image of war as a heroic enterprise remained unclear, however, as Malinowski never actually completed the monograph on 'the heroic enterprise' that he announced in *Coral Gardens*. And for decades to come no other New Guineasts would even try to follow his lead, attempting to fuse the two apparently contrasting images of war. Instead, those who in the following decades would explore the substitution of exchange for war would stay close to Mauss' argument, and stress the image of war as a wasteful disorder over the idea of war as a domain of creative heroism. Following Malinowski's suggestion for further analytical work on the social functions of exchange, anthropologists thus came to write a body of literature that depicts New Guinean warfare as anything but a functional and non-violent custom.

Widely regarded at the time as a fine contribution to the literature by authors as diverse as Roy Wagner (1972) and Paula Brown (1970), Andrew Strathern's 1971 ethnography of the moka ritual of the Melpa well illustrates the point. Just as many other experts on New Guinea did, Strathern depicted life in the island's mountainous interior as a highly competitive struggle for 'a big name' among clans and aspiring leaders. That struggle often generated war, but Strathern saw exchange too: 'Warfare was what decided the ultimate balance or imbalance of physical power between territorial groups,' he wrote; 'but there were, and are, other ways in which competitive spirit and aggressiveness could find expression. Pre-eminent among these "other ways" is ceremonial exchange' (ibid.: 54). In this view, New Guineans were motivated to ceremonial gift exchange because exchange was a means to acquire a big name, but Strathern also suggested that local people understood, and valued, gift exchange as a 'positive alternative to war' (ibid.: 76). According to Strathern, evidence for this understanding could be found in local discourse on big men, which constructed the true big man not as a war leader but as a man of strong *noman* ('social consciousness') who did not suffer from *popokl* ('frustrated anger') and knew how to create exchange relations ('ropes of moka'). Observations made in the colonial era also indicated that the Melpa used exchange to avoid war and foster peace, Strathern (ibid.: 54) continued, choosing words that made it clear that he never considered war to be a functional and harmless ritual:

In many cases in the Highlands (e.g. the Siane, Salisbury 1962) exchange institutions effloresced and developed to a larger scale when Europeans banned warfare. This was not simply a result of a blockage on [fighting]; rather, it was an expression of the interest of big-men in pursuing an avenue of self-aggrandisement which was more effective and less hazardous than warfare itself. Thus it is that we find Kyaka big-men (Bulmer 1960) urging groups which were still fighting to give up warfare and join in the massive cyclical exchange ceremony, the tee, instead. In Hagen also one still hears frequently statements of the type: 'before we fought and killed each other, and this was bad; now a good time has come, and we can pay for killings and make moka'.[3]

The 'total war' before the return
to the war of the gift

As I will argue, the approach to war and the gift that Andrew Strathern practiced around 1970 solved one of the key problems in the anthropological debate on New Guinea of his time. About a decade later, it became the object of a serious critique among specialists on New Guinea, however.[4] After Papua New Guinea gained independence in 1975, spokesmen for various tribes from the highlands started to claim that warfare was a richly rewarding 'custom' that they had inherited from their ancestors and that a truly postcolonial, rather than neocolonial, 'national' state should thus not oppose this. In this changing context, Marilyn Strathern (1985: 122) came to see that the interpretation of gift exchange that had seemed so illuminating around 1970 'pre-judges the nature of violent confrontations, as they occur in the Papua New Guinea Highlands'. In her view, time had come for anthropologist to rethink the work on the gift and war that they had authored in the past. Along with this rethinking, anthropologists should then also reflect upon their own analytic practices, for, from Marilyn Strathern's point of view, these too often seemed to reproduce too much of Western culture. The result of this double rethinking was an interesting new body of ethnographic writing that well fits my claim that anthropologists had certainly not just denied war or downplayed violence. From a historiographic point of view, the reflections on the anthropology of war in New Guinea presented within the context of this literature are not too convincing, however.

When Marilyn Strathern (1985: 122) wrote that '[i]f we do not pre-judge the nature of "violent" behaviour, then we need not pre-judge the nature of "peaceable" behaviour either', that opened the way for a new view of exchange and warfare. It enabled Strathern to think that big men, calling upon combatants to stop fighting and exchange wealth, did not necessarily express the normative aversion to war that ethnographers had formerly recognised. What now seemed to matter to local actors was the spirited efficaciousness of these men, who, calling for the exchange of wealth, converted the exchange of blows and arrows into an exchange of wealth. People valued this power, rather than the consequences – peaceful conduct – that the established anthropology had valued. This was not to say that New Guineans did not recognise war's harmful

effects, but that to these people such harm was not the most salient thing about war. From a New Guinean perspective, warfare occurs within a ritual space, and, as Simon Harrison (1989; 1993) observed, the men who operate within this space are expected to sidestep the morality that structures domestic practice. The value of their acts is measured in terms of the amount of ancestral force that they demonstrate. This perspective suggests that men deserve a big name for knowing how to converse with the ancestral spirits. What matters is the ability to get in touch with the 'wildness' of the spirit world; and this capacity can be demonstrated equally well by a violent exchange of blows as by an impressive exchange of gifts (cf. O'Hanlon 1995; LiPuma 2000).[5] To some extent, anthropology thus returned once more to Malinowski, who, when he wrote of the possibility of writing of war and exchange in 1935, suggested that both were seen in New Guinea as forms of heroic action. For M. Strathern and Harrison the domain of the 'wildness' was not to be represented, however, in the romantic idiom that Malinowski had used, but in Marxist terms. To them, the symbolism of the wild presented an ideology, which ensured that only some (males) benefited from politics, while all suffered the hard work of wealth production and the distress and deaths of combat.

Though some have suggested so (Jolly 1992; Josephides 1991; Keesing 1992; MacIntyre 1995), this ethnography was thus not the site of romantic fantasies of 'noble savagery'. But if this new ethnography of war and exchange was critical of 'wildness', it was even more critical of the older ethnographic literature that attributed a preference for peace to its New Guinean subjects. Apparently, Marilyn Strathern (1985) suggested, anthropologists had been less concerned with the concerns of the New Guineans they met in the field, than with reconfirming an ethnocentric idea they knew from 'western' philosophy, i.e. that all humans of sound reason understand war as problem that they have to overcome. Supported by references to Sahlins (1972: 176), who in the 1960s observed that Mauss's work on war and the gift elaborated an anthropological vision 'brilliantly anticipated' already by Thomas Hobbes, this critique fits my argument. For to suggest that the argument concerning the gift as a pacifying institution is Hobbesian, is to make it more difficult to convincingly associate this argument with a fiction of noble savagery. Noble savagery is, after all, not what Hobbes is known for.

Still, I object. Sahlins' commentary regarding Hobbes' reasoning and Mauss' analysis of the gift as a way to pass from threatening war to peaceful sociality aids the understanding of Mauss' text, but it also isolates this text from its immediate historic context, and thereby eclipses the concerns that guided Mauss towards his neo-Hobbesian argument.[6] Recapturing Sahlins' decontextualising analysis, the critique that Marilyn Strathern and Simon Harrison presented of the ethnographic tradition that authors such as Andrew Strathern extended around 1970 has a similar effect. It helps to understand the argument of the pacifying gift, but obscures the contemporary challenge that motivated the arguments for the gift as a means of social control made around 1970. That many ethnographers of New Guinea around 1970 wrote of a war that was controlled by gift exchange was not because Hobbesian principles dominated their Western imagination. These authors adopted (and creatively reformulated) the neo-Hobbesian idea of the pacifying gift they knew from Mauss and Malinowski, because they recognised that idea as a valuable means to improve the ethnography of New Guinea of their time. At the time, much of New Guineast anthropology had come to emphasise a specific image of war, and writing of the gift an author like A. Strathern sought to undermine the stress on that image of war. And this image of war was not really Hobbes' image of war, for the war that Hobbes wrote about posed a problem that people could handle, whereas the war that spurred Andrew Strathern's writing of around 1970 was a 'total war' that effectively impaired the human capacity to overcome the state of war. Behind this rendering of warfare as total war lay a sense of despair, whereas the Hobbesian vision reflects a strong faith in human rationality. Like the 'anarchic war' that was said to be overcome by the gift in the neo-Hobbesian writings of Malinowski and Mauss, this 'total war' is also a representation that must be situated historically in relation to a public response to the warfare 1914-1918, but this war was even more radically different from the image of 'primitive war' that anthropology is unjustly criticised for.

At the time when Malinowski and Mauss first stressed the relevance of New Guinean exchange to the condition of the modern world of nation-states, in the United States public debate on World War I and international politics had turned into a debate on American identity. Fearful that the vengefulness of the French and British that was expressed in the Versailles Treaty had turned the Allied victory of 1918 into the cause for another war, many 'Americans' regretted that the United States had intervened in 'the European war'. In this context, people felt that time had come for the citizens of the United States to develop a 'genuine nationalistic self-consciousness', and to 'speak the truth about American civilization' (Stearns 1922: vii, iv). Several anthropologists joined this project, among them Ruth Benedict and Margaret Mead. Both of them felt that it was most important to redefine American identity in terms of culture, rather than race, and set out to make their compatriots think about what true American culture should look like. The representation of New Guinea became involved in this project when Benedict recognised that the study on Dobu by (Mead's partner) Reo Fortune (1932) offered much she could use for making her readers 'culture conscious'. Comparing Dobu life to ways of two native American tribes, Benedict made Dobu practice appear so different as to make the impression that culture mattered inescapable, and that it also mattered in what direction people decided to develop the pattern of their culture. As interpreted by Benedict, the ethnography of the Dobu became a call to consider what American culture should not become, and served to alert readers to the attractiveness of Benedict's favourite model for American society, the 'Apollonian' pattern of culture of the Native American Zuni. As rendered by Benedict, the culture of the Dobuan formed a 'paranoid' pattern that ensured that 'all existence appears to him as a cut-throat struggle' (1946: 159). Introducing Dobu, she wrote:

They are said to be magicians who have diabolic power and warriors who halt at no treachery. A couple of generations ago, before white intervention, they were cannibals, and that in an area where many peoples eat no human flesh. They are the feared and distrusted savages of the islands surrounding them. The Dobuans amply deserve the character they are given by their neighbours. They are lawless and treacherous. (ibid.: 120-21)

In Benedict's account this people did 'lack the smoothly working organization of the Trobriands, headed by honored high chiefs and maintaining peaceful and continual reciprocal exchanges of goods and privileges' (ibid.: 121).[7]

Even though these statements made Fortune (1939) protest that Benedict's search for patterns of cultures had resulted in a poor caricature of Dobu life, Mead followed the same procedure in analysing the material that, with Fortune's aid, she had collected along the River Sepik. Whereas Benedict had written of Apollonian and paranoid cultures, Mead wrote of feminine and masculine cultures, for to her the most important questions regarding American culture all had to do with femininity and masculinity. Regarding these questions, Mead asserted that

[t]he tradition in this country has been changing so rapidly that the term 'sissy', which ten years ago meant a boy who showed personality traits regarded as feminine, can now be applied with scathing emphasis by one girl to another. (1935: 212).

Mead wanted her readers to reflect upon the desirability of this cultural change. Men became insecure about their gender role, and what, Mead asked, would happen if people abandoned 'the assumption that women are more opposed to war than men, that any outspoken approval of war is more horrible, more revolting, in women than in men' (ibid.: 213). To Mead, it was clear that this 'meant a loss', since 'the belief that women are naturally more interested in peace … at least puts a slight drag upon agitation for war, prevents a blanket enthusiasm for war being thrust upon the entire younger generation' (ibid.: 212, 213). Reading about three New Guinean peoples, Mead's readers should come to recognise this. Just as ambivalent about the meaning of masculinity and femininity as the people of the United States, Tchambuli society showed the harmful consequences of men's confusion. Arapesh culture had the advantage that it offered a clear model for action, but it was entirely feminine in its orientation, and thus – like Soviet Communism, Mead intimated – repressive of much that Americans had always valued in their true men. Americans should be happy that their culture was different, but, then, they also should not want their culture to become like that of the Mundugumor, entirely masculine. The Mundugumor culture that Mead depicted made for a life much like that of the Dobuans of Benedict's account. As Mead wrote: 'both men and women are expected to be violent, competitive, aggressively sexed, jealous and ready to see and avenge insult, delighting in display, in action, in fighting' (ibid.:

158). With the members of this 'cannibal tribe', the most respected men were those who were known to be 'really bad man', Mead reported, and feelings of solidarity could be observed only when

head-hunting raids are planned, and the whole male community is temporarily united in the raid and the victory-feasts that conclude them. At these feasts a frank and boisterous cannibalism is practiced, each man rejoicing at having a piece of the hated enemy between his teeth. (ibid.: 134)

A horror that Americans should keep at a firm distance, the warfare thus ascribed to the Dobu and the Mundugumor was more like the 'total war' that Malinowski associated with modernity than the 'savage war' or 'anarchic war' that he associated with New Guinea. Nevertheless, one of the ethnographic authorities of the 1950s, Kenneth Read, called upon all students of New Guinea's Central Highlands to recognise Benedict and Mead's work on war and New Guinean culture as paradigmatic for future work. In his own work, he reproduced all of their claims concerning the 'cultural correlates' of what he called 'raids and organized, concerted attacks among [groups] which only a short time ago were, and in some cases still are, so extremely warlike' (1954: 22). On Read's account,

physical aggression is not merely a corollary of intergroup hostility. It is a more fundamental trait, the obverse of a more far-reaching insecurity in interpersonal and group relations. Physical violence and antagonism are the warp of the cultural pattern; present to some extent in most important relationships, they receive innumerable forms of symbolic and institutionalized expressions. (ibid.)

Stopping short of actually using the expression 'total war', Read thus attributed to war the same kind of impact that Malinowski had associated with total war. With his writing, war in New Guinea became a social force that greatly impacted each and every aspect of New Guinea culture. As if following Malinowski's account of total war in modern Europe, Read even suggested that many of New Guinea's cultural institutions were no longer able to perform their regular function, writing, for example:

At the conclusion of initiation ceremonies crowds of women armed with bows and arrows, sticks and stones, dressed in male decorations attack the returning procession of men and boys. Similar fights are staged at marriages when, characteristically enough, a man is required to shoot an arrow into the thigh of his wife. (ibid.: 23)

Inspired by such statements, one of Read's students invoked war as the context that made sense of all the ethnographic details that had for long puzzled students of Highland New Guinea. 'In Bena Bena', Lewis Langness wrote,

the stated aims of warfare were the *complete and total* destruction of the enemy, if possible. This included every man, women, and child, whether old, infirm, or pregnant. Although it is true that most raids resulted in only one, or few deaths, cases are known in which entire groups were destroyed. (1964: 174, emphasis added)

Hence, the presence of nonagnates in putatively agnatic clans in New Guinea:

Groups which constantly find it necessary to scatter and regroup, which are decimated by casualties, which must take refuge with friends (which are willing to accept them for the same reason they want to be accepted) cannot, I submit, maintain lineage purity. ... if expedient, one need not be too particular about someone else's genealogy, so long as he can fight. Perhaps, indeed, strict unilineal descent was too costly. (ibid.)

Likewise, to Langness, war determined the antagonistic relation between men and women:

Living in an hostile environment, and faced with the almost constant threat of annihilation by enemy groups, has resulted in, or is related to, a distinctive pattern of male solidarity which offers what the Bena Bena perceive as a better change for survival. Male solidarity involves the residential separation of the sexes and a complex of beliefs and sanctions designed to insure such separation, as well as a minimal amount of contact between males and females in general. These beliefs and sanctions that exist to buttress the social distance between males and females, although they are functional in terms of group survival in a dangerous and warlike environment, are so only at some cost to the sex-and-dependency needs of individuals and thus ultimately promote hostility and antagonism between the sexes. (1967: 163)

Just like the modern war of Malinowski's 1940 account, the war that Langness imagined was 'total' in both its aims and its paralysing impact on the culture it deformed.[8]

By the time Andrew Strathern first arrived in New Guinea, so many other New Guineasts had rendered the warfare waged in the island's high valleys in such terms that this had provoked advocates of the 'modern British school' (Kuper 1983) in anthropology to put a premium on the creation of an alternative account that would demonstrate how New Guinea Guinea Highlanders 'achieve a kind of social articulation or order that outlasts the bursts of conflict' (Glasse 1959: 289). John Barnes' widely discussed 1962 contribution to *Man* was most important in this respect. Authors like Read and Langness could be right that 'the disorder and irregularity of social life in the Highlands [...] is due in part to the high value placed on killing', but, Barnes (1962: 9) continued, an important observation pointed to a fact that was not yet fully recognised:

the pre-contact population was large and often densely settled; indigenous social institutions preventing excess violence and destruction must necessarily have been effective, for otherwise the population would not have survived. (ibid.)

Articulated in reaction to the anthropology of unrestrained warfare and the militarisation of culture, such writing spurred a search for effective means towards ensuring social order. In that context, colleagues of Barnes and their students recapitulated the anthropology of war and the gift that Malinowski and Mauss had developed in the 1920 and 1930s. Although the 1980s critics of Hobbesianism would suggest that it was their own presuppositions that guided authors like Strathern towards the image of big men who preferred gift exchange over war, these authors re-turned to the exchange of gifts only in order to demonstrate that New Guinea was not the site of total war that many of their colleagues reported it to be.

The composition of Andrew Strathern's *The Rope of Moka* should leave no doubt about this, for Strathern (1971: 53) not only discussed big men and their gift exchange, he also discussed the anthropology of war in New Guinea.[9] Much anthropological work had characterised interior New Guinea by violent warfare, Strathern wrote, continuing that indeed people there frequently went to war. Still, he insisted that the anthropology that had made violent warfare the key to the highland regions in interior New Guinea 'needs correcting in numbers of ways' (ibid.). Whereas it had become convenient for anthropologists to emphasise the destructive violence of New Guinean warfare and to write of a culture that valued war, 'the stress on warlike prowess varies in intensity throughout the Highlands', Strathern (ibid.: 53-54) wrote. Around 1970 Strathern was not yet sufficiently well established in academe to challenge colleagues such as Langness with respect to their own field sites, but by his choice of words Strathern suggested that the sites of total war these authors depicted were only places of marginal importance – 'fringe' or 'eastern' rather than 'central':

[Stress on warlike prowess] is very strong in some of the fringe Highland societies and in Central Highland societies of West Irian (e.g. the Hewa, Steadman n.d., and the Mbogoga Ndani, Ploeg 1965). It is strong also in some of the Eastern Highland societies, for example the Kamano (Berndt 1962), the Bena Bena (Langness 1964), and the Tairora (Watson 1971). Men of violence – whom Salisbury (1964) has dubbed 'despots' – seem to have arisen sporadically in a number of other Highland societies also. But in many of the large, central Highland areas, where population density is heavy, men of violence were not necessarily the important political leaders. And this is correlated with the fact that in these areas there were well-developed inter-group alliances, gradations of enemy relationships, controls on the escalation of fighting. (ibid.: 53)

Only at that point did Strathern turn to the big man and gift exchange as the social type and the institution that guaranteed that central New Guinea was not a place of uncontrolled warfare. The neo-Malinowskian, neo-Maussian argument of exchange rather than war that Strathern presented was thus neither a conventional primitivist denial of violent war, nor a straightforward Hobbesian claim. Instead, as his text itself suggests, his argument on gift giving rather than war was born from an attempt to overcome an anthropology that depicted New Guinea as a site of 'total war'.

Conclusion: possible futures

Elaborating on an image of war – 'total war' – that has not so far been recollected in historicising reflections on the anthropology of war, and tracing the debate the use of this image generated, I have attempted to be more faithful to the complexities of anthropology's historic dealings with warfare than those whose representations I have criticised. Hence, I wrote about the complexities of Malinowski's work, which contain passages differentiating savage from total war occur as well as passages suggesting that New Guineans and modern nations face a similar problem of war and international anarchy. New Guineast anthropology by and large bypassed the first, to elaborate instead upon the second of these suggestions. This happened only because much of the anthropology of New Guinea of the twentieth century depicted New Guinea as a site of total war. That suggestion made it meaningful for others to revive the image of New Guinea as a place where people knew an alternative to war that allowed them an escape from the state of war. My aim was not 'historicist', however – at least not in the sense in which Stocking (1965) has taught historians of anthropology to understand this term. With Stocking, I agree that the anthropology of the past deserves to be studied in its own right, but if Stocking once considered it possible to attempt this in writing on Boas, classical evolutionism, and early cultural anthropology, at present a historicist perspective on anthropology and war is out of question. Due to the work of such authors on the past of the anthropology of war as I have referred to, this subject is too much infused with contemporary interests to attempt to study the anthropology of war for purely historicist reasons. As I sought explicitly to relate my understanding of the material I have found stored in our discipline's archives of anthropology to both the regionalist claims about Hobbesianism and the generalised critique of the alleged attempt to save the mythical Noble Savage by imagining primitive war, I did so because I hope to also speak to the future of the anthropology of war.[10]

For the New Guineasts who first made the critique of Hobbesianism, this facilitated the thick description necessary to make sense of the evaluation of war in terms of ancestral power. This was something they had to do in order to get beyond the arguments on cultural paranoia that had first provoked the counter argument of the pacifying gift. By focussing on Hobbes, these authors could draw attention to the need to relate understandings of war to understandings of the person. After all, Hobbes (1651) had made it exceptionally clear that his view of war and peace was intimately related to a particular view of the human being, and he had first devoted a long series of chapters to the nature of

Man before he turned to war and its resolution. To recall Hobbes was a way to call upon ethnographers of New Guinea to do the same. It offered an introduction to the observation that New Guineans are so interested in the ability to make violence since they normally experience themselves as responding to the moral obligations that follow from the inborn ties that bind them to their maternal relatives, and long to 'produce admiration, fear, desire or other types of affect in others' (Harrison 1985: 117). Whereas for Hobbes, and his followers, the problem is to morally restrain each individual's innate aggressiveness, for New Guineans the problem would be to overcome morality. As has been repeatedly observed already, this argument presents a highly promising contribution to the ethnographic interpretation of New Guinea that goes well beyond both the discourse building upon Benedict and Mead, and the counterargument on the peace of the gift (Knauft 1994; Maclean 1998). On the other hand, anthropological analysis of the practice of warfare in New Guinea should recognise that all recent and current warfare involved, or involves, a changing environment, which have long since been affected by the activities and ideas of Europeans (Görlich 1999; Knauft 1999; Macintyre 1983; 1995). By equating the relevant historic discoursing on war in the West to one particular, radically individualist, contribution from philosophy, the critique of Hobbesianism eclipses the need for such recognition. By demonstrating that ethnographic writings are dialogically interanimated creations, rather than products of such isolated individuals as the reflective New Guineast of the 1980s and 1990s attributed to the West, and by underscoring that ethnographic representation relates to distinct historic situations, this chapter should, in contrast, reveal that need.

At the same time, this chapter should be a warning not to follow the lead of Keeley and Otterbein, who, just as the critics of Hobbesianism, had a distinct future in mind when they presented their view of anthropology's past. Whereas people like Harrison urged for a reconstruction of notions of personhood impacting the understanding of war, and thus the practice of warfare, the historical narrative about Rousseauian myth-making and mystifying the true character of warfare among tribal peoples was presented to arouse a desire for a 'realistic view of all warfare' (Keeley 1996: 24). On this account, the anthropology of war is to be brought in

line with the discipline's 'painfully accumulated facts', which, according to Keeley, 'indicate unequivocally that primitive and prehistoric warfare was just as terrible and effective as the historic and civilized version' (ibid.: 174). It is suggested that any future anthropology should recognise that '[p]rimitive warfare is simply total war conducted with very limited means' (ibid.: 175). Writing so, Keeley suggested that anthropology would make a distinct contribution to military history when it would start constructing the warfare of tribesmen as total war:

The discovery that war is total – that is, between peoples or whole societies, not just the armed forces who represent them – is credited by historians to recent times. … [T]his 'discovery' is comparable to the European discovery of the Far East, Africa, or the Americas. The East Asians, sub-Saharan Africans, and Native Americans always knew where they were; it was the Europeans who were so confused or ignorant. So it is with total war. (ibid.: 175-76)

What Keeley wants anthropologists to do, producing ethnographic reports on non-western 'total war', they have already been doing, however, for decades. Historically, by doing so they made a distinct contribution to the formulation of American culture and identity. The anthropology of New Guinea paid a high price, however, for the anthropology of 'total war' burdened regionalist discourse with an image of war that, while certainly belonging to a specific culture, was foreign to the setting studied in fieldwork. As I have demonstrated, it took New Guineasts many decades to pass beyond the resulting confusion.

Moreover, even if 'an anthropology of total war' could ever have made a distinct contribution to military history, the time for making such a contribution seems over now. While Keeley urged anthropologists to recognise the reality of total war, by now an increasing number of military historians treat 'total war' as a mythologising interpretation of war; not a 'discovery' of the nature of (modern) warfare to be celebrated, but one of those master narratives that, put to use by modern actors, have shaped the structure and texture of social action in the twentieth century. Such a master narrative needs to be ethnographically studied, rather than reproduced. Ethnographers of New Guinea may see a task here, as they move on from the critique of Hobbesianism and the thick description of New

Guinean evaluations of war, to the analysis of the practice of warfare in a historical setting that eventually came to include also western agents. Given that New Guinea was frequently rendered as a place of total war in ethnography, this modern myth to all likelihood also affected the action and speech of such agents of change as Malinowski addressed in 1922 with a claim about warfare in New Guinea being different.

N O T E S

1 For a discussion of relevant historiography, see Chikering (1999).

2 Brandt (2000) offers more detail on Rivers, Wedgwood, Malinowski, and the anthropology that Westermarck promoted at the London School of Economics. On the analysis of World War I, in terms of 'international anarchy' and the discourse on International Law, see Wallace (1988).

3 Brandt (2002: esp. 61-64, 71-77) provides further detail on the discourse on exchange to which A. Strathern's writing of around 1970 belongs.

4 For a detailed discussion of the literature about exchange as a pacifying social practice of the 1970s, see Brandt (2002: 97-106).

5 For further detail, see Brandt (2002: esp. 121-131).

6 Of course, rendering Hobbes simply as a modern Western philosopher also eclipses the historic warfare that he was responding to when writing *Leviathan*, which, as Barker (1993: 135) observes, was the work of a 'philosopher in terror' hoping to safe his country from further bloodshed by the creation of a Cartesian science of politics.

7 I formed my ideas on Americanism in anthropology after reading Stocking (1992: esp. 284-290) and Michaels (1995).

8 The work of Read and Langness, as well as others offering related representations, is discussed in more detail in Brandt (2002: esp. 50-56, 64-71).

9 Preparing for fieldwork in New Guinea in the early 1960s, Strathern spent time also at the Australian National University where he worked together with Paula Brown, who strongly supported the search for social order in New Guinea (see Brandt 2002: 63-64, 72-73).

10 It is to be noted that 'after the fact', Stocking (1992) has recognised the epistemological limitations of his original position on 'historicism' and 'presentism'.

B I B L I O G R A P H Y

- Appadurai, A. 1986. 'Theory in anthropology: Center and periphery'. *Comparative Studies in Society and History*, 28, pp. 356-61.
- Barker, F. 1993. *The Culture of Violence: Tragedy and History*. Manchester: Manchester University Press.
- Barnes, J. 1962. 'African models in the New Guinea Highlands'. *Man*, 62, pp. 5-9.
- Benedict, R. 1946(1934). *Patterns of Culture*. New York: Mentor Books.
- Berndt, R. 1962. *Excess and Restraint: Social Control among a New Guinean Mountain People*. Chicago: University of Chicago Press.
- Brandt, E. 2000. 'Images of War and Savagery: Thinking Anthropologically about Warfare and Civilization, 1871-1930'. *History and Anthropology*, 12, pp. 1-36.
- Brandt, E. 2002. *On War and Anthropology: a History of Debates concerning the New Guinea Highlands and the Balkans*. Amsterdam: Rozenberg Publishers.
- Brown, P. 1970. 'Cimbu transactions'. *Man*, 5, pp. 99-117.
- Bulmer, R. 1960. *Leadership and Social Structure among the Kyaka People of the Western Highlands District of New Guinea*. Ph.D. dissertation, Australian National University, Canberra.
- Chikering, R. 1999. 'Total war: The use and abuse of a concept'. In: M. Boemeke, R. Chickering and S. Föster (eds.): *Anticipating Total War: The German and American Experiences, 1871-1914*. Cambridge: Cambridge University Press. pp. 13-28.
- Clausewitz, C. von 1832. *Vom Kriege*. Berlin: Ferdinand Dümmler.
- Dawson, D. 1996. *The Origins of Western Warfare: Militarism and Morality in the Ancient World*. Boulder (CO): Westview Press.
- Dentan, R. 1988. 'Band-level Eden: A mystifying chimera'. *Cultural Anthropology*, 3, pp. 276-84.
- Dickinson, G. 1916. *The European Anarchy*. London: George Allen and Unwin.
- *Dickinson, G. 1926. The International Anarchy, 1904-1914*. London: The Century.
- Ember, C. 1978. 'Myths about hunter-gatherers'. *Ethnology*, 17, pp. 439-48.
- Fortune, R. 1932. *Sorcerers of Dobu: The Social Anthropology of the Dobu Islanders of the Western Pacific*. London: Dutton.
- Fortune, R. 1939. 'Arapesh warfare'. *American Anthropologist*, 40, pp. 22-41.
- Glasse, R. 1959. 'The Huli descent system. A preliminary account'. *Oceania*, 29, pp. 171-83.
- Görlich, J. 1999. 'The transformation of violence in the colonial encounter: Intercultural discourses and practices in Papua New Guinea'. *Ethnology*, 38, pp. 151-62.
- Harrison, S. 1985. 'Concepts of the person in Avatip religious thought'. *Man*, 20, pp. 115-30.
- Harrison, S. 1989. 'The Symbolic construction of aggression and war in a Sepik River society'. *Man*, 24, pp. 583-99.

- Harrison, S. 1993. *The Mask of War: Violence, Ritual and the Self in Melanesia*. Manchester: Manchester University Press.
- Hobbes, T. 1651. *Leviathan*. London: Green Dragon.
- Howard, M. 1994. 'Constraints on warfare'. In: M. Howard, G. Andreopoulos and M. Shulman (eds.): *The Laws of War: Constraints on Warfare in the Western World*. New Haven: Yale University Press, pp. 1-11.
- Jolly, M. 1992. 'Partible persons and multiple authors, Contribution to Book Review Forum on M. Strathern's (1987) *The Gender of the Gift: Problems with Women and Problems with Society in Melanesia*'. *Pacific Studies*, 15, pp. 137-49.
- Josephides, L. 1991. 'Metaphors, metathemes, and the construction of sociality: A critique of the new Melanesian Ethnography'. *Man*, 26, pp. 145-61.
- Keeley, L. 1996. *War before Civilization: The Myth of the Peaceful Savage*. Oxford: Oxford University Press.
- Keesing, R. 1992. 'Review, Contribution to Book Review Forum on M. Strathern's *The Gender of the Gift: Problems with Women and Problems with Society in Melanesia*'. *Pacific Studies*, 15, pp. 129-37.
- Knauft, B. 1994. 'Review of S. Harrison's (1993) *The Mask of War: Violence, Ritual and the Self in Melanesia*'. *Anthropos*, 89, pp. 612-13.
- Knauft, B. 1999. *From Primitive to Postcolonial in Melanesia and Anthropology*. Ann Arbor: The University of Michigan Press.
- Kuklick, H. 1991. *The Savage Within: The Social History of British Anthropology, 1885-1945*. Cambridge: Cambridge University Press.
- Kuper, A. 1983. *Anthropology and Anthropologists: The Modern British School*. London: Routledge.
- Langness, L. 1964. 'Some problems in the conceptualization of Highlands social structures in New Guinea'. *American Anthropologist*, 66, pp. 162-82.
- Langness, L. 1967. 'Sexual antagonism in the New Guinea Highlands: A Bena Bena example'. *Oceania*, 37, pp. 161-77.
- LiPuma, E. 2000. *Encompassing Others: The Magic of Modernity in Melanesia*. Ann Arbor: The University of Michigan Press.
- Macintyre, M. 1983. 'Warfare and the changing context of "Kune" on Tubetube'. *The Journal of Pacific History*, 18, pp. 11-34.
- MacIntyre, M. 1995. 'Violent bodies and vicious exchanges: Personification and objectivation in the Massim'. *Social Analysis*, 37, pp. 29-43.
- Maclean, N. 1998. 'Mimesis and pacification: The colonial legacy in Papua New Guinea'. *History and Anthropology*, 11, pp. 75-118.
- Malinowski, B. 1920. 'War and weapons among the natives of the Trobriand Islands'. *Man*, 20, pp. 10-12.
- Malinowski, B. 1922. 'Ethnology and the study of society'. *Economica*, 2, pp. 208-19.
- Malinowski, B. 1961(1922). *Argonauts of the Western Pacific*. New York: Dutton.
- Malinowski, B. 1933. 'The work and magic of prosperity in the Trobriand Islands'. *Mensch en Maatschappij*, 9, pp. 154-74.
- Malinowski, B. 1935. *Coral Gardens and their Magic*. London: Allen and Unwin.
- Malinowski, B. 1941. 'An anthropological analysis of war'. *American Journal of Sociology*, 46, pp. 521-50.
- Marwick, A. 1967. *Britain in the Century of Total War: War, Peace and Social Change, 1900-1967*. London: Bodley Head.
- Mauss, M. 1954(1923-24). *The Gift*. London: Cohen and West.
- Mead, M. 1935. *Sex and Temperament in Three Primitive Societies*. New York: William Morrow.
- Michaels, W. 1995. *Our America: Nativism, Modernism, and Pluralism*. Durham: Duke University Press.
- O'Hanlon, M. 1995. 'Modernity and the "graphicalization" of meaning. New Guinea Highland shield design in historical perspective'. *The Journal of the Royal Anthropological Institute, incorporating Man*, 6, pp. 469-94.
- Otterbein, K. 2000. 'A history of research on warfare in anthropology'. *American Anthropologist*, 101, pp. 794-805.
- Pearson, K. 1901. *National Life from the Standpoint of Science*. London: Cambridge University Press.
- Ploeg, A. 1965. *Government in Wanggulam*, Ph.D. dissertation, Australian National University.
- Read, K. 1954. 'Cultures of the Central Highlands, New Guinea'. *Southwestern Journal of Anthropology*, 10, pp. 1-43.
- Rivers, W. 1920. 'The dying-out of native races'. *The Lancet*, 98, pp. 42-44 and 109-11.
- Sahlins, M. 1972. 'The Spirit of the Gift'. In: *Stone Age Economics*. Chicago: Aldine, pp. 149-83.
- Salisbury, R. 1962. *From Stone to Steel: Economic Consequences of a Technological Change in New Guinea*. Melbourne: Melbourne University Press.
- Salisbury, R. 1964. 'Despotism and Australian Administration in the New Guinea Highlands'. *American Anthropologist*, 66, pp. 225-39.
- Simons, A. 1999. 'War: Back to the future'. *Annual Review of Anthropology*, 28, pp. 73-108.
- Steadman, L. n.d. Unpublished seminar paper on the Hewa. Australian National University.
- Stearns, H. (ed.) 1922. *Civilization in the United States: An Inquiry by thirty Americans*. New York: Harcourt, Brace and Company.
- Stocking, G. 1965. 'On the limits of "presentism" and "historicism" in the historiography of the behavioral sciences'. *Journal of the History of the Behavioral Sciences*, 1, pp. 211-18.
- Stocking, G. 1992. *The Ethnographer's Magic and Other Essays in the History of Anthropology*. Madison: The University of Wisconsin Press.
- Strathern, A. 1971. *The Rope of Moka: Big Men and Ceremonial Exchange in Mount Hagen, New Guinea*. Cambridge: Cambridge University Press.
- Strathern, M. 1985. 'Discovering "social control"'. *Journal of Law and Society*, 12, pp. 111-34.

- Turnwald, R. 1912. *Forschungen auf den Salomo-Inseln und dem Bismarck-Archipel*. Berlin: Dietrich Reimer.
- Wagner, R. 1972. 'Review of A. Strathern (1971)'. *American Anthropologist*, 74(6), pp. 1415-16.
- Wallace, S. 1988. *War and the Image of Germany: British Academics 1914-1918*. Edinburgh: John Donald Publishers.
- Watson, J. 1971. 'Tairora: The politics of despotism in a small society'. In: R. Berndt and P. Lawrence (eds.): *Politics in New Guinea*. Nedlands: University of Western Australia. pp. 224-75.
- Wedgwood, C. 1930. 'Some aspects of warfare in Melanesia'. *Oceania*, 1, pp. 5-33.
- Westermarck, E. 1910. 'Prefatory note'. In G. Wheeler (ed.): *The Tribe, And Intertribal Relations in Australia*. London: John Murray Ltd., pp. v-vi.
- Wright, Q. 1942. *A Study of War*. Chicago: Chicago University Press.

War as Practice, Power, and Processor:
A Framework for the Analysis of War
and Social Structural Change

/7

CLAUS BOSSEN

Introduction

This is an inquiry into the relation between war and social change and an attempt at constructing a framework for analysing this relation. On the most general level, it is based on a simple question: How do war and military organisation contribute to social structural change? I use the term 'structural' in order to stress that I do not aim to consider how wars or military organisations enabled an actor or a group of actors topple and replace one power-holder and become the new person(s) in power. Nor do I aim to examine how, for example, conquest enabled the incorporation of new or re-incorporation of old territory, or how war and conquest led to famine and devastation for one group and conspicuous living for another. These are matters of life and death to the people involved, but in the perspective taken here they will only feed into the discussion if the new power-holder, conquest or acquisition of new resources, for instance, led to new social structures.

Whilst there is no doubt that wars have been fought throughout human history (Keeley 1996; Martin and Frayer 1997; O'Connell 1995), and that their consequences for people and their societies can be vast and devastating, it is a matter of dispute whether war should be seen as a having a role or being a factor in the development of human history.

The old evolutionist Herbert Spencer (1967) and his more contemporary heir Robert Carneiro (1970) argue that war is *the* driving force in human evolution, and the renown historian John Keegan asserts that '... all civilizations owe their origin to warriors...' (Keegan 1993: vi), even though he also emphasises how warfare is embedded in the societies of which it is part. On the other hand, the anthropologist Henri Claessen (2000) argues, very much like the sociologist Bruce Porter (1994), that war is a derived phenomenon and thus cannot be a 'factor' in human history, though both acknowledge that the consequences of war for humans and societies are immense. I will sidestep this discussion first of all by rejecting any evolutionary schemata in human history. While I think that it does make sense to establish categories for different kinds of society – such as 'tribe', 'chiefdom' or 'state' – and ask how one kind of society could develop into another, I do not presuppose any historical or systemic logic in these processes. Secondly, I do not think it is constructive to discuss war as a unitary phenomenon; instead, I propose to subdivide war into different elements – e.g., physical violence, military organisation, conquest – and look at war at three different levels: that of practice, that of society and that of process.

The interrelations and feedback between war, military organisation and other sections of society are many and complicated, and in this article I take on the task of constructing a framework for analysing war in an attempt to provide an overview. In the existing literature analysis tends to become bogged down in historical detail or limited to specific geographical areas. Historians like Timothy McNeill (1983) and John Keegan (1993) have produced impressive accounts of the historically changing relationships between war, military organisation, technology and society, but they do not offer a framework for conceptualising these relations in general. Grand Theory sociologists Anthony Giddens (1985b) and Charles Tilly (1990) provide such frameworks, but suffer from their focus on European history and developments after the rise of the European state: while they analyse the concurrent developments of war, military organisation and the European state, they pay little attention to the transition from pre-state to state societies. On the other hand, the analysis of Michael Mann (1993), another Grand Theory sociologist, conceptualises military organisation as one of four sources of social power (the other three being politics, economy and ideology), which seems to me to be one of the best approaches to a historical and theoretically informed understanding of the role of war in the development of societies. My debt to his work will be clear in the following. However, war has more aspects than military organisation and the historical contingent interactions with politics, economy and ideology that Michael Mann outlines. I will add to his theoretical framework the dimensions of practice and process.

Building a framework for the analysis of war in the perspective of social change would be an immense task if it had to take into consideration all the existing literature on war (for overviews, see Modell and Haggerty 1991; Nagengast 1994; Simons 1999), and this particular attempt must be regarded as but an initial contribution. I start from the premise that change takes place gradually or suddenly (Friedman 1982) as intended or unintended results of actions by humans acting within given contexts. It is necessary to combine social structure with human action, as argued by Sherry Ortner (1984) and elaborated by anthropologists and sociologists (e.g. Bourdieu 1993; Giddens 1985a; Mann 1986). For an analysis of the relation between war and social structural change, the practice of war has to be included. In addition to

practice and structure, there is change. At this processual level it is possible, I will argue, to delineate the particular kinds of roles that war can have. Thus, the framework proposed has three levels (figs. 1-3). At one level, I argue, war is seen as a form of social practice based on violent acts, which are embedded in webs of significance and organised socially and in which technology is usually applied. At the second, societal, level, war is conceived as one of four forms of social power: military, political, economic and ideological (following Michael Mann [1986]). Finally, at the third, processual, level war may have three different roles in processes of social change and these are linked to the manifestations of war as military organisation, conquest, or a general context for the reproduction of society. While each level by itself is rather simple, the overall model becomes more complicated (as can be seen in the final figure, Fig. 4).

A framework for the analysis of war and social change

War as social practice

What is war? A minimalist definition views war as '…organised inter-group homicide involving combat teams of two or more persons' (Divale and Harris 1976: 521). This definition includes battles, sieges and campaigns, which in the era of the modern nation-state are associated with war, but also small-scale war acts like skirmishes, raids or feuds. However, most people would probably add further qualifications to the minimalist definition: the single event of the murder of two persons from one group by two members of another group for idiosyncratic reasons and not implicating their respective groups in general would usually not qualify as war. Somehow, war implies some kind of scale and that the acts carried out are *interpreted* not merely as violence, but as part of inter-group relations. The problem of scale is sometimes solved by incorporating some kind of size and extent into a definition of war. Quincy Wright, for example, in the *International Encyclopedia of the Social Sciences* (1968), writes that 'War in the ordinary sense is a conflict among political groups, especially sovereign states, carried on by armed forces of considerable magnitude for a considerable period of time' (Wright 1968: 453). I would argue, however, that it is crucial to distinguish between the analyst's and the actors' perception of the acts of homicide. While we normally would not interpret a single homicide as war even if the victim and the perpetrators were from different groups, and would demand evidence of more incidents before applying that kind of label, two groups may on the other hand have such strained relations that even a single incident would lead to a declaration of war. It is the interpretation and the meaning given to the incident of homicide that is crucial and not the scale of violence itself. War is thus based on acts of homicide that are seen as part of the inter-group relations and hence as part of the political relations between groups. However, while homicide is most often involved in war, this is not always the case, since wars may be declared, warriors or soldiers raised, and weapons used without anyone being killed. Here it is the threat of homicide that qualifies these incidents to be labelled as war. This is reflected in the definition of war by Simon Harrison in the *Encyclopedia of Social and Cultural Anthropology* (Barnard and Spencer 1996):

... anthropologists usually envision war as a particular type of political relationship between groups, in which the groups use, or threaten to use, lethal force against each other in pursuit of their aims. (Harrison 1996: 561)

For the present purposes I will therefore define war as *the organised use or threat of use of lethal force by a minimum of two or more actors from one group against members of another group, which is interpreted by the actors and/or the analyst as part of the relations between the two groups.*

Such a definition, however, looks at war at the level of society; and while it points to central elements such as organisation, lethal force and interpretation, it does not yet perceive war as a practice. At a basic level war is about causing physical harm to other peoples' bodies and acts of war are thus a subcategory of acts of physical violence. While violence may take on 'symbolic', 'mental', 'emotional' and 'structural' forms, in the case of war it is ultimately the physical aspect that must be at the core. David Riches' discussion of violence is of special interest here, since he defines the universal core meaning of violence as 'the intentional rendering of physical hurt to another human being' (Riches 1986b: 4). Violence, Riches furthermore argues, is always embedded in strategy and meaning. It is used as a means to achieve goals, practical or symbolic, and will always imply a need to legitimise the violent

act. Aiming at a cross-cultural theory of violence, Riches argues that while performers, victims and witnesses (including analysts) may disagree as to the legitimacy of violence, they will nevertheless universally recognise violence in the sense of rendering physical hurt to another person. We may disagree as to whether the beating of pupils and children should be labelled as 'education', the police dispersion of demonstrators as 'restoring peace and order', and the incision of patterns upon the skins of young girls and boys should be seen as 'ritual'. We may discuss whether these are all examples of illegitimate violence or not, but we recognise the element of 'rendering physical hurt' in all of the examples and thus recognise violence in its core meaning.

Violence is a highly potent act, according to Riches, because of four key characteristics: firstly, violence always implies contestations of legitimacy and is especially suited for making statements of significance and meaning; secondly, it is a recognisable act in its key sense of 'rendering physical hurt'; thirdly, violence is highly visible to the senses; and fourthly, violence requires little specialised equipment or knowledge (see also Nagengast 1994: 111-16; Riches 1986a; Riches 1986b: 11):

...as a means of transforming the social environment (instrumental purpose), and dramatizing the importance of key social ideas (expressive purpose), violence can be highly efficacious. So it is that the desire to achieve a very wide variety of goals and ambitions is a *sufficient condition* for acts of violence to be performed. (Riches 1986b:11)

In sum, war as social practice can be seen as having at its core the use of physical violence, which has three aspects. Firstly, it is always embedded in webs of meaning and interpretation. Secondly, while one of the reasons for the ubiquity of violence may, as David Riches argues, be found in the small amount of specialised knowledge, training, and technology required to carry it out, war almost everywhere entails some kind of technology and specialised knowledge such as arms, their use, cooperation and tactics. This logically derives from the fact that the expansion of these elements highly increases the scope and intensity of the possible impact. Thirdly, while war is a subcategory of violence it always involves two or more actors who coordinate their actions, and acts of war therefore always imply organisation. War as practice can be represented graphically as in figure 1.

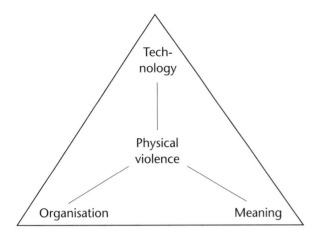

FIG. 1: *War at the level of social practice.*

The 'meaning' aspect of war as social practice underscores the point that violence is not in the arbitrary outcome of biological or psychological drives, but as always embedded in webs of significance. According to Jonathan Spencer,

Anthropology's most useful contribution [to the study of violence] has probably been its documentation of the fact that violence is pre-eminently collective rather than individual, social rather than asocial or anti-social, usually culturally constructed and always culturally interpreted. (Spencer 1996: 559, my insertion)

This applies to the subcategory of violence as well: 'war … is always an expression of culture' (Keegan 1993: 12), and acts of war dramatise key social ideas, create categories of 'us' and 'them' as points of identification, and are embedded in stipulations of legitimate kinds of acts of war. The Nuer restrict intra-village fighting to clubs while spears may be used in inter-village fights; and while the molestation of women and children, the destruction of huts and taking of captives is prohibited in intra-Nuer fights, it is not when the enemy is non-Nuer (Evans-Pritchard 1969: 121, 155). Pre-contact Fijian chiefs led their armies into war, but attacks on the chief himself were strictly prohibited and would lead to cruel revenge upon the transgressor and his family (Clunie 1977). The stipulation of legitimate acts of war varies from group to group and Western observers have often explained these differences by referring to 'ritualised' or 'conventionalised' warfare. However, all societies have stipulations of legitimacy and ritual kinds of skirmishes or fights as precursors to the

total annihilation of the opposing society. This of course does not mean that actors always respect the stipulations of legitimate violence, and often it is not the violence itself but its undue application in a certain context that is criticised in public debate. Nonetheless, history abounds with actors who did not comply with existing limits to violence and who successfully achieved their goals and transformed society. In *Vengeance is their Reply* by Rolf Kuschel (1988), a young warrior on Bellona Island in the Pacific dreams of killing a prominent chief because of the lasting renown he would achieve and the likely incorporation of his name in local mythology, despite the fact that chiefs were absolutely beyond the range of possible legitimate victims.

The 'organisation' aspect of war as social practice emphasises the coordination problems posed by actors engaged in war. More coordination means more efficient attack and defence, whether efforts are invested in large armies, guerrilla warfare, fortresses or evasive resistance. However, the coordination and cohesion of warriors is not always easy. In *The Mask of War*, Simon Harrison (1993) describes the problems of assembling a raid party: recruiters always have to consider whom to include from their own village group, since some of their fellow male villagers are members of different kinship groups and closely related and obliged to people from villages other than their own. In ancient Fiji, village men would readily fight on behalf of the village to which they were related through kinship. However, the large armies of state raised for larger campaigns, *valu ni tu* and *valu rabaraba*, consisted of bands of warriors from different villages and were plagued by mistrust since village chiefs often betrayed their allies. On hearing a rumour of betrayal, the army might dissolve into its different bands (Clunie 1977). The European city states of the 14th and 15th centuries often used hired armies whose cohesion was based on their mutual interest in financial remuneration because their rulers could not count on the loyalty of their subjects, and these hired armies were used to suppress internal rebellion as much as to fight other rulers. When France was attacked after the revolution in 1794, common conscription was invented and the strength of an army of soldiers dedicated to their force was demonstrated by the defeat of the European powers that moved against revolutionary France. An alternative way of achieving cohesion within the military organisation was daily drilling, which in addition to increased efficiency and coordination also created strong psychological bonds between individual soldiers (McNeill 1983: 131-33).

Finally, the 'technology' aspect of war as social practice has significance for the scope and intensity of war. This also has important corollaries for which kinds of war armies can engage in and what kind of coercive rule they can sustain. According to Michael Mann, the logistical limitations upon Mesopotamian armies (they could only march for seven days with a maximum distance of 90 km) had repercussions for rulers who could rule but not govern: they could exact tribute and demand allegiance from people marginal to the centre of power with the threat of military retaliation, which however was costly and took time, so they could not control their subjects on a day-to-day basis (Mann 1986). It is often assumed that the history of war is the history of technological development. The development of ever more efficient technologies and of war in Europe in the 19th and 20th centuries and the concurrent developments in state bureaucracy and taxation can be taken as evidence of this point and indeed make the question of war a central dynamic in social change (see for example McNeill 1983; O'Connell 1995). However, such a perspective is challenged by those who emphasise the importance of local politics, succession struggles and legitimating principles (see Simons 1999). The historian John Keegan (1993) argues that war is always embedded in culture, offering the example of 16th century Japan where rulers first used fire weapons to secure their basis of power. However, after having monopolised fire weapons, they phased them out and military power was subsequently based on the samurai and his sword for the next two centuries. Efficiency is but one concern in war.

The social practice of war should be taken seriously, not only because it approaches war from the perspective of actors, but also because war at this level may constitute social groups and their organisation. Internal cooperation and the division of work, and the experience and interpretation of acts of war can, as is well-known, create a strong sense of 'us' within a group of fighting (wo)men and/or within another group on whose behalf the group fights. In one of the most inspiring books on war in non-state societies, *The Mask of War*, Simon Harrison (1993) argues that the social entity of the 'group' may in a very fundamental way be constituted by war. In Western theories it is often assumed that war is the result of a lack or weakening of social order. This assumption can be traced back to Thomas Hobbes (1991[1651]), who argued that humans in their natural state – i.e., without society – will inevitably become entangled in war against each other. Society, Hobbes argued, arose as a way of establishing an order above individual interests. Harrison, however, argues that to the Avatip of Papua New Guinea acts of war are fundamental acts in the building of society. Their ontological premise is not like that of Hobbes, of man living individually in nature, but instead that of persons inescapably interrelated in networks of kinship to everyone else. Without the social group, the Avatip person would be ontologically dissolved into an indefinitely extendable web of kinship relations and the rights and obligations they entail. War serves to establish non-kinship relations and thus to establish separate identities for individuals and groups. An analysis of war thus cannot assume that social order and groups precede war.

This is nevertheless what we often do in the era of the nation-state, each of which has its own army to the exclusion of all other fighting military organisations. Social order internally and externally is the premise from which exchange-relations are seen as establishing peaceful relations between groups; consequently, war is viewed as the result of a breakdown of these relations. In continuation of this premise, it is discussed whether trade inhibits war (e.g. Mansfield 1994). This is however a normative and culturally biased perception. Based on a review of the literature on war in Melanesia, Bruce Knauft (1990) concludes that while in one instance a group seemed to fight because of a lack of society – i.e. a lack of exchange and kinship relations – other groups fought because of too much society. Marilyn Strathern (1985) argues that the giving of compensation payments and the shaking of hands in Papua New Guinea is not, as a Western approach would assume, an example of the regulatory power of social conflict management. Offensive acts also give the perpetrator esteem and subsequent conciliation, and the giving of compensation and shaking of hands testify to the perpetrators' ability to produce wealth and power to start a violent conflict and end it again. To the Hageners of Papua New Guinea, exchange relations and relations of violent conflict are not normatively different but are rather equally attractive opportunities in the pursuit of male power, exemplified, among other things in the ability to use and control these.

Consequently, we cannot assume that the practice of war takes place in a context of disorder and moral condemnation, just as we cannot assume groups to precede the practices of war. Whether this is the case has to be ascertained on an individual basis.

War at the level society:
the four forms of social power
While the literature on war is extensive, only few sociological theories have been proposed in which war is not a derived but a factor by itself. Among these, the work of Anthony Giddens, (1985b) Michael Mann (1986) and Charles Tilly (1990) stands out. Giddens pays a great deal of attention to military organisation and how it institutionalises domination, but also cautions that

Power may be at its most alarming, and quite often its most horrifying, when applied as a sanction of force. But it is typically at its most intense and durable when running silently through the repetition of institutionalized practices. (Giddens 1985b: 9)

The foci of Giddens' and Tilly's work are nevertheless primarily directed at explaining the emergence of the nation-state in Europe. Michael Mann, on the other hand, has a broader historical and geographical scope, making him theoretically suited for the general perspective of this article. I will therefore rely heavily upon him in the following.

Michael Mann's theory perceives of society as constituted by 'multiple overlapping and intersecting sociospatial networks of power' (Mann 1986: 1). 'Societies', 'groups' and 'states' are not viewed as natural entities but emerge where there is a sociospatial condensation of different kinds networks. The theory therefore does not assume that society precedes war and military organisation and that war is a deviation from social order, but instead views war as one kind of network of power. Power, according to Michael Mann, '...is the ability to pursue and attain goals through mastery of one's environment' (1986: 6). In a Weberian perspective power refers to the ability of one actor to carry through his will despite the resistance of other actors and is seen as something distributed between a number of actors in a zero-sum game. However, power also has, according to Mann, a Parsonian aspect, since a number of actors may cooperate and thus collectively enhance their power over third parties or nature (Mann 1986:

6). Cooperation means social organisation and division of labour and there is, Mann writes, an inherent tendency in collective power also to entail distributive power. Those at the top of a social organisation tend to be more likely able to carry through their will, while those at the bottom are hindered in their resistance by the fact that they can only achieve similar collective benefits if they can establish an alternative division of labour: people with a low position in a social hierarchy are, in Mann's wording, often organisationally outflanked. To Mann, stratification is the central aspect of societies since its dual aspect of distributive and collective power is the means by which humans try to achieve their goals. Both Marxists and neo-Weberians argue, according to Mann, on the premise that stratification is the central aspect of social organisation and that politics, ideology and economy are the main organisations in any society. Mann accepts this but separates military organisation from political organisation. His argument for doing so is historical: unlike the modern nation-state, most states have not had or even claimed a monopoly on organised military force and, furthermore, military groups constitute an independent factor that may act without the consent of the society from which they come (Mann 1986: 11). Mann distinguishes between two dimensions of power which result in four ideal-typical organisations of power. Firstly, power may be assessed according to its extensity and intensity – i.e., its 'ability to organize large numbers of people over far-flung territories in order to engage in minimally stable cooperation', and its 'ability to organize tightly and command a high level of mobilization or commitment from the participants' (Mann 1986: 7). Secondly, power can be assessed according to its authoritative and diffused types, which refers to whether power is 'actually willed by groups and institutions [and]... comprises definite commands and conscious obedience' or whether it is spread in a more 'spontaneous, unconscious, decentred way throughout a population' (Mann 1986: 8, my insertion) (These ideas are, incidentally, very similar to those of Anthony Giddens' [1985b: introduction]).

Putting these four polarities of power together results in four ideal-typical forms of organisational reach:

	Authoritative	*Diffused*
Intensive	Army command structure	A general strike
Extensive	Militaristic empire	Market exchange

(Mann 1986: 9)

The army is an example of concentrated and coercive organisation of power which is intensive, in contrast to the militaristic empire where power is also 'willed' and based on 'definite commands and conscious obedience', but which is likely to get only a low degree of commitment from its subjects. The general strike is Mann's example of a diffuse, intensive organisation of power showing a high degree of commitment and happening more or less spontaneously. Finally, market exchange, which is voluntary and involves transactions that may extend over vast areas, is an example of a diffuse, extensive organisation of power.

Power in its distributive and collective aspects thus has four ideal typical forms of organisation: economic (production and exchange of subsistence needs), ideological (giving meaning, morale and aesthetics to actors), military (providing means of defence and aggression) and political (centralised, institutionalised and territorialised aspects of social relations). Below is a graphic representation of my perception of society as constituted by four networks of power. The broken circle indicates that only the socio-spatial overlap and condensation of these networks lead to what we normally characterise as 'society':

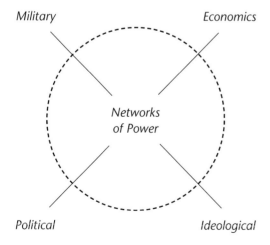

FIG. 2: *War at the level of society.*

All of the four main organisations of power entail a mixture of the above-mentioned aspects of power. Military organisation 'mobilizes violence, the most concentrated, if bluntest, instrument of human power' (Mann 1986: 26). The concentration and tactical use of this form of power is crucial in battles, sieges and skirmishes, so violence that is organised authoritatively, distributively and intensively provides decisive advantages in such situations. However, military organisation also has a more extensive aspect in that raids and punitive actions may be launched over extensive areas:

Thus military power is sociospatially dual: a concentrated core in which positive, coerced controls can be exercised, surrounded by an extensive penumbra in which terrorized population swill not normally step beyond certain niceties of compliance but whose behavior cannot be positively controlled. (Mann 1986: 26)

Likewise, the economic organisation of production, distribution, exchange and consumption has an extensive reach, since distribution and exchange may imply networks crossing vast distances, but it also has an intensive side since, for example, production involves intensive practical, everyday labour.

These four ideal types of social power broadly have the function they indicate, according to Mann, there is no one-to-one relationship between form and function. Economic functions can be handled by states, armies and churches as well as specialised circuits of exchange, just as ideologies can be brandished by economic classes, states and armies. On the one hand, there are obviously important interdependencies between the different kinds of power. Military organisations may, for example, rely on or lean to hierarchies of authority in political networks, be dependent on ideological networks for the creation of solidarity between warrior-soldiers and the endowment of meaning and legitimacy to their task, and be dependent on access to economic networks for campaigns of any size and duration. Such interdependencies are evident in ancient Polynesian chiefdoms like ancient Fiji and Hawaii, where the political status of chief was not only linked closely to divinity but also to the chief's capabilities as a warrior (Kirch 1981; Valeri 1985), and in the case of European state-building in the 18th and 19th centuries (Giddens 1985b; Tilly 1990). On the other hand, military power may be extended into economic, political and ideological networks. The distributed power of military organisation is, according to Mann, amendable to other situations of social cooperation where coercion can be used, such as coerced labour in mines, in plantations and in the building of monuments, fortification or roads. It is, however, less suitable for normal dispersed agriculture, for industry where special skills are required, or for trade, where the costs of coercion exhaust the resources of the military regime (Mann 1986: 26).

The ideal-typical forms of power attain intermittent existence in specific historical situations where they interrelate and give rise to a particular configuration of networks of interaction. In their specific forms, they are interdependent on the development of the other forms of power organisation in their distributive, collective, authoritative, diffuse, extensive and intensive aspects. They are furthermore also dependent in extensity and intensity on the development of technologies of transportation and communication (though Mann argues against technological determinism [1986: 524-26]). As already mentioned, Mann argues that the logistics of pre-industrial societies inhibited unsupported marches of more than 90 kilometres and that military power could thus be exercised at longer distances at high costs. A similar argument is made by Anthony Giddens, who makes the limitations upon European cavalry in the 15th and 16th centuries a pivotal point in the change from the Absolutist to the modern nation-state: in the former, coercion could only be applied temporarily and at costs increasing with the distance from the centre of power, whereas modern technology has enabled the rapid deployment of armies to all of the geographical areas enclosed by the borders of the nation-state. In the former, power diminished with distance; in the latter, power is homogenously present all over (Giddens 1985b, see also Anderson 1991 for a similar point).

What we get from this perception of war at the societal level is the possibility of analysing war as organised physical violence, as defined above, in relation to other kinds of social power. It is often assumed that military organisation is derived from social organisation as such. For example, the 'old' evolutionists such as Morton Fried (1967) and Elman Service (1978) argue that war cannot be a factor in social evolution because it will always rely on political organisation. What they do not take into account is that military organisations might develop independently of other networks of power and/or across established social groups. Obvious examples are the rebel movements in, for instance, Sierra Leone (Richards 1996) and the Al-Qaeda terrorist network. In neither of these cases do we have a military organisation that is part of a 'society' on whose behalf it acts; instead, military organisations fight across such entities. Paul Richards (1996) thus argues that the rebel movement in Sierra Leone takes advantage of young people who drop out of established social networks, forcefully integrates them in a movement that is informed by practices connected with the forest (as opposed to the 'city', both metaphorically and concretely) and makes use of specific kinds of violence (the cutting off of limbs at the wrist or at the elbow) to reach an international audience in order to circumvent the enemy 'group' (the state).

War at the processual level

Since societies are conceived of as overlapping networks of social power, social structural change takes place when the configuration of these networks changes and a more extended and tightly overlapping concentration of networks emerges or more fragmented and loose overlayings result. Michael Mann posits that whereas the four forms of power are 'track-layers' in the world-historical process '… there is no obvious, formulaic, general patterning of the interrelations of power sources.' (Mann 1986: 523). He does point out two different ways in which military power may have a reorganising role in history: in battles military power may decide which kind of society will predominate; and in peace, military organisation may dominate other networks of power and organise societies as such (Mann 1986: 521). I would, however, like to reformulate these two reorganising roles of war in history and add a third.

If we conceive of societies as configurations of overlapping networks of power, war may have a role in changing these configurations in three different ways: firstly, war may as military organisation expand onto the other networks of power of which it is part and from which it feeds; secondly, war may be a means for the forceful integration and subjugation of other configurations of networks of power such as when one group conquers or subjugates another; and, thirdly, as a frequent and recurrent phenomenon war may form a context within which groups develop. War may thus, from the perspective of a single society, have an internal, external or contextual role in social structural change. Graphically, this can be represented as below:

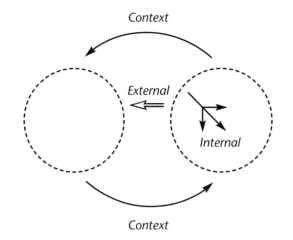

FIG. 3: *War at the level of process.*

These three roles correspond to three basic ways in which war in the literature on the subject is said to change society: war leads to subjugation, the development of a military organisation based on central command and hierarchy which spreads to the rest of society, or provides a context in which political organisation under the threat of annihilation develops into more complex forms.

All three roles are represented in the theory of one of the first scholars to link war and social structural change, Herbert Spencer, who discusses his ideas in his three-volume work Principles in Sociology, published between 1876 and 1896:

Wars originate governmental structures, and strong leadership does not evolve where people live in peace with each other or are scattered and cooperation thus not possible. War between groups, however, furthers the development of leadership by a warrior-chief and, later on through the process of differentiation, by developing the political and the military arms of government as separate organs. Means of communication are eventually developed when the coordination between the different parts of society requires it.... In the process of growth, one group may compound the other, but such incorporation through conquest and subsequent tributary relationship remains unstable until it is 'habituated to combined action against external enemies'. (Spencer 1967: 37)

Wars between groups thus spur the internal development of a society's leadership, and wars lead through conquest to the formation of larger entities,

whose leadership is stabilised through wars with external enemies:

... men who are local rulers while at home and leaders of their respective hands of dependents when fighting a common foe under direction of a general leader, become minor heads disciplined in subordination to the major head and as they carry more or less of this subordination home with them, the military organization developed during war survives as the political organization during peace. (Spencer 1967: 37)

Wars lead, according to Spencer, to the development of centralised government because of the need to coordinate action in order to survive. Subordination in the military is replicated in the political organisation, and gradually more complex forms of society develop. While on the one hand war provided a context in which a society developed internally because of the premium on better management through better integration of a society's various parts, on the other hand war as conquest was a means through which societies grew in size and from which institutionalised inequality arose.

Since Spencer, the linkage between war and evolution has been continued by Franz Oppenheimer (1999, originally 1914) and others who argued that conquest was behind early state formation (see Haas 1982: 63-66; Service 1978: 23-25). A contemporary heir of Spencer is found in Robert Carneiro, who in a number of articles (1970; 1981) argues that

... war has been the principal agent by which human societies, starting as small and simple autonomous communities, have surmounted petty sovereignties and transformed themselves, step by step, into vast and complex states. (Carneiro 1990: 191)

Since people do not voluntarily relinquish their sovereignty, social organisations like the state can only develop through forced subjugation – i.e. conquest – Carneiro argues (1970). Michael Mann's discussion of the emergence of the early state takes the same point of departure, but like many others who discuss the role of war in the formation of early states (Claessen 2000; Claessen and Skalník 1978a; Cohen 1985; Haas 1981; 1982), he argues that war is neither a sufficient nor a necessary factor in state formation. On the other hand, even those who oppose war as a prime mover or as a factor at all because war is derived from other factors acknowledge that the consequences of war can be vast and of immense significance (e.g., Claessen 2000: 110; Claessen and Skalník 1978b: 626; Cohen 1985; Porter 1994: 3). The interrelations and feedback mechanisms that can occur are, judging from the literature, many and complex, but the three roles mentioned above provide the forms.

The first way in which war can lead to change in a society's configuration of power is through the spread of military organisation onto other networks of power. Mann indicated this possibility with the concept of 'compulsory cooperation'. Here the intense, authoritative power of the military is used to intensify the exploitation of concentrated pockets of labour, stabilise systems of value, and protect production and trade, leading to the compulsory diffusion of ideologies (Mann 1986: ch. 5). In Mann's case the assumption is that military organisation is a specific kind of social power that may force itself onto other forms of power. Other authors (e.g., Spencer 1967) argue, however, that because military power is most effective when organised on the basis of a centralised command and a clear-cut hierarchy, frequent war will lead to a higher level of subjugation, centralisation and hierarchy in military organisation than in the rest of the social organisation. If this military organisation spreads to the rest of society in times of peace, the social structure will change towards more hierarchical structures with more centralised control. The crucial task in both cases is to delineate under which circumstances military power would be allowed to spread into other social organisations. Not all economic, political or ideological networks are amendable to the authoritative, intense power of military organisation, and while military leaders may be cherished by their soldiers and the people they defend in times of war, they may not have any power in times of peace. The successful war leaders Geronimo, of the Apache in North America, and Fousive, of the Yanomami in South America, could not transform their military success into civil authority (Chagnon 1974: 177-80). Elman Service (1978) argued that a centralised leadership would be agreed upon by consensus because of the benefits for societal survival that it offers. While this may apply in times of war and explain the emergence of a sophisticated military organisation, it does not explain why such forms of organisation would spread to the rest of society in times of peace. People may relinquish their sovereignty to military command in order to survive, but under which circumstances would they not revoke that decision in times of peace? The process is therefore most likely to take place in circumstances where a society is under constant threat of being annihilated and where military organisation can organise the economic activities, its leadership can achieve political status, or its ideology can tie into the ideology of the group.

The second role of war in processual change is as a means for the coerced integration of other societies into a society's own networks of power – i.e., through conquest or subjugation. The conquest of other groups provides a means by which to acquire access to additional basic resources. The conquest theories have the strength of explaining not only how centralised command and hierarchy emerges, but also how stratification – i.e. privileged access to basic resources in the sense of Morton Fried (1967) – comes about. The conquest theories posit that the subjugation of one group by another, followed up by the extraction of tribute or tax from the conquered group, leads to stratification where previously only egalitarian or ranked political relations prevailed. Over time, military rule is transformed into institutionalised government, laws, and the assimilation of the two groups into one people. Herbert Spencer (1967) argued in this way, and more recently the line of argument has, as mentioned, been taken up by Robert Carneiro (1970), whose theory has become one of the most cited and central contributions to discussions on the relation between war and state formation. Carneiro argues that in a situation where a group's access to land is limited by circumscription (because of ecological conditions, the presence of other groups, or the concentration of scarce resources) a point will arise when it is more feasible to conquer the land of another group than to intensify production by working harder or inventing new technologies of cultivation. Because this point arises before production is maximised, Carneiro argues, the conquering group can achieve higher levels of production by coercing the conquered group. Through such a process, chiefdoms arise by incorporating several villages into one polity. The replication of the process on a larger scale accounts for the rise of states through the incorporation of several regions into one polity. Through a process of 'internal evolution', previous reciprocal or redistributional exchange relations have been transformed, and a society where one powerful stratum taxes a subjugated stratum has appeared.

While neat and elegant, the theory actually only explains why and how entities grow larger through conquest, but not how political organisation changes. Carneiro does not elaborate upon the 'internal evolution' through which political transformation is achieved and instead takes the state for granted at this point (Carneiro 1970: 736). His theory does not actually explain what it intends to do and thus shares the general weakness of the conquest theories. They assumed that the relationship between the conquering and the conquered groups acquires a new quality after conquest. For the theory to work the conquered cannot merely become incorporated into existing exchange relations. A group conquered by a chief and paying their yearly tribute like other groups of the chiefdom would not be part of a state but of a chiefdom. While conquest may explain the subjugation of another group of people, it does not explain the transformation of political power or the social structure as such. The theories offer plausible contexts in which higher levels of social complexity, subjugation and stratification should develop, but not how this would transgress the existing social structure. One possibility is that the war leaders appropriate pieces of land (Webster 1975) or tribute that can provide the initial basis for economic power.

War may, thirdly and finally, provide a context in which societies develop. The argument is that since military effectiveness is enhanced the more its distributive aspects are developed, and that more complex societies enable better organisation of defence and attack, stratification and division of work will develop within societies frequently engaged in war. The evolutionist Herbert Spencer (1967) offers one example of this approach, and Elman Service (1978) similarly argued that war would make people accept the power monopoly of an emerging state since 'good government' would enhance their chances of survival by offering a more powerful organisation for their protection (Service 1978: 270). Michael Mann basically says the same when he states that '[t]hrough battles the logic of destructive military power may decide which form of society will predominate. This is an obvious reorganising role of military power throughout much of history.' (Mann 1986: 521). The crucial task here is to explain when such developments would take place. Herbert Spencer argued that such a development would not occur in the case of extensively dispersed groups, and Carneiro had to invoke a scenario with a resource

shortage and circumscription. Furthermore, just as war and development cannot be equated, neither can peace and stagnation: some societies are continuously at war with each other but do not develop higher complexity. Here war is instead part of the reproduction of social structures (Harrison 1993; Gardner and Heider 1969). The context approach works best as a long-term, survival-of-the-fittest explanation of state formation or to explain development through the threat of annihilation, but we still need to explain when a reconfiguration of the networks of power will be the result.

The three different ways in which war may part of a process of social structural change can of course be closely interlinked. The advantage of separating them is that their distinct ways of operating become clear and that the preconditions that have to be fulfilled also become apparent. In evolutionist scenarios like those of Herbert Spencer and Robert Carneiro war is seen as a unitary phenomenon. They suggest that war may entail a number of developments: better military organisation; better social organisation as such, to the conquest of other group and the development of inequality, to the emergence of the early state; and to the psychological adaptation to submission. They do not, however, specify how these different political changes may occur. War may be a factor in social structural change, but is obviously not always so: while wars proliferate, they mostly only bring about the substitution of one chief with another, but not with a king. One group may defeat another, but still be a chiefdom demanding tribute instead of becoming a state that receives taxes. In the terms of the above framework, social structural change only occurs when the networks of social power are rearranged, and the task for an analysis of the role of war in such a process is to specify the constraints and opportunities available to actors in such processes. When is it possible for a warrior-soldier to transform his status achieved in war into political status, or apply his command of military organisation to other areas of his group? When does conquest and subjugation lead to new forms of political organisation? Under which circumstances does frequent war lead to a process in which social organisation or technologies of violence enter a spiral of refinement and development instead of the reproduction of the status quo?

The dominance of long-distance trade may be an opportunity for leaders to build their own military organisation which is not dependent upon the consent of their own group (Webb 1975), and war may be an opportunity for leaders to acquire small pieces of land from which they can build an economic base and transform the political organisation (Webster 1975). While chiefs, traders, priests and great warriors can be seen as representatives of one kind of social power (i.e. political, economic, ideological and war respectively), most leaders have to use the multiple sources of power in order to uphold their position (Earle 1997). To analyse such processes and discuss the role of war in social change, it is necessary to consider war at both the level of practice and the level of society.

Conclusion

What I have argued above is that war may be considered at three different levels. At the level of practice, war is conceived as acts of physical violence organised by two or more actors that may employ technology to enhance the scope and intensity of violence. The acts are directed at members of another group and are always embedded in a context of meaning which gives significance to the acts and delineates which kinds of violence are deemed legitimate in the actual context and which are not, yet this does not imply that these delineations are always

respected. At the level of society, practices of war are manifestations of one of four social sources of power – economic, political, ideological and military – which are ideal-typically organised in military organisations. All societies consist of overlapping networks of social interaction that are based upon one of these four kinds of social power and which in their specific combination characterise a specific kind of society. At the level of process, war can have three different roles according to whether military organisation spreads into and dominates other networks of power, whether other societies by conquest or subjugation are forcefully integrated in the networks of which the military organisation is also part, or, finally, whether frequent and reoccurring wars provide the context for an ongoing process of development of technology and forms of power, especially social organisation. Processes of change are always induced by actors pursuing interests in a specific context and while some processes of change are the result of conscious choice others result from unintended consequences. Change is spurred by practice and its intended or unintended consequences for the overall configuration of networks of social power. Combined, the three levels of analysis of war at the level of practice, of society and of process can graphically be represented as below:

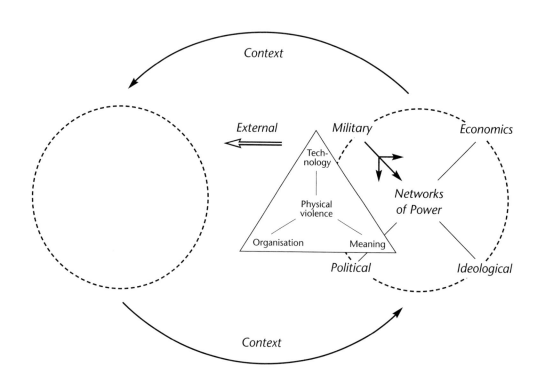

FIG. 4: *The framework for the analysis of war.*

The advantage of this model is its combination of three levels of analysis. War is most often looked upon at the societal level in discussions of war and historical development or social structural change. As I have argued, this is not acceptable for any social theory that aims to take practice seriously, nor is it feasible as an exclusive focus in connection with war. War may change the overall configuration of the social networks of power, but not always. Its ability to do so is dependent on other circumstances. First, the internal spread of military organisation is dependent on the compatibility of military organisation to political organisation, on the linkage of meanings of organised physical violence to ideology and on the possibilities of converting military status into political status. Second, the external application of war leads only to qualitative change to the extent that new forms of subjugation and payments are the result of conquest; and third, for societies that reproduce themselves in a context of frequent war there is the need to specify the dynamic of change that results in a reconfiguration of the networks of power.

BIBLIOGRAPHY

- Anderson, B. 1991. *Imagined Communities. Reflections on the Origin and Spread of Nationalism.* London: Verso.
- Barnard, A. and Spencer, J. (eds.) 1996. *Encyclopedia of Social and Cultural Anthropology.* London: Routledge.
- Bourdieu, P. 1993. *Outline of a Theory of Practice.* Cambridge: Cambridge University Press.
- Carneiro, R.L. 1970. 'A theory of the origin of the state'. *Science*, 169, pp. 733-38.
- Carneiro, R.L. 1981. 'The chiefdom: Precursor of the state'. In: G.D. Jones and R.R. Kautz (eds.): *The Transition to Statehood in the New World.* Cambridge: Cambridge University Press, pp. 37-79.
- Carneiro, R.L. 1990. 'Chiefdom-level warfare as exemplified in Fiji and the Cauca Valley'. In: J. Haas (ed.): *The Anthropology of War.* Cambridge: Cambridge University Press, pp. 190-211.
- Chagnon, N. 1974. *Studying the Yanomamö.* New York: Holt, Rinehart and Winston.
- Claessen, H.J.M. 2000a. *Structural Change. Evolution and Evolutionism in Cultural Anthropology.* Leiden: CNWS.
- Claessen, H.J.M. and Skalník, P. (eds.) 1978a. *The Early State.* The Hague, Paris, New York: Mouton Publishers.
- Claessen, H.J.M. and Skalník, P. 1978b. 'Limits: Beginning and End of the Early State'. In: H.J.M. Claessen and P. Skalník (eds.): *The Early State.* The Hague, Paris, New York: Mouton Publishers, pp. 619-35.
- Clunie, F. 1977. *Fijian Weapons and Warfare.* Suva: Fiji Museum.
- Cohen, R. 1985. 'Warfare and state formation'. In: H.J.M. Claessen, P. van de Velde, and M.E. Smith (eds.): *Development and Decline. The Evolution of Sociopolitical Organisation.* South Hadley: Bergin and Garvey Publishers, pp. 276-89.
- Divale, W.T. and Harris, M. 1976. 'Population, warfare and the male supremacist complex'. *American Anthropologist*, 78, pp. 521-38.
- Earle, T. 1997. *How Chiefs come to Power. The Political Economy in Prehistory.* Stanford: Stanford University Press.
- Evans-Pritchard, E.E. 1969. *The Nuer. A Description of the Modes of Livelihood and Political Institutions of a Nilotic People.* New York: Oxford University Press.
- Fried, Morton H. 1967. *The Evolution of Political Society. An Essay in Political Anthropology.* New York: Random House.
- Friedman, J. 1982. 'Catastrophe and continuity in social evolution'. In: M. Rowlands, C. Renfrew, and B. Seagraves (eds.): *Theory and Explanation in Archaeology.* New York: Academic Press, pp. 175-96.
- Gardner, Robert and Heider, K. 1969. *Gardens of War. Life and Death in the New Guinea Stone Age.* New York: Random House.
- Giddens, A. 1985a. *The Constitution of Society: Outline of the Theory of Structuration.* Cambridge: Polity Press.
- Giddens, A. 1985b. *The Nation-State and Violence.* Cambridge: Polity Press.
- Harrison, S. 1993. *The Mask of War: Violence, Ritual, and the Self in Melanesia.* Manchester: Manchester University Press.
- Harrison, S. 1996. 'War'. In: A. Barnard and J. Spencer (eds.): *Encyclopedia of Social and Cultural Anthropology.* London: Routledge, pp. 561-62.
- Haas, J. 1981. 'Class conflict and the State in the New World'. In: G.D. Jones and R.R. Kautz (eds.): *The Transition to Statehood in the New World.* Cambridge: Cambridge University Press, pp. 80-102.
- Haas, J. 1982. *The Evolution of the Prehistoric State.* New York: Columbia University Press.
- Hobbes, T. 1991(1651). *Leviathan.* Cambridge: Cambridge University Press.
- Keegan, J. 1993. *A History of Warfare.* London: Pimlico.
- Keeley, L.H. 1996. *War Before Civilization.* New York and Oxford: Oxford University Press.
- Kirch, P.V. 1981. *The Evolution of the Polynesian Chiefdoms.* Cambridge: Cambridge University Press.
- Knauft, B. 1990. 'Melanesian warfare: A theoretical history'. *Oceania*, 60, pp. 250-311.
- Kuschel, R. 1988. *Vengeance is their Reply. Blood Feuds and Homicides on Bellona Island.* Copenhagen: Dansk Psykologisk Forlag.
- Mann, M. 1986. *The Sources of Social Power.* Vol. I. *A History of Power from the Beginning to AD 1760.* Cambridge: Cambridge University Press.

- Mann, M. 1993. *The Sources of Social Power*. Vol. II. *The Rise of Classes and Nation States*, 1760-1914. Cambridge: Cambridge University Press.
- Mansfield, E.D. 1994. *Power, Trade, and War*. Princeton: Princeton University Press.
- Martin, D.L, and Frayer, D.W. (eds.) 1997. *Troubled Times. Violence and Warfare in the Past*. Amsterdam: Gordon and Breach Publishers.
- McNeill, W.H. 1983. *The Pursuit of Power: Technology, Armed Force, and Society since A.D. 1000*. Oxford: Blackwell.
- Modell, J. and Haggerty, T. 1991. 'The social impact of war'. *Annual Review of Sociology*, 17, pp. 205-24.
- Nagengast, C. 1994. 'Violence, terror, and the crisis of the state'. *Annual Review of Anthropology* 23, pp. 109-36.
- O'Connell, Robert L. 1995. *Ride of the Second Horseman. The Birth and Death of War*. New York and Oxford: Oxford University Press.
- Ortner, S. 1984. 'Theory in Anthropology since the Sixties'. *Comparative Studies in Society and History*, 28, pp. 368-74.
- Oppenheimer, F. 1999. *The State*. New York: Transaction Publishers.
- Porter, B.D. 1994. *War and the Rise of the State. The Military Foundations of Modern Politics*. New York: The Free Press.
- Richards, P. 1996. *Fighting for the Rain Forest. War, Youth and Resources in Sierra Leone*. London: The International African Institute.
- Riches, D. (ed.) 1986a. *The Anthropology of Violence*. Oxford: Basil Blackwell.
- Riches, D. 1986b. 'The phenomenon of violence'. In: D. Riches (ed.): *The Anthropology of Violence*. Oxford: Basil Blackwell, pp. 1-27.
- Service, E.R. 1978. 'Classical and modern theories of the origins of government'. In: R. Cohen and R.E. Service (eds.): *Origins of the State*. Philadelphia: Institute for the Study of Human Issues, pp. 21-34.
- Simons, A. 1999. 'War: back to the future'. *Annual Review of Anthropology*, 28, pp. 73-108.
- Spencer, H. 1967. *The Evolution of Society. Selections from Herbert Spencer's Principles of Sociology*. Edited and with an introduction by R.L. Carneiro. Chicago and London: University of Chicago Press.
- Spencer, J. 1996. 'Violence'. In: A. Barnard and J. Spencer (eds.): *Encyclopedia of Social and Cultural Anthropology*. London: Routledge, pp. 570-78.
- Strathern, M. 1985. 'Discovering social control'. *Journal of the Law Society*, 12, pp. 111-34.
- Tilly, C. 1990. *Coercion, Capital, and European States*. Oxford: Basil Blackwell.
- Valeri, V. 1985. 'The Conqueror becomes King: A Political Analysis of the Hawaiian Legend of 'Umi'. In: A. Hooper and J. Huntsman (eds.): *Transformations of Polynesian Culture*. Auckland: The Polynesian Society, pp. 79-103.
- Webb, M.C. 1975. 'The flag follows trade: an essay on the necessary interaction of military and commercial factors in state formation'. In: J.A. Sabloff and C.C. Lamberg-Karlovsky (eds.): *Ancient Civilization and Trade*. Albuquerque: University of New Mexico Press, pp. 155-209.
- Webster, D. 1975. 'Warfare and the Evolution of the state: A reconsideration'. *American Antiquity*, 40(2), pp. 464-70.
- Wright, Q. 1968. 'War'. In: D.L. Sills and R.K. Merton (eds.): *International Encyclopedia of the Social Sciences*. New York: The Free Press, pp. 453-68.

Warfare and pre-State Societies

Warfare and pre-State Societies:
An Introduction

HELLE VANDKILDE

/8

This section concerns warfare in non-state or pre-state societies. In our own part of the world social entities without central power no longer exist. To find them we have to reach far back into the prehistory of Europe or to consult ethnographic descriptions of tribal populations outside Europe even today not entirely subordinated to a state power. European prehistory represents excellent research ground for the study of warfare in non-centralised societies outside the sphere of influence of states and, later still, under various forms of impact from such centralised units.

For thousands of years de-centrality was a dominant social principle in Europe and remained so even after the formation of states in certain core regions: around 2000 BC the first state societies emerged in the eastern Mediterranean. Further north, in central Europe and the Balkans, there were attempts to monopolise power early on, but it is not until the 7th and 6th centuries BC – under the influence from Mediterranean city-states – that these efforts turned out more successful. In northern Europe, by comparison, state formation is a late phenomenon; a prolonged process of emulation connected to the expansion and politics of the Roman Empire. Turning to the ethnohistorical and ethnographic record, few, if any of the tribal societies we know of, have maintained themselves independently of the modern world system, and this particular context should be taken into account when assessing their relationship to war.

Research and stereotypic pre-understandings

To venture into the subject of warfare among tribes is intriguing, not least, due to the mythical constructions associated over the years with these societies. The difficulties scholars are facing in avoiding these stereotypic pre-understandings are obvious when looking at the history of archaeological and anthropological research into the subject of warfare and society: we have tended to classify tribal societies as either inherently bellicose or innately peaceful with periodic ups and downs for either view (Otterbein 1999, Vandkilde 2003 and this volume chapter 5). Lately, the former, Hobbesian, view has prevailed in both disciplines.

In prehistoric archaeology, studies devoted specifically to war and violence were few until the mid 1990s, when Lawrence Keeley's book *War before Civilisation* (1996) turned the tide and pushed the opinion from a dove's view on prehistory towards a hawk's view. In the more recent anthropology non-centralised societies are quite often described as being in a constant state of war. This perception of a bellicose other with deep roots in Western thought was in the 1970s reiterated by Pierre Clastres, who argued that warfare should simply be understood as the dominant structure, the essence, of tribal society: warfare, says Clastres, is ubiquitous among tribes, and the very factor that prevents their transformation to state. What Clastres did was to reverse Thomas Hobbes (1958[1651]) famous dictum about the primitive being *'bellum omnium contra omnes'* by arguing that if the state is categorised as the pacifier of that being, then war in primitive society is war against the state; hence the phrase *society against the state* (Clastres 1977; 1994; Bestard and Bidon-Chanal 1979: 225).

There can be no doubt that warfare was a frequent activity among tribal populations in the Americas: Tupi-Guarani, Yanomami, Blackfoot, Apache, Algonquin, Iroquois, and so forth, but also in many other parts of the world, for example in Melanesia and east Africa. Tribes without warfare are definitely few, but Jürg Helbling rightly points out that it is nevertheless significant that they exist (chapter 9). Likewise, one may wonder if omnipresence of war and martial culture really is a fair description of European prehistory – before and after the first states and empires were formed in the east Mediterranean region?

Questions and recent approaches

The more recent debates in anthropology and archaeology bring forth a series of connected questions. Can we really, irrespective of context, talk about a permanent situation of war among tribes and is the often quoted prerogative 'endemic war' really appropriate? Are the use of violence and the waging of war against the other, then, deeply rooted in human biology? Here follows, logically, the question whether the state is a pacifier or creator of that bellicose being? Pacification is one possibility and at the other end of the scale warfare among tribes is considered as a result of state expansion and influence (Ferguson and Whitehead 1992). We may likewise ask whether warfare could possibly be a factor in generating social complexity and hierarchy among tribal societies thus undermining the basic principle of decentralisation? Or is it rather the opposite way around, namely that war merely has a homeostatic function in reproducing these societies as themselves? If the latter is the case, this would support Clastres' view that tribal societies are inevitably structured against change precisely due to the permanency of war (1977; 1994).

The five articles of this section all work towards solving these problems, and all authors share the view of warfare as a violent kind of social action. Jürg Helbling's (chapter 9) contribution is the most general of them, offering important new insights into the theme of warfare in societies without centralised political control – in Polly Wiessner's phrasing acephalous societies (chapter 11). It also assesses critically the prevalent theories of warfare and ventures into the difficult question of the cause, purpose and effects of war in these societies. The remaining four articles are case studies examining warfare in specific non-state societies or regions: Nick Thorpe (chapter 10) examines patterns of warfare in prehistoric Britain and Ireland, whereas Polly Wiessner (chapter 11), Ton Otto (chapter 12) and Chris Gosden (chapter 13) all operate within historical Papua New Guinea, although the actual contexts of Highland and Island New Guinea differ in many aspects.

War was never permanent

What is especially striking – when reading these different accounts from anthropology and archaeology – is the actual variable presence of war, hence strongly implying that war-related violence is culturally and socially constructed rather than rooted in psychology and biology.

Thorpe's examination (chapter 10) demonstrates with clarity that the British evidence of the Neolithic and the Bronze Age cannot support the evolutionary hypothesis of increasing warfare through time. The surprisingly rich record of traumata, weaponry and fortified settlements rather points towards specific horizons of fighting and feuding, especially connected to periods with radical social and economic change. This is much in tune with Wiessner's findings among the notoriously warlike Enga in the Highlands of New Guinea: not even here is war always of the same magnitude and different kinds of war can be distinguished, not all of them equally lethal (chapter 11).

The temporal and regional variation in the presence of war is also very much the focus of Helbling's discussion. He puts emphasis both on the fact that warfare is frequent among tribes and that peaceful tribes exist. He argues further that not all conflicts lead to war and some conflicts are solved peacefully. He shows that hunter-gatherer societies typically have no war, and this argues against biological and psychological explanations of tribal warfare. War is nevertheless a widespread form of social action, which constitutes a particular social environment to which actors and local groups have to adapt for the sake of survival. War does not break out because people are devoted to warfare and killings; it breaks out because it is simply too risky to engage in a strategy of peace: *war is the necessary, even though unintentional and damaging result of the strategic interaction of groups under specific structural conditions* (chapter 9).

The distinct variability in the presence and scale of warfare in time and space points to the conclusion that the predominant understanding of tribal societies as being in an eternal state of war has to be nuanced and contextualised. It is equally thought-provoking that some periods in European prehistory with excessive amounts of prestigious weaponry have surprisingly little evidence of war-inflicted skeletal traumata (cp. Thorpe chapter 10; Robb 1997), and this may suggest that the concepts of cold war and arms race could possibly apply to decentralised societies. This is completely in accordance with Helbling's observation (chapter 9) that among ethnographically described tribes war rarely breaks out when two parties are of equal strength and in a stalemate situation.

War and social change

People caught in the middle of a war certainly experience that conditions for action have changed, but we need to approach the relationship between war and social change in a more sociological manner. Does war produce social change, either smoothly or radically? The answer to the first part of the question is a fairly clear yes. Theories of human practice and interaction have in recent years helped to bring more focus upon the human actor as the instigator of social change, irrespective of whether that action is peaceful or warlike. All kinds of social action produce small-scale change, which we may term 'social reproduction'. Hence, warfare can be said always – whenever it occurs – to contribute to the hardly noticeable change that continuously takes place in all societies.

The answer to the second part of the question is a more hesitant yes, or perhaps. From the point of view of archaeology war seems to occur most frequently in certain periods with radical shifts in the systems of domination in Europe, for example when agriculture and a domesticated way of living were first implemented from the 7th millennium BC onwards, or much later, at the transition to the Middle Bronze Age around 1600 BC when a series of new geo-political figurations formed. Warfare can certainly be attributed some kind of role in the web of causes and effects contained by major macro-regional horizons of social transformation in European prehistory, but at present this role is hard to pinpoint more precisely. Acts of war have often enough been used strategically by groups or individuals to centralise and enhance social power, but how effective is war in this respect? Along side economic power, ideological power and political power, the military source of power is among Michael Mann's four key power sources. It mostly functions as a back-up of one or more of the other three power sources (Mann 1986). Then, in Mann's perception of history, war and military organisation often play roles in the games of power, even if they rarely stand alone.

The articles of this volume allow us to carry the debate about the role of warfare in social transformation further, especially as regards the emergence of strong leadership. Thorpe is critical to the often assumed connection between the enhancement of male status and warfare; at least it is clear from the British evidence that warfare was not only conducted by young males striving for prestige; senior males, and even women, can be seen to have been involved in violent encounters presumably related to war (chapter 10). Outside Britain the prehistoric age-gender pattern is also variable, which should evoke some caution in connecting emphasised masculinity directly with (war-imposed) hierarchy. The (male) warrior companies inhabiting many past and present societies do not inevitably lead to a central leadership even if the potential is present owing to their high standards of internal order and organisation. Rather it seems to me that warrior clubs are quite often placed on the margins of society where they are far from being power fields in themselves. Only grave forms of external pressure seem to be able to change this state of affairs (Vandkilde chapter 5; cp. also Steuer chapter 16).

In the context of historical Papua New Guinea Wiessner points out that egalitarian institutions typically pulled in the opposite direction when ambitious individuals tried to establish themselves as leaders with a following of supporters (chapter 11; also Wiessner 2002). She is also reluctant to accept a direct connection between the rise of hierarchy and the waging of war. True enough, strong leaders eventually rose in Melanesia, but not from conventional warfare, she argues. Among the Highland Enga it was first and foremost war-games and exchange transactions of the so-called Great Ceremonial Wars that played a vital role in this respect, and leadership sometimes passed on from father to son. In the eastern Highlands weak hierarchies developed connected to warfare and male cults, but had little effect on the routines of daily life. It seems likely, according to Wiessner, that it was activities within institutions with a predominantly peaceful purpose which – in her phrasing – paved the road towards changing the role of warfare from a conservative to a progressive force (chapter 11).

This is fairly parallel to the outcome of the debates in the War and Society project group, reported on in the introduction to this book: war is for the most part involved in the formation of hierarchies, but other factors enter into interaction with war. Claus Bossen (cp. chapters 7 and 17) has in particular studied war as a possible processor in state formation. His conclusion is that war can rarely be singled out as the very factor that leads to the emergence of the state: war is almost always present when states are in the making, but a series of contributing factors can be added.

In sum, regularities are difficult to pinpoint apart from the fact that war very often creates more war; most likely as an unintended effect of power-strategic actions. For the moment we can then safely say that warfare is almost always present when societies make the change towards strong leadership and hierarchy, but the role as processor needs more study.

War, tribes and states

It is probably in the nature of states to reproduce themselves by expanding their interests into new territories, such as those inhabited by decentralised societies. Doubtless the latter will be influenced one way or the other, even when the state resides, not in the vicinity, but farther away. Our present knowledge hardly allows us to model the effect of state expansion on decentralised peripheries, with respect to war and peace, in any great detail, and until recently two quite opposite scenarios have quite dominated the research. They originate in the Hobbesian and Rousseaunian worldviews briefly referred to above.

The first standpoint assumes that tribes are by nature peaceful and that warfare amongst them is not an original mode of action, but a result of state contact. Especially Brian Ferguson has advanced this view with primary reference to interaction between colonial authorities and indigenous societies in 19th and 20th century Americas, the so-called 'war-in-the-tribal-zone-theory' (Ferguson 1990; Ferguson and Whitehead 1992). The underlying idea is that a delicate, and predominantly peaceful balance between local tribes was shattered by the state intrusion and as a result widespread acts of war broke out. A similarity may here be noted to Clastres' conceptualisation of tribal warfare as a defence mechanism against the state. New archaeological evidence and analyses, however, seriously undermine this argumentation: war has existed among decentralised societies long before any state was ever present and for that matter continued to exist long after this major watershed (e.g. Thorpe chapter 10; Vandkilde chapter 5; Maschner and Reedy-Maschner 1998; Keeley 1996). This does not necessarily mean that 'the tribal zone' was unaffected, and it is quite likely that the actual level of warfare increased or decreased due to state meddling.

The second standpoint advances the contrasting opinion that state interference has the positive effect of pacifying inherently warlike tribes, but, as argued above, the level of war in decentralised societies has varied across time and space, and cases of peaceful societies can even be cited. This actual variability suggests that state intervention in tribal affairs has effects that logically also must vary from case to case. The configuration of power in itself makes possible different strategic avenues for both parties: state dominance can notably be achieved through coercion and/or persuasion. Likewise, tribal subordination may imply strategies of collaboration and/or resistance that may again have effects on the modes of interaction internally between local groups – whether in a more peaceful or warlike direction (cp. Guha 1997: 20ff).

Monopolising the violence will surely be amongst the primary aims of expanding states, but will they necessarily succeed in pacifying the people they are trying to convert into subjects and by what means? In spite of the ideological construction of Pax Romana, the reactions of Germanic and Gallic tribes to the military and political expansion of the Rome were manifold, varying in a very complex manner between radical militarisation and intimate cooperation, and even emulation of the Roman state organisation. In more recent times there are several examples of immediate or ultimate success at pacification, but this was a two-sided process in which the colonised people acted strategically in their world.

In the Bismarck Archipelago of Papua New Guinea, Gosden points out that pacification was not only the result of colonial efforts to impose a monopoly, but also due to local desires to stop fighting (chapter 13). Otto has in Manus region of Papua New Guinea observed a slightly deviating pattern (chapter 12). He points out that exchange was the crucial glue that ensured the functioning of the social structure and that warfare was traditionally utilised as an alternative resource that contributed to the maintenance of the exchange network and hence in an essential manner to the reproduction of society. In this part of PNG the continued smooth functioning of the traditional exchange systems – rather than peace – was a key issue: a destabilisation of the existing exchange systems occurred first, probably due to impact of the state apparatus, and then as a result an acceptance of colonial authority and its efforts to monopolise violence. Importantly, prior to the decisions to engage in strategies of peace in the Manus province and other regions of Papua New Guinea – the actual level of internal warring was raised – apparently accentuated by the introduction of firearms. The chain of events can be roughly reconstructed as follows: First, in order to obtain firearms attacks were organised on Westerners, who then committed severe retaliations in which villages were destroyed and people killed or driven away. The Western punitive raids and the use of firearms in local warfare disrupted existing exchange relations and the ensuing malaise – probably amplified by epidemic diseases – caused villagers to give up weapons and fighting, and as a consequence the whole social system underwent transformation ending with the firm establishment of colonial authority. A similar pattern of response also involving firearms has been noted among the Yanomami (Clastres 1994), and the adoption of horses among the Abipón of the Grand Chaco, likewise in South America, is known to have increased the level of internal warfare as well as the level of resistance against the Spanish colonisation (Lacroix 1990). In sum, this suggests a varied response to conquest and colonial hegemony and underlines that innovation in fighting technology is a variable that should also be accounted for.

Concluding remarks

It seems to me that the topic of warfare in non-state societies – with or without states in their spheres of interaction – has made some recent advances. The articles of this section testify to this conclusion. To achieve clearer answers about the relationship of non-state societies to violence and warfare archaeological prehistoric sources have to be consulted and compared to anthropological evidence. We could probably learn a lot more by engaging into systematic and context-based comparisons of cases distributed research-strategically across time and space. Attention should be paid to the intricate interplay between action and structure since it is in this configuration that regularity and variability in history are formed.

BIBLIOGRAPHY

- Bestard, J. and Bidon-Chanal, C. 1979. 'Power and war in primitive societies: The work of Pierre Clastres.' *Critique of Anthropology*, 4, pp. 221-27.
- Clastres, P. 1977. *Society against the State*. Oxford: Blackwell.
- Clastres, P. 1994. *Archeology of Violence*. New York: Semiotext(e).
- Ferguson, R.B. 1990. 'Blood of the Leviathan. Western contact and warfare in Amazonia'. *American Ethnologist*, 17(2), pp. 237-57.
- Ferguson, R.B. and Whitehead, N.L. (eds.) 1992. *War in the Tribal Zone. Expanding states and Indigenous Warfare*. Santa Fee: School of American Research.
- Guha, R. 1997. *Dominance without Hegemony. History and Power in Colonial India*. Cambridge (Mass.): Harvard University Press.
- Hobbes, T. 1958(1651). *Leviathan*. Indianapolis: Bobbs-Merrill.
- Keeley, L.H. 1996. *War Before Civilization. The Myth of the Peaceful Savage*. New York and Oxford: Oxford University Press.
- Lacroix, M. 1990. 'Volle Becher sind ihnen lieber als leere Worte. Die Kriegerbünde im Gran Chaco'. In: G. Völger and K. von Welck (eds.): *Männerbande-Männerbünde*. (Ethnologica, Neue Folge, 15). Cologne: Joest-Museum der Stadt Köln, pp. 80-271.
- Mann, M. 1986. *The Social Sources of Power*. Vol. I. *A history of power from the beginning to A.D. 1760*. Cambridge: Cambridge University Press.
- Maschner, H.D.G. and Reedy-Maschner, K.L. 1998. 'Raid, retreat, defend (repeat): The archaeology and ethnohistory of warfare on the North Pacific Rim.' *Journal of Anthropological Archaeology*, 17, pp. 19-51.
- Otterbein, K.F. 1999. 'A history of research on warfare in anthropology.' *American Anthropologist*, 101(4), pp. 794-805.
- Robb, J. 1997. 'Violence and gender in Early Italy'. In: D.L. Martin and D.W. Frayer (eds.): *Troubled Times: Violence and Warfare in the Past*. Langhorne: Gordon and Breach Publishers, pp. 111-44.
- Vandkilde, H. 2003. 'Commemorative tales: archaeological responses to modern myth, politics, and war'. *World Archaeology*, 35(1), pp. 126-44. (Theme: The Social Commemoration of Warfare, edited by Roberta Gilchrist).
- Wiessner, P. 2002. 'The vines of complexity: Egalitarian structures and the institutionalization of inequality among the Enga.' *Current Anthropology*, 43(2), pp. 233-69.

War and Peace in Societies without Central Power: Theories and Perspectives

/9

JÜRG HELBLING

Since the late 1960s – that is, after the decline of structural functionalism and under the impact of the various wars of independence and postcolonial wars in the Third World – anthropology has again become more interested in conflicts and wars (Bohannan 1967; Fried, Harris and Murphy 1968). Various theories of war in societies without a central power have been fiercely debated in the last few decades.[1] In the following article I will examine some aspects of war and peace and discuss those theories of war, which are in the centre of current discussions. I will not deal with civil wars and ethno-political wars, which have also occupied anthropological thinking in recent decades, nor will I discuss the contribution of war to the formation of states. Instead I shall concentrate on war and peace among tribal populations, which are not (no longer, or not yet) completely subordinated to a state power (Ensminger 1992: 143).[2] These so-called tribal wars – as can be observed still today in Amazonia, in the Highlands of New Guinea, in East Africa and elsewhere – are of course not 'modern wars' (i.e. in the sense of wars for secession from a state or for the control of the state apparatus). But they are, nevertheless, wars occurring in the present world of states and in the context of an economic world system – contexts, which have manifold impacts on these wars and modify their character.[3]

Besides discussing some concepts (such as war, conflict, feud and violence) as well as five important theories on war, I will deal with four issues which may be considered important for future research on war in anthropology. First, we should take into account theories of international relations, which may considerably inspire the anthropology of war. These theories are relevant for anthropology because states are political units waging war, as local groups in societies without a state are, and the logic and dynamics of war between states are, despite all the differences between 'primitive' and 'civilised' war (Keeley 1996), comparable to those in war between local groups. Second, the phenomenon of alliance has been neglected by anthropology so far, but this must be taken into consideration because whoever has to wage war also needs allies. Alliance formation influences the regional relation of force between warring local groups, and both victory and defeat may depend on the support of allies. Third, any theory of war also has to explain why in some (but few) tribal societies conflicts between local groups are never carried out by warlike means. Hence, we have to tackle the problem of explaining tribal societies without war. Fourth, the anthropology of war also has to consider the question of pacification. Pacification of warlike tribal groups is not only an interesting historical process as such, but also represents a test field for theories of war in tribal societies. Before addressing these aspects, I will give a brief overview of anthropological theories of war and discuss some conceptual problems.

1. Tribal warfare and theories of war

According to one widely accepted definition, war is a planned and organised armed dispute between political units (Otterbein 1968: 278; 1973: 923ff; 1985: 3; Ember and Ember 1994: 190), or as Ferguson (1984a: 5) puts it: 'an organized, purposeful group action, directed against another group ... involving the actual or potential application of lethal force'.[4] Tribal warfare can take different forms: from ambushes and surprise attacks to open armed clashes on different levels of escalation, ranging from an exchange of insults and the use of long-range weapons, which only cause minor losses, to pitched battles and close combat with spears and axes that cause far more casualties (Turney-High 1949; Hanser 1985). Battles such as these are not common in all societies. For instance, they occur in New Guinea, but not in Amazonia (Hanser 1985). Surprise attacks are by far the most frequent form of tribal warfare and cause the highest proportion of war related casualties. Head-hunting and other forms of conspicuous cruelty are tactics of warfare often used in areas with low population density. By means of such instrumental brutality enemy groups can be terrorised and expelled from an area, which could not be achieved as easily using military force (see Morren 1984 on the Miyanmin; Vayda 1976 on the Iban). Coalitions may be of different size and may differ in stability. In the Highlands of New Guinea, the coalitions described amounted to anything up to 800 or 1000 warriors on each side (see Meggitt 1977 on the Mae Enga; Larson 1987 on the Ilaga Dani), but mostly did not comprise more than about 200 warriors (Hanser 1985: 158ff). In these societies, alliances may be strengthened by gift exchange, alliance feasts and by marriage relationships, all of which are quite costly but render alliances more reliable and long lasting (see Meggitt 1977 and Wiessner and Tumu 1998a; 1998b on the Mae Enga). In other societies, alliances are purely ad-hoc pacts without gift exchange and marriage relations of any importance and are, therefore, far more unstable (see Chagnon 1983 on the Yanomami).

Turney-High (1949), Keegan (1993) and others maintained that the difference between 'primitive' and 'civilised' war is absolute and essential (see Otterbein 1999). According to them, 'primitive warfare' is determined by religious-cultural factors and is mainly a harmless, playful form of fight, causing only minor casualties. In contrast, states rationally calculate the advantages and disadvantages of a war; 'civilised (real) wars' aim at territorial gains or political advantages and cause far more losses than 'primitive wars'. However, this distinction does not make sense against the background of empirical evidence: local groups in tribal societies compare possible gains and losses of a war as well, and they behave strategically and use certain tactics to beat their enemies – as states do. Furthermore, war-related mortality seems to be even higher in 'primitive wars' than in 'civilised wars': war-related mortality as a percentage of total mortality averages between 20% and 30% in tribal societies, whereas it lies below 5% in most state societies (see Keeley 1996: 88ff, 196f). Thus, it is tribal warfare which deserves the description 'total' (in the sense of involving the whole population) rather than wars between states, even though 'unrestricted' and 'restricted' warfare can also be distinguished in tribal societies (Feil 1987: 67f).

Several anthropological theories of war in tribal societies are distinguishable. I will discuss five theories, which take centre stage in current debates on tribal warfare (for a more comprehensive list of theories, see Otterbein 1973 and 1990). But before examining these theories more extensively and presenting yet another new theory of tribal war, we first have to discuss some of the concepts involved, such as war, violence, feud and conflict.

2. War, conflict, feud and violence

War is a planned and organised armed dispute between political units (Otterbein 1973: 923ff; Ferguson 1984a: 5). The political units in a society without a state are local groups (i.e. villages) or coalitions of local groups. Local groups display an internal hierarchy and a leadership structure (elders and juniors, men and women, village headmen, councils of elders etc.), as well as a specific kin composition (such as local kin groups which possibly form political factions). Internal conflicts are usually settled in a peaceful way within the group or – if they cannot be resolved (i.e. if violent self-help prevails) – they may escalate, leading to a splitting of the group. Decisions on war, alliance or truce are usually made through meetings comprising all the adult men in a local group, although group members may differ as to political position ('hawks' and 'doves'), bargaining power and, hence, as to their interests (elders and juniors, competing local kin groups). As a collective decision is finally taken and implemented, local groups can be interpreted as politically autonomous, collective actors with regard to 'foreign policy'.

Furthermore, the analytical difference between 'conflict' and 'war' must be emphasised. In societies without a state conflicts are fundamental features characterising the social interaction within as well as between the groups (Koch 1974a: 16). However, not all conflicts lead to wars, but some are resolved peacefully by means of compensation payments or negotiations, or alternatively violence is contained by limiting it to the direct adversaries and normatively regulated as in a feud (see Greuel 1971 on the Nuer). As serious conflicts may peter out and trivial conflicts escalate into wars, there is no necessary relationship between conflict and war, i.e. between the seriousness of a conflict and the intensity of a war. Thus, war is only one mode of conflict resolution, namely a planned and organised armed conflict between political units.

As for 'violence', this embraces a large number of phenomena ranging from war to hooliganism, from torture to terrorism, from a bull fight to suicide (see Riches 1991: 293f; Aijmer 2000: 1) and it is doubtful whether violence in general is an appropriate object for a theory at all. Violence is defined as intentionally inflicting physical harm on somebody (Riches 1991: 292ff). If we define war as an armed conflict between local groups, violence between individuals (or families) belonging to the same or to different local groups is not war. Violence between individuals often breaks out spontaneously (Knauft 1987) or in a ritualised form, as in the form of fist pounding or stick duels among the Yanomami (see Chagnon 1983). Conflicts within a local group are usually settled peacefully, but they can escalate, leading to a splitting of the group. This is often prevented, however, since a split would weaken the group militarily at a time when it is facing a threat from hostile neighbouring groups. Furthermore, 'war' must be distinguished from 'feud', since a feud consists of violence and counter-violence between individuals and/or families of different local groups. Whereas feud consists of taking revenge for wife stealing, abuse, manslaughter, or sorcery etc. in order to achieve an even score, war usually aims at defeating the enemy by decimating and expelling him (Carneiro 1994: 6).[5] However, the difference between war and feud is often blurred, since feuds may escalate into war between local groups under certain conditions. Some wars – often called 'ritualised warfare' – have much in common with feuds as they are waged with the intention of 'making peace' after a show of force, and end with an exchange of compensation payments.

The difference between violence and war is not merely a terminological exercise, as can be seen in the case of hunter-and-gatherer societies (such as the !Kung San, BaMbuti, Yaghan and the Inuit). In these societies wars usually do not take place, but a high level of inter-personal violence may be observed, with homicide rates even higher than in tribal societies.[6] In contrast to that, the social relations within the groups are largely peaceful in many warlike tribal societies, as among the Dani, the Cheyenne and the Iban (see Kelly 2000: 21). Because war, as an armed conflict between groups, must be distinguished from violence between individuals and families, biological or psychological explanations of tribal war can also be refuted. These theories see war as an extension and accumulation of individual violence and do not distinguish between individual and collective violence, which is a planned and organised endeavour of a local group achieved by a bargaining process within the group. Furthermore, such theories explain violence by some biological potential or psychological mechanism (such as frustration leading to aggression). While nobody has ever contested the proposition of a (biological) capability for aggression, the same also holds true for the ability to behave peacefully, which may also be rooted in our biologically determined behavioural repertoire. The main problem of all theories referring to human universals is that they can explain neither the regional and temporal variation of war within a society, nor its variation between different tribal societies or between different types of societies.

3. Five theories of tribal warfare

In the following, I will discuss the five main theories of tribal war: 1) the biological, 2) the cultural, 3) the ecological and economic, 4) the historical and 5) the political theories of war.

1) Biology

According to sociobiology (see Chagnon 1988; 1990a; 1990b on the Yanomami; Durham 1991 on the Mundurucù; Barash 1981; Gat 2000), tribal wars may ultimately be explained by the competition between men for the scarce resource of women[7]. The theoretical background of this proposition is the assumption that more aggressive men are not only more attractive to women (because they improve their children's chance of survival) but are also able to outdo less aggressive men. Aggressive men are assumed to have a reproductive advantage, because they can control more women and transmit their genes to more (surviving) descendants than their less aggressive competitors. The aggressiveness of men is therefore favoured by sexual selection and ultimately leads to warlike competition between local groups including the abduction of women from enemy groups and allies alike, as Chagnon (1983) has maintained for the Yanomami.[8] Currently two variants of socio-biological theories are discussed (see van der Dennen 1995; 2002). Alexander (1977) has proposed a theory of 'imbalance of power', according to which a group attacks if it is superior in power, and will be rewarded with women and resources (see also Wrangham 1999). This argument, however, already presupposes the existence of inter-group hostility. Hence, an imbalance of power may be a plausible reason for a specific war to break out, but it is neither an explanation for tribal war nor is it a biological theory. The theory of 'male coalitional warfare' (Tooby and Cosmides 1988; van der Dennen 1995; Wrangham 1999) maintains that men in war pursue a high risk, high gain reproductive strategy: the surviving men will gain more women (on average) after a war because war-related female mortality is much lower than male mortality. This theory may explain the (reproductive) interest of men in participating in wars, but it does not explain why war is the dominant mode of interaction between local groups in tribal societies.

To explore the impact of individual and group strategies in a warlike environment on the relative reproductive success (on mortality and fertility) is one matter. To maintain, however, that war is adaptive (in this biological sense) – 'a master adaptation', as Barash (1981: 188) puts it – is not very plausible considering the high costs of war (including the loss of life, the practice of infanticide and the destruction of resources), as even Barash (1981: 181ff) has

to concede. Rather, war constitutes a specific social environment, to which local groups have to adapt in order to survive. Such a warlike environment may have emerged for the first time in world history as an unintended result of becoming sedentary (first among Mesolithic fishers and then among Neolithic farmers), as archaeological data suggest (see Ferrill 1985: 26ff; Gabriel 1990: 31ff; Keeley 1996: 31, 39; Thorpe http://www.hum.au.dk/fark/warfare/thorpe _paper_1.htm; Haas http://www.santafe.edu/sfi/ publications/Working-Papers/ 98-10-088.ps: 6, 8, 10, 13, 18).

The sociobiological explanation of war is also implausible for empirical reasons, as even the example of the Yanomami shows. Although conflicts between men of different villages may emerge because of women (wife stealing and adultery), most of these conflicts are settled in a peaceful way. They may provoke duels, but this only leads to war if the relationships between the villages were already, for other reasons, in a bad state (Lizot 1989: 105f; Alès 1984: 92). Although wife stealing is a welcome side-effect of a successful war campaign (more fertile women, more children and, thus, more influence within a group for a man and more future warriors for the group), it is neither the cause nor the purpose of wars (Alès 1984: 97; Lizot 1989: 106; even Chagnon 1983: 175f).[9] According to Lizot (1989: 104f) aggressive men do not enjoy higher status within their local group, but they do earn greater respect from their enemies: to kill a successful warrior improves one's reputation. Therefore, they become preferred targets in warlike clashes, and their life expectancy is lower than the male average. Excessively aggressive men, who involve their group in unnecessary and unwanted wars, are often killed by their own people, even by close kinsmen (Biocca 1972). Besides that – and this is the crucial point – successful warriors do not have more wives or more children than other men (Lizot 1989: 104f; Albert 1989; 1990; Ferguson 1989b). The proposition of a relative reproductive success of aggressive men is, thus, not confirmed (see also Robarchek and Robarchek 1998: 133ff on the Waorani and Moore 1990 on the Cheyenne). It seems that wife stealing can only be understood against the background of an already existing warlike environment. Men or local groups try to acquire more women and to have more children, in order to improve their political position within the group or to enhance their military strength. Competition for scarce women (an unlimited demand for fertile women) is not a natural phenomenon, but a consequence of warlike competition between local groups (see Harner 1975 on women as the labour force). Therefore, war is the cause, not a consequence, of the scarcity of women.

2) Culture

Cultural theories explain wars by moral ideals and norms, which sanction and value violent behaviour and are reproduced by corresponding modes of socialisation, such as male initiation (Whiting 1965). These motivational dispositions and expectations cause violence and its escalation to war (Robarchek 1989; Robarchek and Robarchek 1992; 1998; Ross 1981; 1986; 1993a; Orywal 1996a; 1998).

It is true that (male) violence (or rather courage and the readiness to be violent) is highly valued in warlike societies, whereas peaceful behaviour and harmonic interaction are the cultural ideals in societies without war. However, these are not causal relations, but correlations, which need to be explained. Norms and ideals which reward violent behaviour only have a selective value (as compared to alternative norms of peacefulness) in an already warlike society. If, among other things, the military success of a group depends on the courage and determination of its warriors, local groups with a higher share of such men will have military advantages in a warlike environment (Peoples 1982). The cultural theory does not, however, explain how and why such a warlike environment comes into existence.

Furthermore, the cultural norms and ideals are one thing, the real attitudes and preferences of actors, however, quite another. It seems astonishing that the proponents of a cultural theory of war have seldom tried to find out the real attitudes of their informants towards war and violence. If they had done so, they would have learned that the facts contradict their proposition that cultural norms and ideals determine human thinking and action. Ethnographic evidence shows that even the most valiant and courageous warriors consider war a bad thing and are afraid of being wounded or killed during fights; many even suffer from war trauma (see Keeley 1996: 395; Knauft 1999: 143ff). Hence, fear seems to be a more important emotion than aggressiveness in the context of war (Gordon and Meggitt 1985: 28, 146; Goldschmidt 1997: 50f). Taking this meta-preference for peace even in warlike societies into account, it is

not surprising that men first have to be motivated to engage in an unavoidable war.

In a society with local groups, entangled in a permanent state of war, there is a high demand for men willing to overcome their fear when time for war has come. Actors will choose that mode of action which is best rewarded and provides the most advantages for men who courageously participate in war. Reputation and prestige, power and booty, slaves and women are the most important advantages a man can get by successfully participating in war. War rituals, prophesying and protective amulets also contribute to overcome fear. And a corresponding mode of socialisation emphasises courage and strength, fearlessness and perseverance (Goldschmidt 1997: 51–56). Since every mode of socialisation produces the kind of individuals which 'society needs', it is not astonishing that the mode of socialisation in these societies produces individuals – through a system of reward and punishment, disapproval and indoctrination – who are prepared to be violent, courageous and aggressive. Such a mode of (male) socialisation is a consequence of war and not the reverse, because it disappears with progressing pacification (Ember and Ember 1994: 192; Goldschmidt 1997: 55, 58f). Aggressiveness must be mobilised and instrumentalised. Aggressiveness is, however, not the dominant feeling in wars, but the quality needed to gain the rewards and advantages which are connected with a successful performance as a warrior (Goldschmidt 1997: 58f). But these rewards and advantages are always weighed against the costs and risks of a war.

Furthermore, cultural norms and values (such as the obligation to take revenge in the case of black magic and manslaughter) never completely determine the actors' behaviour or even their thinking. While the principle of taking revenge is never contested, its application has always to be negotiated and interpreted in each individual case. Whether revenge is taken in a specific case depends on the relations of force between the groups as well as on the bargaining process between 'hawks' and 'doves' within each of the involved groups (see Greuel 1971 on the Nuer; Ferguson 1995 on the Yanomami). Ethnographic evidence shows that a group will only remember an unpaid blood-debt and take revenge if it is stronger than the other group and if there is a good chance of defeating the enemy. The number of conflicts, accusations and grievances will only increase if a group is determined and ready to start a war (see Vayda 1976: 13 on the Maring). Sometimes, a reason for revenge is even intentionally created through a well-calculated provocation in order to have a legitimate reason to attack a weaker group (see Godelier 1982: 155 on the Baruya). If in turn a group is too weak, or interested in an alliance, the accusations of a perpetration are simply 'forgotten' or the death of a group member is attributed to 'natural causes' (see Lizot 1989 on the Yanomami).

Therefore, warlike behavioural ideals and norms, as well as the obligation to take revenge, are not causes of war, but can only be understood against the background of an already existing warlike environment, in which each local group fights for its survival (Helbling 1996a). Even though norms and values rewarding violent behaviour do not explain wars, the description of the cultural dimensions of war remains nevertheless important.[10]

3) Ecology and economy

According to the ecological-economic theory, war is the result of competition due to scarce resources and population pressure (Vayda 1961; 1976; Harris 1977; 1984; Rappaport 1968). The shortage of agricultural land and/or of game, on which local groups are dependent as resources, leads to stress (frustration and aggression) within the groups, as well as to competition between adjacent local groups. These conflicts easily escalate into wars, aiming at appropriating more land from neighbouring enemy groups or at expelling them from their hunting grounds. Rappaport (1968) on the Maring, as well as Harris (1974; 1977) on the Yanomami, reformulated this theory using a functionalist model. According to them, war has the function of lowering population growth (through a reduction of local population by war or, indirectly, through female infanticide) and of preventing the overuse of local resources (by reducing pig population for alliance feasts or by spacing out enemy groups and creating buffer zones where depleted game may recover). But even if war had such ecological functions – which is highly questionable (see Helbling 1991; 1992; 1996a) – it would still have to be explained why local groups – acting according to their interests, not in order to meet the requirements of their ecosystems – decide to wage war. This is even more questionable if one considers the fact that war always entails considerable risks (such as loss of life and the destruction of

resources) and high costs (such as war preparations and recruitment of allies through gifts).

There are examples of warlike tribal societies (especially in the Highlands of New Guinea) with high population densities, where land resources represent a frequent reason for conflicts between adjacent groups, as Meggitt (1977: 14) has argued for the Mae Enga. However, there are several arguments against such a theoretical position. First, conflicts over scarce resources – as other conflicts – do not necessarily lead to war. There are alternatives to a warlike zero-sum struggle over scarce resources, such as relocation of a village or migration of a faction into a thinly populated area, a peaceful exchange of land and trade between local groups, as well as the intensification of agriculture (Ferguson 1989a: 196; Hallpike 1977: 231).[11] Land scarcity may increase conflicts, but conflicts do not have to lead to wars. The same is also true for the Yanomami, who according to Harris (1977) wage war because of scarce game. Lizot (1971: 149-68; 1977: 190-202) and Chagnon (1983: 57, 85f) have shown that the Yanomami consume vegetable and animal protein in sufficient quantities. Many species of wild animals are locally available in high densities and not all species are hunted; the high resource selectivity also weighs against the proposition concerning a general scarcity of hunting game. The hunting territories are sufficiently large and exclusively defined; they are, therefore, never the object of conflicts between local groups (Lizot 1977: 195). Second, there are numerous warlike tribal societies in which population densities are low and resources cannot be said to be scarce at all (Hanser 1985: 269, 285 on the Eastern Highland of New Guinea; for Amazonia see the examples of the Jivaro, Mekranoti, Waorani and the Yanomami). And there are societies with very high population densities, in which hardly any conflicts break out over land or other resources, as among the Ilaga Dani (Larson 1987: 405). Hence, population density is neither a relevant indicator of resource scarcity nor of frequency of conflict or war (Knauft 1999: 124). Third, in many warlike societies with resource scarcity, it is the necessity to prepare for war and to recruit war allies, which forces local groups to pursue an expansive reproductive policy (high birth rates, wife stealing) and to increase production for alliance feasts and gift exchange. As local groups have to compete for allies with multiple alliance options, an inflationary increase in the production of political goods is often the result, as evidenced by the ceremonial exchange of pigs in the Western Highlands of New Guinea (see Meggitt 1974: 198ff and Wiessner and Tumu 1998a; 1998b on the Mae Enga; Helbling 1991 on the Maring). But even in societies where gift exchange and marriage relations only play a minor role (as in Amazonia and the Eastern Highlands), allies must be won by holding feasts and therefore crop production and hunting have to be intensified. Thus the shortage of resources (of land or game) is a consequence rather than a cause of war (Helbling 1991; 1996a; Ferguson 1989a: 185f).[12]

4) History

According to Ferguson and Whitehead (1992: 27f) tribal war is not rooted in the structure of indigenous societies, but first occurred in the course of the expansion of (colonial) states – and the formation of the world economic system. Tribal wars are thus not explained by reference to the internal logic of tribal societies, but as a consequence of the expansion of the state into the 'tribal zone', in which the state interacts with tribal populations. Ferguson and Whitehead maintain that tribal wars break out when local groups start to compete for scarce trading goods (such as iron tools and weapons), for the control of export products (such as slaves) and for a favourable position in regional trading networks (cp. Ferguson 1992 on the Yanomami). Furthermore, the expansion of the colonial state triggered rebellions and wars of resistance in indigenous populations. And the states often supported and supplied groups with arms in order to attack other tribal groups, to punish them for rebellions, or to capture slaves (Ferguson and Whitehead 1992: 19). All these factors have contributed – according to Ferguson and Whitehead – to the emergence of tribal warfare.

It is true that the expansion of colonial states created new constellations of conflict as well as new forms of war, as Ferguson and Whitehead have shown. However, numerous archaeological findings (see extensively Roper 1975; Vencl 1984; 1991; Keeley 1996; Haas http://www.santafe.edu/sfi/publications/Working-Papers/98-10-088.ps, LeBlanc 2003) and ethno-historical data (Knauft 1999: 99ff) indicate that this proposition, according to which tribal wars are caused by the expansion of states into the tribal zone, is wrong.[13] Even the less radical version of this theory – claiming that wars did not emerge for the first time but intensified in the tribal zone

– seems to be one-sided. The interaction of tribal groups with expanding states had different effects. As well as intensifying warfare it also reduced warring in many regions, or even stopped it altogether. As Service (1968) has already shown, defeated populations were forced to retreat into inhospitable areas and to transform themselves into peaceful hunter-and-gatherer societies (see also Dentan 1992; 1994). Even Ferguson (1990a: 242) argued in an earlier article that epidemics decimated indigenous populations (such as the Pemon and the Piaroa) and reduced regional settlement densities to such an extent that local groups were henceforth too far removed from each other to wage war. Furthermore, it should not be forgotten that the politics of all colonial states ultimately aimed at pacifying warlike tribes and at establishing a monopoly of power, which they always achieved sooner or later (see Bodley 1983).

The fact, however, remains uncontested that the states and the world economic system constitute contexts for tribal wars which must be considered in their historical dimensions much more than has been the case up to now, as has been shown by Ferguson (1995) on the Yanomami, Sandin (1967), Pringle (1970) and Wagner (1972) on the Iban, Renato Rosaldo (1980) on the Ilongot, Keesing (1992) on the Kwaio on Malaita, Meggitt (1977), Gordon and Meggitt (1985) and Wiessner and Tumu (1998a; 1998b) on the Mae Enga and others (see also Wolf 1982; 1987).[14] But it is important to analyse not only the wider regional and historical contexts but also the internal logic of indigenous warfare.

5) Politics

According to Koch (1973; 1974a; 1974b; 1976), Spittler (1980a) and Sahlins (1968: 5), war in tribal societies must be explained by the absence of a triadic mode of conflict management (adjudication), i.e. of a superordinate power (such as a state), which can enforce peaceful settlement of conflicts between groups and prevent the escalation of conflicts into wars. If no (efficient) state is present, a 'permanent state of war' will prevail, in which wars may break out at any time. This Hobbesian proposition, however, may hold for tribal societies, but not for hunting-and-gathering societies, which also lack a state, but usually do not wage war.

Marcel Mauss (1926) saw the solution for this Hobbesian problem in gift exchange, presenting an overall alternative to the general state of war.

According to Lévi-Strauss (1967: 78; Clastres 1977: 183ff), every successful exchange prevents war and every war is the result of a failed exchange. It is, however, highly questionable whether gift exchange is an alternative to war. It is true that gift exchange may strengthen an alliance, as Meggitt (1974) has shown for the Mae Enga and Pospisil (1994) for the Ekagi-me. But, as the Mae Enga state, "we first exchange, then we fight", which means that either gifts are exchanged with one group in order to wage war against another, or gift exchange itself creates conflicts, and allied groups become enemies (Gordon and Meggitt 1985: 149). Gift exchange is a means to recruit allies in order to fight and defeat common enemies, but it is neither an alternative to war in general nor a goal in itself (Clastres 1977: 196ff; Meggitt 1974: 170f, 198ff). Gift exchange (and marriage) may even in itself create conflicts between exchange partners, so that allies may turn into enemies (see Tefft and Reinhardt 1974; Tefft 1975; Knauft 1999; Pospisil 1994).

Sillitoe (1978) proposed another political theory of war by referring to the internal power structure of local groups. According to him, political leaders are motivated to instigate wars because ultimately they owe their status to their merits as valiant warriors and organisers of war campaigns. For Godelier (1991) this is especially true for Great Men, but much less for Big Men, who owe their status to their skills as organisers of gift transactions (see also Feil 1987 on the Highlands of New Guinea). However, even an aggressive Great Man must conform to public opinion and cannot permanently act against the interests of the majority of his group. Great Men are often killed by their own people, especially if they have infringed on other group members' interests and have turned into local despots (see Godelier 1982 for the Baruya; Biocca 1972 for the Yanomami; Watson 1971 for the Tairora; Pospisil 1978 for the Ekagi-me).[15]

Otterbein (1985; 1990) put forward yet another political explanation of tribal war, which refers to kinship relations within and between groups. According to him, it is highly probable that wars break out in societies with patrilocal and patrilineal groups (fraternal interest groups), because no relations of kinship amity and loyalty exist between the local groups (see also Murphy 1957, Thoden van Velzen and van Wetering 1960). However, Ember and Ember (1971) and Lang (1977) argued that – if such fraternal interest groups occur – they are the consequences rather than the causes of war. Moreover, rules of descent and locality are cultural norms, which do not even allow us to predict the actual kin composition of local groups (Sahlins 1965). Local groups in a warlike environment most often display a heterogeneous kin composition, comprising defeated allies and immigrants from weaker groups who joined the stronger group in order to enhance their military strength (Hanser 1985: 297ff). Nevertheless, patrilineal kinship rules may form the core of an ideology, which is the normative result of adaptation to a warlike environment, and may enhance the solidarity of co-resident men (Lang 1977; Ember and Ember 1971).

The propositions regarding 'fraternal interest groups' and the lack of an adjudicative power (put forward by Koch 1974a; 1974b), however, converge in the more general approach suggesting that politically autonomous local groups in a multi-centric, anarchic system are an important element for the explanation of tribal warfare. But the emergence of conflicts is still not explained in this way, and social-isation inducing aggressive behaviour (Koch 1974a) or biological dispositions like 'the primate past of man' (Otterbein 1985: 168f) are not convincing either. In contrast, the absence of a superordinate power (such as a state) is able to explain both the conflict ridden relationships between local groups, as well as the high probability that these conflicts will escalate into wars. I have tried to develop such an explanation, which I will sketch in the following (see Helbling 1999).

4. War as a strategic interaction between groups in an anarchic environment

I shall focus on tribal societies in which the state – before pacification – did not play a role, or at least not a decisive one. I will first address the preconditions for the likelihood of war in these societies by referring to both structural conditions and interaction in the form of conflict between local groups. However, those theories which have been refuted as causes of war in the preceding part must also be accounted for in an alternative theory of war. My main proposition is that the cultural, economic, socio-structural and political factors, which these theories hold responsible for tribal war, are merely dependent variables, i.e. the consequences of war.

Let us start with the fact that war – as a purposeful

and planned group action, directed against another group and involving the application of lethal force – was quite frequent in societies without centralised power. Whereas war was hardly fought by nomadic bands, there was a state of permanent war in tribal groups of shifting cultivators, sedentary fishermen and pastoral nomads; war could break out at any time, though without effectively being waged all the time. War has severe consequences: war-related male mortality is as high as 35% on average (N=13) and overall war-related mortality is at 25% on average (N=16).[16] A huge volume of resources is destroyed or misallocated, and a considerable number of the labour force as well. The question then, is how this state of permanent war can be explained. My proposition is that tribal war can be explained by two structural conditions: 1) the anarchic structure of the political system consisting of politically autonomous local groups, and 2) the relative immobility of local groups, i.e. their dependence on locally concentrated resources.

These structural conditions render strategic interaction between local groups warlike, which may be described in terms of game theory.[17] I should add that game theory neither refers to games nor is it a theory. Rather, it is a parsimonious description of different constellations of strategic interaction between social actors, and its logic may be described in purely colloquial terms without any mathematical technicalities. These simplified descriptions should make sense of what we observe, i.e. they are not a priori models into which reality has to fit. Game theory is also a decision theory assuming that actors behave according to their interests and to their evaluation of the advantages and disadvantages of different strategic options. Ultimately this boils down to the assumption that actors always have good reasons for behaving in the way they do. Without this assumption hardly any behaviour could be explained. The analytical aim is to understand the structural property of the environment which may explain why local groups interact the way they do.

Let me first elaborate on the structural conditions of war in tribal societies.

4.1 Structural conditions

The first structural precondition in the explanation of warfare is the anarchic system in which local groups interact, as Thomas Hobbes (1994[1651]), but also Sahlins (1968), Hallpike (1973), Koch (1974a),

Colson (1975), Spittler (1980a; 1980b), Keeley (1996) and others have argued. Conflicts between local groups can be settled either by peaceful or by warlike means. The reason why conflicts between local groups lead to war is that there is no superordinate, centralised power (adjudication) such as a state that could prevent violent settlement of conflicts between local groups and punish those who break agreements for peaceful conflict resolution. This impossibility of precluding violence through bilateral agreements ultimately forces each group to use violence in the first place, in order not to fall victim to the violence of others. However, war – though not interpersonal violence – is extremely rare among nomadic bands of hunters-and-gatherers (see Helbling n.d.b and footnote 6), in spite of the fact that these societies also lack a superordinate power. Hence, the lack of an overarching power or, to put it differently, the political autonomy of local groups is only one structural condition, but not the only one.

The second structural condition responsible for the prevalence of warfare in tribal societies is their dependence on locally concentrated resources, such as fields, herds, pasture or fishing grounds. If local groups depend on locally concentrated resources, they cannot afford to move away and thus avoid armed confrontation with adjacent groups, without incurring high opportunity costs: this would entail losing property, forgoing harvests and risking starvation. In contrast, resources in hunting-and-gathering-societies are usually widely scattered. Hence, mobility is not only a successful production strategy, but also a precondition for evading conflicts and avoiding war (Dyson-Hudson and Smith 1978; Carneiro 1994: 12; Keeley 1996: 31; Haas http://www.santafe.edu/sfi/publications/Working-Papers/ 98-10-088.ps: 8, 10, 18).

Where these two structural conditions exist, wars can break out at any time. They therefore explain the permanent state of war, i.e. constitute the preconditions for the likelihood of war in tribal societies.

4.2 Strategic interaction

A tribal society thus constitutes an anarchic system of autonomous local groups dependent on locally concentrated resources. This structural framework also represents an incentive system in which each local group pursues its own interests in interacting with others, i.e. it causes a specific form of strategic interaction between local groups. The logic of the warlike strategic interaction, resulting from the two structural conditions, may be described as a prisoners' dilemma or a security dilemma.[18]

It may be assumed that politically autonomous local groups would prefer to (co-operate and to) settle their disputes in a peaceful, non-violent way, because they could avoid high losses of human life and resources. (According to the logic of the prisoners' dilemma, co-operation between the groups would provide the highest collective gains for them.) However, because bilateral agreements between local groups aiming at settling conflicts peacefully are neither sanctioned nor enforced by a superordinate power, none of the groups involved can be sure that the other groups will keep such agreements. Hence, it is too risky to pursue a peaceful strategy unilaterally, because a one-sided peace strategy would be interpreted by the other groups as a sign of weakness and this would encourage them to attack. This is because a bellicose strategy not only brings higher gains (by decimating the other groups or expelling them from their territory, and by capturing booty), but it also helps to reduce the highest possible risks, by being prepared for surprise attacks, and so deterring enemies. The adoption of a bellicose strategy is all the more necessary as local groups are dependent on locally concentrated resources and therefore cannot opt for withdrawal, as an alternative to war. The aim of war is thus to deter enemies, to decimate them and weaken them by stealing their women, their animals and their land, in order at the same time to gain strength. To get rid of them – by annihilating them or driving them out into unfertile, disease-stricken areas – is even better.

Thus the two structural conditions create a warlike environment in which local groups have to survive. The mutual mistrust and reciprocal threat of force ultimately compel each group to take steps to ensure its survival. The conflicts, leading to war in an anarchic 'state of warre', are themselves a result of this anarchic system. It is thus not an innate human propensity for aggressiveness (the Hobbesian position has sometimes been misrepresented in that sense) which propels collective violence, but fear. But what about the cultural factors explaining war?

Culture

Although I consider the practical reason of social actors to be the most important aspect, I am not denying the importance of the cultural dimension. It should be stressed that game theory already takes

into account and explains the perceptions and expectations of social actors. It does not ignore them – as some have criticised – but considers them as a part of the game, i.e. controlled aggressiveness is supported by cultural norms and rewarded with prestige, but mistrust and fear are also culturally expressed. The adequacy of cognitive systems and the effectiveness of norms and values may vary. However, if local groups do not realistically perceive their warlike environment and male actors are not motivated to overcome their fear of participating in a war, they will be punished militarily in a selective social environment.[19]

It is thus not astonishing that warlike behavioural ideals, norms and values, as well as corresponding modes of socialisation, aiming at rewarding courageous behaviour and punishing cowardice, correlate with the occurrence of war as we have already seen. However, they only make sense in a social environment which is already warlike; they must therefore – contrary to what a cultural theory of war maintains – be treated as dependent variables, as cultural and behavioral adaptations in a warlike environment. It does not come as a surprise that values rewarding readiness for violence make sense in a warlike environment, because the military success of a local group also depends on the motivation and skill of its adult men in war. It is only by such norms and values, as well as war rituals and protective amulets, that reluctant men are motivated to overcome fear, to participate in a war and to muster the courage and determination to fight, as Harrison (1993) shows for the Manambu (see also Goldschmidt 1997). But despite all these cultural incentives there are still many reasons for a man not to participate in war: bad omens, such as certain birds' song, bad dreams and so on, that allow warriors to stay at home (Goldschmidt 1989). Even staunch tribal warriors dislike war: they fear war-related risks and suffer from war trauma (Knauft 1999; Keeley 1996). They seem to have a meta-preference for peace but see themselves compelled to wage war for reasons of defence. Meggitt (1977: 33) mentions this 'Hobbesian view of war' among the Mae Enga:

Fear is probably a more potent force in shaping human and social destiny than bravery or entrepreneurial skill. ... A climate of suspicion and distrust appears to be a common characteristic of loosely structured or acephalous societies, which like the Enga espouse a fiercely egalitarian ideology. (Gordon and Meggitt 1985: 147)

And one Yanomami is quoted accordingly: 'We are fed up with fighting. We don't want to kill any more. But the others are treacherous, and one cannot trust them' (see also Chagnon 1977: 35f, 129f). This is exactly what the prisoners' dilemma is all about.

Under conditions of mutual mistrust and the reciprocal threat of force, the survival of each group depends on its ability to become larger and stronger and to recruit more allies than its potential adversaries. But as the military superiority of one group inevitably entails the corresponding inferiority of the other groups, these other groups strive to become larger and to win more allies in turn. Local groups are thus trapped in a security dilemma (Jervis 1978; Otterbein 1988), where security for one group leads to insecurity for the others. The logic of this security dilemma causes an 'arms race' between groups, and mandates that each group attacks pre-emptively and tries to decimate, weaken or rout its enemies, because if a group does not attack at a favourable moment it risks being attacked at an unfavourable moment (Waltz 1960: 5).[20] There is no way out of this security dilemma, and groups that behave differently risk being defeated, routed or even annihilated.[21]

4.3 Military strength, group size and alliances

Under the conditions of a security dilemma, the survival of every local group depends on their ability to become stronger, i.e. to become larger and to recruit more allies than their enemies.

Group size

The security dilemma forces each group to enhance its military strength, which basically depends on the number and determination of its warriors. Hence, local groups have – as a consequence – to adopt an 'expansive population policy' (high fertility, positive balance of marriage exchange, wife-stealing) and to encourage immigration. This also explains the relatively high population growth in tribal societies. Each group thus tries to become at least larger in size than its neighbouring rivals.

What about the social organisation of local groups and its relation to war?[22]

Social organisation

The incorporation of refugees (from defeated, allied groups) or regrouping (after a defeat) – as they often occur in warlike tribal societies (see Colson 1975: 29f on the Iroquois; Watson 1983: 231ff on the

Tairora) – may explain why patrilocal, patrilineal male groups are neither common nor necessary in warlike tribal societies. They certainly do not explain war, as the theory of fraternal interest groups maintains (Otterbein 1994). However, an ideology of male solidarity, irrespective of the actual kin composition of a local group, may enhance the unity and solidarity of a group's fighting force, as do rituals and co-residence in a men's house. Such an ideology may well be expressed using (fictive) kinship terms. The crucial elements enhancing the unity of a local group are co-residence and the common threat posed by neighbouring local groups. Adoption and incorporation of refugees into local kin groups does not mean that kinship is a determining factor, but that it is used as a metaphor for amity and co-operation.

Political leaders

As for political organisation, local groups with more efficient leaders will have military advantages (Otterbein 1985: 95; Hallpike 1977: 122-26, 129, 135f). But people will be loyal to a local leader only as long as he is a shrewd organiser of war campaigns, an able mediator in internal disputes and a successful recruiter of allies for his group. The support for leaders within a group is usually stronger in times of war than in peacetime (see Meggitt 1971 on the Mae Enga; Godelier 1982 on the Baruya). This is because unsettled disputes within a local group, or even open violence between group members, may seriously degrade the military strength of a group. The local group is an organisation with a system of sanctioned norms, and it should be a realm of co-operation. A political leader delivers collective goods by organising and co-ordinating war campaigns and alliances, as well as by contributing more than other group members to alliance feasts and compensation payments. For his superior contribution to the military success of the group, he is awarded high status and a good reputation, as long as he delivers. A local leader who turns into a despot, bullying the other group members, is either killed by his own people, or his group masterminds a secret pact with their enemies to have him killed, as among the Tairora (Watson 1971; 1983), the Yanomami (Biocca 1972) or the Baruya (Godelier 1982).

Men and women, seniors and juniors

The most war-prone individuals in a local group are usually young, unmarried men. They stand to gain the most from war by enhancing their status and acquiring women and political goods. This has been reported for the Highlands of New Guinea (Meggitt 1977: 79, 110, 116 on the Mae Enga) and of East African herders (Baxter 1979: 83f; Almagor 1979: 132-41). Elder men, on the other hand, have already gained maximum status and hence stand to lose from war. In many instances, they try to cool down young hotheads and to avoid 'unnecessary wars', although not always successfully.

As for gender relations, it may be said that they also depend – at least partially – on war, but the general picture is not very clear.[23] Almost exclusively, it is men who wage war, and most warlike societies are characterised by a marked asymmetry between men and women which, however, varies widely in tribal societies. This can be shown by examining some examples of warlike tribal societies, although the terms used here are rather vague, such as higher /lower or better/worse to characterise the relative position of (fertile, married) women.[24] The relative position of women among the Yanomami is low (Chagnon 1983; Biocca 1972), perhaps because gift and marriage exchange between groups are not important. Female infanticide and abduction of women are practised, and women appear to have little say in political matters. Nevertheless women often press their men to go to war, because they fear rape or abduction by enemies (Biocca 1972). A similar constellation concerning gift exchange, marriage relationships and the position of women prevails in the Eastern Highlands of New Guinea (see Langness 1967 on the Bena Bena), in contrast to the Western Highland groups. In the latter societies the position of women seems to be better, because women are the objects of marriage exchange controlled by men and they are responsible for raising pigs, which are used in gift exchange between allies. Hence they are highly valued as labour. But women have only minor political influence; in-married women are mistrusted, because quite often they originate from hostile groups (Meggitt 1977). However, women play an important role in opening peace negotiations (Wiessner and Tumu 1998b: 262). Among the Waorani, however, who resemble the Yanomami in many respects, women seem to be in a better position. Their war-related mortality of about 55% is one of the highest in all warlike societies; this is due to the fact that women also participate in raids, and war-related female mortality was at 46% (male

mortality at 63%; Larrick *et al.* 1979). Maybe this is the reason why the relationship between men and women is described as more equal, in contrast to most of the other warlike societies (Robarchek and Robarchek 1998). Kinship and residence may also play a certain role: among the bilaterally organised Iban, the position of women seems to be better and they have more say in choosing their spouse than in other tribal societies (Komanyi 1971; 1990). Among the matrilineal, matrilocal Iroquois – famous for their external warfare – the position of women is high (Colson 1975; Schumacher 1972); however, this is not so among the matrilocal, patrilineal Mundurucu who also wage external wars, but where the in-marrying, unrelated men reside in a men's house (Murphy and Murphy 1974). As this brief overview illustrates, a clear connection between war and gender relations cannot be demonstrated. A lot of systematic and comparative study has still to be done on this topic; but let us now turn to alliance.

Alliance
An expansive population policy will enhance group strength only in the long term. In the short term, the military strength of a local group basically increases with the number and reliability of its allies. Local groups have to form alliances against common enemies, and it is the common enmity against third parties that makes (conditional) co-operation between allies both necessary and possible. Alliance partners can expect more from forging an alliance against third parties than from waging war against each other; however, the modalities of the alliance have to be negotiated (Schelling 1960; Elster 1989).

Anthropological theories of war mostly concentrate on war and its causes, but neglect the formation of alliances. Of course there are ethnographical accounts of alliance formation by kinship, marriage and gift exchange, but hardly any theoretical reflection on this significant phenomenon. However, alliances are a crucial phenomenon in the context of war, as victory or defeat of a group often depends on the number of its allies: whoever has to wage war needs allies. The military strength of a local group depends not only on its size and the number of its warriors, but also on the number and reliability of its allies. And whether a local group will attack or not, will also depend on the number of its own allies, as well as on those of its enemy. Again, anthropology can learn a lot from models, which have been developed by political science (see Riker 1962; Rothstein 1968; Wagner 1986; Walt 1987; Nicholson 1992).

The logic of alliance formation

Alliances are forms of pragmatic co-operation based on (short term) common interests: two groups will do better against an enemy group by forming an alliance. With the help of allies an enemy coalition can be defeated; but allies may also be unreliable, they may withdraw their support at a decisive moment (Vayda 1976 on the Maring) or even commit treason by secretly forming an alliance with a former enemy (see Chagnon 1983 on treacherous feasts among the Yanomami). Hence, the loyalty of allies is always uncertain. They have to be compensated for their losses and be considered in the distribution of booty after a successful war. The mode of co-operation and distribution of gains between allies has to be negotiated accordingly, whereby neither of the allied groups will accept a solution, which is worse than it would be without the alliance. A group looking for allies will try to concede as little as possible, but still enough to avoid losing its ally. Each group will try to profit from its alliance partners' weakness and to get a higher share of the booty. Alliances can thus be seen as games of negotiation: what one group wins, the other one loses, but in contrast to a zero-sum game, both will lose if they do not co-operate (see Schelling 1960; Rapoport 1976). A zero-sum game prevails between warring groups or coalitions (the one wins, what the other loses), but within a coalition the modality of the relations must be negotiated, a game which displays both conflictive and co-operative aspects. The formation of alliances follows the logic of a N-actors-zero-sum game: a zero-sum game between warring coalitions, a bargaining game characterised by both conflictive and co-operative aspects, however, within a coalition (Barth 1959; Riker 1962). Alliances may be reinforced by an exchange of gifts and by marriage relations. But even allied groups may mistrust each other, and often conflicts over the modalities of gift transactions break out between them. A gift may be interpreted as a sign of the sincerity of one's intentions and a means of overcoming mistrust. But the incentive not to deviate from this mode of co-operation is the common interest in an alliance. Exchange will be successful as long as both groups are sufficiently interested in an alliance against common enemies.

Barth (1959) has described the coalition strategies among the Swat Pathan as a zero-sum game with five groups, which either wage war against each other or are allied to each other. The strategy of each local group consists of belonging not only to the stronger coalition (which is large enough to defeat a common enemy) but also to maximise its gains within the coalition (land and cattle, as well as women and an improvement in its strategic position; see Riker 1962). In this way weak groups may become stronger than the strongest group by forming an alliance. But there may be cases where the weaker groups are too weak to form a winning or even a blocking alliance against the strongest group. In this case it makes more sense for a weak group to join the dominant group in order not to be defeated, although it may be exploited by its stronger ally (Nicholson 1992: ch. 11). At any rate, the logic of alliance formation depends on the specific relation of force between the local groups in a region. The relative strength of the groups, the different threats to which they are exposed from other groups, as well as their alternative alliance options, ultimately determine the bargaining power of each group within a coalition (Helbling 1996a; 1999).

Taking the formation of alliances into consideration is a precondition for a more realistic theory of war and of the interface between tribal wars and their regional and national contexts. But what are the economic consequences of alliance formation? What are the consequences of war on the relations between local groups related by kinship and marriage? And to what extent do these relations influence or determine the formation of alliances?

Economy

Expanding group size to enhance military strength and increasing the production of political goods (such as pigs, shells and axes) for allies are military advantages in a warlike environment, but may put considerable stress on local resources. Economically optimal local groups would be much smaller than they are (100-300 persons on average) so that the pressure on local resources would also be smaller (Helbling 1991). Gift exchange, alliance feasts and marriage relationships strengthen alliances, but are quite costly. Production has to be massively intensified: in the Highlands of New Guinea, for instance, this involves the production of pigs and sweet potatoes (Meggitt 1974: 198ff). As a consequence, resources may become scarce. However, these resources are not goods for consumption and their scarcity is not due to population growth. Rather, they are political goods and their inflationary production is due to the requirements of war, i.e. of recruiting allies. Thus, if resources are scarce this is a consequence rather than a cause of war. Besides the inflationary expansion in the production of political goods causing widespread overuse of local resources, war also entails the destruction of property, such as gardens, animals and trees, the killing and misallocation of the labour force, the underuse of land at the border and its overuse in secure areas, and investment in defence works such as palisades. The fact that sometimes an enemy's land is occupied, wealth is stolen or women are abducted does not indicate that these resources are scarce. Rather, it is a means of weakening the enemies (or allies) and of strengthening one's own group at the same time.

Alliance, kinship and marriage

Alliance formation usually follows the criteria of *Realpolitik*, which depends on group interests in a given regional constellation of force (Meggitt 1977: 37). When groups choose allies, kinship relations play only a subordinate role. The Mae Enga wage war as often against related clans as against unrelated ones, but always against neighbouring clans; their allies often – but not always – belong to the same phratry. However, belonging to the same phratry is never reason enough to support a fraternal clan in a war (Meggitt 1977; see also Colson 1975: 12f, 22ff, 29f on the Iroquois). Furthermore, the Mae Enga say that 'we marry the people we fight' (Meggitt 1977: 42). The Yanomami fight against former allies, groups led by close cousins as well as against unrelated groups; often groups of brothers-in-law are allied, but alliances are always brittle and imbued with mistrust and fear of treason. Among the Nuer most of the fighting takes place between neighbouring, closely related local groups belonging to different tertiary sections. Local groups of the same section do not automatically support each other in a fight against a group belonging to the next higher section. And there have been wars where Nuer groups allied to the Dinka fought against other Nuer groups. A careful study of war histories reveals quite a different picture of war and alliance to the one given by the model of the segmentary lineage system (Evans-Pritchard 1940; Kelly 1985). Conflicting loyalties and cross-cutting ties do not seem to affect the policy of entire groups and will not prevent war between them, but only affect individuals, who may not participate in a war in order to avoid clashing with relatives on the opposite side. So we may conclude that kinship relations between local groups will only help to settle conflicts peacefully between allied groups interested in co-operation against common enemies, as Greuel (1971) and Evens (1985) demonstrated for the Nuer. Kinship and marriage, though, are not irrelevant. They are relevant as resources of mobilisation, of which, however, advantage is taken only if *Realpolitik* so requires.

The theory of tribal war which I propose has the advantage of combining the structural with a strategic perspective. Structural elements – such as the political autonomy of local groups in an anarchic system and the dependence of local groups on locally concentrated resources – form the basic conditions for the strategic bellicose of local groups. This strategic interaction includes mutual threat and the adoption of a confrontational strategy, in order to avoid survival risks (as a consequence of a prisoners' dilemma), or striving for military superiority and attacking pre-emptively (as a consequence of a security dilemma).

4.4 Reasons for specific wars

The two structural conditions of war, the absence of an overarching centralised power and high opportunity costs of moving away, account for the warlike interaction between local groups. Where these conditions exist, wars can break out at any time. They explain the permanent state of war, i.e. constitute the preconditions for the likelihood of war in tribal societies. However, a permanent state of war does not imply that war is constantly waged. Whether specific wars break out or not, and whether certain alliances are formed or not, depend on many factors which cannot be fully examined here.[25] One important factor is the regional relations of force between local groups. The relative strength of a local group is – as we have seen – a function of its relative size and solidarity, as well as of the number and reliability of its allies. A local group will only start a war if it thereby expects to improve its present position or to prevent a future deterioration of its situation. Thus the starting point for any analysis is the local groups in a region and the individual factions within these local groups that evaluate the advantages and disadvantages of the different options: waging a war, forming an alliance, concluding a truce or solving a conflict peacefully. Whether war breaks out or not basically depends on the relative regional strength between the local groups and on the decision process within the local groups (between 'doves' and 'hawks'). The relative strength of a local group depends on its relative size and the number of reliable allies, but internal unity and determination are also important, as well as the extent to which it is threatened by other groups. However, we still need more comparative studies in order to explain the variation of wars in different tribal societies.

As Clausewitz (1980[1832-34]: I.1.2–17) has already shown, escalating and de-escalating aspects of war activities can be distinguished. The prisoners' dilemma and, above all, the highly conflictive security dilemma form the background against which local groups decide for or against specific wars and must be considered as escalating forces. Hawkish positions will prevail in every group in a situation of heightened mistrust and conflict. Each group wants to attack at a favourable moment, in order to decimate, expel or at least weaken their enemies (by occupying some of their land, and by stealing property or women), as this is the only means to eliminate the threat from enemy groups and to avoid being attacked by them at an unfavourable moment (i.e. when the enemies are stronger). The relation of force can quickly change, through a variation of group size (by group splitting or immigration) or a shift in alliance politics (loss of allies or recruitment of new ones). Furthermore, the mutual threat between local groups increases with decreasing distance. But there are also times of relative peace, in which no wars break out and conflicts do not escalate to wars. No serious wars will break out as long as two adversaries are of about even strength and in a stalemate situation. In this case only limited fights will occur, in which the adversaries mutually test and demonstrate their strength and their determination, as well as the reliability of their allies. These forms of 'regulated warfare' may be interpreted as an attempt to limit violence between groups and thus to avoid its detrimental consequences for both parties. But if one of the groups has become weaker (because it has lost its allies), the stronger group will immediately escalate the fight and try to rout the other (Vayda 1976 on the Maring). Coalitions may also be so exhausted after a war of attrition with high losses that no further gains can be expected

from continuation of the war. In this case negotiations are started in order to achieve a truce and (in some societies) compensation payments are made, or one of the groups involved will move away and the hostilities ended in that way.

Even war activities can be regulated and are limited by mutual interest. Here we find again the meta-preference of local groups for peace. Agreements as to place and time of a fight, limitations on the kind of weapons used, treatment of the wounded, the protection of women and children, as well as the possibility of truce agreements and compensation payments, are forms of such de-escalating limitations (restricted warfare). However, all such limitations continue to be subject to the logic of the prisoners' dilemma. If it is profitable for one of the coalitions to deviate from the path of de-escalation (if it has gained strength or the enemy has become weaker), it will intensify the fight again and there will be hardly any limitations (unrestricted warfare). The balance of power can change quickly, even in the short term; for instance, if the allies do not show up at a battle or arrive too late. As soon as one of the groups proves to be weaker, the superior group will try to rout or even annihilate it.

To summarise, we have seen that war has highly detrimental consequences for people, property and resources. Nevertheless, local groups in tribal societies cannot avoid waging war under the prevalent structural conditions. There is no centralised power to prevent war between politically autonomous local groups dependent on locally concentrated resources, and pursuing a unilateral peaceful strategy is too risky and may lead to annihilation of the group. These structural factors can change (as with the emergence of a state), with corresponding consequences for the strategic interaction between the local groups (such as, for instance, pacification or an alliance of former enemies against the state). Strategies may also have unintended consequences in the aggregate and may change the structural conditions of interaction between the groups.

5. Pacification

The pacification of warlike groups is not only an interesting historical process as such, but also forms an important field test for theories of war and peace. The factors which were responsible for a successful pacification of formerly warlike societies could also be the causes responsible for the endemic state of war prevailing before their pacification. The ethnographical and historical data on pacification in New Guinea, in Amazonia and in other areas are very detailed (see Rodman and Cooper 1983 on Melanesia; Gordon and Meggitt 1985 on the Mae Enga; Robarchek and Robarchek 1996 on the Waorani; Pringle 1970 and Wagner 1972 on the Iban). But despite this wealth of empirical information, pacification has hardly ever been the object of systematic, theoretical reflection (exceptions are Bodley 1983 and Koch 1983).

Pacification is a process, in the course of which the state enforces its legitimate monopoly of power and brings wars between politically autonomous local groups to an end. Without any doubt, the pacifying states did not pursue philanthropic purposes; the pacification of the 'savages' has never been an end in itself but a precondition for colonial conquest, domination and exploitation and sometimes even elimination of indigenous peoples (see Wolf 1982; Bodley 1983). However, we shall focus on the complex circumstances and conditions of the processes of pacification: the interests and strategies of state actors who try to establish state control, but also the reactions of the politically autonomous local groups, which are entangled in warlike interaction and are put under pressure to stop fighting and to solve their conflicts peacefully. The state monopoly of power is not a constant factor, but the presence and effectiveness of a state varies considerably according to the contexts and phases of colonial expansion (see Bodley 1983). Furthermore, the state is not a homogeneous apparatus, but consists of different actors (such as colonial officials, police and army, local allies but also missionaries) who – together with private actors (such as settlers, traders and entrepreneurs of all kinds) – interact with the different tribal groups. These various actors differ with regard to their interests, capacities of repression and possibilities of reward, and they have influenced (furthered or hindered) pacification in different ways and by different means (Ferguson and Whitehead 1992: 6f, 11). But tribal populations vary in number, military strength and unity as well, and not all local groups in a region may have the same strength and the same interest in stopping warfare.

Many ethnographies report on the relief of members of once warlike societies at the successful cessation of the permanent state of war and the inter-

ruption of the vicious circle of violence and counter-violence (see Colson 1975: 40ff; Knauft 1999: 143ff). This fact shows that local groups are interested in preventing violent clashes with each other. But as we have seen, a unilateral peaceful strategy would be too risky, because a unilateral confrontational strategy provides both higher gains and prevents the utmost losses. A peaceful strategy for groups entangled in a prisoners' dilemma only becomes possible under the influence of a third, superordinate power, which punishes a unilateral confrontational strategy to such an extent that a peaceful strategy becomes the best strategy for each group.

One decisive condition of successful pacification is the enforcement of state power that can force local groups to stop fighting and to enforce peace. The intervention of state actors will be more successful if it is systematic and impartial (as it was in the Highlands of New Guinea, see Gordon and Meggitt 1985) than in cases where the state intervenes unsystematically and partially, as in the case of the Upriver Iban in Borneo (Pringle 1970). However, the effectiveness of a state not only depends on its repressive power, but also on its ability to protect groups who renounce war and reward them selectively with desirable goods (such as iron tools and prestige goods) and the co-option of their leaders. This may explain why, in some cases, even less repressive forms of triadic conflict solution (such as mediation and arbitration) in combination with selective incentives have led to the pacification of warlike groups, such as the Waorani (Robarchek and Robarchek 1998). The example of the Waorani also shows that the higher the costs and risks of war were before, the more the groups are ready to accept peace. A third precondition for enduring pacification – besides selective punishment of warlike groups, selective rewards and protection of groups renouncing war – is the establishment of institutions for the peaceful settlement of conflicts, such as courts, or the acceptance of indigenous forms of mediation in the 'shadow of the Leviathan' (Gordon and Meggitt 1985). However, as long as the state does not punish violent groups systematically (e.g. because police troops are not, or only partially, present), and as long as the state is not able to protect groups willing to stop fighting and fails to establish or support legitimate institutions for the peaceful settlement of conflicts, a confrontational strategy is still the best choice for every local group and the security problem still exists. Local groups will only renounce war if they can be sure of not being attacked by other groups. A general peace can be achieved in a region if the state (or another superordinate instance) systematically punishes a unilateral confrontational strategy (selective repression), efficiently rewards a peaceful strategy (selective reward) and establishes alternative institutions for the peaceful settlement of conflicts.

Future research on the processes of pacification should not only systematise the relevant ethnographical information, but also develop models and concepts reflecting these processes in connection with theories of war and peace in tribal societies.

6. Peace

There are tribal populations, which are not (efficiently) controlled by a state, but where local groups nevertheless do not wage war against each. Every theory of war should also be able to address the problem of peaceful societies. But let us discuss some conceptual problems first.

While war is a collectively planned and organised armed conflict between political units, the definition of peace remains somewhat unclear. Often peace is presented as the normal condition of society and therefore does not need any further explanation. In contrast, war is seen as a deviation from this natural condition, which needs to be explained and, indeed, it has received far more attention in anthropological discussion. However, any theory of tribal war must also address the problem of explaining tribal societies without war and put forward a consistent explanation for both tribal societies with and without war. In the last decades anthropology has become more interested in the causes of peaceful relations between local groups in societies without a state.[26] But not only are there more studies and theories relating to war than to peace, but warlike tribal societies are also far more numerous than peaceful ones. According to Sipes (1973) only five of the 130 societies he investigated are peaceful, and Otterbein (1973) found only four peaceful societies in a sample of 50 (cited in Gregor 1990: 106; Bonta 1993). Most of these peaceful societies are hunters-and-gatherers, hardly any tribal shifting cultivators, pastoral nomads or sedentary fishermen (Sponsel 1996: 103ff; see footnote 2). This points to the fact that tribal societies are warlike, in contrast to hunters-and-gatherers, who usually do not wage war.

Basically, two different conceptions of peace require discussion: positive and negative peace (Dentan 1992: 253f). The first definition – positive peace – not only includes non-violence between local groups but also political and economic equality as well as harmonious interaction within groups (Fabbro 1978).[27] This conception of peace, also implying non-violence between individuals, is far too restrictive, for very few societies would fit into this category: even in societies without war, violence between individuals within a group is not rare (for the BaMbuti, !Kung, Inuit and Yaghan, see Kelly 2000). Therefore this first conception hardly suits the analysis of tribal societies without war. The second conception – negative peace – only entails the absence of collective violence between local groups, whether or not the interaction between individuals is violent. Hence peace is the absence of war, but not of violence between individuals. According to this conception, war and peace may represent two modalities of relationships between groups in a tribal society: peace between allies and simultaneously war against hostile groups. There may be even periods of peace (truce) between hostile groups. In these cases, we may speak of a relative peace in a warlike environment. But there are also societies in which local groups do not wage war against each other (see Dentan 1968 and Gregor and Robarchek 1996 on the Semai; Helbling 1996b; 1998 on the Mangyan). In these societies conflicts are solved through avoidance and retreat, never by means of war, although violence between individuals may occur, as already mentioned. Hence negative peace as an antonym of war refers both to an occasional alternative to war in warlike societies as well as to the general absence of war (but not of violence between individuals).

If peace were the opposite of war, then the explanation for tribal societies without war would just be the reverse of theories of war in warlike tribal societies: that is, the absence of those elements which cause wars, would then also explain the lack of war. However, not all the theories of war listed above are relevant for the explanation of peace, for the same reasons that they are not convincing in explaining war. Thus, Robarchek and Robarchek (1992; 1996; 1998) and Gregor (1990) explained the peacefulness of tribal societies (such as the Mehinaku and the Semai) by cultural values and norms that reward peaceful behaviour and disapprove of violence. But as I have already shown, the peacefulness of tribal societies cannot be explained by such values and norms, because these only make sense in an already peaceful environment. The opposite of the fraternal interest theory is the proposition of conflicting loyalities of men between their local group and kinship group, as for instance in patrilineal and matrilocal societies. But as Hallpike (1973: 463) pointed out, such a constellation does not prevent wars between local groups, but only stops certain individuals from participating in a specific war, as these want to avoid a clash with relatives on the opposite side. Conflicting loyalties may even spread violence by dragging neutral groups into the armed conflict if they offer refuge to a member of one of the main opponents.

A better way of explaining the phenomenon of peaceful societies without central power is to divide them into categories and to explore each of them separately. There are three categories of peaceful societies beyond state control, i.e. indigenous populations not (not yet or no longer) subject to (an effective) power monopoly of a state:

1) Hunter-and-gatherer societies consisting of small and mobile groups of about 25 persons, who live from hunting and gathering (Lee and DeVore 1968: 11). The population density is usually very low, with only very slight population growth, so that contacts between groups are not frequent and, therefore, the conflict potential does remain low. The nomadic economy also allows each group to avoid conflicts with other groups and to prevent violent disputes (Reyna 1994: 37; Sponsel 1996: 103ff). Therefore, hunters-and-gatherers – with few exceptions – hardly ever wage war, and never do so against each other (Helbling n.d.b).[28] In this they differ from tribal societies, which – as can be seen from archaeological findings (Gabriel 1990: 31ff; Kelly 2000: ch. 4) – were predominantly warlike before their pacification. Hence, we have not yet explained the phenomenon of tribal societies without war.

2) Tribal groups, which seem to be peaceful, such as those in the Upper Xingu Basin (Gregor and Robarchek 1996). However, if the wider regional and historical context is taken into consideration, we can discern that the Xingu groups were militarily weakened by wars against powerful adversaries, as well as by epidemics, and had to retreat into inaccessible areas. They found refuge and recovered in a protectorate where they had access to medical care and Western goods. And it was in this 'sanctuary'

(Dentan 1992: 221ff) that the Xingu groups formed a kind of permanent alliance against warlike neighbouring groups outside the protectorate, against which they successfully waged defensive, but also offensive wars (Menget 1993). The alleged peacefulness of the Xinguanos, thus, turns out to be an optical illusion: they do not form a peaceful tribal society, as Gregor and Robarchek maintained, but a permanent alliance between local groups of different ethnic origins in a sanctuary.

3) Nevertheless, there are tribal societies beyond state control in which conflicts between local groups are not settled by means of warfare. These are tribal groups which were forced by militarily superior populations from the lowland to retreat into inaccessible forest and mountain areas, but continue to depend economically on the adjacent dominant population for the provision of goods they cannot manufacture themselves. This situation – described as an 'enclave' by Dentan (1992: 211ff) – can be found among the Semai (Dentan 1968) and the Mangyan (Helbling 1998), who face a lowland population – far superior in numbers and power and far more aggressive – in a tribal zone that is not completely controlled by the state. At the same time they have to work for settlers and to exchange forest products with traders in order to get desired goods such as bush knives, cloth and iron pots. Under such circumstances avoidance of conflicts by retreat and withdrawal or by a peaceful, submissive behaviour are far better survival strategies than armed resistance and sporadic attacks. These local groups must, therefore, adapt to their structural inferiority and factual powerlessness and to their simultaneous economic dependence on their superior neighbours on the social, economical and political levels. Small, mobile groups, a wide network of bilateral kinship and extensive agriculture, combined with hunting and gathering, allow a quick withdrawal and dispersion of the groups, with the possibility of taking refuge in other groups in the case of emergency. It does not come as a surprise that these marginal groups see themselves as timorous and the neighbouring populations in the lowland as violent (McCauley 1990: 14f), for this corresponds to their historic experiences as losers, who always had to withdraw and to retreat. The fear of violence in all its forms (physical and spiritual) is an important regulator of behaviour. This fear makes plausible the norms and behavioural ideals that reward peacefulness and

disapprove of aggressiveness, and, thus, legitimates peaceful behaviour, which matches their actual powerlessness. The emergence of such peaceful norms and ethnic stereotypes, contrasted with the 'violent others', produces the reciprocal expectation that the other local groups also prefer peaceful interaction and, thus, none of the groups has to fear violent conflicts.

Peaceful relations between local groups in a tribal society, only marginally integrated into a state society and not subjected to effective control by the state, can, thus, not be explained by the internal logic of this type of society, but rather by specific circumstances in a tribal zone (i.e. a tribal group facing a far superior population). One important factor explaining the peacefulness of tribal groups is the kind of refuge to which these groups are forced to withdraw. Local groups either sporadically attack neighbouring groups from a secure sanctuary (as the Xingu) or have permanently become peaceful in an enclave (as the Semai and the Mangyan). Both, structural inferiority vis-à-vis a powerful neighbour, which precludes violent resistance and forces retreat, and at the same time economic dependence on settlers and traders (hence an enclave), cause economic, organisational, political and cultural adaptations on the part of the indigenous groups, which make wars between local groups highly improbable (see Helbling 1998). The explanation of peaceful relations between tribal groups emphasises the necessity to explore not only the internal logic of the interaction between tribal groups but also the regional and national context as well the history of tribal societies.

7. Summary and outlook

Although tribal wars always take place in a wider context of expanding or weakening states in a world economic system that influences the course and intensity of wars, it is nevertheless imperative to analyse the internal logic of indigenous war. I have argued that war can be explained by two structural conditions: the political autonomy of local groups in an anarchic system and the dependence of local groups on locally concentrated resources. This political and economic constellation causes a warlike type of strategic interaction between local groups, which may be described in terms of game theory. Despite the high costs and disadvantages of war and despite the fact that a peaceful interaction (co-oper-

ation) yields the highest gain for all groups collectively, each group is compelled to adopt a bellicose strategy. This is because a one-sided peaceful strategy is too risky, since a one-sided bellicose strategy brings the highest gains and a one-sided peaceful one the highest losses. Under these conditions of mutual threat and fear of violence, the survival of each group depends on its ability to become larger and stronger and to recruit more allies than its potential adversaries. The military superiority of one group, however, inevitably constitutes a threat to the others, forcing them to achieve military superiority in turn. And each group tries to attack pre-emptively and to defeat, to weaken or expel its enemy, because if it does not attack at a favourable moment it risks being attacked at an unfavourable moment. Thus, the prisoners' dilemma evolves into an even more conflictive security dilemma. It is against this background that local groups interact strategically and pursue their interests. War is the necessary, even though unintentional and damaging, result of the strategic interaction of groups under specific structural conditions. I have also examined alternative theories of tribal warfare. These theories explain tribal warfare with human aggressiveness rewarded by reproductive success or by cultural norms and values, with a struggle over scarce resources, conflicts between groups of men related by kinship or as caused by politically ambitious group leaders. My conclusion was that these factors are consequences, not causes, of tribal warfare.

The study of tribal wars remains one of the central topics of anthropology. Numerous theories have been put forward in order to explain tribal warfare, and many ethnographic accounts have been published which describe tribal warfare in all its modalities. Archaeology, too, has tackled the problem of warfare, especially since the 1980s and the 1990s (see among others Vencl 1984; 1991; Keeley 1996; Haas and Creamer 1993; 1997; Haas http://www.santafe.edu/sfi/publications/Working-Papers/98-10-088.ps; Martin and Frayer 1997; Thorpe http://www.hum.au.dk/fark/warfare/thorpe_paper_1.htm LeBlanc 2003 and contributors to this volume). But still, *The Oxford Companion to Archaeology* (of 844 pages), edited by Brian Fagan (1996), has no entry on war and war is not even listed in the index (of 24 pages). Our aim should be not only to provide ethnographic and archaeological data on warlike societies but also to explain tribal warfare based on controlled comparison and – where possible – on statistical analysis. Tribal wars are too interesting and too important a theme for anthropology and archaeology to leave their study to sociobiologists, political scientists and military historians. War should again be moved into the centre of anthropological and archaeological research in close co-operation with neighbouring disciplines such as history, political science and sociology.

Acknowledgements

I would like to thank Danilo Geiger, Georg Elwert, Heinzpeter Znoy, Sandra Gysi, Thomas Bernauer, Tobias Schwörer, Polly Wiessner, Helle Vandkilde and Ton Otto for their helpful comments and suggestions.

NOTES

1   For an overview of different theories see Otterbein (1973; 1985; 1990; 1999), Hallpike (1973), Koch (1974a), Nettleship *et al.* (1975), Clastres (1977), Harris (1977), Hanser (1985), Foster and Rubinstein (1986), Wolf (1987), Rubinstein and Foster (1988), Ferguson (1984b; 1990b), Knauft (1990; 1999), McCauley (1990), Haas (1990), Ross (1993a; 1993b), Carneiro (1994), Reyna and Downs (1994), Keeley (1996), Orywal (1996b; 1998), Elwert, Neubert and Feuchtwang (1999) and Simons (1999).

2   I am fully aware of the fact that the term 'tribal' is a contested one (Fried 1975). I use the term 'tribal society' neither in the sense of an developmental stage of human society, as the evolutionists do, nor in the sense of a political organisation of a whole regional population, but in a purely descriptive sense or as an 'ideal type' (Max Weber). The term denotes a regional population of extensive farmers, pastoral nomads or settled fishermen, who live in politically autonomous local groups and entertain relationships of kinship, marriage and gift exchange and who are allied or wage war against each other (Sahlins 1968; Bodley 1997). Tribal societies thus differ from hunters-and-gatherers, who live in small nomadic groups. I do, however, not consider sedentary fishermen such as the Kwakiutl and incipient farmers such as those at the north, east and south-east coast of Australia as hunters-and-gatherers since they – as tribal groups – also depend on locally concentrated resources (Murdock 1968; Lourandos 1997: 44–52, 57–59, 60–69; Helbling n.d.b). Tribal societies also differ from local communities integrated into a state society, controlled by a state with a legitimate monopoly of force, and having to pay tribute and rents.

3   According to a statistical evaluation by Scherrer (2000: 24, 31-35) about 19.6% of all wars between 1985 and 1995 were anti-regime wars and 11.8% were waged between states. In the same period tribal wars represented 15.7% of cases, whereas ethno-national wars amounted to 44.1% of all wars.

4   It does not make sense to define war as organised, collective violence only between states because in that case the proposition of war contributing to the formation of the state would not even be considered.

5   Feud is the legal obligation and right to take revenge for an injustice committed against the perpetrator (or one of his kin). In some cases revenge can be replaced by a compensation payment, which, however, does not always annul the desire to take revenge, but only postpones it (on the Nuer, see Evans-Pritchard 1940 and Greuel 1971; on the Mae Enga see Wiessner and Tumu 1998b).

6   On the absence of war in hunting-and-gathering societies see Johnson and Earle (1987: 19), Steward (1968: 334), Service (1966: 60), Fried (1967: 99-106), Lee and DeVore (1968: 339-40), Wolf (1987: 132, 136), Carneiro (1994: 12ff), Coon (1976: 275), Harris (1977: 42ff), Sponsel (1996: 107), Kelly (2000: 125) and Konner (1982: 204), for archaeological evidence see Sponsel (1996: 103ff), Roper (1969: 330f), Leakey and Lewin (1977: 245), Vencl (1984: 120f), Ferrill (1985: 17), Gabriel (1990: 23), Thorpe (http://www.hum.au.dk/fark/warfare/thorpe_paper_1.htm), Haas (http://www.santafe.edu/sfi/publications/Working-Papers/ 98-10-088.ps) and Keeley (1996: 39). LeBlanc (2003) maintains a contrary view, although he only presents cases of individual violence rather than of war and cases of 'Mesolithic peoples' (such as those at the American Northwest Coast) rather than real hunters-and-gatherers.

7   For older biological theories, focussing on territoriality, cp. Eibl-Eibesfeldt (1984).

8   Durham (1991), Chagnon (1990b) and Gat (2000) also take the competition over scarce food into consideration. The military success in competition over scarce food also leads to (relative) reproductive success. I will address this economic explanation of tribal war, also shared by anthropologists not favouring a biological theory, below.

9   Even in hunting-and-gathering societies, which display a high level of violence between individuals, violence is not predominantly between men and not because of women and sexual rivalry. Adultery and jealousy are only rarely recorded as reasons for violence (Kelly 2000: 31ff).

10 Examples of such descriptions are provided by Harrison (1993) on war rituals among the Manambu (Avatip) of the Sepik, as well as by Michelle Rosaldo (1980) on the Ilongot in the Northern Philippines.

11 Moreover, as Wiessner and Tumu (1998b: 148f) and others have pointed out, not land but rather labour seems to be scarce among the Mae Enga.

12 Ferguson (1990b) made differentiations in the ecological-economic theory of tribal wars, by also taking into account socio-structural and ideological elements. He

claims that infrastructural elements explain why wars are waged; structural elements determine the social patterns of war and why a specific war will break out at a given time; superstructural elements ultimately determine the cognitive conditions of decision processes about war and peace. But even this more sophisticated model continues to uphold the (implausible and empirically not substantiated) premise that economic-ecological elements ultimately cause wars.

13 Even Ferguson (1984b: 271ff) concedes that wars were waged among the Indians of the Northwest coast long before their first contact with representatives of a state; according to Yesner (1994: 161f) a period of about 4000 years.

14 Turton (1979; 1994 on the Mursi) and others focussed especially on the impact that the introduction of firearms had on the frequency and mode of warfare.

15 Clastres (1976) and Harrison (1993) maintained that war is the result of the local groups' endeavour to preserve their sovereignty and autonomy. This is highly implausible, not only because of the functionalist logic of the argument but also because the sovereignty of local groups is only endangered in an already warlike environment. Furthermore, one can ask why warfare, and not more peaceful, less risky and less costly, but nevertheless equivalent forms (ritual, classification, peaceful contest etc.), is used by local groups to mark their identities in the first place.

16 The sample includes the Tauna-Awa, Usurufa, Mae Enga, Kamano, Auyana, Huli, Eipo, Baktaman (Faiwolmin), Dani, Anggor, Abelam, Jivaro, Yanomami, Waorani, Mekranoti and the Blackfoot (see Helbling n.d.a).

17 It is astonishing that, with a few exceptions, anthropological theories of war hardly ever took notice of the relevant theoretical discussions in political science. Anthropology could learn a lot from models of war and alliance between states provided by the theory of international relations, as states are political units of war as are the local groups in tribal societies. Neo-realist theories (see Waltz 1979; Levy 1989; van Evera 1998; 1999) and decision theories, above all game theory (see Rapoport 1974; Brams 1975; Nicholson 1992), are of special interest in understanding the logic of tribal warfare. Levy (1989) gives an overview of the discussions on the causes of war, see also Waltz (1960), Brown (1994), Dougherty and Pfaltzgraff (2001[1990]), Vasquez (1993; 2000), Bueno de Mesquita (1980), Nicholson (1992), Burchill and Linklater (1996).

18 In a prisoners' dilemma co-operation would yield the highest gains collectively for all actors but is too risky for each actor because a one-sided defection brings the highest gains and a one-sided co-operation the highest losses.

19 To be clear I do not adopt a Social Darwinist position but rather follow evolutionary economics (Nelson and Winter 1982; Hodgson 1993; Nelson 1995; for anthropology see Peoples 1982 and Ensminger and Knight 1997). Local groups in a tribal society interact in a warlike environment as firms interact in a capitalist market environment. Those firms pursuing strategies which entail lower costs and higher profits will be more successful and expand, whereas firms with higher costs and lower profits will lose market share and may even become bankrupt.

20 The 'arms race' in tribal societies mainly consists of the enlargement of groups (more warriors) as well as the recruitment of more allies.

21 Under these conditions the adoption of a peaceful strategy by all groups is only possible if a superordinate power effectively punishes those who pursue a unilateral confrontational strategy and rewards those who adopt a peaceful one.

22 It should be stressed that war does not endanger the 'social order', as the structure-functionalists assumed. While there is social order – sanctioned norms, a structure of authority, procedures to settle conflicts peacefully – within a local group as the relevant political unit in tribal societies, the relation between the local groups in a region is a spontaneous order. As I have shown, this order emerges under certain structural conditions as an unintentional result of interaction between local groups.

23 Theoretical attempts to explain war by the higher aggressiveness of men, gender-specific modes of psycho-socialisation (Chodorow 1978) or patriarchal structures (Schmölzer 1996) have failed - why should men be more aggressive in tribal societies than in societies of hunters-and-gatherers? It seems to be more meaningful to study the consequences of a warlike environment on the relations between men and women in local groups (see Harris 1977: ch. 6).

24 For convenience I will use the ethnographic present in the following by thus referring to the date when the field-research had been conducted or the monograph been published.

25 Put in terms of ultimate and proximate causes, it may be said that the structural conditions and the ensuing logic of warlike interaction – and not the maximisation of inclusive fitness, as sociobiology states – are the ultimate causes of tribal warfare. Other causes put forward by alternative theories of tribal warfare may be interpreted as proximate causes.

26 On peaceful societies see Fabbro (1978), Wiberg (1981), Howell and Willis (1989), McCauley (1990), Dentan (1992; 1994), Sponsel and Gregor (1994) and Gregor (1996).

27 According to Fabbro (1978) local groups in peaceful societies display five characteristics: 1) small group size with face-to-face-communication, 2) an egalitarian social organisation, 3) general reciprocity, 4) social control and decision making by consent and 5) non-violent values and socialisation (see also McCauley [1990: 14f]).

28 This is also true for tribal groups which were forced by powerful neighbouring groups to withdraw into inhospitable regions, where they could live in relative security, but – as cultivation was not possible in these areas – had to become hunters-and-gatherers (Service 1968).

BIBLIOGRAPHY

- Aijmer, G. 2000. 'Introduction: The idiom of violence in imagery and discourse'. In: G. Aijmer and J. Abbink (eds.): *Meanings of Violence*. Oxford: Berg, pp. 1-22.
- Albert, B. 1989. 'Yanomami violence. inclusive fitness or ethnographic representation'. *Current Anthropology*, 30(5), pp. 637-40.
- Albert, B. 1990. 'On Yanomami warfare: rejoinder'. *Current Anthropology*, 31(5), pp. 558-63.
- Alès, C. 1984. 'Violence et ordre sociale'. *Etudes Rurales*, 95/96, pp. 89-114.
- Alexander, R. 1977. 'Natural selection and the analysis of human sociality'. In: *Changing Scenes in the Natural Sciences, 1776-1976.* (Special Publication 12). Philadelphia: Academy of Natural Sciences of Philadelphia, pp. 283-337.
- Almagor, U. 1979. 'Raiders and elders: A confrontation of generations among the Dassanetch'. In: K. Fukui and D. Turton (eds.): *Warfare among East African Herders*. Osaka: National Museum of Ethnology Press, pp. 119-45.
- Barash, D. 1981. *Sociobiology: The Whispering Within*. Glasgow: Fontana/Collins.
- Barth, F. 1959. 'Segmentary opposition and the theory of games: A study of Pathan organization'. *The Journal of the Royal Anthropological Institute of Great Britain and Ireland*, 89, pp. 5-22.
- Baxter, P. 1979. 'Boran age sets and warfare'. In: K. Fukui and D. Turton (eds.): *Warfare among East African Herders*. Osaka: National Museum of Ethnology Press, pp. 69-95.
- Biocca, E. 1972. *Yanoama*. Frankfurt: Ullstein.
- Bodley, J. 1983. *Der Weg der Zerstörung. Stammesvölker und die industrielle Zivilisation*. München: Trickster Verlag.
- Bodley, J. 1997. *Cultural Anthropology: Tribes, States, and the Global System*. Mountain View: Mayfield Publishing Company.
- Bohannan, P. (ed.)1967. *Law and Warfare*. New York: The Natural Museum Press.
- Bonta, B. 1993. *Peaceful Peoples: An Annotated Bibliography*. Metuchen: Scarecrow.
- Brams, S. 1975. *Game Theory and Politics*. New York: Free Press.
- Brown, S. 1994. *The Causes and Prevention of War*. New York: St. Martin's Press.
- Bueno de Mesquita, B. 1980. 'Theories of international conflict'. In: T. Gurr (ed.): *Handbook of Political Conflict*. New York: Free Press, pp. 361-98.
- Burchill, S. and Linklater, A. (eds.) 1996. *Theories of International Relations*. New York: St. Martin's Press.
- Carneiro, R. 1994. 'War and peace: alternating realities in human history'. In: S. Reyna and R. Downs (eds.): *Studying War: Anthropological Perspectives*. Langhorn: Gordon and Breaches, pp. 3-27.
- Chagnon, N. 1977. *Yanomamö: The Fierce People*. (2nd edn.). New York: Holt, Rinehart and Winston.
- Chagnon, N. 1983. *Yanomamö: The Fierce People*. (3rd edn.). New York: Holt, Rinehart and Winston.
- Chagnon, N. 1988. 'Life histories, blood revenge, and warfare in a tribal society'. *Science*, 239, pp. 985-92.
- Chagnon, N. 1990a. 'On Yanomamö Violence: Reply to Albert'. *Current Anthropology*, 31(1), pp. 49-53.
- Chagnon, N. 1990b. 'Reproductive and somatic conflicts of interest in the genesis of violence and warfare among the Yanomamö Indians'. In: J. Haas (ed.): *The Anthropology of War*. Cambridge: Cambridge University Press, pp. 77-104.
- Chodorow, N. 1978. *The Reproduction of Mothering*. Berkeley: University of California Press.
- Clastres, P. 1976. 'Die Gesellschaft gegen den Staat'. In: P. Clastres: *Staatsfeinde*. Frankfurt: Suhrkamp.
- Clastres, P. 1977. 'L'archéologie de la violence: La guerre dans les sociétés primitives'. In: P. Clastres: *Recherches d'Anthropologie Politique*. Paris: Seuil, pp. 171-207.
- Clausewitz, Carl von 1980(1832-34). *Vom Kriege*. W. Hahlweg (ed.). Bonn: Dümmlers Verlag.
- Colson, E. 1975. *Tradition and Contact: The Problem of Order*. Chicago: Aldine.
- Coon, C. 1976. *The Hunting Peoples*. Harmondsworth: Penguin.
- Dennen, J. van der 1995. 'Origin and evolution of primitive war'. In: J. van der Dennen and V. Falger (eds.): *Sociobiology and Conflict*. Chicago: Aldine, pp. 149-88.
- Dennen, J. van der 2002. 'Evolutionary theories of warfare in preindustrial (foraging) societies'. *Neuroendocrinological Letters*, 23, pp. 55-65.
- Dentan, R. 1968. *The Semai*. New York: Holt, Rinehart and Winston.
- Dentan, R. 1992. 'The rise, maintenance and destruction of peaceable polity'. In: J. Silverberg and P. Gray (eds.): *To Fight or not to Fight: Violence and Peacefulness in Humans and other Primates*. New York: Oxford University Press, pp. 214-70.
- Dentan, R. 1994. 'Surrendered men: peaceable enclaves in the Post-Enlightenment West'. In: L. Sponsel and T. Gregor (eds.): *The Anthropology of Peace and Nonviolence*. Boulder: Lynne Rienner Publishers, pp. 69-108.
- Dougherty, J. and Pfaltzgraff, R. 1990. *Contending Theories of International Relations*. New York: Harper and Row Publishers.
- Durham, W. 1991. *Coevolution: Genes, Culture and Human Diversity*. Stanford: Stanford University Press.
- Dyson-Hudson, R. and Smith, E. 1978. 'Human territoriality'. *American Anthropologist*, 80(1), pp. 21-41.
- Eibl-Eibesfeldt, I. 1984. *Krieg und Frieden*. München: Piper Verlag.
- Elster, J. 1989. *Nuts and Bolts for the Social Sciences*. Cambridge: Cambridge University Press.
- Elwert, G., Neubert, D. and Feuchtwang, S. 1999. 'The Dynamics of collective violence: An introduction'. In: G. Elwert, D. Neubert and S. Feuchtwang (eds.): *Dynamics of Violence: Processes of Escalation and De-escalation of Violent Group Conflicts*. Berlin: Duncker und Humblot, pp. 9-31.

- Ember, C. and Ember, M. 1971. 'Conditions favoring matrilocal versus patrilocal residence'. *American Anthropologist*, 73, pp. 571-94.
- Ember, C. and Ember, M. 1994. 'Cross-cultural studies of war and peace: recent achievements and future possibilities'. In: S. Reyna and R. Downs (eds.): *Studying War: Anthropological Perspectives*. Langhorn: Gordon and Breach, pp. 185-208.
- Ensminger, J. 1992. *Making a Market. The Institutional Transformation of an African Society*. New York: Cambridge University Press.
- Ensminger, J. and Knight, J. 1997. 'Changing social norms: Common property, bridewealth, and clan exogamy'. *Current Anthropology*, 38(1), pp. 1-24.
- Evans-Pritchard, E. 1940. *The Nuer*. Oxford: Clarendon Press.
- Evens, T. 1985. 'The paradox of the Nuer feud and leopard-skin chief'. *American Ethnologist*, 12, pp. 84-102.
- Evera, Stephen van 1998. 'Offense, defense, and the causes of war'. In: M. Brown, O. Coté, S. Lynn-Jones and S. Miller (eds.): *Theories of War and Peace*. Cambridge (Mass.): MIT Press, pp. 55-93.
- Evera, Stephen van 1999. *Causes of War*. Ithaca: Cornell University Press.
- Fabbro, D. 1978. 'Peaceful societies'. *Journal of Peace Research*, 15, pp. 67-83.
- Fagan, B. (ed.) 1996. *The Oxford Companion to Archaeology*. Oxford: Oxford University Press.
- Feil, D. 1987. *The Evolution of Highland Papua New Guinea Societies*. Cambridge: Cambridge University Press.
- Ferguson, R.B. 1984a. 'Introduction: Studying war'. In: R.B. Ferguson (ed.): *Warfare, Culture and Environment*. Orlando: Academic Press, pp. 1-81.
- Ferguson, R.B. (ed.) 1984b. *Warfare, Culture and Environment*. Orlando: Academic Press.
- Ferguson, R.B. 1989a. 'Game wars? Ecology and conflict in Amazonia'. *Journal of Anthropological Research*, 45, pp. 179-206.
- Ferguson, R.B. 1989b. 'Do Yanomami killers have more kids?' *American Ethnologist*, 16(3), pp. 564-65.
- Ferguson, R.B. 1990a. 'Blood of the Leviathan'. *American Ethnologist*, 17(2), pp. 237-57.
- Ferguson, R.B. 1990b. 'Explaining war'. In: J. Haas (ed.): *The Anthropology of War*. Cambridge: Cambridge University Press, pp. 26-55.
- Ferguson, R.B. 1992. 'A savage encounter'. In: R.B. Ferguson and N. Whitehead (eds.): *War in the Tribal Zone*. Santa Fe: School of American Research Press, pp. 199-227.
- Ferguson, R.B. and Whitehead, N. 1992. 'The violent edge of empire'. In: R.B. Ferguson and N. Whitehead (eds.): *War in the Tribal Zone*. Santa Fe: School of American Research Press, pp. 1-30.
- Ferguson, R.B. 1995. *Yanomami Warfare: A Political History*. Santa Fe: School of American Research Press.
- Ferrill, A. 1985. *The Origin of War*. Boulder: Westview Press.
- Foster, M. and Rubinstein, R. (eds.) 1986. *Peace and War*. New Brunswick: Transaction books.
- Fried, M. 1967. *The Evolution of Political Society*. New York: Random House.
- Fried, M. 1975. *The Notion of Tribe*. Menlo Park: Cummings.
- Fried, M., Harris, M. and Murphy, R. (eds.) 1968. *War: The Anthropology of Armed Conflict*. New York: Natural History Press.
- Gabriel, R. 1990. *The Culture of War*. New York: Greenwood Press.
- Gat, A. 2000. 'The human motivational complex: Evolutionary theory and the causes of hunter-gatherer fighting'. *Anthropological Quarterly*, 73(1, 2): pp. 20-34 (Part 1), pp. 74-88 (Part 2).
- Godelier, M. 1982. *Die Produktion der Grossen Männer. Macht und männliche Vorherrschaft bei den Baruya in Neuguinea*. Frankfurt: Campus.
- Godelier, M. 1991. 'An unfinished attempt at reconstructing the social processes which may have prompted the transformatiuon of great-men societies into big-men societies'. In: M. Godelier and M. Strathern (eds.): *Big Men and Great Men*. Cambridge: Cambridge University Press, pp. 275-304.
- Goldschmidt, W. 1989. 'Personal motivation and institutionalized conflict'. In: M. Foster and R. Rubinstein (eds.): *Peace and War: Cross-cultural Perspectives*. New Brunswick: Transaction Books, pp. 3-14.
- Goldschmidt, W. 1997. 'Inducement to military participation in tribal societies'. In: R. Rubinstein and M. Foster (eds.): *The Social Dynamics of Peace and Conflict*. Boulder: Westview, pp. 47-65.
- Gordon, R. and Meggitt, M. 1985. *Law and Order in the New Guinea Highlands*. Hanover: University Press of New England.
- Gregor, T. 1990 'Uneasy Peace: Intertribal Relations in Brazil's Upper Xingu'. In: J. Haas (ed.): *The Anthropology of War*. Cambridge: Cambridge University Press, pp. 105-24.
- Gregor, T. (ed.) 1996. *A Natural History of Peace*. Nashville: Vanderbilt University Press.
- Gregor, T. and Robarchek, C. 1996. 'Two paths to peace: Semai and Mehinaku nonviolence'. In: T. Gregor (ed.): *A Natural History of Peace*. Nashville: Vanderbilt University Press, pp. 159-88.
- Greuel, P. 1971. 'The Leopard skin chief'. *American Anthropologist*, 73(5), pp. 1115-20.
- Haas, J. (ed.) 1990. *The Anthropology of War*. Cambridge: Cambridge University Press.
- Haas, J. and Creamer, W. 1993. 'Pueblo political organization in 1500: Tinkering with diversity'. http://www.santafe.edu/sfi/publications/Working-Papers/98-11-095.pdf
- Haas, J. and Creamer, W. 1997. 'Warfare among the pueblos: Myth, history, and ethnography'. *Ethnohistory*, 44, pp. 235-61.
- Hallpike, C. 1973. 'Functional interpretation of primitive warfare'. *Man*, 8(3), pp. 451-70.
- Hallpike, C. 1977. *Bloodshed and Vengeance in the Papuan Mountains*. Oxford: Clarendon Press.

- Hanser, P. 1985. *Krieg und Recht*. Berlin: Dietrich Reimer.
- Harner, M. 1975. 'Scarcity, the factors of production, and social evolution'. In: S. Polgar (ed.): *Population, Ecology and Social Evolution*. The Hague: Mouton, pp. 123-38.
- Harris, M. 1974. *Cows, Pigs, Wars, and Witches*. New York: Vintage.
- Harris, M. 1977. *Cannibals and Kings*. Glasgow: Fontana/Collins.
- Harris, M. 1984. 'A cultural materialist theory of band and village warfare: the Yanomamö test'. In: R.B. Ferguson (ed.): *Warfare, Culture, and Environment*. Orlando: Academic Press, pp. 111-40.
- Harrison, S. 1993. *The Mask of War: Violence, Ritual and the Self in Melanesia*. Manchester: Manchester University Press.
- Helbling, J. 1991. 'Reproduktion der Lokalgruppen bei den Maring'. *Zeitschrift für Ethnologie*, 116, pp. 135-65.
- Helbling, J. 1992. 'Ökologie und Politik in nicht-staatlichen Gesellschaften'. *Kölner Zeitschrift für Soziologie und Sozialpsychologie*, 2, pp. 203-25.
- Helbling, J. 1996a. 'Warum bekriegen sich die Yanomami? Versuch einer spieltheoretischen Erklärung'. In: P. Bräunlein and A. Lauser (eds.): *Krieg und Frieden*. Bremen: Keo, pp. 195-223.
- Helbling, J. 1996b. *Macht, Verwandtschaft und Produktion: Die Alangan-Mangyan im Nordosten Mindoros*. Berlin: Reimer.
- Helbling, J. 1998. 'Rückzug und Abhängigkeit: Die Friedfertigkeit der Alangan-Mangyan auf Mindoro (Philippinen)'. *Asiatische Studien*, 2, pp. 383-418.
- Helbling, J. 1999. 'The dynamics of war and alliance among the Yanomami'. In: G. Elwert, D. Neubert and S. Feuchtwang (eds.): *Dynamics of Violence: Processes of Escalation and De-escalation of Violent Group Conflicts*. Berlin: Duncker und Humblot, pp. 103-16.
- Helbling, J. n.d.a. Krieg und Frieden in Gesellschaften ohne Zentralgewalt. Forthcoming.
- Helbling, J. n.d.b. Gewalt und Krieg in der 'Urgesellschaft'? Politische Verhältnisse in Wildbeutergesellschaften. Forthcoming.
- Hobbes, T. 1994(1651). *Leviathan*. W. Euchner (trans. to German). Frankfurt: Suhrkamp.
- Hodgson, G. 1993. *Economics and Evolution*. Cambridge: Polity Press.
- Howell, S. and Willis, R. (eds.) 1989. *Societies at Peace*. London: Routledge.
- Jervis, R. 1978. 'Cooperation under the security dilemma'. *World Politics*, 30, pp. 167-214.
- Johnson, A. and Earle, T. 1987. *The Evolution of Human Societies. From Foraging Group to Agrarian State*. Stanford: Stanford University Press.
- Keegan, J. 1993. *A History of Warfare*. London: Pimlico.
- Keeley, L. 1996. *War Before Civilization*. Oxford: Oxford University Press.
- Keesing, R. 1992. *Custom and Confrontation: The Kwaio Struggle for Cultural Autonomy*. Chicago: University of Chicago Press.
- Kelly, R. 1985. *The Nuer Conquest*. Ann Arbour: University of Michigan Press.
- Kelly, R. 2000. *Warless Societies and the Origin of War*. Ann Arbor: University of Michigan Press.
- Knauft, B. 1987. 'Reconsidering violence in simple human societies'. *Current Anthropology*, 28, pp. 457-500.
- Knauft, B. 1990. 'Melanesian warfare: a theoretical history'. *Oceania*, 60, pp. 250-311.
- Knauft, B. 1999. 'Warfare and history in Melanesia'. In: *From Primitive to Postcolonial in Melanesia and Anthropology*. Ann Arbor: University of Michigan Press, pp. 89-156.
- Koch, K.-F. 1973. 'The Etiology and sociostructural conditions of violent conflict management in Jalé society (New Guinea)'. In: J. De Wit and W. Hartup (eds.): *Determinants and Origins of Aggressive Behavior*. The Hague: Mouton, pp. 461-71.
- Koch, K.-F. 1974a. *The Anthropology of Warfare*. (Addison-Wesley Module in Anthropology 52). Reading: Addison-Wesley Publishing Company.
- Koch, K.-F. 1974b. *War and Peace in Jalémó. The Management of Conflict in Highland New Guinea*. Cambridge (Mass.): Harvard University Press.
- Koch, K.-F. 1976. 'Konfliktmanagement und Rechtsethnologie'. *Sociologus*, 26, pp. 96-129.
- Koch, K.-F. 1983. 'Epilogue. Pacification: Perspectives from conflict theory'. In: M. Rodman and M. Cooper (eds.): *The Pacification of Melanesia*. Lanham (Md.): University Press of America, pp. 199-207.
- Komanyi, M. 1971 'Iban woman's role: A brief summary of observations at Samu on the Paku River'. *Sarawak Museum Journal*, 19, pp. 253-56.
- Komanyi, M. 1990. 'The real and ideal participation in decison-making of Iban women: A study of a longhouse community in Sarawak'. *Acta Ethnbogr.*, 36(1/4), pp. 141-86.
- Konner, M. 1982. *The Tangled Wing: Biological Constraints on the Human Spirit*. New York: Holt, Rinehart and Winston.
- Lang, H. 1977. Exogamie und interner Krieg in Gesellschaften ohne Zentralgewalt. Dissertation, Hamburg.
- Langness, L. 1967. 'Sexual antagonism in the New Guinea Highlands: A Bena Bena example'. *Oceania*, 37, pp. 161-77.
- Larrick, J., Yost, J., Kaplan, J., King, G. and Mayhall, J. 1979. 'Patterns of health and disease among the Waorani Indians'. *Medical Anthropology*, 3, pp. 147-89.
- Larson, G. 1987. *The Structure and Demography of the Cycle of Warfare among the Ilaga Dani of Irian Jaya*. Ann Arbor: University Microfilms International.
- Leakey, R. and Lewin, R. 1977. *Origins*. London: Macdonald and Jane's.
- LeBlanc, S. 2003. *Constant Battles: The Myth of the Peaceful Noble Savage*. New York: St. Martin's Press.
- Lee, R. and DeVore, I. 1968. 'The problems in the study of hunters and gatherers'. In: R. Lee and I. DeVore (eds.): *Man the Hunter*. Chicago: Aldine, pp. 3-12.

- Levy, J. 1989. 'The causes of war: a review of theories and evidence'. In: P. Tetlock, J.L. Husbands, R. Jervis, P. Stern and C. Tilly (eds.): *Behavior, Society, and Nuclear War*. Oxford: Oxford University Press, pp. 209-333.
- Lévi-Strauss, C. 1967. *Les Structures Élémentaires de la Parenté*. Deuxième édition. Paris: Presse Universitaire de France.
- Lizot, J. 1971. 'Economie ou Société'. In: *Les Yanomami Centraux*. (Cahiers de L'Homme, 22). Paris: EHESS, pp. 137-75.
- Lizot, Jacques 1977. 'Population, ressources et guerre'. In: *Les Yanomami Centraux*. (Cahiers de L'Homme, 22). Paris: EHESS, pp. 177-209.
- Lizot, J. 1989. 'A propos de la guerre'. *Journal de la Société des Américanistes*, 75, pp. 91-113.
- Lourandos, H. 1997. *Continent of Hunter-gatherers: New Perspective in Australian Prehistory*. Cambridge: Cambridge University Press.
- Martin, D. and Frayer, D. (eds.) 1997. *Troubled Times: Violence and Warfare in the Past*. Langhorn: Gordon and Breach.
- Mauss, M. 1968(1926). *Die Gabe*. Frankfurt: Suhrkamp.
- McCauley, C. 1990. 'Conference overview'. In: J. Haas (ed.): *The Anthropology of War*. Cambridge: Cambridge University Press, pp. 1-25.
- Meggitt, M. 1971. 'The pattern of leadership among the Mae Enga of New Guinea'. In: R. Berndt and P. Lawrence (eds.): *Politics in New Guinea*. Nedlands: University of Western Australia Press, pp. 191-206.
- Meggitt, M. 1974. 'Pigs are our hearts: The *te* exchange cycle among the Mae Enga of New Guinea'. *Oceania*, 44, pp. 165-203.
- Meggitt, M. 1977. *Blood is their Argument: Warfare among Mae Enga Tribesmen of the New Guinea Highlands*. Palo Alto: Mayfield Publishing Company.
- Menget, P. 1993. 'Les frontières de la chefferie. Remarques sur le système politique du Haut Xingu (Brésil)'. *L'Homme*, 126-128, pp. 59-76.
- Moore, J. 1990. 'The reproductive success of the Cheyenne war chiefs'. *Current Anthropology*, 31(3), pp. 322-30.
- Morren, G. 1984. 'Warfare on the Highland fringe of New Guinea: The case of the mountain Ok'. In: Ferguson, R.B. (ed.): *Warfare, Culture, and Environment*. Orlando: Academic Press, pp. 169-207.
- Murdock, G.P. 1968. 'The current status of the world´s hunting and gathering peoples'. In: R. Lee and I. DeVore (eds.): *Man the Hunter*. Chicago: Aldine, pp. 13-20.
- Murphy, R. 1957. 'Intergroup hostility and social cohesion'. *American Anthropologist*, 59(6), pp. 1018-35.
- Murphy, Y. and Murphy, R. 1974. *Women of the Forest*. New York: Columbia University Press.
- Nelson, R. 1995. 'Recent evolutionary theorizing about economic change'. *Journal of Economic Literature*, XXXIII, pp. 48-90.
- Nelson, R. and Winter, S. 1982. *An Evolutionary Theory of Economic Change*. Cambridge: Harvard University Press.
- Nettleship, M., Givens, R.D. and Nettleship, A. (eds.) 1975. *War, its Causes and Correlates*. The Hague: Mouton.
- Nicholson, M. 1992. *Rationality and the Analysis of International Conflict*. Cambridge: Cambridge University Press.
- Orywal, E. 1996a. 'Krieg in den Köpfen. Ein Vorwort'. In: E. Orywal, A. Rao and M. Bollig (eds.): *Krieg und Kampf. Die Gewalt in den Köpfen*. Berlin: Reimer, pp. 7-12.
- Orywal, E. 1996b. 'Krieg als Konfliktstrategie. Zur Plausibilität von Kriegstheorien aus kognitionsethnologischer Sicht'. *Zeitschrift für Ethnologie*, 121, pp. 1-48.
- Orywal, E. 1998. 'Zur Anthropologie des Krieges – ein interdisziplinärer Überblick'. In: R. Eckert (ed.): *Wiederkehr des 'Volksgeistes'?* Opladen: Leske und Budrich, pp. 83-142.
- Otterbein, K. 1968. 'Internal war: A cross-cultural study'. *American Anthropologist*, 70(2), pp. 277-89.
- Otterbein, K. 1973. 'The anthropology of war'. In: J. Honigmann (ed.): *Handbook of Social and Cultural Anthropology*. Chicago: Rand McNally, pp. 923-58.
- Otterbein, K. 1985. *The Evolution of War*. (3rd edn.). New Haven: HRAF Press.
- Otterbein, K. 1988. 'The dilemma of disarming'. In: K. Otterbein: *Feuding and warfare. Selected works of Keith E. Otterbein*. Langhorn: Gordon and Breach Publishers, pp. 181-94.
- Otterbein, K. 1990. 'Convergence in the anthropological study of warfare'. In: K. Otterbein: *Feuding and warfare. Selected works of Keith E. Otterbein*. Langhorn: Gordon and Breach Publishers, pp. 171-80.
- Otterbein, K. 1994. *Feuding and Warfare. Selected Works of Keith E. Otterbein*. Langhorn: Gordon and Breach Publishers.
- Otterbein, K. 1999. 'A history of research on warfare in anthropology'. *American Anthropologist*, 101(4), pp. 794-805.
- Peoples, J. 1982. 'Individual or Group advantage? A reinterpretation of the Maring ritual cycle'. *Current Anthropology*, 23(3), pp. 291-310.
- Pospisil, L. 1978. *The Kapauku Papuans of West New Guinea*. New York: Holt, Rinehart and Winston.
- Pospisil, L. 1994. 'I am sorry I cannot kill you anymore: War and peace among the Kapauku'. In: S. Reyna and R. Downs (eds.): *Studying Warfare: Anthropological Perspectives*. Amsterdam: Gordon and Breach Publishers, pp. 113-26.
- Pringle, R. 1970. *Rajahs and rebels: The Ibans of Sarawak under Brooke Rule, 1841-1941*. London: Macmillan.
- Rapoport, A. 1974. *Konflikt in einer vom Menschen Gemachten Umwelt*. Darmstadt: Darmstädter Blätter.
- Rapoport, A. 1976. *Kämpfe, Spiele und Debatten*. Darmstadt: Darmstädter Blätter.
- Rappaport, R. 1968. *Pigs for the Ancestors: Ritual in the Ecology of a New Guinea People*. New Haven: Yale University Press.
- Reyna, S. 1994. 'A mode of domination approach to organized violence'. In: S. Reyna and R. Downs (eds.): *Studying Warfare: Anthropological Perspectives*. Amsterdam: Gordon and Breach Publishers, pp. 29-65.

- Reyna, S. and Downs, R. (eds.) 1994. *Studying Warfare: Anthropological Perspectives*. Langhorn: Gordon and Breach.
- Riches, D. 1991. 'Aggression, war, violence: space/time and paradigm'. Man, 26(2), pp. 281-98.
- Riker, W. 1962. *Theory of Political Coalition*. New Haven: Yale University Press.
- Robarchek, C. 1989. 'Hobbesian and Rousseauan images of man'. In: S. Howell and R. Willis (eds.): *Societies at Peace*. London: Routledge, pp. 31-44.
- Robarchek, C. and Robarchek, C. 1992. 'Cultures of war and peace'. In: J. Silverberg and P. Gray (eds.): *To Fight or not to Fight: Violence and Peacefulness in Humans and other Primates*. New York: Oxford University Press, pp. 189-213.
- Robarchek, C. and Robarchek, C. 1996. 'The Aucas, the cannibals, and the missionaries: From warfare to peacefulness among the Waorani'. In: T. Gregor (ed.): *A Natural History of Peace*. Nashville: Vanderbilt University Press, pp. 159-88.
- Robarchek, C. and Robarchek, C. 1998. *Waorani: The Contexts of Violence and War*. Fort Worth: Harcourt Brace College Publishers.
- Rodman, M. and Cooper, M. (eds.) 1983. *The Pacification of Melanesia*. Lanham (Md.): University Press of America.
- Roper, M. 1969. 'A survey of the evidence for intrahuman killing in the pleistocene'. *Current Anthropology*, 10(4,2), pp. 427-59.
- Roper, M. 1975. 'Evidence of warfare in the Near East from 10000-4300 BC'. In: M. Nettleship, R.D. Givens and A. Nettleship (eds.): *War, its Causes and Correlates*. The Hague: Mouton, pp. 299-343.
- Rosaldo, M. 1980. *Knowledge and Passion: Ilongot Notions of Self and Social Life*. Cambridge: Cambridge University Press.
- Rosaldo, R. 1980. *Ilongot Headhunting*. Stanford: Stanford University Press.
- Ross, M. 1981. 'Socioeconomic complexity, socialization, and political differenciation'. *Ethos*, 9(3), pp. 217-47.
- Ross, M. 1986. 'A cross-cultural theory of political conflict and violence'. *Political Psychology*, 7, pp. 427-69.
- Ross, M. 1993a. *The Culture of Conflict*. New Haven: Yale University Press.
- Ross, M. 1993b. *The Management of Conflict*. New Haven: Yale University Press.
- Rothstein, R. 1968. *Alliances and Small Powers*. New York: Columbia University Press.
- Rubinstein, R. and Foster, M. (eds.) 1988. *The Social Dynamics of Peace and Conflict*. Boulder: Westview.
- Sahlins, M. 1965. 'On the ideology and composition of descent groups'. *Man*, 65, pp. 104-107.
- Sahlins, M. 1968. *Tribesmen*. Englewood Cliffs: Prentice-Hall.
- Sandin, B. 1967. *The Sea Dayaks of Borneo before White Rajah Rule*. London: MacMillan.
- Schelling, T. 1960. *The Strategy of Conflict*. Cambridge (Mass.): Harvard University Press.
- Scherrer, C. 2000. 'Ethnonationalismus als globales Phänomen'. In: R. Moser (ed.): *Die Bedeutung des Ethnischen im Zeitalter der Globalisierung*. Bern: Haupt Verlag, pp. 17-90.
- Schmölzer, H. 1996. *Der Krieg ist Männlich; ist der Friede Weiblich?* Wien: Verlag für Gesellschaftskritik.
- Schumacher, I. 1972. *Gesellschaftsstruktur und Rolle der Frau bei den Irokesen*. Berlin: Duncker und Humblot.
- Service, E. 1966. *The Hunters*. Englewood Cliffs: Prentice-Hall.
- Service, E. 1968. 'War and our "contemporary ancestors"'. In: M. Fried, M. Harris and R. Murphy (eds.): *War: The Anthropology of Armed Conflict*. New York: Natural History Press, pp. 160-67.
- Sillitoe, P. 1978. 'Big men and war in New Guinea'. *Man*, 13(2), pp. 252-71.
- Simons, A. 1999. 'War: back to the future'. *Annual Review of Anthropology*, 28, pp. 73-108.
- Sipes, R. 1973. 'War, sports and aggression: An empirical test of two rival theories'. *American Anthropologist*, 75, pp. 64-86.
- Spittler, G. 1980a. 'Konfliktaustragung in akephalen Gesellschaften: Selbsthilfe und Verhandlung'. In: E. Blankenburg (ed.): *Alternative Rechtsformen und Alternativen zum Recht*. (Jahrbuch für Rechtssoziologie und Rechtstheorie, 6). Opladen: Westdeutscher Verlag, pp. 142-64.
- Spittler, G. 1980b. 'Streitschlichtung im Schatten des Leviathan. Eine Darstellung und Kritik rechtsethnologischer Untersuchungen'. *Zeitschrift für Rechtssoziologie*, 1, pp. 4-32.
- Sponsel, L. and Gregor, T. (eds.) 1994. *The Anthropology of Peace and Nonviolence*. Boulder: Lynne Rienner Publishers.
- Sponsel, L. 1996. 'A natural history of peace'. In: T. Gregor (ed.): *A Natural History of Peace*. Nashville: Vanderbilt University Press, pp. 95-125.
- Steward, J. 1968. 'Causal factors and processes in the evolution of prefarming societies'. In: R. Lee and I. DeVore (eds.): *Man the Hunter*. New York: Aldine Publishing Company, pp. 321-34.
- Tefft, S. 1975. 'Warfare regulation: A cross-cultural test of hypotheses'. In: M. Nettleship, D. Givens and A. Nettleship (eds.): *War: Its Causes and Correlates*. Chicago: Aldine, pp. 693-712.
- Tefft, S. and Reinhardt, D. 1974. 'Warfare regulation'. *Behavior Science Research*, 9, pp. 151-72.
- Thoden van Velzen, H.U.E. and van Wetering, W. 1960. 'Residence, power groups and intra-societal aggression'. *International Archives of Ethnography*, 49, pp. 169-200.
- Tooby, J. and Cosmides, L. 1988. 'The evolution of war and its cognitive foundations'. *Institute for Evolutionary Studies Technical Report*, 88(1), pp. 1-15.
- Turney-High, H. 1949. *Primitive War: Its Practices and Concepts*. Columbia: University of South Carolina Press.

- Turton, D. 1979. 'War, peace and Mursi identity'. In: K. Fukui and D. Turton (eds.): *Warfare among East African Herders*. Osaka: National Museum of Ethnology Press, pp. 179-210.
- Turton, D. 1994. 'Mursi political identity and warfare: The survival of an idea'. In: K. Fukui and J. Markakis (eds.): *Ethnicity and Conflict in Africa*. London: James Currey, pp. 14-31.
- Vasquez, J. 1993. *The War Puzzle*. Cambridge: Cambridge University Press.
- Vasquez, J. 2000. 'What do we know about war?' In: J. Vasquez (ed.): *What Do We Know about War?* Lanham (Md.): Rowman and Littlefield, pp. 335-70.
- Vayda, A. 1961. 'Expansion and warfare among swidden agriculturalists'. *American Anthropologist*, 63(2), pp. 346-58.
- Vayda, A. 1976. *War in Ecological Perspective*. New York: Plenum Press.
- Vencl, S. 1984. 'War and warfare in archaeology'. *Journal of Anthropological Archaeology*, 3, pp. 116-32.
- Vencl, S. 1991. 'Interpretation des blessures causées par les armes à l époque du Mésolithique'. *L'Anthropologie*, 95, pp. 219-28.
- Wagner, H. 1986. 'The theory of games and the balance theory'. *World Politics*, 4, pp. 546-76.
- Wagner, U. 1972. *Colonialism and Iban Warfare*. Stockholm: OBE-Tryck.
- Walt, S. 1987. *The Origins of Alliances*. Ithaca: Cornell University Press.
- Waltz, K. 1960. *Man, the State and War*. New York: University of Columbia Press.
- Waltz, K. 1979. *Theory of International Politics*. New York: McCraw-Hill.
- Watson, J. 1971. 'Tairora: The politics of despotism in a small society'. In: R. Berndt and P. Lawrence (eds.): *Politics in New Guinea*. Nedlands: University of Western Australia Press, pp. 224-75.
- Watson, J. 1983. *Tairora Culture*. Seattle: University of Washington Press.
- Whiting, B. 1965. 'Sex identity conflict and physical violence: A comparative study'. *American Anthropologist*, 67(2), pp. 123-40.
- Wiberg, H. 1981. 'What have we learned about peace?' *Journal of Peace Research*, XVIII, pp. 111-48.
- Wiessner, P. and Tumu, A. 1998a. 'The capacity and constraints of kinship in the development of the Enga Tee ceremonial exchange network'. In: T. Schweizer and D. White (eds.): *Kinship, Networks and Exchange*. Cambridge: Cambridge University Press, pp. 277-302.
- Wiessner, P. and Tumu, A. 1998b. *Historical Vines*. Washington: Smithonian Institution Press.
- Wolf, E. 1982. *Europe and the People without History*. Berkeley: University of California Press.
- Wolf, E. 1987. 'Cycles of war'. In: K. More (ed.): *Waymarks: The Notre Dame Inaugural Lectures in Anthropology*. Notre Dame: University of Notre Dame Press, pp. 127-50.
- Wrangham, R. 1999. 'Evolution of coalitionary killing'. *Yearbook of Physical Anthropology*, 42, pp. 1-30.
- Yesner, D. 1994. 'Seasonality and resource stress among hunter-gatherers'. In: E. Burch and L. Ellanna (eds.): *Key Issues in Hunter-gatherer Research*. Oxford: Berg, pp. 151-68.

Fighting and Feuding in Neolithic and Bronze Age Britain and Ireland

I . J . N .   T H O R P E

/10

Although this paper concentrates on evidence from Britain and Ireland, it is inevitably situated within a far wider theoretical debate, which mostly sets the British and Irish material within a universal or at least pan-European framework. The specifics of the data examined here are therefore weighed against these general models as well as being assessed in their own terms.

Origins are always attractive subjects, and the origin of war in early prehistory is no exception, having been considered recently from the viewpoint of biological anthropology (e.g. Wrangham 1999), social anthropology (e.g. Otterbein 1997; Kelly 2000), military history (e.g. Keegan 1993; O'Connell 1995), history (e.g. Dawson 1996) and archaeology (e.g. Keeley 1996).

Of these fields, the most interest in early war has been shown by anthropologists of varying kinds, as it is central to that elusive quality, 'human nature'. Strangely, archaeology has largely been an onlooker in this argument, which has almost entirely been fought out using evidence from contemporary societies (with Ferguson 1997 an important exception).

The most influential of these general theories have been various approaches within evolutionary theory (see Laland and Brown 2002 for a clear guide to the similarities and differences of the different schools),

the materialist approach (e.g. Ferguson 1990) and cultural evolution (Dawson 1999; 2001).

Within evolutionary theory the particular strand which has shown most interest in the origins of human conflict is evolutionary psychology. This is because evolutionary psychologists see humans as shaped by an ancestral environment long past, dubbed the environment of evolutionary adaptation (EEA). The EEA broadly equates to the Palaeolithic and Mesolithic (e.g. Cosmides *et al.* 1992; Pinker 1998: 42), with the development of agriculture marking a crucial break. Thus any pattern to be discerned in Neolithic or Bronze Age warfare has its origins in earlier times and should clearly fall into line with one of the three main competing theories for warfare situated within evolutionary psychology – territorial, reproductive and status competition.

The territorial model originates in modern times with E.O. Wilson, who argued from the sociobiological perspective that ethnocentricity was a product of natural selection (1978: 119):

Our brains do appear to be programmed to the following extent: we are inclined to partition other people into friends and aliens, in the same sense that birds are inclined to learn territorial songs and to navigate by the polar constellations. We tend to fear deeply the actions of strangers and to solve conflicts by aggression. These learning rules are most likely to have evolved during the past hundreds or thousands of years of human evolution ...

More specifically, Wrangham (1999) has argued for continuity of a territorial instinct from the common ancestor of chimpanzees and humans. He argues that a territorial instinct exists in modern chimpanzees, with young male chimpanzee patrols of territorial borders leading to conflicts of extermination with neighbouring groups, improving the victors' access to resources. He compares these with the territorial nature of modern American gang culture, with the link in his argument provided by comparisons with the Yanomami of the Amazon as an example of 'primitive' culture.

The reproductive theory of warfare is based on analogies with primate behaviour in which male-centred competition, over access to females, takes violent form (e.g. Wrangham and Peterson 1996). In human societies this is argued to take the form of successful warriors having more wives and, crucially, more children. The other approach regards warfare as the outcome of violent competition by young males striving for status and prestige (Maschner and Reedy-Maschner 1998), even when there is no prospect of territorial gain; reproductive success is not the primary aim here, even though it may be a consequence of gaining status.

There are a variety of problematic assumptions involved in these theories, which struggle to encompass the ethnographic evidence, let alone the archaeological record (Thorpe 2003b). They also tend to blur almost entirely any distinction between individual acts of violence directed against those nearest to them, conflict within and between groups directed against specific individuals, often in the form of feuds between kin-groups, and actual warfare (by which I mean here organised, premeditated, aggression between autonomous political units, where the individual identity of the victim is less important than their membership of the group being attacked).

In addition there are some specific issues relating to prehistoric archaeology. First, while prestige is clearly a significant factor in the creation of warriors (Clastres 1994: 169-200), just as with the reproductive theory, however, dubbing violence the business of men (e.g. Gilbert 1994; van der Dennen 1995) avoids the considerable ethnographic, historical and

archaeological evidence, especially from Asia (Rolle 1989) and the Americas, of female warriors and even female war chiefs (Koehler 1997; Hollimon 2001). The existence of female warriors in prehistoric Europe is an area which has received extremely little attention.

The other fundamental problem for evolutionary anthropology is that, as various critics have noted (e.g. Knauft 1991; Foley 1996), almost no archaeologists believe in an unchanging environment of evolutionary adaptation until the advent of agriculture. Instead, the pattern of conflict and warfare may well be expected to vary through time and across space.

The other main camp within anthropology is the materialist. Those favouring a universal materialist interpretation start from the standpoint that warfare is completely irrational, and therefore one would only risk one's life in combat when there was a desperate need for land, or more immediately food (e.g. Ferguson 1990). Ferguson (1990) argues that motivations as stated by the participants to ethnographers hide the real motives of achieving basic material goals. Resource shortages are, of course, a staple of archaeological explanations, and one can note some clear difficulties for the materialists here. For example, many models of the transition to agriculture in the Levant involve a supposed shortage of resources, either through over-population (Cohen 1989), climatic decline (Henry 1989) or seasonal shortfalls (Harris 1990), yet the evidence for conflict within the Natufian gatherer-hunter population here is negligible, only a single traumatic injury from some 400 skeletons excavated (Thorpe 2003b). Of course, the significance of material considerations in many conflicts is undeniable, and must have played some part in prehistoric warfare. Material considerations are not, however, convincing as a total or sole explanation.

A recent reappearance in warfare studies is that of cultural evolution (Dawson 1996; 1999; 2001), which harks back to 1960s notions of a simple division of prehistoric societies into bands, tribes, chiefdoms and states. For Dawson a key development was agriculture. In this he follows earlier commentators who see conflict as a consequence of settled agrarian communities (e.g. Leakey and Lewin 1992); although he does propose that in prehistory the pattern of warfare may have been cyclical rather than linear. In such models the walls of Jericho, one of the earliest towns, feature prominently (for an alternative interpretation as a flood defence see Bar-Yosef 1986) along with the attack on the Early Neolithic causewayed enclosure of Hambledon Hill, England (Mercer 1999), considered below, or Schletz in Austria (Windl 1994), an enclosure with multiple burials of individuals who met a violent end in the ditch. The most dramatic case is that of Talheim in Germany (Wahl and König 1987), c. 5000 BC, where a mass grave contained men, women and children, killed by axe and adze blows to the head.

To assess the validity of this proposal we need to consider the evidence from the Mesolithic, as whatever the level of Neolithic violence may have been, we can not simply assume this appeared from nothing. Certainly the evidence for both serious injuries and violent death from Mesolithic skeletons in Europe is steadily growing (Vencl 1999; Grünberg 2000; Thorpe 2003a; Thorpe 2003b). It is clear that conflict is *not* a feature confined to the period after the development of agriculture. If we look at northern Europe, there is plentiful evidence of individuals with healed injuries. One of the best known examples is the Korsør Nor harbour settlement site on the coast of Zealand, dating to the Ertebølle period, where three of the seven burials at the site were of individuals who had suffered traumatic skull injuries (Bennike 1997).

Skeletal material also points to the existence of conflicts occurring on a much larger scale, perhaps actual Mesolithic warfare. At Ofnet cave in Bavaria two pits contained the skulls and vertebrae of 38 individuals, all stained with red ochre, dating to around 6500 BC (Frayer 1997). Most were children; two thirds of the adults were females. Grave goods of deer teeth and shells were associated only with adult females and children. Half the individuals were fatally wounded by blunt mace-like weapons, with females, males and children (even infants) all injured, but males suffering the most wounds (Frayer 1997). The scale of this apparent massacre suggests the destruction of a whole community, followed by the ceremonial burial of 'trophy skulls' (Keeley 1996: 102). Certainly, there are many accounts in the ethnographic record which demonstrate the very careful curation of skulls taken in warfare (e.g. Sterpin 1993). We should also note that none of the conditions of sedentism, territoriality and status competition attributed to the Ertebølle (which therefore fits quite well with the general theories

outlined above) have been claimed for the German Mesolithic (Jochim 1998).

It is also crucial to note that even within relatively small areas there can be significant differences in the level of inter-personal violence revealed by skeletal injuries. In the Iron Gates area of the Balkan Mesolithic the sites of Vlasac and Schela Cladovei show clear evidence of conflict (Radovanovic 1996; Chapman 1999). The level of violence is especially high at Schela Cladovei – among the fifty-six skeletons excavated there are six cases of projectile injuries (four male, one female, one unsexed, mostly from bone points) along with some half a dozen examples of cranial injuries (mostly not healed before death), so that about one third of all adults from the site had traumatic injuries. However, Schela Cladovei and Vlasac are anomalous within the region's Mesolithic cemeteries as a whole – the latest overall figures are eight projectile injuries out of 400 skeletons, and roughly ten individuals with fractures. These two sites thus provide the vast bulk of cases of violence from just one fifth of the total examined burial population. Thus sweeping generalisations arguing for high (or low) levels of conflict in early prehistory may not hold true even for relatively small regions.

Turning to Britain, we do not have the skeletal record possessed by our continental colleagues for the Mesolithic, with the material from the only cemetery of this date (Aveline's Hole in southwest England) largely lost or destroyed, and the remainder almost all single bones, so the existence of conflict can not be documented in any clear way. There are, however, a few exceptions. Cheddar Man, near Aveline's Hole, dated to c. 8000 BC, is a complete skeleton (Tratman 1975) with a variety of traumatic injuries to the skull and collar-bone. Most of these have been attributed to the effects of water (Stringer 1985), but the most recent examination by pathologists suggests that he suffered several blows to the head some time before death, one of these resulting in the formation of an abscess between the eyes, this probably becoming infected and bringing about his death (Wilson 2001: 54). The cranium of a probable adult male, dating to c. 8500 BC, from the Ogof-yr-Ychen cave on Caldey Island off Wales has a healed depressed fracture (Schulting 1998: 277). Finally, a skull from Aveline's Hole appears from examination of a photograph (the skull itself has not been located) to have a perforation of the skull (Schulting 1998: 273).

One can either take these scattered examples to mean that the evidence for larger-scale conflict will one day emerge, or that things were genuinely different in Britain. If we believe the latter, then this may have implications for our interpretation of the Early Neolithic.

Early Neolithic

Beginning with individuals, and perhaps individual acts of violence, there are clear Early Neolithic examples of death through arrowhead injury at the chambered tombs of Ascott-under-Wychwood (Selkirk 1971) in western England (probably an adult male) and Tulloch of Assery Tomb B (Corcoran 1964-6) in Scotland (an adult), in both cases with the arrowhead being embedded in a vertebra. In Ireland, the portal tomb of Poulnabrone (Lynch and Ó Donnabháin 1994), Co. Clare, contained an adult male with the tip of a chert projectile point embedded in his right hip. An important addition to this small group is the rib fragment from a young adult at Penywyrlod in Wales with an embedded arrowhead tip (Wysocki and Whittle 2000). This only penetrated the bone slightly, and was on a downward path at the moment of entry, suggesting a shot from a distance. The wider significance of this discovery is that the Penywyrlod tomb was fully published in the relatively recent past (Britnell and Savory 1984) and thus suggests that many more cases may have been missed where they were not specifically sought for.

Another adult at Ascott-under-Wychwood in the Thames Valley had an arrowhead beneath a rib, which may be the cause of death (Selkirk 1971). This was also the preferred interpretation of the complete leaf-shaped arrowhead found in the throat region of an elderly male burial in the northeast chamber at West Kennet in Wessex (Piggott 1962: 25). He also had a fracture of the radius and a large abscess cavity at the head of the humerus, interpreted as the result of a wound through the muscle. At Wayland's Smithy chambered tomb in the Thames Valley (Whittle 1991) three leaf-shaped arrowheads were found, each on the pelvis of an adult; all had broken points. Significantly, the arrowheads were regarded as having had their points deliberately broken off and discarded before burial by the original excavator in the interim report (Atkinson 1965) but interpreted as either being grave goods or the cause of death in the

final report (Whittle 1991). That these individuals were buried in such important monuments may show that this was a method of death which was approved socially. A similar explanation may be appropriate in the case of the tip of a leaf-shaped arrowhead found in the entrance to the northern passage at the Hazleton chambered tomb in western England, as the excavator, Alan Saville (1990: 264) has argued, given that the rest of the piece was not present. The articulated and disarticulated skeletons from this area of the passage were all male, although it might have entered with one of the bodies which were located in the chamber to which the passage had originally led. Less well recorded examples have come from the burial deposits in several other chambered tombs in western England and Wales, including Notgrove, West Tump, Rodmarton (two examples), Sale's Lot, Adam's Grave and Ty-Isaf (Saville 1990: 264) and at Harborough Rocks in Derbyshire in Northern England (Burl 1981: 95-96). These might be seen as the cause of death rather than as examples of the deliberate breakage of objects as a funeral rite (Grinsell 1961).

Outside chambered tombs, the Fengate multiple burial in eastern England included an adult male aged 25-30 with an arrowhead with a broken tip lodged between his ribs (Pryor 1984: 19-27). There are several examples from Wessex: an adult male burial in the ditch of the Wor Barrow earthen long barrow (Pitt-Rivers 1898: 63) was found with an arrowhead between the ribs; an adult at the Tarrant Launceston round barrow (Piggott and Piggott 1944: 75) had an arrowhead among the ribs; an arrowhead with a broken tip comes from Fyfield long barrow (Thurnam 1869: 194) and the tip of a broken arrowhead from Chute long barrow (Passmore 1942).

In Yorkshire, at the Wold Newton round barrow in Yorkshire (Mortimer 1905: 350-52) an arrowhead was discovered lying on the pelvis of one of the burials, and at Callis Wold 275 round barrow two arrowheads with broken tips were recovered near the knees of an adult (Mortimer 1905: 161-63) and a further arrowhead at the hip of another adult. Another case of an arrowhead with a broken tip comes from the Aldro 88 round barrow (Mortimer 1905: 59).

The causewayed enclosure of Hambledon Hill in Wessex has also famously produced likely evidence of death through arrowshot. A young (about nineteen years old) adult male was buried in the outwork ditch surrounding the enclosure complex at the time that the timber palisade surrounding the site was burnt down and pushed back into the ditch along with chalk rubble (Mercer 1980; 1988). He had been killed by an arrowhead (missing its tip) shot from behind into the lungs. Another young adult male was found nearby in the rubble layer (Mercer 1988), now known to have an arrowhead in the throat (Mercer 1999).

Other Neolithic burial sites have produced dozens of arrowheads (Green 1980), mostly poorly recorded, and the possibility exists that modern excavations might have produced evidence for violence from many of these. The individual examples of death through arrowhead injury are all adult males, where the sex has been identified (although Ascott-under-Wychwood may be an exception if Darvill's description of the skeleton as that of a woman is correct [1987: 68]).

In addition to arrowhead injuries, there are a number of examples of skull and other injuries caused by short range weapons such as antler tines, clubs and axes (as recorded in mainland Europe, e.g. at Talheim, mentioned above). As well as the individuals killed by arrowshot at Hambledon Hill, two skulls (one of a male, the other unsexed) had healed injuries (Roberts and Cox 2003: 73). At the Poulnabrone portal dolmen in Ireland (Lynch and Ó Donnabháin 1994), in addition to the adult male with an arrowhead injury there were individuals with healed fractures to the skull and to a rib, giving a total of three individuals out of twenty-two with traumatic injuries. Tulloch of Assery Tomb A (Corcoran 1964-6), immediately adjacent to Tomb B, with an adult killed by arrowshot, produced an adolescent, aged approximately fourteen years, who had apparently suffered a blow to the skull. Other less clear-cut examples come from the long barrow at Barrow Hills in the Thames Valley (Barclay and Halpin 1999: 29) – an adult (probably male) with a healed forearm fracture; from the Fussell's Lodge long barrow in Wessex (Ashbee 1966) – an adult with a broken forearm; from the Wayland's Smithy, Wessex, chambered tomb (Whittle 1991) – a young adult with a healed forearm fracture and a young adult female with a possible upper arm injury; from the Isbister chambered tomb on Orkney (Hedges 1983: 118-19) – an adult male with a broken and healed forearm and adults with broken ribs; from the Hazleton chambered tomb (Saville 1990: 191) – ribs with healed fractures from at least two adults.

There are similar discoveries from other common locations for burial. Perhaps coming from the very end of the Early Neolithic, the massive round barrow at Duggleby Howe in Yorkshire (Mortimer 1905: 23-42; Loveday 2002), produced a skull part way down one of the burial shafts with a large hole in it. The adult male from Painsthorpe barrow 118 (Mortimer 1905: 125-28), also in Yorkshire, found with a jet slider, who had suffered a heavy blow on the skull causing fracturing and the formation of a large blood clot (Brothwell 1960). Of a similar date is the primary burial of an adult female inside a ring-ditch at Five Knolls, central England (Dunning and Wheeler 1931), with a polished flint knife, who had a healed forearm fracture.

The site of Linkardstown, Co. Carlow, which gives the series of late Early Neolithic Irish round barrows covering cists their name, produced an adult male burial with several healed fractures from blows to the skull, buried with an axe (Raftery 1944). At the Staines causewayed enclosure in the Thames Valley the head (skull and mandible) of an adult male aged twenty-five to thirty years old were found in the primary fill of the outer ditch (Robertson-Mackay 1987). He had two healed head wounds on the right side of the head, and met his death through a series of four blows to the head from a blunt object and was finally decapitated. The published evidence is, however, not entirely convincing (R. Schulting pers. comm.). Another case which probably belongs in the Early Neolithic is from the cist burial under a cairn at Glenquickan in southwest Scotland, discovered when the site was destroyed in 1809. The adult contained within it had had an arm almost severed from the shoulder by a blow from a greenstone axe, a chip of which was found embedded in the bone (Burl 1987: 112).

Outside the norm of Early Neolithic burial places, the adult (possibly male) skull from Haywood Cave in Somerset with a depressed fracture (Everton and Everton 1972) may well date to the Early Neolithic given the radiocarbon date on a vertebra from the same deposit (Hedges *et al.* 1997: 446). One of the earliest British wetland burials, from intertidal peats in Hartlepool Bay on the east coast of northern England, was an adult male who had two old blows on the skull and a healed rib fracture (Tooley 1978).

Undoubtedly the most significant development concerning injuries from short-range weapons is the programme of re-examination of cranial trauma from

Early Neolithic tombs in southern Britain (with northern Britain and Ireland to follow) (Schulting and Wysocki 2002). Of 350 skulls examined, twenty six (7.4%) are high or medium probability traumatic cases, with another five possibles. Of the twenty six, nine are perimortem injuries with no signs of healing, so the majority are old injuries. This provides an interesting match with the Danish Mesolithic material alluded to above. Mostly there is one healed injury per skull, e.g. Norton Bavant long barrow in Wessex, but an adult male from Fussell's Lodge long barrow had three (Ashbee 1966). A slight majority of injuries are on the left side of the cranium, perhaps suggesting right-handed attackers, as is the case with the Danish Mesolithic material (Bennike 1985). Some injuries appear from their size to have been inflicted with antler tines, or possibly pointed wooden clubs. Of the perimortem injuries, examples include a probable axe blow on an adult (probably female) from Coldrum, a chambered tomb in Southeast England and a possible club blow on an adolescent from Belas Knap chambered tomb in western England. At both sites there were also individuals with healed depressions in the skull: at Coldrum an adult male and at Belas Knap an adult female (Schulting *et al.* n.d.). Overall, both women and men suffered injuries, but women apparently in slightly higher numbers (Schulting *et al.* n.d.).

This re-examination prompts a reconsideration concerning the many claims in the antiquarian literature, previously passed over as over-interpretation (Schulting and Wysocki 2002). According to Thurnam (1869: 185), 'In a large proportion of the long barrows which I have opened, many of the skulls exhumed have been found to be cleft, apparently by a blunt weapon, such as a club or stone axe.' Turning to specific cases, the majority of Thurnam's observations came from Wessex. At the long barrow of Boles' Barrow near Stonehenge, the barrow diggers Colt Hoare and Cunnington encountered a skull in a primary burial location which 'appeared to have been cut in two with a sword' (Colt Hoare 1812: 88). Thurnam reopened the mound in 1864, concluding from his examination that several of the skulls had been hacked apart (Thurnam 1869). He also suggested that the apparent level of violent death was so high that dismemberment of corpses as a funerary ritual was a more likely explanation. A later generation of Cunningtons in 1886 re-excavated the burial deposit once more, noting that 'with one exception,

the blows were inflicted on the *left* side of the cranium' (Cunnington 1889: 107). They also argued that one of the dead had been decapitated. At Tilshead Lodge long barrow one of two individuals had a fractured skull (Thurnam 1869: 186). In the northern chamber at West Kennet chambered tomb the skull of two adult males were claimed to be fractured before death (Thurnam 1869: 227). In western England Thurnam identified one skull as fractured at the Littleton Drew chambered tomb, and at the Rodmarton chambered tomb no less than four of the thirteen skulls, those of men from twenty to fifty years old, were interpreted as showing signs of violence (Thurnam 1869: 227-28). From Belas Knap chambered tomb, Crawford (1925: 67-79) identified three cases of cranial trauma (an adult male of about thirty, another adult and a child) from the western chamber and two cases (both elderly males). At Lanhill chambered long barrow in Wessex an adult male seems to have suffered a crippling injury to the left elbow in childhood, possibly resulting from a penetrating wound (Keiller and Piggott 1938). Although some of these antiquarian observations have not been confirmed by re-examination, for example Tilshead (R. Schulting pers. comm.), they do merit serious consideration.

Moving from burials to settlements, the focus of attention here has been the causewayed enclosures mostly found in southern England, but with examples also now known from Wales, the Isle of Man and Ireland (Oswald *et al.* 2001). The interpretation of these has swung between the poles of ceremonial centres and defended settlements for some seven decades, as general explanations then as interpretations of individual sites. The first archaeological investigation of a causewayed enclosure, by another of the Wessex Cunningtons, produced some confusion from the observation of the interrupted nature of the ditches, broken by frequent causeways (Cunnington 1912: 48):

It is impracticable to regard these breaks in the entrenchment as due to an unfinished undertaking, or as entrances in any ordinary sense, and the only other feasible theory seems to be that they had some distinct purpose in the scheme of defence; that they were, indeed, a strengthening and not a weakening factor in this seemingly not very strongly defended place.

The general notion had long existed that with the adoption of an agricultural economy would come a greater attachment to land and a need to protect people and animals, and these 'Neolithic camps' (Curwen 1930) fitted neatly into such a picture (Oswald *et al.* 2001).

With the results of a further ninety years of survey and excavation, what general observations may we make now? Beginning in southwest Britain, one of the most intensively investigated causewayed enclosures is that at Hambledon Hill in Dorset (Mercer 1980; 1988; 1990: 38-44; 1999). Within this major complex there are a number of separate enclosures. The Stepleton Enclosure has a rather different lithic and ceramic assemblage to the Main Enclosure, which seems to be the ritual heart of the site, where bodies were perhaps exposed and pits filled with exotic offerings to the dead. Mercer also believes that a third enclosure existed, below the Iron Age hillfort. This would have lain in the most naturally defensible location. All three enclosures were themselves enclosed by a series of outworks running for over 3000 m, less causewayed and more rampart-like, held together by timber uprights and cross-beams. This massive undertaking would have required both the clearance of a significant area of forest to produce the 10,000 oak beams required and the mobilisation of a substantial labour force. Gateways through the outworks were lined with massive oak posts. The purpose of this remarkable earthwork system was possibly to be the protection of herds of cattle on which the wealth of the community was built (Mercer 1990: 42-43). It is important, however, to note that the Earlier Neolithic date of the outworks has not been established by excavation (R. Mercer pers. comm.).

Then around 3500 BC this high status site was eventually attacked and destroyed. The two bodies in the ditch along with destruction debris mentioned above seem conclusive evidence of an attack by a force of archers. (However, Saville [2002] does note that in one case the arrowhead tip is missing and that the arrowhead is of unusually fine quality for the site, so it could have been a personal possession intended for reworking). In addition to the two young males in the ditch, another young man was apparently abandoned on the outer edge of the ditch, where his body was dismembered by animals; the partly intact skeleton of an older adult female was found in the upper fill of the old Stepleton enclosure ditch where it had been dragged by dogs or wolves. Of a different character (possibly one of

the attacking force – Mercer 1990: 41) is the burial of another young adult male in a pit with pottery fragments and a quern-stone which was backfilled with scorched chalk rubble for which the only known source is the destroyed rampart. The evidence from Hambledon Hill for warfare seems as clear as a dryland and plough-damaged site could hope to produce. The only element which one might expect to be there which is not is a concentration of arrowheads, but Mercer (1988: 105) suggests that the excavated evidence represents only 'a preliminary skirmish in an altogether more serious encounter', with the proposed enclosure below the Iron Age hillfort as the primary target of the attack. Unfortunately, excavation of the Neolithic levels here would necessitate the removal of the protected Iron Age site, so this possibility remains tantalisingly out of reach.

A related interpretation of events has been proposed for Crickley Hill in Gloucestershire (Dixon 1988). Here a causewayed enclosure with a double line of banks and ditches may have been burnt down, with buildings inside the enclosure destroyed and burnt material thrown into the ditch. Following this possible catastrophe, a single line of enclosing bank and ditch was constructed just beyond the line of the earlier enclosure. This was a more substantial rampart, with just two or three entrances and a palisade at the back of the rampart. Then the palisade was burnt and some 400 arrowheads were found around the entrances, in and around the stockade and along the fenced route leading in to the site. Following this the site was abandoned for settlement. However, some of the arrowheads may well relate to an attack on the earlier enclosure, so this figure could be an overestimate.

A similar defensive work may have been present at the Hembury enclosure (Liddell 1930; 1931; 1932; 1935; Todd 1984), where the inner ditch contains extensive traces of in-situ burning of the timber palisade and stones. Liddell's excavations recovered over 120 arrowheads, most of them burnt and broken, many of them located in the vicinity of a burnt gateway in the line of the inner enclosure. Immediately inside the enclosure behind the gateway was a substantial building, also burnt down apparently in the same episode of destruction.

We must also consider here the related class of monuments known as 'tor enclosures' (Silvester 1979). They are defined as sites with irregular stone banks with narrow entrances at fairly frequent intervals surrounding tors (natural granite outcrops); where they mostly differ from causewayed enclosures is in lacking a ditch, although this would have been extremely difficult to cut out of the rock. Of the fourteen certain or possible examples of this group (Oswald *et al.* 2001) only two have been excavated and thus confirmed as Early Neolithic in date – Carn Brea (Mercer 1981) and Helman Tor (Mercer 1997), both in Cornwall. At Carn Brea the hilltop enclosure was surrounded by a massive wall originally well over two metres high and some two metres wide at the base, with further walls outside this. For part of the circuit there was also a causewayed ditch outside the wall, which was not a quarry for the stone to construct the wall. There are over 750 arrowheads from Mercer's excavations, some found among the stones of the enclosure wall, a record for any Early Neolithic British site in both absolute number and as a proportion of the retouched lithic assemblage (35.7%). Several hundred more were found in antiquarian diggings and as surface finds. There was widespread burning in the buildings on the eastern summit of the tor and many of the arrowheads were broken and burnt. Mercer sees Carn Brea as a high status settlement, which was eventually attacked and destroyed, possibly by the inhabitants of the hilltop enclosure of Helman Tor some 40 kms away.

Helman Tor is a somewhat smaller enclosure, but again defined by a massive boulder wall. Limited excavation within this revealed substantial midden deposits and a possible layer of burning, again with a high proportion of arrowheads among the lithic assemblage, but not in such dramatic numbers as at Carn Brea.

The Maiden Castle causewayed enclosure underlying the famous hillfort (Wheeler 1943; Sharples 1991) also produced several broken arrowheads leading to the recent excavator (Sharples 1991: 255) entertaining the possibility of violent attack there too. However, as he himself notes, the area excavated is too small for such a pattern to emerge, and certainly the proportion of arrowheads in the retouched lithic assemblage is nowhere near that from Carn Brea.

In Southeast England the causewayed enclosures of Sussex have been divided into two groups – one of fortified settlement enclosures and another of exposure burial (Drewett 1977) or unfortified ceremonial/ritual enclosures (Drewett *et al.* 1988: 34-44; Drewett 1994). The fortified enclosures (Whitehawk

and The Trundle) are located on hill tops, have views in all directions, have multiple ditch circuits (two at the Trundle, at least four and possibly six at Whitehawk) have some evidence for internal and defensive features, have environmental evidence suggesting mixed farming in the locality, and flintwork suggesting a wide range of tools and craft activities. However, apart from the location, the evidence is not especially clear-cut. Whitehawk has one possible internal structure and a number of pits; and The Trundle just a single post-hole, which may be part of an entrance structure. Whitehawk has a possible gateway and palisades behind the third and fourth (counting outwards) ditches. Craft activities are not confined to Whitehawk and The Trundle (Russell 2002: 87), while ritual deposits in the form of human and animal burials occur at both these sites. Saville (2002) has also questioned the validity of drawing such a sharp distinction between the lithic assemblages from the various enclosures.

In Essex, in eastern England, the Orsett enclosure (Hedges and Buckley 1978) has strong similarities to Whitehawk, with three ditch circuits and a palisade line within the middle ditch. There may have been a single bank between the two outer ditches created from the soil from both ditches, suggesting that they were contemporary. However, aerial photographs indicate that the ditch circuits were not complete, negating the defensive value of the substantial bank and palisade.

Moving to the Thames Valley, the Abingdon enclosure (Avery 1982) has two ditch circuits, the outer succeeding the inner. Bradley (1986) has argued that the outer ditch had an accompanying palisade and that it is not clear that the ditch was broken by regular causeways, so that this phase of the 'whole earthwork seems to assume defensible proportions' (1986: 183-84). However, there is little trace of settlement activity in the area newly enclosed by the larger ditch circuit, so a defended settlement seems unlikely. Unfortunately, the extent of gravel extraction here is such that a resolution of the question is now impossible.

Turning to the north and west, a concentration of complete arrowheads (some 130, mostly from the enclosure ditch) indicates the possibility of an attack on the large D-shaped Billown enclosure on the Isle of Man (Darvill 2001). Conflict has also been proposed for the only definite causewayed enclosure in Ireland, at Donegore Hill in Country Antrim (Mallory and Hartwell 1984). Located on a hilltop with a double line of ditch and palisade circuits surrounding a ploughed area from which some 45,000 sherds from 1,500 vessels of Early Neolithic pottery have come (Sheridan 2001), the site certainly has claims for being a defended settlement. However, plough damage made the recovery of evidence for internal structures very difficult, the flint assemblage contains rather few arrowheads, which are mostly scattered across the site (Nelis 2003), the palisade appears to be rather slight and lacking traces of burning and the ditches are wide and flat. All these factors suggest that Donegore was probably not a defended site, contrary to early thoughts (e.g. Sheridan 1991).

The palisade enclosures at Knowth, C. Meath (Eogan and Roche 1997), predate the famous Middle Neolithic passage graves. Inside there are a number of pits but no clear settlement traces. The lack of evidence for destruction and of associated arrowheads and the slight nature of the palisade again suggest a non-defensive enclosure.

From preliminary reports (e.g. Logue 2003) the Thornhill, Co. Londonderry, site has a palisade line with evidence of rebuilding after destruction by fire, with seven burnt and unburnt leaf-shaped arrowheads (of a total of 21 from the excavations as a whole) associated with the destruction phase. Inside the enclosure, which was rebuilt and seems to have grown over time, were several rectangular houses and outbuildings associated with Earlier Neolithic pottery. Logue (2003) has suggested that the palisade enclosure was located so as to secure a prominent position in exchange networks of flint and stone.

Also in Ireland, the Ballynagilly, Co. Tyrone, long house has been argued to be a scene of conflict, with the house being burnt down and six leaf-shaped arrowheads in and around the house but only three discovered elsewhere in the extensive excavations (ApSimon 1971). This is certainly possible, but the precise stratigraphic relationship between the arrowheads and the house would need to be established. There appears to be a similar case at Islandmagee, Co. Antrim (Moore 2003), where the rectangular House 1 was burnt down c. 3800 BC, an action associated with some fifteen projectile points.

From this brief summary it should be clear that there are plausible defended, and sometimes attacked, Early Neolithic enclosures from across south-western Britain and Ireland. However, this does not mean that I agree with Lawrence Keeley's recent criticism of British archaeologists for attempting to 'pacify the past' (1996: 18). Ironically, he contrasts Alasdair Whittle with 'the archaeologists who have conducted extensive excavations of some of these enclosures', when Whittle has actually carried out a major excavation at the most famous of all causewayed enclosures – Windmill Hill (Whittle *et al.* 1999). The ditches at Windmill Hill contain not only deliberately dumped mounds of feasting debris and fragments of items which arrived there through long distance exchange, but also burials, such as that of a child. In the case of Windmill Hill, a wide range of activities was carried out at this major site, but not settlement and not defence. There are also sites in eastern England which have been intensively investigated but present no evidence for defence or attack and little of possible settlement – Briar Hill (Bamford 1985), Etton (Pryor 1998) and Haddenham (Evans 1988; Hodder 1992), along with those Sussex examples identified by Drewett (1994) as non-defensive. Saville (2002) notes that in terms of numbers of arrowheads only Carn Brea, Crickley Hill and Hembury stand out, with other enclosures having a consistently minor presence of the type. So on this specific criterion there is very limited evidence to suggest a consistent pattern of attacks on enclosures.

Roger Mercer himself, one of the excavators cited by Keeley, has recently stated clearly that he 'does not argue that all Neolithic enclosures were defensive – many were clearly not, and indeed components of the site at Hambledon Hill were not' (1999: 156). We must not forget that sites such as Hambledon Hill and Crickley Hill began life as ceremonial enclosures, but only later in their history became defensive enclosures, reducing the number of entrances and constructing banks and palisades. As Mercer stresses, the ritual element is central to any understanding of causewayed enclosures. At Hambledon Hill the Main Enclosure, with skulls in the ditch along with the remains of feasting debris, possibly funerary feasts, continued in use in tandem with the Stepleton Enclosure.

The ritual authority residing in such centres may have been crucial to their transformation into defensive settlements, given that they existed on the edge of the settled landscape, rather than at its centre, where one would expect the pressure on resources to be at its greatest (Thorpe 1996: 177). The other major factor may have been the potential control over exchange processes which permanent occupation of an enclosure could bring, were exotic items to become a source of social power. Such developments involved only a minority of sites, in western Britain (Oswald *et al.* 2001: 128-29) in a horizon around 3500 BC, and where we have dating evidence for the individual arrowhead deaths, these occur in the same period. The broader context is the emergence of regional pottery styles, and greater sedentism in the form of the first fields and large houses, perhaps resulting in attempts at political centralisation, although they did not succeed. It is likely that this evidence for large scale conflict relates to the more general pattern of the emergence of the individual within society. Towards the end of the Earlier Neolithic individual burials take place under long barrows in Wessex and western England (a class of monument which had previously covered disarticulated and tangled masses of bones [Thorpe 1984]), in the north earthen round barrows with individual interments appear, as at Duggleby Howe in Yorkshire (Loveday 2002), and in Ireland the Linkardstown cist burials develop (Cooney 2000: 97-99). These share features of the burial of recognisable individual skeletons, mostly of single males where the skeletal material has been studied, sometimes placed in separate graves, and with the provision of definite grave goods. However, we must not forget that some of the individual cases of trauma, especially the skull injuries, come from the communal monuments dating to the very beginning of the Neolithic, such as Fussell's Lodge. At least small scale conflict was present, therefore, from the origins of agriculture in Britain, and probably before, so a simple materialist model of conflict appearing along with pressure on land does not work.

Later Neolithic

With this evidence from the latter part of the Early Neolithic one might expect the Late Neolithic to be an even more intense period of conflict. Yet the main successor monument type to causewayed enclosures, henges, have long been accepted as non-defensive (Wainwright 1989), because of the existence of an external bank allowing an attacking force to fire down on the defenders. (Burl 1987: 129 is an extremely rare exception to this unanimity, and he does not address the issue of the external bank.) Indeed, there is remarkably little evidence for conflict of any kind

from the early part of the Later Neolithic (from c. 3300-2700 BC). There may be two factors contributing to this: the burial record is generally sparse, especially in terms of complete bodies in protected locations such as below the mounds of round barrows, and a change in the form of arrowheads, from leaf-shaped to the asymmetrical petit-tranchet form (Edmonds and Thomas 1987). Such arrowheads would be more suitable as a short-range weapon and would tend to inflict wide bleeding wounds. However, Edmonds and Thomas note that such arrowheads can penetrate bone, so a complete absence of arrowhead injuries should not be expected.

But after this possible period of peace, the general assumption has been that a period of conflict arose in the later part of the Later Neolithic (after 2700 BC) with the appearance of Beakers. Certainly this was the traditional view of an invading group of Beaker warriors arriving from the continent. Wessex archaeologist Charles Warne saw the Beaker invaders of Britain as wielding 'the gory battle-axe, reeking and fresh from deadly strife' (1866: 63), while for Lord Abercromby (1912: 64) the Beaker Folk 'presented an appearance of great ferocity and brutality', and indeed their faces betrayed to his keen eye clear criminal features. Even trepanned skulls were interpreted as gruesome drinking vessels made from victims – these are now thought to be the result of medical operations (James and Thorpe 1994: 24-33), although these may in some cases result from attempts to relieve problems produced by skull fractures arising from conflicts.

The reasons behind this lie, of course, in the Beaker-associated set of equipment including metal daggers and copies in flint, battle-axes, barbed and tanged arrowheads and stone wristguards plausibly symbolising archery and warring (Replogle 1980), and argued to be associated with males (Clarke 1970: 264-65) in warrior graves (Ashbee 1978: ch. 7).

Both the association of archery with males and the pan-European nature of the Beaker phenomenon have recently been confirmed by the burial from Boscombe Down near Stonehenge (Fitzpatrick 2002): this adult male (aged 35-45) was buried with five Beakers, gold hair tresses, three copper knives, a slate wristguard on his left forearm, and another wristguard by his knee, and a cache of flints (possibly in a bag); in the grave fill were fifteen barbed-and-tanged arrowheads. Isotopic analysis of his tooth enamel points to a birthplace in central Europe.

The invasionist hypothesis has largely been abandoned today (although some element of population movement is plausible [Brodie 1994], as the Boscombe Down burial suggests), but the impression nevertheless remains of the Late Neolithic as a period of conflict, due to the warfare symbolism of the artefact repertoire.

But what is the actual evidence for this? In contrast to the Early Neolithic, only two plausibly defensible Late Neolithic enclosures have been claimed. One is at Mount Pleasant in Wessex, where around 2200 BC a substantial timber palisade some 800 m long was constructed to cut off an area of over four hectares with just two narrow entrances established by excavation. The palisade, which stood up to six metres high, was partially destroyed by fire, partially dismantled and partially left to rot in situ. Wainwright (1979: ch. 4 and 237-45) felt that the palisade was defensive, but the only potentially contemporary activities related to the enclosure are an internal stone setting (a three-sided cove) and some pottery, flintwork and animal bone in ditch deposits. Although Wainwright interpreted the ditch material as 'domestic refuse' (1979: 241) which arrived in the ditches after the demolition but derived from settlement activity contemporary with the enclosure, reconsideration of the finds suggests that this is the result of deliberate deposition not the dumping of rubbish (Thorpe 1989: 322-25; Thomas 1996: 214-22). The siting of the enclosure on a spur with a steep fall in the ground on one side shows the natural contours of the site could be considered to be taken into the account for defensive purposes. However, the relatively low-lying setting of the enclosures still makes this an unlikely location for a defended settlement, with far more suitable locations such as Maiden Castle (with an earlier causewayed enclosure) in the vicinity.

In Scotland, at Meldon Bridge there was a rather earlier palisade enclosure built around 3000 BC. The palisade here cuts off a large area (c. seven hectares) between two rivers and is again an impressive construction. This was at first thought to be a defended settlement (Burgess 1976), but it is now clear that the settlement debris found inside the enclosure is not contemporary with the palisade. It is now thought that although the perimeter palisade had a defensive potential the site is best seen as one of a group of enclosures with timber avenues which are believed to be primarily ceremonial in nature (Speak and Burgess 1999).

These reinterpretations of the Mount Pleasant and Meldon Bridge palisade enclosures come against the background of the later discovery and excavation of other palisade enclosures with little claim to a defensive purpose (Gibson 2002). These include Greyhound Yard, Dorchester (Woodward et al. 1993), only some two kilometres from Mount Pleasant, Hindwell in Wales (Gibson 1999) and two enclosures at West Kennet in the river valley below the chambered tomb (Whittle 1997).

Unfortunately for the theory of blood-soaked 'Beaker Folk' examination of copper and early bronze daggers has revealed no traces of violent conflict. Many battle-axes are elaborate objects, sometimes decorated, and frequently have blunted edges. This need not, of course, mean that battle-axes were never used as weapons, for there are cultures, such as the Maori, in which famous weapons became highly decorated, but almost no skeletal evidence exists in the form of battle-axe wounds from Britain. Similarly, examination of copper daggers failed to provide traces of their use in combat (Wall 1987). Given the background of nineteenth century exaggeration it is not surprising that some archaeologists have concluded that the Beaker period was peaceful.

What we do see in the Late Neolithic are individuals who appear to have suffered traumatic injuries. At Smeeton Westerby, central England, a disturbed adolescent burial from a Beaker barrow has a possible injury to a chest vertebra (Clay and Stirland 1981). At Chilbolton, Wessex (Russell 1990) the forearm of a young adult male accompanied by a Beaker (radiocarbon dated to c. 2200 BC) had a parry fracture. The Liffs Low, northern England, stone cairn contained a young adult male with a healed fracture of the upper arm; this burial appears to have been associated with a Beaker (Barnatt 1996). In Southeast England, the primary burial in the Pyecombe barrow (Butler 1991) was a mature adult male with healed fractures to the forearm and collarbone, associated with a Beaker, bronze dagger and stone wristguard. A number of possible traumatic injuries are recorded for the Barnack barrow, eastern England (Donaldson 1977). These are a probable older adult male with two depressions on the skull which could result from blows with a hammer; an adult male with a healed linear wound on the cranium, and an older adult male with a healed fracture of the forearm.

More significant are those individuals who seem to have died violent deaths. At the Oxford University

Parks Science Area in the Thames Valley a round barrow ditch enclosed four inhumation burials in graves (Boston *et al.* n.d.). The earliest of these burials (radiocarbon dated to c. 2200 BC) was a woman aged over fifty, who had suffered a blow from an axe to the back of the head, which had begun to heal before death. She was accompanied by a late-style Beaker and a used flint toolkit. Although an unlikely warrior this woman was buried with respect – being the primary burial and the only one with a Beaker.

At Stonehenge (Evans 1984) the burial of a young man, aged twenty five to thirty, was found in the ditch, accompanied by a stone wristguard and three barbed and tanged arrowheads; the tip of one of these was embedded in one of the ribs, while the tip of a fourth arrowhead was found in the sternum, probably having passed through the heart, and another rib has a groove cut in it which was also interpreted as the result of an arrow injury. All the arrow shots appear to have been from close range. Does this represent a ritual killing (as argued by Gibson 1994), the execution of a criminal, or alternatively a victim of conflict? Certainly the notion of an execution seems highly implausible given the setting – someone who had broken social rules is unlikely to have been seen as a suitable person to bury in such prestigious surroundings.

Also in Wessex there is the burial at Barrow Hills (Barclay and Halpin 1999). A young adult male, aged twenty to thirty, was found with variety of artefacts including a Beaker, a bronze awl and a set of five fine barbed and tanged arrowheads. When the skeleton was lifted another arrowhead was discovered lying against the spine – this crude example has an impact fracture at the tip and both barbs are broken and was probably the cause of death.

In central Wales, excavations at the Sarn-y-Bryn Caled timber circle (Gibson 1994) produced the cremation of a young adult buried in a pit at the centre of the circle, which dates to the very end of the Late Neolithic or the start of the Earlier Bronze Age. (The radiocarbon date on charcoal is earlier than the typological associations of the arrowheads). The burial was accompanied by four fine barbed-and-tanged arrowheads discovered amongst the mass of bone. The arrowheads were burnt, but they had not disintegrated completely, so Gibson concludes that they must have been protected from the hottest flames. Moreover, they had clearly been used, and two had lost their tips as a result of breakage on

impact. Gibson thus argues that the arrowheads were in the body during the cremation, thus preserving them. Moreover, he notes the fine quality of the arrowheads, which contrast with the death-dealing example at Barrow Hills, and suggests that the Sarn-y-Bryn Caled death was a sacrifice.

Similar finds are identified by Green (in Gibson 1994) from the cairn of Twr Gwyn Mawr in Wales (Davies 1857) and Grandtully in Scotland (Simpson and Coles 1990). At Twr Gwyn Mawr one of the two barbed-and-tanged arrowheads found with a cremation appears to have an impact fracture at the tip according to the published illustration – unfortunately the finds are lost. At Grandtully the cremation of an adult, possibly female, was found with five finely made barbed-and-tanged arrowheads which were burnt, but had not completely disintegrated. Again Gibson suggests they may have been in the body during the cremation. However, if they were in a body cavity, this would become the hottest part of the body during the cremation process due to the body fats (McKinley 1989), so we may instead be dealing with chance survival as a result of the variability of wind, consistency of the pyre heat (lower towards the edges) or material falling down through the pyre and being collected after the cremation was over (S. Leach pers. comm.).

A further possible example comes from the Fordington Farm barrow in Wessex (Bellamy 1991). A young adult male with a direct radiocarbon date of c. 2350 BC was found in a grave with a complete barbed-and-tanged arrowhead in the stomach; the young man had a healed forearm fracture. However, the particular form of barbed-and-tanged arrowhead is found elsewhere in association with Earlier Bronze Age artefact types, and the grave containing the burial cut through the mound from the first phase of burials at the site, covering burials with direct radiocarbon dates of c. 2100 BC but no grave goods.

Rather than interpreting these discoveries in terms of sacrifice I would argue instead that we may see here the remains of those killed in the course of small scale conflicts, whose bravery was then recognised by a prestigious burial. This interpretation may be strengthened by a discovery made during the re-analysis of the Beaker burials from the small flat grave cemetery at Staxton in Yorkshire, northern England (Stead 1959; S. Leach pers. comm.). Burial 11, an adult male found with a well-made Beaker, had a major weapon injury to the left shoulder, a (battle-axe?) blade fracturing the clavicle and embedding itself into the scapula (S. Leach pers. comm.). This is the most likely cause of death. A healed fracture on the side of the left hand may possibly represent a defensive injury resulting from an earlier episode of violence.

It is crucial to note that these injuries were not noted in the original publication and it is therefore possible that further evidence of traumatic injury may come to light. In particular, several of the cave burials of northern England appear to have suffered sudden deaths. However, direct radiocarbon dates on a number of these from Yorkshire have produced Roman period dates (S. Leach pers. comm.), even for examples apparently associated with Beaker pottery, so caution is necessary in assigning a date to these without evidence from radiocarbon dates.

While the Beaker grave-goods of daggers and battle-axes symbolised warfare, the lack of clear evidence for conflict, particularly given the substantial number of surviving skeletons from the period, makes it more likely that raiding not warfare was typical of Beaker period Britain. This would perhaps result in relatively low levels of casualties, but might have been a vital source of status acquisition. It may be that the European-scale Beaker network of prestige goods competition saw a period of suppression of warfare proper (R. Mercer pers. comm.).

Earlier Bronze Age

Turning to the Earlier Bronze Age, after 2000 BC, it is important to note the differences between Britain and continental Europe. Daggers remain the dominant form in burial (Gerloff 1975). Although halberds and spearheads do appear, they are quite rare, except for halberds in Ireland (Waddell 2000: 129-31). Although halberds are undoubtedly an unwieldy looking weapon, frequent use damage to the back of the hafting plate suggests that they may have been used in a similar way to medieval pole arms, that is mostly using the wooden staff and only striking with the metal head to deliver a *coup de grace* (O'Flaherty *et al.* 2002). Where reliable skeletal reports exist, daggers are associated with males, and have thus long been seen as warrior equipment, following Beaker traditions. As with the Later Neolithic copper daggers, however, there are few traces of combat on the daggers themselves (Wall 1987), except for some examples from the River Thames

(York 2002). Some daggers also seem inappropriate as weapons, as they are too small, have highly polished and unworn blades, have very wide blades, or rounded tips (Gerloff 1975: 46 and 55). Swords or rapiers are rare.

Moreover, we also have a considerable skeletal record from the period, and this is relatively silent when it comes to the victims of conflict. In terms of individual episodes of combat these seem to be relatively few, as other surveys of the evidence have suggested (Osgood 1998: 19).

A similar case of death through arrowhead injury to those noted from the Later Neolithic may occur at Ballymacaldrack, Co. Antrim (Tomb and Davies 1938). Here the cremation of an adult, possibly female, in an Early Bronze Age Collared Urn was accompanied by a rough barbed-and-tanged arrowhead with a broken tip.

In western England, the Court Hill round barrow covered the primary burial of a young adult male with his left upper arm chopped through, probably the cause of death (Grinsell 1971: 120; Bristow 1998: Vol. II, 72). The barrow at Withington, northwest England (Wilson 1981), contained as a primary burial a cremated young adult female (radiocarbon dated to c. 1700 BC) who had a head injury in the process of healing. At Cnip, Isle of Lewis, Scotland (Dunwell *et al.* 1995), an older adult male (also dated to c. 1700 BC) buried with an undecorated pot had extensive but healed facial trauma.

A prehistoric bog burial which probably dates to the Earlier Bronze Age was found at Pilling in northwest England in 1864. A decapitated female skull was discovered wrapped in cloth, together with two strings of jet beads, one with a large amber bead at the centre (Edwards 1969). A probable dryland decapitation burial directly dated by radiocarbon to the beginning of the Earlier Bronze Age has been discovered at the foot of the Gog Magog Hills just outside Cambridge in eastern England (Hinman 2001). Following a possible decapitation, the remainder of an adult male was buried in a pit which was later reopened to remove further portions of the body. In neither of these cases, however, is there any particular reason to suggest that the decapitation took place as an act of war. We may instead be looking at the result of the execution of socially defined outcasts such as witches and their subsequent violent treatment to prevent their return. Certainly, the large sample of Earlier Bronze Age bog bodies from the Wissey Embayment area of eastern England contains no cases of traumatic injury (Healy 1996), except for a possible case of an older adult (probably male) missing some teeth.

A skeleton of an adult male dating to the Earlier Bronze Age (accompanied by a food vessel and a battle-axe) from Callis Wold 23 barrow in Yorkshire (Mortimer 1905: 153-56) had received an extensive wound to the left wrist, causing the hand bones to fuse with those of the arm (Brothwell 1959-60). Another possible case is that of an older adult male from the Tallington round barrow in eastern England (Simpson 1976) with a possibly injured upper arm. From the very limited evidence available, two round barrows in Amesbury near Stonehenge have produced cases of Earlier Bronze Age victims of conflict (Thomas 1956; Ashbee 1960: 79). Under one round barrow was an unaccompanied skeleton with the skull and mandible removed, the right arm missing and the left severed by a cut through the forearm. A burial below a nearby disc barrow 'also showed signs of such treatment', although whether this is dismemberment or violence is not made clear. The dangers of relying on old and incompletely published accounts of traumatic injuries are clear from the case of Stillorgan in Ireland, where a 1955 excavation uncovered a female skeleton in a stone cist with an oyster shell and a flint flake; a blow to the head was stated to be the cause of death (noted by Waddell 1990: 86). However, a recent re-examination of the skeleton suggests that this is post-mortem damage (B. Molloy pers. comm.). The recently rediscovered and extensively injured Sonna Demesne man (Sikora and Buckley 2003) was also not, as believed, a Bronze Age victim but a medieval one as revealed by radiocarbon dating (B. Molloy pers. comm.).

The disputed Sutton Veny bell barrow 'warrior' (Johnston 1980; Osgood 1999) from Wessex may also belong here. The barrow covered a central burial in a wooden coffin with Earlier Bronze Age pottery and a bronze dagger. On the edge of the mound, apparently below the surviving remnants of the mound, was a grave containing the burial of an adult aged twenty-four to twenty-eight years, accompanied only by a shark tooth, 'the victim of a particularly violent head-wound, probably from a sword' (Johnston 1980: 38). Because of this interpretation of the traumatic injury, the burial was assumed to date to the Later Bronze Age, when swords were in use, and thus the burial had to be a subsequent burial

(after the mound was constructed) in a grave which was then carefully backfilled and packed down with chalk. Recently, Osgood (1999) has suggested that the absence of Later Bronze Age burial in Britain makes this more likely to be an Anglo-Saxon secondary burial – a well established type. However, the illustrations of the wounds (unfortunately the skull itself has been mislaid) are also consistent with an interpretation of blunt force trauma (S. Leach pers. comm.), especially as the fracture does not cross sutures, so this could indeed be an Earlier Bronze Age violent death.

Although the relative paucity of the artefactual and skeletal record for conflict in the Earlier Bronze Age is clear, one indirect piece of evidence does point in the direction of warfare. The Irish wooden shield mould from Kilmahamogue used for making V-notched leather shields dates (from direct radiocarbon dating – Hedges *et al.* 1991) to around 1900-1600 BC. This is the earliest known so far of the group, but few of the others have been directly dated. As Osgood notes (1998: 10), this would preclude any idea of dagger combat. Perhaps the concentration of swords noted by Harding in Ireland (2000: 279) may have a beginning in the Early Bronze Age, although shields do not automatically imply the presence of swords.

Although the beauty of the male warrior (Treherne 1995) seems to be present in the Earlier Bronze Age of Britain, we can see few archaeological traces of their actions. Settlements in general are difficult to locate, so we may be dealing with a fairly mobile population, therefore static defences such as enclosures may have been inappropriate, and they are indeed entirely absent.

Overall, the Later Neolithic and Earlier Bronze Age record of conflict is remarkably thin. This raises the question of how long the 'warrior aristocracy' commonly proposed for Late Neolithic-Early Bronze Age Britain could have sustained itself without warring. Interestingly, Robb (1997) has noted a similar pattern for Italy, in which the frequency of cranial trauma is highest in the Early and Middle Neolithic but declines in the Late Neolithic even though weapons made of imported metal appear and weaponry becomes a central theme in rock art. Perhaps the pan-European prestige good networks (the most obvious of which are the movement of metal and amber) acted to suppress warfare, although this appears not to be the case elsewhere, e.g. in Norway (Fyllingen this volume).

Later Bronze Age

Certainly when it comes to the Middle and Late Bronze Age, after c. 1400 BC, together making up the Later Bronze Age period, there is far more evidence of conflict, although most is circumstantial. British and Irish dirks and rapiers (Burgess and Gerloff 1981) are longer and thus have more serious potential as weapons than the Earlier Bronze Age daggers from which they developed, with dirks for stabbing and rapiers for thrusting. However,

the extreme length, narrowness and general fragility of many rapier blades, especially of the longer and finer ... examples ... combined with the inherent weakness of the butt attachment method, leaves little doubt that such weapons could not have been successfully used in combat. (Burgess and Gerloff 1981: 5)

Despite this, York (2002) notes that 86% of the dirks and rapiers from the River Thames show signs of use in the form of edge damage, while torn rivet-holes suggest that rapiers were used as slashing weapons.

At this time we see the appearance in far greater numbers of shields, swords, and spearheads. Examination of shields and the far more common swords and spearheads shows clear traces of damage – piercings by spearheads in the case of the shields, edge notching in the case of the swords, and broken tips in the case of the spearheads.

The damage to shields has been considered by Osgood (1998: 8-11), who proposes that it may mostly be ritual in nature ('killing' the now sacred object) rather than the traces of their use in conflicts. The shields from Beith in Scotland (stabbed by a blade), Long Wittenham in the Thames Valley (stabbed by a spear) and one from Country Antrim in Ireland (slashed with a sword) could all result from combat incidents, but the recent discovery from South Cadbury in Southwest England (Coles *et al.* 1999) does not fit this scenario. Three holes were punched through the shield, probably using a sharpened stake, after it was placed in the top of an ancient enclosure ditch. This strongly favours ritual rather than combat violence, although as Osgood notes (1998:9) the specific form taken by one may have been influenced by the other.

Moreover, it seems clear that the most common British shield type ('Yetholm type') is too thin to resist a determined blow (Coles 1962; Coles *et al.* 1999). There are only two examples of the thicker 'Nipperweise' type known from Britain (Needham 1979).

The edge notching on swords is argued by Bridgford (1997; 2000) to be the result of direct impact on their edges and is most likely to have occurred during the use of swords as weapons. She distinguishes these edge-damaged from heavily damaged and hacked swords, which are interpreted as having been deliberately destroyed. Some clear regional patterns emerge from her re-examination of British and Irish swords. Over 90% of the Irish swords exhibited this edge notching (Bridgford 1997: 106), and about 75% of Scottish examples (Bridgford 2000: Table 4.2.4), while in Southeast England the figure falls to around a half (Bridgford 2000: Table 4.2.4). However, this is exaggerated by including small fragments of sword blades from 'scrap hoards' in the figures for Southeast England (S. Bridgford pers. comm.), and York's specific study of bronzes from the River Thames (2002) has a figure of 84% used swords. Combining this with the unusually large number of swords in Ireland by comparison with other areas of Europe (Harding 2000: 279) it might seem that Ireland was the scene of constant conflict. Against this, the different recording methods used in Bridgford 1997 and 2000 mean that the Irish figure given above is an over-estimate and the true figure may be close to 75% (S. Bridgford pers. comm.).

However, both Osgood and Harding have questioned the interpretation of the notching. Osgood (1998: 13) suggests that misuse of the swords could have caused notching, and points to the English Civil War of the seventeenth century when officers feared that common soldiers would break any swords issued to them by chopping up firewood. Harding suggests that Bronze Age swords would have been ineffective in a slashing role, producing only bruising, and were much more suited to stabbing (1999a: 166). Both stress that we are not dealing with medieval-style hand-to-hand combat or modern fencing, with the sword used as much to parry the opponent's blows as to land them oneself, as the shortness of the Later Bronze Age swords would mean that warriors were very close to each other (Harding 1999a: 166). Osgood is right to urge a degree of caution, but the level of edge damage seems excessive for misuse, while the social importance of swords seen in later periods (with swords being named and magical powers attributed to them) makes it likely that they were highly valued in the Bronze Age too (Kristiansen 2002). As Bridgford (2002: 127) notes, producing the sharp hardened edges of swords 'required time, knowledge and consummate skill'. This would be a clear contrast to the utilitarian swords of the seventeenth century and make misuse far less plausible. She also argues (S. Bridgford pers. comm.) that the nature of the notching on Later Bronze Age swords is unlike known cases of misuse.

Against Harding's point, Bridgford (1997) notes that the early Irish Ballintober type swords had a weakness in the attachment of the hilt to the tang which would result in breakage if used as a slashing weapon. Many of the Ballintober type swords have torn rivet holes suggesting that there was a problem which was recognised. This was then solved by redesigning later types. Moreover, Kristiansen (2002) notes that later swords have a wider blade and a balance point further down the blade, more suitable for a slashing weapon.

Spearheads of the Later Bronze Age have generally been divided into lighter throwing spears and heavier thrusting spears. The development of hollow-cast spearheads may have been to lighten them so larger examples could be thrown (Bridgford 2000: 203). Both York (2002) and Bridgford (2000) have identified large numbers of spearheads with damage to the tip. Bridgford's more wide-ranging survey identified both typological and geographical variations in the frequency of damage, with more elaborate and smaller spearheads and those from Southeast England being less commonly damaged. There has as yet been no challenge to the view that tip damage to spearheads is the result of combat, except that spearheads are of course known at a later date as hunting weapons.

The broad-bladed spearhead with pegs in the socket found at North Ferriby in north east England could have been used like a harpoon, with the spearhead breaking off and encumbering the shield-bearer by becoming stuck in the shield (Bartlett and Hawkes 1965), in a similar fashion to the Roman *pilum*. As Osgood (1998: 15) notes, this is only a single example and further examples need to be found to consider this a significant part of Later Bronze Age warfare.

Despite the general lack of formal burials from the British Later Bronze Age (Brück 1995) there are a number of cases which point to the existence of conflict. Early in the nineteenth century a rare socketed bronze dagger of Later Bronze Age date was discovered in Drumman More Lake, Co. Armagh,

Ireland, embedded in the skull of what was apparently a complete inhumation burial (Waddell 1984). An equally old account is provided by Colt Hoare (1812: 181-82) of a cremation in a large apparently plain urn (which 'fell to pieces the following day') with an unburnt jaw and femur above this. The jaw 'had evident marks of a contusion' and Colt Hoare believed that this was saved from the cremation pyre to mark the cause of death. Although Burl (1987: 197) regards this as an Earlier Bronze Age burial, the urn sounds as though it might be Later Bronze Age, while Grinsell (1957: 167) argues that the unburnt bones were intrusive and thus unconnected to the urn. Since the bones have disappeared along with the urn, this has to remain an open case.

The burial deposits in The Sculptor's Cave in Scotland (Benton 1931) may belong here, as several decapitated individuals (and some evidence that skulls were suspended from the walls or ceiling of the cave – Brück 1995: 276) were found at the site, as well as Bronze Age objects. However, there was also late Roman (and Viking) material in an upper layer from which most of the bones were recovered. It is not clear from the publication whether any of the decapitated individuals were found in the secure Bronze Age layer, and it is now clear that many victims of violent death are found in British caves in the Roman period (S. Leach pers. comm.). A more reliable case comes from Antofts Windypit in Yorkshire, where a possible female of middle to old age was killed by a sharp blade injury to the head (S. Leach pers. comm.); there is a direct radiocarbon date of c. 1350 BC for the skull.

The decapitated skull of an adult male from the burnt mound on the bank of the River Soar, Leicestershire (Beamish and Ripper 2000), is better dated (by radiocarbon to c. 900 BC), but it can not be determined whether he was a victim of conflict.

At Dorchester-on-Thames (Knight *et al.* 1972) a spearhead broke off in the victim's pelvis as it was being pulled out, suggesting the use of great force. The date is around 1100 BC (Osgood 1998: 21). The most direct evidence of mass violent death comes from Tormarton in the west of England (Knight *et al.* 1972), where two young adult males had been killed from behind in a spear attack. One had fragments of spearpoints in the vertebrae and the pelvis and had also suffered a blow to the head. The other had a spear wound in the pelvis. The radiocarbon date for this event is around 1400 BC, at the beginning of the Later Bronze Age. Recent re-excavation of the site (Osgood this volume) demonstrated that the bodies (along with three others without visible wounds) had been thrown into a ditch which was then swiftly backfilled.

These cases not only demonstrate that spearheads were definitely used to kill people, they also provide a link with one of the major developments of Later Bronze Age Britain. The Tormarton ditch was a linear ditch, one of many dividing up the landscape in the Later Bronze Age. What may be an analogous case linking ditches and violence, albeit indirectly, comes from Middle Farm in Wessex (Smith *et al.* 1997: 75-79, 157). In the fill of a linear ditch containing Middle Bronze Age pottery were the skeletons of four adults (two males, a possible female and one unsexed). Of these, an adult male, with a direct radiocarbon date of c. 1400 BC, had a healed fracture of the forearm.

The long-distance boundaries and field systems which can still be seen in many upland areas (e.g. Wessex – McOmish *et al.* 2002, Cunliffe 2004; Dartmoor – Fleming 1988; North Yorkshire – Spratt 1989; and Wales – Murphy 2001) appear in great numbers from the beginning of the Later Bronze Age. Excavations in advance of destruction of sites have added greatly to this picture in the British lowlands in areas such as East Anglia (Pryor 1996; Malim 2001) and most spectacularly in the Thames Valley, where large areas of field systems, probably for stock management, can now be identified (Yates 2001).

The longer boundaries often seem to be placed to maximise visibility (e.g. Cunliffe and Poole 2000; McOmish *et al.* 2002: 64) and given the considerable investment of labour involved it is widely agreed that these make statements about land ownership (Bradley *et al.* 1994: 152), with land and its possibility of surplus production becoming far more important at this time.

Among the extensive field systems of the Thames Valley are a number of enclosed settlements which have been dubbed 'ringforts' (Needham 1992). Excavated examples include two at Mucking (Bond 1988; Clark 1993), Springfield Lyons (Brown 2001), South Hornchurch (Guttmann and Last 2000) and Carshalton (Adkins and Needham 1985).

These ringworks vary in size from only thirty metres in diameter to a few over two hundred metres across. They may have one or several entrances with a main entrance facing East with a gateway. The case

for being the residences of high-status families (e.g. Needham 1992) is not confirmed by the general lack of prestige goods (as Needham himself notes), and as Guttmann and Last (2000) suggest, it may well be that similarity of form should not be equated with similarity of function. Deliberate deposits in the enclosing ditches (including the clay moulds for swords at Springfield Lyons) do not imply an imperative to maintain defensive structures, while the number of entrances in some cases implies that military considerations were not a priority.

Other examples of ringworks occur in Central England at Thrapston (Hull 2000-2001) and in northern England at Thwing (Manby 1980), and the potentially related stone-built ringforts of the western coast of Ireland (Cotter 2000), but they appear to be a rarity outside the Thames Valley and Southeast England.

The main roles of ringwork enclosures seem to be in overlooking and overseeing agricultural production and in monitoring movement along river valleys (Bridgford 2000: 207; Guttmann and Last 2000; Brown 2001; Yates 2001) and changes in the importance of emphasising the distinction between 'insiders' and 'outsiders' by creating clear boundaries between important spaces and the world beyond, a development which fits clearly with the new emphasis on land boundaries (Thomas 1997). A role for ringworks as the residences of high-status families may be limited to areas of Britain where they are rarer. This may be connected with the relative lack of damage to swords and spearheads recorded by Bridgford (2000): perhaps the ringworks acted as a deterrent to raiders.

It is now well established that the earliest hillforts are of Bronze Age date. These appear to be defended settlements with substantial earthen banks and timber ramparts (Avery 1993). The dating of hillforts has taken a number of swings, with early enthusiasm for a large number of hillforts in the Later bronze Age being tempered by the possibility that many hilltops saw activity in the Later Bronze Age but not necessarily enclosure. Thus sites such as Mam Tor (Coombs and Thompson 1979) in northern England and Hog Cliff Hill in Wessex (Ellison and Rahtz 1987) can not be taken to be enclosed at that time. Nevertheless, there are examples with clear dating evidence in the form of radiocarbon dates or artefact assemblages from the ramparts and ditches from areas including Southeast England (Hamilton and

Manley 1997), the Thames Valley (Miles 1997), Wessex (Needham and Ambers 1994), the Southwest (Ellis 1989), western England (Ellis 1993), Wales (Musson 1991), northern England (Stead 1968) and Ireland (Mallory 1995).

The only Later Bronze Age hillfort with apparent direct evidence of conflict is Dinorben in Wales, with defences dated to around 800 BC by radiocarbon (Guilbert 1981). 'In the bottom of the ditch there were three fragmentary male skeletons, one with its skull cleft in two', according to Gardner and Savory (1964: 45). This need not, of course, represent an episode of conflict between groups. Many of these hillforts appear to have relatively slight defences, at least compared with Iron Age hillforts, but this need not mean that their wall-and-fill ramparts (Avery 1993: 122-27) were of negligible defensive value. At The Breiddin, also in Wales (Musson 1991) the extent of burning of the rampart and the subsequent abandonment of the site for several centuries suggests to the excavator that destruction by fire is a possibility here. So a threat may well have existed. The threat may not, of course, have been to the hillfort itself much of the time, and Hamilton and Manley (1997) argue that in Sussex in Southeast England the first hillforts were like the ringworks (not found here) predominantly concerned with looking out over the landscape below to watch people and livestock. In Wessex, ridge-end hillforts (Cunliffe 2004) were also situated to oversee the landscape. Similarly, the Beeston hillfort (Ellis 1993) in western England and The Breiddin in Wales (Musson 1991) were ideally suited to monitor movement in the landscape below.

The other element to consider in the Later Bronze Age landscape is the presence of horses in much larger numbers than hitherto (R. Bendrey pers. comm.) at sites such as Potterne in Wessex (Lawson 2000). As others have noted this makes the possibility of mobile raiders in war-bands, perhaps engaged in cattle theft for prestige, a strong likelihood (Harding 1999b; Osgood et al. 2000: 34). The slashing sword and the spear would be the weapons of choice of such raiders, while the ringworks and the hillforts (slightly defended because a siege was highly unlikely to be undertaken) acted as a deterrent. The apparent emphasis on land and the inheritance of land may have been more to do with controlling the livestock on the land, with the stock at more risk than the land itself.

Conclusion

First it should be clear that the British evidence does not show an increasing trend to a more warlike society through time, as a simple cultural evolutionary model would imply. Instead, it seems that there are two main horizons of conflict: the Early Neolithic and the Later Bronze Age, with a long period of relative lack of conflict between them. This runs contrary to traditional models of Beaker invaders and Wessex Culture warrior chiefs, but the sample of defensible sites and skeletal remains is sufficiently large to point strongly in this direction. There is generally a good agreement between the different classes of evidence – where there are skeletal remains showing trauma there are also plausible weapons and defensible sites. This sequence of cycles of intensity of conflict also contrasts with the implications of theories based on evolutionary psychology, which provide no basis for the interpretation of lower levels of conflict.

As far as the limited skeletal evidence is concerned, war in prehistoric Britain was mostly the business of men, although there are cases of females suffering traumatic injuries from the Early Neolithic onwards, and we should certainly not assume that they were passive victims. It is certainly not clear that warfare was the product of young men striving for status. There are sufficient older males among the violently killed to demonstrate that this can not be an absolute pattern. Unfortunately it is not possible at present to determine how old healed injuries were.

Except perhaps at the end of the Early Neolithic and in certain geographical regions in the Later Bronze Age the materialist theory seems not to fit either, as resources are not the apparent source of conflict. Instead, less tangible notions of prestige and honour seem to be at stake; even in the Early Neolithic of Ireland, when sites may be destroyed in conflicts over the movement of axes, it is extremely unlikely that they were essential to economic production, as there were no significant differences between the various rocks used for axe production in terms of hardness or durability.

Thus none of the generalised theories of human conflict are particularly successful in providing a model of the development of violence, conflict and warfare in British prehistory, implying that we need to develop more nuanced accounts of conflict which situate it in its specific historical and social context.

Acknowledgements

I should like to thank the following for discussions, the provision of information, and helpful comments on a draft of this text: members of the War and Society research group and the editors of this volume, Robin Bendrey (King Alfred's College), Dr Sue Bridgford, Stephany Leach (King Alfred's College), Dr Roger Mercer (Royal Commission on Ancient and Historical Monuments of Scotland), Barry Molloy (University College Dublin), Dr Richard Osgood (South Gloucestershire County Council), and Dr Rick Schulting (Queen's University, Belfast).

BIBLIOGRAPHY

- Abercromby, J. 1912. *A Study of the Bronze Age Pottery in Great Britain and Ireland and its Associated Grave-Goods.* Oxford: Clarendon Press.
- Adkins, L. and Needham, S. 1985. 'New research on a Late Bronze Age enclosure at Queen Mary's Hospital, Carshalton'. *Surrey Archaeological Collections*, 76, pp. 11-50.
- ApSimon, A. 1971. 'Donegore'. *Current Archaeology*, 24, pp. 11-13.
- Ashbee, P. 1960. *The Bronze Age Round Barrow in Britain.* London: Phoenix House.
- Ashbee, P. 1966. 'The Fussell's Lodge long barrow excavations 1957'. *Archaeologia*, 100, pp. 1-80.
- Ashbee, P. 1978. *The Ancient British.* Norwich: Geo Abstracts.
- Atkinson, R.J.C. 1965. 'Wayland's Smithy'. *Antiquity*, 39, pp. 126-33.
- Avery, M. 1982. 'The Neolithic causewayed enclosure, Abingdon'. In: H. Case and A. Whittle (eds.): *Settlement Patterns in the Oxford Region: Excavations at the Abingdon Causewayed Enclosure and other Sites.* (Council for British Archaeology Research Report, 44). London: Council for British Archaeology, pp. 10-50.
- Avery, M. 1993. *Hillfort Defences of Southern Britain.* (British Archaeological Reports, 231). Oxford: British Archaeological Reports.
- Bamford, H.N. 1985. *Briar Hill.* (Archaeological Monograph, 3). Northampton: Northamptonshire Development Corporation.
- Barclay, A and Halpin, C. 1999. *Excavations at Barrow Hills, Radley, Oxfordshire.* Vol. I. *The Neolithic and Bronze Age Monument Complex.* Oxford: Oxford Archaeological Unit.
- Barnatt, J. 1996. 'A multiphased barrow at Liffs Low, near Biggin, Derbyshire'. In: J. Barnatt and J. Collis (eds.): *Barrows in the Peak District: Recent Research.* Sheffield: J.R. Collis Publications, pp. 95-136.
- Bar-Yosef, O. 1986. 'The walls of Jericho: An alternative explanation'. *Current Anthropology*, 27, pp. 157-62.
- Bartlett, J.E. and Hawkes, C.F.C. 1965. 'A barbed bronze spearhead from North Ferriby, Yorkshire, England'. *Proceedings of the Prehistoric Society*, 31, pp. 370-73.
- Beamish, M. and Ripper, S. 2000 'Burnt mounds in the East Midlands'. *Antiquity*, 74, pp. 37-38.
- Bellamy, P.S. 1991. 'The excavation of Fordington Farm round barrow'. *Proceedings of the Dorset Natural History and Archaeological Society*, 113, pp. 107-132.
- Bennike, P. 1985. *Palaeopathology of Danish Skeletons.* Copenhagen: Akademisk Forlag.
- Bennike, P. 1997. 'Death in the Mesolithic. Two old men from Korsør Nor'. In: L. Pedersen, A. Fischer and B. Aaby (eds.): *The Danish Storebælt since the Ice Age.* Copenhagen: A/S Storebælt Fixed Link, pp. 99-105.
- Benton, S. 1931. 'The Excavation of the sculptor's cave, Covesea, Morayshire'. *Proceedings of the Society of Antiquaries of Scotland*, 65, pp. 177-216.
- Bond, D. 1988. *Excavation at the North Ring, Mucking, Essex: a Late Bronze Age Enclosure.* (East Anglian Archaeology Report, 43). Chelmsford: Essex Planning Department.
- Boston, C., Bowater, C. and Boyle, A. n.d. *Excavations at the Proposed Centre for Gene Function, University Parks, Oxford.*
- Bradley, R. 1986. 'A reinterpretation of the Abingdon causewayed enclosure'. *Oxoniensia*, 51, pp. 183-87.
- Bradley, R., Entwistle, R. and Raymond, F. 1994. *Prehistoric Land Divisions on Salisbury Plain.* (English Heritage Archaeological Report, 2). London: English Heritage.
- Bridgford, S.D. 1997. 'Mightier than the pen? (an edgewise look at Irish Bronze Age swords)'. In: J. Carman (ed.): *Material Harm: archaeological studies of war and violence.* Glasgow: Cruithne Press, pp. 95-115.
- Bridgford, S.D. 2000. Weapons, Warfare and Society in Britain 1250-750 BC. Unpublished Ph.D thesis, University of Sheffield, Sheffield.
- Bridgford, S.D. 2002. 'Bronze and the first arms race – cause, effect or coincidence?'. In: B.S. Ottaway and E.C. Wager (eds.): *Metals and Society.* (British Archaeological Reports International Series, 1061). Oxford: Archaeopress, pp. 123-32.
- Bristow, P.H.W. 1998. *Attitudes to Disposal of the Dead in Southern Britain 3500 BC-43 AD.* Vol. I-III. (British Archaeological Reports, 274). Oxford: J and E Hedges.
- Britnell, W.J. and Savory, H.N. 1984. *Gwernvale and Penywyrlod: two Neolithic long cairns in Brecknock.* Cardiff: Cambrian Archaeological Association.
- Brodie, N. 1994. *The Neolithic-Bronze Age Transition in Britain.* (British Archaeological Reports British Series, 238). Oxford: Tempvs Reparatvm.
- Brothwell, D. 1959-60. 'The Bronze Age people of Yorkshire: A general survey'. *The Advancement of Science*, 16, pp. 311-19.
- Brothwell, D. 1960. 'The palaeopathology of Early British Man: An essay on the problems of diagnosis and analysis'. *Journal of the Royal Anthropological Institute*, 91, pp. 318-44.

- Brown, N. 2001. 'The Late Bronze Age enclosure at Springfield Lyons in its landscape context'. *Essex Archaeology and History*, 32, pp. 92-101.
- Brück, J. 1995. 'A place for the dead: the role of human remains in Late Bronze Age Britain'. *Proceedings of the Prehistoric Society*, 61, pp. 245-77.
- Burgess, C. B. 1976. 'Meldon Bridge: a Neolithic defended promontory complex near Peebles'. In: C.B. Burgess and R. Miket (eds.): *Settlement and Economy in the Third and Second Millennia BC*. (British Archaeological Reports, 33). Oxford: British Archaeological Reports, pp. 151-79.
- Burgess, C.B. and Gerloff, S. 1981. *The Dirks and Rapiers of Great Britain and Ireland*. (Prähistorische Bronzefunde IV, 7). Munich: Beck.
- Burl, A. 1981. *Rites of the Gods*. London: Dent.
- Burl, A. 1987. *The Stonehenge People*. London: Dent.
- Butler, C. 1991. 'The Excavation of a beaker bowl barrow at Pyecombe, West Sussex'. *Sussex Archaeological Collections*, 129, pp. 1-28.
- Chapman, J. 1999. 'The origins of warfare in the prehistory of Central and Eastern Europe'. In: J. Carman and A. Harding (eds.): *Ancient Warfare*. Stroud: Sutton, pp. 101-42.
- Clark, A. 1993. *Excavations at Mucking*. Vol. I: *The Site Atlas*. London: English Heritage.
- Clarke, D.L. 1970. *Beaker Pottery of Great Britain and Ireland*. Cambridge: Cambridge University Press.
- Clastres, P. 1994. *Archeology of Violence*. New York: Semiotext(e).
- Clay, P. and Stirland, A. 1981. 'The Smeeton Westerby Beaker burial – some additional information'. *Transactions of the Leicestershire Archaeological and Historical Society*, 56, pp. 97-99.
- Cohen, M. H. 1989. *Health and the Rise of Civilization*. New Haven: Yale University Press.
- Coles, J. M. 1962. 'European Bronze Age shields'. *Proceedings of the Prehistoric Society*, 28, pp. 156-90.
- Coles, J.M., Leach, P., Minnitt, S.C., Tabor R. and Wilson, A.S. 1999. 'A Later Bronze Age shield from South Cadbury, Somerset, England'. *Antiquity*, 73, pp. 33-48.
- Colt Hoare, R. 1812. *The Ancient History of Wiltshire*. London: Miller.
- Coombs, D.G. and Thompson, F.H. 1979. 'Excavation of the hill fort of Mam Tor, Derbyshire 1965-69'. *Derbyshire Archaeological Journal*, 99, pp. 7-51.
- Cooney, G. 2000. *Landscapes of Neolithic Ireland*. London: Routledge.
- Corcoran, J.W.X.P. 1964-6. 'The excavation of three chambered cairns at Loch Calder, Caithness'. *Proceedings of the Society of Antiquaries of Scotland*, 98, pp. 1-75.
- Cosmides, L., Tooby, J. and Barkow, J. 1992. 'Introduction: Evolutionary psychology and conceptual integration'. In: J. Barkow, L. Cosmides and J. Tooby (eds.): *The Adapted Mind*. Oxford: Oxford University Press, pp. 3-15.
- Cotter, C. 2000. 'The chronology and affinities of the stone forts along the Atlantic Coast of Ireland'. In: J.C.
- Henderson (ed.): *The Prehistory and Early History of Atlantic Europe*. (British Archaeological Reports International Series, 861). Oxford: Archaeopress, pp.171-79.
- Crawford, O.G.S. 1925. *The Long Barrows of the Cotswolds*. Gloucester: John Bellows.
- Cunliffe, B. 2004. 'Wessex cowboys?' *Oxford Journal of Archaeology*, 23, pp. 61-81.
- Cunliffe, B. and Poole, C. 2000. *The Danebury Environs Programme. The Prehistory of a Wessex Landscape: Vol. II, part 7. Windy Dido, Cholderton, Hants, 1995*. (Oxford University Committee for Archaeology Monograph, 49). Oxford: Oxford University Committee for Archaeology.
- Cunnington, M.E. 1912. 'Knap Hill Camp'. *Wiltshire Archaeological and Natural History Magazine*, 37, pp. 42-65.
- Cunnington, W. 1889. 'Notes on Bowl's Barrow'. *Wiltshire Archaeological Magazine*, 24, pp. 104-17.
- Curwen, E.C. 1930. 'Neolithic Camps'. *Antiquity*, 4, pp. 22-54.
- Darvill, T.C. 1987. *Prehistoric Britain*. London: Batsford.
- Darvill, T.C. 2001. 'Neolithic enclosures in the Isle of Man'. In: T.C. Darvill and J. Thomas (eds.): *Neolithic Enclosures in Atlantic Northwest Europe*. (Neolithic Studies Group Seminar Papers, 6). Oxford: Oxbow, pp. 155-70.
- Davies, D. 1857. 'Celtic sepulture on the mountains of Carno, Montgomeryshire'. *Archaeologia Cambrensis*, 3rd series, 3, pp. 301-305.
- Dawson, D. 1996. 'The origins of war: Biological and anthropological theories'. *History and Theory*, 35, pp. 1-28.
- Dawson, D. 1999. 'Evolutionary theory and group selection: The question of warfare'. *History and Theory*, 38, pp. 79-100.
- Dawson, D. 2001. *The First Armies*. London: Cassell.
- Dennen, J. van der 1995. *The Origin of War: the Evolution of a Male-Coalitional Reproductive Strategy*. Groningen: Origin Press.
- Dixon, P. 1988. 'The Neolithic settlements on Crickley Hill'. In: C. Burgess, P. Topping, C. Mordant and M. Maddison (eds.): *Enclosures and Defences in the Neolithic of Western Europe*. (British Archaeological Reports International Series, 403). Oxford: British Archaeological Reports, pp. 75-87.
- Donaldson, P. 1977. 'The excavation of a multiple round barrow at Barnack, Cambridgeshire'. *The Antiquaries Journal*, 57, pp. 197-231.
- Drewett, P. 1977. 'The excavation of a Neolithic cause-wayed enclosure on Offham Hill, East Sussex, 1976'. *Proceedings of the Prehistoric Society*, 43, pp. 201-41.
- Drewett, P., Rudling, D. and Gardiner, M. 1988. *The South-East to AD 1000*. London: Longman.
- Drewett, P. 1994. 'Dr V. Seton Williams' excavations at Combe Hill, 1962, and the role of Neolithic causewayed enclosures in Sussex'. *Sussex Archaeological Collections*, 132, pp. 7-24.
- Dunning, G.C. and Wheeler, R.E.M. 1931. 'A barrow at Dunstable, Bedfordshire'. *Archaeological Journal*, 88, pp. 193-217.

- Dunwell, A.J., Neighbour, T. and Cowie, T.G. 1995. 'A cist burial adjacent to the Bronze Age cairn at Cnip, Uig, Isle of Lewis'. *Proceedings of the Society of Antiquaries of Scotland*, 125, pp. 279-88.
- Edmonds, M. and Thomas, J. 1987. 'The archers: An everyday story of country folk'. In: A.G. Brown and M.R. Edmonds (eds.): *Lithic Analysis and Later British Prehistory*. (British Archaeological Reports, 162). Oxford: British Archaeological Reports, pp. 187-99.
- Edwards, B.J.N. 1969. 'Lancashire archaeological notes: prehistoric and Roman'. *Transactions of the Historical Society of Lancashire and Cheshire*, 121, pp. 99-106.
- Ellis, P. 1989. 'Norton Fitzwarren Hillfort: a report on the excavations by Nancy and Philip Langmaid between 1968 and 1971'. *Somerset Archaeology and Natural History*, 133, pp. 1-74.
- Ellis, P. 1993. *Beeston Castle, Cheshire: a Report on the Excavations 1968-85*. (English Heritage Archaeological Report, 23). London: English Heritage.
- Ellison, A. and Rahtz, P. 1987. 'Excavations at Hog Cliff Hill, Maiden Newton, Dorset'. *Proceedings of the Prehistoric Society*, 51, pp. 223-69.
- Eogan, G. and Roche, H. 1997. *Excavations at Knowth*. Vol. II. Dublin: Royal Irish Academy.
- Evans, C. 1988. 'Excavations at Haddenham, Cambridgeshire: a "planned" enclosure and its regional affinities'. In: C. Burgess, P. Topping, C. Mordant and M. Maddison (eds.): *Enclosures and Defences in the Neolithic of Western Europe*. (British Archaeological Reports International Series, 403). Oxford: British Archaeological Reports, pp. 47-73.
- Evans, J.G. 1984. 'Stonehenge – The environment in the Late Neolithic *and* Early Bronze Age and A Beaker-Age burial'. *Wiltshire Archaeological and Natural History Magazine*, 78, pp. 7-30.
- Everton, A. and Everton, R. 1972. 'Hay Wood Cave burials, Mendip Hills, Somerset'. *Proceedings of the University of Bristol Spelaeological Society*, 13, pp. 5-29.
- Ferguson, R.B. 1990. 'Explaining war'. In: J. Haas (ed.): *The Anthropology of War*. Cambridge: Cambridge University Press, pp. 26-55.
- Ferguson, R.B. 1997. 'Violence and war in prehistory'. In: D.L. Martin and D.W. Frayer (eds.): *Troubled Times*. New York: Gordon and Breach, pp. 321-55.
- Fitzpatrick, A.P. 2002. '"The Amesbury Archer": a well-furnished Early Bronze Age burial in southern England'. *Antiquity*, 76, pp. 629-30.
- Fleming, A. 1988. *The Dartmoor Reaves: Investigating Prehistoric Land Divisions*. London: Batsford.
- Foley, R.A. 1996. 'An evolutionary and chronological framework for human social behaviour'. *Proceedings of the British Academy*, 88, pp. 95-117.
- Frayer, D.W. 1997. 'Ofnet: Evidence of a Mesolithic massacre'. In: D.L. Martin and D.W. Frayer (eds): *Troubled Times*. New York: Gordon and Breach, pp. 181-216.
- Gardner, W. and Savory, H.N. 1964. *Dinorben: a Hillfort Occupied in Early Iron Age and Roman Times*. Cardiff: National Museum of Wales.
- Gerloff, S. 1975. *The Early Bronze Age Daggers in Great Britain and a Reconsideration of the Wessex Culture*. (Prähistorische Bronzefunde VI, 2). Munich: Beck.
- Gibson, A. 1994. 'Excavations at the Sarn-y-bryn-caled cursus complex, Welshpool, Powys, and the timber circles of Great Britain and Ireland'. *Proceedings of the Prehistoric Society*, 60, pp. 143-223.
- Gibson, A. 1999. *The Walton Basin Project: Excavation and Survey in a Prehistoric Landscape 1993-7*. (Council for British Archaeology Research Report, 118). York: Council for British Archaeology.
- Gibson, A. 2002. 'The Later Neolithic palisaded enclosures of the United Kingdom'. In: A. Gibson (ed.): *Behind Wooden Walls: Neolithic Palisaded Enclosures in Europe*. (British Archaeological Reports International Series, 1013). Oxford: Archaeopress, pp. 5-23.
- Gilbert, P. 1994. 'Male violence: Towards an integration'. In: J. Archer (ed.): *Male Violence*. London: Routledge, pp. 352-89.
- Green, H.S. 1980. *The Flint Arrowheads of the British Isles*. (British Archaeological Reports, 75). Oxford: British Archaeological Reports.
- Grinsell, L.V. 1957. 'Archaeological gazetteer'. In: E. Critall (ed.): *A History of Wiltshire*. (Victoria County History Vol. I, part 1). London, pp. 21-279.
- Grinsell, L.V. 1961. 'The breaking of objects as a funerary rite'. *Folklore*, 72: pp. 475-91.
- Grinsell, L.V. 1971. 'Someset Barrows, Part II: north and east'. *Proceedings of the Somerset Archaeological and Natural History Society*, 115, supplement, pp. 44-137.
- Grünberg, J.M. 2000. *Mesolithische Bestattungen in Europa*. Rahden: Verlag Marie Leidorf.
- Guilbert, G. 1981. 'Dinorben'. *Archaeology in Wales*, 21, p. 39.
- Guttmann, E.B.A. and Last, J. 2000. 'A Late Bronze Age landscape at South Hornchurch, Essex'. *Proceedings of the Prehistoric Society*, 66, pp. 319-59.
- Hamilton, S. and Manley, J. 1997. 'Points of view: Prominent enclosures in 1st Millennium BC Sussex'. *Sussex Archaeological Collections*, 135, pp. 93-112.
- Harding, A.F. 1999a. 'Warfare: a defining characteristic of Bronze Age Europe?' In: J. Carman and A. Harding (eds.): *Ancient Warfare*. Stroud: Sutton, pp. 157-73.
- Harding, A.F. 1999b. 'Swords, shields and scholars: Bronze Age warfare, past and present'. In: A. F. Harding (ed.): *Experiment and Design: Archaeological Studies in Honour of John Coles*. Oxford: Oxbow, pp. 87-93.
- Harding, A.F. 2000. *European Societies in the Bronze Age*. Cambridge: Cambridge University Press.
- Harris, D.R. 1990. *Settling Down and Breaking Ground: the Neolithic Revolution*. Amsterdam: Twaalfde Kroon-Voordracht.
- Healy, F. 1996. *The Fenland Project, Number 11: The Wissey Embayment: Evidence for Pre-Iron Age Occupation Accumulated Prior to the Fenland Project*. (East Anglian Archaeology Report, 76). Norwich: Norfolk Museums Service.

- Hedges, J.W. 1983. *Isbister: A chambered tomb in Orkney.* (British Archaeological Reports British Series, 115). Oxford: British Archaeological Reports.
- Hedges, J. and Buckley, D. 1978. 'Excavations at a Neolithic causewayed enclosure, Orsett, Essex, 1975'. *Proceedings of the Prehistoric Society,* 44, pp. 219-308.
- Hedges, R.E.M., Housley, R.A., Bronk, C., and van Klinken, G.J. 1991. 'Radiocarbon dates from the Oxford AMS System: *Archaeometry* datelist 12'. *Archaeometry,* 33, pp. 121-34.
- Hedges, R.E.M., Pettitt, P.B., Bronk Ramsay, C., and van Klinken, G.J. 1997. 'Radiocarbon dates from the Oxford AMS System: *Archaeometry* datelist 24'. *Archaeometry,* 39, pp. 445-71.
- Henry, D.O. 1989. *From Foraging to Agriculture: the Levant at the End of the Ice Age.* Philadelphia: University of Pennsylvania Press.
- Hinman, M. 2001. 'Ritual activity at the foot of the Gog Magog Hills, Cambridge'. *In: Bronze Age Landscapes: Tradition and Transformation.* Oxford: Oxbow, pp. 33-40.
- Hodder, I. 1992. 'The Haddenham causewayed enclosure – a hermeneutic circle'. In: I. Hodder (ed.): *Theory and Practice in Archaeology.* London: Routledge, pp. 213-40.
- Hollimon, S. 2001. 'Warfare and gender in the northern plains: Osteological evidence of trauma reconsidered'. In: B. Arnold and N. Wicker (eds.): *Gender and the Archaeology of Death.* Walnut Creek: Altamira Press, pp. 179-93.
- Hull, G. 2000-2001. 'A Late Bronze Age ringwork, pits and later features at Thrapston, Northamptonshire'. *Northamptonshire Archaeology,* 29, pp. 73-92.
- James, P.J. and Thorpe, I.J.N. 1994. *Ancient Inventions.* London: Michael O'Mara.
- Jochim, M.A. 1998. *A Hunter-Gatherer Landscape: southwest Germany in the late Paleolithic and Mesolithic.* New York: Plenum.
- Johnston, D.E. 1980. 'The excavation of a bell-barrow at Sutton Veny, Wilts'. *Wiltshire Archaeological Magazine,* 72-73, pp. 29-50.
- Keegan, J. 1993. *A History of Warfare.* London: Hutchinson.
- Keeley, L.H. 1996. *War Before Civilization: the Myth of the Peaceful Savage.* Oxford: Oxford University Press.
- Keiller, A. and Piggott, S. 1938. 'Excavation of an untouched chamber in the Lanhill long barrow'. *Proceedings of the Prehistoric Society,* 4, pp. 122-50.
- Kelly, R.C. 2000. *Warless Societies and the Origin of War.* Ann Arbor: University of Michigan Press.
- Koehler, L. 1997. 'Earth mothers, warriors, horticulturalists, artists, and chiefs: Women among the Mississippian and Mississippian-Oneota peoples, AD 1000 to 1750'. In: C. Claasen and R.A. Joyce (eds.): *Women in Prehistory: North America and Mesoamerica.* Philadelphia: University of Pennsylvania Press, pp. 211-26.
- Knauft, B.M. 1991. 'Violence and sociality in human evolution'. *Current Anthropology,* 32, pp. 391-428.
- Knight, R .W., Browne, C., and Grinsell, L.V. 1972. 'Prehistoric skeletons from Tormarton'. *Transactions of the Bristol and Gloucester Archaeological Society,* 91, pp. 14-17.
- Kristiansen, K. 2002. 'The Tale of the Sword – Swords and Swordfighters in Bronze Age Europe'. *Oxford Journal of Archaeology,* 21, pp. 319-32.
- Laland, K.N. and Brown, G.R. 2002. *Sense and Nonsense: Evolutionary Perspectives on Human Behaviour.* Oxford: Oxford University Press.
- Lawson, A.J. 2000. *Potterne 1982-5: Animal Husbandry in Later Prehistoric Wiltshire.* Salisbury: Trust for Wessex Archaeology.
- Leakey, R.B. and Lewin, R. 1992. *Origins Reconsidered.* New York: Doubleday.
- Liddell, D.M. 1930. 'Report on the excavations at Hembury Fort, Devon, 1930'. *Proceedings of the Devon Archaeological Exploration Society,* 1, pp. 40-63.
- Liddell, D.M. 1931. 'Report on the excavations at Hembury Fort, Devon, 1931'. *Proceedings of the Devon Archaeological Exploration Society,* 1, pp. 90-119.
- Liddell, D.M. 1932. 'Report on the excavations at Hembury Fort, Devon'. *Proceedings of the Devon Archaeological Exploration Society,* 1, pp. 162-90.
- Liddell, D.M. 1935. 'Report on the excavations at Hembury Fort, Devon'. *Proceedings of the Devon Archaeological Exploration Society,* 2, pp. 135-75.
- Logue, P. 2003. 'Excavations at Thornhill, Co. Londonderry'. In: I. Armit, E. Murphy, E. Nelis and D. Simpson (eds.): *Neolithic Settlement in Ireland and Western Britain.* Oxford: Oxbow, pp. 149-55.
- Loveday, R. 2002. 'Duggleby Howe revisited'. *Oxford Journal of Archaeology,* 21, pp. 135-46.
- Lynch, A. and Ó Donnabháin, B. 1994. 'Poulnabrone portal tomb'. *The Other Clare,* 18, pp. 5-7.
- McKinley, J.I. 1989. 'Cremations: expectations, methodologies and realities'. In: C.A. Roberts, F. Lee and J. Bintliff (eds.): *Burial Archaeology: current research, methods and developments.* (British Archaeological Reports British Series, 211). Oxford: British Archaeological Reports, pp. 65-76.
- McOmish, D., Field, D. and Brown, G. 2002. *The Field Archaeology of the Salisbury Plain Training Area.* London: English Heritage.
- Malim, T. 2001. 'Place and space in the Cambridgeshire Bronze Age'. In: J. Brück (ed.): *Bronze Age Landscapes: Tradition and Transformation.* Oxford: Oxbow, pp. 9-22.
- Mallory, J.P. 1995. 'Haughey's Fort and the Navan Complex in the Late Bronze Age'. In: J. Waddell and E. Shee Twohig (eds.): *Ireland in the Bronze Age.* Dublin: The Stationery Office, pp. 73-86.
- Mallory, J.P. and Hartwell, B. 1984. 'Donegore'. *Current Archaeology,* 92, pp. 271-75.
- Manby, T.G. 1980. 'Bronze Age settlement in Eastern Yorkshire'. In: J. Barrett and R. Bradley (eds.): *Settlement and Society in the British Later Bronze Age.* (British Archaeological Reports, 83). Oxford: British Archaeological Reports, pp. 307-44.
- Maschner, H.D.G. and Reedy-Maschner, K.L. 1998. 'Raid, retreat, defend (repeat): The archaeology and ethnohistory of warfare on the North Pacific Rim'. *Journal of Anthropological Archaeology,* 17, pp. 19-51.

- Mercer, R.J. 1980. *Hambledon Hill: a Neolithic Landscape.* Edinburgh: Edinburgh University Press.
- Mercer, R.J. 1981. 'Excavations at Carn Brea, Illogan, Cornwall, 1970-73'. *Cornish Archaeology*, 20, pp. 1-204.
- Mercer, R.J. 1988. 'Hambledon Hill, Dorset, England'. In: C. Burgess, P. Topping, C. Mordant and M. Maddison (eds.): *Enclosures and Defences in the Neolithic of Western Europe.* (British Archaeological Reports International Series, 403). Oxford: British Archaeological Reports, pp. 89-106.
- Mercer, R.J. 1990. *Causewayed Enclosures.* Aylesbury: Shire.
- Mercer, R.J. 1997. 'The excavation of a Neolithic enclosure complex at Helman Tor, Lostwithiel, Cornwall'. *Cornish Archaeology*, 36, pp. 5-63.
- Mercer, R.J. 1999. 'The origins of warfare in the British Isles'. In: J. Carman and A. Harding (eds.): *Ancient Warfare.* Stroud: Sutton, pp. 143-56.
- Miles, D. 1997. 'Conflict and complexity: the later prehistory of the Oxford Region'. *Oxoniensia*, 62, pp. 1-19.
- Moore, D.G. 2003. 'Neolithic houses in Ballyharry Townland, Islandmagee, Co. Antrim'. In: I. Armit, E. Murphy, E. Nelis and D. Simpson (eds.): *Neolithic Settlement in Ireland and Western Britain.* Oxford: Oxbow, pp. 156-63.
- Mortimer, J.R. 1905. *Forty Years' Researches in British and Saxon Burial-Mounds of East Yorkshire.* London: Brown.
- Murphy, K. 2001. 'A prehistoric field system and related monuments on St David's Head and Carn Llidi, Pembrokeshire'. *Proceedings of the Prehistoric Society*, 67, pp. 85-99.
- Musson, C. 1991. *The Breiddin Hillfort: a later prehistoric settlement in the Welsh Marches.* (British Archaeology Research Report, 76). London: Council for British Archaeology.
- Needham, S.P. 1979. 'Two recent British shield finds and their continental parallels'. *Proceedings of the Prehistoric Society*, 45, pp. 111-34.
- Needham, S.P. 1992. 'The structure of settlement and ritual in the Late Bronze Age of South-East Britain'. In: C. Mordant and A. Richard (eds.): *L'Habitat et L'Occupation du Sol a L'Age du Bronze en Europe.* (Documents Prehistoriques, 4). Paris: Edition du Comite des Travaux Historiques et Scientifiques, pp. 49-69.
- Needham, S.P. and Ambers, J. 1994. 'Redating Rams Hill and reconsidering Bronze Age enclosure'. *Proceedings of the Prehistoric Society*, 60, pp. 225-43.
- Nelis, E. 2003. 'Donegore Hill and Lyles Hill, Neolithic enclosed sites in Co. Antrim: the lithic assemblages'. In: I. Armit, E. Murphy, E. Nelis and D. Simpson (eds.): *Neolithic Settlement in Ireland and Western Britain.* Oxford: Oxbow, pp. 203-17.
- O'Connell, R.L. 1995. *Ride of the Second Horseman.* Oxford University Press.
- O'Flaherty, R., Rankin, B. and Williams, L. 2002. 'Reconstructing an Early Bronze Age halberd'. *Archaeology Ireland*, 16(3), pp. 30-34.
- Osgood, R. 1998. *Warfare in the Late Bronze Age of North Europe.* (British Archaeological Reports International Series, 694). Oxford: Archaeopress.
- Osgood, R. 1999. 'The unknown warrior? The re-evaluation of a skeleton from a bell barrow at Sutton Veny, Wiltshire'. *Wiltshire Archaeological Magazine*, 92, pp. 120-32.
- Osgood, R. and Monks, S. with Toms, J. 2000. *Bronze Age Warfare.* Stroud: Sutton.
- Oswald, A., Dyer, C and Barber, M. 2001. *The Creation of Monuments: Neolithic Causewayed Enclosures in the British Isles.* London: English Heritage.
- Otterbein, K.F. 1997. 'The origins of war'. *Critical Review*, 11, pp. 251-77.
- Passmore, A.D. 1942. 'Chute, Barrow 1'. *Wiltshire Archaeological and Natural History Magazine*, 50, pp. 100-101.
- Piggott, S. 1962. *The West Kennet Long Barrow.* London: HMSO.
- Piggott, S. and Piggott, C.M. 1944. 'Excavation of barrows on Crichel and Launceston Down, Dorset'. *Archaeologia*, 90, pp. 48-80.
- Pinker, S. 1998. *How the Mind Works.* London: Penguin.
- Pitt-Rivers, A.H.L.F. 1898. *Excavations in Cranborne Chase*: Volume IV. London: privately printed.
- Pryor, F. 1984. *Excavation at Fengate, Peterborough, England: the Fourth Report.* (Northamptonshire Archaeological Society Monograph, 2). Northampton: Northamptonshire Archaeological Society.
- Pryor, F. 1996. 'Sheep, stocklands and farm systems: Bronze Age livestock poulations in the Fenlands of eastern England'. *Antiquity*, 70, pp. 313-24.
- Pryor, F. 1998. *Etton: excavations at a Neolithic causewayed enclosure near Maxey, Cambridgeshire.* (English Heritage Archaeological Report, 18). London: English Heritage.
- Radovanovic, I. 1996. *The Iron Gates Mesolithic.* (International Monographs in Prehistory, 11). Ann Arbor: International Monographs in Prehistory.
- Raftery, J. 1944. 'A Neolithic burial in Co. Carlow'. *Journal of the Royal Society of Antiquaries of Ireland*, 74, pp. 61-62.
- Replogle, B.A. 1980. 'Social dimensions of British and German bell-beaker burials: An exploratory study'. *Journal of Indo-European Studies*, 8, pp. 165-99.
- Robb, J. 1997. 'Violence and gender in Early Italy'. In: D.L. Martin and D.W. Frayer (eds.): *Troubled Times.* New York: Gordon and Breach, pp. 111-44.
- Roberts, C. and Cox, M. 2003. *Health and Disease in Britain from Prehistory to the Present Day.* Stroud: Sutton.
- Robertson-Mackay, R. 1987. 'The Neolithic causewayed enclosure at Staines, Surrey: Excavations 1961-63'. *Proceedings of the Prehistoric Society*, 53, pp. 23-128 and microfiche.
- Rolle, R. 1989. *The World of the Scythians.* London: Batsford.
- Russell, M. 2002. *Monuments of the British Neolithic: the Roots of Architecture.* Stroud: Tempus.
- Russell, A.D. 1990. 'Two beaker burials from Chilbolton, Hampshire'. *Proceedings of the Prehistoric Society*, 56, pp. 153-72.

- Saville, A. 1990. *Hazleton North, Gloucestershire, 1979-1982: the Excavation of a Neolithic Long Cairn of the Cotswold-Severn Group.* (English Heritage Archaeological Report, 13). London: English Heritage.
- Saville, A. 2002. 'Lithic artefacts from Neolithic causewayed enclosures: Character and meaning'. In: G. Varndell and P. Topping (eds.): *Enclosures in Neolithic Europe.* Oxford: Oxbow, pp. 91-105.
- Schulting, R. 1998. Slighting the Sea: the Mesolithic-Neolithic Transition in Northwest Europe. Unpublished PhD thesis, University of Reading, Reading.
- Schulting, R. and Wysocki, M. 2002. 'Cranial trauma in the British Earlier Neolithic'. *Past*, 41, pp. 4-6.
- Schulting, R., Wysocki, M., Gonzalez, S. and Turner, A. n.d. *Neolithic clubbing: new evidence for inter-personal violence in the British Neolithic.*
- Selkirk, A. 1971. 'Ascott-under-Wychwood'. *Current Archaeology*, 24, pp. 7-10.
- Sharples, N.M. 1991. *Maiden Castle: Excavations and Field Survey 1985-6.* London: Historic Buildings and Monuments Commission for England.
- Sheridan, A. 1991. 'The first farmers'. In: M. Ryan (ed.): *The Illustrated Archaeology of Ireland.* Dublin: Country House, pp. 47-52.
- Sheridan, A. 2001. 'Donegore Hill and other Irish Neolithic enclosures: a view from outside'. In: T. C. Darvill and J. Thomas (eds.): *Neolithic Enclosures in Atlantic Northwest Europe.* (Neolithic Studies Group Seminar Papers, 6). Oxford: Oxbow, pp. 171-89.
- Sikora, M. and Buckley, L. 2003. 'Casting new light on old excavations'. *Archaeology Ireland*, 17(1), pp. 17.
- Silvester, R.J. 1979. 'The relationship of First Millennium settlement to the upland areas of the South-West'. *Proceedings of the Devon Archaeological Society*, 37, pp. 176-90.
- Simpson, D.D.A. and Coles, J.M. 1990. 'Excavations at Grandtully, Perthshire'. *Proceedings of the Society of Antiquaries of Scotland*, 120, pp. 33-44.
- Simpson, W.G. 1976. 'A barrow cemetery of the Second Millennium BC at Tallington, Lincolnshire'. *Proceedings of the Prehistoric Society*, 42, pp. 215-39.
- Smith, R.J.C., Healy, F., Allen, M.J., Morris, E.L., Barnes, I. and Woodward, P.J. 1997. *Excavations Along the Route of the Dorchester By-pass, Dorset, 1986-8.* (Wessex Archaeology Report, 11). Salisbury. Trust for Wessex Archaeology.
- Speak, S. and Burgess, C.B. 1999. 'Meldon Bridge: A centre of the Third Millennium BC in Peeblesshire'. *Proceedings of the Society of Antiquaries of Scotland*, 129, pp. 1-118.
- Spratt, D.A. 1989. *Linear Earthworks of the Tabular Hills, north-east Yorkshire.* Sheffield: J.R. Collis Publications.
- Stead, I.M. 1959. 'The excavation of Beaker burials at Staxton, East Riding, 1957'. *Yorkshire Archaeological Journal*, 40, pp. 129-44.
- Stead, I.M. 1968. 'A Iron Age hillfort at Grimthorpe, Yorkshire'. *Proceedings of the Prehistoric Society*, 34, pp. 173-89.
- Sterpin, A. 1993. 'La chasse aux scalps chez les nivacle du Gran Chaco'. *Journal de la Société des Americanistes*, 79, pp. 33-66.
- Stringer, C.B. 1985. 'The hominid remains from Gough's Cave'. *Proceedings of the University of Bristol Spelaeological Society*, 17, pp. 145-52.
- Thomas, N. 1956. 'Excavation and fieldwork in Wiltshire: 1956'. *Wiltshire Archaeological Magazine*, 56, pp. 231-52.
- Thomas, R. 1997. 'Land, kinship relations and the rise of enclosed settlement in First Millennium BC Britain'. *Oxford Journal of Archaeology*, 16, pp. 211-18.
- Thomas, J. 1996. *Time, Culture and Identity.* London: Routledge.
- Thorpe, I.J.N. 1984. 'Ritual, power and ideology: A reconstruction of Earlier Neolithic rituals in Wessex.' In: R. Bradley and J. Gardiner (eds.): *Neolithic Studies: a review of Some Current Research.* (British Archaeological Reports, 133). Oxford: British Archaeological Reports, pp. 41-60.
- Thorpe, I.J.N. 1989. Neolithic and Earlier Bronze Age Wessex and Yorkshire: A Comparative Study. Ph.D. thesis, University College London, London.
- Thorpe, I.J.N. 1996. *The Origins of Agriculture in Europe.* London: Routledge.
- Thorpe, I.J.N. 2003a. 'Death and violence – the Later Mesolithic of Southern Scandinavia'. In: L. Bevan and J. Moore (eds.): *Peopling the Mesolithic in a Northern Environment.* (British Archaeological Reports International Series, 1157). Oxford: Archaeopress, pp. 171-80.
- Thorpe, I.J.N. 2003b. 'Anthropology, Archaeology and the origin of Warfare'. *World Archaeology*, 35, pp. 145-65.
- Tomb, J.J. and Davies, O. 1938. 'Urns from Ballymacaldrack'. *Ulster Journal of Archaeology*, 1, pp. 219-21.
- Thurnam, J. 1869. 'On Ancient British Barrows, especially those of Wiltshire and the adjoining counties. (Part I, Long Barrows)'. *Archaeologia*, 42, pp. 161-244.
- Todd, M. 1984. 'Excavations at Hembury (Devon), 1980-83: a summary report'. *The Antiquaries Journal*, 64, pp. 251-68.
- Tooley, M.J. 1978. 'The history of Hartlepool Bay'. *The International Journal of Nautical Archaeology and Underwater Exploration*, 7, pp. 71-75.
- Tratman, E.K. 1975. 'Problems of "The Cheddar Man", Gough's Cave, Somerset'. *Proceedings of the University of Bristol Spelaeological Society*, 14, pp. 7-23.
- Treherne, P. 1995. 'The warrior's beauty: The masculine body and self-identity in Bronze-Age Europe'. *Journal of European Archaeology*, 3, pp. 105-44.
- Vencl, S. 1999. 'Stone Age warfare'. In: J. Carman and A. Harding (eds.): *Ancient Warfare.* Stroud: Sutton, pp. 57-72.
- Waddell, J. 1984. 'Bronzes and bones'. *The Journal of Irish Archaeology*, 2, pp. 71-72.
- Waddell, J. 1990. *The Bronze Age Burials of Ireland.* Galway: Galway University Press.
- Waddell, J. 2000. *The Prehistoric Archaeology of Ireland.* Dublin: Wordwell.

Wahl, J. and König, H.G. 1987. 'Anthropologisch-trauma-tologische Untersuchung der menschlichen Skelettreste aus dem Bandkeramischen Massengrab bei Talheim, Kreis Heilbronn'. *Fundberichte aus Baden-Württemberg*, 12, pp. 65–193.

Wainwright, G.J. 1979. *Mount Pleasant, Dorset: Excavations 1970-1971*. (Research Report of the Society of Antiquaries of London, 37). London: Society of Antiquaries of London.

Wainwright, G.J. 1989. *The Henge Monuments*. London: Thames and Hudson.

Wall, J. 1987. 'The role of daggers in Early Bronze Age Britain: the evidence of wear analysis'. *Oxford Archaeological Journal*, 6, pp. 115-18.

Warne, C. 1866. *The Celtic Tumuli of Dorset*. London: Russell Smith.

Wheeler, R.E.M. 1943. *Maiden Castle, Dorset*. (Research Report of the Society of Antiquaries of London, 12). London: Society of Antiquaries of London.

Whittle, A.W.R. 1991. 'Wayland's Smithy, Oxfordshire: Excavations at the Neolithic Tomb in 1962-63 by R.J.C. Atkinson and S. Piggott'. *Proceedings of the Prehistoric Society*, 57(2), pp. 61-101.

Whittle, A.W.R. 1997. *Sacred Mound, Holy Rings: Silbury Hill and the West Kennet palisade enclosures: a Later Neolithic complex in North, Wiltshire*. Oxford: Oxbow.

Whittle, A.W.R., Pollard, J. and Grigson, C. 1999. *The Harmony of Symbols: the Windmill Hill causewayed enclosure, Wiltshire*. Oxford: Oxbow.

Wilson, D. 1981. 'Withington'. *Current Archaeology*, 7, pp. 155-57.

Wilson, E.O. 1978. *On Human Nature*. Cambridge (Mass.): Harvard University Press.

Wilson, I. 2001. *Past Lives*. London: Cassell.

Windl, H. 1994. 'Zehn Jahre Grabung Schletz, VB Mistelbach, NÖ'. *Archäologie Österreichs*, 5, pp. 11-18.

Woodward, P.J., Davies, S.M. and Graham, A.H. 1993. *Excavations at The Old Methodist Chapel and Greyhound Yard, Dorchester*. (Dorset Natural History and Archaeological Society Monograph, 12). Dorchester: Dorset Natural History and Archaeological Society.

Wrangham, R.W. 1999. 'The evolution of coalitionary killing'. *Yearbook of Physical Anthropology*, 42, pp. 1-30.

Wrangham, R. and Peterson, D. 1996. *Demonic Males: Apes and the Origins of Human Violence*. London: Bloomsbury.

Wysocki, M. and Whittle, A.W.R. 2000. 'Diversity, lifestyles and rites: new biological and archaeological evidence from British Earlier Neolithic mortuary assem-blages'. *Antiquity*, 74, pp. 591-601.

Yates, D. 2001. 'Bronze Age agricultural intensification in the Thames Valley and Estuary'. In: J. Brück (ed.): *Bronze Age Landscapes: Tradition and Transformation*. Oxford: Oxbow, pp. 65-82.

York, J. 2002. 'The Life Cycle of Bronze Age metalwork from the Thames'. *Oxford Journal of Archaeology*, 21, pp. 77-92.

The Impact of Egalitarian Institutions on Warfare among the Enga: An Ethnohistorical Perspective

Akali taiyoko ongo kunao napenge.
The blood of a man does not wash off easily.

POLLY WIESSNER

/11

The importance of war as 'prime mover' towards hierarchical political complexity has long been a matter of debate. While there is little doubt that warfare was a significant force in the political dynamics of centralised societies and in the rise of the state, its role in acephalous societies remains unclear. Some argue that war as an agent of cultural selection weeds out less adaptive cultural institutions and technology, thereby selecting for efficiency and complexity (Carniero 1970). Moreover, because individuals will not willingly give up sovereignty without coercion, warfare is one likely mechanism for increasing social hierarchy (Ferguson 1990: 11). Others contest the role of warfare as a 'progressive' force in acephalous societies on the grounds that personal involvement, weak organisational structures, shifting alliances, diverse motivations of participants, and inefficient technology make warfare 'unproductive' for political evolution (Brown 1978; van der Dennen 1995; Fried 1967; Meyer 1990; Montagu 1976; Naroll and Divale 1976; Otterbein 1970; 1994; Turney-High 1949; Wright 1942).

From existing evidence, it is not easy to determine if and how warfare contributes to generating complexity in acephalous societies. Archaeological indicators such as skeletal damage, weaponry, fortifications, or depictions of combat in art are usually sparse for tribal societies, sometimes indicating little more than the presence or absence of war. Most ethnographic studies of war, excellent though they are, have been conducted in tribal societies that were undergoing rapid change owing to direct and indirect effects of contact with western cultures (Ferguson and Whitehead 1992). Moreover, ethnographic studies of war lack time depth, lending them to the conclusion that war has a homeostatic function (Rappaport 1968; Hallpike 1973; 1987). Ethnohistorical studies of tribal warfare have their own difficulties, for example, interpretation of oral records and biases of narrators. However, they still hold considerable potential for exploring the relation of war to political developments because they contain far more information on the objectives, courses, and outcomes of war than do archaeological studies, and yet have the time depth not available in most ethnographic works.

The objective of this paper will be to use ethnohistorical data from the Enga of Papua New Guinea to explore the role of warfare in pre-contact sociopolitical change over a period of some 250 years following the introduction of the sweet potato and prior to first contact with Europeans. Three questions will be addressed: (1) In acephalous societies like Enga, what is the impact of egalitarian institutions on warfare in contrast to exchange? (2) Given egalitarian constraints operative in Enga history, what role did warfare play in generating the hierarchical complexity that unfolded during the period considered? (3) Did the Enga experiment with warfare to try to circumvent egalitarian constraints and use warfare more productively in their strategies?

Background

The Enga are a highland horticultural population in Papua New Guinea (fig. 1) who live at altitudes of 1500-2500 m above sea level cultivating sweet potato and other crops in an intensive system of mulch mounding to feed large human and pig populations. They are well known in the anthropological literature through the works of Feil (1984), Gordon and Meggitt (1985), Lacey (1975; 1979), Meggitt (1965; 1972; 1974; 1977), Talyaga (1982), Waddell (1972), and Wohlt (1978), amongst many others. The Enga population, which numbers approximately 230,000 today, is divided into a segmentary lineage system of phratries or tribes composed of some 1000 to 6,000 members, and their constituent exogamous clans, sub-clans, and lineages (Meggitt 1965). Patrilineally inherited clan membership furnishes a pool of people who cooperate in agricultural enterprises, defence, procurement of spouses, raising wealth for a variety of payments, and in the past, communicating with the spirit world. Affinal and maternal ties established by exogamous marriage and maintained by reciprocal exchange provide access to resources and

assistance outside the clan. Except in times of warfare when affines may be on the enemy side, there is little conflict between agnatic and affinal loyalties – the clan sees the wide range of affinal ties held by individual members as enhancing the clan's strength (Wiessner and Tumu 1998: 172). Approximately 10% of men in eastern Enga and 30% in western Enga join the clans of maternal or affinal relatives.

Enga women devote themselves primarily to family, gardening, and pig husbandry, while the politics of warfare, exchange, and pursuit of 'name' or reputation, occupies much of men's time and effort. 'Name' can be obtained through several channels: warfare, ritual expertise, mediation, organisation, and the management of wealth. However, great warriors have little say outside of the context of battle, the power of ritual experts is limited to their realms of practice, while big men, the masters of mediation and exchange, are the most revered leaders who exerted strong influence in many contexts.

The Enga hold a rich body of historical traditions (*atome pii*) that are held distinct from myth (*tindii pii*) in that they are said to have originated in eyewitness accounts. Historical traditions contain information on subsistence, wars, migrations, agriculture, the development of cults and ceremonial exchange networks, leadership, trade, environmental disasters, and fashions in song and dress. They cover a period of some 250-400 years that begins just prior to the introduction of the sweet potato along local trade routes and continue into the present. Accompanying genealogies allow events to be placed in a chronological framework (fig. 2).

Between 1985 and 1995, Akii Tumu, Nitze Pupu, and I collected and analysed the historical traditions of 110 tribes (phratries) of Enga (Wiessner and Tumu 1998).[1] The results of our studies indicate that the period between the introduction of the sweet potato and first contact with Europeans was one of rapid change. The earliest of historical traditions recording events prior to the introduction of the sweet potato describe the population of Enga as diverse with people practicing subsistence strategies that varied by altitudinal zone. Eastern Enga (1500-1900 m above sea level) was occupied by horticulturalists who cultivated taro, yams, bananas, sugarcane, and other crops in the wide, fertile valleys of the Lai and Saka. In central Enga (1900 m-2100 m) roughly equal emphasis was placed on gardening, hunting, and gathering. In the high country of

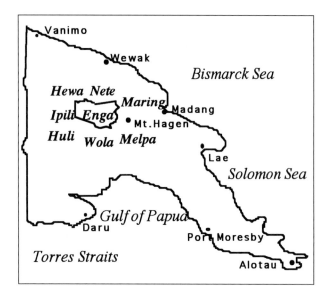

FIG. 1: *Map of Enga territory, Papua New Guinea.*

western Enga above 2100 m lived scattered mobile groups who depended heavily on hunting and gathering, while shifting horticulturalists lived a precarious existence in the steep and infertile valleys below. Marriage and exchange, on the one hand, and tension and misunderstanding, on the other characterised relations between the two. In these early traditions, the thriving exchange economy is described as revolving around the circulation of stone axes, cosmetic oil, salt, plumes and shells; pigs receive little note.

The introduction of the sweet potato released many constraints on production and made it possible to produce a substantial surplus of pigs for the first time (Watson 1965a; 1965b; 1977). Immediate reactions to the arrival of sweet potato differed by area, however, historical traditions from all areas report substantial shifts in population distribution, population growth, and the expansion of ceremonial exchange and religious ritual in response to mounting social and political complexities. Three large networks arose that can all be counted as systems of ceremonial exchange (fig. 3). The first of these was the Kepele cult network of western Enga that linked more than fifty tribes of western Enga in a ritual network. Kepele ceremonies drew hundreds, and, in later generations, thousands of participants to initiate boys, express the equality of male tribal members, communicate with the ancestors, and host guests from other tribes for the massive Kepele feasts. The second was the Great Ceremonial Wars of central Enga, tournament fights fought recurrently between entire tribes or pairs of tribes to demonstrate strength and brew the feasts and exchanges of enormous proportions that followed. The exchanges following the four Great Ceremonial Wars forged links between the inhabitants' four major valley systems and adjacent outlying areas. The third major exchange network was the Tee Ceremonial Exchange Cycle, a three-phase cycle of enchained exchange festivals that encompassed the majority of clans in eastern Enga and many clans of central Enga by first contact. As the Great Ceremonial Wars and Tee Cycle expanded in eastern and central Enga, bachelors' cults, female spirit cults, and ancestral cults were developed and circulated within Enga or imported from neighboring linguistic groups. The circulation of these cults did much to standardise values between areas and specify what was valued (Wiessner and Tumu 1999).

FIG. 2: *Chronological scheme of events discussed in text.*

Generations before present:

c. 9-12	• Introduction of sweet potato to Enga and beginning of Enga historical traditions. (c. 250-400 B.P.)
8	• Population shift from high altitudes to lower valleys • Beginning of early Tee cycle
7	• Kepele cult first practiced by horticulturalists of western Enga
6	• First Great Ceremonial Wars fought
5 (c.1855-1885?)	• Kepele cult imported into central Enga, called Aeatee • War reparations initiated for peace making
4 (c.1885-1915)	• Tee cycle expanded to finance Great Ceremonial Wars • Aeatee (Kepele) cult linked to the Tee Cycle and Great Ceremonial Wars
3 (c.1915-1945)	• Tee cycle begins to subsume Great Ceremonial Wars • Aeatee/Kepele cult used to organise the Tee • First contact with Europeans in early 1930s • 1938-41 last Great Ceremonial War fought • Tee cycle subsumes Great War exchange routes
2 (c.1945-1975)	• Tee cycle continues to expand
1 (c. 1975-2005)	• 1975 Papua New Guinea's Independence

Note: We have calculated a generation to be 30 years, though certainly for the earliest generations time distortions such as telescoping may occur. In view of this, events that occurred in the second to fourth generation before present were roughly dated in relation to known occurrences; from the fifth to eight generation before present, they were sequenced by genealogy but no attempts were made at dating. Prior to the eight generation, they can be neither dated nor sequenced. It is reassuring to note that trends such as the spread of the Tee cycle or major cults do show temporal consistency within and between areas.

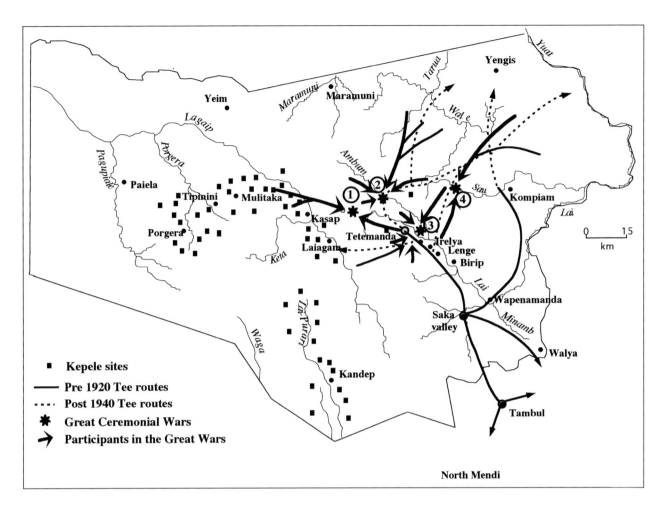

FIG. 3: *The Tee cycle, Great Wars, and Kepele Cult networks.*

Around the fourth to fifth generation before present, as the Great Ceremonial Wars expanded under the forces of dramatic inter-group competition, leaders of central Enga constructed complex alliances of exchange to effectively tap into the wealth of the Tee Cycle and to reinvest the great quantities of wealth that flowed out of the Great War exchanges. The cost, conflicts, and complexity of organisation of the Great Ceremonial Wars and Tee Cycle became formidable. In response, leaders imported the integrative Kepele cult from western Enga and used it as an occasion to unite clans and tribes for exchange and organise the timing and flow of wealth within and between the Great Ceremonial Wars and Tee Cycle. The three networks became linked. In the late 1930s the Great Wars were discontinued and their networks replaced by the Tee cycle. The full-blown Tee was vast, involving a population of some 40,000 people in eastern and central Enga at first contact.

The impact of egalitarian institutions on Enga warfare and exchange

Like many acephalous societies, Enga is governed by the constraints of strong egalitarian institutions stipulating potential equality. I define institutions following the New Institutional Economics as "the rules of the game" (North 1990) and their accompanying ideologies that set values, priorities, and world views (Ensminger 1992). I will refer to egalitarian institutions rather than considering Enga an egalitarian society, because pronounced inequalities or differences in spheres of influence do exist, for example, in relations between the sexes. Rules and ideologies of equality apply within the sexes – all Enga married men hold equal rights to be granted land, allocate household labour and its products as they see fit, receive support by group members in procuring spouses, be protected by the clan, have a voice in decision making, and pursue status.[2] Egalitarian institutions and ideals apply to potential equality but not equality of outcome (Robbins 1994). Though initially defined as equals, men are challenged to excel and achieve status via words and actions that are perceived to benefit the clan or sub-clan. However, name and fame are transient and the demise of big men is rapid when their actions are not perceived as furthering clan interests or when they infringe on the rights or resources of group members. Ideals of potential equality are extended to parallel units in the segmentary lineage system – clans are considered equal to other clans, lineages to other lineages, and so on.

Given the tension between potential equality and encouraged inequality of outcome in the arenas of status and influence, potential equality has to be continually maintained. This task falls first in the hands of individuals and then in the hands of lineages, sub-clans, or clans when the individual cannot manage alone. Each man is expected to exert and defend his equality and the rights of his household. Consequently, assertiveness is admired, promoting strong self-reliant individuals who make things happen to increase the prosperity of family and clan. Nonetheless, men who are weak still hold basic rights to land, spouses, voice, and protection.[3] Individual assertiveness is only expected to go so far – when a man is assaulted physically or verbally in the process of defending his rights, or when he faces hardship, group members are obliged to come to his assistance. Thus, contrasting moralities exist, the one espousing individual assertiveness, and the other, sociality, empathy, solidarity, loyalty, and sacrifice for the group.[4]

Egalitarian institutions differentially affect exchange and warfare. For social and economic exchange, they greatly facilitate cooperation by reducing transaction costs, that is, assessing value, protecting rights, and enforcing agreements (North 1990). For example, equality standardises information for most forms of exchange by stipulating that individuals are equals and that exchange should therefore be balanced over the long run. Within the clan or sub-clan, individuals offer help to others knowing that as equals they will be able to request and receive assistance when assistance is needed without the fear that assistance given to others will be used to build position and subordinate them. With individuals outside the group, where the most important exchanges involve significant delays and temporary imbalance, equality of exchange partners

is crucial to foster the trust that wealth given will be repaid. Under conditions of inequality and mistrust, delayed exchange is quickly eroded by fears of exploitation. Finally, to be received as an equal facilitates the mobility that was so crucial for men who moved from clan to clan to organise exchange. As Kelly (2000) has pointed out, hierarchies do not mesh easily. It is unlikely that the great exchange networks that developed in Enga history could have done so outside of the matrix of balance and trust fostered by equality. And big men appeared to have sensed this – in negotiating the import of cults, they sought those which underwrote potential equality between all men during periods of history when very real inequalities were emerging.

While egalitarian institutions fostered developments in exchange, they inhibited the same in warfare. Three dimensions of potential equality exerted profound constraints on warfare: rights over the allocation of one's own labour and the products of labour, voice in decision making, and equal rights to be defended by the clan members. Each of these merit discussion.

Rights over labour and its products. Perhaps the most strongly held right in Enga, as in other societies promoting equality within the sexes, is the right of individuals over their own labour. Though fellow clan members, affinal kin, or maternal kin choose to allocate a portion of their labour or wealth to assisting one another in a variety of activities, attempts to appropriate the labour of others meet with strong resistance. Men and women with established households never exchange their labour in return for payment – even allies in warfare are not compensated for their efforts, only for lives lost.

An individual's right to control over his or her own labour and its products has two ramifications for warfare. The first is that neither big men nor war leaders can command or buy the services of warriors[5], nor can they subjugate the losers and appropriate their labour. In all cases recorded in historical traditions, the defeated in warfare retreated or disbanded to settle elsewhere rather than forfeit their equality and independence. Second, because Enga cannot appropriate the labour of others, it is labour rather than land that is in short supply. Given a shortage of labour, the only way by which to increase the amount of wealth for distribution, other than by augmenting the work force via polygamy[6], is to construct individual networks of exchange outside the clan and thereby obtain access to the produces of the labour of others. Configurations of external exchange ties follow marriage links and thus differ for each household in a clan. When war erupts, it differentially affects the economic status of each household, generating conflicts of interest within the clan. Households having strong ties in the enemy clan seek to re-establish peace as soon as possible, while those with weak ties may opt to continue fighting. War can be particularly damaging to the households of big men who hold the broadest exchange ties.

Participation in decision-making. A major forum for maintaining equality amongst men is the clan meeting held to discuss issues of corporate group concern (Sackschewsky *et al.* 1970). In clan meetings, each mature individual is expected to express his own opinion, introduce topics, or ask questions to elicit the thoughts of others. Information and opinions are often phrased in symbolic speech that reveals only part of a man's knowledge or position at one time. Other men listen attentively, trying to analyse the speaker's words and gestures for intent and new information. Personal interests of the speaker are then evaluated in view of clan interests. Finally, opinions are synthesised so that individual positions are aligned with those of other members of the group and a decision is reached. Big men provide insightful information, warn of the dangers of certain actions, and seek to coordinate consensus, but do not assert any jurisdiction over clan brothers nor engage in blatantly persuasive speech, particularly when the issue is warfare. They have more influence than others if and only if they can convince their 'brothers' that their plans will optimally benefit the group. Men who aggressively persuade others to fight may risk blame when lives of men from other sub-clans are lost. Required consensus for warfare makes it difficult for any man to launch a war to pursue his own Machiavellian interests; competitors within the clan are extremely wary of each other's motives.

Right to be protected and defended by the group. As equals, men in Enga are neither judged positively or negatively by clan members nor given official recognition or punishment for their deeds. Certain insignia or items of paraphernalia may be worn on public occasions to advertise a man's achievements, however, it is the individual himself who chooses to wear these at the risk of mockery should others not judge his accomplishments as positively. The clan confers no honours. Likewise if a man has committed reckless acts or crimes, he will not be judged and punished by the clan. For internal cases, the offender and his kin must negotiate reparations for his deeds or risk retaliation by the kin of the offended party. In external relations, the entire clan backs a man, for the clan's reputation depends on the defence of members. Women receive protection from their husbands' clans and their natal clans alike.

Guaranteed vengeance or affronts to group members can be executed against any member of the enemy group, following the principle of social substitution which Kelly (2000) considers to be the hallmark of true war. It assures backing for each individual and signals the strength of the group relative to surrounding ones. Unavenged wrongs generate intense feelings of humiliation and insecurity. In a very real sense the pain of violence or shame are shared by clan members as expressed in the following quote:

Now I will talk about warfare. This is what our forefathers said: When a man was killed, the clan of the killers sang songs of bravery and victory. They would shout, '*Auu*' ('*Hurray*' or '*Well done*') to announce the death of an enemy. Then their land would be like a high mountain (*manda singi*) and that is how it was down through the generations. The members of the deceased's clan would become small (*koo injingi*). They would be nothing. But, when they had avenged the death of their clansman then they would be all right. Their hearts would be open (*mona lyangenge*). In other words, when one fights and takes revenge for the death of a fellow clansman, then one gets even and back on equal footing. (Tengene Teyao, Yakani Kalia clan of Wakumale, Wabag)

The assurance of clan backing, right or wrong, makes men willing to undertake acts of verbal or physical aggression that they might not otherwise contemplate. It also means that the rash actions of one man or a small group of men could derail the course of war or peace making. As a result, men hoping for a short surgical strike often find their clans embroiled in a enduring struggle fuelled by individual retributive action. It is perhaps for this reason that education for warfare was tuned to make young men ready and eager to defend their clans when attacked, but cautious about initiating violence. Moreover there were no initiations, cults, or other institutions designed to valorise warfare and organise men into a fighting force. Boys learned the skills of warfare at an early age in mock wars with grass arrows and by accompanying their fathers to wars. Stories of wars told and retold in men's houses recounted feats and defeats of past battles in rather factual terms and did not praise war heroes. Sessions for dream interpretation made men assess clan position in warfare and exchange. Proverbs and other teachings warned men against reckless murder or seeking to join the wars of allied clans (Gibbs 2002). Bachelors' cults bonded clan brothers for a number of activities, including agricultural cooperation, exchange, and warfare, but praise poetry detailing ideal men did not mention prowess in warfare (Wiessner and Tumu 1988).

In summary, the above egalitarian constraints made it difficult for enterprising men to pursue individual goals through warfare, assemble a fighting force larger than the clan and allied volunteers, and steer the course of war in a predictable manner. Below I will illustrate some of these constraints in action by briefly describing the course of a typical Enga war from about 1940-1990 drawing on the work of Meggitt (1977), Wormsley and Toke (1985), Lakau (1994), Allen and Giddings (1982), and two wars chronicled in my fieldwork of the 1980s.[7] This discussion will provide a baseline from which to address the second question of this paper: the role warfare in earlier generations of Enga history.

Enga warfare from c. 1940-1990

In considering Enga warfare three points should be noted. First, like 94% of societies surveyed by Kelly (2000), Enga clans are not always at war. Meggitt (1977) recorded eighty-six wars recalled by elders for fourteen clans of central Enga from approximately 1900-1950, suggesting that during this period Enga clans may go to war every 5-10 years.[8] Some clans go through periods in which they fight frequently and others may go for two or more decades without warfare.[9] Second, the Enga have no permanent enemies and engage regularly in marriage and exchange with all surrounding clans during times of peace, though they can specify neighbouring clans with whom they fight more or less often. Third, Enga life involves much verbal and physical violence, however, the majority of conflicts are solved through mediation (Talyaga 1982). Why a few incidents escalate into armed conflict depends on historical relations, tensions of the time, success of immediate attempts at mediation, and ulterior motivations of those involved, including interest in taking land, revenge, or pursuit of other political goals.

Most Enga wars are the outgrowth of disputes between individuals (Wormsley and Toke 1985:30). Whether fighting breaks out spontaneously or members of one clan wish to launch an attack on another, a secret clan meeting (*kambuingi*) is held excluding members with mixed loyalties (Sackschewsky *et al.* 1970). In such meetings, each man expresses his opinions and then efforts are made to coalesce conflicting opinions into consensus. Internal conflicts of interest may be considerable. Even if some men are interested in gaining land, these desires are not usually voiced, though regaining land lost in former battles may be. This is because only one lineage is likely to benefit from small areas of land gained, the outcome of war is unpredictable, losses may be high, and land taken is usually contested militarily for generations. Big men must be careful in these circumstances to summarise the situation, warn of the pros and cons of warfare, but not try to coerce fellow clansmen. Once made, however, decisions to go to war are binding, though decisions to seek peace may be thwarted by a small contingent of angry men who decide to take violent action. Men in a single clan go to war for very different reasons: in response to the triggering incident, to make a name in battle, for the excitement and brotherhood, to gain a small piece of land, to make a name through peace negotiations, or to fight out old grudges in a new war. There is little that leaders can do to prevent individuals from pursuing private agendas. What unites all men is the desire to uphold clan honour in the face of provocation by another clan, whether this is verbal insult, physical aggression, or the destruction of property.

As opposing clans are composed of acquaintances, relatives, and 'yesterday's friends', numerous rites and activities are performed to unite the clan and distance the enemy. At the onset of the fight dehumanising songs are sung and insults are hurled across the border zone to taunt the enemy. Pigs are slaughtered to elicit the help of the ancestors and rites of divination are held to determine what the fighting will bring. Few men can resist the call of brotherhood and the excitement of battle; the recalcitrant are goaded with insults until they join. Individuals draw on a repertoire of fight magic purchased from ritual experts for skill in warfare and protection from enemy arrows (*telya lakoe nemongo*), while women observe practices believed to safeguard their men. Should the events of battle take a turn for the worse, a secret clan gathering (*yanda aiputi*) followed by rites of absolution is held to confess past grudges and deeds against clan members. At any point during a war if clan members feel the war is going badly and attribute their losses to rifts between brothers, a clan gathering (*yanda aiputi*) may be held in a secret place during which each man confesses past bad feelings, thoughts, or deeds against fellow clan members followed by rites of absolution.

By day volleys of arrows are exchanged in an atmosphere that may initially resemble a sports match with intoxicating team spirit. However once a man has been killed, 'sport' turns to rage. At this point women, who have no say in decisions regarding warfare, retreat with their children, possessions, and pigs to stay with relatives for the duration of the war (Kyakas and Wiessner 1992). As a war intensifies, small groups of men undertake lethal ambushes and night raids to burn houses, greatly increasing the probability of fatalities. With each man killed or wounded, each house burned, or piece of property destroyed, fear and anger take control and the possibilities for peaceful settlement become more remote. At such times men describe physiological and emotional changes stimulated by fear, anger and, very importantly, sleep deprivation; the effects of these factors on judgment should not be underestimated.[10]

Units of conflict extend beyond the political and territorial group of the clan, and for any given war, some external ties are suppressed and other activated when allies, usually men with affinal ties, come to join the fight. Allies are a mixed blessing. On the one hand the number of allies has an important impact on the outcome of the fight. On the other, allied deaths incur great costs to the host clan, particularly if the enemy tries to incite tension within an alliance by targeting and killing allied men (see also Strathern and Stewart 2000). Moreover, allies may have their own motives for assisting, for example, the desire to fight out their own grudges on the land of another or to weaken a threatening clan in somebody else's fight. For these reasons, the help of allies can be rejected. If allies are accepted, a ceremony is performed which commits the host clan to paying reparations for allied warriors killed. As a war progresses, so do the various motivations of everybody involved.

An end to a war may be called after days, weeks, or months, for a variety of reasons: when losses on both sides are roughly equal, when the offended party feels they have taken revenge and balance of power is restored, when either side fears that debts incurred from allied deaths are too high to pay, or when both sides tire of fighting and wish to resume exchange. Whereas big men may have been unable to exert much influence over the course of the war, when reconciliation is desired, they take the lead. The months that follow are devoted to re-establishing balance and respect. First allies have to be appeased and compensated for men lost, else war will break out between allies and hosts. Next big men have to initiate peace with the enemy. The process of reconciliation is painful and fraught with problems caused by internal disputes and the question of who should pay compensation first. In wars of short duration, land overrun by victors may be returned in the interest of peace; in long and bitter ones as much land as possible may be taken to punish the losers. Land is the means of sustenance and space in which individuals and clans assert themselves and realise their ambitions. It therefore represents both pride and independence. To take the land of another is the ultimate humiliation.

Enga say that 'wars start from the tongue', accordingly people are careful with their words and allow only skilled orators to negotiate peace. Meanwhile, the bed and possessions of the victim are removed so that his clan brothers will not be reminded of him and seek revenge. When payment of reparations is agreed upon, big men call a clan meeting in which each male member is urged to contribute pigs and cash to the war reparation exchanges. Clan meetings are called again and again until the total sum of wealth to be given is considered sufficient. War reparations exchanges take place in three stages over a period of two years or longer: (1) *kepa singi*, the first payment of steamed pork to the clan of the deceased in public; (2) *saandi pingi* or *yangenge*, initiatory gifts of pigs/piglets, goods and valuables given in private by the clan of the deceased to the clan of the killer to oblige the final payment of live pigs (3) *akali buingi*, the formal payment on the clan's ceremonial ground of live pigs, cassowary, axes, salt, or other goods and valuables to the clan of the victim by the clan of the killer. Individual families give payments to their kin or exchange partners in the enemy clan with larger sums directed to the immediate kin of the victim. Returns go directly to the donor household; wealth is not pooled and redistributed by big men, though payments are given during a coordinated distribution on the clan's ceremonial grounds. The three-phase reparation exchanges conducted over a period of approximately two years deters aggression with the promise of wealth to come and plays on the healing hands of time. The clan of the killer seeks to surprise the clan of the victim by giving more than expected and thereby trying to express respect and good will. Individuals, particularly big men, salvage some of the losses incurred in warfare by enhancing their reputations as peacemakers and forming strategic alliances.

Once reparations are paid it is said that 'the spear is broken', *yandate lakenge*, and that people 'can stay with good hearts', *mona epe palenge* (Young 2002: 24). War reparations and balance of power restored reactivate profitable exchanges between households of the two clans. Nonetheless, new tensions can be generated if any disgruntled men seek revenge through payback murders. If land is gained by one side, it is often left as barren no-man's-land for years or even decades until the victor begins to occupy it, first by planting trees or grazing pigs and only later by building gardens and houses. Meggitt (1977) argues that the Enga rely on uninterrupted use of land gained over time to effect social amnesia in which *de facto* possession is converted into *de jure* title. However, our historical studies, those of Wormsley and Toke (1985), and even remarks made by Meggitt (1977) indicate that social amnesia is not prevalent among the Enga, nor is land taken in warfare regarded as the legitimate property of the victors unless one party disbands or migrates to another area. Many wars have deep historical roots.

In summary, owing to egalitarian constraints and colonial influence, there is no indication that warfare was a moving force towards generating political complexity between 1939 when warfare began to be suppressed by Australian patrols and independence in 1975. At best, it helped redistribute the population over resources and establish balance; at worst it was a negative force inhibiting developments. But was this also the case for earlier generations? If so, or if not, why? To address these questions, I will turn to historical studies carried out by Akii Tumu and myself.

An historical perspective on Enga warfare

Historical traditions describing Enga wars of the past abound. In our studies we systematically collected migration histories, but not warfare histories nor details of wars after c. 1920-30. To centre on recent wars would have eclipsed other important issues and made elders suspicious of the motives behind our research (Wiessner and Tumu 1998). However, in the course of recounting their tribal histories elders did tell of wars that had a lasting impact on politics in that they led to the separation of clans of a tribe, generated lasting antagonisms, or set off migrations. Because we have thorough and systematic data on wars that led to migrations, but only sporadic data on wars that had only short-term impact, we can give no concrete figures on the frequency of war through time.

Most Enga historical traditions recounting wars are surprising in that they do not glorify war, valorise the deeds of heroes, nor seek to establish right or wrong. Avenging verbal or physical insults to the clan or its members, whether deserved or not, is taken as a just reason to go to war. In most war traditions, names of clans are substituted for the names of individuals by two generations after the war occurred; thus they can be used to justify interests of clans but not of individuals. Some accounts of war are mere sketches, while others are told in considerable detail. For most it is possible to determine: (1) in which generation a war took place, (2) what was the triggering incident, (3) who participated, (4) whether war was premeditated or the product of unfortunate misunderstandings, (5) if motivations for the fighting were altered significantly during the course of a war, and (6) what was the outcome. The data on wars presented in Tables 1-2 are divided into three periods. The first period covers events that occurred before the eighth generation and is of unknown length. We only have one fixed date from this period, the mid-1600s when a volcano on Long Island erupted and covered the PNG Highlands in ash. The second period covers the 7-6th generations before present, a time in which events can be sequenced but for which we have no fixed dates. If generations during this period are approximately 30 years, this may cover a sixty-year period (c. 1795-1855); however collapsing of generations may have occurred. On the assumption that a generation is 30 years (Lacey 1975), the third period includes wars fought during the 5-4th generations before present and can be dated to approximately 1855-1915 from known events (Wiessner and Tumu 1998: 33). We do not have comparable, systematically collected data on wars from approximately 1920 on. Unfortunately we cannot use Meggitt's data for this time period owing to problems with dating and compatible methodology.[11]

Wars prior to the eighth generation

Prior to approximately the eighth generation, migrations were common and appear to have been set off by new opportunities provided by the sweet potato. Population density is described as sparse, life as lonely, and spouses hard to find. Clans often sought new members amongst maternal and affinal kin to build communities (see also Wohlt 1978). Of 84 clans in eastern and central Enga who migrated in this period, 59 (70%) sought to join relatives residing in the lower valleys while only 25 (30%) migrated as a result of conflict (Table 1).[12] Nonetheless, prior to the eighth generation life was by no means peaceful. Accounts of internal warfare abound, telling how brother clans fought and redistributed themselves over tribal land or how clans from different tribes fought to space troublesome neighbours. A wide spectrum of wars is described, ranging from skirmishes among hunters in the high altitudes to full-blown wars involving entire tribes and their allies. Virtually all wars from

this period began with personal disputes; some are described as escalating from brawls, to combat with sticks and clubs, and on into full-blown wars fought with spears and bows and arrows. The most common causes of disputes (84% of all cases) were over sharing work, possessions, or meat, and the majority of wars were internal, that is, between clans of a tribe (Table 2). For 12 out of 13 wars that ended in migration, the losing party left voluntarily. For six of these (46%), historical traditions mention the regret of the victors at the loss of a brother or friend. In nearly half the migrations, relatives living elsewhere encouraged the losers to leave the area of conflict and offered them good land.

In the early generations, then, war appears to have solved problems by dividing groups that had grown too large to cooperate or by spacing troublesome neighbours. The outcome of most of these wars was dispersal within tribal lands and re-establishment of balance of power and exchange relations.

Table 1. Number and average distance of migrations by generation for eastern, central Enga and the western Lagaip up to the Keta river

Generation	8+	7th-6th	5th-4th
number of migrant clans	59	22	9
average distance of migration	29 km	32 km	14 km
after warfare: no. of migrant clans	25	71	29
average distance of migration	27 km	26 km	17 km

– These data include only clans who were totally displaced from the central valleys of Enga, not immigration into Enga nor secondary migrations once groups had established new residences in fringe areas.

– Total number of clans whose history was covered in these areas is approximately 300.

Table 2. Incidents that set off tribal wars by generation

Generation events	8+ sweet potato population shifts	7th-6th migrations rise of Tee, Great Wars, Kepele cult	5th-4th Tee, Great Wars and Kepele flourish
triggering incidents:			
hunting/meat sharing	12 (39%)	4 (19%)	4 (13%)
possessions or work sharing	14 (45%)	5 (24%)	3 (9%)
pigs	1 (3%)	2 (9.5%)	3 (9%)
pandanus	1 (3%)	0 (0%)	2 (6%)
political/ homicide**	1 (3%)	5 (24%)	7 (22%)
rape/adultery	2 (7%)	3 (14%)	4 (13%)
garden/land disputes	0 (0)%	2 (9.5%)	9 (28%)
total	31 (100%)	21 (100%)	32 (100%)

* 8+ encompasses of some 150+ years around the time of the introduction of the sweet potato or shortly after.
** This category includes: homicide, disrupting a funeral, refusal to help as allies, refusal pay war reparations, hindering exchange, etc.

If groups were too small to be viable after dispersal, they recruited new members via affinal or maternal ties. In the case of routing, land abandoned after conflicts often could not be filled by the 'victor' and was given to allies with the hope of replacing bad neighbours with good ones.

Wars of the 7th to 6th generation

The population shifts that occurred 8+ generations following the introduction of the sweet potato had their repercussions in the 7-6th generations. Immigrant groups so eagerly recruited in the 8+ generations often did not provide the supportive neighbours their hosts had hoped for. Within a generation or two the hosts were at war with the hosted, and sometimes it was the hosts who were then displaced. At this time the number of voluntary migrations on the part of those seeking new land decreased (Table 1). Though sharing and cooperation still played a role in generating conflict, pigs, homicide, political disputes, rape and garden disputes became more common causes for armed conflict. For wars leading to migrations, on which we have systematically collected data for all generations, external wars became as frequent as internal ones. Incidents of voluntary departure of an offended party accompanied by expressions of regret on the part of the offender declined.

Wars of the 7th-6th generations appear to have been larger in scale and led to larger population movements in terms of numbers of clans displaced than at any other time in Enga history. Some involved entire tribes, for instance two wars of this period together led to the replacement of the entire population in the Ambum valley and the acquisition of approximately 160 sq km of good agricultural land. One of these wars was likely to have been motivated by desire on the part of clans in the high country for fertile garden land in the Ambum valley; the other grew out of a personal dispute between two men whose tribes already had more good land than they could fill. In both cases the victors were hard pressed to fill the land vacated and had to call on allies from other tribes to occupy a share.

Not only were many of the wars of the 7th-6th generations large in scale but also extremely destructive. To give one description recorded by Roderic Lacey:

...In those days warfare was prevalent and it was in one of these wars that the sons of Yoponda and Nenaini became involved. The war by which they were forced out of their territory was against the Tendepa people and one other group whose name I cannot recall. The war broke out after a quarrel about a stolen boar and a tussle over a piece of land through which one man was building a garden fence.

The war that followed was one of the longest that was ever fought in the area. It went on and on until there was not more food left and all the pig stock was destroyed too. When the war was over, compensation payments (to allies) still had to be made, but Nenaini and Yoponda had no pigs with which to pay and so they were faced with another problem besides war. (The problem was solved by the losers giving their land to their allies as payment and migrating out of the area upon the invitation of relatives.) (From R. Lacey, 1975: 259-60; Narrated by Kale, Yoponda, Walya in November 1972, transl. Nut Koleala)

Nonetheless, most wars that led to migrations for the 7th-6th generations appear to have developed out of smaller disputes, escalated, and changed in intent as the war progressed, just as did fights of more recent decades. The following historical tradition from the Lagaip valley of western Enga provides a good example.

Maipu, a 'son of Diuatini' (i.e. from the Monaini Diuatini clan), lived at Poko near Mulitaka a long time ago (sixth to seventh generation before present). A man from the Diuatini Maipa sub-clan killed a man from the Kaia tribe and a war broke out. During the war no Diuatini man was killed to avenge the death of the Kaia man, and so when the war came to an end, Kaia decided to avenge the death in a payback killing rather than in a tribal war. Kaia offered a Diuatini man one of their women in payment if he killed a fellow Diuatini clansman. He accepted the offer and murdered a young man from the Diuatini Kakaipu clan by sneaking into his house and hacking him up with an axe while he was sleeping. The man who was murdered had just gotten married.

The Diuatini Kakaipu sub-clan blamed the matter on the Diuatini Maipu sub-clan who had started the war in the first place. They told the Maipu sub-clan that they would later help them in their war against Kaia, but that first they needed to avenge the death of their own man. They did so one day when a man from another Diuatini subclan was building a house. A Maipu man came over to help him thatch the roof of his house, and while they were working, a Kakaipu man, who had been hiding in the bushes, jumped out and split the Maipu's man's head open with his axe. Shortly after when a Maipu man met a Kakaipu man, he murdered him in payback. By this time the situation had gotten very tense and a tribal war broke out within Diuatini. Many people were killed. Maipu clansmen and their wives and children fled to Maramuni by the Molyoko track. Kakaipu and other Diuatini clansmen fled by the Pawapi track. Most of them later settled near Pasalakusa. (Apele Ipai, Diuatini Kakaipu, Maramuni)

Other wars motivated by political intent produced a domino effect:

The story begins with four sons of Londeale, Wapo, Mangalia, Koepe and Kalipa. One day they decided to hold a traditional dance (sing-sing) at Tupisamanda, so Wapo and Mangalya clansmen went to Kandepe to get some mambo oil to which was used for traditional body decoration during dances. They arrived back from Kandepe with the containers of oil and were upset to find that their brothers had held the sing-sing in their absence. Since they had missed the first, they staged their own the next day. (Sing-sings were used as occasions to gather people for exchange.)

Yambetane was annoyed with Sikini's troublesome sons since a second sing-sing would attract more people and lead to more damage of gardens, theft, and so on. They fought with Sikini and drove them back to Kapetemanda and then on to Takenemanda where they took refuge. While they were at Takenamanda, a man from the Itokone Lundopa clan betrayed them. The Itokone Lundopa man made arrangements with Yambetane for a rope to be lowered down the cliff below the men's house where the Sikini men were sleeping so that Yambetane could climb up and attack at night. The plan worked and two Sikini men were killed.

Later the Anjo clansmen from the Kandawolini tribe told Yambetane that they had been betrayed by a Lundopa man and that they should kill a Lundopa man in revenge. Anjo was fighting Lundopa at that time and thus offered Sikini some of their land at Mapemanda in the Pina area of the Lai valley if they killed a Lundopa man. The Sikini Londeale men then did as Anjo asked and killed a Lundopa man both to take revenge and to obtain the land offered to them by Anjo. They then moved to their new land at Mapemanda where Koepa and Kalipa are today. Once settled at Mapemanda, Sikini Londeale clans got into a fight with Lyongeni and Waingini over a bird stolen from a trap, drove the latter out to Kompiama and took over their land. They also drove out parts of Tinilapini and remaining clansmen from Aiamane to Kompiama and took their land.

Sikini Wapo and some of the Sikini Lakai clans remained on the land in the Saka although their brother clans moved to the Lai. (Gabriele Konge and Joseph Kambao of Sikini Koepa clan, Pina)

In the last case the underlying impetus for the initial war was almost certainly politics of Tee exchange which were often negotiated during traditional dances. For the second bout of fighting, revenge appears to have been the motive, and for the final one, conflict that arose when an immigrant group disrupted the balance of power in their new area of settlement. Note that here, as in most historical narratives of the time, land was described as plentiful and offered to immigrant groups who promised to be good allies and to open opportunities for exchange. In many wars of the 7-6th generations, if land was vacated by the losers, it was shared with allies of the victors or in some cases given to allies of the defeated before their departure. As in former times, warfare temporarily solved problems of difficult and uncooperative neighbours by spacing conflicting groups and in some cases replacing them by allies anticipated to be better neighbours.

As a result of the turmoil of the 7-6th generations, three large ceremonial exchange networks mentioned earlier were formed: the Tee ceremonial exchange cycle of eastern Enga, the Great Ceremonial Wars of central Enga, and the Kepele Cult network of western Enga. By the 5th to 4th generations these systems had expanded to form broader networks of enchained exchange throughout Enga. All three systems put demands on pig production and external finance. Access to the products of labour of people in neighbouring groups became more important than ever before.

War in the 5th-4th generations

By the 5th-4th generations garden disputes, homicide, and other political issues became the predominant triggering incidents for wars (Table 2). As in other generations most wars arose out of individual disputes. Seventy percent of wars leading to migration were internal ones; voluntary departure of an aggrieved party is unusual. However, the number of clans who migrated after warfare decreased radically in the 5th-4th generations (Table 1), and by the 3rd-2nd generations we only recorded five migrations of entire clans for eastern and central Enga after warfare, though numerous sub-clans disbanded and resettled with affinal and maternal kin. Where the real difference lies between the 5th-4th generations and preceding ones is in the outcome of wars. For the first time in Enga history, traditions tell of war reparations, which were formerly paid to allies only, being extended to enemies in attempt to restore peace without resorting to spatial separation. The changing outcomes of wars can be attributed to two factors. First, with a growing population, relatives were not as eager to welcome new immigrants. Second, and very importantly, opponents did not want to expel neighbouring groups who provided valuable partners in networks of enchained exchange; they only sought to establish balance so that exchange could flow. Emphasis shifts from battle descriptions to efforts to restore balance peacefully. The following tradition tells of how such efforts were defeated by runaway aggression. Nonetheless, it is the peacemakers who emerge as heroes.

Yoelya and Petakini were brothers. They were descendants of peacemakers. Yokone was the son of Yoelya from the Wakemane sub-clan of the Apulini Talyulu clan. Yokone was a married adult. One day it happened that Yokone was passing by a garden where the daughter of Piuku was working. Piuku was a big-man of the Wataipa sub-clan of the Apulini Talyulu clan. Piuku's daughter was young and extremely beautiful and Yokone was sexually aroused by the sight of her. He dragged her to a secluded place and raped her.

After hearing of the incident, Piuku and his supporters went over to the house of Yokone. They opened the door and entered, taking several pigs which were inside. They did this because Piuku was the leader of the Wataipa sub-clan and what had happened to his daughter was a serious offense to him. When they broke into Yokone's house, they were fully armed with spears, bows and arrows.

Petakini who was the cousin of Yokone heard of the rape and immediately took one female and one male pig from his herd and promised them to Piuku. He also said that he would give him another pig which had not yet returned from the bush to try to prevent the incident from escalating into further violence. He told Piuku and his men to return home with their weapons. Then Petakini went to Yokone and handed over the pigs to him and Yokone took them and gave them to Piuku and his men. Petakini who had come to witness the giving of the pigs went back to his house in Wandi. Shortly after he left, Piuku and his men who had accepted the pigs planned to kill Yokone. They soon forgot the promise that Petakini had made earlier that meant in effect that Petakini as a big-man, would be able to increase the number of pigs paid in compensation to Piuku for the rape of his daughter.

Yokone and his men were fully aware of the tense situation. They were armed and when Piuku and his supporters came to attack Yokone, they made a counter attack. The two groups

fought and before long Piuku was fatally wounded with a spear and died. The Apulini Kapeali clan took part in the fight and supported Wakemane. Unaware that the fight had broken out, Petakini, Tamati and a couple of other Wakemane men were at Wandi when they heard a cry from the fight area announcing Piuku's death and then a song of victory sung by Wakemane: 'The leader of Wataipa is now gone. The eaters-of-liver will mourn for him.' The eaters-of-liver refers to poor men.

On the same day Eneakali, the brother of Kanapatoakali, was killed by the Wataipa. He was from the Yakani Kalia clan of Wakumale and was visiting his cousin Petakini before his death and so fought as an ally. The skirmish then escalated into a full-blown war. Neighboring clans took sides and the Yakani clans of Paluia and Sane came to help Wakumane in the fight. The war went on for two weeks and then both sides agreed to pause for two weeks to recover from their exhaustion. After the break the war continued. It was a terrible war. It is said that even the mourning house for Piuku was burned down. After another two weeks of fighting, the warring clans sent messages to one another and withdrew from the war zone.

The narrative goes on in some detail to tell of peacemaking, exchange of reparations, the renown achieved by the peace-makers and the profitable exchange that was resumed. No land was gained or lost.

Summary

Historical traditions indicate that war was always prevalent in Enga life; whereas Enga say that long ago life was quite different with respect to such matters as population, ritual, and exchange, we heard no claim that there was a time before warfare. Although population grew substantially during the period considered, the presence, scale, organisation and technology of war changed little. Some wars were fought over land as Meggitt (1977) suggests and others may have been motivated by the strategies of Machiavellian big men as Sillitoe (1978) proposes, however, there is little in historical traditions that suggest these were the primary forces behind most wars.[13] Even though triggering incidents for wars changed with the interests of the time, the overall intent of most wars regardless of generation seems to have been to solve problems with troublesome neighbours after insult or injury and recreate balance of power, relations of equality, and respect. This goal could be achieved by spacing conflicting groups and then re-establishing equality, routing neighbours with whom problems seemed insoluble, or, in later generations through the exchange of wealth in war reparations. Given a balance of power, wealth could flow along exchange networks and people could transcend the bounds of household labour in assembling wealth for distribution. As Tuzin (1996) has proposed, many wars were fought in the ultimate interest of peace. That is not to say that balance and equality were achieved with ease through warfare – as often as not weak leadership, diverse interests of fighters, unanticipated events of battle, and the need for revenge propelled wars into vicious revenge cycles.

Whereas the constraints exerted by egalitarian institutions on warfare relegated warfare to a conservative, levelling force throughout Enga history, it did have important indirect effects on the development of ritual, exchange, and political complexity. First it redistributed people over land, greatly expanding the territory occupied by the Enga. Second, it continually re-established the political matrix necessary for exchange to expand and flourish – a balance of power and equality. Third, while conventional warfare itself underwent little change, great strides were made in restoring order through the payment of reparations with enemy. The negotiation and payment of war provided forum for the emergence of leaders who had influence beyond the borders of their clans and furnished an important impetus for developing enchained finance.

The Great Ceremonial Wars

The above results beg the third and final question of this paper: 'Did Enga experiment with warfare to try to circumvent egalitarian constraints in order to use warfare more productively in their strategies?' Material contained in historical traditions indicates that the answer to this question is a definite 'yes'. Throughout Enga history, enterprising men took advantage of any means available to them to enhance the prosperity of their households and clans. Warfare was no exception. From the eighth generation on there are historical narratives describing how big men tampered with wars to construct formats that would provide the benefits of war without the disadvantages, for example by trying to construct more 'sportive' wars of short duration to generate the exchanges that followed. These efforts had either limited success or were one-shot affairs that could not be duplicated, however, there is one notable exception: the Great Ceremonial Wars of central Enga (Wiessner and Tumu 1998: ch. 10).

The Great Ceremonial Wars are said to have commenced in approximately the sixth generation. The development of the Great Ceremonial Wars is not well documented in historical traditions; however, we do know that four Great Wars between different pairs of tribes of central Enga were fought in repeated episodes at 10-30 year intervals beginning in the mid-1800s until the early to mid-1900s. The last great ceremonial war fought c. 1940. The Great Wars developed out of large vicious wars between entire tribes or pairs of tribes in the 8th-6th generations, two of which appear to have been over resources and two over seemingly minor issues. The Great Wars were gradually constructed by altering the rules of conventional war in order to preserve some of the benefits of warfare without the destruction. That is to say, they were designed to create larger units of cooperation and profit from the exchange that ensued between hosts, owners of the fight, and allies.

The purpose of the great Wars was summed up by two of the elders who participated in the last Great Ceremonial war:

The Great Wars were planned and planted like a garden for the exchange that would follow. They were arranged when goods and valuables were plentiful and when there were so many pigs that women complained about their workloads. Everybody knew what they were in for, how reparations were paid for deaths, and what the results would be. They were designed to open up new areas, further existing exchange relations, foster tribal unity, and provide a competitive, but structured environment in which young men could strive for leadership. These qualities of the Great Wars made them differ from conventional wars, which disrupted relationships of trade and exchange, causing havoc and sometimes irreparable damage. The distributions of wealth that took place after the Great Wars brought trade goods from outlying areas into the Wabag area on the trade paths initially established by the salt trade. (Ambone Mati, Itapuni Nemani clan, Kopena [Wabag])

The underlying purpose of these wars was to bring people together – they were formal and ceremonial. They were fought to show the numerical strength and solidarity of a tribe and the physical build and wealth of the warriors; figuratively it is said that in the wars, 'They exposed themselves to the sun'. The Great Wars were events for socializing. After getting to know each other, they would kill many pigs and hold feasts [Great War exchanges] (Depoane of the Yakani Timali tribe, Lenge [Wabag]).

The Great Wars were fought between entire tribes or pairs of tribes who were 'the owners of the war' and intermediary tribes who hosted men from the respective sides. The hosts provided their guests with food, water, entertainment, and front line fighters for the duration of the war. The timing and location of Great War episodes was first negotiated by fight leaders (*watenge*), big men chosen from clans of the 'owners of the fight' for their ability to plan, put on spectacular public performances, and organise exchange. Then a mock attack was staged to spark the war. For weeks before the battles began, people from the hosting tribes on both sides received warriors in their own houses and began to make preparations, sing, dance and brew the fighting spirit. Meanwhile fight leaders (*watenge*), drew up plans for battle and the exchanges to follow.

On an appointed day hundreds of warriors, or in later generations a couple of thousand, appeared on the battlefield in full ceremonial regalia. Fight leaders who engaged in flamboyant ritualised competition announced a formal beginning. Each side had one or two fight leaders from each participating clan, who were big men renowned for their public performance and ability to organise exchange. Fight leaders were considered fair targets for humiliation, for example, warriors sought to capture or to steal their plumes, but it was considered foul play to kill them for they were the men who would orchestrate the Great War exchanges.[14] Fighting took place in a designated zone on the land of the hosts so that no land could be gained or lost. By day warriors fought in front of hundreds or thousands of spectators, while the women sang and danced on the sidelines. By night they ate, drank, talked with their hosts, and courted their daughters. The immunity extended to war leaders did not apply to ordinary men who did indeed die in battle, sometimes at the hand of 'friendly fire' when mistaken for enemy. But death rates were generally low because lethal tactics such as night raids and ambushes were frowned upon. For example, in the last Great War, which lasted for several weeks and involved over a thousand men, three men were killed on one side and four on the other. The battles continued for weeks or months until fight leaders decided to hold a closing ceremony and cast their arms into the river. In these 'fights without anger', no land could be gained or lost, no damage was inflicted on property, and the men who died were said to have given

their lives for a worthy cause. Their deaths were not avenged.

Next a series of massive and festive exchanges was initiated; they would continue for two to four years. Essentially, they transformed the close relationships between hosts and hosted that had formed during their weeks or months together into exchange partnerships. First, the hosts paid war reparations for allied deaths and then held a feast to mark the end of the war and initiate the host payments from the 'owners of the fight'. Pigs and cassowary were dramatically and ceremoniously presented for this event. After some months had elapsed the owners of the fight reciprocated with large distributions of raw pork in which individuals gave to the families who had hosted them. In the last great fight more than 20 participant clans slaughtered some 1000-2000 pigs on one day. Hosts and allies got up at dawn and travelled over hill and vale from clan to clan collecting pork from families they had hosted. It was said that even the dogs could eat no more. During the months that followed, hosting families gave another round of initiatory gifts which were reciprocated by the hosted in the form of live pigs decorated for the occasion, marsupials, cassowary, goods and valuables. In this way bonds that had been formed during fighting were turned into exchange partnerships. Meanwhile the Great War courtship parties generated post-war marriages.

After the closing feast, interaction between opposing sides that had been forbidden during the Great Wars could be resumed immediately without tension, for the Great Wars were said to be without anger. Furnished with new exchange ties, opponents became desirable exchange partners for one another. However, no food distributed during Great War exchanges was to cross enemy lines, a prohibition believed to be enforced by the 'sky beings' who punished transgressors.[15]

The Great Wars were fought repeatedly at 7-20 year intervals from the early to mid 1800s on, peaking in the early 20th century. Around the 5th-4th generations big men tapped into the emerging Tee cycle in order to bring pigs from eastern Enga to provision the exchanges and send pigs flowing out of the exchanges eastward to repay creditors. With this additional influx of wealth, both the Tee Cycle and Great Wars expanded greatly, creating networks which linked the Lagaip, Lai, Ambum, and Sau valleys (see Fig. 3). The most recent Great Wars involved up to 3000 warriors, the exchange of between 6,000 and 10,000 pigs and many trade goods, by contrast to most conventional wars that drew a couple of hundred warriors and the exchange of 60-300 pigs. Starting around 1900 the Great Wars became formidable to organise. As a result, three of the Great Wars were discontinued well before Australian patrols entered the area when their organisers decided it would be more profitable to subsume Great War networks with those of the Tee Cycle. The last episode of the fourth Great War was fought around 1940. We do not know if this one would have continued had Colonial intervention not occurred. Thereafter only smaller, conventional wars that had always been fought in the intervening years between the Great War episodes persisted.

The Great Wars were ingeniously constructed to circumvent the constraints of egalitarian institutions and allow warfare to yield productive outcomes. First they alleviated the need for revenge, so that the Great Wars would not escalate or take an unforeseen course, but proceed uninhibited to the exchange phase. Second, exchange ties were disrupted with the enemy only during the period of fighting; meanwhile, strong ties were forged with hosts during the war. In some Great War episodes, reparations were paid to enemies as well as hosts and allies. Third, internal conflicts were few when land, property and exchange ties were not at stake. With minimal internal conflict and shared motivations on the part of most men – to be able to participate in the exciting battles, demonstrate prowess, take part in social events, and profit from exchange – the Great Wars grew to much greater proportions than did any other events in Enga history. The Great Wars accomplished what was intended but often not accomplished in conventional wars – to display strength and establish relations with surrounding groups conducive to profitable trade and exchange. And such large-scale events provided unique opportunities for strong leaders to arise. By the second generation that some Great Wars were fought, historical traditions and genealogies indicate the position of Great War leader was passed from father to son or nephew in response to public demand for continuity in leadership. Names of the Great War leaders were known throughout Enga.

Concluding remarks

Egalitarian institutions that facilitate exchange vary greatly between societies, as does the context in which they operate. Their impact on warfare changes accordingly. In other Highland societies such as the Huli (Glasse 1968) and the Chimbu (Brown 1964; 1978) where men were potential equals, where emphasis was placed on managing wealth, and where substantial amounts of wealth were obtained through external ties, warfare appears to be a leveling force with more indirect than direct effects on the development of political complexity just as it was in Enga. In the eastern Highlands groups where marriage did not foster such strong exchange ties and emphasis was not on management of wealth (Du Toit 1975; Lindenbaum 1979; Godelier 1982a; 1982b; Robbins 1982; Watson 1983: 114), fewer internal dilemmas were generated by war. Consequently, unrestricted warfare caused continual fear of annihilation, conflicts were settled only by dispersal, and the weak hierarchies that did develop applied only to warfare and ritual (Du Toit 1975; Godelier 1982a; 1982b; Robbins 1982; Watson 1983: 93). Despite such variation, when egalitarian institutions preclude the exploitation of the labour and every individual has a right to be defended and avenged by the group, warfare may serve to vent anger, redistribute groups over the landscape, promote the solidarity, or re-establish balance, however, it will not be a moving force towards hierarchical complexity.

That said, New Guinea is a land of intrigue and experimentation in all realms of life; warfare is no exception. Enterprising men did indeed find paths to circumvent egalitarian constraints on warfare. The Enga achieved this by the ritualisation of war without invoking hierarchical structures, but yet allowing inheritance of leadership to arise out of public demand for continuity of leadership for the organisation of these popular events. By contrast, some Sepik societies imported hierarchical ritual structures for contexts linked to war, while secular equality prevailed in day-to-day life. As Tuzin (1976; 2001) has shown for the Ilahita Arapesh of the Middle Sepik, the complex ritual organisation of the Tamboran cult was imported from the neighboring Abelam to provide a hierarchical structure that was activated in certain contexts in order to to counteract tendencies towards fission. Military strength was thereby maintained. The upshot was the formation of a community of 1500 people in society formerly made up of small hamlets composed of but a few extended families (Tuzin 2001). Harrison (1985; 1993) describes a similar situation for the Atavip in which ritual hierarchy was an alternative form of social action that temporarily altered the secular equality of daily life. Hierarchical values celebrated in the male cult were responses to the real and perceived exigencies of war but had few repercussions on daily life. They existed side-by-side with secular institutions of equality and were only activated for unity or defence. It is perhaps in institutions imported or generated to side step the constraints of warfare in societies with strong egalitarian institutions that the transition from warfare as a conservative to a progressive force may be found.

Acknowledgments

I thank Akii Tumu, Pesone Munini, Alome Kyakas and Nitze Pupu for their collabouration over more than a decade. The Max Planck Society and Enga Provincial Government generously provided funding for this project. We are particularly grateful to the many Enga who gave us their time and their knowledge. I would also like to thank Jim Roscoe and participants in the Seminar on 'Warfare in the South Pacific: Strategies, Histories and Politics' for helpful comments and criticisms.

NOTES

1   See Wiessner and Tumu (1998) and Wiessner (2002: electronic appendix) for an in depth discussion of oral traditions as history and of our methodology.

2   Women also hold equal rights vis-a-vis other women – rights to garden land, support from their husbands, payments to their kin for child growth, and protection vis-a-vis other women.

3   For example, studies by Meggitt (1974: 191, n. 43) indicate that in the 1960s big men did not have significantly larger land holdings than their fellow clan members, though they had larger households and thus more family

4   The impact of contrasting moralities on individual personality has been noted in other areas of New Guinea (Brison 1991; Harrison 1985; 1993; Robbins 1998; Strathern and Stewart 2000: 65-66).

5   An exception exists when men are unable to execute a pay back murder in a distant clan and hire a man from an adjacent clan to do so. Such paid killers are despised and can be expelled from their own clan to live in the clan that hired them.

6   Polygyny is limited to some 10-15% of men who can provide land, support, and equitable payments to affines for all wives. Prior to contact, genealogies indicate that men rarely had more than 2-3 wives.

7   I am grateful to Pesone Munini who collected detailed notes on events of these wars in which I could not participate.

8   Our historical studies conducted in the same clans as Meggitt's (1977) studies indicate that a number of the fights recorded by Meggitt did not take place between 1900 and 1955 but well before 1900, including all four of the clan routs given by Meggitt. Lakau (1994) has reached a similar conclusion. Frequency of warfare and rates of displacement of clans after warfare may thus be overestimated in Meggitt's studies.

9   The title of Young's (2002) thesis, 'Our Land Is Green and Black', reflects these extended periods of peace. As one young man explained: clans can remain at peace for long periods of time and when they do their casuarina trees flourish so that their leaves are indeed green and their trunks dark.

10 Men in 28 out of 49 (57%) households visited by Pesone Munini during a war in 1985 complained of serious sleep deprivation.

11 From our work and that of Lakau (1994), it is evident that the wars which displaced entire clans that Meggitt placed between 1900 and 1955 actually took place two or more generations earlier. Because Meggitt's data has been destroyed, we were not able to check on the dating of wars in groups that were not displaced. Moreover, our methodology for collecting 'causes' of wars differed from Meggitt's. We recorded 'triggering incidents' as described in historical traditions. As far as we can determine, Meggitt asked retrospectively about the causes of war. Both methods are valid but yield different results. Reasons for a war given after a fight is over are often different from the triggering incident, simply because different motivations come into play once a war is underway.

12 Western Enga is not included here because, unlike in eastern Enga, many small migrations of lineages or sub-clans took place during this period.

13 The following considerations question Meggitt's thesis that the Enga fight over land: (1) frequency and severity of warfare does not appear to have increased with population growth, (2) it is labour not land that is short in Enga, (3) war continues to rage today when land can no longer be gained or lost, (4) land shortage or desire to take the land of another does not enter into a single historical tradition as an explanation for war, though people did have disputes over gardens that triggered wars, just as they had disputes of a wide variety of other issues. Even Meggitt suggests in his final chapter that land shortage is perceived rather than real and that Enga explanations may represent a culture-bound definition of relative scarcity. Since perceived land shortage apparently only becomes an explanation for war after contact with

Europeans, it is possible that land conquest as an explanation is bound to inter-cultural communication – the only explanation for warfare that the Colonial administration accepted as a 'rational' one for warfare.

14 We were told of one incident in which young men ambushed and killed a fight leader; they were praised in song by the women but scolded harshly by the older men. The opponents accepted the death as a man lost in the Great Wars and did not take revenge.

15 Interestingly the sky beings, unlike the ancestors, were not associated with specific tribes but were believed to punish or protect all human beings depending on their behaviour.

BIBLIOGRAPHY

- Allen, B. and Giddings, R. 1982. 'Land disputes and violence in Enga'. In: B. Carrad, D. Lea and K. Talyaga (eds.): *Enga: Foundations for Development*. (Enga Yaaka Lasemana 3). Armidale: Dept. of Geography, Univ. of New England, pp. 179-97.
- Brison, K. 1991. 'Changing constructions of masculinity in a Sepik society'. *Ethnology*, 34(3), pp.155-75.
- Brown, P. 1964. 'Enemies and affines'. *Ethnology*, 3(4), pp. 335-56.
- Brown, P. 1978. *Highland Peoples of New Guinea*. London: Cambridge University Press.
- Carneiro, R. 1970. 'A theory of the origin of the state'. *Science*, 169, pp. 733-38.
- Ensminger, J. 1992. *Making a Market: Transformation of an African Society*. Cambridge: Cambridge University Press.
- Dennen, J.M.G. van der. 1995. *The Origin of War*. Vol. I and II. Groningen: Origin Press.
- Du Toit, B. 1975. *Akuna, a New Guinea Village Community*. Rotterdam: A.A. Balkema.
- Feil, D.K. 1984. *Ways of Exchange: The Enga Tee of Papua New Guinea*. St. Lucia: University of Queensland Press.
- Ferguson, R.B. and Whitehead, N.L. 1992. *War in the Tribal zone: Expanding States and Indigenous Warfare*. Santa Fe (New Mexico): School of American Research Press.
- Ferguson, R.B. 1990. 'Explaining War'. In: J. Haas (ed.): *The Anthropology of War*. Cambridge: Cambridge University Press, pp. 26-55.
- Fried, M. 1967. *The Evolution of Political Society: An Essay in Political Anthropology*. New York: Random House.
- Gibbs, P. 2002. Missionary intervention and Melanesian values in Papua New Guinea. Paper delivered at the conference on Christianity and Native Cultures, University of Notre Dame, September 2002.
- Glasse, R. 1968. *The Huli of Papua: a Cognatic Descent System*. Paris: Mouton.
- Godelier, M. 1982a. 'Social Hierarchies among the Baruya of New Guinea'. In: A. Strathern (ed.): *Inequality in New Guinea Highlands Societies*. Cambridge: Cambridge University Press, pp. 3-34.
- Godelier, M. 1982b. *La Production des Grands Hommes*. Paris: Fayard.
- Gordon, R. and Meggitt, M. 1985. *Law and Order in the New Guinea Highlands*. Hanover: University Press of New England.
- Hallpike C.R. 1973. 'Functionalist interpretations of primitive warfare'. *Man*, 8(3), pp. 451-70.
- Hallpike, C.R. 1987. *Principles of Social Evolution*. New York: Clarendon and Oxford University Press.
- Harrison, S.J. 1985. 'Ritual hierarchy and secular equality in a Sepik River village'. *American Ethnologist*, 12(3), pp. 413-36.
- Harrison, S.J. 1993. *The Mask of War: Violence, Ritual and the Self in Melanesia*. Manchester: Manchester University Press.
- Kelly, R.C. 2000. *Warless Societies and the Origin of War*. Ann Arbor: University of Michigan Press.
- Kyakas, A. and Wiessner, P. 1992. *From Inside the Women's House: Enga Women's Lives and Traditions*. Robert Brown: Brisbane.
- Lacey, R. 1975. Oral Traditions as History: An Exploration of Oral Sources among the Enga of the New Guinea Highlands. Unpublished Ph.D. thesis. University of Wisconsin.
- Lacey, R. 1979. 'Holders of the way: A study of precolonial socio-economic history in Papua New Guinea'. *Journal of the Polynesian Society*, 88, pp. 277-325.
- Lakau, A. 1994. Customary Land Tenure and Alienation of Customary Land rights among the Kaina, Enga Province, Papua New Guinea. Ph.D. dissertation, Department of Geographical Studies, University of Queensland.
- Lindenbaum, S. 1979. *Kuru Sorcery*. Palo Alto: Mayfield Publishers.
- Meggitt, M. 1965. *The Lineage System of the Mae-Enga of New Guinea*. New York: Barnes and Noble.
- Meggitt, M. 1972. 'System and sub-system: the "Te" exchange cycle among the Mae Enga'. *Human Ecology*, 1, pp. 111-23.
- Meggitt, M. 1974. '"Pigs are our hearts!" The Te exchange cycle among the Mae Enga of New Guinea'. *Oceania*, 44(3), pp. 165-203.
- Meggitt, M. 1977. *Blood Is Their Argument*. Palo Alto: Mayfield.
- Meyer, P. 1990. 'Human nature and the function of war in social evolution: A critical review of a recent form of the naturalistic fallacy'. In: J.M.G. van der Dennen and V.S.E. Falger (eds.): *Sociobiology and Conflict: Evolutionary Perspectives on competition, Cooperation, Violence and Warfare*. London: Chapman and Hall, pp. 227-40.
- Montagu, M.F.A. 1976. *The Nature of Human Aggression*. New York: Oxford University Press.
- Naroll, R. and Divale, W.T. 1976. 'Natural selection in cultural evolution: Warfare versus peaceful diffusion'. *American Ethnologist*, 3(1), pp. 97-128.
- North, D.C. 1990. *Institutions, Institutional Change and Economic Performance*. Cambridge: Cambridge University Press.

- Otterbein, K.F. 1970. *The Evolution of War: A Cross-cultural Study*. New Haven: HRAF Press.
- Otterbein, K.F. 1994. *Feuding and Warfare: Selected Works of Keith F. Otterbein*. Langhorne: Gordon and Breach.
- Rappaport, R. 1968. *Pigs for the Ancestors. Ritual in the Ecology of a New Guinea People*. New Haven (Conn.): Yale University Press.
- Robbins, J. 1994. 'Equality as Value: Ideology in Dumont, Melanesia and the West'. *Social Analysis*, 36, pp. 21-70.
- Robbins, J. 1998. 'Becoming sinners: Christianity and desire among the Urapmin of Papua New Guinea'. *Ethnology*, 37(4), pp. 299-316.
- Robbins, S. 1982. *Auyana: Those Who Held onto Home*. Seattle: University of Washington Press.
- Sillitoe, P. 1978. 'Big men and war in New Guinea'. *Man*, 13(2), pp. 352-72.
- Strathern, A.J. and Stewart, P. 2000. *Arrow Talk: Transaction, Transition, and Contradiction in New Guinea Highlands History*. Kent, Ohio: Kent State University Press.
- Sackschewsky, M., Gruenahgen, D. and Ingebritson, J. 1970. 'The clan meeting in Enga society'. In: P. Brennan (ed.): *Exploring Enga Culture: Studies in Missionary Anthropology*. Wapenamanda: Kristen Press, pp. 51-101.
- Talyaga, K. 1982. 'The Enga yesterday and today: A personal account'. In: B. Carrad, D. Lea and K. Talyaga (eds.): *Enga: Foundations for Development*. (Enga Yaaka Lasemana 3). Armidale: Dept. of Geography, Univ. of New England, pp. 59-75.
- Turney-High, H. 1949. *Primitive War: Its Practice and Concepts*. Columbia: University of South Carolina Press.
- Tuzin, D.F. 1976. *The Ilahita Arapesh: Dimensions of Unity*. Los Angeles: University of California Press.
- Tuzin, D.F. 1996. 'The spectre of peace in unlikely places: Concept and paradox in the anthropology of peace'. In: T. Gregor (ed.): *A Natural History of Peace*. Nashville: Vanderbilt University Press, pp. 3-33.
- Tuzin, D.F. 2001. *Social Complexity in the Making: A Case Study among the Arapesh of New Guinea*. New York: Routledge.
- Waddell, E. 1972. *The Mound Builders: Agricultural Practices, Environment and Society in the Central Highlands of New Guinea*. Seattle: University of Washington Press.
- Watson, J.B. 1965a. 'The significance of recent ecological change in the Central Highlands of New Guinea'. *Journal of the Polynesian Society*, 74, pp. 438-50.
- Watson, J.B. 1965b. 'From hunting to horticulture in the New Guinea Highlands'. *Ethnology*, 4(3), pp. 295-309.
- Watson, J.B. 1977. 'Pigs, fodder and the Jones effect in post-Ipomoean New Guinea'. Ethnology, 16(1), pp. 57-70.
- Watson, J.B. 1983. *Tairora Culture*. Seattle: University of Washington Press.
- Wiessner, P. 2002. 'The vines of complexity: Egalitarian structures and the institutionalization of inequality among the Enga'. *Current Anthropology*, 43(2), pp. 233-69.
- Wiessner, P. and Tumu, A. 1998. *Historical Vines: Enga Networks of Exchange, Ritual, and Warfare in Papua New Guinea*. Washington, D.C.: Smithsonian Institution Press.
- Wiessner, P. and Tumu, A. 1999. 'A collage of cults'. *Canberra Anthropology*, 22, pp. 34-65.
- Wohlt, P. 1978. *Ecology, Agriculture and Social Organization: The Dynamics of Group Composition in the Highlands of New Guinea*. Ann Arbor: University Microfilms.
- Wormsley, W. and Toke, M. 1985. *The Enga Law and Order Project*. Report to the Enga Provincial Government, Wabag, Papua New Guinea.
- Wright, Q. 1942. *A Study of War*. Vol. I. Chicago: University of Chicago Press.
- Young, D.W. 2002. *'Our Land is Green and Black'. Traditional and Modern Methods for Sustaining Peaceful Intergroup Relations among the Enga of Papua New Guinea*. Goroka: Melanesian Institute Press.

Warfare and Exchange in a Melanesian Society before Colonial Pacification: The Case of Manus, Papua New Guinea[1]

Wohl nirgends ist der Kriegszustand ein so permanenter wie bei den Moanus, und eine Folge davon ist, dass der Stamm, der sonst alle Bedingungen in sich vereint, um sich zu vermehren und zu gedeihen, so verschwindend klein ist. An Veranlassungen zum Kriege fehlt es [...] niemals, aber auch ohne Veranlassung allein aus Kampflust zieht man in den Krieg. Das Töten eines Feindes ist die Hauptsache; die Eroberung des Gebietes ist Nebensache, tritt aber ein, wenn der Feind gänzlich vernichtet und aus seinen Wohnsitzen vertrieben wird. Kriegsbeute, bestehend aus Kähnen mit Zubehör, Muschelgeld und sonstigem Eigentum, wird nicht verschmäht; Häuser werden in Brand gesteckt und Kochgeschirre zerschlagen. Was an Menschen lebend in die Hände des Siegers fällt, wird als Sklaven fortgeführt, wer sich nicht flüchtet, wird erschlagen, sei es Mann oder Weib, jung oder alt. Dabei werden die schauerlichsten Greueltaten verübt und die Leute nicht selten zu Tode gemartert. Hat man Zeit, so nimmt man auch wohl die Leichen der Gefallenen mit und verkauft sie an die Usiai. (Parkinson 1911: 400-401)[2]

/12

TON OTTO

Warfare in Melanesia and Manus

Just before the arrival of Western colonisers the Admiralty Islands – now Manus Province in Papua New Guinea (Fig. 1) – were a warlike environment. This picture is confirmed both by the local oral tradition and by observations from the early colonisers. The ubiquity of warfare had a deep impact on the local social system, of which it was a part, and it is primarily this system that I wish to describe in this chapter. Melanesian anthropology has contributed an important body of ethnographic material as well as theoretical sophistication to the study of tribal warfare (see Knauft 1999; Brandt chapter 6; Helbling chapter 9). Because warfare ended in Manus at the beginning of the 20th century, the Manus material has not really been integrated into the study of Melanesian tribal warfare. Of the modern anthropologists working in Manus, only Theodore Schwartz (1963) pays due attention to precolonial warfare in his major article on 'systems of areal integration', which has been an important source of inspiration for the present chapter. The basis for the following reconstruction is, apart from extensive historical material written in English, Russian and particularly German (Otto 1994b), a body of oral histories which I collected primarily during my first fieldwork in Manus, mainly on Baluan Island, from March 1986 to March 1988. This material was not collected with the aim of studying precolonial warfare, but out of a general interest in social and cultural change. Its extent and historical depth can by no means match that of the oral histories collected by Wiessner and Tumu among the Enga (see Wiessner chapter 11), but it still provides an important additional perspective on precolonial Manus warfare.

The study of warfare in Melanesia has provided a range of theoretical frameworks for explanation, which can be roughly divided into those with a structural and/or functional focus, those with an ecological focus, and those with a cultural focus. In this chapter I will propose a framework in which exchange and network relations have a central place. Even though ecological division and specialisation were part of and gave shape to the exchange relations, they cannot be seen as a proper explanation for the prevalence of warfare. The same applies to the unmistaken cultural focus on warrior prowess and warlike aggressiveness. I think my focus on networks and exchange can best be seen as a variation of a structural and functional framework, even though precolonial warfare in Manus did not function to create larger groups by intern control and external conflict. Warfare can be seen as a crucial and sustaining element of the precolonial regional

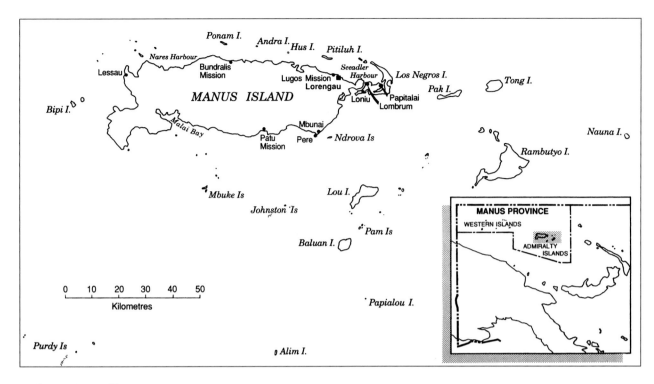

FIG. 1: *Map of Manus.*

system of exchange, comprising small autonomous groups that depended greatly on their exchange and network relations with each other. Once the sustainability of the exchange and warring relations was undermined by colonial military technology and organisation, the system collapsed and was superseded by a new system with colonial administrative centres supported by the colonial military.

Manus at the time of colonisation (1885)

The people of the Admiralty Islands live in a varied landscape. The main island is hilly and even partly mountainous with steep river valleys and sharp ridges with the highest peak rising to 718 metres. The coast of this island consists for the most part of mangrove swamps. To the west, north and northeast of the main island are flat sandy islands with large coral reefs around them, while to the south, southeast and east are a number of higher islands of volcanic origin. When Western colonisers reached the archipelago in the latter part of the 19th century, they found a population that had diversified and specialised. The various groups of inhabitants had adapted to different ecological niches and had developed specialised forms of livelihood. In the interior of the main island people lived from their gardens

with taro as the main crop. They inhabited small villages of less than 100 people mostly situated on hilltops. Those with access to the coastal areas produced sago flour from the sago palm. People on the small western, northern and north-eastern islands had specialised in fishing and depended on exchange with the inland people to obtain starch, sago leaves and timber. The higher volcanic islands had fertile soil and their inhabitants produced taro, yams, different kinds of fruit and nuts. These people lived spread over their islands in small hamlets. Close to these islands as well as along the south coast of the main island, a fourth type of people lived in large villages of 100 to 200 inhabitants with houses built on poles in the shallow lagoons. These people specialised in fishing and trade and obtained most of their food and building materials from exchange with islanders and mainland people. These lagoon dwellers are nowadays known as Titans (referring to their language) but at the time of the first contacts with Westerners they were mostly called 'Manus true', as they dominated the initial exchanges and confrontations. Whereas the Titans shared a language and a common history, the other groups were linguistically more diverse. Linguists have identified around twenty-seven languages which can be divided in four main groups.

In precolonial Manus villages were the largest social units. A well-defined village leadership did not exist, but in special situations such as during a major feast the leader of the largest clan could act as such. In the case of warfare the whole village would often unite under one war leader. A village – typically called 'place' (for living) in Manus languages – was often divided in a number of 'little places', which were the residential units of clans: groups of people who consider themselves related through descent from a common ancestor. Both the larger 'place' and the 'little place' had proper names. A (little) place consisted of several 'houses' which carried the name of an ancestor. These named houses referred to patrilineal descent groups which owned land and other kinds of property. Patrilineages were the smallest social units with clearly defined leadership. Seniority was an important principle of hierarchical classification and the leader of a lineage was in principle its most senior male member, just as the leading lineage of a 'place' often descended from the eldest son of the common ancestor. In practice, however, the rules of descent were not always strictly observed and could be subordinated to other considerations – for instance, if the eldest son was unfit to be a leader, one of his brothers could take over.

In most Manus villages, clans or 'little places' would have their own 'men's house' in which the unmarried men of the clan would sleep. In front of the men's house was a great open space for ceremonial purposes, displays and dances. Like the smaller houses for married couples which would be clustered around the ceremonial space, men's houses were constructed from tree poles, bush ropes and sago leafs. Some Manus villages, for example on Baluan and Pam, did not have men's houses. These, mostly small, villages consisted of a number of large houses divided in two parts: one half for the men, the other for women and children. The women's side (um) had fireplaces for cooking and beds along the walls, the men's side (lui) also had beds and a large entrance with a high doorstep on which the men would sit watching the dancing space in front of the house (kulului). If the house belonged to an important leader, the front facade would be painted (on sheath leafs of betel nut palm). Like men's houses in other villages, the importance of a leader could also be expressed by hanging up a great number of cowry shells (Ovula ovum) from the top of the entrance.

Manus societies were characterised by a dual rank system. A person could either belong to a 'house' of high status which provided clan leaders or to a 'house' of low status. High status groups were called lapan, which was also the name of the leader of such a group. The followers of a lapan, which included the low-status groups that supported a leading 'house', were collectively called lau. Lapan rank was hereditary just like lineage leadership but could be lost if a leading family failed to live up to expectations. In precolonial Manus there were two main areas in which the continuing vitality and power of a leading family had to be demonstrated, namely, warfare and the organisation of large, so-called lapan feasts. The latter were a kind of mortuary feast to commemorate a deceased lapan. They were organised by the lapan's successor and were highly competitive: the name of the new lapan depended to a large degree on his organisational and oratory skills and the size and the success of the lapan feast. In addition to inheritance, the renown that could be gained from these two types of enterprise, feasting and fighting, was a crucial element in maintaining the lapan status (Otto 1994a).

Exchange and networks

In the absence of strong and lasting political units, exchange was the central socially binding mechanism in the Admiralty Islands, both within villages and between them. Exchanges had many different forms: from the spectacular and large-scale distributions during lapan feasts, which occurred relatively seldom, to the daily exchanges of limited amounts of food and work between close relatives. Exchange was a necessity for the islanders in order to obtain the various items and foodstuff needed for survival. As mentioned, various groups in the archipelago had specialised in exploiting different ecological niches. Therefore, the fishermen living on small low islands and in lagoon villages had to exchange their fish and shells for garden food and bush materials with the people from the mainland and the larger islands. This happened mostly during regular markets in which two groups met formally and traded on the basis of fixed barter rates. In addition one could obtain necessary food supplies and other items by visiting a trade friend who would be able to give credit.

Apart from the direct exchange of food and raw materials the economic system in Manus was also based on the specialised production of utensils and luxury items such as shell money, wooden bowls, soup ladles, carved beds and combs (Mead 1930; 1963[1930]: 221-22). Sometimes the production of such items was considered a monopoly, which would be defended with force if necessary, and sometimes it was connected with privileged access to special ecological resources such as clay for pots, produced on Hus and Mbuke, or obsidian for spear points made on Lou. These special items were often acquired from trade friends or during lapan feasts but they were also exchanged during smaller ceremonial exchanges between kin. Thus exchange between people living in different villages and islands was conditioned both by ecological differences and specialisation in production. But exchange also permeated daily life within villages between near and more distantly related kinsfolk. Here exchange was not primarily a means to acquire the necessities for living; rather it was a way to mark individual and group identities and to confirm social relationships.

The important transitional events in an individual's life – birth, marriage and death – were celebrated with a ceremony to which various groups contributed with specific types of goods. What was exchanged and by whom depended on the type of event (Carrier and Carrier 1991; Schwarz 1963; Otto 1991). A marriage ceremony, for example, consisted of a series of exchanges in which the family of the bride contributed mostly food and cloth, while the groom's family provided primarily wealth items: dogs' teeth and strings of shell beads. During such a ceremony the bride and groom were joined in marriage, while at the same time a relationship was cemented between their respective 'houses' which would last for several generations; the children (and grandchildren) of the bride would namely maintain a relationship to the 'house' of their (grand)mother, in relation to which they were classified as 'son of the woman' (narumpein in the Baluan language), while the male line was called 'son of the man' (narumwen). Whereas rights to land were primarily inherited in the male line, the 'children of the woman' would keep secondary rights on the land of their mother's 'house' as well as the obligation to assist this descent group in large enterprises, such as lapan feasts. On the other hand, at the death of a person, his or her nearest patrilineal kin would still have to pay a major gift to the 'house' of the deceased's mother as a final recognition of the debt to this group. Thus

exchanges not only established a person's transition from one status to another – from unnamed to named, from unmarried to married and from living to dead – they also defined alliances between various 'houses'. Two types of relationships were particularly important in establishing such alliances: the affinal relationship created by marriage and the relation between the children of a brother and sister created by birth.

The centrality of exchange as a key element in Manus culture has been observed by previous ethnographers of the region. Margaret Mead emphasised the role of the individual in establishing exchange networks:

In Manus we have seen that we are dealing with individuals living in clusters with different degrees of common identifying features, that all activity was initiated by individuals and was carried out, not by lineages, or clans or villages as members of the groups, but because the individuals initiating the proposed affinal exchange, trading voyage or raiding party, invoked lineage or clan or village membership as a reason for temporary co-operation by other individuals (Mead cited in Schwartz 1963: 59)

If instead of asking a Manus 'to what do you belong?' one asks, 'what are your roads?' the answer is a network of various degrees of intimacy covering all adjacent villages of the Manus linguistic group, and a series of trade friends in faraway villages. (ibid.)

Even though Mead's description may be considered extremely individualistic – not all choices of collaboration are equally free and there exist quite explicit and binding expectations of mutual obligation between members of kin groups – she is certainly right in stating that individual actors play an important role in establishing and maintaining their own personal networks of exchange.

Also Theodore Schwartz emphasises the role of individuals in establishing integrative networks. He prefers to see the local leader or '"big man" as an entrepreneur, integrator, node in a network rather than as leader and representative of a group' (Schwartz 1963: 68). Being such a node in an exchange network allows for the temporary accumulation of enormous wealth, which gives lasting prestige to the individual.

Part of the fascination in the psychology of ceremonial exchange derives from the temporary accumulation and distribution of magnitudes of wealth far beyond the means of any individual participant. Exchange allows large numbers of people to participate in a travelling wave of wealth that accumulates now on this spot, disperses, and then accumulates on another spot in a complex social network, allowing great scope for improvisational and entrepreneurial skills. The principals of the exchange derive prestige from the temporary amassing of wealth, but so also, to a lesser degree, do all participants in the exchange who at another time will themselves be principles. This pattern is familiar and well described from many parts of Papua New Guinea. For the people of Manus it provides perhaps the central focus of life interest and enhancement of self. (Schwartz 1982: 395-96)

Schwartz (1963) rightly points out that genealogies are important in Manus, not as a biological record but rather as a valuable topological record of possible 'roads' that can be invoked according to circumstance. The widespread occurrence of adoption also points in the same direction: kin connections are not only determined by biological descent but are constructed according to need and occasion. Interestingly, Schwartz develops the network perspective to encompass the whole archipelago in what he calls an 'areal culture': 'That is, not only are the cultures historically related but they are so interdependent culturally, and otherwise ecologically, that each must be considered as part-culture not sufficient or fully understandable in itself' (1963: 89). The type of integration thus established he characterises as a 'particularistic, entrepreneur-centred, dispersed network'.

Other anthropologists working in Manus after Mead and Schwartz have also emphasised the importance of exchange. In particular Achsah and James Carrier have analysed the structural and processual aspects of exchange relationships (Carrier and Carrier 1989; 1991). Of course this is a well-known theme in Melanesian anthropology more generally, which has led to interesting theoretical developments. Marilyn Strathern in particular has done much to develop our understanding of exchange relationships in Melanesia (see Strathern 1988; 1996; 2004). Whether or not we conceive of Melanesian persons as exchanging parts of their selves (thus becoming partible persons as Strathern suggests), it is commonly accepted knowledge that groups exchange individual group members (especially women, but also children, captives and other dependents), and persons exchange things but also immaterial property such as rituals, magic and knowledge. Exchange and the concomitant logic of reciprocity are central to understanding Melanesian personhood and sociality. Acts of violence can enter such reciprocity and disrupt existing links but they may also establish new social relationships. The complementary relationship between warfare and exchange has often been emphasised (Corbey chapter 3), but in this chapter I want to highlight their possible integration within the same social system instead of seeing exchange as the solution to and conclusion of warfare. Contra Mauss I see warfare as much as a social act as exchange (cp. Harrison 1993: 21, 149). Precolonial Manus was a type of regional system in which social groups were small and could easily split and fuse (Schwartz 1963); therefore, exchange should be considered as the crucial social glue that allowed the regional system to function. I will argue that warfare entered exchange relationships as another resource and helped maintain the system of interdependent exchange partners. But before developing this argument, we have to find out what kind of war was waged in the archipelago.

Precolonial warfare

The first agents of Western colonialism considered the Admiralty Islands as a wild and dangerous place, even in comparison with other coastal Melanesian societies that were colonised in the latter quarter of the 19th century. Parkinson (1911) writes that the state of war was more permanent here than elsewhere (see quotation above) and Schnee (1904: 195) compares the character of the Manus warriors with that of his own soldiers from other places in New Guinea in the following way:

Vielleicht dämmert auch in den braven Neumecklenburgern und Bukas das Bewusstsein, dass sie es hier mit einem von Natur intelligenteren, hinterlistigeren und in jeder hinsicht gefährlicheren Gegner zu tun haben, als sie selbst es ihrer natürlichen Anlage nach sind.[3]

Similar observations abound and there is no doubt that warfare played an important role in the Admiralties from the perspective of outsiders. The earliest relevant sources stem from the short visit by the English ship 'The Challenger' in 1875 and two longer visits by the Russian explorer Mikloucho-Maclay in 1877 (two weeks) and 1879 (four weeks). The Challenger expedition resulted in a number of reports (Spry 1877; Swire 1938), the most detailed ones ethnographically coming from the naturalist H.N. Moseley (1876; 1879). Moseley observed a fortified village on one of the islands and noted that each small island appeared to have its own 'chief', whose 'power seemed to depend on his fighting qualities' (1876: 414). Even though one of the chiefs obviously had considerable power – for example, he was able to take goods given to some of the men away from them for redistribution in the group – Moseley noted that no ceremonious respect was paid to him at all. The inhabitants were most eager to trade for (hoop) iron and must have known about and obtained some of that material through previous contact and exchange (1879: 451).

Mikloucho-Maclay's descriptions are extremely interesting as an early form of field anthropology: he stayed and slept in the villages instead of returning to the ship for safety. He tells about his close escape from an ambush and relates the killing of a trader who was set ashore during his first trip to Manus, while he was able to free another who was robbed but survived thanks to the mercy of an old Manus man. Mikloucho-Maclay provides detailed observations and also receives information from a Malay man who had been held captive in a village and whom he freed by paying a ransom. Among other things he tells about a recent native raid on a village where seven men were killed (1879: 174), but also about an attack on a Western trading vessel by a great number of warriors on canoes from the village where one of the traders was killed. The captain and crew were able to ward off the attack and apparently killed or wounded more then fifty native warriors, who appeared unfamiliar with the lethal power of Western firearms (1879: 156-57). Mikloucho-Maclay asserts with certainty that cannibalism occurred on the islands. He observed a young woman tearing off and eating pieces of flesh from a human bone and he received corroboratory accounts from the men who were held captive on the islands that it actually happened frequently (1879: 147, 176). Concerning the issue of leadership, Mikloucho-Maclay doubts whether there actually were chiefs on the islands. Some men clearly had more authority than others but he ascribes this to personal qualities and character rather than to position (1877: 79; 1879: 176).

In 1885 the Admiralties were declared a German protectorate, thus formally becoming a colony of the German Empire. Subsequently, the stream of Westerners visiting the islands increased considerably; this stream included administrators, missionaries, and other people with an interest in native affairs. In the colonial journal *Deutsches Kolonialblatt* (DKB) a lot of information can be found about the population, but the most comprehensive descriptions concerning the last part of the 19th century and the first decade of the 20th are provided in Parkinson 1911, Schnee 1904 and Nevermann 1934. The latter provides a compilation of all the available sources (including DKB), most of which are in German. One could rightly criticise some of these descriptions as coloured; for example, Schnee shows great admiration for the fierceness of the Manus warriors. But on the other hand his descriptions are so detailed and informative that it was possible for me to compare his story of the murder of a German trader and the subsequent punitive expedition with oral evidence I collected on the islands, allowing me, by combining and interpreting evidence, to give an alternative explanation of events, alliances and motives (Otto 1991: 89-98).

What picture emerges of warfare in Manus from the combined evidence of these historical sources? The main weapon used in fighting was the spear, which had a point of obsidian or hard wood. There were no bows and arrows and no shields. Warriors would throw their spears and dodge the ones thrown back at them. Another weapon was the dagger, but this was mainly used for ambush. Warfare ranged from open combat to surprise raids and ambushes. Peaceful gatherings, such as markets and feasts, could turn into violent clashes, and participants normally carried their weapons along. If a whole village was attacked, the defeated enemies would be killed, their pottery and other property destroyed and the houses set afire. Useful booty was taken and the dead bodies eaten or exchanged with other villages. Some people were kept alive: women would be taken along to serve as prostitutes, especially for the young men (see also Mead 1963[1930]: 221), and men were sometimes taken as a kind of slave. Parkinson describes the occurrence of sea battles between warriors throwing their spears from canoes. The main strategy was to kill the steersman and capsize the canoe, and thereafter spear the enemies lying in the water. The Western sources mention frequent conflicts and villagers were apparently on guard all the time. Villages were built on small islands or on hill tops or surrounded by palisades for defence purposes, and paths to the

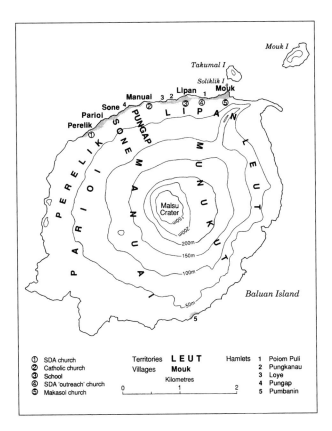

FIG. 2: *Map of Baluan.*

villages often had booby traps with hidden spears and points. We get a picture of a time in which war in its different forms was a common occurrence, permeating daily life and the organisation of society.

Stories about leadership and war

Another important source for our knowledge of precolonial warfare is the indigenous stories about lapans in the past. During my fieldwork on Baluan (Fig. 2) from 1986 to 1988 I collected a considerable number of these stories, many of them on tape. These stories are the property of the houses that descend from the lapans concerned and are rarely told in public.[4] They are handed down from generation to generation within the lineage. However, as they often go back six or more generations, during which groups split up, moved and became autonomous, there are generally several houses that guard the memory of the same lapan. My collection has therefore allowed me to cross-check central events and names and has revealed both consistency and interesting differences in perspective.

A common theme of these stories is how lapans received their status and power, because it appears

that many lapans in fact were not of noble birth, as the cultural rule would have it, but obtained their status through other means. One way was adoption. The forefather of the two strongest clans of the largest village on Baluan drifted as a boy from mainland Manus to the island. Here he was found and later adopted by two lapans who did not have a son themselves. Another founder of a lapan line was a war captive who was bought by a Baluan leader and later given to another lapan for adoption. These two boys became strong warriors and protectors of their houses and clans according to the oral traditions that guard their memory (Otto 1991: 59-60). In addition to adoption, marriage was also an important strategy to link a strong person to a house and it could be used as a means of transferring the lapan status. In the following I will follow one such lapan story in some detail as an example of the kind of information it contains. I have five tape-recorded versions of the story but will use only one, as the differences are not relevant in the present context. The chosen version was told by the undisputed leader of the house and direct descendant of the founding father, whose story is told. Unlike some other stories, which involve contemporary disputes, it appears unproblematic to present this story in written form. The narrator is Kalai Poraken, lapan of the Yongkul house and the acknowledged traditional leader of the village of Parioi at the time of my fieldwork.

Kalai begins the story by relating how two brothers got into a quarrel about the fruits of a breadfruit tree. One of the brothers, Marankopat, decides to leave and goes to the lineage of his mother. His mother's brother takes him in, raises him and organises his marriage. What follows is a verbatim translation of the rest of the story:

In those days the side of Lipan would often attack the side of Mun, which are the people of Sone, Parioi, Perelik, and Manuai.[5] The men of Lipan would capture some people of Mun and sell them to the islanders of Lou for obsidian and wooden bowls. The people of Lou would then eat the Baluan captives.

All right, Marankopat used to hear this when he still lived in Munukut.[6] He was married now and he asked his uncles to build a house. They made a house for him. They erected the houseposts and put the rest on it. While they were still working on it, the men of Lipan went up the mountain to Manuai village. They chased away the people[7] and caught one boy. They put him on a bed and four men carried him. They went down along the road in Pumaliok on their way to Lipan.

The mother of the boy was crying and called the name of Marankopat. Marankopat took all his spears and ran to Pumaliok. When he met the men of Lipan carrying the boy, he killed one with his spear. They left the boy and fled, taking the dead body with them. Marankopat untied the boy and called the mother to come from Manuai and get her child back. The mother came and received the child. Marankopat went back to his uncles who were building his house. He said to them: 'The name of this house you are making is Yongkul'. Then he told them the name of the spear with which he killed a man of Lipan: 'The name of this spear is Kumkilamut'. The uncles finished the house, which was now called Yongkul. The meaning of Yongkul, leaf of the breadfruit tree, is that it protects you from sun and rain. Therefore, the house was called Yongkul. If the men of Lipan would come again the people of Mun could find protection under this house. The meaning of the name of the spear, Kumkilamut, is that the east side (Kum) could not come to fight any more and kill the people of Mun because there was a strong man to chase them away.

Kalai goes on to relate his line of descent from Marankopat (see Fig. 3). Another version tells how the latter was married to the daughter of a lapan of an important house in Parioi, who transferred lapan status to Marankopat after he had demonstrated prowess in fighting. The next episode of the story describes how a descendant of Marankopat moved to the village of Manuai to found another lapan house, called PulianPaluai. The name literally means 'mountain of Baluan' and was intended to claim the highest status for this house. Kalai recounts that Marankopat's son Keket had two sons, Popol and Ngi Kuian. The wives of the two brothers had an argument (apparently a common theme in lapan stories) and Kuian decided to leave the village of Parioi. He went to Manuai, taking his lapan status with him. His line of descendants has provided the leading lapan of Manuai ever since. The third episode concerns a fight between the houses of Yongkul and PulianPaluai. Kalai's verbatim account continues below.

Version 1 (Kalai)

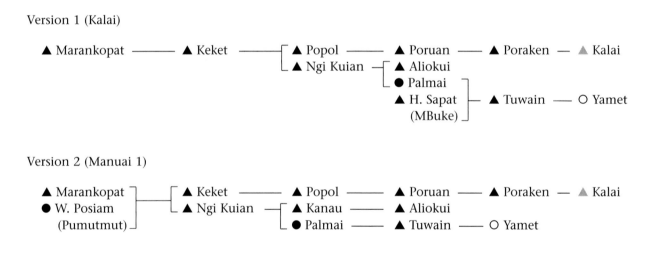

Version 2 (Manuai 1)

Version 3 (Manuai 1 and Poipoi 2)

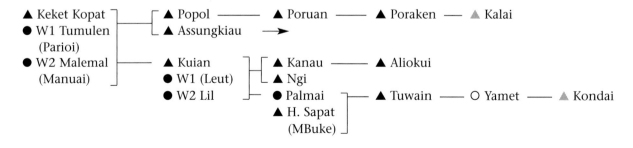

FIG. 3: *Genealogy of Yongkul.*

All right, this clan split into two parts. One part went to Manuai and was called Umtan Kuian and the other stayed in Parioi and was called Umtan Popol.[8] I will now talk about the time of Aliokui and Poruan. The two had the following dispute. Aliokui wanted to be the head lapan and Poruan said: 'No, I am in the original lineage of the lapan and I remain the head'. This dispute came up and Aliokui looked for a way to kill Poruan. Poruan went to the bush to work in his garden. Poruan had two wives. One was called Posiam and the other Sapou. In the afternoon they returned from their work and they were ambushed by Aliokui and his line. They fought and Poruan was killed. Poruan was carrying his son Poraken. When his father died Poraken fell onto the ground and a stone wall broke down and covered him. The group of Aliokui did not kill Poraken. Poraken stayed alive.

All right, the people of Parioi wanted to take revenge on Umtan Kuian for the murder of Poruan. When they were mourning for Poruan and killed pigs for this purpose, they took a prime part of a pig and put an obsidian dagger inside it. They sent this to Umtan Palasip because a woman of Yongkul had been married to this house. They followed this woman when they sent the dagger to Umtan Palasip.[9] Palasip cut the piece of pig and found the dagger. He called his people and said: 'We have a big thing at hand now because they have sent us this dagger'. They wondered who would be fit to eat the piece of pig. One person was very skilled in throwing the spear. His name was Kanau. He has descendants who live in Manuai now. They told him: 'You must eat this piece of pig and kill Aliokui'. All right, they went to the lapang, which is the place where they practised fighting. Palasip told Kanau now: 'If you do not shoot at Aliokui, I will shoot you'. When they were ready to go, Aliokui called Kanau and said: 'Let us go to the fight'. They went and some others stayed behind, because it was a practice ground and many people went there to practise or to watch. Aliokui threw his spears up the hill and the people there dodged them. Then the people uphill shot their spears and the group of Aliokui moved backwards. When they wanted to go uphill again, Kanau hit Aliokui with his spear. One of the people on the hillside called out that they had killed a man. The people thought that someone from above had killed Aliokui, but this was a trick.[10] Someone from below had shot him. This was to pay back the murder of Poruan by the line of Aliokui.

This short story about the house of Yongkul contains a wealth of relevant information about practices of violent conflict and warfare. In the first place one can discern different types of warfare. The villages of the east (Kum) were in a permanent relation of enmity with the villages of the west (Mun). This relation would often take a formalised or ritualised form, as on the practice ground, or could lead to incidental raids on enemy territory leading to killings, destruction of property, and the taking of captives. From other oral sources we know that periods of (relative) peace alternated with periods of war. At the time of the German punitive raid on the island in 1900 (see above) the pressure from the west on the east was so great that one clan had found it necessary to build a palisade around its settlement. In the German raid this palisade as well as the whole village was destroyed (oral information). Also the murdering of whole villages is kept in memory. There were always some people who escaped and found shelter with other clans. Their descendants were called *puanin*, the fruits of war. The villagers of Baluan lived uphill and away from the coast because they were afraid of raids from the Titan people, who would come in their big seagoing canoes. In remembered history several groups of Titan people settled very close to the Baluan coast, obviously with permission from the locals with whom they entered relations of exchange, in particular trading fish for garden produce. It is likely that these Titan groups also provided a kind of protection against raids from other Titan groups. Clearly, Titan groups were in competition (and sometimes at war) with each other for privileged access to agricultural villages.[11]

Violent conflict could also arise between closely related descent groups. In the story related above we have an example of an ambush in which close relatives were killed by the aggressors. Deceit and ruse were commonplace. One story relates how one group (Luibuai) was invited by another (Muiou) to a gathering with the stated purpose of commemorating a dead relative. Once the guests were seated, all in pairs, one guest next to one host, they were attacked and killed with daggers which the hosts had hidden in their curly hair. The reason given for this is that the Muiou wanted to take revenge on the Luibuai for using a kind of decoration that was considered as the Muiou's property. The dominant Muiou were thus prepared to defend one of their privileges in this bloody way. It is perhaps doubtful whether these kinds of events should be called war, but they certainly often resulted in one. For example, the killing of Aliokui at the practice ground (see above) led to a ferocious war between the people of Mun and the falsely accused Munukut, who were severely beaten.

A central theme that is well illustrated by the Yongkul story is the close connection between leadership and prowess in battle. Marankopat became the leader of a lapan house because of his warring qualities and similar explanations permeate stories about other lapans. It appears that at a time when warfare was ubiquitous, being a good warrior was the decisive quality for a leader. Even though my informants emphasised that lapanship was hereditary, the historical narratives they told me point out that hereditary leaders used the means at their disposal to get strong and aggressive men into the family, namely, adoption and marriage. I was also told that strong and warlike leaders would fill their followers with fear and therefore were able to exact more obedience and support than other leaders. My investigation into the use of lapan titles (Otto 1994a) supports the present argument about the centrality of warfare in issues of leadership: the leaders of dominant clans all had titles relating to warfare, whereas lapans who had only titles referring to peace belonged to less powerful groups.[12]

An important aspect of leadership is the strong competition for status between leaders of descent groups. In the story above, this competition between closely related leaders leads to killing and warfare. Status competition appears to be a very strong motive in Manus societies (see Schwartz 1963). It often caused groups to split as one leader would not accept the dominance of another. This competition not only engendered frequent violent conflicts but also caused the groups involved in these conflicts to remain relatively small. Fission of descent groups was the order of the day and fusion happened only when defeated groups were too small to defend themselves and had to join another, larger group for survival.

Clearly the quest for status was an important motive for going into war and the demonstration of prowess was probably a means to prevent others from attacking one's own group. But warfare also provided other, more material spoils. There was of course booty, but also dead bodies that could be eaten or traded for coveted materials. In the story it is specifically mentioned that human meat was exchanged for obsidian, necessary for warfare, and wooden bowls from Lou. This was a common theme and explanation in many other Baluan stories. Finally, warfare could provide human labour. On Baluan I did not hear any stories about keeping captive women as prostitutes, but there is frequent mention

of adoption or the keeping of captives as dependents. In some cases this would eventually lead to exchange relations with the groups from which the captives originated.

A final and crucial point concerns the nature of relationships and networks in the case of warfare. In the story of the murder of Aliokui, the people of Parioi used an ally in the clan of the victim to execute the revenge. In this case an affinal relationship was obviously considered as more important than clan allegiance and the obligation established through marriage was thus able to override the collaboration and loyalty required between members of the same clan. Because the identity of the real murderer was kept hidden and another clan got the blame, a war within the clan was avoided. But the episode clearly supports Mead's observation that individuals on Manus could invoke different kinds of relationships in the pursuit of their strategic aims (Mead 1961[1937]; Schwartz 1963).

Warfare, exchange, and areal integration

The prevailing logic underlying warfare on Manus was the necessity to retaliate. A strike always invoked a counter-strike and every insult or offence called for payback. If a group was not able to react, it was in danger of losing its place as a centre of action and agency, because it would become easy prey for others. The maintenance of an aggressive and violent image was therefore an asset in the relations between groups. Alliances would be made but were unstable, as they could easily be overturned by other interests. In principle, no one could be trusted and therefore, following the logic of the prisoners' dilemma (see Helbling chapter 9), it was better to attack and strike first than to be caught off one's guard.

At an abstract level one can consider the logic of payback as a modality of reciprocity, albeit a negative form of reciprocity. Many have underlined the centrality of reciprocity in Melanesian societies following Mauss' logic of the gift, but it is less common to extend this reciprocity to include forms of payback (cp. Trompf 1991: 51-77, Harrison 1993: 16ff). However, the gift versus counter-gift, strike versus counter-strike, and action versus reaction rationality appears to be a central theme both in practice and in myth (Pouwer 1975). Having basically the same logic, can negative reciprocity also be seen as having the same effect as positive reciprocity, namely, the establishment of links and the integration of society? I would argue it does. In the first instance an act of violence or warfare of course disrupts a relationship (as between the two lineages of Yongkul), but if one takes a wider perspective it is possible to see that it in fact contributed to the integration of small units on Manus.

Firstly, warfare forced groups to find allies, thus strengthening some relationships while disrupting others. Warfare also produced spoils in the form of booty, human flesh and captives, which entered into existing exchange relationships and sometimes even established new ones. Because of the rule of negative reciprocity there was plenty of opportunity to engage in war, as it was easy to find insult (evidence from oral history). The famous 'Kampflust' of the people of Manus (see Parkinson 1911 quoted above) can be explained by the opportunities for prestige provided by warfare, both for a war leader and his young warriors. For them warfare could be a profitable affair, as long as they remained on the winning side; waging war was therefore one of the options an entrepreneurial leader had. The other main arena for prestige and status competition was the organisation of big lapan feasts. This was by no means an easier road, and possibly not even a safer one, as oral history has it that a number of potential lapans died while they were preparing their big feast. This high mortality is ascribed to the use of powerful magic by opponents and the lack of strength of the would-be leaders, but in a Western frame of explanation one could speculate about causes like exhaustion and stress-related diseases.

The strong motivation for status led to fierce competition, which in turn caused groups to split up and enter into relations of conflict (and exchange) with each other. This kept the groups as centres of action and agency relatively small, making them dependent on allies and thus on exchange relationships with others. In other words, status competition and warfare had a conservative and integrative effect on the system of regional exchange and interdependence. War was also used to defend local specialisations of production and privileged access to trading partners, thus augmenting the necessity for exchange in addition to ecological differences (see Schwartz 1963: 83-85). In short, warfare was one of

the resources at the disposal of entrepreneurial leaders and a functional part of the regional system of integration.

With the arrival and permanent settlement of colonists this system changed fundamentally. It will take another article to describe the changes in some detail, but here I will briefly mention the main elements of the transformation. Firstly, the archipelago saw the introduction of Western firearms, which had greater destructive power than the native spears and could reach further. There was fierce competition concerning the acquisition of these new weapons, which at first could only be obtained by attacks on Western traders, many of whom were murdered. However, the new warring relations established with this new military technology proved to be very disruptive also for the victors, as more people were killed and more groups eradicated than before (Nevermann 1934). In addition, the same period saw the spread of major epidemics which killed many inhabitants and in many places had disastrous effects on the number of the population (cp. Gosden chapter 13).

This general disruption was aggravated and extended by the increasing number of punitive raids organised by the colonists who wished to punish native groups for murdering white traders. Warring native groups with their new powerful weapons went on long raiding tours, but had little or no possibility to return to a more stable settlement. Their villages were burned down by the white punitive raiders and they had to remain on the run for extended periods of time. In the long run this was not a sustainable situation and after the establishment of a permanent police station on Manus in 1911, a relatively fast and overwhelming transformation of the regional system occurred: the local warriors gave up their weapons, both the captured guns and their own spears, and transferred the power to retaliate and execute pay-back to the colonial authority. In my view this transformation was as much an act of rational choice and consent within dramatically changed circumstances as it was forced upon the natives by the superior military power of the colonisers. Pax germanica was established both through force and consent as a new form of collaboration between indigenous populations and German colonists.

NOTES

1 I thank Jürg Helbling and Helle Vandkilde for useful comments on drafts of this article.

2 'The condition of war is probably nowhere as permanent as among the Manus, and a result of this is that the tribe, which otherwise harbours all conditions needed to grow and thrive, is so vanishingly small. Causes for war are never absent, but also without cause, merely out of lust for battle, war is initiated. The killing of an enemy is the main objective; the conquest of territory is of secondary importance, but takes place when the enemy is completely destroyed and is driven off his places of residence. Loot, consisting of boats with equipment, shell money and other kinds of property is not disdained; houses are set on fire and kitchenware smashed. The people who fall into the hands of the victor alive are taken as slaves; those, who do not flee, are killed, be it man or woman, young or old. In that connection the most dreadful and horrible deeds are committed and people are not rarely tortured to death. Is there time, one also takes along the corpses of the fallen and sells them to the Usiai.' (my translation).

3 'Perhaps the sterling New Irelanders and Bukas begin to realize that they are dealing with an adversary who is naturally more intelligent, more cunning, and in every respect more dangerous than they are themselves in their natural disposition.' (my translation).

4 The stories are used as evidence in the local land court if a lineage's status or property is under dispute.

5 Lipan is on the east side of the island, also called 'Kum', and 'Mun' refers to the west side.

6 This is the name of the clan of his father, which lived uphill and belonged to the east side of Baluan.

7 Another version relates the warriors' destructive activities: putting fire to the houses, killing the pigs, and breaking the clay pots.

8 Umtan Popol means 'house of Popol' and this is the lineage of Yongkul, who remained in the village of Parioi. Umtan Kuian means 'house of Kuian' and this is the new lineage that moved to Manuai.

9 Umtan Palasip is a house in the village/clan of Manuai!

10 This event is also part of another clan's history, namely the Munukut, who were the victims of the deceit because they were wrongly accused of killing Aliokui and had to endure a severe pay-back.

11 The murder of the German trader on Kumuli near Baluan and the subsequent raids are good examples of this (see Otto 1991: 89-98).

12 A number of leading lapans had both warlike and peace-related titles. The latter were often connected to original inhabitants and owners of the land, whereas war titles were obtained through fighting prowess and were often in the possession of newcomers to the island.

BIBLIOGRAPHY

- Carrier, J.G. and Carrier A.H. 1989. *Wage, Trade and Exchange in Melanesia: A Manus society in the modern state*. Berkeley: University of California Press.

- Carrier, A.H. and Carrier, J.G. 1991. *Structure and Process in a Melanesian Society*. Chur: Harwood Academic Publishers.

- Harrison, S. 1993. *The Mask of War: Violence, ritual and the self in Melanesia*. Manchester and New York: Manchester University Press.

- Knauft, B. 1999. *From Primitive to Postcolonial in Melanesia and Anthropology*. Ann Arbor: The University of Michigan Press.

- Mead, M. 1930. 'Melanesian middlemen'. *Natural History*, 30, pp. 115-30.

- Mead, M. 1961(1937). 'The Manus of the Admiralty Islands'. In: M. Mead (ed.): *Cooperation and Competition among Primitive Peoples*. Boston: Beacon Press, pp. 210-39.

- Mead, M. 1963(1930). *Growing up in New Guinea*. London: Penguin Books.

- Mikloucho-Maclay, N. 1877. 'Admiralty Islands: Sketches from travels in Western Micronesia and Northern Melanesia'. In: Mikloucho-Maclay's journeys in the islands in the Pacific, October 1873 – August 1881: Diaries, letters, and reports translated by C.L. Sentinella from the Sobranie Sochinenii Mikloucho-Maklaia, vol. 2. MLMSS 2913, item 2. Sydney: Mitchell Library, pp. 61-99.

- Mikloucho-Maclay, N. 1879. 'The islands of Andra and Sorri (from the 1879 diary)'. In: Mikloucho-Maclay's journeys in the islands in the Pacific, October 1873 – August 1881: Diaries, letters, and reports translated by C.L. Sentinella from the Sobranie Sochinenii Mikloucho-Maklaia, vol. 2. MLMSS 2913, item 2. Sydney: Mitchell Library, pp. 140-82.

- Moseley, H.N. 1876. 'On the inhabitants of the Admiralty Islands'. *Journal of the Anthropological Institute of Great Britain and Ireland*, 6, pp. 379-429.

- Moseley, H.N. 1879. *Notes by a Naturalist on the 'Challenger'*. London: Macmillan and Co.

- Neverman, H. 1934. 'Admiralitäts-Inseln'. In: Thilenius, G. (ed.): *Ergebnisse der Südsee-Expedition 1908-1910, II Ethnographie: A. Melanesien*. Vol. III. Hamburg: Friederichsen, De Gruyter and Co.

- Otto, T. 1991. *The Politics of Tradition in Baluan: Social Change and the Construction of the Past in a Manus Society*. PhD thesis, The Australian National University, Canberra and Nijmegen: Centre for Pacific Studies.

- Otto, T. 1994a. 'Feasting and fighting: rank and power in precolonial Baluan'. In M. Jolly and M.S. Mosko (eds.): Transformations of hierarchy. Structure, history and horizon in the Austronesian world. Special Volume of *History and Anthropology*, 7(1-4), pp. 223-39.

- Otto, T. 1994b. 'A bibliography of Manus Province'. *Research in Melanesia*, 18, pp. 113-77.

- Parkinson, R. 1911. *Dreissig Jahre in der Südsee: Land und Leute, Sitten und Gebräuche im Bismarckarchipel und auf den deutschen Salomoinseln.* Stuttgart: Verlag von Strecker and Schröder.

- Pouwer, J. 1975. 'Structural history: A New Guinea case study'. In: W. van Beek and J. Scherer (eds.): *Explorations in the Anthropology of Religion.* The Hague: Nijhoff, pp. 80-112.

- Schnee, H. 1904. *Bilder aus der Südsee: Unter den kannibalischen Stämmen des Bismarck-Archipels.* Berlin: Dietrich Reimer.

- Schwartz, T. 1963. 'Systems of areal integration: Some considerations based on the Admiralty Islands of Northern Melanesia'. *Anthropological Forum,* 1, pp. 56-97.

- Schwartz, T. 1982. 'Alcohol use in Manus villages'. In: M. Marshall (ed.): *Through a Glass Darkly: Beer and Modernization in Papua New Guinea.* (Institute of Applied Social and Economic Research Monograph, 18). Boroko: Institute of Applied Social and Economic Research, pp. 391-403.

- Spry, W.J.J. 1877. *The Cruise of HMS 'Challenger': Voyages over many seas, scenes in many lands.* London: Sampson Low, Marston, Searle, and Rivington.

- Strathern, M. 1988. *The Gender of the Gift.* Berkeley: University of California Press.

- Strathern, M. 1996. 'Cutting the network'. *The Journal of the Royal Anthropological Institute* (N.S.), 2(3), pp. 517-35.

- Strathern, M. 2004. 'Transactions: an analytical foray'. In E. Hirsch and M. Strathern (eds.): *Transactions and Creations: Property debates and the stimulus of Melanesia.* New York and Oxford: Berghahn, pp. 85-109.

- Swire, H. 1938. *The Voyage of the Challenger: A Personal Narrative of the Historic Circumnavigation of the Globe in the Years 1872-1876.* Vol. I and II. London: The Golden Cockerel Press.

- Trompf, G.W. 1991. *Melanesian Religion.* Cambridge: Cambridge University Press.

Warfare and Colonialism in the Bismarck Archipelago, Papua New Guinea

CHRIS GOSDEN

/13

War is a simple three-letter word, but is not a singular category. It represents a spectrum of behaviour from aggression contained within social rules to that which erupts despite social regulation and is then difficult to contain. Von Clausewitz's famous definition of war as politics by other means, might in Papua New Guinea, be formulated as exchange by other means. Groups which are most closely locked together in exchange relationships are also most likely to fight, as the competitive spirit found in exchanges spills over into more naked aggression. This form of warfare is not benign, as Wiessner (chapter 11) shows, but it is contained within rules and can be brought to an end by mutually understood forms of mediation. But occasionally war breaks free from all social regulation, taking a form which is hard to contain and thus threatens the very basis for society. I want to look at one such case of unconstrained warfare, brought about by early colonial relations in Papua New Guinea and I shall consider the lessons that unregulated warfare holds for an understanding of war as a whole.

Colonialism is basic to an understanding of war in Papua New Guinea and war is vital to understanding colonialism. War was the point at which the prestige of the colonist confronted the prestige of local men. In many areas of New Guinea war and

exchange were the basis for male standing, so that success in one or both of these arenas was vital to gain a name in the region. In the Australian colonial period (1914-1975) territory within New Guinea fell under four different classifications: uncontacted, contacted, pacified and under full control (similar classifications were used by the Germans between 1884-1914). Pacification meant an acknowledgement that the sole legitimate use of force lay in the hands of the colonial state, which went rather further than giving up warfare, also including local forms of punishment and pay-back. Nevertheless, stopping fighting between groups was the topic most mentioned in patrol officers' reports and this is what was foremost in their minds when thinking of pacification. In most areas of the Bismarck Archipelago people gave up warfare very readily in the early twentieth century. Did this represent an easy victory for the colonial forces and an acknowledgement by local people of the overwhelming power of the new police and military forces at colonial command? Things are not as simple as that and in order to understand 'pacification' we need to consider the type of warfare that was given up, which was itself an outcome of colonialism.

To understand the warfare that ceased we need to embed it within colonial structures. The formal colonial period started in 1884 when Germany took over New Guinea and Britain made Papua a colony. But this formal starting point is somewhat illusory in terms of the nature and genesis of colonial forces. The creation of the colony of New South Wales in 1788 put New Ireland, Manus and, to some extent, New Britain on novel sets of shipping routes between Australia and Asia, so that ships' crews came ashore for water and fresh food in coastal areas of the Bismarck Archipelago, swapping European goods as barter. Whaling in the region also increased through the nineteenth century, creating further contacts between outsiders and New Guineans (Gray 1999). Relations with outsiders changed fundamentally in the 1860s and 1870s when the first recruiters and traders became active in the Bismarck Archipelago. The demands of the plantation economies elsewhere in the Pacific (such as Samoa and the New Hebrides) and the Queensland cane fields meant that reserves of labour were regularly sought and the islands of the Bismarck Archipelago represented a major supply of 'recruits' (Firth 1973; 1977; 1986; Hempenstall 1978). Recruitment was largely unregulated and the setting up of the German colonial government on the Gazelle Peninsula in 1884 did little to change this before 1900, when Governor Hahl took over. A basic part of the trader's and recruiter's means of operation was the rifle. Relations with local people were often poor and deaths of whites and locals were common, so that the rifle was the basis for self-protection. Unlike in British Papua trading in guns was not illegal in German New Guinea, so that the rifle was part of the basic inducement to trade or allow people to be recruited. Figures do not exist for the numbers of rifles traded, but there are accounts of whalers in the 1860s confronting people on the shores of New Ireland, all of whom were armed. Guns were regularly used in violence between whites and locals but also became a part of local warfare.

The firm of Godeffroy und Sohn set up their first trading station at Matupi in Blanche Bay near the present-day town of Rabaul in 1874, and were soon followed by the smaller firm of Robertson and Hernsheim in 1875. They found the locals well-accustomed to white traders and already speaking pidgin (Hempenstall 1978: 119). The missionary George Brown was also in the first trickle of settlement, and he set up mission stations and made converts between 1877 and 1881. These early traders lived very isolated lives, isolated that is from other Europeans, but they were plunged into a mass of relations with local people, which were not always that happy. Ten of Godeffroy's twelve agents died in the late 1870s, and the firearms which were a part of early trade exacerbated conflicts. There were said to be 700 guns in Tolai hands by 1887 (Hempenstall 1978: 123).

In the Arawe Islands on the south coast of New Britain I have collected oral historical accounts dating from the late nineteenth century, mapped old village sites and taken geneaologies in an attempt to understand the confused situation in the last decades of the nineteenth century. These accounts tell of whole villages being massacred partly due to the use of rifles and of long-distance raids taking place from villages on the south coast who attacked those at the western end of New Britain, in the Cape Gloucester area. The levels of death and the geographical scale of the warfare at this time were unprecedented and alarming to people now, even in retrospect. At this distance in time it is hard to reconstruct in any detail quite what happened, but it seems clear that the areas, like the Arawe Islands, with a relatively high population and more access to traders and recruiters than other groups suddenly had a huge advantage in inter-group violence which they used in the short term. If present-day accounts do reflect some of the sentiments of over a century ago, even the victors were nervous of the newly destabilised situation. If war in New Guinea is exchange pursued by other means, to kill your exchange partners in large numbers is not a sustainable strategy.

The new, unbridled forms of warfare took place against a background of other forms of disruption and dislocation. People, particularly younger men, left the villages in large numbers to work on plantations, mission stations and in towns. We have no figures for levels of recruitment in the nineteenth century, but can gain some indication of the situation in the earlier period by looking at the statistics of the early twentieth century. Between 1912 and the start of 1914 there was a ban on recruitment in the Sulka and Mengen areas and the whole coastline from Cape Gloucester to Montagu Harbour. It was noted that ruthless recruiting by the Forsayths (a plantation-owning family) had made villagers flee inland and one of their recruiters, Karl Münster, was sentenced to three months jail, which was extremely unusual (Firth 1973: 224, 228). In 1913 a Forsayth recruiter broke the prohibition by recruiting labourers for the company's Arawe plantation from Rauto on the adjacent coast. In 1913 recruitment was reaching its limit in coastal areas: it is estimated that nearly every unmarried man in the villages of the north-west coast of New Britain was a recruit that year or had been one recently (Firth 1973: 173), and much the same might have been true of the south coast.

The process of recruiting was obviously a violent and destructive one, which had multiple effects. The Germans estimated that there were 152,075 people living on the Bismarck Archipelago in January 1914, and although we should not let the spurious accuracy of this figure mislead us, it does provide an estimate with some factual backing (Firth 1973: 142). Of these 17,529 were working for Europeans or Chinese in some capacity – that is getting on for 12% of the population at one time.

The other set of devastating changes that came about in the German period was epidemics of disease. Smallpox and flu epidemics spread right through New Guinea in 1893-4 and had major effects in the Bismarck Archipelago (Parkinson 1999[1907]: 90-91). Estimates of how many people died are difficult to come by. In Witu plantations were set up on land left vacant after 50% of the population died from small-pox and the suppression of two uprisings in 1901 and 1903 (Firth 1973: 136). Witu was densely populated so it might have suffered worse than some areas from density-dependent diseases. However, the Arawe Islands also represent high levels of population, and according to local testimony many died at this time. A further devastating event which had a major effect on the west end of New Britain and its south coast was the collapse of the Ritter volcano in 1888. Ritter is out in the Vitiaz Strait and its collapse created tidal waves (tsunamis) which wiped out villages in the Cape Gloucester area, along the coast of Umboi and along the south coast, where local people have stories of waves several metres high. Loss of life would have been high on Gloucester and Umboi, but less on the south coast where destruction of villages and gardens would have been the main effects. This geological event, essentially random when viewed in terms of historical process, would have added an extra dimension to the confusion and disruption of the times.

Up until 1900 was a period of chaos and mass-death, now hard to document in detail, partly because so many died that oral histories were wiped out and because the unprecedented nature of the events made narratives hard to construct. The last three or

FIG. 1: *Two Arawe men (names not recorded) displaying shields, one wearing a pig's tusk ornament as if for a fight. Taken by Beatrice Blackwood in Kandrian, between June and August 1937; PRM BB.P.13.222. Photo courtesy of Pitt Rivers Museum, University of Oxford.*

four decades of the nineteenth century created a radical break with the past and a newly destructive warfare was central to that break. After about 1900 the new German administration under Governor Hahl (1980) brought a new shape and structure to life. From 1896 onwards Hahl also appointed a series of village headmen, the *luluais* (a corruption of a Tolai word for leader) and their deputies, the *tultuls*. Their initial roles were to supervise road construction and adjudicate in local disputes, such that they could impose fines of up to 25 marks (10 fathoms of *tambu* [shell money]). By 1900 there were forty-four luluais on the Gazelle Peninsula and twenty-three in the Duke of Yorks, which were the only areas of Hahl's authority. Government influence gradually spread over the next fourteen years as new government stations and sub-stations were set up. The administrative structure that the Australians took over in 1914 consisted of district offices in Kavieng, Madang and Rabaul, plus second-class stations at Namatanai, Kieta, Morobe, Aitape and Lorengau (Rowley 1958: 16). The administrative staff was not huge and was mainly concentrated in Rabaul: in 1913 there were eighty-four administrative staff altogether in New Guinea, seventy-two in the Bismarck Archipelago and the Solomons and twelve on mainland New Guinea. Each station had at least one German officer, and these officers laid and maintained roads, levied the head tax, appointed luluais and tultuls, kept law and order, and attempted some mild regulation of recruiting.

In the early twentieth century warfare was given up suddenly and has never reappeared as an institutionalised element of life. This might be seen as the colonial power pacifying the natives and this is certainly how the Germans viewed the matter. But the small numbers of German patrol officers who visited most areas once or twice a year could not have forced heavily armed local populations to give up war. The response was a voluntary one on the part of local people, perhaps using the colonial power as an excuse, an external justification for what they wanted to do anyway.

War was given up with some relief, but also at a considerable cost and that cost was paid in basic social re-organisation. On the Arawe Islands where people were previously spread among a number of small hamlets on the south coast of New Britain and a number of islands, they now cluster in big villages on five islands and one area of the mainland (Gosden and Pavlides 1994). In the present, people in many parts of the province and particularly on the coast live in villages of several hundred inhabitants, but these were formed often around the beginning of this century from a number of smaller hamlets, each of which was defended and centred around a men's house (Counts and Counts 1970: 92; Zelenietz and Grant 1986: 204). Chinnery (1925; 1926) has reported an identical form of organisation for the Kaulong and other areas of the south coast.

These small hamlets seem to have been the basic settlement pattern throughout West New Britain and the Siassi Islands (Freedman 1970), being given up on the coast fairly early in the history of European colonialism and later (if at all) in the inland areas. But this was not just a settlement pattern but a form of social organisation based on cognatic ties which allowed changing rights to land and other resources.

This single form of settlement pattern and social structure is an ancient one dating back around 1000 years (Gosden and Pavlides 1994). It allowed for both mobility of settlement and open links between

groups. What happened from the early twentieth century onwards is that the settlement pattern changed from small hamlets to large villages, but the open network of connections continued and expanded. People congregated into larger villages, but were still able to move regularly from one area to another on trading expeditions or to take up gardening land in a new place through kin connections. New larger settlements formed a labour pool for larger gardens to feed the settlements themselves, but also the new needs of plantations and traders for food. The new villages were also focal points in the movement of people through the area utilising their kin connections to travel far and wide. The large villages at first only existed on the coast, where government influence is greatest, and settlement agglomeration happened later inland; in some areas, predominantly inland ones, it has never happened. All these changes brought into question the nature of the community and these questions of community have considerable implications for the exchange and use of material culture.

Museum collections and colonial histories

Studying colonial histories in New Guinea is not easy and there is no single source which provides an adequate understanding of the complexity of colonial change. In addition to the oral and written histories drawn on above, museum collections and their attendant documentation are an important source of evidence of how the types of material culture and the uses to which things have been put have altered over the colonial period. Recently Chantal Knowles and I have organised a project analysing material in museum collections from the south coast of New Britain in order to chart the changes brought about by the colonial period in this region (Gosden and Knowles 2001), important amongst which is the cessation of warfare. The material data used in this study consist of four collections that were made between 1909 and 1937, and the results of fieldwork which was carried out by Gosden and Pavlides between 1985 and 1992. This latter work involved a survey of material culture then in use on the Arawe Islands, including items used in exchange systems, as well as those bought from stores. Initial results from this work were presented in Pavlides (1988). Information of recent movements and uses of material culture can provide a point of comparison and contrast with the situation earlier in the twentieth century.

Each of the collections was a result not just of the material culture in use and circulation at the time, but also of the interests of the collector, which needs to be taken into account when analysing the structure of the collections. The first collector, A.B. Lewis, had a special interest in connectivity and hence items used in trade and exchange, taking care to discover where things were made and how they had been traded. Felix Speiser had general culture historical interests and he tried to chart the movement of materials across Melanesia in a fashion that might indicate migrations of peoples and links between them. We know least about J.A. Todd's work, as he was a Ph.D. student at Sydney who never finished his thesis and published very little pertaining to the collections he made. As far as we can tell, he was interested in totemism and objects that might reflect the varying social groups people belonged to, as well as the sets of symbols and beliefs lying behind these groups. Beatrice Blackwood concentrated on technologies, especially stone tools and was generally concerned to document the processes people used when making things. The other factor is what local people were prepared to sell, as any collection is the outcome of negotiations. Despite their individual biases, each collected a good range of materials which were generally representative of the broader changes in material culture through the earlier part of the twentieth century.

The first of the collections was made by A.B. Lewis who did fieldwork on the south coast of New Britain from December 1909 to February 1910, as part of a total of four years in the Pacific. As the first curator of anthropology at the Field Museum in Chicago, a major part of his interest was in collecting artefacts which were well documented in terms of their production, use and exchange. He has extensive field notes which refer to the hundreds of objects he collected (the exact number is not known at present, but 324 have been counted for New Britain), plus some 250 photographs of the area and of items in use. The second collector – Felix Speiser – was in New Britain in 1929 as part of a larger trip to different parts of Melanesia in search of comparative material designed to throw light on Melanesian unity and origins. He collected large numbers of artefacts (111), well documented as to provenance, as well as pho-

tographs and a short film of initiation rites, in which he had a special interest. All Speiser's material is held in the Museum für Völkerkunde in Basel. J.A. Todd made two field trips to south-western New Britain in April 1933 to April 1934, and in 1935 to 1936 while he was a Ph.D. student at the University of Sydney. He made a collection of 185 items on his first visit. This collection is now in the Australian Museum, Sydney with some attendant documentation, although the bulk of his field notes, sound recordings and photographs have been lost. The final collector considered in the analysis was Beatrice Blackwood of the Pitt Rivers Museum. She was a curator and made an extensive collection of artefacts (271 pieces) from the south coast of New Britain during a seven month stay in 1937, together with meticulous documentation of many aspects of life, including genealogies, word lists and some 350 photographs. She also made a short movie film. She was especially interested in production, technology and skull deformation. A comparison and contrast of the material in these collections provides vital insights into the processes of colonial change.

In Table 1 percentages are given and these refer to the number of potential classes of item in each collection which falls into each of the categories. For example, Lewis' collection has examples of all the types of containers that are known from the collections as a whole (100%), while he has only 44% of the different types of musical instruments. Given that each collection represents a particular 'time-slice', we are attempting to look at change through time by focusing on the increases and decreases in the different types of objects within each collection, bearing in mind the biasing effects of the interests of the collectors. Let us compare Lewis' collection (1910) and Pavlides' observations (1992): over a period of 82 years, little change is discernible in the collections in the categories of hunting and fishing, food preparation and craft production. However, over the same period, a major decline can be seen in the availability and collection of objects associated with warfare, stone tools, containers and material culture pertaining to music.

Warfare and music were linked, and the decline in the collected objects that are connected with both of these aspects of life was associated with the sharp decline in exclusively male artefacts collected since Lewis' visit in 1910 (see Table 1). Lewis arrived

Table 1. Changes in the percentages of artefact types in each category between the four collections, plus Pavlides' observations

	Lewis 1910	Speiser 1929	Todd 1935	Blackwood 1937	Pavlides 1992	Decline
Hunting and fishing	69%	62%	23%	62%	54%	10%
Warfare	75%	100%	50%	75%	0%	100%
Craft production	50%	17%	17%	83%	83%	None
Axes and obsidian	33%	40%	10%	70%	0%	100%
Food prodn and eating	56%	44%	44%	33%	56%	None
Containers	100%	50%	83%	50%	30%	70%
Ornaments and clothes	68%	58%	42%	47%	32%	30%
Valuables	53%	35%	59%	53%	Increase	Increase
Music	44%	78%	11%	78%	22%	50%

at end of the old order on the south coast of New Britain; there was still some fighting, and although some large villages were to be found there were also smaller defended settlements located away from the coast. Men's shields were decorated and the decoration on the inside of the shield, nearest to the man's body, was his own and was repeated on the bark-cloth belt worn around his waist (Fig. 1). Some types of ornaments, such as pig's-tusk mouth ornaments could only be worn after killing someone in battle. By the 1980s and 1990s when Pavlides carried out a survey of material culture in the Arawe Islands, shields and spears were absent there (although they remained as items of ritual elsewhere in the region) and pig's-tusk ornaments had been revalued as items of wealth any man of some standing could wear.

Ritual itself saw a shift from male-only forms to those that involve the whole community. Circumcision, marriage and death reflect entry into the community, a change of state within it, and exit from it. Rituals surrounding these changes predominate today and probably have done since just after the First World War. The earliest collection, that of Lewis, is the only one in which masks relating to the male-only *warku* cult are found. Five examples were collected by Lewis, and derived from *warku* ceremonies, which were men-only and centred on the men's house, but these ceremonies might have been in decline even when Lewis was there. Speiser discussed *warku* with Aliwa, a local bigman. On 21 April 1929 Aliwa said that there were two *warku* masks, named *sala* and *einsparna*, left on Pililo, all the others had been destroyed by the missions or bought by whites (including Lewis presumably). *Warku* masks were

used during circumcisions which were carried out in the men's house. Women were allowed to glimpse them under certain circumstances, unlike in Siassi or New Guinea, where *warku* were also found and where it was completely forbidden for women to see the masks. *Warku* masks are not found in any later collections, and this accords with the testimony of people in the Arawes, who said that *warku* ceremonies had not been run for most of the last century. *Warku* were also connected with bull-roarers, which were still present when Blackwood was collecting (she has two in her collection) but have subsequently disappeared.

The sale of the *warku* masks and ceremony is part of a much larger movement of objects and ritual between areas in New Britain that were earlier more distinct in their material culture and ritual. There are complicated reasons behind these movements of ideas and rituals, partly to do with the opening up of exchange networks and partly with the mixing of people on plantations and missions which enabled people from one area to learn about life in another (Chowning 1969: 29). The earlier importation of male-only ceremonies shifted to an emphasis on, and acquisition of, community-wide ceremony.

Other sets of distinction which broke down were those within the local communities relating to the ownership or control of objects. At the beginning of the twentieth century on the coast and later in inland areas, only big men and women could own and control ritually powerful objects, such as *mokmok*, *singa* and named pearl shells. Such items were dangerous and could have ill effects on the community at large if in the hands of the inexperienced or unskilled individual (Goodale 1995: 89). Now all men and women own such stones, although it may be true that the most powerful objects are restricted to the most powerful individuals. The same was also true of certain forms of taro and other plants, which at one time could only be planted by people with sufficient personal power to control the magic associated with their growth. Communities in western New Britain have allowed the erosion of such distinctions over the last century, which has opened exchange networks still further. Although the evidence is not clear, it seems that people of Aliwa's standing and generation were the last who had the power to enforce such distinctions in ownership and use. Arawe communities were much more internally fractured and distinct from each other a century ago, but the breakdown of a series of differences has led to a massive expansion of the social universe that people now inhabit.

One indication of the greater size and intensity of links between people has been an inflation in payments of various kinds, especially marriage payments. The one category of items which shows an increase in Table 1 is that of valuables. This increase is made up of both a greater number of different objects exchanged and a greater frequency of exchanges for existing objects. These increases are difficult to quantify, but it is certain that there has been an inflation in certain types of payment, such as bride price, with far more objects needed to complete a bride price now than earlier in the century (Gosden and Pavlides 1994: Table 2). Many objects are used in a variety of forms of ritual, such as initiation or death rituals, and it is likely that payments of all types have increased. Such rituals and forms of exchange, involving the kin group as a whole, are now central to peoples' lives and have replaced the male-centred rituals of the nineteenth century. We feel that it is no coincidence that most such rituals concern life cycles and initiation of children into the group. Such an emphasis on the group is partly as a result of the many changes to the nature of group life, through alterations of settlement patterns and movements of people as a result of labour recruitment, as well as shifts in the relationships between men and women. The group as a whole was re-thought, and people have used existing cultural means, constructed in novel ways, to carry out this re-thinking.

People's motivation to engage in exchange expanded as other opportunities for social advancement declined. For men exchanges helped compensate for the lack of opportunities once found in warfare and which for women had existed in much more restricted form while warfare was constant. Today throughout the region peoples' social reputations, their names, are associated with the ability to host song festivals, feasts and ceremonies and to engage in exchanges. Both Maschio (1994) and Goodale (1995) emphasise the two-way process whereby people in western New Britain develop their individuality through social links with others. Relations with other people and groups necessitate the husbanding of material resources, trade objects, the creation and maintenance of links with others, and the retention of knowledge about the social world in general. Knowledge about objects and people and the ability to use this knowledge constructively was the basis for social success, and this is brought about by both travel and a wide circle of friends.

Conclusions

On the south coast of New Britain, the end of the nineteenth century and the beginning of the twentieth saw basic re-organisations of the community, gender relations and some of the values attached to objects. War, part of the basis for male prestige, was abandoned and the shields and spears used in fighting disappeared or became accoutrements of dance. The supply of guns dried up as recruiting was both curtailed and more strongly regulated. In the movements from hamlets to large villages gender relations were re-thought. The fact that people could now live in undefended settlements, changed the settlement pattern and all the social relations within and between settelements. The earlier separation of men and women broke down and instead of sleeping only in the men's house, men with wives and children slept in family houses, giving more prominence to the nuclear family. Men-only rituals, like the *warku* cult, were let go and the masks sold off. In their place was an efflorescence of cult activity concerning the nature of community and community membership. The exchange networks which supplied valuables for community-orientated rituals also expanded greatly. Prestige, for women and for men, derived from sponsoring rituals and the exchanges which funded them. Exchange became war by other means, but with more pacific aims.

Death through warfare, disease and tsunami and absence of many due to high levels of recruitment all put the community under threat. Local responses were hugely innovative, re-ordering many of the key relationships composing local cultural forms. Maleness was rethought and became less to do with the use of force, concentrating rather on using extensive cognatic relations to engage in exchange.

The colonial changes observed in New Britain are not found everywhere in Papua New Guinea. In the Highlands, for instance, as Wiessner brings out well in her contribution, there is a much shorter and more regulated colonial history which has unfolded since the 1930s. Violence by colonial officers was at much lower levels than on the coast, as by this time New Guinea was a Mandated Territory under the League of Nations and all local deaths at the hands of whites had to be reported and explained. The sale of guns to local people was banned. Local warfare has been harder to eradicate and has seen a considerable resurgence in recent years unlike the situation on the coast. The rather lesser traumas of the colonial period may partly explain why western goods are regularly incorporated into socially important exchanges in the Highlands, whereas in many areas of New Britain goods of western origin are rigorously excluded from bride prices or initiation payments. Although exchange has expanded this has not included western goods. The dangers of colonial contact, stemming from an earlier period of trauma are still felt today.

War is not one thing, but many. Because war is often part of the basic social norms of society it will vary as these norms vary. Occasionally war breaks the rules and has to be contained, as is the case here for New Britain and other areas of the Bismarck Archipelago. In either case, war is a crucial diagnostic aspect of social forms, their core values and history. Colonial change is often hard to track in detail, but can be done using multiple sources of evidence from oral histories kept by people from the region in question, historical records and museum collections. The combination of disease and unregulated war at the end of the nineteenth century still stays in people's memories today in a general sense. Many present ways of life also derive from the alterations that these traumas brought about. These were, in turn, due to a combination of local desires to stop fighting and colonial needs to impose a monopoly of state violence as the ultimate basis of colonial power. Local and colonial desires combined to refashion life fundamentally once war ceased.

BIBLIOGRAPHY

- Chinnery, E.W.P. 1925. 'Notes of the natives of certain villages of the Mandated Territory of New Guinea'. *Territory of New Guinea, Anthropological Reports Nos. 1 and 2*. Melbourne: Government Printer.
- Chinnery, E.W.P. 1926. 'Certain natives of south New Britain and Dampier Straits'. *Territory of New Guinea, Anthropological Reports No 3*. Melbourne: Government Printer.
- Chowning, A. 1969. 'Recent acculturation between tribes in Papua-New Guinea'. *The Journal of Pacific History*, 4, pp. 27-40.
- Counts, D.E.A. and Counts, D. 1970. 'The *Vula* of Kaliai: a primitive currency with commercial use'. *Oceania*, 41, pp. 90-105.
- Firth, S.G. 1973. German Recruitment and Employment of Labourers in the West Pacific before the First World War. Unpublished D.Phil. thesis, University of Oxford, Oxford.
- Firth, S.G. 1977. 'German firms in the Pacific Islands 1857-1914'. In: J.A. Moses and P.M. Kennedy (eds.): *Germany in the Pacific and the Far East 1870-1914*. Brisbane: University of Queensland Press, pp. 3-25.
- Firth, S.G. 1986. *New Guinea under the Germans*. Port Moresby: Web Books.
- Freedman, M.P. 1970. 'Social organization of a Siassi Island community'. In: T.G. Harding and B.J. Wallace (eds.): *Cultures of the Pacific*. New York: The Free Press, pp. 159-79.
- Goodale, J. 1995. *To Sing with Pigs is Human: the Concept of Person in Papua New Guinea*. Seattle: University of Washington Press.
- Gosden, C. and Knowles, C. 2001. *Collecting Colonialism. Material culture and Colonial Change in Papua New Guinea*. Oxford: Berg.
- Gosden, C. and Pavlides, C. 1994. 'Are islands insular? Landscape vs. seascape in the case of the Arawe Islands, Papua New Guinea'. *Archaeology in Oceania*, 29, pp. 162-71.
- Gray, A.C. 1999. 'Trading contacts in the Bismarck Archipelago during the whaling era, 1799-1844'. *Journal of Pacific History*, 34, pp. 23-43.
- Hahl, A. 1980(1937). *Governor in New Guinea*. P. Sack and D. Clark (eds. and trans.). Canberrra: Australian National University Press.
- Hempenstall, P.J. 1978. *Pacific Islanders under German Rule: A Study in the Meaning of Colonial Resistance*. Canberra: Australian National University Press.
- Maschio, T. 1994. *To Remember the Faces of the Dead: the Plenitude of Memory in southwestern New Britain*. Madison: University of Wisconsin Press.
- Parkinson, R. 1999(1907). *Thirty Years in the South Seas*. J. Dennison (trans.). Bathurst: Crawford House Press.
- Pavlides, C. 1988. Trade and Exchange in the Arawe Islands, West New Britain, Papua New Guinea. Unpublished Honours thesis, La Trobe University, Melbourne.
- Rowley, C.D. 1958. *The Australians in German New Guinea*, Melbourne: Melbourne University Press.
- Zelenietz, M. and Grant, J. 1986. 'The problem with *Pisins*'. Parts I and II. *Oceania*, 56, pp. 199-214, 264-74.

Warfare and the State

Warfare and the State:
An Introduction

HENRIK THRANE

/14

This section spans a wide field in space and time, nevertheless covering the same subject, namely the relationship of warfare to the state. It ranges from Europe to Africa and the Pacific. Claessen's article (chapter 15) serves as introduction with its general theoretical discussion of one of the main issues of the War & Society project. War as part of the state concept is a commonly accepted idea. Indeed it has been maintained time and again that the logistics and organisation of warfare were crucial elements in the rise of states, the question being the identity of the prime mover. War in pre-state or non-state societies was sometimes even seen as a methodological misunderstanding.

We still grapple with the proper way of distinguishing war from other organised forms of violence. Several definitions have been suggested (Turney-High 1971; Keegan 1996: 89ff). For archaeologists these are somewhat academic discussions. We only have scarce and sometimes (most often) very particular evidence of the scale of violence in prehistoric societies. The evidence needed to define the level of violence as one proper for war – in Claessen's definition (chapter 15): 'legitimised and organised deadly violence between centralised polities (states and paramount chiefdoms)' – is hardly ever conclusive before we reach proto-history. While legitimisation is not accessible in archaeological evidence, the deadly consequences sometimes are – cp. the section 'Warfare, Rituals and Mass Graves'. Considering the difficulties we have in defining when chiefdoms arose and what kind they were – not to mention paramount chiefdoms or early states – we are left with general statements based upon precisely such deliberations as those presented in this section. We have to take our clues from the ethnographical and historical interpretations of, to us very late, phenomena such as the (early) states and similar advanced social systems. This leaves out 99% of the history of mankind. Our difficulties as prehistorians transpire from several sections in this book.

The concept

Henri Claessen demonstrates the reasons and ideas behind the assumption of a causal interconnection between war and state focused upon war's creative role in this dualism (chapter 15).

A distinction has often been made between primitive and civilised (i.e. serious, genuinely lethal) warfare (Keeley 1996). Primitive warfare was seen as limited in space and duration with perhaps as high a social involvement as 'proper' war, but to a certain extent ritualised, always brief and with strictly limited physical consequences for the participants – the warriors. One gets the impression that the fights were organised for the fun of it and as a way of letting off steam in societies where conflicts might otherwise simmer too long and become too detrimental (Otterbein 1970).

Cause and effect chains are inevitable elements of any historical research not satisfied with establishing the pure facts. In any non-historical context it is highly unlikely that we can obtain knowledge of all the possible causal elements which led to the existence of a phenomenon, and surely we all realise that we live in a complicated world where causes are normally multitudinous and the decisive ones not so easy to distinguish. A priori it seems simplistic to say that 'war makes states'.

Access to resources and pressure on land are of universal importance to any society and must be classic factors in social development and differentiation. Warfare was certainly ever present as an alternative solution, no doubt linked to the perception of 'group identity'. Claessen sees the creative force of conflict and war as of little value (chapter 15). Anyone not associated with the technology of arms production will surely agree with him on general principles. That war and the military in states could become an economic power of sometimes destructive dimensions, e.g. in the later Roman Empire, is seen as connected with developed forms of states. It was most likely caused by the enormity of the state rather than vice versa.

The Betsileo case – apart from the distinction of skirmishes from war – raises an interesting question: how can archaeology understand the chain of events such as that leading to the differential development of neighbours and the formation of hillforts? If such limited war activity in a certain historic context was enough to lead to a hillfort system and an early state, was there proper war in Europe before the establishment of the Roman Empire? Were the hundreds of hillforts in Bronze and Iron Age Europe the results of a similar situation and can they be taken as signs of state formation?

Five factors leading to (early) state formation partly coincide with the four factors behind conflict. They reflect reactions to the demands on society caused by stress from increased population density. As Claessen warns us, even these factors do not automatically lead to an early state. An independent action is needed to trigger that development. His cautionary article presents a salutary warning against oversimplified thinking (chapter 15)

Robert Carneiro (1970) has managed to lay a trail that has absorbed much energy from his fellow anthropologists. Claessen (chapter 15) pinpoints the ambiguity of Carneiro's models, but his distinction between derivative and causal factors in the cause-effect chain is useful. That chains of events differ across the world can hardly surprise in view of the enormous diversity in anthropological experience – which makes it so hard for archaeologists to behave correctly in the world of analogy: it is amazing how we have loved to infer from living society to dead society over 200 years of research and how often we have erred. The usual problem with models pops up – if they are to be universal they become so elementary that they become useless.

Scales of conflict

Claessen finds that the relatively high number of casualties at the Dutch Middle Bronze Age site of Wassenaar (chapter 15; cp. Thorpe chapter 10 and Osgood chapter 23) does not justify the term war. This is where we may disagree. Is war not wholesale murder and must it leave a certain number of victims? Do efficiency and ruthlessness not constitute elements of war?

Of course the slaughter of small groups can be carried out by equally small groups, but that does not exclude the possibility that these were parts of a larger force practising proper war (e.g. at Crow Creek: Keeley 1996). So once again, the evidence is not as easily interpreted beyond the simple statement of facts. If war differs from skirmish or simple conflict – just as a matter of scale – we can only speak of war when societies have become complicated or large enough to be states.

Africa

Jan Abbink (chapter 18) changes the perspective to state and 'culture' in sub-Saharan Africa, examining recent attempts to see 'culture' as a decisive factor in the recent state of affairs. Ritualised violence was here traditionally often exercised to avoid larger bloodshed, and violent practice has led to 'cultures of violence'. Abbink's definition of war wants to propagate an interactionist view and is non-material and therefore difficult to apply to prehistoric 'cultures' where the elements of his definition are a matter of interpretation of mute sources. One may ask whether Africa is a good general example because of the strong and varied influence of colonisation and the reactions to it, not least the artificial colonial and postcolonial forced groupings of heterogeneous, independent 'cultures' where conflict seems predictable and inevitable – because of the rupture of old obligations? This reminder of reality would be good reading for well meaning politicians. Six elements which are (were) important to the study of conflict, resistance and warfare are listed by Abbink (chapter 18). Most of them, if not all, occur on other continents.

The conflict with western ideals is here, as elsewhere, a central problem in recent world affairs. We are back at the causal problem: can 'cultural factors' alone explain cultural change?

Whether cultural values have autonomous causal force is a matter of belief, says Abbink. 'Cultures of violence' as the term goes, are seen as transitory (chapter 18), which seems to be the only optimistic aspect of this devastatingly wide spread phenomenon.

Ethiopia provides an example of a region where non-state warfare has been known for more than a hundred years and apparently remains unchanged by recent attempts at state formations.

Ecological crises and the present level of weapon technology are lethal elements in any ever so slightly unstable society (culture). The horrible efficiency of modern weaponry is one of the far too many elements available to trigger a circle of violence in Africa as in other regions (cp. Kapferer 2004). The old time predictability has vanished, as the old people lament, and as any politician would surely agree. Our successors may be able to recognise, in the future archaeology of our era, the gun in its development as the material expression – and part-cause of this development. The African landscapes must be littered with thousands of spent cartridges and other rubbish of violence as archaeological sources; far outnumbering the handaxes of many millennia!

Cultural mediation seems to be the inevitable loser in face of the international economy selling its fast, long distance destructive guns to anyone. What would happen if the weapons import to Africa could be stopped? (I know it is an academic question). The modern states in Africa crosscut all the traditional cultural identities and one wonders how long they can survive – now with the additional burden of AIDS.

The Pacific

Claus Bossen takes us to the Pacific (chapter 17) – as do several other articles in this volume. His cases are historically documented; Fiji and Hawaii are both classic cases of chieftains, wars and early states.

That states make war is rather clear. Whether war makes states is of course the central theme of this section. Bossen stresses the individual factors in any specific historical development. A monopoly of violence presupposes laws backed up by law enforcement; however, individual and gang violence survives in even the most developed states. Pre-modern states were surely saturated by violence at home, in institutions, in the streets etc., completely outside the control of the armed forces of a state apparatus.

Bosssen's stance is that while leaders make war, war was not enough to make a state, which seems a sound statement. He divides the discourse on how war caused states to be formed into 1. the context of social organisation, 2. military organisation, 3. the means to enlarge political power. They are all aspects of social organisation. It is interesting that the motivation for state formation has been seen as either rather egoistic (Fried) or altruistic (Service) which presumably more than anything else reflects how we as individuals, anthropologists or others, are oriented in our own social world.

Bossen examines how war contributed to a change of social organisation (chapter 17), which leads us to think of the *Iliad*. Homer's early city states with their kingly hierarchy would be an example of early states with warriors recruited through the '(Indo) European' model of '*Gefolgschaft*' which is not quite the same as a paid military force (cp. Vandkilde chapters 26 and 34). When is a society in a constant state of war? Were the earliest Greek cities proper states? War without a strict social organisation is just a shambles.

Can a centralised military command structure appear by itself, isolated from social structures? One thing is efficiency but will such a system be adopted by people who are engaged in different social relations in their ordinary lives? War leaders (generals) were temporary, *ad hoc*, positions in Europe's pre-states and '*Gefolgschaft*' is a temporary and personal reciprocal obligation (Timpe 1998). War organisation is only crucial in situations of constant outside threat and with a suitable subsistence production.

Conquest is of course a result of war, but not an indispensable element of war. Elements of warfare may be conducive to increased centralisation and hierarchisation in already centralised and hierarchised societies, nonetheless it is not war itself, but the opportunities it may offer and the combination with other domains of power that may lead the way to state formation. It is never a simple process. Fiji is a good example with historical sources on warfare and archaeological evidence in the shape of fortified sites. The fluid transition from small-scale ambushes to the later large-scale operations is a good illustration of our terminological despair. The effects of large-scale operations were serious for individuals and social groups (villages). The conquest of a village involved wholesale killing, which presumably would leave archaeological traces like the mass graves in European prehistory.

In large-scale chiefdoms war was used to change eco-political conditions by subjugation.

Malcolm Webb is quoted by Bossen (chapter 17) as pointing out the role of long distance trade. This raises another issue of common interest – the role of peaceful external relations in the transformation of societies. With the steady interest in this topic this looks like a fruitful alternative to violent relations. Hawaii is another classic in anthropology with a very detailed historical record, expressing enough power to deter attacks. In Fiji warring prevented a more complex society based on exchange relations and the use of limited external resources for the decisive victories – a special case one may ask? Only a combination of several areas of power led to the formation of states by war.

Protohistoric Europe

Heiko Steuer's article (chapter 16) expands the chronological range of the other contribution in this volume to the protohistoric Migration Period. He employs historical sources, but the emphasis is archaeological. The general aim is to produce a model of warfare and state formation.

The fluid state of Germanic and Gallic tribal units, at least as recorded by their names, has become much clearer with recent research. Territorially they were only defined by their relation to the Roman Empire; a case of periphery reacting to its centre in fundamental aspects of social organisation. There is hardly any doubt that the Barbarian societies were profoundly influenced by the existence of the Roman state from Caesar's time onwards. The migrations of the period are now interpreted as military campaigns rather than the wholesale movements of entire populations that we all believed in a generation ago – viz. the Goths. Such tribal names are now taken to cover specific warrior bands, although the Latin writers mention women as part of the phenomenon. Here war is seen as indispensable to the tribal constitution and development of native states. The warrior identity again appears in a continuous 'state of warre' – annually recurring campaigns at least are suggested.

Symbiosis with great states seems indispensable to Barbarian social development beyond the *limes*, already before the rise of the Roman Empire. The discrepancy between the detailed historical information on geography and political processes and the lack of corresponding support from archaeology is a sad reminder of the absence of clear indications of these processes in mute prehistoric material evidence, in spite of our occupation with ethnic and migratory problems for more than 100 years. Settlement data promise new ways out of the methodical quagmire.

We are taken through a series of stages of increasing contact with the Romans (cp. Green chapter 20) and their successors as dominant powers in Europe through to the Viking Age. All the Barbarians, including Goths and Huns, were out for gold, having seen its value during their service in the Roman forces. They knew how to press the Romans who, after all, found it cheaper to buy peace than to raise new armies. A crucial element in this symbiosis was milking the cow without ruining the udder. So in a war period the warlords thrived, enabling them to improve their standing and adapting Roman standards in art and other material spheres, as archaeology clearly demonstrates. The spiral of warring, and gaining by it, certainly assisted or even led to many changes in Barbarian Europe.

Loyalty, apart from that of warriors and their leader, seems to have been a very relative quality – very much like what we see now in modern tribal societies in the East. Steuer's model, though based upon concrete European historic (and archaeological evidence), is suggested to have a wider use in regions where similar symbioses existed between states/empires and their unruly, greedy, and mobile tribal neighbours (chapter 16).

The interpretation of the most ubiquitous of archaeological sources, namely graves with weapons, remains crucial to any archaeological assessment of war in pre- and protohistoric societies. Were they only for professionals or for the members of the 'Gefolgschaft' (or Roman units)? The answer will lead back to the Bronze Age situations for which it is currently debated whether a 'Gefolgschaft' may be assumed without any historical evidence. It seems debatable whether the motifs behind warrior bands were defence or attack and whether raids were not part of the social set-up much earlier than this period – indeed the early Neolithic site of Talheim and the Middle Bronze Age site of Wassenaar with mass graves could be interpreted in this manner.

Without states the major component in the symbiosis would be missing. They were at the same time instigators – by recruiting Barbarian warrior bands to fight each other – and the object of the envy of all the other warrior bands. So the close interrelationship would have led nowhere without the constant presence and needs of the great State. This seems equally true of the specific Roman case and in general. The value of Steuer's study (chapter 16) lies in the long series of imperial records of situations which enable him to present this outline of the history and the ensuing model.

An elaboration and extension of it to other regions and periods is a tempting prospect.

We must realise that there are no easy solutions. An increased awareness of qualities and weaknesses is the only way to better models and safer conclusions. That is what the articles forming this section have endeavoured to provide.

REFERENCES

- Carneiro, R.L. 1970. 'A Theory of the Origin of the State'. *Science,* 169, pp. 733-38.
- Kapferer, B. (ed.) 2004. *State, Sovereignty, War. Civil Violence in Emerging Global Realities.* Oxford: Berghahn.
- Keegan, J.1993. *A History of Warfare.* London: Pimlico.
- Keeley, L. H. 1996. *War Before Civilization. The myth of the peaceful savage.* New York and Oxford: Oxford University Press.
- Otterbein, K. 1970. *The Evolution of War.* (3rd edn.). New Haven: HRAF Press.
- Timpe, D. 1998. 'Gefolgschaft'. *Hoops Reallexicon der germanischen Altertumskunde* X. Berlin. pp. 537-46.
- Turney-High, H.H. 1971(1949). *Primitive War: Its practices and concepts* (2nd edition). Columbia: University of South Carolina Press.

War and State Formation: What is the Connection?

HENRI J.M. CLAESSEN

/15

In this article I will investigate if and if so, to what extent, there is a connection between war and state formation. This connection is often more assumed than demonstrated, and it can be doubted if war is a creative force behind state formation at all. I will first make a distinction between war, and other types of conflict, and present some views on the concept of causality. Then I will discuss some theories and statements on the connection between war and state formation. Finally I will try to answer the question to what extent war has played a role in the formation of states – or, for that matter, in the development of hierarchy, chiefdoms or stratified societies.

On the definition of war

A definition of war may resemble at first sight a mere terminological exercise, but in view of the importance that this concept is accorded in the construction of theories such an exercise is warranted. War is usually associated with large-scale, organised violence between societies. When the term 'war' is used, one thinks of extensive, well-organised chiefdoms and states. In the voluminous *War, its Causes and Correlates*, Martin Nettleship states that 'war is a civilized phenomenon, different from primitive

fighting' (1975: 86). He does not draw, however, a line between 'fighting' and 'war'. In his vision, one should rather think of a continuum than in terms of a sharp division.

The French anthropologists, Bazin and Terray, keep all possibilities open when they call war 'une pratique sociale particulière', which might be considered a poor definition (1982: 10). There are several more rather general statements on war, such as the one by Marvin Harris (1977: 33), that war is 'organized inter-group suicide'. It is, however, Ronald Cohen who casts the most light on the phenomenon of war, by defining it as: 'publicly legitimised and organized offensive and/or defensive deadly violence between polities' (1985: 276-77). When the violence is restricted to groups within a political unit, or between small political units, fighting for revenge, feuds, need for booty, or on the ground of considerations of social and political prestige, he employs terms such as 'conflict', 'raid', 'attack' and the like. I shall follow him in this approach and limit the term 'war' to legitimised and organised deadly violence between centralised polities such as early states and paramount chiefdoms.

About causality

To establish causality in the social sciences or prehistory is a tricky matter (Köbben 1970). Stated baldly, one speaks of cause and effect when Phenomenon B follows Phenomenon A after a certain period of time and there is every reason to suppose that B would not have been able to exist without A. The problem is that in our sciences, the question of cause and effect is a complex, often recalcitrant phenomenon, however. After all, what is the cause of something? How far should we penetrate back in time to find 'the' origin of something? One can easily conceive of a situation in which Phenomenon A has apparently been caused by Phenomenon B, but behind B lurks an 'instigator' C – and in that case is B or C the cause of A? (Claessen 2000: 8).

Philippe van Parijs (1981: 4-20) has devoted a great deal of attention to the question of cause and effect in social evolution. In his view the scholar should first establish the time sequence. Secondly he or she should point out patterns that have been established earlier, and finally, he or she should be able to make the assumed relation comprehensible. Thus there is absolutely no question of any 'black-box' reasoning, only the beginning and end of which are known. To give an example: imagine I raise my hand and directly afterwards a small dog dies in the street. Is this then a question of *post hoc, ergo propter hoc*? Naturally, it is highly unlikely; imagine, however, that I raise my hand ten times and ten times a small dog dies. Is there then any question of cause and effect? At this point we must take recourse to the assistance of van Parijs' criteria: have there been earlier confirmed regularities in this area? Is it possible to make the assumed relationship comprehensible? Because neither of these requirements can be fulfilled, we are not yet able to register the hand-dog relationship among the causalities.

We have to be very cautious when we want to establish a cause and effect relationship – even when many people suggest that such a relation exists. This caution is especially necessary in the field of the war-and-state formation theories, as we will see presently.

Pre-state conflict

War and conflict are quite common phenomena in human cultures. There is no reason to overestimate the degree of peacefulness in the bands of hunters and gatherers; there have been recorded many cases of conflict and violence among these peoples (see e.g. Lee and DeVore 1968), though full-fledged war seems hardly to occur among such groups (Steward 1968: 333-34). Matters could also greatly escalate in the prehistoric period, which is demonstrated by a Bronze Age grave excavated in Wassenaar in the western Netherlands in 1986 (Louwe Kooijmans 1993). This contained twelve bodies buried together. One whole group – men, women, and children – was killed at one and the same time; the bodies show obvious traces of violence. This must have been an organised attack, of which the purpose was the destruction of the opponents (Claessen 2000: 104). It is, the number of casualties notwithstanding, not a case of war in the sense of the definition adopted here. This view is confirmed by the findings in the recently defended doctoral thesis of David Fontijn (2003).

A slaughter on this scale easily outstrips the relatively small number of victims who fall in the notorious, endemic conflicts among the Yanomami. Here the number of victims is restricted to but a few each time (Chagnon 1968). It must be kept in mind that the Yanomami, though being a small scale society, base their subsistence mainly on horticulture. The Yanomami can be considered a good example of a society, living in a 'State of Warre' – a characterisation formulated in 1642 by the English philosopher Hobbes. It is true that Hobbes did not claim that warfare would be permanent in such societies, but that a permanent 'State of Warre' would prevail, because there were no institutions to prevent the conflict. People would therefore have to live in permanent uncertainty and insecurity (Sahlins 1968: 4-6; Claessen 2000: 102).

Such a 'State of Warre' was not only found in relatively simple societies, such as the Yanomami, but also in the quite complex League of the Iroquois and among their neighbours, the Huron (Claessen 2000: 105; Morgan 1851: 58; Trigger 1969: 42-53). With regard to the Huron, Heidenreich (1978: 385) even states that 'Theoretically any man could plan and organize a war party if he got enough support, but in most cases this task was assumed by the experienced war chiefs'. Neither among the Iroquois nor among the Huron was there any question of professional soldiers, or a military organisation, an affair of the Nation, or of any military action legitimised by the Nation as a whole. The many conflicts, raids, attacks, and violence found with the Iroquois and the Huron thus cannot be called 'war' in the sense of the definition developed by Cohen.

On the connection between war and state formation

In his well-known theory of the origin of the state (1970), Carneiro does not distinguish between war and conflict; he uses the terms indiscriminately. This theory deals with a population which is increasing in size and lives in a limited area blessed with resources. The growing number of people forces a struggle for existence and the defeated groups are faced with the choice of accepting subjugation (and exploitation) or heading off into the desert. The organisation which the victors develop to keep the defeated groups in subjugation he calls the state.

Unfortunately, Carneiro does not tell how such a state organisation was formed, and neither, how such state organisation was supposed to function. Famous and elegant though the model is, it is ambiguous. In some publications it is used to defend the evolutionary role of population growth and in others to demonstrate that conflict lies at the root of the development of more complex forms of political organisation. Of course, the two influences are not mutually exclusive, but this ambiguity does tend to undermine the soundness of the proposed reasoning. In fact, it is impossible to confirm much more than that in this model conflict was the result of population growth. This makes war or conflict a *derivative factor* (Claessen 2000: 103). The state formation, which eventually takes place in the circumscribed area, does not solve the initial problems, because the population pressure is not removed and it may even increase. The development of remedies by the leaders of the polity – apart from a forced migration of those who lost the battle – may vary from irrigation works to trade. Undeniably these measures increase the carrying capacity of the area and lessen the relative population pressure. At the same time they make it in their interest for the population to remain in the area in which they are settled and make it worthwhile for them to defend it.

The most significant objection to Carneiro's circumscription model is the absolute character, which he attributes to it: state formation has always and exclusively taken place by these means (Claessen 2000: 95). In later publications Carneiro also invari-

ably makes circumscription, war and violence the causal factors in the formation of the state, or in the development of chiefdoms (Carneiro 1981; 1987). In one of his most recent publications (2002: 90ff) the same views come to the fore again. Here it is stated that chiefdoms (the precursors of the state), develop only in the Neolithic, when

several factors led Neolithic villages to transcend local autonomy and create the multi-village political units we call chiefdoms. These conditions were, essentially, the presence of agriculture, the existence of environmental or social circumscription, population pressure, and warfare. Together, they formed the necessary and sufficient conditions that triggered the process. Almost irresistibly they led to the rise of chiefdoms, and then, in more limited areas, to the emergence of the state.

Here he added more explicitly the presence of agriculture, but in the original formulations agriculture already played a role. For the rest, the same views are expressed: neighbouring villages have to be defeated by force of arms, incorporation of the defeated polities in their own polity, prisoners of war must work as slaves, close supporters are used to administer the conquered territory, their own subjects are required to pay tribute, and have to serve as fighting men in times of war. It is the same story all over again. Cases are known, however, in which war – in a circumscribed area – did not produce a state or a more complex sociopolitical organisation, as appears from the following examples.

– Martin van Bakel (1989) describes the situation in the small, densely populated Polynesian island of Rapa which with about 2000 inhabitants in an area of 36 km² in the beginning of the 19th century, certainly can be considered a densely populated 'circumscribed area'. The population was scattered here over about twenty villages, each having some arable land and access to the sea. Between the villages great tensions existed; there was found a permanent 'State of Warre'. In the course of time some chiefs succeed sometimes in subjugating a few villages for a short period of time, but up until the colonial era the island population remained divided over a number of virtually impregnable fortresses. There were numerous raids and attacks, but never did one of the groups succeeded in uniting the whole of the island under one rule. All the attacks and violence, combined with serious population pressure, did not produce a state. Only when, with the arrival of the

Europeans, firearms were imported, did the conquest of some of the forts became possible (Van Bakel 1989: 140-60, 209-12; see also Hanson 1970).

– Timothy Earle (1997) presents data on the Mantaro Valley in Peru over the period AD 500-1534. He points out that 'In the Mantaro Valley, warfare merely resulted in a political standoff between hill-fort chiefdoms. Here both detailed documentary evidence and well-preserved fortified communities testify to the limited significance of warfare as a political strategy for the expansion of chiefdoms' (1997: 111). Interestingly, it was specifically the Peruvian valleys that were used by Carneiro to explain the evolution of states by circumscription and war. But, as Earle states: 'The important point is that burgeoning warfare among these comparatively simple societies corresponded with the collapse of regionally integrated polities and does not appear to have facilitated political integration' (1997: 113).

One cannot but conclude that the circumscription-and-war-or-violence theory does not provide a general explanation of the phenomenon of state formation. The cases mentioned above fulfil all requirements of Carneiro's theory, but no state did emerge here. Apparently more factors are needed to make state formation possible than just war and circumscription.

This conclusion finds confirmation in Tymowski's article on 'The Army and State Formation in West Africa' (1981). In it he describes how in two cases (Kenedugu and the state of Samori) the formation of a state was greatly promoted by military activities. First, however, he points to a number of conditions that the formation of these states actually made possible:

In both cases, we have to do with communities which began producing an economic surplus capable of maintaining a ruling group much earlier than the states actually developed. In both cases, too, the communities had knowledge of state organization. It proved, however, neither to be the economic surplus nor the readily available organizational models that automatically caused states to emerge. (1981: 428)

There were factors hampering this process such as periods of economic stagnation, and a lack of external threat that would force these societies to organise in self-defence. Promoting state formation were, according to Tymowski, such factors as population growth in the early nineteenth century, and the development of trade. 'The need to ensure security on trading routes was one of the stimuli for the development of the state' (1981: 429). Trade led to growing social differentiation, and growing commercial opportunities for the ruling class. The interests of the trading and the ruling groups began to merge. To better exploit the population another sociopolitical organisation was necessary, and to enforce the new system an army was needed: 'the victory in the power struggle went to those who more quickly and efficiently created an army' (1981: 430). Thus in Tymowski's analysis, the interaction of an increasing surplus production, population growth, trade, and military activities led in the end to the emergence of these two West African states. The army was used in these polities to protect trade, and to keep their own population in check. Though there certainly was violence, there was no war (in the sense of Ronald Cohen's definition) connected with the formation of these states; the army acted as a state police force.

There are also cases known in which increasing political centralisation occurred, but where neither population growth nor war has played a demonstrable part. For example, Leslie Gunawardana (1981; 1985) describes how in medieval Sri Lanka differences in the quality of land and the presence of irrigation water led to differences in status and wealth between the villagers and also to differences in prestige and power among the villages concerned. In this situation a stratified society evolved without either population pressure or subjugation having had any effect at all. The differences between the owners and the non-owners, or those with very little land, were reinforced by the development of endogamous marriage networks between the prominent members in the various villages. The presentation of pieces of land and the transfer of small-scale irrigation works to Buddhist monasteries won the donors a wealth of *karma*, which could legitimate their material advantage over others. The situation Gunawardana describes cannot be squared easily with Carneiro's model, for here it was mainly ecological, economical, and ideological factors that led to a hierarchisation, which in due course gave rise to the emergence of chiefdoms (Gunawardana 1981: 136, 138). Only when ambitious chiefs tried to merge a number of chiefdoms into one large polity – or an early state – did military activities began to play a role.

The same type of development is found with the Kachin of Highland Burma (Leach 1954). The Kachin had an economy in which shifting cultivation was the principal means of livelihood. They lived in villages, the inhabitants of which were related to each other by a complex marriage structure in which the brides went in one direction and the bride prices in the other one. Under the prevailing ideology the giver was more highly placed than the receiver. The agricultural system presented one possible source of disturbance of this balanced situation. Generally speaking, the small fields did not give a high yield, but every so often there were ripe crops to be harvested. At that particular moment the owners of that field had an abundant surplus, and because the crop could not be stored, it was the custom to organise a feast for all the members of the village. The host of the feast derived prestige from his action. However, because every so often every family was in a position to give such a feast, this caused little structural change. But it might just so happen that one of the families was able to give several feasts one after the other. This did change the egalitarian structure and the family concerned accrued more permanent prestige to itself. The daughters of the prestigious family became considerably more expensive and the higher bride prices increased its prosperity – and thus the prestige of the family – even more. If this trend continued, the girls became too expensive for the boys of the village. They then 'constructed' a new marriage circuit, and the prestigious family now entered a bride exchange circuit in which the notables of a wider area participated. If the prosperity of the family continued the less fortunate villagers sought an explanation for these uncommon developments. This was found in religious terms: the more fortunate fellow-villagers apparently had better access to the ancestors or to the spirits than the more ordinary mortals. Now the development reaches a crucial phase. Up to that point the position of the notable family had been based on distribution: the giving of feasts and gifts. But, once the villagers understood how matters stood, the stream of gifts changed direction. The villagers began to offer the well-to-do family small gifts with the request that they put in a good word for them with their ancestors. Naturally, this request was acceded to and within the shortest possible time material goods flowed in to the notable and he reciprocated this with immaterial matters – a veritable realisation of Marx' Asiatic mode of production (Claessen 2000: 60; Friedman 1979).

These developments were not limited to the Kachin; in fact the same events occurred in many parts of Southeast Asia. The consequence of these developments was in the end the emergence of a number of rather unstable early states in the area, a process described in detail by Renée Hagesteijn (1989). As the mountainous terrain made effective war and conquest practically impossible, ambitious chiefs sought to enlarge their influence by strategic marriages, and the concluding of treaties – which never held longer than the life time of those who concluded them. The whole complex of cause and effect here has nothing to do with environmental or social circumscription and just as little with war or population pressure. In this case the qualifying factors were incidental overproduction, the existence of an ideology which accorded the giver a higher status than the receiver, and the circumstance that after a prolonged period of prosperity the villagers were inclined to give the fortunate farmer presents in return for blessings. Ideological factors clearly played the crucial role here.

In other places relatively peaceful developments can also be found. Conrad Kottak (1972) takes population growth as his point of departure. He describes how, in the distant past, a limited number of people lived on the shores of Lake Victoria in Africa. The climate was good, the land fertile, and the lake provided fish. Under such favourable conditions, the population grew and as time passed the whole shore of the lake was inhabited. The unabated population growth forced the societies involved to look for more areas of settlement in the hinterland, at some distance of the shore of the lake. Those who were 'sent' to the hinterland were the younger sons of younger sons and their dependents. Those who ended up in the hinterland were certainly not banished to a wilderness; the land there was fertile and the climate was good. Their great disadvantage was that they had no direct access to the lake (and consequently to fish and trade). If they were able to share in any of the benefits offered by the lake, they were dependent on the generosity of the dwellers of the lake shore – their older brothers and cousins. In this case a situation arose in which not everybody of the same age and the same sex had equal access to the means of livelihood; that is to say there arose what Fried (1967) defines as a stratified society. Kottak's model reveals how virtually without a ripple a society can glide into a situation in which terms like 'rank' and 'stratified' are applicable; the whole process happens just as silently as it did in Kachin society (cp. Service 1975: 75-78 for a comparable model).

Another phenomenon likewise contributed greatly to the development of ranking and stratification – and the formation of chiefdoms and early states – in Africa: the ideology of the firstcomer's primacy. Leadership in many societies was usually allocated to the leader of the lineage which had settled the area first and, by 'opening up' the land, had entered into a 'contract' with the earth spirit(s), which continued to give him access to fertility in exchange for offerings (Claessen and Oosten 1996: 368-70, and the chapters on West Africa in that volume; Kopytoff 1999; Bay 1998). Small groups of cultivators who later wanted to settle in this area and desired to make a claim on the fertility magic of the ritual leader, the earth priest, had to ask his permission and display a certain degree of obedience. Treading this peaceful path, gradually not inconsiderable territorial units or villages emerged. This system of peaceful enlargement is still found in many places in Africa. Zuiderwijk, who recently investigated the situation in Cameroon, informs us that:

The clan that first occupied the land and founded a village is called the clan of the chiefs. In principle and often in practice the chief is indeed a member of this clan. Clans that arrived later were warmly welcomed, and even given virgin land to cultivate, but they came to be under the symbolic authority of those who arrived first. (1998: 92)

It is against this background that Igor Kopytoff states that chiefs here do not so much 'rise above their neighbours as they were so to speak, 'levitated' upwards as more immigrants arrived and inserted more layers at the bottom of the hierarchy' (1999: 88). In the formation of most of the African early states, especially those in West Africa, these ideological principles lay at their foundation. In particular, the assumed ritual influence of the ruler on fertility provided a strong form of legitimation to the government (Claessen and Oosten 1996).

Our investigations thus far have shown that war, even in a circumscribed area, does not always produce more complex sociopolitical structures; the stagnating developments in Rapa and the Mantaro Valley are cases in point. Moreover, there were found instances in which stratified societies emerged without war (or circumscription) playing a role at all – as was demonstrated by the Kachin, the people living on the shores of Lake Victoria, the villages in Sri Lanka, and the political structures in Africa based on the firstcomer-ideology. The article by Tymowski showed that in West Africa not only an armed police force, but also favourable economic and demographic circumstances were needed for the emergence of states. War thus far does not seem a very promising factor in explaining social evolution – and yet mankind has fought numerous wars, and committed numerous acts of violence. Why so many wars?

Reasons for war and conflict

As was stated in the discussion of causality, there often lurks behind Phenomenon B, another Phenomenon C. In other words, causalities behind causalities. Regarding war it can be safely stated that in the majority of cases war was not waged for the sake of war itself, nor was there conflict for conflict's sake. There was always some reason or a complex of reasons lying at the base of it. In short, wars between 'civilised peoples', as well as raids and fights between societies which are classified as 'lower' are derived events, instigated by other, more compelling, factors. They thus cannot be included in the explanation of evolution as independent factors, even though they sometimes had enormous consequences; still less can war or conflict be considered to be a 'first cause', neither a 'necessary' nor a 'sufficient condition' in the development of more complex societies. The reasons for conflict and war can be summed up globally in the following categories:

- Demographic causes (population pressure, exiguous means of subsistence, shortage of land);
- Economic causes (shortage of food and/or raw materials; the competition to control markets, trade routes, or ports; the search for prestige goods to bind people, etc.);
- Ideological causes (religious oppositions; questions of honor and shame; reciprocal obligations; feelings of vengeance; struggles for succession);
- Politico-strategic causes (security reasons; elimination of threatening neighbors; claims to influential positions).

It seems probable that in many instances more than one of these causes simultaneously played a role and the motivation for one participant in the conflict will have differed from that of another (Claessen 2000: 109-10).

The role of war

Little value should be attached to the creative force of conflict and war as such. Hallpike (1986: 233-35) has argued that military activities (under which he includes war as well as conflict) generally have few constructive consequences. Admittedly, some planning, coordination and leadership are required if anything above a simple brawl is to be successful, but that is as far as it goes. Above this, Hallpike thinks the strengthening of leadership or authority is more likely to occur if some form of leadership or authority already existed before this strife.

Mere violence, however, cannot lead to permanent institutions of political authority, and to my knowledge there is no instance of military activity *by itself* ever having led to the emergence of even chiefly authority. (Hallpike 1986: 235, italics in original)

Earle, who devoted a great deal of attention to conflict and war in his book about the evolution of chiefdoms is very cautious in his pronouncements about them. He has this to say about war (1997: 106):

Inherently warfare is limited in its effectiveness as a power for central control. Although military forces may create a broadly integrated polity, it can as well dissolve it by intrigue, coup, and rebellion. The power of force rips at the social fabric, the institutions of society. To be effective as a power of centrality, coercive force must itself be controlled, a difficult task that is achieved by binding the military with economic and ideological tethers.

Apart from the difficulty of political leaders controlling the military, valuable developments can indeed take place in the aftermath of conflict and war. This holds especially when these are looked at from an organisational point of view. In *The Early State* (Claessen and Skalník 1978: 626) it has already been argued that the origins of the state cannot be attributed to war, but can indeed be largely encouraged by war and tension. Generally speaking, it can be said that in conflict or war, regulations and provisions have to be laid down. This requirement is just as valid for the Yanomami as for the Incas or the Carolingians. Scouts have to be selected and sent out; their reports have to be processed. Warriors have to be assembled. Rewards have to be organised, even when the army had to forage from the land, and the rewards for loyal servants have to be honoured, no matter how the war has been concluded.

Political leaders who have numerous, loyal, followers at their disposal sometimes experience great difficulty in meeting their obligations in peacetime – the Carolingians are a prime example of this. It was above all in their empire where *Gefolgschaft* obligations played a great role that the leaders often found themselves in great difficulties. Loyalty remained intact as long as the leader looked after his followers generously. This sometimes meant that new wars and conquests were unavoidable, but could also lead to pressure being brought to bear on farmers and citizens to pay ever higher taxes. Better administrative organisation was an absolute prerequisite and we may assume that the waging of war or the threat of war was partly responsible for the development of this.

Elsewhere, in the empire of the Incas, the development of a road system was a corollary of the never ending wars; the other impulse for road-building came from the necessity to transport great quantities of food and valuables. The seasonal wars in the African kingdom of Dahomey were first caused by the necessity of procuring people to sacrifice, and later by the interesting possibility of selling them as slaves – a development that resembles greatly the repeated wars of the Aztecs to make a sufficient number of prisoners for the necessary human sacrifices (Claessen 2000: passim). Here Cohen's dictum that 'states make war' is fully realised (Cohen 1985).

Emergence of states with or without war

Thus far it has been demonstrated that for the emergence of rank, stratification, or chiefdoms war and/or circumscription was neither a necessary nor a sufficient cause. One might argue, however, that for the formation of states war or violence (or eventually circumscription) was a necessary factor, or that in cases where a state emerged, war and/or violence (and/or circumscription) played a crucial role. Before going into this matter, I will first define the early state: *an early state is a three-tier (national, regional, local level) sociopolitical organisation for the regulation of social relations in a complex, stratified society, characterised by a shared ideology of which reciprocity between the strata is the basic principle; the ruler (the central government) has the legitimised power to enforce decisions* (adapted from Claessen and Skalník 1978: 640). How then did this type of sociopolitical organisation develop? Let us take as an example the formation of the (early) state of the Betsileo (based on Kottak 1980).

In the sixteenth and seventeenth century the Betsileo lived in the eastern part of the southern highlands of Madagascar. They pursued irrigated rice cultivation. In the seventeenth century they were threatened by a group of inhabitants from the coast who were in search of plunder and slaves. This threat drove them to seek help from one another and they retreated to the hilltops, around which they built fortifications. It should be clear that no war (as defined above), ever took place here. The worst that happened were some skirmishes. As time passed, increasing numbers of people found their way to these relatively safe places and the need for more clear-cut leadership made itself felt. The existing traditional forms of leadership (family and clan chiefs) gradually ceded place to a more formal hierarchy and the originally somewhat amorphous sacred qualities which were ascribed to clan chiefs were now transposed to the most prominent of them, those who had won themselves status and power in the administrative apparatus, which was beginning to take shape. This is the way in which a process was set in motion, which after a good century or more carried the Betsileo to the stage of an early state. The development took a completely different course among the Isandra who lived close by. The Isandra were cattle-herders, who were used to leading a nomadic life with their herds; their sociopolitical organisation was quite limited. When the threat of the coastal people began to be palpable in their area they chose the most obvious solution: they withdrew to safer places with their herds and only returned after the raiders had left the area. During those tumultuous years virtually nothing changed among the Isandra; whereas among the Betsileo a state structure came into existence.

This case makes several interesting observations possible. The first is that the development of the state was not planned beforehand; it was the unexpected outcome of living together in the hilltop forts. Second, it is a clear example of Gregory Johnson's thesis that a growing number of people, living together for a longer period of time, need to develop stronger forms of leadership – or fall apart (Johnson 1982; Claessen 2000). The Betsileo had to choose between stronger leadership or slavery. One might say that these hilltop forts were a kind of circumscribed area. That is correct – but contrary to Carneiro's theory there was no fighting among the Betsileo in the forts; as far as there was fighting, they fought foreign raiders. Third, the previous existence of an ideology of sacred leadership appeared to be very important: with the help of this ideology a stronger type of leadership could be explained and legitimised (a phenomenon we found also to be crucial among the Kachin, and the villages in Sri Lanka). Fourth, the survival of the Betsileo demanded a lot of management on the side of the rulers. The 'servant of the community' by his management activities became its master (Friedrich Engels, quoted in Claessen 2000: 29). And, finally, the whole exercise was set in motion to preserve the irrigated rice fields – and thus their food supplies. There was the danger of plunder and slave hunts, and to defend themselves the Betsileo organised the hilltop forts and were successful in their defense. Violence against the enemy certainly was found here; the term 'war' seems not to be warranted to characterise the skirmishes, however.

If we try to generalise the conclusions above, combining them with the data from the cases presented earlier in this chapter, it seems that the following factors – in a complex interaction – are conducive to the formation of an early state (cp. Claessen and Oosten 1996: 5; Claessen 2000: 188; Claessen 2002: 107-11):

- Population growth. This leads, when the society in question stays together, to a more complex sociopolitical organisation, and makes an increase in food production inevitable.
- An ideology, which explains and legitimates (stronger) forms of leadership. This ideology is usually an elaboration of already existing ideas.
- The domination of the economy. A larger population and a more developed form of leadership demand an increase in production. This demands more management, and more land has to be brought under cultivation. Better transport and better storage facilities have to be developed.

- A more complex sociopolitical organisation demands more functionaries, more juridical measures and more stratification. Over time a complex administrative apparatus – or bureaucracy – will emerge.
- A limited military apparatus or police force. This is necessary to maintain external independence and internal safety. This apparatus in its turn makes possible a stronger administrative organisation and a more developed economy.

Even when these factors are found together, an early state will not emerge automatically. According to Patricia Shifferd (1987: 47ff):

Continued centralization, although clearly observable taking human civilization as a whole, is certainly not inevitable for individual cases. In fact such continued centralization was the least common outcome in the sample at hand.

Some action or event – internal or external – is needed to trigger the development. Such action or event may even have taken place long ago, not directed especially towards this goal. When such a process is set into motion, it tends to enforce itself, and the various factors influence each other. The process can be compared with a snowball: once it comes into motion it grows faster and faster (Claessen and Skalník 1978: 624). In the case of the Betsileo the 'event' (Claessen 2002: 110-12) was the threat posed by slave hunters; in the two African cases it was the need to protect trade; in some of the Southeast Asian polities it was the introduction of new religious ideas with the help of which the ruler could develop a stronger form of legitimation (Hagesteijn 1989). It is certainly possible to distinguish these factors, but it is also clear that they are closely intertwined.

Sometimes war was involved in the development towards the state. In the cases presented here the role of war and conflict as an evolutionary factor was limited, or secondary only. Neither war nor conflict (or for that matter, circumscription) should be considered as a sufficient or a necessary factor. Even less so could war be considered causal to the development of the state, for in that case there should have been first war, and then, as a consequence of it, the formation of a state.

Concluding summary

We are now in the position to formulate an answer to the question posed at the beginning of this paper: is there a connection between war and state formation? The answer must be that there is some connection, indeed, for several times war, fighting, or violence, was connected with the formation of states, chiefdoms or other types of more complex sociopolitical organisations. This connection, however, should not be considered a causality; at best it was a corollary of the developments. In all cases presented a number of interacting factors played a role in the formation of a chiefdom or state. Neither war, nor other types of violence, appeared to be a necessary or sufficient condition in their development. Moreover, instances were presented in which neither war, nor circumscription, played any role in the formation of a more complex sociopolitical organisation.

War, however, is not wholly without influence. There have been developments triggered by war, such as the erection of defensive structures, the development of a better administrative organisation, etc., developments that were conducive to stronger or better types of sociopolitical organisation. However, when leaders did not succeed in 'taming' the military, in tying the warriors to the government, the effect of war and strife was in the end only destructive; greater units did not develop; never ending fighting came in its place (cp. Earle 1997). One might argue that destruction is also an evolutionary force – and I will not deny it – but the aim of this article was an inquiry into the connection between war and state formation. That connection appeared to be a limited one.

BIBLIOGRAPHY

- Bakel, M.A. van 1989. Samen leven in gebondenheid en vrijheid. Evolutie en ontwikkeling in Polynesië. Unpublished Ph.D. thesis, Leiden University, Leiden.
- Bay, E.G. 1998. *Wives of the Leopard. Gender, Politics, and Culture in the Kingdom of Dahomey.* Charlottesville: University of Virginia Press.
- Bazin, J. and Terray, E. (eds.) 1982. *Guerres de Lignages et Guerres d'États en Afrique.* Paris: Editions des Archives contemporaines.
- Carneiro, R.L. 1970. 'A theory of the origin of the state'. *Science,* 169, pp. 733-38.

- Carneiro, R.L. 1981. 'The chiefdom: precursor of the state'. In: G.D. Jones and R.R. Kautz (eds.): *The Transition to Statehood in the New World*. Cambridge: Cambridge University Press, pp. 37-79.
- Carneiro, R.L. 1987. 'Cross-currents in the theory of state formation'. *American Ethnologist*, 14, pp. 756-70.
- Carneiro, R.L. 2002. 'Was the chiefdom a congelation of ideas?' *Social Evolution and History*, 1, pp. 80-100.
- Chagnon, N. 1968. *Yanomamö, the Fierce People*. New York: Holt, Rinehart and Winston.
- Claessen, H.J.M. 2000. *Structural change; Evolution and Evolutionism in Cultural Anthropology*. Leiden: CNWS Publications.
- Claessen, H.J.M. 2002. 'Was the state inevitable?' *Social Evolution and History*, 1, pp. 101-17.
- Claessen, H.J.M. and Oosten, J.G. 1996. 'Introduction' and 'Discussion and considerations'. In: H.J.M. Claessen and J.G. Oosten (eds.): *Ideology and the Formation of Early States*. Leiden: Brill, pp. 1-23 and 359-405.
- Claessen, H.J.M. and Skalník, P. 1978. 'Limits: Beginning and end of the earl state'. In: H.J.M. Claessen and P. Skalník (eds.): *The Early State*. The Hague: Mouton, pp. 619-38.
- Cohen, R. 1985. 'Warfare and state formation'. In: H.J.M. Claessen, P. van de Velde and M.E. Smith, (eds.): *Development and Decline*. South Hadley: Bergin and Garvey, pp. 276-89.
- Earle, T. 1997. *How Chiefs Come to Power*. Stanford (Cal.): Stanford University Press.
- Fontijn, D. 2003. *Sacrificial Landscapes; Cultural Biographies of Persons, Objects and 'Natural Places' in the Bronze Age of the Southern Netherlands, c. 2300-600 BC*. Ph.D. thesis, Leiden University.
- Fried, M.H. 1967. *The Evolution of Political Society*. New York: Random House.
- Friedman, J. 1979. *System, Structure and Contradiction; The Evolution of Asiatic Social Formations*. Copenhagen: National Museum of Denmark.
- Gunawardana, R.A.H.L. 1981. 'Social function and political power. A case study of state formation in irrigation society'. In: H.J.M. Claessen and P. Skalník (eds.): *The Study of the State*. The Hague: Mouton, pp. 133-54.
- Gunawardana, R.A.H.L. 1985. 'Total power or shared power? A study of the hydraulic state and its transformations in Sri Lanka from the third to the ninth century A.D.' In: H.J.M. Claessen, P. van de Velde and M.E. Smith (eds.): *Development and Decline*. South Hadley: Bergin and Garvey, pp. 219-45.
- Hagesteijn, R.R. 1989. *Circles of Kings; Political Dynamics in Early Continental Southeast Asia*. (Verhandelingen van het KITLV, 138). Dordrecht: Foris.
- Hallpike, C.R. 1986. *The Principles of Social Evolution*. Oxford: Clarendon.
- Hanson, F.A. 1970. *Rapan Lifeways; Society and History on a Polynesian Island*. Boston: Little, Brown and Company.
- Harris, M. 1977. *Cannibals and Kings; The Origins of Cultures*. New York: Random House.
- Heidenreich, C.F. 1978. 'Huron'. In: B.C. Trigger (ed.): *Handbook of North American Indians*. 15. *Northeast*. Washington D.C.: Smithsonian Institution, pp. 368-89.
- Johnson, G.A. 1982. 'Organizational structure and scalar stress'. In: C. Renfrew, M. Rowlands and B. Seagraves (eds.): *Theory and Explanation in Archaeology*, New York: Academic Press, pp. 389-421.
- Köbben, A.J.F. 1970. 'Cause and intention'. In: R. Naroll and R. Cohen (eds.): *A Handbook of Method in Cultural Anthropology*. New York: Columbia University Press, pp. 89-98.
- Kopytoff, I. 1999. 'Permutations in patrimonialism and populism: The Aghem chiefdoms of Western Cameroon'. In: S. Keech McIntosh (ed.): *Beyond Chiefdoms; Pathways to Complexity in Africa*. Cambridge: Cambridge University Press, pp. 88-96.
- Kottak, C.Ph. 1972. 'Ecological variables in the origin and evolution of the African states'. *Comparative Studies in Society and History*, 14, pp. 351-80.
- Kottak, C.Ph. 1980. *The Past in the Present. History, Ecology and Cultural Variation in Highland Madagascar*. Ann Arbor: University of Michigan Press.
- Leach, E. 1954. *Political Systems of Highland Burma*. London: Athlone.
- Lee, R.B. and DeVore, I. (eds.) 1968. *Man the Hunter*. New York: Aldine.
- Louwe Kooijmans, L.P. 1993. 'An Early/Middle Bronze Age multiple burial at Wassenaar, the Netherlands'. *Analecta Praehistorica Leidensia*, 26, pp.1-20.
- Morgan, L.H. 1851. *League of the Ho-de-no-sau-nee, Iroquois*. Rochester: Sage and Brother.
- Nettleship, M.A. 1975. 'Definitions'. In: M.A. Nettleship, R.D. Givens, and A. Nettleship (eds.): *War, its Causes and Correlates*. The Hague: Mouton, pp. 73-92.
- Parijs, Ph. van 1981. *Evolutionary Explanation in the Social Sciences*. Totowa (N.J.): Rowman and Littlefield.
- Sahlins, M.D. 1968. *Tribesmen*. Englewood Cliffs (N.J.): Prentice Hall.
- Service, E.R. 1975. *Origins of the State and Civilization*. New York: Norton.
- Shifferd, P.A. 1987. 'Aztecs and Africans: Political processes in twenty-two early states'. In: H.J.M. Claessen and P. van de Velde (eds.): *Early State Dynamics*. Leiden: Brill, pp. 39-53.
- Steward, J.H. 1968. 'Causal factors and processes in the evolution of pre-farming societies'. In: R.B. Lee and I. DeVore (eds.): *Man the Hunter*. New York: Aldine, pp. 321-34.
- Trigger, B.G. 1969. *The Huron: Farmers of the North*. New York: Holt, Rinehart and Winston.
- Tymowski, M. 1981. 'The army and the formation of the states of West Africa in the nineteenth century. The cases of Kenedugu and Samori'. In: H.J.M. Claessen and P. Skalník (eds.): *The Study of the State*. The Hague: Mouton, pp. 427-42.
- Zuiderwijk, A. 1998. *Farming Gently, Farming Fast. Migration, Incorporation and Agricultural Change in the Mandara Mountains of Northern Cameroon*. Leiden: Centre of Environmental Science, Leiden University.

Warrior Bands, War Lords, and the Birth of Tribes and States in the First Millennium AD in Middle Europe

/16

HEIKO STEUER

The subject: the background

It is my intention to formulate a model on war. But it is not warfare itself that will be considered in all its aspects (Steuer 2001; Jørgensen and Clausen 1997), but rather the causes and effects of wars conducted by war lords and their warrior bands. Wars from the 4th century BC to the 10th/11th century AD will be considered in devising the model – wars spanning a period of more than a thousand years, from the ancient Celts to the Normans. Based upon the reports in the written sources, characteristic phenomena (*Erscheinungen*) will be singled out as criteria for defining this type of warfare. Not all criteria will be obtainable for every epoch, but in the overall view a varied number of criteria can confirm the comparability of the armed conflicts and therefore also their socio-political backgrounds. (This model is not a newly formulated thesis. Most of it can be found in Wells 1999). The question of the expression of these events in the archaeological sources will only be raised in the second place, and it will be shown that, contrary to general opinion, warrior bands can never, or rarely, be recognised using archaeological methods. On the contrary, following the conclusion of the socio-political process – the birth of tribes and states – the later occupation of land as a result of these wars would leave marks on the archaeological remains.

The Celtic and Germanic societies of central Europe of the first millennium were permanently changing. These variations appeared regularly, to some extent in waves or phases. We can abstract the rules; but history does not repeat itself, that is why the various epochs are not completely identical. The different reports in the written sources describe this change in an indirect manner. The recorded names of the active groups – we will call them tribes following Caesar, Tacitus and the later Ammianus Marcellinus – emerge and disappear again. The names of tribes during the time of Caesar (100-44 BC) (*De Bello Gallico*, 58-52 BC) differ from those of the time of Tacitus (55-116/120 AD) (*Germania*, 98 AD) or those of Ammianus Marcellinus (second half of the 4th century), Gregory of Tours (ca. 540-594) (*Historia Francorum*), or the time of the Carolingian and Ottonian historians. 150 to 250 years lie between each of these reports, the equivalent of five to eight generations.

Various ethnogeneses (tribalisations) are reflected in the change of the names, but even in the preservation of an old name the old tribe does not continue. A new ethnogenesis creates other political entities. There are Ariovistusus' *Suebi*, the Suebi of the age of Tacitus, the *Suebi* (*Suevi*) of the migration period in Spain, etc. The same applies to the *Marcomanni* during the times of Ariovistusus or Maroboduus and those of the Marcomannic Wars 166-180 AD, which are not the same tribe. The *Alamanni*, mentioned for the first time around 300 AD, were made up of various groups of Germanic people, then constituted themselves in a second ethnogenesis (tribalisation) in the Merovingian Empire around 500 and then again for a third ethnogenesis (tribalisation) in the dukedom/duchy of the 8th/9th century.

What are these tribes? Are they the population of territorial units, the unification of several settled landscapes with large numbers of villages (*Siedlungs-kammern*)? It is a matter of discussion whether the tribes which are mentioned by name in the ancient sources emerged only as territorially bound units in reaction to the overly powerful threat of neighbouring empires or states; as the open ranked society based on clan-like organisation with its numerous inner dependencies had become obsolete. The Celtic tribes with their central oppida mentioned by Caesar are a reaction to the pressure of the Mediterranean states and only appear in the Middle and Late La Tène period with the aid of returning mercenaries. The Germanic tribes may only have constituted themselves due to the pressure of the Celtic oppidum-civilisation on the Germanic clans, to which Germanic mercenaries in the service (army) of Celtic nobility may have contributed on their return home.

Ethnology has pointed out this development for Africa during the colonial age of the 19th century: before the intrusion of Europeans there would appear to have been none of the tribal structures that later on became apparent and which up to today provide a continuous cause for conflict and wars (Vail 1989; Sigrist 1994; Lentz 1998). Today the political and social organisation of rule under Shaka (Marx 1998), the founder of the Zulu tribe or realm, is under discussion in this way, and likewise the deadly conflict between Hutu and Tutsi, today two different tribes, but before colonisation two different economic structures: farming and cattle breeding.

The idea of tribes (*gentes*), of a people (*Volk*), and of a nation developed as part of the nation-state thinking of the 19th century. Events in the present (in the Balkans, in Africa etc.) show that ethnic clashes may also be conflicts among rival bands of thieves (warrior bands).

But it is impossible to speak about these terms (*termini*) in all their complexity in this paper. In recent decades there has been a very controversial and many-sided discussion concerning the terms 'tribe/*Stamm*' or 'ethnogenesis/tribalisation/*Stammes-bildung*' (compare developments for instance from Wenskus 1977 to Pohl 1998 and 2002, or also Bowlus 2002; Barth 1969). In numerous *lemmata* the *Hoops Reallexikon der Germanischen Altertumskunde* (Pohl 1998) tried to specify this discussion in its many aspects. For some other terms like '*Gefolgschaft*' (*comitatus*), i.e. followers or retinue, problems in the translation from Latin into German or the English language exist as well. For the structuralist attempt – an old one, not a poststructuralist or contextual attempt – made in this paper it may be enough to formulate a model as an idea.

'*Völkerwanderungen*', translated as migrations of whole peoples, were in fact not migrations of peoples but rather military campaigns. These campaigns only had temporary camps as stations, not permanent settlements. Campaigns of warrior bands could be undertaken without difficulty, as the landscape was relatively empty. Although the landscapes were populated completely and systematically with settlements like a network in distances not more than 5 kilometres as a rule, there was still enough space to allow larger military units with their horses to pass through, either unnoticed or officially sanctioned.

The groups called in the sources by names such as *Suebi* or *Langobardi* were warrior bands and not neighbouring societies living in villages with families or kinship groups based upon relationships among relatives, and with a clan-system. Perhaps only the leading clans went by these names. Through their concepts of order, ancient writers equated the names of the mobile and multi-ethnic warrior bands with inhabitants of the territories from which some of the warriors were drawn. The authors, with their background in the Roman Empire, did not recognise the principal difference between warrior bands and rural populations with the same names. The names randomly appear next to one another, not because parts of tribes had migrated to other countries, but because a military unit had been divided and deployed at various places.

The *Völkerwanderungen* of that period therefore are not the migrations of tribes with the whole family, their mobile belongings, cattle and all, as was believed in earlier research, but rather campaigns of warrior bands whose wars only much later led to the occupation of land. Their emigration did not leave the native lands depleted of their population.

Depopulation had completely different reasons and backgrounds. For instance:

– The Germanii disappear from the territories east of the river Elbe long after the end of the so-called migration-period in the 6th century, and Slavic tribes immigrate.

– The landscape of Anglia (Angeln) in modern Jutland, and other areas along the coastline of the North Sea, only appears to become depopulated more than a century after the emigration to England around the middle of the 5th century; it would therefore seem that only latecomers were concerned.

– People, i.e. warrior bands, left the densely populated areas. Even when these warrior bands settled in new countries, no major changes came into effect in the native areas, a fact that can be attested in the archaeological features – the structures of the villages show no change. At the very least no decline in the density of the population can be registered.

Wars are an immanent part of social change. Ethnogenesis or tribal constitutions (tribalisations) and the development of states were not possible without wars. One therefore has to consider the effect of the feedback. Because higher organised societies (empires) influenced their surroundings, tribes developed, taking the place of clan-societies, in order to defend themselves. Warrior bands split off after the constitution of these tribes. Because warrior bands appear and influence the higher organised societies (kingdoms or empires) through mercenary services or as organised bands of thieves (warrior bands) – threatening them militarily and through raids – they in turn attempt to influence the tribal communities through paying mercenaries or drawing up contracts with these groups in order to use the tribes and warrior bands for their own ends.

War

War is the armed conflict between groups of men, i.e. armies of various sizes (Steuer 2001: 347ff).

a) The armies or warrior bands or '*Gefolgschaften*' as military units can be levied from the villages of a territory in order to protect them. Out of the clan-based society – due to the necessity of organisation – a tribe will arise, possessing its own territory. This tribe can then even erect special fortifications to defend the settlement area as a whole, as can be observed during the later period of the Roman Iron Age when constructions consisting of ramparts,

palisades and ditches were erected along the borders on land or in the sea. The territories were up to 30 to 50 kilometres wide. These areas of roughly 2500 square kilometres would incorporate about 100 villages with areas of 25 square kilometres per village, each village consisting of 10 households with 10 inhabitants per household which in turn leads to a total population of 10,000 people. Up to 20% of this total number could go to war, which means that such an area could raise an army of up to 2000 warriors. Another method of reckoning is based on the number of weapons in the find from Illerup: based on the ratio of excavated shield-bosses made of silver, bronze and iron which are 6 to 30 to 350, a leader with a silver shield commanded a warrior band with the strength of a Roman '*centuria*' (60 to 80 soldiers). The entire force could easily have reached the size of a Roman '*auxilia*' unit with 1500 to 2000 men, corresponding to the number of warriors from the area of one tribe in the Germanic territories. The warrior in the burial of Gommern in Saxony-Anhalt with his precious grave-goods and silver fitted shield corresponds to the leaders in the military unit from Illerup in Jutland, and he would have been one of the leaders of the warrior-contingent from the 30 kilometre wide settlement landscape around Gommern.

b) The military units (army) can be made up of warriors and a leader (king, *rex* or *dux*) who completely separate themselves from the structures of the clan or tribe and move about in an 'unattached' manner in order to plunder and pillage (gaining the spoils of war-booty '*Kriegsbeute*'). Warriors from various tribes came together and the old ethnic affiliations lost their meaning.

c) These military units can be recruited and offer their services to a higher order of state (an empire) as auxiliary units, or as groups of mercenaries.

Warfare is a lifestyle, combined with a certain mentality of the warrior (Bodmer 1957). This is a behaviour which accepts war as a way of life and struggle of existence (besides the rural life), from which a warrior or noble caste emerges. War was waged almost every year in the early historic societies of the first millennium. The function of a military unit varied, depending on its role in the formation of a tribe or the constitution of a state. The organised states or empires, as in the cases of the Roman Empire or the Merovingian Empire, waged war annually along various borders, sometimes in defence, but mostly in order to expand their political power and to obtain war booty. The warrior bands in turn waged war every year in order to 'earn' their living, either against other armies or, more often, against the settled communities of tribal societies, or they waged war against the more organised states in the form of raids.

The historical phases

Early and Middle La Tène

Celtic mercenaries appear during the 4th century in the area of the Mediterranean, for example in the wars of the tyrant Dionysius (c. 430-367 BC) of Syracuse against Carthage. The invasions of Celtic warrior bands into northern Italy commenced soon after 400 BC, culminating in the plundering of Rome under Brennus and the battle at the river Allia in 387. The withdrawal from Rome was achieved with the payment of 1000 pounds of gold. The Celtic occupation of land began in the *ager Gallicus* along the Adriatic coast. The Celtic groups named in this migration were the Senones and the Boii, as well as the Insubres. None of this can be proved archaeologically; only solitary assemblages with weapons and belt ornaments are to be found among the cemeteries of local character from the 5th to the 3rd century BC; settlements remain unknown. The Celtic immigrants would appear to have adapted immediately to the local culture, to have acculturated themselves. But there is a cultural feedback to the areas from the north of the Alps to the river Marne, mirrored in the decorative style of the jewellery.

A general feature of importance to this discussion is that mobile warrior bands allow themselves, as a first step, to be recruited as mercenaries. They return to whence they came with their pay. The early Celtic minting of coins in gold and silver imitates the coins of Phillip (*382, king 359-336) and Alexander the Great (*356, king 336-323). These mercenaries come from the cultures on the fringes (*Randkulturen* – cultures in peripheral or border regions) of the more state-like higher organised societies. In the next phase, they resort to wars of aggression under the leadership of warrior-kings (*Heerkönige* – war leaders, commanders). The name *Brennus* is an occupational name and relates to the Breton word *brennin* – which means king. The names of the 'tribes' in the written sources are not valid for whole peoples, but rather for the warrior bands. The occupation of land only begins later and is not reflected in the archaeological evidence.

Late La Tène

Around 72/71 BC Ariovistus, *rex Sueborum* (Pliny *Historia Naturalis* 2,170) or *rex Germanorum* (Caesar *de Bello Gallici* 1,31,10), appears among the *Sequani* as a leader of mercenaries. They had employed him and his 15,000 warriors for the war against the Haedui. The number grew to 120,000 as time passed on. In the decisive battle against Caesar 58 BC 24,000 Harudes, descendants of the warriors that had once roamed Central Europe with the Cimbri, as well as Marcomanni, Triboki, Vangiones, Nemeti, Eudusii and Suebi – totalling at least 7 different groups – could be found amongst the armies. That this coalition was made up from various individual warrior bands is a remarkable fact. Where Ariovistus came from is still unknown – he may have been a Tribokian. But these names again do not describe peoples, but rather warrior bands that may indeed mainly have been recruited from just one tribe.

Ariovistus remained as a mercenary leader in Gaul and controlled one third of the territories of the *Sequani*. The campaign ended in the year 58 BC with his defeat, after which the troops moved about for 14 years 'without a roof over their heads' (*qui inter annos XIIII non subissent* / Caesar *de Bello Gallici* 1,36,7). There are no archaeological traces of Ariovistus' campaigns. But archaeologists think they may be able to prove an occupation of land through the 'Germanic' finds in the Wetterau and from the Rhine-Main area to the northern part of the upper Rhine in the Late La Tène period. There can be no doubt that there are correspondences in the inventories of pottery and weapons between the Polish Przeworsk Culture and cultural groups in central Germany, Thuringia and the river Elbe region. There is a cultural link between the groups along the Rhine and the cultures further east. It cannot yet be decided if this is an extension or expansion of an archaeological culture group (*Kultur-* or *Formenkreis*), or if it is an immigration of people towards the Rhine, as the archaeological finds in both regions are contemporaneous. But there is no connection between the mobile and very mixed warrior bands of Ariovistus and the rural settlement areas along the Rhine and Main rivers with influences from the areas of the Przeworsk Culture, as Suebi from distant territories only made up a small part of the troops.

The age of Arminius

Ariovistus was not only the enemy of Caesar, but was also *amicus populi Romani* in the year 59 BC. And the Cheruscan with the name of Arminius from the stirps regia of the tribe was a Roman citizen and belonged to the class of *eques*, he was a Roman knight: he had been to Rome and led a band of mercenaries, an auxiliary unit, in the Roman army. In the famous Varus-Schlacht (*clades Variana*), or '*Schlacht im Teutoburger Wald*', some 'tribes', the *Marsi*, *Bructeri*, probably *Chatti* and later on *Langobardi* and *Semnones*, could be found fighting alongside the *Cherusci* under his command. The term *Cherusci sociique* is used. Again a coalition of various warrior bands came together, some groups may earlier have been in action as *auxilia*, or other groups that had not been in Roman military service. The so-called 'Germanic Battle of Freedom' under the leadership of Arminius was – according to the historian Dieter Timpe (Timpe 1970) – no popular uprising, but rather the mutiny or rebellion of Roman mercenaries, of a regular or normal *auxilia*.

One other war lord of this period was Maroboduus, *genere nobilis* (Gaius Velleius Paterculus 2,108,2), a Marcoman who had also been to Rome before he made himself a king. He then assembled a coalition of various warrior bands, made up of *Lugii*, *Zumi* (Zumern), *Butoni*, *Sibinii* (Sibiner), *Semnones* and *Langobardi*, a force of 70,000 foot soldiers and 4000 horsemen. The areas of settlement of the *Marcomanni* in Bohemia, to where, following his differences with Arminius, Maroboduus had retreated, and the emigration of the *Marcomanni* after the death of Maroboduus into Slovakia are said to be recognisable archaeologically. These are, however, patterns of rural settlements which cannot have had any direct connection with the mobile warrior bands.

The age of the Marcomannic Wars 166-180 AD

The wars of Marcus Aurelius and his son Commodus were conflicts involving the Roman army in defence against the warrior bands that were threatening the Roman provinces from the interior of the Germanic territory. In the beginning, the *Victuales* and *Marcomanni*, as well as others, are mentioned (Marcus Aurelius 14.1: *Victuales et Marcomannis cuncta turbantibus aliis gentibus*), who demand the allocation of land as a tribute following their raids. An army of 6000 allied *Marcomanni*, *Langobardi*, *Obii* and others is mentioned in the year 166-167. Later on, the Marcomannic king Ballomarius is mentioned as conducting negotiations and as speaking for 11 legations (Cassius Dio 71.3.1a), therefore as a leader of a coalition with ten other war lords. The traces of the Marcomannic Wars are said to be clearly recognisable in the archaeological record, for example in the presence of Roman weapons – swords with ring pommels and coats of ring-mail – as well as spurs (*Stuhlsporen*) of the special Mušov-type in a corridor of traffic between Jutland and Bohemia via the river Elbe. A feedback into the areas whence the Germanic armies came could thereby be recognised.

The so-called Völkerwanderungszeit,
or Migration Period in the 3rd and 4th centuries

The invasions of Alamans, Franks and Saxons into the Roman Empire began as raids of warrior bands who wanted to pillage. Some of the warrior bands are directly employed as mercenaries, not least as garrisons of the Roman forts along the imperial borders of the Rhine and Danube in late antiquity.

The Roman commander – and later emperor – Julian won the battle of Argentoratum (Strasbourg, France) over an Alamannic army in the year 357. The Alamannic leaders were king Chnodomar and his nephew Serapio/Agenarich, who had assembled a coalition of five kings (Vestralp, Urius, Ursicinus, Suomer and Hortar) as well as ten *regales* (on horseback) and a respectable number of nobles (*optimates*), in total 35,000 armed men (*armatores*) on the side of the Alamanni. There were 17 war lords for 35,000 men, which meant that every one of them had 2000 to 3000 warriors under his command. Other Alamannic kings are known from this period, as found in the descriptions of Ammianus Marcellinus; for example, the brothers Gundomad and Vadomar, and the son of Vadomar with the name of Vithikap, were leaders of war bands – which at times waged war upon Rome, and at other times were in the service of Rome as leaders of mercenaries. In the Roman military handbook, the *Notitia Dignitatum* from the time of around 400 AD, there are mentions of Germanic units, the *Brisigavi iuniores* and the *Brisigavi seniores* from the Breisgau, the *Lentienses*, the warriors from the Linzgau, and others. The Romans recruited warriors from regions in which Alamannic groups had begun to settle. A phase of occupation of land ensued, following the initial period of raids and service as mercenaries, constituting the basis for the first ethnogenesis (tribalisation) of the Alamanni.

Archaeological traces from the beginnings are very rare. There are only a few burials, the total number being out of all proportion to the strength and numbers of the armies of which we know. It may therefore be assumed that at first these war bands returned to their homelands on both banks of the river Elbe some hundreds of kilometres from the Limes. The discoveries of enormous lost booty from the Rhine, as in the cases of Neupotz and Hagenbach, are proofs of this thesis. In the areas of origin beyond the rivers Elbe and Saale, in Thuringia, the so-called princely graves of the Hassleben-Leuna group are interpreted as the burials of leaders of mercenaries and their families. A number of years ago Joachim Werner (1973) pointed out the rich furnishings with vessels made of bronze and glass, with Aurei as Charon's-pennies and with knobbed bow fibulae (*Bügelknopffibeln*) – Roman officers' fibulae – to mark mercenaries from the '*Gallisches Sonderreich*', the separated Empire of Gaul under emperor Postumus.

It appears that from the 3rd to the 5th centuries warrior bands moved to the south from beyond the river Elbe and later made way for the ethnogenesis (tribalisation) of the Alamanni. Warrior bands from northern Germany headed into northern Gaul and led to the ethnogenesis of the Franks, and the warrior bands of the Angles and Saxons from the coastal area of Schleswig-Holstein contributed to an ethnogenesis in England.

The density of settlement after the acquisition/-occupation of land only began to increase during the 5th century in the Alamannic area. Military hill-forts and rural settlements belonged to this pattern. Only during this phase does an increase in the population become archaeologically evident, which maintains close cultural relations to the areas of origin beyond the Elbe. A steady flow of people towards the south is also recognisable, but so is a cultural feedback which influenced material culture, from fibulae to ceramic inventories.

It remains unanswered how far the Roman military-belts of the 4th/5th century – which can be found in grave furnishings in Germany – should to be seen as indicators of returned mercenaries. Germanic craftsmen themselves began to produce belt mounts of that type even quite far from the border to the Roman Empire as a result of a fashion trend. It is rather the deposits of gold coins, ring-jewellery and fibulae which were probably the possessions of former high-ranking leaders of mercenaries.

The Vikings in the west during the 9th and 10th centuries

The number of raids of the Danish and Norwegian Vikings into the Carolingian Empire was on the increase from 840. War lords with war bands came in order to plunder with their ships that could transport 30 to 60 oarsmen and warriors. In the beginning they returned to the north, but later they began to spend the winters in enemy territory. With 100 ships easily 3000 to 6000 men were involved. The Danish king and war lord Harald Klak offered mercenary services, and Louis I ('The Pious' [died 840]) granted him a fief in Friesia, so that he would fight other Vikings. The Danish king Horik I offered Louis his support for the same reasons. He also demanded Friesia, and furthermore the lands of the Abodrites as fiefs. In order to give his demands more weight, he sent several hundred ships against Hamburg. As the Germanic kings and war lords had offered mercenary services and 'foeratii contracts' after their raids in Roman times, so the Viking kings were now attempting the same.

Under the reign of Karl II ('The Bald', *dem Kahlen* [838-877]) the payment of 'Danegeld' began. It was the time of the Viking irregulars under independent chiefs, who in turn were linked in alliances. A chief or war lord by the name of Weland let himself and his followers be recruited by Karl for payment, and fought other Viking irregulars/bands. He had access to 200 ships and wanted 5000 pounds of silver as well as provisions for driving other Vikings away. Such alliances between Vikings and local rulers were also to be found in England and Ireland before the Vikings openly resorted to raiding. A great band or army of Vikings gathered in the year 865 in Kent and remained together until 879; the army conquered several Anglo-Saxon kingdoms. Parts of this army began to settle in England from 876 onwards; others moved on against France in 879 and besieged Paris in 885 and 887, but returned to England in 892 and were defeated by Alfred the Great. This so-called Great Army, a coalition of numerous war lords and their armies, was continuously strengthened through fresh units. Attempts at colonisation only began after 900 (see Coupland 1998).

The Normans and the Mediterranean during the 11th century

The Viking leader Rollo spent the first ten years of the 10th century fighting for booty in the area of the river Seine. He was granted a fiefdom in the year 911 after converting to Christianity – land at the estuary of the Seine, later known as Normandy. He thereby became the first duke of Normandy. The descendants of Rollo tried to strengthen the fiefdom and to incorporate the 'Viking' nobility in a tighter fashion, but parts of this group rebelled and moved on to new raids in the Mediterranean.

The first group of Norman knights appeared in Salerno in southern Italy in the year 999: They were a group of pilgrims returning from Jerusalem and had stopped over in Salerno when the city was attacked by Saracens. They procured weapons and horses and chased the Saracens away. The lord of Salerno, Gaimar IV (999-1027), wanted to employ them as mercenaries, but they returned to Normandy accompanied by envoys who in turn were to employ others as mercenaries. Other versions mention that warriors from Normandy appeared in southern Italy at the latest in 1015-16. They were in the military service of the local nobles and therefore in the service of the emperor against the Muslims or Saracens and Byzantine enemies. 250 Normans are listed as defeated mercenaries in a battle against Byzantine forces; various later sources mention contingents of 300 knights under the command of leaders who were granted Sicilian towns as fiefdoms.

The Norman Robert Guiscard followed his brothers to southern Italy in 1046-47, conquered Calabria and became duke of Apulia and Calabria as well as Sicily. He sent his younger brother Roger (26 years old) with just 60 knights to hold the area of Calabria. The conquest of Sicily followed in 1061 through a first wave of attack with 13 ships and 270 knights which was then joined by a group of another 170 knights in the second wave of attack in order to conquer Messina. The battle and victory of 700 knights over 15,000 Muslims followed. Sicily, which had become Muslim in the 9th century, had been reconquered for the empire in the later part of the 11th century; Palermo was taken in 1072, the last Muslim outpost fell in 1091.

The rules of the war lords and war(rior) band warfare

Criteria for this special kind of warfare can be given with the help of the following scheme of development:

1 In the beginning it was mercenaries from the central European clan and tribal communities who offered their services to the armies of the more advanced organised states. They returned home with gold and silver which they had received as pay, as well as with concepts and ideas of the more advanced forms of societies.

2 The formation of warrior bands under the leadership of war lords – who gathered entourages of warriors around them – followed next; the war lords and their warriors emerged from the old tribal society, separated from it and became individual entities with their own names that had less and less involvement with their communities and lands of origin.

3 In order to keep these warrior bands together, the war lords had to see to it that a steady income was ensured, which in turn was gained through warfare, although sometimes the mere threat of war was sufficient to trigger payments of tributes. These raids were aimed at areas outside the home territories, often enough (in terms of civilisation) more organised – and therefore promising more booty – political entities, empires or states.

4 Successful leaders attempted to strengthen their armies, for which they needed more income, which in turn meant that they had to wage wars more frequently, in the end almost continuously. This phase is characterised by a lack of fortifications, as territory was of no importance to warrior bands that were constantly on the move.

5 In order to be successful against growing military resistance from the affected areas a number of leaders joined together and formed military coalitions.

6 After a series of raids the warrior bands began to occupy and at the same time to settle in the territories they had fought in, instead of returning to their lands of origin. Or the most powerful leader attempted to build his own empire in enemy territory. He thereby had to assert himself against his opponents. The threatened countries solved the problem through the inclusion of war-band leaders into their own system of government (foederati or fiefdom contracts). They then tried to use these forces against further threats from without. In the case of a lack of a local government organisation, the leaders formed their own empires on the territories of the state which they had occupied, or where they could settle as mercenaries.

All these proceedings refer to Central or temperate Europe – from this part of the world I have provided examples, but it may also be possible to apply this sequence to other districts and cultures in the world – with individual features dependent on the special character of the epoch. Generalisation is only possible in a limited manner, but will be attempted here for model-forming purposes. Comparisons are possible with other parts of the world and other epochs, for instance in the special circumstances of neighbouring traditional and nomadic empires, as in China (Barfield 1989).

The movement from warrior band to empire or territorial government always follows a pattern of which the individual stages cannot – or at best in some stages only – be archaeologically recognised. It is only the final goal, the completed occupation of land and new settlements complete with an ethnogenesis (tribalisation), which leaves its traces in the archaeological material. The signs of the constitution of a tribe are (1) the filling of the area of settlement with latecoming immigrants and (2) the connection to the various lands of origin (the Saxons in England and lower Saxony; Franks and northern Germany; Alamanni and the area between Mecklenburg and Bohemia beyond the river Elbe; Normans in Italy and in Normandy).

It is an unanswered question, then, what the furnishing of male burials with weapons in Early La Tène, in the Roman Iron Age, in Merovingian times and during the Viking Ages may imply.

Are weapons symbols of social status and rank, or indeed signs of actual participation in combat, or even in wars? In all cases only a small proportion of the total numbers of male burials from each period include weapons in their inventories; these weapon assemblages often depend on the age of the dead man. There is a difference between younger and older warriors. For the early period there is also the possibility that at first only members of warrior bands or auxiliary units of the Roman Empire were buried together with their weapons.

Pattern of interpretation

The function of the warriors in the constitution(s) and development(s) of tribes (tribalisation) and states (empires, territorial governments) is recognisable through the described stages.

In conclusion it can be said, that (Fig. 1): Warfare forces clan societies to organise themselves in tribes. Tribes develop, which have to set up military units in order to fend off threats and to defend themselves. A part of the warrior group joins together to form warrior bands that offer services as mercenaries or undertake raids. A cyclical development creates war: The contrast between states at different levels of development and between states and clan-based societies leads to warfare, which in turn leads to the constitution of tribes. As tribes developed, warfare between tribes followed. As warrior bands, which searched for and needed 'work', developed on another level (higher than tribes and above and also independent of the tribes), the raid was invented and mercenary services offered to state-organised societies. These always had reasons for waging war against their neighbours.

The question must be asked, why did the social form of organisation called the warrior band and its behaviour – the annual raid – develop? They must have had either a regular or specific function in European history; the regularity with which this behaviour can be observed cannot be overlooked. Possible explanations could be:

1 The warrior bands always existed, as they were necessary due to certain inner structures of their societies. A part of the young warriors could not live in the rural surroundings of the settlements at home because, for example, of laws regulating inheritances which stood against them. They separated themselves from the clan and tribal societies. The wars necessary for their income were mostly internal, as can be seen in the Celtic world (sacrificial areas as in the case of Gournay in France) or in the Germanic world (the great weapon deposit of Illerup in Jutland). They only appear sporadically in historic sources – mainly when the raids were conducted against more organised states.

2 The co-existence of states – with varied levels of organisation – and open ranked societies, clan and tribal societies, forces the cultures along the fringes (border lines) to adopt the new organisational form of the warrior bands which make profits by mercenary services or raids on the of more advanced political entities.

The cyclic developments can thus be explained by the fact that there are always new structures of states emerging, which in turn influence their neighbourhood. That is why my repeated model

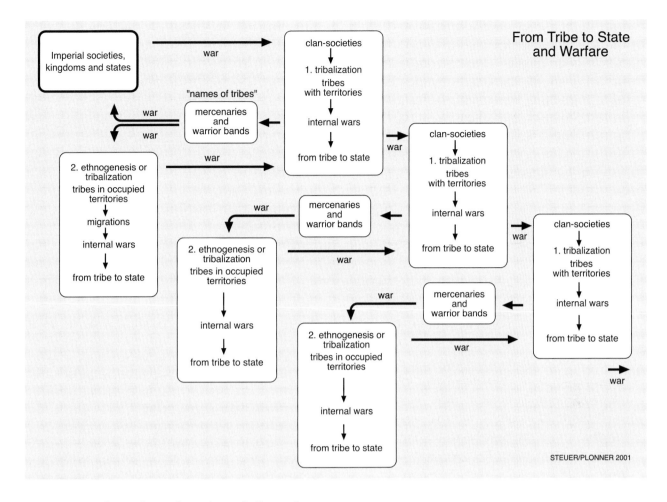

STEUER/PLONNER 2001

FIG. 1: *Warfare and transformations of tribes and states.*

can be used – and not only for the times beginning with the Celts and the La Tène period, but equally for the Hallstatt period and earlier epochs of temperate Europe.

– The Mediterranean world with Greece, Rome and Carthage influenced the clan-based societies of the La Tène period – the first recognisable response being the Celtic mercenaries in the area of the Mediterranean.

– The secondary structuring development of tribes and areas of rule (oppidum-civilisation) in the Celtic world in turn influenced the early Germanic world of beyond the Rhine. The first recognisable response was the *Suebi* mercenaries of Ariovistus employed by the Celtic lords of the *Sequani*.

– The Roman Empire in its varied stages of expansion influenced the Germanic world, after integrating and assimilating the Celtic world in the Empire as the Gallic provinces. During the Roman Iron Age far inside the Germanic countries areas of

rule developed which defended themselves with ramparts and trenches. How many of these basic units merged into a tribe still remains unknown. 100 tribal districts of the Suebi are mentioned by Caesar (*de Bello Gallici* 4.1). These sent out 1000 armed men every year on raids, altogether 100,000 men – perhaps a reflex of this situation. (Caesar goes on to report that the rest of the warriors stayed at home in order to do their farming so that their families would have an income, but were sent out the following year when it was their turn). The '*Suebenknoten*' (knot of the *Suebi*), the characteristic knot of hair worn on the side of the head of the warriors, is spread across the whole of Germania, and Tacitus points out that it was an honour for a warrior to wear this knot (*Germania* 38).

The first response outside Germany was the bands of mercenaries serving as *auxilia*-units in the Roman army since the times of Arminius. A later response was the groups of mercenaries made up of the so-called Alamanni and Franks who served during the 3rd century.

– The Frankish Empires of the Merovingian and, even more so, the Carolingian dynasty had the same stimulating effect on the Scandinavian peoples. Scandinavians can be traced back to the Merovingian or Vendel period where they had been employed as mercenaries of the Franks (so-called ring-swords may be signs of such alliances: Steuer 1987).

– Raiding bands of Vikings are employed as mercenaries against other bands of Vikings, the situation later being resolved through the granting of fiefdoms.

– Mercenaries from Normandy conquer the Muslim south of Italy and found a kingdom.

The feedback such activities provided to the places of origin only become apparent when written sources report on it. Otherwise, the development from clan- and family-based societies to tribal societies and from that stage – with the key word 'centre of wealth' – to early areas of rule (empires, states) can only be deduced in an indirect manner (compare: Mortensen and Rasmussen 1988; 1991).

BIBLIOGRAPHY

• Barfield, T.J. 1989. *The Perilous Frontier: Nomadic Empires and China*. (Studies in Social Discontinuity). Oxford: Blackwell.

• Barth, F. 1969. 'Introduction'. In: F. Barth (ed.): *Ethnic Groups and Boundaries. The Social Organization of Culture Difference*. Oslo: Universitetsforlaget, pp. 9-38.

• Bodmer, J.P. 1957. *Kriegertum als Form der menschlichen Existenz im Frühmittelalter*. Zürich: Fretz und Wasmuth.

• Bowlus, Ch.R. 2002. 'Ethnogenesis: The tyranny of a concept'. In: A. Gillett (ed.): *On Barbarian Identity. Critical Approaches to Ethnicity in the Early Middle Ages*. (Studies in the Early Middle Ages, 4). Turnhout: Brepols, pp. 241-56.

• Coupland, S. 1998. 'From poachers to gamekeepers: Scandinavian war lords and Carolingian kings'. *Early Medieval Europe*, 7(1), pp. 85-114.

• Jørgensen, A.N. and Clausen, B.L. (eds.) 1997. *Military Aspects of Scandinavian Society in a European Perspective, AD 1-1300*. Copenhagen: Nationalmuseet.

• Lentz, C. 1998. *Die Konstruktion von Ethnizität. Eine politische Geschichte Nord-West Ghanas 1870-1990*. Cologne.

• Marx, Chr. 1998. 'Die "Militärmonarchie" des Shaka Zulu - koloniale Mythen und neue Forschungen'. *Arbeitskreis Militärgeschichte e.V. Newsletter* 6, pp. 6-9.

• Mortensen, P. and Rasmussen, B.M. (eds.) 1988. *Jernalderens Stammesamfund. Fra Stamme til Stat i Danmark 1*. Aarhus: Aarhus University Press.

• Mortensen, P. and Rasmussen, B.M. (eds.) 1991. *Høvdingesamfund og Kongemagt. Fra Stamme til Stat i Danmark 2*. Aarhus: Aarhus University Press.

• Pohl, W. 1998. 'Gentilismus'. In: *Hoops Reallexikon der Germanischen Altertumskunde*. Vol. XI. Berlin and New York: Walter de Gruyter, pp. 91-101.

• Pohl, W. 2002. 'Ethnicity, theory, and tradition: A response'. In: A. Gillett (ed.): *On Barbarian Identity. Critical Approaches to Ethnicity in the Early Middle Ages*. (Studies in the Early Middle Ages, 4). Turnhout: Brepols, pp. 221-39.

• Sigrist, Chr. 1994. *Regulierte Anarchie: Untersuchungen zum Fehlen und zur Entstehung politischer Herrschaft in segmnentären Gesellschaften Afrikas*. (3rd edn.). Hamburg.

• Steuer, H. 1987. 'Helm und Ringschwert – Prunkbewaffnung und Rangabzeichen germanischer Krieger'. *Studien zur Sachsenforschung*, 6, pp. 189-236.

• Steuer, H. 2001. 'Kriegswesen. III. Kulturgeschichtliches und Archäologisches'. In: *Hoops Reallexikon der Germanischen Altertumskunde*. Vol. XVII. Berlin and New York: Walter de Gruyter, pp. 347-73.

• Vail, L. (ed.) 1989. *The Invention of Tribalism*. London.

• Timpe, D. 1970. *Arminius-Studien*. Heidelberg: Carl Winter.

• Wells, P.S. 1999. *The Barbarians Speak. How the Conquered Peoples Shaped Roman Europe*, Princeton: Princeton University Press.

• Wenskus, R. 1977. *Stammesbildung und Verfassung*. (2nd edn.). Cologne and Vienna: Böhlau.

• Werner, J. 1973. 'Bemerkungen zur mitteldeutschen Skelettgräbergruppe Hassleben-Leuna. Zur Herkunft der ingentia auxilia Germanorum des gallischen Sonderreichs in den Jahren 259-274 n. Chr.'. In: H. Beumann (ed.): *Festschrift für Walter Schlesinger*. (Mitteldeutsche Forschungen, 74/1). Cologne and Vienna: Böhlau Verlag, pp. 1-30.

Chiefs Made War and War Made States?
War and Early State Formation in
Ancient Fiji and Hawaii

CLAUS BOSSEN

/17

The epigram 'War made state and state made war' seems to emblematically condense recent discussions of the emergence of the European nation-state.[1] Various works indicate the close inter-relations between war, military organisation, and the development of medieval bureaucratic organisation, and the significance of these connections to the emergence of the territorially based nation-state with its (claimed) monopoly of violence (e.g., Giddens 1985; McNeill 1982; Porter 1994; Tilly 1990). That '...state made (and makes) war' is beyond doubt when looking at European history. More generally, once an elite group has gained privileged access to basic resources and thus a position of power, it can employ military organisation to protect its interests internally and externally. But what about 'War made state ...'? Is there really a connection between the organisation for defence and attack and the emergence of the state? Does the search for protection against external threats result in complex organisation with internal subjugation, either by consensus or through warring leaders who use military organisation to suppress the people they were meant to protect? The coincidence of frequent, often prolonged wars and the rise of modern nation-states in Europe in the last millennium offers a good case through which to argue this in the affirmative. But the *long-durée* of

this development also suggests that it was not the logical outcome of pre-set conditions. Various actors had to play their cards right in the right social structural conditions in the right historical context.

The dictum 'War made state' has a long history in European thought concerning the emergence of the early ('archaic') and European states.[2] However, according to archaeological and anthropological evidence,[3] while wars have been fought throughout human history, states remain rare occurrences in world historical terms. Depending on how one chooses to define a state, such political entities may only have emerged ten to twenty times before the European nation-state and its subsequent establishment of a global system of nation-states through commercial and military enterprise during the last two centuries (Mann 1986). War, conquest and military organisation are apparently not sufficient to cause state formation, and pre-state societies seem to have in-built anti-state features that discourage stratification and military leadership from developing into class division, a monopoly of means of violence, and a bureaucratic, political organisation.

In the following, I pursue the problematic of war and early state formation by focussing on what are now the two island groups of Fiji and Hawaii at the time of European contact around 1800. Fiji and

Hawaii seem especially suited for the analysis of this question, since their highly hierarchical societies frequently engaged in wars, conquests and quests for political domination. Furthermore, ancient Hawaii was close to or already an early state, while stratification and central government was less institutionalised in Fiji. They therefore offer good cases through which to pursue a comparative discussion of the dynamics of war, leadership, and the institutionalisation of power. After a short theoretical overview of the discussion of war and early state formation, I analyse the cases from Fiji and Hawaii, respectively, and conclude with a comparative discussion of the dynamics of war and early state formation. I argue that while leaders did make war, war was not sufficient to make state. To approach this problem, I adopt two conceptual strategies. Firstly, I circumvene the question of definitions, and instead of trying to define, for instance, 'chiefdom' or 'early state', I ask how war contributed to changes in or the reproduction of social organisation. I assume that early state formation was not a sudden, definitive break with previous social structures, but instead an ongoing process of construction and deconstruction that only sometimes resulted in a stable, early state of some permanence. While definitions are useful devices to indicate what we are talking about, a strict focus upon these in cases of state *formation* – i.e., cases that are somewhere between 'chiefdom' and 'early state' – would make the argument about war and early state formation depend upon definitions instead of throwing light upon the processes involved. Secondly, I divide the very different arguments of how war causes early state formation into three groups according to whether they treat war as a context, as military organisation or as a means. By these two conceptual strategies I hope to achieve a framework within which processes of war and early state formation can be discussed in a comprehensive way.

War and state formation

Early state formation has been the subject of a long debate. In addition to war and conquest, the following causes for its occurrence have been suggested: the emergence of private property, the central management of irrigation systems, the development of social classes, population pressure, and long-distance trade.[4] None of these have been established as a sufficient or necessary cause by itself in state formation,

and 'single-factor explanations belong to the kindergarten state of state theory because origins are extremely diverse' (Mann 1986: 50). According to contemporary consensus, the formation of early states is a complex process involving an (indeterminate?) number of factors. Considerable effort has been made to provide extensive overviews of these factors and their varying constellations in different contexts of state formation (Claessen and Skalník 1978; Cohen 1978; Haas 1982; Mann 1986). An inherent problem in this discussion has been the lack of a common definition of the state and of the kind of society that preceded it.[5] There is furthermore no agreement on what kind of process was involved in the transition from non-state to state. Different approaches and definitions direct attention to different interconnections. A classic case is the long-time debate between a conflict and an integration position, exemplified by Morton Fried (1967) and Elman Service (1975). Is the state primarily formed to safeguard the interests of an economically dominant class (Fried's position), or is the state formed as a social organisation, which has superior capacities to protect and integrate its members (Service's position)?[6] Those who argue for the prominence of war in state formation actually argue that war leads to stratification as well as a better form of social organisation and integration. In order not to get into the egg-and-hen problem of deciding which traits of the state came first, I adopt a broad perspective and ask how war may contribute to either of these: i.e., through centralisation, stratification, a putative monopoly of violence, or the institutionalisation of government on a 'basis superior to kinship' (Fried 1967).

War as the motor behind social evolution was proposed by Herbert Spencer in *Principles of Sociology* (1967), originally published between 1876 and 1896. In more recent times, Spencer's intellectual heir, Robert Carneiro, has argued for 'the centrality of war in political evolution' (Carneiro 1990: 190). People only accept limitations of their sovereignty through force, according to Carneiro, and war is thus the basis for the formation of chiefdoms as well as states:

Given the universal disinclination of human groups to relinquish their sovereignty, the surmounting of village autonomy could not have occurred peacefully or voluntarily. It could – and did – occur only by force of arms. (Carneiro 1990: 191)

As a mono-causal theory, Carneiro's position has been refuted by cases where statehood has been achieved without evidence of warfare (Claessen and Skalník 1978; Haas 1982; Mann 1986). However, his remark upon the difficulties in establishing coercive authority is central. It remains to be explained, 'how authority was converted into power that could be used *either* coercively against the people who granted authority in the first place *or* to deprive people of the rights of material subsistence' (Mann 1986: 49). Coercive rule presupposes an organised body of men of enforcement, which presupposes the compulsory extraction of surplus production for its sustainability. Somehow people have to be forced by circumstances to accept coercive rule and not withdraw their products or disperse outside the reach of government. Non-state societies nevertheless seem to have in-built structural features which work against the establishment of coercive control by one part of society over the other. Malcolm Webb states the problem thus:

The problem is the specification of mechanisms whose operation would in time generate a monopoly of coercive force in the hands of the tribal leadership, in a situation where the previous lack of such a monopoly would of itself seem to preclude the possibility of ever securing just such a monopoly. In other words, the full emergence of state would appear to require eventually the final overthrow of a previously existing traditional system of authority, social controls, and resource allocation whose inelasticity and whose decentralized and localized organization would indeed have very largely inhibited even the initial concentration of power and resources required for further significant steps in this direction. Despite the areas of continuity, or seeming continuity, between advanced tribal and incipient state systems of governance, the shift from the former to the latter entails a basic and total alteration in the manner in which the authority of the leadership is ultimately enforced and upheld. (Webb 1975: 157)

As mentioned above, there are multiple roads to statehood and therefore multiple 'mechanisms' that may lead to such a shift in the mode of power. Malcolm Webb suggests that long-distance trade enables privileged groups to establish themselves as a separate class that has the economic resources for a paid military. Michael Mann, who emphasises a multi-causal theory, argues that the 'mechanism' is a gradual condensation of various networks of relations that effectuates a 'caging' of people into society

(Mann 1986: ch. 3). In the present context I only consider the theories that propose war as a state-creating mechanism and briefly outline the strengths and weaknesses of these theories below. Particular attention is given to Robert Carneiro's theory (1970) because of its centrality in the discussion.

The role of war

War as a road to state formation is attractive because it apparently offers a plausible explanation for the appearance of subjugation, centralisation of government, and greater social complexity. In general, it has been argued that war promotes social change in three basic ways. First, war is considered a context for social organisations, and the argument is that in a constant environment of war more complex societies have a higher chance of survival because the centralised co-ordination of large groups of warriors provides more concentrated military power. Second, the focus is on military organisation: the premium on effective military organisation is seen as leading to the more rapid development of state-like traits such as centralisation and subjugation in the military during war, spreading to the rest of society even when the war is over. Third, war is regarded as a means to increase political power. The conquest of other groups, it is argued, leads to subjugative stratification and superseding forms of government. Robert Carneiro's theory falls into this third category.

The first approach sees war as a context within which social organisation develops. During times of war, the argument goes, organisation of defence and attack may be developed into higher degrees of complexity that enable higher efficiency, and hence more complex societies emerge. The evolutionist Herbert Spencer (1967) provides an early example of this approach, and Elman Service (1975) argued that war would make people accept the power monopoly of an emerging state since 'good government' would enhance their chances of survival by offering a more powerful organisation for their protection (Service 1975: 270). While the logic of how war and social development might be linked is well constructed in this approach, it leaves two questions unanswered. Firstly, how is the transformation of social organisation achieved? Even granted the questionable assumption that all members of the group agree upon the desirability of such a change, reconstruction of society is a difficult task,[7] and the approach leaves the political transformation involved in a black box between war and higher complexity. Secondly, some societies apparently do not develop into larger entities of higher complexity. Herbert Spencer explained this by the extensive dispersion of these groups or by the fact that they lived in peace with each other (Spencer 1967: 33). However, war and development cannot be equated and neither can peace and stagnation. Some societies are continuously at war with each other but do not develop higher complexity; war is instead part of the reproduction of social structures (Harrison 1993; Gardner and Heider 1968). The context approach works best as a long-term, survival-of-the-fittest explanation of state formation or to explain development through the threat of annihilation, but otherwise it leaves the process of change in the black box.

The second approach argues that military organisation spreads to the rest of society. The assumption is that since military power is most effective when organised on the basis of centralised command and a clear-cut hierarchy, frequent war will lead to a higher level of subjugation, centralisation, and hierarchy in military organisation than in the rest of the social organisation. If this military organisation is adopted in times of peace by the rest of society, the social structure shifts towards more hierarchical structures with more centralised control. Whereas the first approach assumes a close connection between social and military organisation, the second approach assumes their separation and reintegration. First, military organisation develops at least semi-independently of social organisation, which is then transformed into the model of the military. The crucial points, however, are whether a military organisation could become semi-independent of social organisation and whether military organisational capabilities could spread into other areas of social life. Military leaders may be cherished by their soldiers and the people they defend in times of war, but have no power in times of peace. The successful war leaders Geronimo of the Apache in North America and Fousive of the Yanomami in South America could not transform their military success into 'civil' authority (see Chagnon 1974: 177-80). Elman Service (1975) argued that a centralised leadership is agreed upon by consensus because of the benefits for societal survival that it offers. While this may apply in times of war and explain the emergence of a sophisticated military organisation, it does not explain why such forms of organisation would spread to the rest of society in times of peace. People may relent their sovereignty to military command in order to survive, but under which circumstances would they not revoke that decision in times of peace. Furthermore, centralised, hierarchical command may not at all be a suitable way to organise agricultural production, trade, political organisation or religious life. The thesis thus seems most plausible in circumstances where a society is under constant threat of being annihilated and where especially economic production may suitably be conducted under centralised, hierarchical command – such as large irrigation systems.

The third and final approach sees war as a means to increase and transform political power. The conquest of other groups provides a means to acquire access to additional basic resources. The conquest theories have the strength of explaining not only how centralised command and hierarchy emerges, but also how stratification, i.e. privileged access to basic resources in the sense of Morton Fried (1967), comes about. The conquest theories posit that the subjugation of one group by another and followed up by the extraction of tribute or tax from the conquered group leads to stratification where previously only egalitarian or ranked political relations prevailed. Over time military rule is transformed into institutionalised government, laws, and the assimilation of the two groups into one people. Herbert Spencer (1967) argued in this way, and more recently Robert Carneiro (1970) has taken up this line of argument, his theory having become one of the most cited and central contributions to discussions on the relation between war and state formation. Carneiro argues that in a situation where a group's access to land is limited by circumscription (because of ecological conditions, the presence of other groups, or the concentration of scarce resources) a point will arise when it is more feasible to conquer the land of another group than to intensify production by working harder or to invent new technologies of cultivation. Because this point arises before production is maximised, Carneiro argues, the conquering group can achieve higher levels of production by coercing the conquered group. Through such a process chiefdoms arise through the incorporation of several villages into one polity. The replication of the process on a larger scale accounts for the rise of states through the incorporation of several regions into one polity. Through a process of internal evolution, previous reciprocal or redistributional exchange relations are transformed, and a society appears where one powerful stratum taxes a subjugated stratum appears.

While neat and elegant the theory actually only explains why and how entities grow larger through conquest, but not how political organisation changes. Carneiro does not elaborate upon the 'internal evolution' through which political transformation is achieved and instead takes the state for granted at this point (Carneiro 1970: 736). So his theory does not actually explain what it intends to do and thus shares the general weakness of the conquest theories, which assume that the relationship between the conquering and the conquered groups acquires a new quality after conquest. For the theory to work the conquered cannot merely become incorporated into existing exchange relations. A group conquered by a chief and paying their yearly tribute like other groups of the chiefdom would not be part of a state but of a chiefdom. While conquest may explain the subjugation of another group of people, it does not explain the transformation of political power or the social structure as such. The theories offer plausible contexts in which higher levels of social complexity, subjugation and stratification should develop, but not how this would transgress the existing social structure. They assume but do not explain how and why development of political organisation occurs.

The different approaches invoke different causalities to explain state formation. The context approach assumes that social and military organisations reflect each other and thus change together in a context of frequent war. The focus on military organisation assumes instead that the military can develop prior to social organisation and then by usurpation or consensus be reintegrated on a higher level. Finally, the focus on war as a means proposes conquest as a means to assure basic resources and increase political power. They can of course be combined into a scenario that makes a strong case for the centrality of war in social change and state formation. But their distinct weaknesses should be remembered. We cannot assume that war means development and peace stagnation, since war might be part of social reproduction, and even if war is an impetus to social change we need to specify when war transgresses existing social structure and how political change takes place. On the other hand, we cannot

assume that military organisation is independent of social organisation and as a result can develop in advance, but need to specify under which circumstances that independent military development and the subsequent adoption of military organisation by other sections of society would take place. Finally, while war may be a means to gain access to basic resources and to enlarge political entities, it does not by necessity entail the transformation of political organisation. Hence, ultimately state formation as political transformation is not explained, only made probable.

David Webster (1975) is one of the few to have addressed the problem of political change within the context of war. He bases his argument on Carneiro's theory (1970) and goes on to consider the problem of structural change in terms similar to those of Malcolm Webb (1975) mentioned above:

On one hand, the structure of ranked society, with its hierarchical organization of political authority, centralization of some forms of economic activity, and at least incipient forms of wealth accumulation and differential consumption, prefigures basic characteristics of the state – concentration of wealth and coercive force in the hands of a small segment of society. On the other hand, the kinship idiom which integrates ranked societies would seem to frustrate the evolution of state-type institutions in several fundamental ways. In such societies wealth accumulation is limited by the reciprocal relations of the chief to his producer-kinsmen. Effective monopolisation of coercive force is difficult because what limited access to force the chief possesses is largely derived from his redistributional activities. (Webster 1975: 460)

According to Webster, chiefs tread a fine line between receiving gifts and blatantly and coercively mobilising labour and taxation. If chiefs demand too much, their people will abandon them. In a quest to monopolise power the system would therefore become unstable. Another source of instability is the competition between almost equals in the hierarchy for the position of chiefs (Webster 1975: 466).

Warfare can, however, Webster argues, overcome the hindrances to social structural change under certain circumstances, and he invokes in the process all three approaches presented above. Continued warfare would put a premium of higher chances of survival on those chiefdoms with a stable military leadership and a stable political leadership (thesis one above), thus dampening internal competition and conflict.

Furthermore, territorial conquest would allow the incorporation of small pieces of land outside of normal kinship obligations (thesis three above). This land could be monopolised by military leaders whose privileged access to wealth would enable them to augment their power, either in order to become heads of state or to employ their own military to enforce their rule (thesis two).

Constant warfare not only provided an important and highly adaptive managerial function for emergent elite segments of society, but it also stimulated the acquisition of small amounts of 'wealth' (i.e., basic resources) which were external to the traditional system and could be manipulated in various self-serving ways by these same groups to dampen internal dissension and attract supporters. Out of this milieu developed political and economic special-interest groups, which ultimately provided the basis for social stratification. (Webster 1975: 469)

The way out of the structural constraints of non-state (chiefdom) societies is thus privileged access to basic resources through war, which also stiffens competition between would-be leaders and thus allows political power to stabilise. Essentially this means that it is the opportunities provided by war for tying the concentrated, coercive power of the military into other areas of society that may enable social structural change. The extra-military factor for Webb was long-distance trade, while the factor for Webster was acquisition of land outside the sphere of redistribution by chiefs.

The above overview of theories on war and state formation can be summarised in the following way: while war may indeed spur the development of social complexity and military organisation and lead to subjugation, by itself it does not explain the processes that lead to changes in the social structure as such. The processes leading to changes in political organisation in connection with war require further description and analysis, and the arguments of Webb and Webster suggest that it is the linkage of the power of the war leader with other domains of power that may effectuate such a change. War between the Fijian groups in the first half of the 19th century and the Hawaiian groups at the end of the 18th century provide evidence to support such an analysis: war was frequent within these groups, led to the conquest of people and land, and was a means for warring leaders to expand their power and domain. An examination

of war in these island groups will thus address all three of the above theses of the connection between war and social structural change. Additionally, it offers the opportunity to further assess the arguments of Robert Carneiro (1990).

War and conquest in Fiji in the early 19th century

Though Abel Tasman passed by the Fiji islands in 1643, the first information on the inhabitants acquired by Europeans was given to James Cook in July 1777 when he visited nearby Tonga:

Fidgee lays three days sail from Tongatabu in the direction of NWBW, they describe it as a high but very fruitfull island abounding with Hogs, Dogs, Fowls and all the kinds of fruit and roots that are found on any of the others. It is not subject to Tongatabu on the Contrary they frequently make War on each other and it appeared from several circumstances that these people stand in much fear of those of Fidgee, and no wonder since the one is [a] Humane and peacable Nation, whereas the other is said to be Canibals, brave, Savage and Cruel. (Beaglehole 1967: 163, my insertion)

The Tongans here probably gave Cook one side of a local version of the stereotypic Noble Savage-Barbarian dichotomy used by European discoverers and travellers to categorise people outside Europe. Cook was obviously endeared by the islands of present-day Tonga, which he called the 'Friendly Islands', and susceptible to such a stereotypic characterisation. Accounts by travellers coming to Fiji later on would however add nuances to this depiction of the inhabitants of 'Fidgee' island, though they confirmed that these people did fight wars. Captain Cook, however, sailed on to discover Hawaii without visiting the Fiji islands.

The first wave of Europeans came after sandalwood had been discovered on Vanua Levu, the second biggest island in the group, in 1801. Accordingly, European ships regularly visited Vanua Levu from 1804 to 1813, at which point most of the sandalwood had been cut and the number of European ships decreased. Trade in beche-de-mer, a sea slug of great value in China, caused them to resume sailing to Fiji in the 1830s, and a little later the first missionaries arrived. A short-term boom in cotton led to an influx of Europeans from Australia, and in 1855 they started to dominate local politics. The various accounts of these people provide useful information about war and its role in the societies of this group of islands on its own terms.[8]

Wars were frequent on these islands in the first half of the 19th century and archaeological evidence testifies to the existence of fortifications for several centuries (Frost 1979; Palmer 1969; Rowlands and Best 1980). Wars could range from ambushes and raids involving just a few warriors to large-scale campaigns, battles and sieges with thousands of warriors. Bows and arrows, clubs, slings and spears were used with casualties ranging from nil to hundreds of dead men, women and children. In the wake of European trade, muskets were introduced and strengthened those few parties that disposed of them, but the advantage wore off as they became more widespread in the 1850s. While they tended to replace the bow and arrow, their impact on the way that wars were fought was minor: village fortifications were provided with higher earthen moulds as protection against enemy bullets. On top of the moulds, the usual palisades and entrances ports were built around the village (Clunie 1977).

The reason for the relatively frequent occurrence of war was, in the words of the missionary Thomas Williams, the 'many independent governments, each of which seeks aggrandisement at the expense of the rest' (Williams 1982[1858]: 43). However, these 'governments' were not all independent. In addition to the independent or ruling groups, there were three statuses: *bati* (border), *qali* (tributary) or *kaisi* (slave). In *Kaisi* villages lived those who had suffered total defeat in war and were relegated to the status of slaves. *Qali* villages had surrendered or acknowledged the superiority of the leading group, to which they were paying an annual tribute of first fruits consisting of staples, woven mats, women and other valuables. *Bati* villages were more independent, non-tribute paying villages often situated between neighbouring confederacies and thus those first involved in war. There was considerable variation in the internal organisation of each group, especially between the more egalitarian western groups and the more hierarchical groups in the east, but a general pattern of basic kinship-based organisation on the basis of which political alliances were made was generally common (See France 1969: 9-18; Routledge 1985: 28-29). Since state formation (or lack thereof) is in focus here, I shall concentrate on the most hierarchical societies of Eastern Fiji, which were also the

ones that had developed into the largest entities. The dominating confederacies here were those of Bau, Rewa, Verata, Cakaudrove, Bua and Lakemba; and the basic social entity was the *yavusa* (clan), in which all members claimed a common descent and which was sub-divided into *mataqalis* (sub-clans) and *itokatoka*'s (extended households).[9] Several *yavusa* might gather together into a *vanua* (tribe) and several *vanua* could in turn form a *matanitu* (confederation). Within a *yavusa* each *mataqali* would be assigned particular services in everyday life and on ceremonial occasions in return for the protection of the community and the right to cultivate land (France 1969: 15). A *yavusa* usually formed a village and was based on kinship and common descent. The latter could be extended to the level of *vanua*, which, however, like the *matanitu*, mainly consisted of political alliances. Though in general applicable, this ideal model was never present, since the growth or decrease of a population and a victory or defeat in war might compel groups to separate from their own village or to migrate to become part of other villages.

Each entity was led by a ruling group from which the chief was chosen on the basis of descent and aptness as a leader and warrior (Routledge 1985: 35). While chiefs in the western part had to rely on the consent and approval by their kin, chiefs in the eastern part had absolute power and were shown the utmost respect (see, for example, Lockerby 1922: 21; Williams 1982[1858]: 22). In these highly hierarchical confederacies where chiefs had absolute power, wars were frequently fought for all kinds of reasons: perceived sleights of honour, revenge, to acquire 'long pigs' (humans) for consumption, to procure women as concubines, or in a quest for power. Most wars were small, involving only ambushes, raids and skirmishes in which single or small parties of warriors attacked other warriors or unsuspecting men, women or children in the gardens, out fishing or the like. This required skills in handling weapons, knowing how to dodge arrows or stones or throw clubs, and in making hidden attacks on various terrains, but by individual warriors or small groups with only small organisational demands. In the following, I concentrate on the large-scale wars, *valu ni tu*, which were wars between *matanitu* and required more organisation, co-ordination and command over numerous contingents of warriors over extended periods of time. Though sometimes fought because of some small insult or breach of convention, they eventually led to the inclusion of new tribute-paying *qali* or *kaisi* slave villages and the acquisition of new territory by conquest.

The wars

Wars in Fiji in the first half of the 18th century were part of social life. Boys were taught the skills of war from an early age and for adult men war was an occasion for them to prove their worth (Clunie 1977: 42-44). In parts of Fiji, it was customary to honour a man who had killed with his club with the title of *koroi* in a four-day ceremony. A warrior who had slain ten enemies was awarded the title of *koli* and the right to wear a cone shell, *tora*, on his arm. Killing another ten, twenty or thirty enemies was awarded with new titles and armshells. Before entering a large-scale war the gods were consulted and offerings made. Individual warriors would bless their weapons or go through rituals of invulnerability before going to war, and on behalf of the whole group the chief would present major offerings. In the case of victory, the warriors would present the bodies of slain enemies to the gods before eating them. But if the campaign was unsuccessful the warriors and chiefs might let their disappointment and anger out on the gods, to whom they related as if they had human personalities (Erskine 1853: 247).

Large-scale wars, *Valu ni tu*, tended to be openly conducted, formally declared, and involved armies marching to attack fortified towns. A council of chiefs, priests and elders would weigh the pros and cons, consider the loyalty of *qali* and *bati* chiefdoms, consult the gods, and if things looked favourable formally declare war. After the decision had been made, considerable diplomatic negotiation and communication ensued in order to secure support.

Orders are sent by the Chief to all under his rule to be in readiness, and application is made to friendly powers for help. A flat refusal to comply with the summons of the Chief, by any place on which he had a claim, would, sooner or later, be visited by the destruction of the offenders. Efforts are made to neutralise each other's influence. *A* sends a whale's tooth to *B*, entreating his aid against *C*, who, hearing of this, sends a larger tooth to *B*, to *bika* -'press down'-the present from *A*; and thus *B* joins neither party. Sometimes two hostile Chiefs will each make a superior Chief the stay of their hopes; he, for his own interest, trims between the two, and often aids the weaker party,

that he may damage the stronger, yet professing, all the time, a deep interest in his welfare. (Williams 1982[1858]: 44-45)

The paramount chief or a close relative renown for his capabilities in war would be appointed 'war chief', *turaga-ni-valu*, to lead the army into battle. While heralds were sent to allies to secure support, non-involved villages were warned of the imminent fighting in order for them to stay within their premises, since 'the warriors once unleashed tending to slay anyone they strayed across, not pausing to confirm whether he or she was friend or foe' (Clunie 1977: 13). All men who could use a weapon would be drawn into the irregular army, which could number thousands of warriors. Before going into battle, a review was held where the chiefs and their contingents of warriors were presented with whale teeth, gifts of food, and promised further gifts in the case of victory. The chiefs and warriors would in return pledge their loyalty to the paramount chief and boast of their deeds of bravery and the damage they were going to do to the enemy (see Williams 1982[1858]: 46-48; Lockerby 1922: 31-32). Sometimes paramount chiefs would give away weapons and promise more weapons and possibly women in return for success. Though villages were most often inter-related, the only loyalty that counted to a warrior at the time of war was that to this tribe and his chief (Willams 1982[1858]: 45). In the case of large-scale war, allegiance was paid to the paramount chief, though the army was still organised on the basis of the village, or *yavusa*.

A period of ambushes and raids, sometimes lasting for months, often preceded the large-scale war, which might not even eventuate. When it did, the attackers advanced on the fortifications of their opponents, with scouts and friendly villages reporting enemy movements and the defenders slowly retreating to their fortifications if the advance could not be halted. Fortifications were either built on hilltops or on the ground and testify to their considerable experience with war. Villages had palisades all around with a few gates that led the attackers through winding corridors into the village. The corridors had small holes through which the defenders could shoot at the attackers while they were winding their way through, trying not to step on the spikes on the ground and avoiding spears aimed at them from the platforms on top of the corridors. On the outside, the villages were enclosed by one, sometimes two, three or four ditches broken by a few earthen causeways or bridges that could be withdrawn at times of war (Clunie 1977: 14-19; Lockerby 1922; Jackson 1853). The attackers usually did not storm the fortifications immediately after the defenders' retreated into their fort; instead they took time to clear the area for snipers, and if the fort seemed strong and resistance fierce the attackers erected their own defences, lines of communication and escape routes. The attackers might continue to be highly motivated but were sometimes demoralised and gave up the attack. If the attack was to be made, the allied contingents were briefed by the *turaga-ni-valu* and they would attack with the battle-flag of *masi* (bark cloth) tied to the end of each of their spears (Clunie 1977: 26). Forts were seldom taken by a head-on attack, and if they were at all defeated it was only after a prolonged series of small attacks, false retreats and sometimes treachery. If taken, the palisades were broken down, the defending warriors, women and children killed, and the village burnt down.

Before this happened it might, however, have been possible to negotiate a surrender, which was ratified by the presentation of a basket of earth with a reed stuck into it. This symbolised the surrender of tribal land to the winning chief. Whale teeth, *tabua*, other valuable property, and young women of high rank often accompanied the presentation. Depending on the conditions of surrender, the attackers might burn down the town and take weapons, food and other property as spoils of war. They might force the defeated tribe into slavery, force them to pay tribute regularly, or drive them from their land for a period of time or forever. Landless tribes might then resettle with a friendly tribe to which they nonetheless would have to pay tribute.

Large double-hulled canoes, *drua*, which could be more than thirty metres in length and carry up to 250 warriors, were essential for transporting armies and to extend influence and power geographically. Chiefs had these *drua* built by their own people or commissioned from Fijian or Tongan specialists. The construction would take five to seven years and was a costly affair only to be carried out by a powerful chief. Battles were also fought at sea with flotillas of small canoes and large double-hulled canoes. Allied fleets could number more than a hundred canoes and thousands of men (Clunie 1977: 21-23). Usually the tactic would be to ram into the other canoe, possibly after having launched waves of

arrows or bullets at it, and then try to finish the enemy off by hand.

There were of course difficulties in launching *valu-ni-tu*, large-scale wars. Gathering together the different contingents of warriors, which sometimes numbered thousands, and possibly transporting them on board canoes across the sea to the enemy fort demanded high levels of co-ordination. Supplying food provisions for the army was a problem and limited the duration of campaigns to a few weeks or months. Plundering enemy fields, fruit trees and coconut palms was a means to enlarge the army's own supplies, in case the enemy had not already harvested as much as possible and destroyed the rest. Much skirmishing took place between foraging parties from both sides, and warriors began to eat dogs, cats, lizards and snakes to display toughness and endurance (Clunie 1977: 31).

On top of this, forging an alliance in the first place was a major problem, and because of this intensive diplomatic communications followed a decision to launch a major war as described above. While the support of qali villages and chiefdoms could usually be relied upon, those that had bati status were more independent and their support had to be secured and further allies sought. Even though allegiance was secured before engaging in war, the situation was never totally in hand. Treachery and deception played a major part in Fijian warfare, and villages and chiefdoms might change sides during the parade or even during the fighting. Allies might pretend to support one side but launch a surprise attack and kill the *turaga-ni-valu* and his warriors, or join the siege of a fort just to lead the attackers into ambush. As a result, contingents of warriors were always alert and quick to retreat at the cry of treachery. The sources on Fiji in the first half of the 19th century are full of accounts of the plots, treacheries and deceits employed during war. The position of power that chiefs and their generals had was therefore not unchallenged or at any time totally secure, which brings us back to the question of the position of chiefs in general.

The power of chiefs

Generally, there were six classes of people in eastern societies: 1. kings and queens, 2. chiefs of large islands or districts, 3. chiefs of towns, priests and *mata-ni-vanua* (heralds), 4. distinguished warriors of low birth and chiefs of clans (*mataqali*), 5. common people and, finally, 6. slaves by war (*kaisi*) (Williams 1982[1858]: 32). More particularly, the *yavusa* of Bau, situated on a small island of the same name, was led by the *Roko Tui Bau* (Sacred lord of Bau) whose person was sacred, never took part in wars, and concerned himself with spiritual matters. The temporal chief, the *Vunivalu* (root of war), was commander-in-chief and executive sovereign, who almost had as many privileges as the *Roko Tui Bau*. Naulivou, who successfully engaged in a series of conquests to enhance Bau's position in the late 18th and early 19th century, curtailed the powers of the *Roko Tui Bau* in the process (Routledge 1985: 43). The *Roko Tui* and the *Vunivalu* were from different *mataqalis*, just as the *mata ni vanua* (heralds) and the *Bete* (priests), who were next in the hierarchy. The *mataqali* Lasakau provided the warriors, while *mataqali* Butoni were fishermen (Thomson 1968[1908]: 61-62). Just below the *Roko Tui* and the *Vunivalu* were the chiefs of *bati* and *qali* villages, while the *kaisi* (slave) villages were at the absolute bottom. The latter provided the chief with the daily food and were employed to maintain houses and boats or to carry out whatever tasks were needed. The competing confederacy of Rewa similarly had a *Roko Tui* and a *Vunivalu*, but the former had not relinquished its executive powers, nor had the latter reduced these to secondary status. Below these two ruling families were six *mataqali*, which supplied the heralds, *mata-ni-vanua*, and had no obligations toward the chiefs other than leading the army into battle and making ambushes (ibid.: 366). Next in the hierarchy were chiefs of aligned yavusa, some of which were qali or bati. At the bottom again were the *kaisi* villages, of which one chief, Cakonauto, claimed to have seventy which provided him and his village with provisions on a daily basis. Cakonauto would occasionally amuse himself by shooting at the people of Drekete and Noco when they came with food. Samuel Jackson gives the following account:

...the reason he [Cakonauto] gave for this almost unaccountable cruelty to the slave people was, that they were a very rebellious and formidable people some forty or fifty years ago, and almost got the upper hand of Rewa, and that Rewa had rather a hard struggle to extricate herself from the threatening and disagreeable situation she was placed in. Of all the slave people Drekete was the worst and the 'vura' (or ringleader); and at the time of their hardest struggle, the Rewa people swore that, if they did succeed in quelling and get the upper

hand of them they would for ever after keep them in the most abject slavery. (Jackson 1853: 463, my insertion)

As mentioned above, the person of a chief was sacred in Bau and Rewa, and some chiefs even claimed divine origins. They were thought to impart a degree of sacredness on everything they touched and were surrounded by a considerable amount of tabus – i.e., special rules of behaviour sanctioned by religion. They ate alone, sometimes fed by attendants since they were not allowed to touch food, and commoners were expected to show profound deference when meeting a chief. A special royal dialect existed in which 'not a member of a chief's body, or the commonest acts of his life, are mentioned in ordinary phraseology, but are all hyperbolized' (Williams 1982[1858]: 37). The entourage of a chief consisted of two or three personal attendants, two or three priests and a number of wives (ibid.: 24-26). Amongst the men of official importance were the *mata-ni-vanua*, heralds, whose function it was to bring messages from the chief and to associate chiefdoms, each of which would have their own heralds stationed. Bau would thus have a whole group of heralds from other chiefdoms present as a kind of diplomatic corps. Mnemonic sticks or reeds were employed to deliver the exact wording of messages (ibid.: 27).

Chiefs were chosen from the chiefly *mataqali*, in principle on the basis of seniority and primogeniture so that the eldest brother would replace a dead chief or, if no brothers were living, the eldest son. This principle, however, was complicated by the fact that rank was inherited from the mother and most often would reflect the kind of support a son would be able to get from his mother's clan (*yavusa*). Also of importance was the acknowledgement of aptness, which would be judged according to the personal abilities of the candidate. Aptness must be assumed to have included a combination of political acumen in manoeuvring safely within the troubled waters of the intrigues of the chiefly court and prowess in warfare to assure and enhance the standing of the chiefdom. The position of chief was sacred almost to the level of the lower gods, and he had absolute power over commoners and slaves including choice over life and death and eviction of whole villages from the land they occupied. Land rights were basically of three kinds: the most exclusive was the right to the house site which was vested in the occupant of the house and which usually was not terminated by the death of the head of the household, but passed one of his sons. Rights to cultivated lands were held by extended families, *tokatoka*, subject to the provision of services assigned to the *tokatoka's mataqali* and payment of an annual tribute, *sevu* or first fruits, to the chief who granted the right to use it and the provision of services. Services and tribute were just as much recognition of the chief's authority as of his superior rights to land, and unmet obligations would be met by eviction (France 1969: 15). The group would then seek the protection of another chief to whom they would have to provide services and tribute. Chiefs could give personal land rights as gifts to acknowledge bravery in battle, for defending a corpse in battle (since it was the ultimate humiliation to the person and his family to be eaten by the enemy), or for having had the role of mourner of a chief. Basically the extent of the land to which a chief could grant rights corresponded to the area he could defend. Powerful chiefs could thus accommodate more escaping, evicted and defecting groups than lesser ones and hence by granting rights of cultivation they accumulated more tribute and more services. Tribute would either be presented at the chief's place or collected by the chief, who would come with his entourage, which would then have to be hosted (Thomson 1968[1908]: 59). Most of the tribute would be redistributed and a good chief was generally considered to be a generous chief.

The considerable powers of a chief were however restricted by several factors. Since there were usually several candidates for the title, a chief continuously had to face the possibility of treachery from his rivals and the possibility of small groups or even *vanua* defecting to other *matanitu*. Chiefly councils furthermore restricted the authority of the chief. Decisions to make war, alliances or judgements of law would be made by councils that would not allow the chief to arbitrarily push through decisions. Finally, the predicament of the chief was enhanced by the institution of *vasu* (literally, 'nephew'), according to which a man could appropriate anything he might want from his maternal uncle or those under his uncle's power except the wives, home and land of a chief. *Vasus* provided, according to Thomas Williams 'the high-pressure power of Fijian despotism' (Williams 1982[1858]: 34); ultimately no chief whose sister had a son was his own master. Under the direction of a chief a *vasu* was a powerful political instrument in Fiji, and thus chiefs sought to marry high-ranking women of other chiefdoms since through their sons they would therefore have considerable claims on that chiefdom.

Conquest and social change in Fiji

Warfare in Fiji in the first half of the 19th century depended upon and was part of the prevailing social relations outlined above. It was an avenue to greater power for ambitious chiefs, part of male identity, and its enduring threat of war led to the fortification of villages. Though some clans were given the role of warriors, as in the case of Rewa, where the six *mataqali* of Sauturaga had no other obligations than to lead warriors in war, the army consisted of all men capable of carrying weapons. Armies were raised on the basis of alliance or the status of *bati* or *qali* and organised on a tribal basis under the command of the *Vunivalu*. The question is, however, the role of war in the transformation of social structure. There seems to have been a kind of self-reinforcing dynamic that could effectuate a conglomeration of single villages and chiefdoms into larger entities, since a chief could enhance his annual tribute and amount of services by subjugating other villages and chiefdoms. The more villages he could hold in a *qali* relation or defeat totally and thus reduce to the status of slaves, the more whale teeth, food, services and women he would be able to claim. These could be used to commission the production of more weapons and more double-hulled canoes (*drua*) to be employed or given away to allies, which likewise could be raised in larger numbers with more wealth. Thus bigger armies could be raised to conquer yet more villages. However, two observations caution against embracing a 'chiefs made war and war made state' position. Firstly, alliances and larger political entities were not only the result of conquest, but also of the conscious use of *vasu* relationships. Also, the successful chiefs of rising Bau, Naulivou and Cakobau required sons of tributary and friendly chiefs to live on the island of Bau to acquire a distinct Bauan point of view (Routledge 1985: 43). While there is no doubt that conquest played a major part in the formation of *matanitu*, it had to be supported by political alliances through marriages, reciprocity of gifts and mutual support. Secondly, while conquest was an important dynamic in the political life of the *matanitu* of Fiji, it did transform basic social relations. Depending on the definition we may consider confederacies like

Bau and Rewa to be either complex chiefdoms or inchoate early states.[10]

Carneiro argues in an examination of the case of Fiji that 'From the point of political evolution, the salient feature of chiefdom-level warfare is, of course that it led to conquest' (Carneiro 1990: 207). However, according to Carneiro, states were not formed in Fiji because the motive for war was not only conquest but also revenge. It seems questionable to base the cause of evolution on the motives of actors, and Carneiro wisely strengthens his argument by arguing that the wrong motives led to consequences that did not imply evolution or change in political organisation:

The aim of much of the fighting in Fiji ... was the destruction, even annihilation, of the enemy, not only his person but his property. Thus, chiefdom-level warfare often exhibits the 'regressive' element of destruction and dispersion rather than the 'progressive' one of amalgamation and consolidation. (Carneiro 1990: 207)

Warfare in Fiji did however lead to amalgamation and consolidation. Carneiro argues that there was not enough of it, which is nevertheless a quantitative argument. The crux of state formation, however, is not quantitative aggrandisement of power, but qualitative change in political organisation. He does not explain how less revenge and more amalgamation and consolidation would amount to this.

War and conquest in Hawaii in the late 18th century

Having sailed past the Fiji islands in 1777, Captain James Cook arrived half a year later, in January 1778, at a group of islands that he first named after Lord Sandwich and which later would be called the Hawaiian Islands.[11] Cook's journal of his third and fatal journey into the Pacific Ocean was published in 1784, and a brief comment mentioned the favourable prizes some of his seamen had obtained in China for a few furs from North America. Accordingly, in 1786 the second European ship to visit the islands inaugurated a fur trade between North America and China that would last almost a decade. The Hawaiian Islands became important for the provisioning of salt pork, yams, fresh food and firewood, which the islanders traded for iron tools and before long also for muskets and cannons. When Captain George Vancouver, who was with Captain Cook in 1778, came back to Hawaii in 1790, he remarked the following:

The alteration which has taken place in the several governments of these islands since their first discovery by Captain Cook, has arisen from incessant war, instigated both at home and abroad by ambitious and enterprizing chieftains; which the commerce for European arms and ammunition cannot fail of encouraging to the most deplorable extent. (Vancouver 1984: 477)

During the 1790s, muskets and cannons spread in the islands, local islanders worked on the trading ships, and the Hawaiian chiefs started adopting foreigners into their courts to train their men in the use of muskets and cannons (Kuykendall 1968: chs. 2 and 3; Sahlins 1992: 36-54). Though arms and cannons quickly became important in the quest for power by different chiefs they did not substantially alter the form of warfare (Howe 1984: 158). However, in 1795 the chief of Hawaii, Kamehameha, conquered the neighbouring chiefdom, which included the islands of Maui, Oahu, Molokai, Lanai and Kahoolawe. Kamehameha could now dominate the trade with Europeans and establish privileged access to weapons, which led to the formation of the first polity encompassing all the Hawaiian islands under the rule of Kamehameha. The privileged access to outside sources was a major factor in Kamehameha's consolidation of his rule and the transformation of political structures, and in the following the focus is therefore on war and social structure before 1795.

In the 1780s, the eight Hawaiian Islands (Hawaii, Maui, Oahu, Kauai, Molokai, Lanai, Niihau and Kahoolawe) were divided into four major polities that fought for supremacy under the leadership of their respective high chiefs, ali'i nui. Each polity was in turn subdivided into districts, moku, administered by lesser chiefs, ali'i ai moku, appointed by the ali'i nui. Districts were then subdivided into sub-districts, ahuapua'a, which were led by the konihiki appointed by the ali'i ai moku. There was a distinct difference between the class of ali'i, from which the chiefs and their administrative nobility came, and the class of commoners who cultivated the land, the maka'ainana. The commoners were not organised into kinship units, but consisted of an 'assemblage of ego-based bilateral kindred (interrelated to varying degrees by marriage) rather than corporate kinship units' (Hommon 1986: 57. See also Sahlins 1992: 31-33).

Upon the conquest or succession of a new chief, new chiefs of the districts and sub-districts were appointed. Whereas the class of *ali'i* were cosmopolitan, or rather pan-archipelagic, in their orientation, the commoners lived in self-sufficient, endogamous groups within the sub-districts, though they might migrate if their chiefs became too oppressive (Malo 1903: 92). In addition to the *ali'i* and *maka'ainana* there was a group of people without status, the *kauwa*, who were slaves, probably either because they had broken important prohibitions or as the result of conquest (Ellis 1979: 106; Kamakau 1964: 8; Malo 1903: 96-101). This social structure appears to have been relatively homogenous throughout the islands and to have been established centuries before contact.[12] Archaeology and mythical history suggest that a cleavage between commoner, *maka'ainana*, and noble, *ali'i*, had developed by 1600 (Hommon 1986; Kirch 1985).

At the time when Western ships made contact with Hawaii, war was frequent (Fornander 1880; Kamakau 1992: 230). Warfare would typically be occasioned either by the death of the high chief and the ensuing quest for power between successors and their allies or by ambitious chiefs attempting to enlarge their domain. According to Hawaiian myths, the attainment of political power through warfare goes, as far back as the 16th century and the high chief of Hawaii, 'Umi (Hommon 1986: 67). Patrick Kirch (1985) writes as follows:

The political history not only of Hawaii but of the other major islands as well, during the final two centuries prior to European intrusion, was one of constant attempts by ruling chiefs to extend their domains through conquest and annexation of lands. Campaigns extended beyond the borders of individual islands, and Moloka'i was a frequent prize of both the Maui and O'ahu chiefdoms, ... The expansion of a chiefdom was generally short-lived, followed within a generation or two by collapse and retrenchment, frequently precipitated by usurpation of the paramountship by a junior collateral able to enlist the aid of other mal-content chiefs and warriors. Thus the late political history of the Islands was cyclical. (Kirch 1985: 307)

The wars

The chiefs declared and led the great wars of ancient Hawaii. There was no standing army or separate group of men trained exclusively for war. Instead, all able-bodied men were potential warriors that the chiefs could call upon. Wars were fought with slings, spears, javelins, daggers and tripping devices, and while clubs were used they did not have the same significance or high status as in the rest of Polynesia and Fiji. Bows and arrows were not used (Hiroa 1964). In general, all men learned fighting skills when growing up through stone-throwing, slinging, wrestling, throwing the javelin, and engaging in sham battles. Hawaiian warriors seem to have developed admirable skills in throwing and dodging stones and spears. Commoners learned the skills of war from each other, while the men of the *ali'i* were supposed to be especially capable warriors: 'All the chiefs in the government were trained in military exercises until they had attained greater skill than was possessed by any of the common people' (Malo 1903: 261). Included in the chief's court were specialists, *kalaimoku*, who 'were well versed in the principles of warfare. They knew how to set a battle in order, how to conduct it aright, how to adapt the order of battle to the ground' (Malo 1903: 259).

There were two great reasons why a kalaimoku had superior ability as a councillor to others. In the first place, they were instructed in the traditional wisdom of former kalaimokus, and in the second place their whole lives were spent with kings. When one king died, they lived with his successor until his death, and so on. Thus they became well acquainted with the methods adopted by different kings, also with those used by the kings of ancient times. (Malo 1903: 261)

Chiefs and their courts were highly skilled in warfare and the male commoners constituted 'in times of peace a reserve army, each individual of which kept his weapons in readiness in his house' (Emory 1970: 233). The Hawaiian polities seem to a have been highly effective in mobilising their troops and conducting war. They might transport themselves on foot across an island or in small and large double-hulled canoes that could reach all the other islands in the group. The passage between Oahu and Kauai, however, had such strong a current and the sea there was so rough that Kamehameha's first attempt, in 1796, to subjugate this last polity under his rule failed (Kuykendall 1968: 47-48). By 1810, however, Kamehameha's power and naval capabilities were so great, enhanced also by the acquisition of European ships, and the chief of Kauai (and Niihau), that Kaumualii accepted Kamehameha as his superior without conquest.

Sham battles were part of being militarily alert and were frequently held. George Vancouver witnessed such a battle on Monday the 4th of March 1793:

The party consisted of about an hundred and fifty men armed with spears; these were divided into three parties nearly in equal numbers, two were placed at a little distance from each other; that on our right was to represent the armies of *Teteeree* and *Taio* [Kahekili, chief of Maui, and his younger brother Kaeo. Two of Kamehameha's main opponents at the time], that on the left the army of *Tamaahmaah* [Kamehameha]. Their spears on this occasion were blunt pointed sticks, ... the battle began by throwing their sham spears at each other. These were parried in most instances with great dexterity, but such as were thrown with effect produced contusions and wounds, which, through fortunately of no dangerous tendency, were yet very considerable, and it was admirable to observe the great good humour and evenness of temper that was preserved by those who were thus injured. This battle was a mere skirmish, neither party being supported, nor advancing in any order but such as the fancy of the individuals directed. (Vancouver 1984: 832-33, my insertions)

Warriors that wanted to prove their military skills would walk up in front of the line of opponents and challenge them. With a spear in their left hand they would parry some of the spears thrown at them, while with their right hand they would catch spears in flight and return them. Vancouver was much impressed:

In this exercise no one seemed to excel his Owhyhean majesty [Kamehameha], who entered the lists for a short time, and defended himself with the greatest dexterity, much to our surprize and admiration; in one instance particularly, against six spears that were hurled at him nearly at the same instant; three he caught as they were flying, with one hand, two he broke by parrying them with his spear in the other, and the sixth, by a trifling inclination of his body, passed harmless. (Vancouver 1984: 833)

After a short intermission, the first skirmish-like battle was followed by another, which was quite different:

The warriers who were armed with the *pallaloos* [long spears], now advanced with a considerable degree of order, and a scene of very different exploits commenced; presenting, in comparison to what before had been exhibited, a wonderful degree of improved knowledge in military evolutions. This body of men, composing several ranks, formed in close and regular

order, constituted a firm and compact phalanx, which in actual service, I was informed, was not easily to be broken. (Vancouver 1984: 834)

In consultancy with the priests the chief decided upon actual warfare and war strategies. The chief sent out messengers to the districts and villages under the chief's command to summon a number of men in proportion to the kind of war at hand. Rising to the task of being warriors, men would bring their own weapons, candlenuts for torches, calabashes for water, dried fish, or other portable provisions. Sometimes, another officer from the king's court would check whether the men summoned came along: 'if he found any lingering behind who ought to have been with the army, he cut or slit one of their ears, tied a rope around their body, and in this manner led them to the camp' (Ellis 1979: 99). The shame and humiliation of staying at home, however, often made it unnecessary to send this officer around. They sometimes launched attacks after only brief preparations, while at other times they spent up to a year building war canoes, crafting weapons and securing provisions (Kamakau 1992: 28, 74). Meanwhile, spies could be sent out to assess the strength of the enemy (Kamakau 1992: 55ff).

The form of warfare could range from surprise attacks upon the unsuspecting opponent to formalised battles where time, place and mode of fighting had been agreed upon previously (Handy 1970: 234). The former most likely occurred in quests for more land, whereas the latter would seem to war as the last means in disputes arising from slanders, minor injustices and revenge that were solved through negotiations. Conventional warfare took place when two renowned warriors, one from each side, would start the battle by fighting between themselves while the rest of the army looked on from the sidelines (Ellis 1979: 104). Another form of conventional warfare was the *kukulu*, whereby the two armies would face each other in two single lines and await the cry for the battle to begin (Malo 1903: 259, 268). In less conventional battles, the armies would be organised into different formations suited to the landscape in which the fighting took place. In the *kahului* formation, soldiers formed a crescent, whereas the army was organised into the *makawalu* formation – groups of soldiers spread irregularly in the terrain – when the grounds was level with scrub (see Malo 1903: 259-60, 268-69; Emory 1970: 237-38).

Commoners were led by their own chief, who in turn took orders from a superior chief, forming a hierarchy that replicated the hierarchy of the social structure, with the paramount chief, the *ali'i nui*, at the apex (Ellis 1979: 101).

The nobles were supposed to be especially skilled in warfare, and great warriors could earn fame and prominence. Kamakau 1992 describes a celebrated warrior:

...a chief from Kolokai [was] celebrated for his skill and strength. He would toss an antagonist in mid-air and tear at him, so that he was helplessly mangled when he reached the earth. Ka-makau-ki'i, a famous fighter from Hawaii, was a swift runner, so strong and sure with his spear thrust or other death-dealing weapon that he never missed a hair of even a louse on a man's head, and so quick was his grasp that he could catch a bird before it took wing. (Kamakau 1992: 80, my insertion)

One example of the warrior ethos was the defiance of eight famous warriors of Oahu who on their own went to counter a surprise attack by Kahekili, chief of Maui, on their island: 'It was a chivalrous undertaking, a forlorn hope ... but fully within the spirit of the time for personal valour, audacity, and total disregard of consequences' (Fornander 1880: 223).

Wars would sometimes last only one day, with the armies getting no further than marching up in front of each other before a truce was achieved, or they could consist of skirmishes lasting months (Fornander 1880: 139). At other times wars would develop into battles lasting several days (five seems the typical number. I'i 1959: 70; Fornander 1880: 139) or into sieges of the fortifications into which the enemy had retreated after having lost a battle. Fortifications could either be natural points of defence or constructed forts, and some were renown for being impregnable (e.g. Fornander 1880: 98). Sieges might be long and unsuccessful (Fornander 1880: 147), and in some cases victory was only achieved by cutting off water supplies (I'i 1959: 116). Victorious armies would kill enemy warriors, women, old people and children. The conquerors made the survivors their slaves. If the defeated chief was captured alive the conquering chief might sacrifice him to the gods as a final confirmation of the new holder of the office of power (see Valeri 1985). Chiefs of the winning party were appointed as holders of the offices of chief of district and sub-districts and in this way land was redistributed amongst the winning party.

Warring chiefs of Hawaii

War in ancient Hawaii was an important means for getting power and a social field in which *ali'i* and *maka'ainana* could achieve fame as great warriors. The economy of these societies and the rules of succession to office after the death of a paramount chief made warfare an attractive option for chiefs and their rivals.

Ideally, succession to the office of *ali'i nui* was determined by rank, which was inherited from the father as well as the mother. Determination of rank, however, was often difficult since a chief might have several wives, and the eldest son of the first wife might thus have a lower rank than a younger brother whose mother was of higher rank. The system encouraged marriage to close kin, since the highest possible rank within a domain would be achieved by a child of a son and daughter of the *ali'i nui* and his highest-ranking wife. Slightly lower rank would result from marriage between a brother and half-sister or between cousins (Malo 1903: 80-82; Kamakau 1964: 4-5). Genealogical relationships were much entangled and as a result assessing rank was often extremely difficult, leading to contestations between several equal or near-equal successors to the office of chief, so a war between contenders who sought the support of factions within the polity often followed the death of a chief (Valeri 1985: 159-60). The stakes could be high since the new *ali'i nui* would re-appoint the lower offices of chief and the status quo might thus be in danger. If a group of chiefs felt themselves at a disadvantage with the new division of the land, they could contact a rival to try to get him to lead a revolt and have him installed as paramount chief. Accordingly, for example, the chief of Hawaii, Kalaiopuu, appointed his son Kiwalao as his successor, but a group of chiefs disappointed by Kiwalao's redistribution of land succeeded in having Kamehameha, cousin to Kiwaloa, lead a successful revolt and become the new paramount chief of Hawaii (Howe 1984: 155; Fornander 1880: 299-310). So the *ali'i nui* derived his position from a combination of rank and the ability to obtain support from the most powerful aggregation of interest groups. The latter claim would often have to be proven through internal warfare and subsequent conquest of the *ali'i nui* position. Once in power, a chief would have to look out for and maybe kill potential usurpers by such methods as poisoning (Fornander 1880: 142, 218).

Ultimately, the *ali'i nu* owned the land, granting the lesser chiefs rights to a district of land on the condition of an annual tribute. In the same manner, these chiefs in turn subdivided their districts off to lesser chiefs, who eventually granted the *maka'ainana* land to cultivate. The annual tribute on all levels would be paid in connection with the Makahiki festival in the form of first fruits to the immediate superior chief, who on his part would present a tribute to his superior (Sahlins 1992: 25-27; Valeri 1985: 156-57). On the lower levels of this hierarchy most of the tribute would consist of taro and other daily food staples, but on the higher levels, where tribute would come from more distant lands, the tribute would be paid in the form of pigs, dogs, salted fish, bark cloth, canoes, nets, mats and feathers (Sahlins 1992: 28). Either because they were more storable and incorporated more labour or because of their symbolic value. The accumulated goods of tribute would then be redistributed, with the greatest shares going to those closest to the *ali'i nui*, while the *maka'ainana* would get nothing (Valeri 1985: 204). In ancient Hawaii, the chiefly economy was thus characterised by differential redistribution.

An *ali'i nui* basically had three ways in which to enhance his status. Firstly, he could marry as many high-ranking wives as possible, thereby ensuring that his successor and children have a high level of sanctity. In this way the *mana* of his successor and children would be high and therefore they would deserve a high degree of reverence from lesser chiefs and *maka-ainana* in particular. Social life was restricted by a system of religious prohibitions, *kapus*, that especially related to relations between men and women and entailed rules of abject behaviour towards the paramount chief (I'i 1959: 58-59; Malo 1903: 50, 83; Valeri 1985: 90-95). The position of the chief was supported by religion, which, in its royalist version, depicted the paramount chief as the embodiment of the society as a whole and as the apex of a hierarchy that gradually approximated the divine status of the gods. The position was enacted in the *luakini* ritual, in which the king sacrificed a human, who might be a defeated chief or the first killed in a battle (Valeri 1985). The strength of a chief's sanctity was not just a given essence, but implied efficacy (ibid.: 90-105), and thus had to be enacted and shown through the volume of tribute and valuables, the building of temples, or conquest.

Secondly, a chief could try to raise the productivity of his polity to the highest possible limits. Depending on the strategy of the chief this could become manifest either in larger amounts of tribute presented at the *makahini* festival and the ensuing higher amounts of valuables ultimately presented to him as *ali'i nui* or it could become manifest in the building of temples, which required large numbers, sometimes thousands, of people investing their labour in construction while being fed by those still remaining in agriculture. In some cases such a demonstration of power would be enough to discourage enemies to attack. 'It can thus be said that kings fight by building temples and not only by arms' (ibid.: 235).

The strategy of productivity, however, had the inherent danger of alienating the commoner people, the *maka ai'nana*. An *ali'i nui* had to balance his enactment of his absolute powers with the risk of rebellion by lesser chiefs or commoners if they considered his rule too oppressive or exploitative. According to Marshal Sahlins, the threshold of popular resentment was rather low, and since the commoners were neither bound to the soil nor consanguineally connected to their chiefs, they could chose to migrate or revolt (Sahlins 1992: 30). David Malo lists the chiefs who were killed by the commoners (Malo 1903: 258), and Kamakau (1992), for example, tells the story of the two sons of the ancient king of Hawaii, 'Umi-a-Liloa, who inherited the Hilo and Kona parts of the island, respectively. In the end, however, the younger brother, Keawa-nui-a-'Umi, became ruler of all of Hawaii, since the elder brother, Ke-li'i-o-kaloa, exploited his people too harshly.

[Ke-li'i-o-kaloa] deserted the advice of the wise, he paid attention to that of fools, thus forsaking the teachings of his father and the learned men of his kingdoms. He deserted the god and oppressed the people. These were his oppressive deeds: he seized the property of the chiefs and that of the *konohiki* of the chiefs, the food of the commoners, their pigs, dogs, chickens, and other property. The coconut trees that were planted were hewn down, so were the people's *kou* trees. Their canoes and fish were seized; and people were compelled to do burdensome tasks such as diving for *'ina* sea urchins, *wana* sea urchins, and sea weeds at night. Many were the oppressive deeds committed by this chief Ke-li'i-okaloa. Therefore some of the chiefs and commoners went to Hilo, to Keawa-nui-a-'Umi, and offered him the kingdom of Kona. (Kamakau 1992: 35)

The third strategy for enhancing status was to acquire more (cultivated) land, which meant more tribute and more prestigious valuables, red feathers and bark cloth, and thus higher status for the *ali'i nui* and, to a lesser degree, his court. So there was a premium on acquiring more land, which could only be achieved by conquest, since all of the Hawaiian islands were already divided into different polities, and successfully conducted conquest simultaneously meant more tribute and the diminishing of the status of the defeated chief. This avenue to greater status and power had its self-reinforcing dynamic, since the ability to build double-hulled canoes, commission weapons, and raise big armies increased with the number of people and amount of tribute and services the *ali'i nui* could command. Warfare as a strategy to keep and enlarge power can be neatly summarised through the example given by Fornander (1880) of the ancient chief of Hawaii, Kalaniopuu:

… at the usual redistribution of lands at his accession, apparently all were satisfied or none dared to resist. For several years afterwards he occupied himself diligently in reorganising the affairs of the state, augmenting the warlike resources of the island, building war-canoes, collecting arms, &c, and his own and the neighbouring islands enjoyed a season of rest from foreign and domestic strife and warfare.

But *Kalaniopuu* was ambitious for fame in his island world by warlike exploits and by enlarging his domain with the acquisition of neighbouring territory. Possibly also he may have been moved by reasons of policy, such as finding occupation abroad for the yong and restless chiefs with who every district abounded. Suddenly, therefore, he concentrated his forces and war-canoes at Kohala, and, without previous rupture of peace or declaration of war, he invaded Maui… (Fornander 1880: 146)

As with the other strategies for increasing status, war also entailed dangers. In addition to the obvious risk of badly losing a war campaign, a warring chief risked rebellion at home while conquering abroad. Accordingly, after having conquered Maui in 1790, Kamehameha had to return to Hawaii quickly and abandon further conquest because of a revolt against him (Fornander 1880: 235-41). To be a successful chief of ancient Hawaii thus required careful and skilful manoeuvring between different factions amongst the nobles, pushing levels of tribute to high levels of productivity without causing a rebellion, and enhancing *mana* through marriage strategies, which also might pave the way for alliances.

In sum, the paramount chief of the Hawaiian Islands was a position that was religiously sanctioned, partly inherited, and at the apex of a differentiated redistributive economy. The might of the chief grew with the domain over which he could claim paramountcy, since more lands meant more tribute, more status, and the ability to enlist more labour. The power of the chief might be invested in the building of *luakini* temples, which in some instances involved thousands of people that had to be fed and organised. In other instances, however, the accumulated goods would be used to build canoes, request more weapons, and raise armies. These could then be employed to attack and hopefully conquer more land and people, which in turn would enable yet another round of fighting through actual conquest or temple building.

War and conquest in Fiji and Hawaii

In a comparative perspective, Fiji and Hawaii are quite similar in various respects. In Fiji as well as in Hawaii, chiefs made war frequently and, according to archaeological evidence and oral history, had been doing so for centuries. The position of chief was religiously sanctioned and occupied an intermediate place between gods and humans. Access to the position of chief was gained through a combination of inherited rank (implying degrees of sanctity), military skills, and the ability to gain support from other groups. In Fiji and Hawaii, the warring chiefs had a small retinue of warriors based mainly on close kin and relied upon alliances with other chiefs and upon subordinate chiefs when they wanted to raise armies. Though motivations for war might vary from perceived sleights or questions of honour to outright quests for chiefly paramountcy, large-scale warfare took place and required the co-ordination and command of thousands of warriors, access to large canoes and weapons, and considerable amounts of food.

However, there were also distinct differences. In war, Fijians relied on bows and arrows for long-distance hostility and upon clubs for close combat, while Hawaiians used spears and javelins for combat at a distance and swords with shark teeth for close encounters. Fijian warfare seemed to centre on fortifications and sieges; Fijians had developed considerable skills in building and organising their defence through fortifications and, inversely, in conducting sieges. In Hawaii, outright battles were more common, and Hawaiians had developed different battle formations for encounters in the open. The closed kind of phalanxes that Vancouver witnessed does not seem to have had any parallel in Fiji, just as the significant role of treachery and shifting alliances during battles and sieges does not seem to have been of significance in Hawaii. The prevalence of battles instead of sieges and the ability to form different battle formations seems to indicate that war in Hawaii was based on more stable and complex military organisation.

This difference in military organisation seems in turn to be based on a difference in social organisation as such. While the distinction between the chiefly elite and commoners was distinct and implied the absolute, despotic right of chiefs over commoners in both Fiji and Hawaii, the internal political relations between various chiefs were different. Paramount chiefs in Fiji could through intrigue, conspiracy and conquest influence who was to acquire lesser chiefly positions, but they could not by right of their position decide this by public decree as in Hawaii. The *turaga ni matanitu* of Fiji could demand tribute and request warriors and services from subjugated *qali* villages, but apparently not from the same position as the *ali'i ai mokus* of Hawaii. The ability of Hawaiian chiefs to commission the construction of irrigation and fishponds, and thus invest their power in economic relations, also does not seem to have had an equivalent in Fiji. Furthermore, while Hawaiian religion had developed a hierarchy amongst its myriad of gods, at the pinnacle of which were a few central gods who were the object of central rituals and temples, Fijian religion seems to have been less focused. In contrast to Hawaii, in Fiji there was no overall ritual including all groups within the territory of the paramount chiefs where tribute was paid. Buildings for worship existed in both cases, though the Fijian *Bure Kalous* propably could not match the Hawaiian *lukakinis* in degree of work-investment. The difference in military organisation between Fiji and Hawaii thus seems related to a parallel difference in the degree of centralisation.

In both Fiji and Hawaii, a self-reinforcing dynamic existed in which the successful conquest of another polity meant access to more resources in the form of tribute, people, food and wealth. This in turn could be invested in raising yet more warriors and weapons for new conquests, and in this way larger political entities were built, leading to the formation of large polities in Fiji. In these confederacies, however, political relations were inherently unstable since internal revolt or defection might challenge the power of the leading chief. In neither case was the power of the chief based upon military prowess or success alone. The right alliance between chiefly families in cases of marriage was necessary to procure a suitable, high-ranking chief, who had to achieve a leading position within the field of qualified heirs or usurpers through military prowess and political acumen to forge alliances. There is no evidence of a non-chiefly war leader who revolts against his chief by using his command of military power. A basic difficulty of such a task was that armies were mobilised from various sections of the political entity and only forged into a unit as long as the political hierarchy was stable. If a paramount chief was challenged, the army was most likely to break down into smaller unities according to political allegiance.

In a comparative perspective, a higher level of organised warfare in Hawaii seems to parallel higher levels of centralisation in politics, religion and economy. It is a matter of definitional choice whether this difference should be considered one of degree or of quality. The travel accounts of early Europeans speak of 'chiefs' and 'kings' in both cases, and while more recent anthropological literature tends to regard the groups in Fiji as 'chiefdoms', those of Hawaii are referred to as 'chiefdoms' (e.g., Earle 1997; Cordy 1974), 'early states' (e.g., Claessen and Skalnik 1978) or something in between (Hommon 1986). Instead of trying to solve the question of definition, it seems more fruitful to pay attention to the role of war in reproducing or changing social organisation. In both cases, war was obviously a means to achieve and enlarge political power. While this also implied a centralisation of rule as long as the conqueror was successful, there is no evidence that further stratification, more bureaucratisation, or a higher degree of monopolised violence was the result. Rather, in Fiji and Hawaii, war did not surpass existing social structures, but might be regarded as part of it, since the ability to launch wars was integral to the power of political leaders.

The central problem was that while war could enlarge political power, it took more to use this power to transform existing structures. Naulivou and Cakobau of Bau, Fiji, consciously used the *vasu* relationship to convert their military success into political power, just as the *ali'i nui* of Hawaii acquired high-ranking wives to enhance their sons' *mana*. However, in both cases these attempts at conversion were soon neutralised by competing, equal-ranking claims to *vasu* or *mana*. The result was several competing successors whose rivalries often reduced the built-up power base.

Chiefly rule in Fiji and Hawaii relied upon different forms of power: economic, political, ideological and military. While war and conquest offered a self-reinforcing dynamic through which chiefs could acquire more power, more was required to stabilise and make this a lasting, qualitative difference. Instead we must envisage a dynamic combination of existing historical conditions, actions of one or more actors, and a reconfiguration of forms of power to make warring leaders, deliberately or unintentionally, come out of warfare as heads of state instead of warring chiefs. The different fates of Fiji's Cakobau and Hawaii's Kamehameha are interesting in this respect. Kamehameha of Hawaii won a decisive battle against Oahu in 1795. Under normal circumstances, the shifting fates of Hawaiian chiefs might also have been Kamehameha's, and his supreme dominance only temporary. Kamehameha, however, could use his temporary supremacy to secure a dominance of trade in firearms and other goods with Europeans and thus further enhance his power and advantage over rival chiefs. Furthermore, his heir, Kamehameha II, under the strong influence of his mother and Kamehameha's favorite wife, Kaahumanu, abolished the *kapu* system, which amongst other things meant that rivals could not legitimise claims to his position by reference to *mana*. The fate of Cakobau in Fiji was quite different. He gained temporary supremacy over Rewa in 1845 when the town of the latter was burned down and a chief loyal to Bau, Cakonauto, was subsequently installed in Rewa. Cakonauto, however, died in 1851, and in defiance of Bau, the Rewans installed Qaraniqio, whose aim was a new, independent and powerful Rewa. The Bauans staying at Rewa were ejected and an important *bati* chiefdom, Kaba, defected to Rewa. By 1855, Cakobau and Bau were in severe difficulty, and only by converting to Christianity and thereby securing the support of the Tongan King and his army could Cakobau remain the dominant chief of Fiji. In a position of dominance, Kamehameha could engage upon major political reform and further institutionalise his power after 1795; however, Cakobau became dependent on local European immigrants and the

Tongan warriors. The already existing degree of institutionalised hierarchy as well as incidental historical circumstances seem to have made the difference (see also Routledge 1973).

'...and war made state'?

Hawaiian and Fijian chiefs went to war to aggrandise their power and status, and chiefs like Kamehameha and Cakobau, who could achieve a self-reinforcing dynamic whereby war meant more tribute and larger armies, which in turn meant greater military power, could build up large confederacies. None of them relied on war and military alone, but they had their power enhanced by their divine status and had to build alliances to remain chiefs. As outlined above, however, war did not make the state. To achieve this, the configuration of different sources of power had to be right, and even then the process relied on contingent factors. What implications does this have for the three approaches to war and state formation presented initially?

These approaches focused on war as context, military organisation and means, respectively. The context approach posited that a situation of constant war would promote the development of social complexity because of the advantages gained in respect to protection against outside attack. At first glance, this approach would seem to be confirmed by the war practices of Fiji and Hawaii, since wars were frequent and the people invested considerable co-operative efforts in building fortifications and carrying out large-scale war campaigns. The political units in Hawaii were already larger than the kinship units, which were only relevant at a local level, *ahuapua'a*, and while in Fiji kinship and ancestral descent was important for constituting clans and sub-clans, the *vanua* and *mataqali*, they already practiced the associative incorporation of groups that fled from suppression or had been expelled. However, the local kinship groups were also involved in the exchange of goods, valuables and women on a scale that surpassed these units, and that might arguably be considered the basis for extra-local cooperation and association. One might even argue that had it not been for the frequency of war, these exchange relations might have developed to an extent in scale and intensity that would have effectuated the formation of a complex society. In Fiji, establishing alliances with other groups was essential in war-making, and the outcome of campaigns might decide whether one group would be required to pay tribute to another, so in the short term war did effectuate larger-than-local, subjugating networks. In a long-term perspective, however, the problem was to stabilise these. The fact that authority had not yet developed into a form of stabilised state organisation despite centuries of warfare and conquest testifies to this.

In a short-term perspective, we may find support for the propositions of Herbert Spencer (1967), who argued that frequent war, stratification and despotic rule are (evolutionarily) related, and those of Elman Service (1975), who argued that political leadership and larger social entities arose because of the advantages this provided in questions of attack and defence. In a long-term perspective, the question arises why even larger, more centralised and bureaucratic entities did not develop. There might be two answers to this: either war itself is not enough to transform social relations and has to be tied into other forms of power to enable this or it might have fragmenting as well as centralising effects. The combined use of different sources of power by chiefs in Fiji and Hawaii, as described above, supports the former proposition. In support of latter proposition, I argue that subjugation by force and the creation of *vasu* claims by acquiring high-ranking women as war tribute create strong motivations for defection and an unwillingness to meet moral obligations. Hence, the violence of war disrupts the construction of vast, lasting exchange networks based on reciprocity. While these might also create stratification, the absence of violence makes it easier to tie them in with moral and therefore integrative obligations.

The second approach to war posited that the premium on effectiveness in warfare would result in military organisation becoming more hierarchical, centralised and subjugating than other sections of a group's social life. It was objected that this would only occur if military organisation and status could be transferred from structures of war into structures of peace. The comparison of Fiji and Hawaii suggested that their military organisations relied on other social relations, and there were no indications that subjugation or centralisation existed to a higher degree in the military than in other social hierarchies. There is therefore little to suggest that military organisation might separate and develop along an autonomous trajectory, and the question of re-integrating the military into the other social areas is

thus void. War and military organisation were certainly sources of fame and status. Any exceptionally skilled warrior would gain fame in Fiji and Hawaii, but by already being part of the chiefly clan or the *ali'i* he could become a paramount chief. So, the fact that the political leaders owed their status to religious sanctity as well as military prowess does make military skill a source of power, but it simultaneously limits its potency to the elite groups. In Fiji, the ritual for acknowledging great warriors after war consisted in the warriors being isolated for four days before receiving the honours from the chief. This ritual neutralised attempts by a great warrior to convert his feats of war into political power, unless he initiated a revolt immediately upon return.

The third thesis posited that the state would form as the result of conquest. The basic weakness of this thesis was that while conquest may result in larger units it does not per se result in social structural change. In Fiji as well as Hawaii, war resulted in subjugation and sometimes even in the relegation of whole groups to the status of slaves. But that did not imply a basic change in the social structure of these organisations. Conquest was an important means in the struggle of chiefs to become more powerful, but even for the most successful, such as Cakobau and Kamehameha, this did not amount to transforming the social structure as such. The *matanitus* of Fiji were inherently unstable, a stable hierarchy of power never eventuated, and a routine and rule-regulated kind of bureaucracy never developed. Over time, of course, this might have developed. In Fiji, Bau's founder, Naulivou, was the head of a more hierarchical group than those of Western Fiji, and a local flotsam of Europeans and their muskets reinforced his military power. But even though the coincidence of more hierarchy and muskets did give Bau a certain advantage for some years, this did not lead to a permanent position of power. In Hawaii, the paramount chiefs of the main islands had been at war for centuries, apparently within the same basic social structures. Only at a contingent time in history did this change, when Kamehameha secured dominance over the archipelago and used a near-monopoly on trade with Westerners to secure this position and subsequently change the political organisation. Even then, converting this position of power into a new social structure required the establishment of new practices, such as the abolition of the *kapu* system by Kamehameha's heir.

Hawaii and Fiji are extraordinarily well-suited to help illustrate the relation between war and social structure: wars were frequent, war was a means to enlarge the power of chiefs, and there was no relegation of war to the margins of society, which would have inhibited higher levels of subjugation and central command in military organisation from spreading to the rest of society. War was of central importance but nevertheless did not result in state formation. The central problem is that whereas organised violence is a concentrated, coercive form of power, it is difficult to extend spatially over large areas and over extended lengths of time. It therefore has to tie in with other forms of power – for instance, political, religious, ideological and economic. This is, of course, what the chroniclers of European state formation also show in addition to the 'war made state and state made war' epigram (Giddens 1985; Mann 1986; McNeill 1982; Porter 1994; Tilly 1990). War alone did not make state, and this explains why chiefs who made war only made state when military power and success could combine with other social areas of power under the right historical circumstances.

NOTES

1   The epigram is usually accredited to Charles Tilly (1975: 42), whose work *The Formation of the Nation State in Western Europe* (1975) was a main impetus for this discussion.

2   On war and early states, see Claessen and Skalnìk (1978: 7-14), Haas (1982: 133-40), Fried (1967: 213-16), Service (1975: 271-73) and Claessen (2000: 103-12).

3   See Martin and Frayer (1997), Keeley (1996) and O'Connell (1995).

4   For overviews see Claessen and Skalnìk (1978) and Cohen and Service (1978).

5   On the problem of definitions, see Claessen and Skalnik (1978), Cohen (1978) and Jones and Kautz (1981).

6   For a thorough discussion of these positions, see Haas (1982).

7   On conflict versus integration approaches to state formation, see Haas (1982).

8   For instance, the sandalwood-trader Lockerby (1922), the absconding seaman Jackson (1853), the missionaries Williams (1982) and Waterhouse (1997), the wife of a beche-de-mer trader Wallis (1983[1851]), and the exploring expeditions of Erskine (1853) and Wilkes (1845).

9   These are the standardised terms later elevated to orthodox tradition by the colonial administration and reflect a Bauan perspective. See France (1969: 9-18), Clammer (1973) and Routledge (1985: 28-29).

10 Cordy (1977) defines a complex chiefdom as 'stratified or incipient stratified societies with two or more chiefly rank or status levels and two or more chiefly redistribution levels. The upper levels of these societies are free from subsistence work and the paramount decision-making level has control of some force in his sanctions' (Cordy 1977: 92).

The inchoate early state exists, according to Henri Claessen and Peter Skalník, where 'Kinship, family, and community ties still dominate relations in the political field; where full-time specialists are rare; where taxation systems are only primitive and *ad hoc* taxes are frequent; and where social differences are offset by reciprocity and close contacts between rulers and ruled' (Claessen and Skalník 1978: 23).

Bau and Rewa furthermore seem to fit into Robert Carneiro's definition of a state as 'an autonomous political unit, encompassing many communities within its territory and having a centralised government with the power to collect taxes, draft men for work or war, and decree and enforce laws' (Carneiro 1970: 733).

11 In the following I use 'Hawaiian' to indicate the whole group of islands while 'Hawaii' is used for that particular island.

12 However, Kamakau writes the following: 'The chiefs did not rule alike on all the islands. It is said that on Oahu and Kauai the chiefs did not oppress the common people they did not tax them heavily and they gave the people land where they could live at peace and in a settled fashion' (Kamakau 1964: 231).

BIBLIOGRAPHY

- Beaglehole, J.C. (ed.) 1967. *The Journals of Captain James Cook on his Voyages of Discovery*. Vol. III: *The Voyage of the Resolution and Discovery 1776-1780*. Cambridge: Published for the Hakluyt Society at the University Press.

- Carneiro, R.L. 1970. 'A theory of the origin of the state'. *Science*, 169, pp. 733-38.

- Carneiro, R.L. 1990. 'Chiefdom-level warfare as exemplified in Fiji and the Cauca Valley'. In: J. Haas (ed.): *The Anthropology of War*. Cambridge: Cambridge University Press, pp. 190-211.

- Chagnon, N.A. 1974. *Studying the Yanomamö*. New York: Holt, Rinehart and Winston.

- Claessen, H.J.M. 2000. *Structural Change. Evolution and Evolutionism in Cultural Anthropology*. Leiden: CNWS.

- Claessen, H.J.M. and Skalník, P. (eds.) 1978. *The Early State*. The Hague, Paris and New York: Mouton Publishers.

- Clammer, J. 1973. 'Colonialism and the Perception of Tradition in Fiji'. In: T. Asad (ed.): *Anthropology and the Colonial Encounter*. London: Ithaca Press.

- Clunie, F. 1977. *Fijian Weapons and Warfare*. Suva: Fiji Museum.

- Cohen, R. 1978. 'State origins: a reappraisal'. In: H.J.M. Claessen and P. Skalník (eds.): *The Early State*. The Hague, Paris and New York: Mouton Publishers, pp. 31-75

- Cohen, R., and Service, E.R. (eds.) 1978. *Origins of the State: the Anthropology of Political Evolution*. Philadelphia: Institute for the Study of Human Issues.

- Cordy, R.H. 1977. 'Complex Rank Cultural Systems in the Hawaiian Islands: Suggested Explanations for their Origin'. *Archaeology and Physical Anthropology in Oceania*, IX, pp. 89-109.

- Earle, T. 1997. *How Chiefs come to Power. The Political Economy in Prehistory*. Stanford: Stanford University Press.

- Ellis, W. 1979. *Journal of William Ellis. Narrative of a Tour of Hawaii, or Owhyhee; with Remarks on the History, Traditions, Manners, Customs, and Language of the Inhabitants of the Sandwhich Islands*. Vermont and Tokyo: Charles E. Tuttle Company.

- Emory, K.P. 1970. 'Warfare'. In: E.S.C. Handy, K.P. Emory, E.H. Bryan, P.S. Buck, and J.H. Wise: *Ancient Hawaiian Civilization*. Rutland (Vt.) and Tokyo: Charles E. Tuttle Co., pp. 233-40.

- Erskine, J.E. 1853. *Journal of a Cruise among the Islands of the Western Pacific: Including The Feejees and Others Inhabited by the Polynesian Race in Her Majesty's Ship Havannah*. London: John Murray.

- Fornander, A. 1880. *An Account of the Polynesian Race. Its Origin and Migrations and the Ancient History of the Hawaiian People to the Times of Kamehameha I*. Vol. II. Rutland (Vt.): Charles E. Tuttle.

- France, P. 1969. *The Charter of the Land. Custom and Colonization in Fiji*. Melbourne: Oxford University Press.

- Fried, M.H. 1967. *The Evolution of Political Society. An Essay in Political Anthropology*. New York: Random House.

- Frost, E.L. 1979. 'Fiji'. In: J.D. Jennings (ed.): *The Prehistory of Polynesia*. Cambridge: Harvard University Press, pp. 61-81.

- Gardner, R. and Heider, K.G. 1968. *Gardens of War*. New York: Random House.

- Giddens, A. 1985. *The Nation-State and Violence. Volume Two of a Contemporary Critique of Historical Materialism*. Cornwall: Polity Press.

- Haas, J. 1982. *The Evolution of the Prehistoric State*. New York: Columbia University Press.

- Handy, E.S.C., Emory, K.P., Bryan, E.H., Buck, P.S. and Wise, J.H. 1970. *Ancient Hawaiian Civilization*. Rutland (Vt.) and Tokyo: Charles E. Tuttle Company.

- Harrison, S. 1993. *The Mask of War: Violence, Ritual, and the Self in Melanesia*. Manchester: Manchester University Press.

- Hiroa, T.R. 1964. *Arts and Crafts of Hawaii. Section X: War and Weapons*. (Bernice P. Bishop Museum Special Publication, 45). Honolulu: Bishop Museum Press.

- Hommon, R.J. 1986. 'Social evolution in ancient Hawaii'. In: P.V. Kirch (ed.): *Island Societies. Archaeological Approaches to Evolution and Transformation*. Cambridge: Cambridge University Press, pp. 55-68.

- Howe, K.R. 1984. *Where the Waves Fall*. Honolulu: University of Hawaii Press.

- Jackson, S. 1853. 'Jackson's narrative'. In: J.E. Erskine (ed.): *Journal of a Cruise among the Islands of the Western Pacific: Including The Feejees and Others Inhabited by the Polynesian Race in Her Majesty's Ship Havannah*. London: John Murray, pp. 412-77.

- Jones, G.D. and Kautz, R.R. (eds.) 1981. *The Transition to Statehood in the New World*. Cambridge: Cambridge University Press.

- Kamakau, S.M. 1964. *Ka Po'e Kahiko – The People of Old*. Honolulu: Bishop Museum Press.

- Kamakau, S.M. 1992. *Ruling Chiefs of Hawaii*. Honolulu: Kamehameha Schools Press.

- Keeley, L.H. 1996. *War Before Civilization*. New York and Oxford: Oxford University Press.

- Kirch, P.V. 1985. *Feathered Gods and Fishhooks. An Introduction to Hawaiian Archaeology and Prehistory*. Honolulu: University of Hawaii Press.

- Kuykendall, R.S. 1968. *The Hawaiian Kingdom*. Vol. I: *1778-1854 Foundation and Transformation*. Honolulu: University of Hawaii Press.

- I'i, J.P. 1959. *Fragments of Hawaiian History*. Honolulu: Bishop Museum Press.

- Lockerby, W. 1922. *The Journal of William Lockerby. Sandalwood Trader in the Fijian Islands during the Years 1808-1809*. London: Hakluyt Society.

- Malo, D. 1903. *Hawaiian Antiquities*. Honolulu: Hawaiian Gazette.

- Mann, M. 1986. *The Sources of Social Power. A History of Power from the Beginning to A.D. 1760*. Vol. I. Cambridge: Cambridge University Press.

- Martin, D.L. and Frayer, D.W. (eds.) 1997. *Troubled Times. Violence and Warfare in the Past*. New York: Gordon and Breach Publishers.

- McNeill, W.H. 1982. *The Pursuit of Power: Technology, Armed Force, and Society since A.D. 1000*. Chicago: University of Chicago Press.

- O'Connell, R.L. 1995. *Ride of the Second Horseman. The Birth and Death of War*. New York and Oxford: Oxford University Press.

- Palmer, B. 1969. 'Ring-ditch fortifications in Windward Viti Levu, Fiji'. *Archaeology and Physical Anthropology in Oceania*, 4, pp. 181-97.

- Porter, B.D. 1994. *War and the Rise of the State: the Military Foundations of Modern Politics*. New York: Free Press.

- Routledge, D. 1973. 'The failure of Cakobau, Chief of Bau, to become King of Fiji'. In: G.A. Wood and O'Connor (eds.): *W.P. Morrell: A Tribute*. University of Otago Press: Dunedin, pp. 125-40.

- Routledge, D. 1985. *Matanitu. The Struggle for Power in Early Fiji*. Suva: University of the South Pacific.

- Rowlands, M.J. and Best, S. 1980. 'Survey and Excavation on the Kedekede Hillfort, Lakeba Island, Lau Group, Fiji'. *Archaeology and Physical Anthropology in Oceania*, XV, pp. 29-49.

- Sahlins, M. 1992. *Historical Ethnography*. Vol. I. *The Anthropology of History in the Kingdom of Hawaii*. Chicago and London: University of Chicago Press.
- Service, E. 1975. *Origins of the State and Civilization*. New York: W.W. Norton and Company.
- Spencer, H. 1967. *Principles of Sociology*. London: Macmillan.
- Thomson, B. 1968(1908). *The Fijians. A Study in the Decay of Custom*. London: Dawsons of Pall Mall.
- Tilly, C. (ed.) 1975. *The Formation of National States in Western Europe*. Princeton: Princeton University Press.
- Tilly, C. 1990. *Coercion, Capital, and European States, A.D. 990-1992*. Oxford: Basil Blackwell.
- Valeri, V. 1985. 'The Conqueror becomes King: a political analysis of the Hawaiian legend of 'Umi'. In: A. Hooper and J. Huntsman (eds.): *Transformations of Polynesian Culture*. (The Polynesian Society Memoir, 45). Auckland: The Polynesian Society, pp. 79-103.
- Vancouver, G. 1984. *A Voyage of Discovery to the North Pacific Ocean and Round the World 1791-1795*. London: The Hakluyt Society.
- Wallis, M. 1983(1851). *Life in Feejee, or, Five Years Among the Cannibals*. Suva: Fiji Museum.
- Waterhouse, J. 1997. *The King and People of Fiji*. Honolulu: University of Hawaii Press.
- Webb, M.C. 1975. 'The flag follows trade: an essay on the necessary interaction of military and commercial factors in state formation'. In: J.A. Sabloff and C.C. Lamberg-Karlovsky (eds.): *Ancient Civilization and Trade*. Albuquerque: University of New Mexico Press, pp. 155-209.
- Webster, D. 1975. 'Warfare and the evolution of the state: a reconsideration'. *American Antiquity*, 40, pp. 464-70.
- Wilkes, C. 1845. *Narrative of the United States Exploring Expedition during the Years 1838-1842*. Vol. III. Philadelphia: Lea and Blanchard.
- Williams, T. 1982(1858). *Fiji and the Fijians*. Vol I: *The Islands and Their Inhabitants*. Suva: Fiji Museum.

Warfare in Africa:
Reframing State and 'Culture'
as Factors of Violent Conflict

/**18**

JAN ABBINK

Introduction:
relating warfare, politics, and culture

This chapter explores the connections between culture, politics and warfare in Africa. Warfare, as the organised use of massive violence by states, insurgent groups or other collectivities for various political or predatory aims, is still frequent in contemporary Africa. Violence, defined as the human practice of intentional and contested rendering of harm or lethal force, aimed at intimidation or enforcing dominance, is a universal in all societies at some point or other. But some societies are alleged to be more prone to it than others. State societies with the so-called monopoly on the legitimate use of force do not necessarily evince less inter-personal violence or warfare than stateless or 'tribal' societies known from the anthropological record. Indeed, processes of state formation show phases of intense violent action aimed to establish hegemony by newly emerging elites claiming the monopoly of the means of violence and the extraction of surplus and other resources. While other chapters in this book evaluate the more strictly anthropological, comparative or archaeological aspects of warfare, in this chapter I will discuss a certain geographical area – Sub-Saharan Africa – from a broader political anthropology perspective.

In popular discourse and media images, postcolonial Africa often figures as a continent of ceaseless internecine wars and ethno-regional conflict. Many countries have or had either an internal armed conflict or a high level of communal violence. The terms 'low intensity warfare' (van Creveld 1991), 'new wars' (Kaldor 1999) or 'anarchy' (Kaplan 1994) are often seen as applying particularly to the African situation, especially after the end of the Cold War. Congo, Somalia, Sudan, Angola, Liberia, Sierra Leone, Guinea, or the Saharan Republic are notorious examples, in which hundreds of thousands of people died in the past decade alone. It is, however, difficult to consider the continent as showing unitary political or historical traditions that would explain the mayhem. Only factors like the low level of socio-economic development, steady ecological decline, inequalities and a great lack of legitimacy of regimes and state elites are common across the region. What seems sure is that indigenous African traditions of political culture, community mediation and leadership accountability do not articulate well with the state structures that were imposed in the colonial period or with those that emerged in the African post-colonies. In many accounts of state decline and persistent every-day violence in Africa, explanatory recourse is taken to what we might call the cultural argument: not only are political and economic inequalities or elite misrule seen as responsible but also 'culture', held to be the complex of values, norms and inherited 'traditional' ways of doing things. Culture is seen as involved in (re)producing enmity, rivalry and violence, or as creating certain preconditions to it (cp. Ferme 2001) In the recent debate on differing historical paths of development (cp. Landes 1998; Harrison and Huntington 2000) it is also contended that cultural elements are often decisive.

Without disputing neither the dismal realities of contemporary violence and warfare in Africa (see Mkandawire 2002; Richards 1996; Braathen *et al.* 2000) nor their puzzling supernatural aspects (Ellis 1999), nor the force of cultural traditions (cp. MacGaffey 2000; Ferme 2001), this chapter seeks to critically assess the cultural argument on the basis of a discussion of some African examples. In this context, some reflection on culture in general as well as on African political culture – with the necessary caveats about its complexity and diversity – is necessary, to estimate the role of states and political elites in (re)shaping violent representations and practices.

One interesting point is that in debates on violence in Africa that started with the genre of resistance studies in the 1950s (cp. van Walraven and Abbink 2003), the explanatory arguments were not so much cultural but political and economic in nature, stressing the context of colonial oppression, inequality in the global system of power, and the emergence of justified resistance. In the early 21st century, there is a moment of resigned despair about ongoing violence and bloodshed without clear political ideologies or aims. This makes analysts look for cultural-historical factors, now coupled with a sense of pessimism about the spread and impact of AIDS on the continent (cp. de Waal 2003). It is remarkable that comparatively little attention was paid to modes of 'traditional' conflict resolution and why they failed. These were part of African political culture (cp. Zartman 1999), but were largely bypassed or neutralised by colonial rule and postcolonial state power in a discourse emphasising the 'civilisational shortcomings' of African societies. With examples of the mutilation and torture policies of, e.g., RUF in Sierra Leone and Renamo in Mozambique, the internal wars in Sudan, Liberia and in Congo after Mobutu's demise, the dominant image is that the kind of wars and practices of violence in Africa – while not unknown in other parts of the world – are particularly 'uncivil', evoking a deep sense of shame among those involved and among more distanced observers powerless to prevent them (cp. Keane 1996: 95).

Violence and African political culture

Violence can be analysed in a variety of forms, among them ritual and political. African political culture and power structures had a high degree of ritualised violence (MacGaffey 2000), both expressive of internal tension and political rivalry as well as canalising them. Indeed, as MacGaffey contends (ibid.), the ritual enactment of violence frequently served to avoid real bloodshed. A supernatural connection was often evident in the exercise and control of violence itself, as in the institution of 'divine kings' or of substitutive sacrifices (Simonse 1992; de Heusch 1985). Much of Africa's violence and warfare is easily explained politically – related to indigenous state formation, political rivalry, resource competition and conquest. In this there is similarity with other continents, not the least Europe (see Tilly 1985). Its political cultures were, however, quite dissimilar. In the early 20th century, African resistance against colonialism and national oppression became important. Initially, in the late 19th century, Africans resisted the imposition of colonial rule in a fragmented, localised manner, contesting the defeat of indigenous rulers and political units, the introduction of new labour regimes and production systems, and the undermining of culture and religion. Later, movements emerged that became ideologically motivated rebellions, especially during and after the Second World War. Contemporary movements are less clear in their aims or programs (Mkandawire 2002). In the current age of new wars (Kaldor 1999) the challenge is to find out what indeed are the motivations and ideals for which people fight, and, if any, how to explain them. It is easy to see that there are manifold bases for resentment and protest, but most contemporary insurgencies lose their original aims in the ongoing practices of warfare and violence. Instances are the SPLM in southern Sudan (which has been quite oppressive of smaller ethnic groups in the South), or even more, the Somali armed factions, and the various rebel movements in the DR Congo.[1] There is a need to understand the generalisation of violent practice into what has been called 'cultures of violence' (cp. Abbink 1998: 273).

Culture

If we speak about culture and its role both as a 'source' of violent expression as well as a resource or framework for conflict resolution, two problems come up. The first is: what is culture? Can it be defined in a workable manner after the critiques of the 1980s, is there a shared and transmitted framework of thought, symbols and behaviour or a *habitus* that is normative for a group of people, and if so, do all who are reckoned to it subscribe to the culture's assumed or normative values? I here advance a concept of culture that recognises cognitive and symbolic coherence[2] on the one hand, and fluidity and malleability of collective identity references and ways of life on the other, as the latter are embroiled in and constituted by practice – patterns of enduring material engagement with natural conditions, and of commodity production and division of labour. Culture is a complex of (cosmological) ideas, value orientations and practices of a group that, while dynamic, evinces some historical durability, sharedness, and a thematic profile or 'style'. Ideas of kinship and relatedness, and beliefs about humans and the supernatural, such as witchcraft and 'hidden forces', often play a great role, especially in African cultural traditions. Thus, an *interactionist* view, based on the articulation of material and political processes on the one hand, and of socio-cognitive internalisation of lived differences on the other, is pleaded for.

The second issue is the evaluative dimension: are cultures equivalent and to be respected as such? In view of the great differences in the degrees of ethnocentrism and enmity towards others,[3] one is inclined to say no, unfortunately not.[4] The culture concept – in the sense of referring to the shared, inherited traditions of a collective group – has had a chequered history and is tied up with the history and identity of 'nations', of which in Germany the counter-Enlightenment thinker J.G. Herder was one of the most prominent ideologues. Although Herder wrote at a specific juncture of European history and aimed at reinstating the role of the fragmented German-speaking nation against the large, more unified imperial structures of Britain and especially France (with their emerging universalist-revolutionary Enlightenment discourse), he recognised one important thing: the problematic relationship between existing cultural difference and national, or nowadays predominantly ethnic, identity, and the political form this would take. His approach implied that

there is a latent (and often not so latent) ethnocentrism prevalent in virtually every cultural/national group. This fact makes consensus politics based on the recognition of shared aims and values very difficult. And even then we assume that some groups already *possess* the cognitive or social value of recognising that people have legitimate differences rooted in their own history and in principle the right to express them. But it is easy to see the quite varying degrees in which value is attached to either violence or mediation and accommodation across cultural traditions. The idea of mediating conflict is itself a cultural value not cherished by all.

This leads to another relevant issue in the debate: in the appeal to 'cultures' as factors in warfare or as resources in conflict resolution it is be recognised that they are not homogeneous wholes easily compared. We cannot be sure that all these so-called cultures are warlike or, on the other hand, appreciate debate, dialogue and exchange. What institutions or social strata or organisations exist *within* those nations or cultures that exhibit either warlike or conciliatory discourse? What formations of power and what 'cultural requirements' show a predisposition to warlike activities? It has, for example, been claimed that the culturally crucial bridewealth system of the Nuer pastoralists in the Sudan had an inbuilt tendency for violent expansion (Kelly 1985). But the culturally very similar Dinka did not have this predisposition.

The state in Africa

In Africa, levels of collective, political, violence seem higher because of the problematic nature of state formation and dismal economic development in the wake of colonialism. In the light of quite different cultural backgrounds and systems of political control within postcolonial state boundaries, African politics are a challenge to general political theory and philosophy. Major political thinkers like Machiavelli, Hobbes, Locke, Spinoza and, in the early 20th century, M. Weber and G. Sorel wrote about violence and politics in Europe. How would their ideas and theories relate to African realities, to 'culture' and to models of the person and of the political in Africa? In the decentralised and largely non-literate societies of pre-colonial Africa a discourse on politics and political man similar to that in Europe did not emerge.[5] But there were complex local ideologies or cosmologies of power, usually related to religious notions (as in the Aksumite empire, the Ashanti kingdom, imperial Christian Ethiopia, or the Islamic empires and emirates in various parts of Africa), and the relation between religious norms and political authority was often tense in Africa as well.

In the colonial period, European powers forged authoritarian states in Africa with often racist overtones that were imposed on dispersed and small-scale polities. The new borders forcibly brought together groups that were often strangers and that had not been based primarily on a territorial concept but on personal-political loyalties and cultural affinity. The colonial state thus sat uneasily on a wide variety of socio-cultural and political traditions. The new elites fostered by the colonial authorities were partly artificial, with little grass-roots support, and their staying on in postcolonial times was predicated on networks of patron-client relations and on a balancing act between ethnic and regional interests. Patterns of mimetic rivalry between elites or aspiring elites from various ethnic or regional origins contributed to violent responses. Postcolonial states were not successful in promoting socio-economic development, internal peace or legitimacy. According to Chabal and Daloz (1999) many now seem to thrive on disorder, made into an instrument of power.

Diversity

To appreciate the role of cultural factors in conflict, warfare and conflict prevention, the idea of deep-seated diversity must be recognised. Cultural diversity in Africa is reproduced in everyday contexts and does not go away, not even in the march of globalisation. It is constructed in representations, notions of relatedness (e.g., kinship), socialisation patterns and behavioural routines, and has an *embodied* character. The expectation that people in developing areas will give up their own life ways and cultural representations in the pursuit of 'development', wealth, etc. is of course incorrect. For example, while many people in the ex-colonial, developing world are drawn to modernity, they have little interest in what they often see as the empty culture of Westerners, without religion, without ethical codes, without clear family norms of mutual assistance, reciprocity, etc.[6] This implies that when debating conflict solutions and restoration of 'normality', an

initial position of respect should be taken towards existing diversity – although this is not to say that all cultures are equal or equivalent. I do not plead for cultural relativism, which I find logically untenable and morally unsound.[7] But as people with such divergent backgrounds, representations and political aims clash in zones of conflict, the search for shared moral criteria and autonomy requires taking these divergences seriously and exploring their scope and relevance. It is not only a question of uncovering the 'material interests' and 'resources' over which people are held to fight. Even if these are at stake, the 'language of violence' is clothed in cultural, often ethnic, terms.[8] Indeed, it is hard to see how cultural factors *cannot* be involved.

When looking for features of African cultures that might shape responses to conflict, resistance and warfare, the following elements emerge:

a) In most African societies, the cultural commitments of people are not to large, overarching structures, but with clan- and kin-groups, territorial groups, or occupational groups: a more collective not an individual frame of reference.

b) Age grading and authority based on seniority or (status) hierarchy had political relevance. Senior elders had ritual and other obligations to other sections of the population, however.

c) Legitimate authority was buttressed by ritual-religious mediation figures, e.g. local priests, earth-shrine holders, sacrificers, diviners, or divine chiefs/kings.

d) Social life was pervaded by notions of reciprocity, with expected rights and duties, and compensation for tort.

e) Political culture was primarily shaped by lines of personal loyalty, not of territorially-based authority.

f) Personal status and social adulthood was often shaped by non-civic, non-political mechanisms like initiation and ritual transitions, which created primary social identities for people.

It should be noted that although conflict and warfare were frequent in Africa, especially in dramatic phases of state building and expansion through conquest, there was also a cultural stratum of organised ritual and corporate group relations that allowed mediation and peacemaking and in fact permitted groups to live in relative peace in *non-state* conditions. For instance, pastoral societies like Turkana, Dinka, Samburu, Maasai or Boran were largely successful in 'acephalous' self-regulation.

Recognising the above characteristics, it can be concluded that Western political theory, which emerged from the political philosophers cited above, and was based on ideas of individual rights and duties before king or state, does not find automatic application in African societies. Even if they contained ideals of freedom, personal dignity, equality before the law or the ruler, and justice and equity based on reciprocity, they structured them in different ways, not easily translatable into a formal 'political system' as we know it.

The role of culture in conflict: causal lines?

A recent theoretical approach in conflict studies had tried to shed new light on the role of culture. Authors like S. Huntington, L. Harrison and D. Landes, political scientists and historians,[9] have come to correlate certain cultures or 'civilisations' with values and behaviour that make them relatively more prone to certain social and other processes than others (e.g., economic and social development, material progress, institutional improvement, or on the other hand stagnation, conflict, levelling, the recourse to violence in political life). Basically they stand in the tradition of Max Weber with his classic work on the Protestant ethic, and rehearse, in a more sophisticated form, ideas derived from the modernisation theories of the 1950s.

Indeed, culture has a role to play: people act on ideas and representations, not on unmediated material interests. Perceptions of relative deprivation, ideals of a better future, equality, group solidarity, religious purity or unity, etc. drive collective movements. What the above school of thought has achieved is the reopening of an important debate, and a renewed attention to the interaction of cultural representations and material progress. They have underlined that there *are* affinities that matter. This position has also indicated the limits of historical-institutional approaches to conflict, violence, lack of development, etc. But the question remains: can 'cultural factors' in themselves be *explanatory* in accounting for these problems? Such a reductionist view would deny that cultural traditions are prone to change and suggest cultural change is to be explained by culture.

However, a properly historical view of culture recognises that people are situated actors, drawing upon historically inherited bodies of meaning and ranges of choice. The analysis of how cultural elements shape conflict, warfare and mediation remains a running task for social science because a huge amount of factual, empirical developments in local societies is simply not known or is glossed over. Neither are the cultural dialectics of *interpersonal* relations and the generative mechanisms producing humiliation and resentment (cp. Miller 1993; Bailey 1991) sufficiently taken into account. But, while cultural values play a great role, it seems to be a statement of faith to say that they have autonomous causal force.

'Cultures of violence' as an outcome of warfare

The term 'cultures of violence' is regularly used in recent studies on war and conflict. When not clearly defined and made testable in a comparative perspective, it is a very problematic concept. If cultures of violence[10] do exist, one can imagine the difficulties of ever reaching workable peace and reconciliation deals. My opinion is that first of all the term is a metaphor with limitations. But 'cultures of violence' – as social formations where power is based on unmediated violent performance, and where public discourse is shaped by symbols and acts of violence and intimidation – can exist, usually for short periods. They may appear as a semi-institutionalised, objectified domain of generalised violent action. They could emerge in times of dramatic transitions – systemic political change, a power vacuum, economic upheaval, and decline of a traditional social order, often resulting in youth revolt and in the establishing of a regime based on policing and power enforcement. Many revolutions, regime changes and violent collective movements can be correlated with processes of social exclusion and blocked mobility that hit younger generations, who then search for alternatives: either migration, crime or armed revolt. The latter can go wrong and drag on for much longer than expected – fighting and resource extraction become a 'way of life' with an economic logic, and this is why an entire society can be drawn into a state of disruption and violence as a result. Somalia (*moriyaan* or war lord gangs), Sierra Leone (RUF) and ex-Zaire (various militias and rebel groups after the disappearance of Mobutu in 1997) are telling examples. A culture of violence in the more ethnographic sense, as a *habitus* of violence, where acts of intimidation, killing and impunity become sources of prestige-building and achievement among peers, is also frequent. Examples are the drug trader-dominated sub-cultures in the slums of Brazilian cities, the rule of the Lord's Resistance Army in northern Uganda, and the highly militarised pastoral groups like the Karimojong, where a man without a Kalashnikov and a public status of killer is no longer respected (Mirzeler and Young 2000).

It is likely that in the coming decades, despite all efforts to the contrary, violent conflicts in Africa, either political or more diffusely social or criminal, will become more general. Conditions of external dominance, continued economic exploitation, political inequality and incompetence, decline of the monopolies of the state, and un-reflected globalisation processes contribute to this and will have maximum impact on fragmented local societies, that have lost their moral fibre due to the decline of age graded authority structures, economic survival problems, the growing subjection of women, and a decline of traditional religious-ritual mediation practices.

The example of southern Ethiopia

Ethiopia is one of the oldest states in Africa, dating its precursors to the 4th century BC. The Christian mediaeval Ethiopian highland state of Aksum that emerged in the first century AD expanded gradually in rivalry with traditionalist societies and later with small Islamic emirates on the Red Sea coast. Ethiopian history was thus marked by almost continuous warfare and violence. This tradition of warfare and identity forged in war allowed it to resist European colonialism in 1896, when an Italian colonial invasion army was beaten at Adua. Internally, Ethiopia was a strongly hierarchical country and never attained any solid unity in the face of its culturally and religiously heterogeneous character. The imperial era ended in 1975 with the deposition and secret killing of Emperor Haile Sellassie. A socialist-Marxist dictatorship reigned until 1991 and installed a regime of fear and intimidation (cp. Abbink 1995), testing the country's civilisational basis and very survival. Tens of thousands of young people were killed in the 'Red Terror' period and

several hundred thousand in civil wars in the north and east. On the state level, one could speak of a culture of violence being instituted. A new political system was established after a military victory in 1991 of an ethno-regional guerrilla movement from a minority region (Tigray). It instituted a system of ethnic self-government, creating ethno-regions and de-emphasising national identity. This ideologically novel approach led to a decentralisation and 'localisation' of conflicts, this time all in 'ethnic' terms, but did not offer institutional means and mechanisms for solving them (cp. also Dereje 2001; Tronvoll 2001). A combination of factors inherent in the ethnicised political system and wider, global, developments seemed to lead to the continued presence of the sub-text of violence in the political system and discourse of Ethiopia, to which thousands have fallen victim.

As an illustration of the connection of warfare, culture and politics, I focus briefly on a case-study of some developments in a multi-ethnic setting in southern Ethiopia (Abbink 1998; 2000). It is similar to a few dozen other cases occurring in Ethiopia and elsewhere in Africa. On such a local scale, universal processes of group differentiation, the role of culture in violent conflict, the laboriousness of conciliation, and the mechanisms of social reproduction of conflict can be recognised quite clearly. It is in these local contexts that everyday perceptions of conflict and the options or seductions of violence and war are reproduced.

Southwestern Ethiopia is marked by heterogeneity. Small-scale ethnic group or 'tribal' warfare was frequent, mostly between groups occupying different niches in a wider politico-ecological system of relations. Highland farming groups opposed lowland pastoralists, with smaller groups of hunter-gatherers in between and choosing sides according to their best survival chances at the moment. Southwest Ethiopia and the adjacent areas of Kenya and Sudan have in past two decades have seen a notable transformation if not acceleration of violence and warfare between 'ethnic' communities,[11] both among each other and between some of them and the Ethiopian state. The region is home to a dozen agro-pastoralist groups (e.g., Suri, Baale, Mursi, Kara, Dassanetch, Nyangatom, Toposa, Hamar, Bashada, Bodi) and a smaller number of farming peoples (Bench, Tishana-Me'en, Dizi, Aari, Dime) and hunter-gatherers (Kwegu, Band, Menja), each with different languages and cultural backgrounds. The area has in fact seen regular patterns of non-state warfare over a period of more than a hundred years (cp. Fukui and Turton 1979; Alvarsson 1989). While the nature of violent confrontation is changing, the violence and warlike battles in this region have shown no sign of relenting in the past decades, despite (or due to?) an encroaching central state.

A linkage that I studied in particular was that of the *Dizi* sedentary peasants and the *Suri* agro-pastoralists (and to a lesser extent with the people in the small, mixed towns) in the Maji area (Abbink 1998; 2000). Conditions of recurring ecological crisis, population movement and growth, the rapid spread of modern semi-automatic weapons since the late 1980s, and certain state policies form the backdrop for recent developments in this area: a near permanent state of armed incidents, cattle raids, and killings. These conditions articulate with cultural factors that generate both diversity as well as cross-cutting mechanisms for mediation and co-operation. As said above, the current violence relates to older forms of so-called 'tribal' warfare (there was no golden era of tribal peace in this area) but is essentially changed by recent political developments in Ethiopia, such as new resource competition over land, cattle and alluvial gold, and the ethnicisation of politics and local administration under the guise of a new federal policy of 'ethnic self-determination'. From Sudan there is the ongoing impact of the devastating civil war between the Islamist government and the insurgent movement SPLA fighting for an autonomous Southern Sudan just across the border.

Violent conflict between Suri and Dizi is marked by two aspects: intensification (more frequent recourse to violent acts, more cruelty and more dead and wounded per encounter), and its cyclical nature. A truce is rarely achieved, and if one is made it is soon broken. This seems to reveal a crisis in customary mediation practices between the groups. There is a view among the local people that '...it never used to be so bad like this in the past'. They feel their society to be in disarray. This is not a predictable view that one might expect elders in general to give, because oral traditions and life histories seem to indicate that the past was indeed of a different order. Historically, the construction and discourse of difference between the ethno-cultural groups, through the maintenance of ritual codes and recognised symbiosis, served to manage and 'restrain violence' and

create predictability. The current transformation of a state of organised group conflict, that was interspersed with ritualised mediation establishing peace between collective groups, into a pattern of generalised, 'low intensity' warfare practised by members of these groups in a dispersed and unsanctioned manner is fuelled by material factors of resource pressure and demographic growth, by changes in arms technology, and by the new discourse of 'ethnic claims' made in the Ethiopian political arena. The process visible here is also well-known among the Nuer (Hutchinson 2001) and the Karimojong (Mirzeler and Young 2000), for example, and reveals that the balance of power between groups – and even their very construction as 'ethnic groups' – is caught in widening flows of goods, new power configurations and ideologies that reshape ideas of politics, group interest, cultural belonging and identity, as well as the value of 'mediation'.

Among the Suri and Dizi (as with neighbouring groups), this recent escalation of violence is thus related to a devaluation of organised 'traditional' conflict resolution, formerly defining or constraining the relations between them. Discursive mediation was provided by the ritual code, as when representatives of the contending parties came together to perform a ceremony of reconciliation, with a joint sacrificial livestock killing and ritual use of the animal parts. The new state authorities, while claiming a monopoly over the regulation of violence, did *not* substitute their own ways of mediation or conflict resolution. They bypassed local leaders, devalued mediation meetings, and imposed state 'solutions' based on police and army force (cp. Abbink 2000). In a significant move, the state authorities even forbade ceremonial stick fighting among the Suri as being 'too violent'.

At present, the ethnic groups convert their systems of ritual mediation and conciliation into practices of short-term violent 'settling' of disputes. This is largely based on the perception of what group at that particular moment is strongest in terms of fire-power, economic position, or in its relations to the state. The essential point seems to be that the construction of power between groups is shifting, and that inter-cultural values of mediation are 'put on hold'. It can be easily recognised that the use of semi-automatic rifles as a tool in violent confrontations (since the mid-1980s) has played a specific role, enabling warfare on a 'higher' level, with more dead and wounded than under the previous technological regime of spears, clubs, knives, and old three-shot reloading rifles.[12] New weapons have a far-reaching social effect, finding new 'targets', creating a rupture in self-perception and social experience among local people, and lead to a reconfiguration of gender relations. The gun also becomes the object of a 'cult' of violent performance.

This process is accompanied by an attempt at boundary construction: cultural or 'ethnic' difference is asserted, either on the basis of material or imagined concerns, and is being brought in as a conflict-generating element. Suri and Dizi hardly inter-marry any more, and trade has gone down. In its turn, the 'boundary' becomes an ideological, 'primordial' reason to fight out differences instead of discursively mediating them. In this respect, the conflict between the groups in the Maji area – an area *not* notably transformed by forces of economic or cultural 'globalisation' – is no different from 'ethnic' group conflicts in post-modern industrial or modernising societies in, for instance, Eastern Europe or ex-Yugoslavia. One also sees that boundaries are created not on the basis of pre-existing, fixed ethnic or tribal groups but the other way around: collectivities based on presumed ethnic markers are defined due to conflict over material issues. The conversion of the ritual codes and social patterns of co-operation into violence has predictable regularities based on political ecology (state versus local society, emerging perceptions of resource competition, and new weapons technology) and its ideational reflection, or better, its cultural appropriation, in the various local societies. 'Culture' or cultural difference in this sense is not a self-propelling cause of conflict.

The present-day violence in southern Ethiopia, however, seems to defy initiatives to revive reconciliation (cp. Abbink 2000). The reason is that the actual performance of violence has significantly transgressed the accepted cultural or ritual bounds (e.g. killing women and children in the fields or on the road, killing elders not involved in any fighting, shooting captured cattle from a distance). Usually, instead of reconciliation revenge is sought, seemingly based on the idea, also according to Suri elders who reject the violence of the younger age grade, that 'the gun solves problems' and 'saves time'. Mediation is rejected when retribution is so easy. With men possessing at least one automatic rifle per person and it having become a necessary symbol of

'manhood' and even of social identity of the junior age grade members, a Suri ethos of violent self-assertion was reinforced, at the expense of notions of balance and conciliation. This might be identified as the beginning of a 'culture of violence', and it can be recognised in many emergent violent conflicts. A barrier to dialogue and ritually negotiating difference has been created. We could extrapolate the crucial elements of this case to other social settings, even in large-scale societies. Cultural mediation is extremely vulnerable in the face of dramatic shifts of power.

Conclusions

African state regimes, modelled on a political modernity that does not fit the ethnic and cultural realities of African societies, pursue a totalitarian, imposing logic of governmentality that undermines or manipulates cultural bases of belonging, and enhances antagonistic responses to its exercise of power – even in the federal forms as in Ethiopia. 'Seeing like a state', in James Scott's felicitous phrase (1998), does not only refer to the ill-guided social engineering attempts like villagisation, huge infrastructure projects, or collectivisation of agriculture, but extends to schemes of guided decentralisation and local rule, ideological and cultural policies, and ethnic divide-and-rule. In these domains, there is an ill-understood, or just plainly neglected, vernacular of cultural understandings of what also constitutes authority, value, and legitimacy that eludes state politicians and planners, and subverts their schemes (cp. Donham 1999).

'Cultures of violence', as societies that come to evince a pattern of institutionalised violence defining the political and civic order, are still rare and usually temporary. In Africa, classic examples are the terror rule of the 19th century Zulu king Shaka (Walter 1969) or, in modern times, Renamo rule in its territories (Seibert 2003): in both political regimes state practices of intimidation caused fear and terror to trickle down to the level of the common people, making violent practice contagious and general. Nevertheless, developments in conflict and warfare in the current globalising world suggest that the emergence of cultures of violence will become more widespread. Contributory factors here are the degeneration of armed insurgencies into movements of forced extraction of resources and the exploitation of the common people, and a world-wide spread of organised crime (and private business deals shading into crime, e.g., the arms trade). Criminal activity tends to become socially rooted and thus gives rise to new forms of political authority and exploitation, blurring the boundaries between a legitimate *constitutional state* and informal formations of power (incidentally, Europe is not excluded from such processes).

In Africa one notes not only the decline of the state as an accepted mechanism of social redistribution and furthering of the common good, but also the recent emergence of violent youth movements showing a mixture of street politics, crime-fighting, cultural revivalism and illegal activities of their own. Salient examples are the *Mungiki* in Kenya, the Bakassi Boys and the Odua People's Congress in Nigeria (see Wamue 2001; Harnischfeger 2003; Akinyele 2001). These movements have ideas and programs of reform, social recovery or 'cleansing', but also tend to institute a culture of self-righteous violence and urban warfare that makes itself the local norm. While the state, as an idea and an effective institution, is reclining or crumbling in much of Africa, the current phase of social transition shows new political and cultural forms of organisation that redefine collectivities. In the interstitial spaces of failing states and emerging informal global networks, new human collectivities claim agency on the basis of ethnic, cultural and religious elements, which are combined to forge new units and frameworks of power. Through their informality, hybridity and their connections to hidden and often mystical dimensions of meaning, these groups are transforming politics, culture and the exercise of violence in Africa. States in Africa only rarely provide a joint discursive space, and people are 'retrieved' by non-state identities or loyalties emanating from local cultural assumptions.

In reframing the role of state and culture in analyses of African warfare a few general conclusions come up. First it seems obvious that a denial of the relevance of cultural notions and images in present-day struggles is unwarranted. Culture, seen as an unbounded but more or less coherent reservoir and repertoire of inherited meanings internalised by a group of people, shapes their *habitus* and behavioural patterns in situations of conflict and war. As boundaries emerge in conflict situations, people draw together on presumed 'identity markers'. These are of a cultural nature. Apart from that, *practices* of violence and warfare – their 'style' – go back to patterns

of the pre-state past. This does not logically entail that some 'cultures' follow a war logic or are by definition 'more violent' than others. This depends on social and political conditions, not least the impact of emerging states and political regimes that establish themselves with force over communities based on kinship or traditional forms of authority. The modern state in Africa, barely a hundred years old, has decisively reshaped conflict and warfare. This usually led to increasing antagonisms and evoked contestation and revolt. The postcolonial state of the last forty years has shown itself to be a predictable mechanism of elite resource extraction and ethno-regional patronage, subverting the indigenous notions of equity and reciprocity. In the political life of the postcolony, ethno-cultural differences were made the bone of contention rather than the bridge for accommodation.

Hence, while a cultural dimension in warfare and conflict is ever-present and is often underestimated, a theory of culture and agency is needed that reflects the dialectic of cultural tradition and political agency in wider structural conditions. Popular and political science approaches that seek a direct explanatory link are therefore unsatisfactory. The cultural argument to explain warfare is weak. A theoretical approach that reflects the constitution and workings of cultural elements in shaping the 'commitments' of people in social and political interaction is useful in the analysis of local constructions and constitutions of politics, conflict and warfare, but only if these processes are set in a broader dynamics of environmental and political forces. 'Cultural' factors change and respond to these, but do not determine their outcome.

NOTES

1   While editing this paper in mid-2003, messages about 'tribal massacres' in north-western Congo kept coming out, e.g., one in April whereby about 200 to 300 people were killed , and one in October, when 65 people were massacred, including 45 children (see http://news.bbc.co.uk/2/hi/africa/3169860.stm). See also the BBC report on Congo by F. Keane, 'Africa's forgotten and ignored war', at http://news.bbc.co.uk/2/hi/programmes/from_our_own_correspondent/3201770.stm.
2   To say 'unity' would be too strong.
3   See for instance, Schlee 1999.
4   See also the debates in UNESCO's *World Culture Report 2000* (Paris, 2001), and Edgerton 2000.

5 But since the late 19th century, African traditions of political thought have emerged, cp. Boele van Hensbroek 1999; Kiros 2001.

6 A similar attitude is often seen towards rich Asians and Arab-Islamic people. Their arrogance is resented.

7 Cp. Spiro 1986 and 1996.

8 See Braathen *et al.* 2000.

9 Cp. Huntington 1997; Harrison and Huntington 2000; Landes 1998.

10 I discussed this earlier in J. Abbink 1993.

11 Here distinguished on the basis of differences in language, different histories of origins, and separate political identities, despite partly overlapping ecological niches and even cultural traits.

12 When I arrived for the first time among the Suri in southern Ethiopia in 1988, I still saw many men armed with spears. A decade later, no more spears were to be found, not even after an intensive search: they were discarded or reforged into agricultural tools. In their stead, however, automatic rifles have now become part of the personal equipment of any adult man.

BIBLIOGRAPHY

(Note: Ethiopian authors are by custom cited by their first name)

- Abbink, J. 1993. Cultures of Violence. A Comparative Study of Cultural Forms of Violence. Nijmegen: ICSA (NWO research proposal 1993).
- Abbink, J. 1995. 'Transformations of violence in twentieth-century Ethiopia: cultural roots, political conjunctures'. *Focaal*, 25, pp. 57-77.
- Abbink, J. 1998. 'Ritual and political forms of violent practice among the Suri of southern Ethiopia'. *Cahiers d'Etudes Africaines*, 38(150-152), pp. 271-96.
- Abbink, J. 2000. 'Violence and the crisis of conciliation: Suri, Dizi and the state in southwest Ethiopia'. *Africa*, 70, pp. 527-50.
- Akinyele, R.T. 2001. 'Ethnic militancy and national stability in Nigeria: a case study of the Odua People's Congress'. *African Affairs*, 100, pp. 623-40.
- Alvarsson, J.-A. 1989. *Starvation and Peace or Food and War? Aspects of armed conflict in the Lower Omo Valley, Ethiopia.* Uppsala: Department of Anthropology, Uppsala University.
- Bailey, F.G. 1991. *The Prevalence of Deceit.* Ithaca (NY) and London: Cornell University Press.
- Boele van Hensbroek, P.B. 1999. *Political Discourses in African Thought, 1860 to the Present.* Westport (Conn.): Praeger.
- Braathen, E., Bøås, M. and Sæther, G. 2000. *Ethnicity Kills? The Politics of War, Peace and Ethnicity in Subsaharan Africa.* Basingstoke and London: Macmillan.
- Chabal, P. and Daloz, J.-P. 1999. *Africa Works. Disorder as Political Instrument.* Oxford: James Currey.
- Creveld, M. van 1991. *The Transformation of War.* New York: MacMillan.
- Dereje Feyissa 2001. State, conflict and identity politics: the case of Anywaa – Nuer relations in Gambela, Western Ethiopia. Paper for the Conference 'Changing Identifications and Alliances in Northeastern Africa'. (June 5-9, 2001), Max Planck Institute for Social Anthropology, Halle, Germany.
- Donham, D.L. 1999. *Marxist Modern. An Ethnographic History of the Ethiopian Revolution.* Berkeley, Los Angeles and London: University of California Press; Oxford: James Currey.
- Edgerton, R.B. 2000. 'Traditional beliefs and practices: are some better than others?' In: L. Harrison and S. Huntington (eds.): *Culture Matters. How Values Shape Human Progress.* New York: Basic Books, pp. 126-40.
- Ellis, S.E. 1999. *The Mask of Anarchy: the Destruction of Liberia and the Religious Dimension of an African Civil War.* London and New York: C. Hurst.
- Ferme, M.C. 2001. *The Underneath of Things: Violence, History and the Everyday in Sierra Leone.* Berkeley, Los Angeles and London: University of California Press.
- Fukui, K. and Turton, D. (eds.) 1979. *Warfare among East African Herders.* Osaka: National Museum of Ethnology.
- Harnischfeger, J. 2003. 'The Bakassi boys: fighting crime in Nigeria'. *Journal of Modern African Studies*, 41, pp. 23-49.
- Harrison, L. and Huntington, S. (eds.) 2000. *Culture Matters. How Values Shape Human Progress.* New York: Basic Books.
- Heusch, L. de 1985. *Sacrifice in Africa. A Structuralist Approach.* Manchester: Manchester University Press.
- Huntington, S.P. 1997. *The Clash of Civilizations and the Remaking of World Order.* London: Touchstone Books.
- Hutchinson, S.E. 2001. 'A curse from God? Religious and political dimensions of the post-1991 rise of ethnic violence in South Sudan'. *Journal of Modern African Studies*, 39, pp. 307-31.
- Kaldor, M. 1999. *New and Old Wars. Organized Violence in the Global Era.* Cambridge: Polity Press.
- Kaplan, R. 1994. 'The coming anarchy'. *The Atlantic Monthly*, 273, pp. 44-76.
- Keane, J. 1996. *Reflections of Violence.* London and New York: Verso.
- Kelly, R.W. 1985. *The Nuer Conquest.* Ann Arbor: University of Michigan Press.
- Kiros, T. (ed.) 2001. *Explorations in African Political Thought.* New York and London: Routledge.
- Landes, D.S. 1998. *The Wealth and Poverty of Nations.* London: Little, Brown and Co.
- MacGaffey, W. 2000. 'Aesthetics and politics of violence in Central Africa'. *Journal of African Cultural Studies*, 13, pp. 63-75.
- Miller, W.I. 1993. *Humiliation and Other Essays on Honor, Social Discomfort and Violence.* Ithaca (NY) and London: Cornell University Press.
- Mirzeler, M.K. and Young, C. 2000. 'Pastoral politics in the northeast periphery in Uganda: AK-47 as change agent'. *Journal of Modern African Studies*, 38, pp. 407-30.

- Mkandawire, T. 2002. 'The terrible toll of post-colonial "rebel movements" in Africa: towards an explanation of the violence towards the peasantry'. *Journal of Modern African Studies*, 40, pp. 181-215.
- Richards, P. 1996. *Fighting for the Rainforest. War, Youth and Resources in Sierra Leone*. London and New York: Routledge.
- Schlee, G. (ed.) 1999. *Imagined Differences. Hatred and the Construction of Identity*. Münster, Hamburg and London: Lit Verlag.
- Scott, J.C. 1998. *Seeing Like a State. How Certain Schemes to Improve the Human Condition Have Failed*. New Haven and London: Yale University Press.
- Seibert, G. 2003. 'The vagaries of violence and power in post-colonial Mozambique'. In: J. Abbink, M. de Bruijn and K. van Walraven (eds.): *Rethinking Resistance: Revolt and Violence in African History*. Leiden: Brill, pp. 253-76.
- Simonse, S. 1992. *Kings of Disaster*. Leiden: Brill.
- Spiro, M.E. 1986. 'Cultural relativism and the future of anthropology'. In: *Cultural Anthropology*, 1, pp. 259-86.
- Spiro, M.E. 1996. 'Postmodernist anthropology, subjectivity and science: a modernist critique'. In: *Comparative Studies in Society and History*, 38, pp. 759-80.
- Tilly, Ch. 1985. 'War making and state making as organized crime'. In: P.B. Evans, D. Rueschemeyer and Th. Skocpol (eds.): *Bringing the State Back In*. Cambridge: Cambridge University Press, pp. 169-91.
- Tronvoll, K. 2001. 'Voting, violence and violation: peasant voices on the flawed elections in Hadiya, southern Ethiopia'. *Journal of Modern African Studies*, 39, pp. 697-716.
- Waal, A. de 2003. 'How will HIV/AIDS transform African governance?' *African Affairs*, 102, pp. 1-23.
- Walraven, K. van and Abbink, J. 2003. 'Rethinking resistance in Africa: an introduction'. In: J. Abbink, M. de Bruijn and K. van Walraven (eds.): *Rethinking Resistance: Revolt and Violence in African History*. Leiden: Brill, pp. 1-40.
- Walter, E.V. 1969. *Terror and Resistance*. New York: Oxford University Press.
- Wamue, G.N. 2001. 'Revisiting our indigenous shrines through Mungiki'. *African Affairs*, 100, pp. 453-67.
- Zartman, I.W. (ed.) 1999. *Traditional Cures for Modern Conflicts: African Conflict 'Medicine'. Boulder* (Co.): Lynne Rienner.

Warfare, Rituals, and Mass Graves /19-24

Warfare, Rituals, and Mass Graves:
An Introduction

HENRIK THRANE

/19

This section comprises three articles presenting archaeological and osteological material plus a broader article on the treatment of prisoners of war or what are presumed to be such persons. They may not be POWs in the proper sense of the word – i.e., male fighters – but rather hostages or persons captured in connection with other conflicts (ancient POWs; cp. Gelb 1973).

The notion that the physical act of violence leaves traces such as traumata which may still be recognised centuries after the lethal event is relevant to all of these persons. This in itself is not as simple as it may sound given that there are many ways of maltreating and killing people and animals without leaving traces on the skeletal parts of the corpse. Recent atrocities may be listed that involved inflicting maximal harm on the soft parts of the body, and the well-preserved bog bodies of the north European Bronze and Early Iron Age were normally strangled rather than killed by having their throat cut. Other gory details may be found in Miranda Green's article. Even well-preserved bones do not lend themselves immediately to the study of pre-mortal lesions. The pathologist has to know what to look for and be able to distinguish between the kind of violent treatment that interests us here and other marks such as from the dissection of a corpse in connection with culturally required complex burial rites – for instance, in the Neolithic – or from conditions preventing a timely and normal burial in sacred ground such as with Norse Christians under Arctic conditions (Jørgensen 2001), among whom high-ranking persons were even salted or boiled before being transported to their final resting place. Elsewhere similar cases may have existed in contexts which we do not understand (Hansen 1995: 125ff, 162ff, 254ff).

The use of analogies

The articles illustrate the very important archaeological issue of the extent to which we are permitted to transfer analogies from other fields of individual and social behaviour to our prehistoric past, where nearly everything is different except perhaps the tool kits – stone or iron?

In Miranda Green's article the transfer is as direct as possible within a roughly contemporary Europe, but there are nonetheless source-critical, methodological problems which are not easily transcended. The most spectacular archaeological evidence represents actrocities frequently not covered by the Romans, presumably because they were not allowed anywhere near sites like Gournay, and informing the occupying force may have been worse than letting your war leader down. One may also ask whether the protohistoric cases are close enough to purely prehistoric ones as regards the simple matter of scale to be genuinely applicable.

Green treats the situation in Early Rome and its Barbarian European semi-circle to the north and west as perceived and described by the Romans. Her article may be read in conjunction with Steuer's to compare this aspect to the general trend in warfare during those Iron Age centuries. She compares these descriptions with the archaeological evidence that has recently significantly increased and improved in respect to our objectives here. While practically all the individual pieces of evidence certainly speak of violence and humiliation, without contemporary written evidence the war context could be hard to establish for all cases. In this respect the Celtic (Gaulic) sites of Gournay and Ribemont with their masses of destroyed iron weapons and male bones provide evidence of large-scale conflicts which deserve to be called wars, similar to the later weapon deposits in Danish bogs.

Degrading your opponents

The phenomenon of humiliation is much overlooked but is surely elementary to all societies. It may be assumed in the cases mentioned even if human bones are absent from the Scandinavian weapon sacrifices. Humiliation includes archtypical elements that may be assumed to be universal and others that are culturally specific and therefore not easy to infer by analogy to historical or ethnographical cases.

Revenge

Treating the sexes differently seems to be a universal phenomenon – for example, the selling of women and children and the killing of males. We interpret the past in our own image. If we are normal citizens in a civilised country we tend to assume that people of the past behaved rationally and decently too in spite of all the evidence to the contrary. Placing the enemy under the victor's foot must at all times have been a very degrading and insulting act, as proven by its continuous use in royal iconology – from Narmer's palette to Achaemenid Persia and Imperial Rome (e.g., Porada 1965). It is easy to forget that going berserk is ancient Norse behaviour and that other drug-induced behaviour no doubt guided people in many situations – without making them shamans.

War is a collective effort involving psychological excitement in order to overcome the individual's fear and exhaustion. In situations of stress, normal behaviour, no matter how rational it may have been, tends to be forgotten or disregarded (Camus 1947). Even in our own era, gentle, decent men may return from the hell of the trenches (Graves 1929) to a VC earned in a moment of extreme duress – the apogee of a fine warrior (Dinesen 1929). This must have been all the more manifest in an ancient society where culture was saturated by the warrior's ethos – as the Latin writers imply and describe for their contemporary Barbarian neighbours and they themselves were, too.

Hallucinogens are not easy to discern, but I think we have to count on their presence when looking at behaviour that seems unnecessarily gruesome. They may explain something. No doubt stress (read warfare) aided in the transgression of thresholds of behaviour, and the common excitement carried the violence much further, as numerous ethnographic cases inform us. This is the sort of behaviour that we may transfer to the other silent cultures without too many reservations. Human resilience and the human will to survive is amazing and nearly infinite, with cultural limitations (van der Post 1970; Levi 1957).

Forensic analysis of traumata

Poor 5000 year-old Ötzi, apparently no longer a humble shepherd but a man of status, has now, after the initial clumsy and degrading de-icing, received the full attention of a battery of forensic specialists (*Discovery*, 7 Dec. 2003). He apparently fought several opponents at close range during his last forty-eight hours. The fatal flint arrow in his back was fired from c. 30 m by a person from Ötzi's home valley on the south side of the Alps who followed (or stalked) him all the way up the very difficult slope where he expired. That is the present story, but even the exemplary preservation and application of the most modern forensic methods do not allow more than a tentative reconstruction. We shall never know what lay behind the slaying – a feud, a quarrel over status, a local war?

His traumata were ignored for years because the otherwise scrupulous scientists thought that the damages to skin and so forth were caused by the brutal way in which the frozen corpse was recovered. We have here another classic case of how difficult it is even for scientists to realise what they see; they only examine what they imagine should be there to examine – so much for objectivity.

The pathologists who present their respective analyses of the much less well-preserved Bronze Age and Medieval evidence of mass killings from Scandinavia take pains to inform us of the methodological problems involved in dealing with ancient skeletal material. This is a necessary reminder that when using the results of other disciplines we should consider them on their terms rather than ours. Both cases also illustrate that the archaeological context is not always as simple as non-archaeologists would like it to be – and as we ourselves would hope. The excavators of the Sund skeletons were far from clear about what they had actually found (Farbregd *et al.* 1974), an uncertainty which cannot but be projected onto the interpretation of the bones and their relevance for wider speculations. The Sandbjerg mass grave had to be emptied so fast that elementary details are not available and some observations on the bones which might have been made in situ could not be made. On the other hand, Pia Bennike had the opportunity to handle the exhumation, which no doubt has improved the quality of the find greatly in this respect.

Both cases show how old and new excavations are needed to construct a fuller picture of the individual case. The difference in method and the loss of information are rather severe sometimes – such as at Tormarton, Sandbjerg and, indeed, Sund. The Tormarton case is presented by the (re-)excavator, who has a different approach to the two speared men and their mates in the ditch. The context is territorial through its association with the ditch and as a still not crystal clear, element of the demarcation. The slain men are regarded as the defenders destroyed and thrown in their own ditch. It is a good story, but not a closed case – archaeological cases rarely are. The twenty-five plus Sund skeletons are notable among the cases presented here in representing a section of a (complete?) local population. This is very different from the all-male populations of the Sandbjerg pit and the Tormarton ditch. The ages represented at Sund range from infant to mature adult, and both sexes are among them – eight children, ten women and eight men (Torgersen 1974); without a doubt, this is one of the few Bronze Age skeleton samples from Scandinavia where children are reasonably represented, whereas in the burials they are heavily under-represented. The Sund population is in fact the largest from a single site in Scandinavia. The small number of contemporary skeletons from southern Scandinavia where the full Bronze Age culture flourished contrasts with the situation at Sund. This may sound odd considering the wonderful preservation conditions of the Middle Bronze Age oak coffins with their clothes and wooden artefacts (Jensen 1998); however, that preservation only too rarely coincides with a corresponding preservation of the deceased, apart from their hair. The few skeletons that have been somehow preserved have been neglected too often, as the bones from Sund certainly were for years. The discrepancies in post-burial preservation and post-excavation conditions manifest themselves here, as it so often happens. Therefore, the Sund material unfortunately stands rather alone. It is no wonder that the excavators were not quite at ease with the interpretation (Farbregd *et al.* 1974). We have to look toward Central Europe at a much earlier stage for decent populations which repay intensive studies of health and so on which are known from extensive cemetery material from Early Bronze Age Austria and neighbouring regions (e.g. Neugebauer 1991; Hårde chapter 24).

Context of violence

The main problem here, as in any purely archaeological source, is what the social context of this violence was. The Sandbjerg case is no different in the absence of relevant historical sources. Sund reminds us of other, very rare mass graves from Talheim and Wassenaar (cp. Thorpe chapter 10, and consider the time span), and we take these at face value as representing an entire local community that had received a treatment similar to that measured out to the villagers of Fiji when a village was stormed (Bossen chapter 17). Only at Sund there were no villages, but rather dispersed single farms. Is this an unfortunate farmstead's inhabitants who were killed? Why? Was this just a raid that took a group of them by surprise so the raiders could abduct the livestock and crops? Did any escape?

These collective murders are a rather recently observed phenomenon and present a range of theoretical issues which deserve closer scrutiny. We will do well to realise that mass graves are a phenomenon with many facets deserving individual consideration and definitions. Collective tombs are a well-known phenomenon at different times and places, some representing simultaneous interment in one and the same tomb or grave pit and others not serving as mass graves in the proper sense of the term but rather as collective recipients for successively buried corpses placed in tombs or pits which were accessible for reuse (cp. Thrane 1978). Neither of these groups necessarily have anything to do with violent conflicts. The first group could rather be the result of epidemic diseases, while the second group may have no more than a cultural explanation. The immediate interment of a group of people in a common burial pit, well ordered or haphazard, may have different explanations depending upon the nature of the occasion and the care of the group that buried the corpses. Without traumata we cannot assume that such burials reflect war or violence.

Even when written sources inform of mass graves, archaeological evidence is sometimes not available, as in Anglo-Saxon England (Härke 1978: 91).

Massed skeletons, associated with clear destruction levels, are known from Near Eastern cities and forts like Tepe Gawra, Hasanlu or similar sites (Tobler 1950: 25f; Muscarella 1989), from Roman Europe – for instance, the famous clades variana, the battlefield in the Teutoburg Forest (von Carnap-Bornheim 1999) or Gelduba/Krefeld Gellep (Pirling 1996), and of course elsewhere; these are just random samples. Here the corpses were left as they fell in the last battle, so they are not graves, properly speaking. At Wassenaar they were neatly arranged in anatomical order according to sex and age – a regular burial which we must attribute to the survivors or neighbours who felt obliged to follow the rites of passage (Kooijmans 1996). The accumulated multiple (collective) graves of the late Aunjetitz culture are different in such aspects as their accumulation over a certain amount of time and the strange distribution of bones of individuals in different levels (cp. Hårde chapter 24: Kettlasbrunn in eastern Lower Austria and Nižna Myšľa in eastern Slovakia). Similar situations suddenly appear later (cp. Rittershofer 1997). Buried outside the regular cemeteries these mass graves must illustrate a different social context. Whether the individuals were natives or outsiders from other communities is an open question. Treating one's own people this way would only seem meaningful if their status justified this humiliating treatment. The locality of Sund could be consistent with a collective, rather disrespectful interment of a group which still deserved a burial – within a circular ditch, as one might expect in a proper burial enclosure, with or without a covering cairn.

Quite different are the Tormarton and Sandbjerg men who we assume represent the warrior element of local society, or possibly not that local at Sandbjerg. Here we meet men who died together, or at least they were interred as a group, and thus give us insight into the size of warrior units – if only minimal. Unfortunately, we have no direct way of asserting how large a segment (if at all) of the local force is represented by these deceased persons. We do not know how large a fragment of the population of Sandbjerg (and Sund) was made up of the total fighting force or local population. How professional – i.e., non-local – were they? In that specific period a mixture of natives and foreigners – the latter mainly professional soldiers – might be expected to join in battle. In a city like Næstved there would have been people around to take care of the inhumation, whatever the motivation may have been.

Who were the defenders and who were the intruders? These are questions that we cannot answer yet but that may be answerable when the use of DNA analysis is so universal that we are able to distinguish local populations. In small-scale warfare (typically at the raid level in the Bronze Age?), the raiders presumably would have come from such a similar environment that not even this scientific approach could distinguish aggressor from aggressed.

So, even under optimal conditions, the truth seems more elusive than it appears to be when present-day methods of pathology are applied to murder cases. This is a constant and vital difference between 'historical' and contemporary evidence. Rather than being a pessimistic view this is a conservative one that is slightly more sceptical about holes in the arguments which could be refuted in court. Before concluding it is sound procedure to examine not only the bits of evidence but also the chains of thought linking them together into an argument as well as the logic behind that thinking. Of course, it makes for less exciting material but may be more challenging for the reader.

BIBLIOGRAPHY

- Camus, A. 1947. *La Peste*. Paris: Gallimard.
- Carnap-Bornheim, C. von 1999. 'Archäologisch-historische Überlegungen zum Fundplats Kalkrieser-Niewedder Senke in den Jahren zwischen 9 und 15 n. Chr.' In: W. Schlüter and R. Wiegels (eds.): *Rom, Germanien und die Ausgrabungen von Kalkriese*. Osnabrück, pp. 495-508.
- Dinesen, T. 1929. *No Man's Land*. Copenhagen.
- Farbregd, O., Marstrander, S. and Torgersen, J. 1974. *Bronsealders skjelettfunn på Sund, Inderøy, Nord-Trøndelag*. (Rapport arkeologisk serie, 1974:3). Trondheim.
- Gelb, I.J. 1973. 'Prisoners of War in Early Mesopotamia'. *JNES 32*, pp. 70-98.
- Graves, R. 1929. *Goodbye to all that*. London: Jonathon Cape.
- Hansen, U.L. 1995. *Himlingøje – Seeland – Europa*. Copenhagen: Det Kongelige Nordiske Oldskriftselskab.
- Härke, H. 1998. 'Briten und Angelsachsen in nachrömischen England'. *Studien zur Sachsenforschung, 11*. Hildesheim, pp. 89-119.
- Jørgensen, J.B. 2001. 'Nordbogravene ved Brattahlid'. *Tidsskriftet Grønland*, 3, pp. 81-99.
- Künzl, E. 1993. 'Schlösser und Fesseln'. In: *Das Alamannenbeute aus den Rhein bei Neupotz 1*. Mainz, pp. 365-91.
- Levi, P. 1957. *Se questo e un uomo*. Torino: Einaudi
- Louwe Kooijmans, L.P. 1993. 'An Early/Middle Bronze Age multiple burial at Wassenaar, the Netherlands'. *Analecta Praehistorica Leidensiana*, 26, pp. 1-20.
- Muscarella, O.W., 1989. 'Warfare at Hasanlu in the Late 9th Cent. BC'. *Expedition*, 31, pp. 24-36.
- Neugebauer, J. 1991. *Die Nekropole von Gemeinlebarn F*. (Römisch-Germanische Forschungen, 49). Berlin.
- Pirling, R. 1996. 'Krefeld-Gellep in der Spätantike'. In: *Die Franken, Wegbereiter Europas I*. Mainz: von Zabern, pp. 81–84.
- Porada, E. 1965. *Ancient Iran: The Art of Pre-Islamic Times*. London: Methuen.
- Post, L. van der 1970. *The Night of the New Moon*. London.
- Rittershofer, K.-F. (ed.) 1997. *Special Burials in the Bronze Age of Eastern Central Europe. Proceedings of the Conference of the Bronze Age Study-Group at Pottenstein 1990*. Espelkamp: Verlag Marie Leidorf.
- Thrane, H. 1978. *Sukas IV, A Middle Bronze Age Collective Grave on Tall Sukas*. Copenhagen.
- Tobler, W. 1950. *Excavations at Tepe Gawra II*. Philadelphia: University Museum Monographs.
- Torgersen, J. 1974. 'Osteologisk undersøkelse'. In: O. Farbregd, S. Marstrander and J. Torgersen (eds.): *Bronsealders skjelettfunn på Sund, Inderøy, Nord-Trøndelag*. Trondheim, pp. 19-23.

Semiologies of Subjugation: The Ritualisation of War-Prisoners in Later European Antiquity

> Nine dogs had the prince, that fed beneath his table,
> and of these Achilles cut the throats of twain, and
> cast them upon the pyre and twelve valiant sons of
> the great-souled Trojans slew he with the bronze –
> and grim was the work he purposed in his heart –
> and thereto he set the iron might of fire, to range at
> large. Then he uttered a groan, and called on his dear
> comrade by name: 'Hail, I bid thee, O Patroclus,
> even in the house of Hades, for now am I bringing
> all to pass, which aforetime I promised thee. Twelve
> valiant sons of the great-souled Trojans, lo all these
> together with thee the flame devoureth.
> (Homer *Iliad* XXIII: lines 175-84; trans. Murray 1963: 507-509)

/20

MIRANDA ALDHOUSE-GREEN

Ancient Greek literature is rich in references to the treatment of high-ranking prisoners-of-war, treatment that on occasion took the form of sacrificial killing in the context of aversion or reprisal rituals. The quotation above relates to a composite act of honour and revenge by Achilles in the extremity of his grief at the death of Patroclus, his close comrade-in-arms. He was a warrior of high status whose death could only be avenged by the retaliatory killing of equally noble prisoners. Furthermore, honour could be satisfied and compensation deemed acceptable only if the reprisal killing involved several deaths in payment for the one. The passage illustrates well the essential ambiguity with which foreign prisoners-of-war could be regarded in antiquity. Their high rank might be acknowledged in their selection as appropriate offerings in ritual acts of reciprocity and substitution. However, their lesser worth, as outsiders, foreigners and vanquished enemies, is reflected in the perceived necessity of sacrificing several lives as compensation for one.

Issues concerning attitudes to prisoners-of-war possess a strong pulse of contemporary resonance in the emotive images of Taliban prisoners captured during the war in Afghanistan, following the terrorist attacks on the US on September 11, 2001, and disseminated in media coverage throughout the world.

These captives, suspected of Al Qaeda involvement in international terrorism, have been treated in a manner that combines the perceived need for high security with a heavily ritualised attitude, on the part of the US government and military. Such treatment is associated with what is presented as deliberate humiliation, de-humanisation and denial of identity. In addition to their transportation from Afghanistan on a twenty-six-hour flight to the military base at Guantanamo Bay in Cuba, if we are to believe the reports available to us (for example in the BBC 10.00 news January 15, 2002; Purves 2002: 14), the detention of these prisoners involves shackling and solitary confinement in 'cages' with open-mesh walls, so that they can be seen by their captors at all times. Graphic pictures display kneeling captives, hooded or blindfolded, wearing red uniforms and ear-protectors. The description of the removal of their beards (this last a direct contravention of Afghani Islamic religious tradition), seems to reflect attitudes of extreme hostility, fear, contempt and denial of basic human rights that resonate alarmingly with past military approaches to foreign prisoners.

Similarly, records of experiences during the Second world War speak of the totality of the prisoner-of-war experience: capture, the journey towards the prison-camp, interrogation, were all perceived as rites of passage in a weird, surreal world of dis-identity and helplessness as foreign captives (Liddle and McKenzie 2000: 310-28). Gratuitous violence, degradation and deprivation were regularly meted out to prisoners-of-war both in German (op. cit.: 322) and Japanese camps, and to foreign civilians in the latter (Cooper 2000: 42-48; Cooper 2001: 49-52; Cliff 1998).

This paper seeks to investigate the symbolism and ritualisation associated with warfare and, in particular, the symbolic treatment of war-captives in later European antiquity (Fig. 1). Issues concerning detention, restraint, foreignness, deprivation of status, shame, humiliation and loss of identity are of primary interest. While most prominent consideration is given to archaeological data, the evidence of Classical literature is also consulted. These ancient texts provide a perspective on ancient attitudes to warfare, defeat and imprisonment which, to a degree, may be used to complement and contextualise material culture, despite the oft-discussed biases and distortions present in the Greek and Roman texts dealing with 'barbarian' practices.

Conceptualising the defeated

So ended the battle by which the tribe of the Nervii, and even their name, were virtually wiped out...In describing the disaster their tribe had suffered, they said that from their council of 600, only three men had survived, and barely 500 from their fighting force of 60,000. Wishing it to be seen that I treated unfortunate suppliants mercifully, I took the greatest care to keep them safe. (Caesar *De Bello Gallico* II: 28; trans. Wiseman and Wiseman 1980: 54)

This is a chilling account by Julius Caesar of the genocide that befell the fiercely intractable Gallo-Belgic polity of the Nervii at the hands of the late republican Roman army in 57 BC, during the early campaigns in Gaul. It serves as a stark reminder that ethnic cleansing is not new and that war in antiquity could annihilate entire tribes. Caesar's somewhat detached description of the Nervians' fate contains the telling phrase 'Wishing it to be seen that..'; moral responsibility thus gave place to expediency and a perceived need for a particular image of balanced clemency to be projected back to the Roman Senate (and, of course, relayed to other Gaulish tribes) lest Caesar appear to have let his bloodlust run amok.

The prisoner-of-war is, by definition, a subjugated being, the victim of belonging to the losing side in battle. From the perspective of conquerors in any society, past or present, attitudes to captured forces are likely to be complex and to include conceit, relief, anger and fear. Such composite perceptions are exemplified by recent and current conflicts throughout the world, most notably in the treatment of Taliban terrorist suspects (above). In modern warfare, war-captives may be the only means of confronting enemies face-to-face, and they may thus represent the entire hostility of the opposing forces. In both ancient and modern conflicts, issues of strangeness, foreignness and 'otherness' are powerful images of defeat. We must remember that for both past and present-day communities, a foreigner may not belong to a faraway place but may simply be someone not of one's community or even non-kin.

Imprisonment in the context of war or peace is associated with disempowerment, defunctionalisation, de-humanisation, shame, isolation, detention,

FIG. 1: *Human body placed in disused grain storage pit at the Iron Age hillfort of Danebury, Hampshire, southern England. © The Danebury Trust.*

FIG. 2: *Cardiff University students wearing one of the late Iron Age iron slave gang chains from the watery deposit at Llyn Cerrig Bach, Anglesey, North Wales. © National Museums & Galleries of Wales.*

reclusion, deprivation of liberty and privacy, and coercion. Feelings of anonymity and group-identity by the imprisoned might be reinforced by a uniform, by being roped or chained together (Fig. 2) or even by nakedness (Hill 2000: 317-26). In her study of Peruvian Moche material culture of the 1st millennium AD Erica Hill has examined issues of corporeality, in the context of war-captivity and sacrifice, arguing that images of naked, roped-together sacrificial prisoners can be read as a means of denying individual identity (Hill 2000: 317).

If we follow Foucault's argument (1977: 231-38), captivity additionally involves both actual and symbolic transformation of being, not only from free to incarcerated but also from individual to corporate personhood. Indeed, if we apply such a model to prisoners-of-war, their transformative state might include the change from high to low status. In antiquity (as, unhappily, in certain more recent contexts) the war-captive's sudden alteration in fortune would most likely be associated with enslavement, a change from free warrior to 'owned person' (Freeman 1996: 172). For prisoners-of-war, the sense of humiliation and

isolation are reinforced: they are detained in an alien world, by those whom they have been taught to despise. Their deprivation of their power to fight is represented by being bound or placed in a chaingang. Their weapons might be symbolically destroyed and/or set up as trophies by the victors. Additional shame might accrue to war-captives were they forced to act as informers on their own people: Caesar tells us (*de Bello Gallico* II, 16-17) that he regularly used Gaulish prisoners to gain information about the enemy.

In the ancient world, most prisoners-of-war would have ended up being executed or sold as slaves, sometimes in huge numbers. Linear B tablets relating to the Late Bronze Age Aegean record lists of slaves, either captured in warfare or bought, consisting mainly of women and children from Asia Minor. Homeric epics depict a Mycenaean world in which defeat in battle resulted in death for adult men and the enslavement of non-combatants (Nikolaidou and Kokkinidou 1997: 194-97). Two examples from Roman wars serve to illustrate the extent of this trade in vanquished foes. In the mid-3rd century

BC, during the First Punic War, Rome captured the Sicilian city of Acragas and enslaved the entire Greek population of 25,000 (Freeman 1996: 320). A century later, Roman campaigns against the Iberian Lusitani resulted in the enslavement of 20,000 prisoners (Cunliffe 2001: 372).

In seeking to explore symbolic and ritualistic attitudes to war-captives in the ancient world, it is necessary to engage with the phenomenon of warfare itself. Fighting an enemy involves risk (of defeat, injury or death), collectivity, issues of identity, confrontation of other worlds and power:

A first principle of successful campaigning in war, once active hostilities have begun, is to concentrate all the forces that you can muster and make for your enemy's source of power as speedily as possible, there to deal him the decisive blow which in turn will place your secondary objectives within your power.
(Frere and Fulford 2001: 45)

Thus, warfare and war-captivity are each associated with group activity, with issues of power, with the tension of enmity and the necessary hierarchy of winning and losing, superiority and inferiority. Such asymmetrical relationships between victor and vanquished are graphically illustrated by the war-iconography of ancient Egypt, where the conquering ruler may be depicted as a huge standing individual, brandishing a weapon in one hand, while in the other he grasps the hair of much smaller, kneeling, often bound captives, sometimes shown in groups (Welsby 1996: 25, fig. 6; 44). The clutching of the prisoners' hair is a well-known *leitmotif* of insult, contempt and indignity (see below), as are the discrepant sizes of the images and the contrast between empowered upright stance and the disempowered degradation of kneeling, which both diminishes and incapacitates.

Imaging the captured: a comparative case-study from the Nile Valley

Although far away from our study-area, it is useful to take a sideways glance at ancient Nile Valley victory-iconography since it provides an evocative insight into attitudes towards prisoners-of-war and the manner in which defeated foreign enemies were presented on royal conquest-imagery. There is a grammar of representation whose origins can be specifically traced back as far as the New Kingdom of the mid

second millennium BC. But elements of this tradition belong to a much earlier period, to before 3000 BC, when a tomb was built at Hierakonopolis in Upper Egypt, its walls carved with a scene depicting the conquering ruler, brandishing a weapon as the symbol of his might. He towers above three diminutive war-captives, who kneel bound at his feet, tethered together by a rope he holds in his hand (Filer 1997: 57). Perhaps the most powerful image of conquest appears carved on the walls of the New Kingdom temple of Rameses II at Abu Simbel, near Aswan (Shinnie 1996: 83, pl. 19b). Nubian prisoners are depicted, roped together at the neck, with their arms tied behind them; their necks jut out and their heads are all bent forward in the identical attitude forced by the neck-rope, and are thus robbed of any individuality. Most interesting, though, is the way these captured people are portrayed: their negroid features are emphasised, in a manner designed to stress their foreignness, their difference from their Egyptian conquerors. As Peter Shinnie points out, this is all the more significant in so far as skeletal remains of Nubians belonging to this period do not show markedly negroid characteristics.

The 'grammar' of Nile Valley war-prisoner imagery was maintained in Nubia throughout the rule of the Kushites (from the 8th century BC until the 4th century AD). The Lion Temple at Naqa again depicts conquerors, king Netekamani and queen Amanitare (who reigned at the end of the 1st century BC/early 1st century AD). They are portrayed as huge in proportion to the hapless captives whose heads they smite as they crouch in large numbers at their feet, bound at wrist and neck (Shinnie 1967: fig. 24). Other imagery of foreign war-prisoners sends out similar messages of utter subjugation: to the same period as Naqa belongs a small bronze figurine, from Meroe, of a captive, lying helpless on his face, naked apart from a round, feathered cap, his arms bound behind him at the elbows and his ankles tied together, his feet bent up and back until they rest against the bent forearms (Welsby 1996: 60-61, fig. 20). A large stone statue of a similarly trussed prisoner comes from Basa, but this one is being devoured by a lion. Kushite rulers had such images strategically placed at gathering-points in the landscape, particularly at *hafirs* (reservoirs), as a means of reminding people of their power (Welsby 1996: 37-38, fig. 10). One such ruler, king Shorkaror, who reigned during the earlier 1st century AD, had himself represented on a rock carving at Jebel Qeil standing upon a group of fettered war-prisoners (Welsby 1996: 60). Even more evocative of humiliation (to modern western sensibilities at least) is the carving from the Lion Temple at Musawwarat es Sufra, dating to c. 220 BC, depicting three prisoners kneeling one behind the other, linked by a neck-rope the end of which is grasped by the trunk of an elephant which appears to be in charge of them (Welsby 1996: 44, fig. 12).

Scrutiny of ancient Egyptian and Kushite iconography reveals the extent to which certain ideas concerning conquest are emphasised and reinforced: prisoners-of-war are presented as small, less than human, often unclothed, their foreignness exaggerated and their helplessness emphasised. But at the same time, the level of physical restraint applied appears to depict them as worthy opponents, dangerous adversaries whose savagery demands their bondage hand, foot and neck. Similar apparent paradoxes in attitudes to foreign war-captives resonate throughout antiquity and, indeed, occur in modern contexts. It is as though the might of the conqueror who, needless to say, has control over the propaganda of war-imagery, has to be expressed not only in terms of relative status but also in the presentation of risk faced in battle and the consonant valour of the victors.

Bondage in Iron Age Britain and Gaul

Steeds too he adds, and darts from foemen...
And captives he had bound, hands lashed behind,
To send as offerings to the shade, and, slain,
Dash with their blood the fire...
(Virgil *Aeneid* X: lines 59-93; trans. Rhoades 1957: 243)

In constructing his epic poem in praise of Augustus, Virgil deliberately modelled his imperial eulogy on the Homeric tradition of the *Iliad* and the *Odyssey*, in which suprahuman heroes consorted with the gods in a world where people and divine beings shared a numinous landscape. This passage presents the eponymous Trojan war-lord Aeneas in mourning for the death of Pallas, his comrade-in-arms slain by his enemy Turnus, Prince of the Latins. The episode appears to be a conscious imitation of the episode in the *Iliad* in which Achilles presides over the funeral of Patroclus (*Iliad* XXIII) with which this paper begins. Like Achilles, Aeneas conducted reprisal-sacrifices in revenge and honour for his friend, rituals

in which several Latins were killed in compensation for the fallen hero of the winning side. Virgil specifically alludes to the binding of these Latian captives prior to their sacrificial murder and the consignment of their corpses to the funeral pyre (Green 2001: 142).

Bound bodies

The sacrificial war-victims described in both the *Iliad* and the *Aeneid* are each foreign and noble but only the Latin captives are mentioned as being fettered. This difference between the two heroic texts is probably not significant; there is abundant evidence, from material culture, iconography and ancient literature, that the Romans and their adversaries in western Europe bound or chained their prisoners-of-war. But skeletal remains from Aegean communities of the early 1st millennium BC, too, demonstrate that the binding of sacrificial war-captives was also practised in the Greek world at around the time Homer wrote his two great epic poems. At Eleutherna in Crete (Stampholidis 1996: 164-89), a young nobleman was cremated and his remains placed in the central area of his tomb. At the edge of a second, unlit pyre was the inhumed corpse of a man of similar age and of equally robust physique, suggesting that he, too, may have been a warrior. The limbs of this second individual had been tightly bound and show signs of deliberate mutilation and he had been decapitated; he clearly met his death within the context of the first man's cremation. It is possible that the inhumed body belonged to a slave, killed as an 'attendant' sacrifice, to accompany his master to the underworld, but it is perhaps more likely that the site of Eleutherna provides archaeological testimony to a reprisal sacrifice similar to that enacted by Achilles to avenge Patroclus or Aeneas in compensation for Pallas. Such a killing of a high-ranking hero might, at one and the same time, represent retaliation and sacrifice. The ritual victim's limbs might have been abused to dishonour him as a defeated foreigner, and as a vengeance-killing, and his decapitation might also be thus explained. As Stampholidis has observed, reprisal killing in Homeric literature was sometimes carried out in this manner (*Odyssey* XXIII-XXIV; *Iliad* XI: 145ff; XII: 202ff), perhaps to signify dishonour, ritual insult or trophy-taking. The broader issue of head-ritual in the context of warfare is discussed later in this paper.

FIG. 3: *Multiple human bodies, placed in flexed position in a cleared grain silo at Danebury, Hampshire, southern England. © The Danebury Trust.*

As at Cretan Eleutherna, skeletal evidence from western Iron Age Europe sometimes suggests the binding of war-captives in a quasi-punitive, quasi-ceremonial context. Indeed, Graeco-Roman literature on ancient Gaul and Britain provides testimony to the ritualistic aspects of detention and punishment. The fate of the Arvernian war-leader Vercingetorix is

a good example: after the fall of Alesia in 52 BC, the freedom-fighter was kept, a chained prisoner, in Rome for five years before he contributed to the spectacle of Julius Caesar's Triumph in 46 BC and his public and highly ritualistic execution by strangulation (Freeman 1996: 364). There is a small but significant body of Gallo-British evidence for the binding of individuals tentatively identified as prisoners-of-war who may have been victims of human sacrifice. Two Iron Age sites – Danebury in southern England and Acy-Romance in the Ardennes region of northern Gaul – provide a fascinating insight into the way such individuals may have been treated. On both sites, there seems to have been an emphasis on physical restraint, including binding and confinement in small spaces.

The 'pit-tradition' at Danebury endured for at least the first half of the 1st millennium BC. A pattern of ritual behaviour observed at the site involved the clearing-out of disused grain silos dug into the chalk and the subsequent deposition of whole or partial human bodies (Fig. 3), those of animals, ironwork and other material in what seems to have been votive activity. This could, perhaps, be associated with the final phases of an ongoing negotiation with the spirit-world in respect of fertility and crop-growth, a discourse that began with the sinking of the corn-storage pits, and continued with their use, closure and clearance. The complex rituals perhaps concluded with the emplacement of thank-offerings (Cunliffe 1992: 69-83) in the pits that had been guarded and blessed by the infernal powers into whose territory they had encroached. Several of the human bodies (for example Green 2001: figs. 54, 56, 57) were crammed into tiny spaces (Fig. 4) and/or were so tightly flexed as to suggest that their limbs were originally bound. Some bodies were weighted down or crushed by great blocks of chalk or flint (figs. 5-6), again as if to express the symbolism of restraint and, maybe, to suggest their burial alive, whether or not such a fate was actually meted out to these victims. It can only be conjectural to identify these persons as prisoners-of-war, but some had sustained head or body injuries consistent with warfare and the vast majority of the pit-burials comprised young adult males, in the prime of their fighting lives. They were special burials, in so far as they represented only about 6% of the estimated population at Danebury, or perhaps one interment every few years, and the absence of grave-goods could be interpreted as significant in terms of contempt, low status and denial of identity. All of these might be indicators of appropriate attitudes to foreign, defeated war-dead though, of course, there may be many other possible interpretations.

Danebury was by no means the only southern British Iron Age fortified site to have produced bound

FIG. 4: *Human body folded into small space at edge of grain storage pit at Danebury.* © *The Danebury Trust.*

'pit-bodies'. An evocative burial, again of a young man, was excavated at South Cadbury, Somerset; he – significantly perhaps – had been bound and placed in a tightly-crouched position upside-down in a pit cut into the rear of the late Iron Age earthen rampart (Alcock 1972: 103, pl. 31). It could be argued that the emplacement of this individual within the hillfort defences reflects honourable status, as befitted a war-hero whose valour lived on after his death and imbued the rampart with spiritual strength (Green 2001: 146). Conversely, his binding and his position head-down may instead be associated with insult and low rank. Indeed, such apparent ambiguity might reflect the paradoxical status of a valiant prisoner-of-war, an enemy whose fighting-prowess was symbolically harnessed to enhance the defences of those who captured and executed him. The inversion of his body is interesting: it may be associated with facilitating his entry into the otherworld or with symbolic contempt. Sacrificial prisoners-of-war are represented upside-down in Moche ceramic iconography (Hill 2000: 317-26): the Bronze Age rock carving at Hamn in Sweden depicts a line of inverted armless prisoners tethered by their feet (Coles 1990: 61, fig. 48).

FIG. 5: *Body of a man, probably with hands bound, in grain storage pit at Danebury. © The Danebury Trust.*

FIG. 6: *Deliberately crushed human skeleton from a Danebury pit. © The Danebury Trust.*

The deposition of bodies face-down may, in certain specific circumstances, equally serve to make statements concerning status, contempt or other attitudes to special dead that might be linked to the despatch of captives: it is interesting that the young male bog body from Lindow Moss in Cheshire (north-west England), who was probably the victim of human sacrifice during the 1st century AD, received serious head-injuries sustained as he knelt before his assailant. He was garrotted and his throat slit before being kneed in the back and pushed face-down into a marsh-pool (Stead *et al.* 1986). Equally enigmatic is the burial of a mature woman, who died in the middle Iron Age near Northampton, in the Midlands: she, likewise, was interred face-down, but with her head upraised as though she had been buried alive; oddest of all was the lead torc round her neck, broken in two and placed back-to-front, with the terminals on her cervical vertebrae (Jenkinson 2002; Chapman 2001: 1-42). Her strange burial, wearing a neckring uniquely of lead, suggests she was special; in this instance, the torc may indicate symbolic bondage, and the use of a heavy base metal could signify the contempt meted out to a foreign prisoner, hostage or war-slave. Continuance of this

kind of 'humiliation' treatment of the dead in Roman Britain is attested by evidence from a recently-cemetery at Southwark in London in which the remains of an adult male were discovered, his legs bound together and a spear driven through both ankles (Denison 2002: 6).

Acy-Romance (Ardennes) was the site of a settlement including a public central ritual space and a cluster of cemeteries (Lambot 1998; 2000; Green 2001: 130-31, 145). To the early-mid 2nd century BC belong a series of curious interments, of some twenty or so young men, in a seated position, around the edge of the central sacred place; three others were aligned towards the rising sun. All these bodies had been subjected to an idiosyncratic 'biography' prior to their eventual burial. Evidence suggested to the excavators that each person was squeezed into a small wooden crate and lowered into a deep 'dessication-pit' until the flesh had dried and then re-interred in its final resting-place. The deliberate confinement of the corpses in the wooden containers appears analogous to the small spaces into which some of the southern British Iron Age pit-bodies were inserted, as if symbolic imprisonment were being effected. Like those interred at Danebury and elsewhere in central southern England, the Acy-Romance seated men were buried without grave-goods, as if to stress their alien, low-ranking status or the denial of their individual identity. At Acy, the box-burials are in stark contrast to a series of rich cremations found in the cemeteries, which included elaborate grave-furniture containing liturgical equipment, some of which seems to identify the deceased as religious practitioners. Bernard Lambot (1998) is of the opinion that the seated burials at Acy represent victims of human sacrifice. This interpretation has to remain speculative but one further body from the site supports the enactment of ritual murder here: this also belonged to a young adult man but his body lay extended and in such a position as to testify that he died with his hands tied behind his back at the wrists. His skull provides clues as to the violent manner of his death: he was killed by a savage blow to his head from an axe; of great interest was the discovery, in one of the rich cremation-graves, of an implement whose blade closely matches the head-wound of the fettered victim.

The site of Fesques (Seine-Maritime), excavated by Etienne Mantel, has revealed significant evidence for the abnormal treatment of human behaviour: the remains of twenty-six people were found, including some in a seated position. The foot-bones of some of them show signs that the bodies had been suspended, alive or dead: if alive, then they would inevitably have been bound. The human remains were accompanied by huge quantities of animal bones, all occurring on a large site, tentatively interpreted as a place of assembly. Jean-Louis Brunaux (2000: 16-18) has compared Fesques with the public gathering-place mentioned by Caesar (*de Bello Gallico* I: 4) as belonging to the Helvetian chieftain Orgetorix. If the bodies were strung up, they would appear to be analogous to the treatment of people at places like Ribemont (below), but some were found deliberately positioned in a seated attitude, highly reminiscent of the Acy-Romance burials. Brunaux's view is that the bodies at Fesques exhibit a Gallic community's response to the need for both divine and earthly justice and that the suspended bodies were those of transgressors, offered to the gods in retribution either after battle or for flouting rules of social conduct. In making this argument, he points to iconography on painted Greek vases, in which bound, seated victims are depicted (Brunaux 2000: 16).

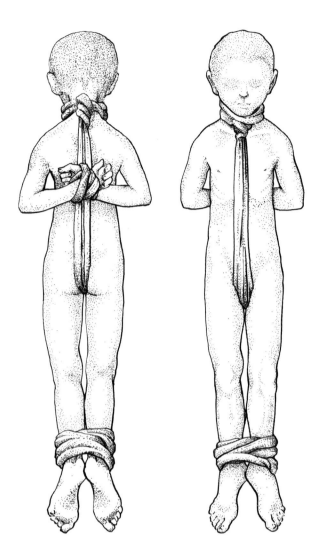

FIG. 7: *An adolescent boy placed, bound, in a bog at Kayhausen, Schleswig-Holstein, North Germany, in 2nd-1st century BC. © Anne Leaver (after van der Sanden 1996).*

Bondage associated with ritual behaviour is well-attested in Iron Age Europe. Some of the northern bog bodies show evidence of such physical restraint in the context of what appears to represent sacrificial killing. A good illustration of such practice is the body of a teenage boy, cast – arguably as a sacrificial act - into a peat-marsh at Kayhausen in Schleswig-Holstein (Fig. 7) during the second-first century BC (van der Sanden 1996: 93, 141, pl. 117; Green 2001: 122, 158, fig. 59). His hands had been tied behind his back with strips of woollen cloth, his feet bound with a folded cloak; another length of cloth had been passed between his legs, pulled taut and wound tightly round his throat, with the effect that any attempt at struggling would throttle him. It is most unlikely that this young boy was himself a combatant prisoner-of-war for he suffered from a sufficiently severe congenital hip-defect as to cause him considerable difficulties in walking, but he could nonetheless have been a casualty of warfare or even a hostage. In the context of symbolic bondage during the north European Iron Age, a comment by Tacitus may be highly significant. In his treatise the Germania, the Roman imperial commentator noted a singular religious custom among the Semnones, a tribe belonging to the powerful *Germanic* confederation of the Suebi:

At a set time all the peoples of this blood gather,
in their embassies, in a wood hallowed by the
auguries of their ancestors and the awe of ages.
The sacrifice in public of a human victim marks
the grisly opening of their savage ritual.
In another way, too, reverence is paid to the grove.
No one may enter it unless he is bound with a cord.
By this he acknowledges his own inferiority and the
power of the deity. Should he chance to fall, he
must not get up on his feet again. He must roll
out over the ground. All this complex of
superstition reflects the belief that in that grove
the nation had its birth. and that there dwells the
god who rules over all, while the rest of the world
is subject to his sway.
(Tacitus *Germania* XXXIX; trans. Mattingly 1948: 132-33)

An image is thus presented of the grove deity's devotees, playing out the role of subservient prisoners or sacrificial victims, fettered to the spirit of the wood and helpless in the supernatural domain. The reference to falling and rolling over suggests to me that the worshipping visitors to the sacred grove behaved as though they, like the Kayhausen boy, were bound hand and foot.

Manacles and chain-gangs: the material culture of restraint

What furnace in town,
What anvil isn't forging their heavy
chains? That's how
most of our iron is used – in fetters!
Well may you fear
That none will be left for plowshares,
hoes or spades in a year'
(Juvenal *Satires* III; trans. Creekmore 1963: 62)

The poet Juvenal published his first book of Satires (I-V) in c. AD 110, when in his mid-forties. He begins the first poem by explaining how he was spurred on to write satire by the moral corruption and hypocrisy he witnessed all around him in Rome. A self-confessed homophobe, xenophobe and mysogynist, his work is nonetheless an amusing and informative commentary on the social values obtaining in the city during the high Roman imperial period. The focus of Satire III is the urban crime-rate, hence his remark about the enforced waste of iron resources on the manufacture of fetters.

Manacles and chains were not reserved for law-breaking criminals but were often used to control large numbers of owned persons, of whom ancient slave-keeping societies were rightly apprehensive. Thus, the unfree workers at the silver mines of Laurion in Attica were kept chained together (Murray 1988: 218), so that they could neither escape nor overpower their captors.

Classical writers on their 'barbarian' neighbours in Gaul and Britain describe how high-ranking prisoners-of-war were chained, probably more as a symbolic gesture than because it was necessary. Tacitus presents a vivid picture of the British freedom-fighter Caratacus, who had led the south Wales polity of the Silures in a tough campaign against the Romans in the mid 1st century AD. When British defeat seemed inevitable, Caratacus fled for sanctuary to Cartimandua, ruler of the powerful northern hegemony of the Brigantes, only to be delivered as a captive to the Romans by the quisling queen. Tacitus (*Annals* XII: 33-35) describes Caratacus's display at the emperor Claudius's triumph in Rome:

For the people were summoned as though for a fine spectacle, while the Guard stood in arms on the parade ground before their camp. Then there was a march past, with Caratacus' petty vassals, and the decorations and neck-chains and spoils of his foreign wars. Next were displayed his brothers, wife and daughter. Last came the king himself.

Using his favourite medium of *oratio obliqua* (reported speech), Tacitus gives Caratacus a proud soliloquy which includes the noble phrase 'humiliation is my lot, glory yours' and a persuasive presentation of arguments as to why Rome and Claudius should be magnanimous to defeated enemies. 'Claudius responded by pardoning him and his wife and brothers', releasing them from their chains.

Contemporary material culture supports the image of the chained prisoner described by Tacitus and others; imagery depicts captives bound or in chains, and sometimes the fetters themselves survive in the archaeological record. The site at Quidney Farm, Saham Toney (Norfolk) lies in the environs of what appears to have been a major focus of symbolic activity, attested by the series of ritual pits at Ashill nearby. Metal detectorists at Quidney Farm discovered an assemblage of late Iron Age and Roman-period metalwork, including sets of horse-gear and a pair of iron manacles, similar to those found at Silchester, dated between the 1st and 3rd centuries AD (Bates 2000: 230). If, as suggested, Saham Toney was the site of a significant Icenian tribal centre, then it is not unlikely that assemblies and other gatherings took place here, perhaps involving the judgement of criminals or war-prisoners who may well have been fettered while on public display. The presence of horse-harness implies that some kind of military display may have occurred here, in which case it is all the more likely that the manacles may have restrained a battle-captive.

Another British site producing wrist-shackles is Read's Cavern, Churhill in Somerset (south-west England). Here, a cave-deposit, dated to between about 50 BC and AD 43, produced the skeletons of four people, including a child together with animal bones, pottery and metal artifacts, including a pair of iron manacles (Bristow 1998: 325). The notion that the assemblage may have other than mundane connotations is supported both by the cave-situation and the identification of cut-marks on the human bone, as if some kind of ritual activity was taking place, involving defleshing or even cannibalism, both of which have been noted as occurring elsewhere in late Iron Age and Romano-British contexts (Aldhouse-Green 2001: 56-61).

To the Roman period belongs a small copper-alloy amulet (Fig. 8), in the form of a naked prisoner bound at the wrist with a rope that also encircles his neck (Green 1978: 48, pl. 138; 2001: col. pl. 20). Like the Kayhausen boy (see above), the double binding seems to have been so arranged that any attempt at struggling to free the hands would result in self-strangulation. The figurine comes from Brough-under-Stainmore in Cumbria, and it is tempting to interpret the image as that of a prisoner-of-war, maybe even a human sacrificial victim. Tim Taylor has suggested (2002: 146) that death by throttling

FIG. 8: *Romano-British copper-alloy amulet in the form of a bound prisoner; Brough-under-Stainmore, Cumbria, north-west England.* © *The British Museum.*

may have had specific rationales, associated with denial of freedom for the dead soul. Strangling or hanging arrests expiry of the last breath, and it is the dying exhalation that, for many societies, signifies true death and the ability of the soul to migrate to the Otherworld. The deliberate intention to leave 'vexed ghosts' wandering in limbo may have been seen as an exquisite form of punishment for an abnormal life or as a means of rendering the sacrificial victim as an especially potent divine gift.

Late Iron Age sites in Britain and Gaul have produced rare but unequivocal examples of gang-chains, made of wrought iron and used to restrain five or six people by means of hinged collars connected to coarse-linked chains. Two massive gang-chains (Fig. 9) were found in the rich metalwork assemblage ritually deposited at Llyn Cerrig Bach on Anglesey, off the north-west coast of Wales (Macdonald 1996: 32-33; Parker Pearson 2000: 8-11). Other British gang-chains include those from a rich late Iron Age

FIG. 9: *Late Iron Age iron slave gang chain (one of a pair), from the watery deposit at Llyn Cerrig Bach, Anglesey, North Wales. © National Museum of Wales.*

chieftain's grave at Barton (Cambridgeshire) and a gang-chain designed for six people from the *oppidum* at Bigberry in Kent, one of the fortified sites quite possibly attacked by Julius Caesar in the mid-1st century BC (Wiseman and Wiseman 1980: 95). A fragmentary example, of 2nd century BC date, was found in 1999 at a site known as Le Petit Chauvort, Verdun-sur-le-Doubs (Saône-et-Loire). Vincent Guichard, Director of the Centre Archéologique Européen du Mont Beuvray, argues convincingly (pers. comm.) that Gallo-British iron gang-chains were ritualised objects, used in the transport of prisoners-of-war and/or sacrificial victims rather than merely as a means of moving large numbers of slaves around. He points out that these instruments of restraint almost always come from special contexts: ritual desposits in water, sacred sites, graves or in association with martial equipment. Furthermore, the incredible investment of time, resources and craftsmanship in producing these chains has to be considered. The manufacture of these objects demanded great black-smithing skills (Green 2002) and the chains themselves were probably heavily invested with value and meaning.

One of the significant, but under-recognised, factors in the use of gang-chains, whether in prehistoric or more modern contexts, is the notion of spectacle, theatre, performance, high visibility and public humiliation. Chaining people together by the neck forces each captive to stand or walk stooped, with his head down: this has already been noted in the depiction of Nubian prisoners in Egyptian war-iconography (above). An experiment was done in about 1990 by the National Museum of Wales, using Cardiff University students and one of the Llyn Cerrig gang-chains (Green 2001: col. pl. 21). Each individual, however much he varies in height from his fellow-prisoners, has to assume a corporate stance and, in fact, to surrender his personhood to the anonymous collective of the group, who all behave as one (*see* Fig. 2). Associated with this is the protracted process of putting people into the chains and taking them out. Placing several people in a gang-chain is a complicated procedure, and the first person in will wait a considerable time for all his fellow miscreants to be chained up with him. Likewise the first one in will inevitably be the last one out and again will wait perhaps for some time to be freed. The implication is that the most important, dangerous or despised captive will be enchained first (S. Aldhouse-Green, pers. comm.)

In order to appreciate the gang-chain as a symbolic piece of punitive theatre, we should turn to Michel Foucault's description of the French penal system in the earlier 19th century (Foucault 1977: 257-61). The chain-gang as a means of transporting convicted French felons only ceased in 1836. In what Foucault termed 'penitentiary science', it had a multiple role providing ritual, spectacle, shock and deterrent value. The 'ritual' commenced with the fastening of the iron collars, each about an inch thick, at the Bicêtre prison, where the head warder, known as the *artoupan* acted as temporary black-smith. It took three men to fasten the ring: one to hold the block, the second to hold together the two halves of the collar round the captive's neck, and the third to strike home the bolt and thus close the circle. As the ring was fastened, the hammer-blows shook the prisoner physically and there was always the *frisson* of risk that the hammer would strike his head rather than the iron bolt. Once the chain-gang was in place, it would then move through the French countryside in a ritualised performance, bringing villagers from their homes to marvel at the spectacle. Foucault describes the way the crowd hurled insults at prisoners or gaolers as they passed, depending on their sympathies, and how 'in every town it passed through, the chain-gang brought its festival with it' (1977: 261). The departing prison-gang was perceived almost in the manner of a scapegoat, taking away with it all the ills of the community and cleansing the regions through which it passed (ibid.: 259; Green 2001: 144-45).

We can learn a great deal from scrutiny of the French chain-gang, in terms of the ritual and symbolism associated with its production and use. It was clearly by no means a simply functional contraption, designed solely to detain and control. The highly visible, public manner in which individuals were chained up, by gaoler-smiths, the noise, heat and danger, followed by the movement of the chain from village to village across France, created a spectacle attracting crowds and inducing an almost religious air of festival and collective catharsis. One other dimension to the French chain-gang is its potential as a weapon. Foucault refers (1977: 261) to the risk run by prison-warders, once the chain-gang was set up, for it was then possible for prisoners to act together and engulf their guards in the chain, that could be used to bludgeon or to strangle them.

In *King Leopold's Ghost*, a highly evocative account of colonialist rule in the Belgian Congo of the 1890s, Hochschild (1998: 119) describes the use of the slave gang-chain. In the words of a Congo state official at the time 'A file of poor devils, chained by the neck, carried my trunks and boxes towards the dock'. It was reported that the death-rate in the Congolese chain-gangs was so high that light steel chains were substituted for the heavy iron collars previously employed, and that there was a constant demand for new supplies. There was also an economic problem in linking people together in this way, for if the gang passed across a bridge over a river and one person lost his footing and fell in, the whole gang was lost (Hochschild 1998: 130). But at the end of the 19th century, the horrific spectacle of lines of children in chains, orphan-victims of colonial raids on villages, could still be seen, on their way to Catholic mission stations (ibid.: 134-35). The history of the Belgian Congo serves as a stark reminder that not only adult prisoners or miscreants were chained. The Congolese children, though, were – in a real and tragic sense – prisoners-of-war. Seen from the perspective of such 19th century chain-gangs, in France and the Congo, the ritualistic aspects of Iron Age gang-chains take on a grim and startling focus.

Iconographies of humiliation

On an engraved Bronze Age rock-panel at Hamn, east of Kville in Sweden, is depicted a martial scene including warriors and boats, perhaps involved in a coastal raiding party. But the imagery of this panel also includes what appears to represent a group of prisoners-of-war, hanging upside down by their feet from a horizontal bar or timber (Coles 1990: 61, fig. 48). These small figures are distinctive, not only in their inverted position but also in the way they are depicted armless. Like a chain-gang, the group is tethered together but they are unable to walk and the absence of their upper limbs carries a powerful message of disempowerment: they have, quite literally, been disarmed and are thus shown in an attitude of utter, dehumanised subjugation.

There is a body of iconography from late Iron Age and Roman Europe depicting defeated captives chained, bound and humiliated. A sarcophagus of Roman date in the Mansell Collection at Rome depicts a group of conquered Gauls, wearing torcs, kneeling or sitting on the ground while heavily-armed Roman footsoldiers and horsemen bludgeon them into submission; one victim is shown with his hands tied tightly behind him (Wiseman and Wiseman 1980: 52). The triumphal arch set up in the 1st century BC at Carpentras (Vaucluse) in southern Gaul again depicts native prisoners, this time chained together with their hands tethered behind them (ibid.: 197). One of the captives is clearly of high status for he wears a tunic of chain-mail, an expensive and highly-prized late Iron Age piece of military hardware, probably designed more for show than defence. The juxtaposition of chain-mail and chain-fetters lends a piquant irony to the carving, in so far as the former reflects power and status and the latter their reverse.

A group of victory-sculptures from the northern frontier regions of northern Britain exhibits a complex grammar of humiliation-imagery, wherein conquered Britons are presented according to stereotypic *topoi* of insult, disempowerment and foreignness, usually bound and seated. On these stones, barbarians are depicted in attitudes of subjugation relative to their triumphal Roman opponents. One distance-slab built on the Antonine Wall, from Bridgness, West Lothian is of especial interest for included in the discrepant symbolism of conqueror and vanquished are the issues of number and relative position. A single Roman cavalryman is depicted at an aggressive gallop; below him are four captives, one beneath the hooves, another fatally injured in the back by a spear, a third bound and seated on the ground, facing the viewer and the fourth, also bound and seated but also decapitated (Ferris 1994: 25; Keppie and Arnold 1984: no. 68, pl. 21).

Trajan's Column was built as a piece of flamboyant imperial propaganda to celebrate and commemorate Roman triumph over Dacian barbarism in the earlier 1st century AD. It presents a fine assemblage of iconographic evidence for the humiliation and physical restraint of foreign prisoners-of-war. Several scenes depict Dacian captives, hands bound behind them (Le Bohec 1994: fig. 29; Settis et al. 1988: 315, pl. 57). A grammar of foreignness is very clear: frequently the enemy is depicted by Roman sculptors as expressly different from Roman soldiers. Thus the Dacians are shaggy-haired, bearded, half-naked or wearing civilian dress: flowing robes and heavy cloaks (Settis et al. 1988: 279, pl. 21). Sometimes they wear oriental-style 'Phrygian' caps, to emphasise their barbarity and otherness (Le Bohec 1994: fig. 29). One interesting feature is the way that Dacian prisoners are repeatedly depicted as held by the elbows or with a Roman soldier grasping them by the hair as they are presented to the emperor (Settis et al. 1988: 431, pl. 173). Both the elbow-restraint and the grabbing of hair have been mentioned earlier in this paper within the context of Egyptian conquest-imagery; the latter seems definitely to have been a gesture of contempt, insult, a means of both physical and spiritual subjugation. Furthermore, just as Egyptian war-imagery exaggerates the negroid foreignness of Nubians (see above), so do some carvings on Trajan's Column: Numidian cavalry, auxiliary-fighters attached to the Roman army, are represented with crinkly hair in dreadlocks and negroid features, engaging with Dacian infantrymen (Settis et al. 1988: 356, pl. 98).

One curious scene (Settis et al. 1988: 326, pl. 68; Le Bohec 1994: fig. 32) depicts a group of three naked prisoners in a tower-like edifice, hands bound behind them, being tortured by elderly, grim-faced women who grasp tufts of their hair and threaten them with firebrands and daggers. The presence of the women would seem to suggest that these victims are Roman soldiers, suffering at the hands of Dacian women, who may even – perhaps – be identified as priestesses. The scene is highly reminiscent of a ritual enacted by the north European tribe of the Cimbri, as recorded by the 1st century Greek geographer Strabo (Geography VII: 2, 3), wherein prisoners-of-war were sacrificed by aged female priests (see below).

In a further repeated pattern of submission, Trajan's Column shows frequent images of Dacians in attitudes of supplication. Many are depicted kneeling before Trajan on one knee, hands outstretched to implore mercy from the emperor or prostrated before him (Settis et al. 1988: 386, pl. 128; 349, pl. 91; 387, pl. 129). It is clear that many of these are noblemen; some are very elderly (Settis et al. 1988: 501, pl. 243). One carving depicts a group of 'barbarians', kneeling in a plea for clemency, being threatened by Roman soldiers. Dacian prisoners may sit shackled at the feet of their captors (Settis et al. 1988: 315, pl. 57); some defeated chieftains are even depicted committing suicide rather than surrender their freedom (Le Bohec 1994: fig. 104). One panel depicts a group of Dacian ambassadors, bearded, semi-naked and wearing Phrygian caps (Settis 1988: 439, pl. 181). Another shows what appears to represent the wholesale deportation of a Dacian community – old men, women and children – from a captured city (Le Bohec 1994: figs. 113-14), in precisely the manner described by Julius Caesar as being inflicted on the defeated inhabitants of Gaulish settlements.

So Trajan's Column may be interpreted as the embodiment of conquest and submission, as presented by the conquerors. The Dacians are barbarians, individuals rather than part of a disciplined professional army; they wear effete, uncivilised dress, unkempt hair and beards and both women, children and the elderly are part of the native scene. Defeat is shown by bondage, humiliation, deportation, enforced supplication; the conquering Roman forces, led by the towering figure of the soldier-emperor Trajan, are depicted in attitudes of contemptuous triumph over Dacian commoners and noblemen alike.

Julius Caesar's commentary on the Gallic Wars (de Bello Gallico), compiled in the mid-1st century BC, contains accounts of subjugation that accord well with the conquest-imagery of the Antonine Wall and Trajan's Column. Caesar's testimony paints a grim picture of the fate that befell the defeated, particularly those chiefs and communities that fought hard for their freedom. As mentioned at the beginning of the paper, capture in the context of war against Rome usually resulted in enslavement or execution.

(But sometimes, when the prisoner was of high status, he was first exhibited in Rome.) Defeat meant enforced movement of whole groups from their ancestral homelands or the use of captives as bargaining-counters. Caesar describes the fate of Gaulish freedom-fighters and their leader Vercingetorix after the siege of Alesia in 52 BC:

I [Caesar] took my place on the fortifications in front of
the camp and the chiefs were brought to me there.
Vercingetorix was surrendered, and the weapons were laid
down before me. I kept the Aeduan and Arvernian
prisoners back, hoping to use them to regain the loyalty
of their tribes. The rest I distributed as booty among the
entire army, giving one prisoner to each of my men.
(*DBG* VII: 89; trans. Wiseman and Wiseman 1980: 176).

Plutarch presents a powerful image of the defeated chieftain:

Vercingetorix, the supreme leader of the whole war,
put on his most beautiful armour, had his horse carefully
groomed, and rode out through the gates [of Alesia].
Caesar was sitting down and Vercingetorix, after riding
round him in a circle, leaped down from his horse, stripped
off his armour, and sat at Caesar's feet silent and motionless
until he was taken away under arrest, a prisoner reserved for
his triumph
(Plutarch *Life of Caesar*; trans. Warner 1958: 240-41).

War-captive iconography even belonging to peoples living across the world from Iron Age Europe can, nonetheless, resonate with the kind of attitudes and presentation we have discussed above. (We have seen examples already in the royal triumph-images of New Kingdom Egypt.) The Moche communities inhabiting the northern coast of Peru between about AD 100 and 800 left behind remnants of a material culture in which conquest, capture and enslavement were habitually employed as subjects for iconographic representation on ceramic vessels. Erica Hill (2000: 317-26) argues with conviction that the Moche used the human body to explore perceptions and meanings associated with their social structure, tensions and change within that structure. One of the themes of Moche art is human sacrifice, the victims apparently selected from conquered communities. This representation of enslaved prisoners-of-war is highly reminiscent both of Egyptian conquest-imagery and that of Iron Age Europe and the

western Roman empire, in so far as *topoi* of degradation and physical restraint are repeatedly expressed. Moche sacrificial art includes scenes wherein lines of prisoners are bound together, sometimes by neck-ropes very similar to gang-chains and sometimes tethered by horizontal poles (Hill 2000: 320, figs. 1, 2). A guard or priest-warder may be depicted holding the tethering-rope and walking behind or in front of the captives. The discrepant manner in which prisoners and warders are treated is highly significant: the prisoner is naked, the captor fully-clad, often armed; the captive may be represented upside-down (ibid.: fig. 3), an iconographic device designed to be read as a 'docile', submissive and biddable body as opposed to the vigorous life-force exhibited by the gaolers. Hill's figure 3 shows a naked, inverted prisoner-of-war in company with two heavily-armed warriors.

Interestingly, vulnerability and humiliation may be depicted by means of clear, often exaggerated representation of genitalia, between splayed legs, whilst denial of individual identity is attested by the generalised blandness of facial features. As has been identified in Egyptian and Roman imagery (above), the Moche captive's hair may be grasped by his captor, partly to represent defeat and partly to secure the head upright for decapitation. (We shall see later that the taking of trophy-heads is depicted on some of the scenes of Dacian defeat on Trajan's Column). On Moche prisoner-art, the captive's hair may be shown cut short and disarranged, a feature that Hill (2000: 320-23) has interpreted as signifying the altered state of the war-prisoner, disempowerment and abrupt loss of status.

The bodies of Moche captives may also be depicted mutilated, dismembered, with disaggregated body-parts, perhaps in symbolism of sacrifice, degradation and both somatic and spiritual dissolution. The dismemberment of the Bronze Age prisoners on the Swedish rock carving at Hamn – also depicted upside-down – thus springs sharply into focus. It may even be possible to apply such a perception to the way in which human bog bodies were sometimes apparently mutilated pre-mortem and to the disarticulation of some of the Iron Age bodies from British sites such as Danebury.

In summary, a range of themes associated with bondage, humiliation, contempt and submission can be identified in Iron Age and Roman Europe, themes that contain strong resonances with the triumph-imagery of other ancient communities, such as those of Egypt and Peru. Witness to such attitudes to inferior, captured status is borne by literature, iconography and the artefactual evidence for restraining devices. The whole assemblage of data is indicative of complex symbolism in which contesting tensions of rank and submission are played out on a public stage. As Foucault reminds us, we should not underestimate the power of theatre, of performance and the potency of the messages such displays conveyed from their initiators to their consumers.

War, sacrifice, and status

Meanwhile, Themistocles was offering sacrifice alongside the admiral's trireme. Here, three remarkably handsome prisoners were brought before him, magnificently dressed and wearing gold ornaments. they were reported to be the sons of Sandauce, the king's sister and Artayctus. At the very moment that Euphrantides the prophet saw them, a great bright flame shot up from the [animal] victims awaiting sacrifice at the altar and a sneeze was heard on the right, which is a good omen. At this, Euphrantides clasped Themistocles by the right hand and commanded him to dedicate the young men by cutting off their forelocks and then to offer up a prayer and sacrifice them all to Dionysus, the Eater of Flesh, for if this were done, it would bring deliverance and victory to the Greeks. Themistocles was appalled at this terrible and monstrous command from the prophet, as it seemed to him. But the people, as so often happens at moments of crisis, were ready to find salvation in the miraculous rather than in a rational course of action. And so they called upon the name of the god with one voice, dragged the prisoners to the altar and compelled the sacrifice to be carried out as the prophet had demanded.'
(Plutarch *Life of Themistocles* XIII, 2; trans. Scott Kilvert 1960: 90-91)

Plutarch's piece refers to an episode during the Persian Wars, just before the Battle of Salamis in 480 BC, led by Themistocles and the Athenian fleet. The passage contains a great deal of interest in so far as a direct connection is made between battle-captives and human sacrifice within the context of great crisis. The Persian prisoners-of-war are young, male, good-looking and noble and the account demonstrates the importance of acknowledging the juxtaposition of high status and captivity in antiquity. Like Homer's 'valiant sons of the great-souled Trojans'

(see opening quotation), these young, high-ranking prisoners were clearly considered as valuable royal hostages, and were deemed by the Athenian seer Euphrantides as appropriate sacrificial victims to Dionysus, even though ritual murder was repugnant to Themistocles.

One significant detail about the Salamis episode concerns the treatment of the sacrificial victims' hair: we have seen that the conquest-iconography of both ancient Egypt and imperial Rome (Trajan's Column) includes a seemingly aggressive image of subjugation, namely the grabbing of the captive's hair by the victor. Plutarch presents us with a variation on this theme that may help develop the way we read this action. By cutting off the Persian prisoners' forelocks, Themistocles is initiating the sacrificial process, claiming the victims as the property of the gods and robbing them of their identity. Hair is a significant part of the persona; in many societies, rites of passage involve cutting or growing hair and beards. In recording the rebellion of the Rhenish tribes against Roman imperial rule in AD 70, Tacitus describes a ritual action by the freedom-fighter Civilis, saying that 'After his first military action against the Romans, Civilis had sworn an oath, like the primitive savage that he was, to dye his hair red and let it grow until such time as he had annihilated the legions. Now that the vow was fulfilled, he shaved off his long beard' (Tacitus *Histories* IV: 61; trans. Wellesley 1964: 247). Interfering with other people's hair against their will constitutes a physical assault and a symbolic violation: cutting or shaving hair (Fig. 10) changes appearance and may signify an attempt to alter perceptions associated with age, identity, gender and status.

It is significant that so many northern European bog bodies, suspected as being sacrificial victims or otherwise marginalised individuals, show signs of their heads being shaven shortly before death. This happened to the young girl drowned and pinned down in a marsh at Windeby in Schleswig-Holstein sometime in the first few centuries AD, and to another adolescent girl, strangled and placed in a peat-bog at Yde in the Netherlands: in both cases, half their hair was shaved shorter than the rest. A male bog-victim at Windeby also had had his hair shaved close; and a mature Danish female ritual murder victim from Huldremose had her hair cut off and placed beside her, one strand woven round her neck (Green 2001: 117-18; van der Sanden 1996;

Glob 1969: 114-16). It is interesting that, in discussing the customs and *mores* of certain Germanic tribes, the Roman author Tacitus specifically mentions the infliction of head-shaving as a shaming, punitive act, along with public stripping, flogging and ejection from the marital home, on women caught in adultery (*Germania* XII), treatment which – like the more modern practice of tarring and feathering – was designed to degrade and dehumanise transgressors. It is worth mentioning a modern analogy, to which my attention has been drawn by Helle Vandkilde (pers. comm.): after the Germans capitulated in Denmark at the end of the Second World War, vindictive Danes showed their desire to humiliate and wreak vengeance on the enemy by forcibly cutting off the hair of girls identified as German.

One of the panels on Trajan's Column appears to depict a group of naked Roman prisoners-of-war in the charge of elderly Dacian women (see above), who threaten them with fire and sword. Such an image resonates strongly with a passage in Strabo's *Geography*, where he describes ancient Cimbrian priestesses engaged in the sacrifice of battle-captives:

They were grey with age, and wore white tunics and over these, cloaks of finest linen and girdles of bronze. Their feet were bare. These women would enter the [army] camp, sword in hand, and go up to the prisoners, crown them, and lead them up to a bronze vessel which might hold some twenty measures. One of the women would mount a step and, leaning over the cauldron, cut the throat of a prisoner [of war], who was held up over the vessel's rim. Others cut open the body and, after inspecting the entrails, would foretell victory for their countrymen.

(Strabo *Geography* VII: 2, 3; trans. Jones 1924)

Like the young Persians at Salamis, described by Plutarch, Strabo's Cimbrian victims were human sacrifices, ritually killed in the context of warfare for the purpose of ensuring victory. The Persians were princes; we have no information as to the social status of the Cimbrians, but the reference to crowning is interesting. Such adornment implies that the prisoners were decorated for their ritual murder, in much the same way as domestic animals were adorned prior to their sacrifice in Classical cult-practice. Similarly, we have references to human scapegoat sacrifices taking place at the southern Gaulish city of Massilia, founded as a Phonecian Greek colony in 600 BC and maintaining a mixed Gallo-Greek culture centuries later. The scapegoat or *pharmakos*

FIG. 10: *Reconstructed stone statue of a seated war-deity or chieftain, his hand resting on a shaven severed head (perhaps symbolic of conquest); from a shrine at Entremont in southern France. 3rd-2nd century BC. © Author (after Benoit 1981).*

was a familiar phenomenon in Greek city-states: an individual of low status was chosen from the community, treated as a pampered and revered person for a time and then driven out of the city, heaped with curses and carrying away the townsfolk's diseases and misfortunes with him, thus purifying the community (Hughes 1991: 139-65). According to Roman literary sources, the Massilian scapegoat ceremony differed from the Greek in so far as the victim was actually killed, by stoning or being cast into the sea. But the pre-slaughter ritual of cosseting and dressing-up, including his adornment with a leaf-crown, resonates with Strabo's account of the hapless prisoner-of-war victims quoted above. His description of sacrificial blood-letting provides a link with a British ritual murder ritual recorded by Tacitus who, in the context of the Roman destruction of the sacred grove on Anglesey in AD 60, alludes to the horrors to be seen in the holy wood:

It was their religion to drench their altars in the
blood of prisoners and consult their gods by
means of human entrails.
(Tacitus *Annals* XIV: 30-31; trans. Grant 1956: 317)

Tacitus does not make it clear as to whether the
British victims were war-captives, rather than criminals or those guilty of sacrilege. We must be mindful,
too, that – even if the blood and innards decorating
the island shrine did belong to defeated combatants,
they may have been killed on the battlefield and
their bodies then used in victory-rituals.

So it appears that, in European antiquity, capture
and imprisonment were charged with multiple layers
of meaning. The abruptness of transition between
high and low status, the paradox of high native rank
and low captured rank and the link between prisoners-of-war and sacrifice may all be identified in the
ancient literature. The expediency of executing
troublesome enemies is juxtaposed with notions of
vengeance, catharsis and the selection of scapegoats:
lesser or foreign persons could be legitimate sacrifices to the gods of the victorious homeland. The
treatment of hair is especially fascinating, for hair
and interference with hair possess sophisticated
metaphoric intricacies associated with belonging,
exclusion, gender, denial of identity and shame. Hair
is a liminal bodily substance, growing from within
but appearing outside the confines of the body; it is
physically painless to cut it but doing so severs links
with the past and changes the persona of its owner.
Its continued growth beyond death perhaps gave
hair a magical, supernatural quality which sometimes had to be disempowered in the intended
humiliation of a victim.

Heads and arms: decapitation, trophy-taking, and insult

When they have decided to fight a battle, it is to Mars that they
usually dedicate the spoils they hope to win: and if they are
successful, they sacrifice the captured animals and collect all
the rest of the spoils in one place. Among many of the tribes it
is possible to see piles of these objects on consecrated ground.
It is most unusual for anyone to dare to go against the
religious law and hide his booty at home, or remove any of the
objects that have been placed on such piles. The punishment
laid down for that crime is death by the most terrible torture'
(Caesar *de Bello Gallico* VI: 17; trans. Wiseman and Wiseman
1980: 123)

The final section of my paper looks at post-battle rituals involving the treatment of things belonging to
the vanquished, including weapons and the severed
heads of the fallen. All these items may be broadly
categorised as 'trophies', symbols of triumph and of
insult to the enemy. Caesar's comment (above) is
important for it stresses the connection between
trophy-taking and religion in late Iron Age Gaul.
Furthermore, we have Plutarch's description (above)
of Vercingetorix's behaviour when surrendering,
namely the Gaulish war-leader's deliberation removal
of his martial accoutrements in front of Caesar, an
act that signified not only defeat but the relinquish-
ment of status.

There is a solid body of archaeological evidence
for the practice of weapons-trophy taking in later
Iron Age Europe, both in the form of iconography
and material culture. The so-called 'war monument'
in the sanctuary of Athene Nikephoros at Pergamum
was set up in the early 2nd century BC to commemorate the victory of the Attalid kingdom over the
marauding Galatians (Szabó 1991: 335). The monument is carved with mountains of captured Gaulish
weapons, the trophies of war and is a poignant, permanent reminder of the defeated. This kind of highly-
visible statement can be identified on public iconography in many parts of the Graeco-Roman world. The
triumphal arches of the 1st century BC at Orange and
Carpentras in southern France exemplify this tradition; they show images of battle-scenes, conquered
Gaulish prisoners and piles of weapons captured from
the enemy (Wiseman and Wiseman 1980: 197).

The northern European Iron Age has produced a
rich array of 'sacrificed' weapons. Particularly evocative are the boat-offerings, and here I will cite two
Danish finds: the Hjortspring and Nydam boats. Both
vessels were deliberately sunk in peat-bogs, the former
in the 3rd century BC, the latter six centuries later.
Each boat was heaped up with weapons and armour,
some of it ritually broken, some showing battle-
wear (Randsborg 1995; Gebűhr 2001). The Nydam
boat was probably never used but fashioned especially for sacrifice in acts of planned deposition. It is
presumed that the the battle-scarred weapons were
booty taken from the enemy, though the victorious
might also have sacrificed their arms. Its purposeful
destruction finds resonance in a text written by the
Roman writer Orosius in the 5th century AD, com-
menting on a Germanic victory-custom apparently
driven by notions of ritual closure or riddance:

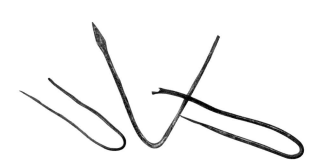

FIG. 11: *Part of a group of six ritually-damaged copper-alloy model spears, from a Romano-British sanctuary at Woodeaton, Oxfordshire, central-southern England. © † Betty Naggar.*

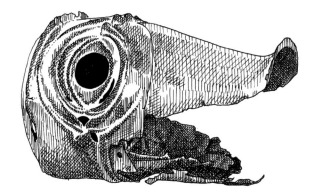

FIG. 12: *Ritually-decapitated copper-alloy boar-head mouth of a carnyx, from a watery deposit at Deskford, Scotland. © Paul Jenkins.*

The enemies who had stormed both camps and had gained an enormous booty destroyed, as a result of an unknown and unusual oath, everything that had fallen into their hands: garments were torn and trodden into the mire, gold and silver were thrown into the river, armour was smashed, the decorative horse equipment destroyed, the horses themselves were drowned in the whirling current, the people strung up on trees with ropes around their necks so that the victors retained no booty of any sort and the defeated experienced no mercy.
(Orosius *Historiae adversum Paganos* V: 16, 1-7, after Gebűhr 2001: 10).

There is a group of 'war sanctuaries' in northern Gaul belonging to the middle and later Iron Age, whose investigation has revealed a complex kaleidoscope of ritual including animal sacrifice and trophy-offerings that took the form both of enemy weapon-collection and display of dead battle-prisoners. At both Gournay-sur-Aronde (Oise) and Ribemont-sur-Ancre (Somme), literally thousands of weapons were heaped up within the sacred precincts, many of them deliberately destroyed or damaged first (Cadoux 1984; Brunaux 1996: 69-77; Brunaux *et al.* 1985; Brunaux 1991: 364-65; Smith 2000: 152-63). The weapons appear to have come from several different regions of Gaul and Germany, suggesting that they belonged to outsiders. The ritual breakage of war-gear (Fig. 11) is often interpreted as an act of sacred violence, associated with the severance of martial offerings from earthworld and their transference to the otherworld (Bradley 1990; Green 2001: 50-51). But it may make sense to turn the argument on its head

and suggest that such destructive action may be part of 'insult' ritual, a gesture of contempt towards the enemy whose swords and shields have been captured, and a message of triumph to the gods of victory to whom the spoils of successful warfare were dedicated. Recent re-examination of the bronze carnyx from Deskford in Scotland (Fig. 12) shows that it was ritually 'decapitated' before being deposited as an offering in a peat-bog (Hunter 2001: 77-108), perhaps as an act of contempt for an enemy object. It is certainly possible to point to modern analogies, wherein the arms of vanquished enemies are ceremonially broken in front of them to signify their utter humiliation and subjugation: this occurred after the Second World War, when the Japanese Emperor surrendered to the Americans in 1945 (Leaver, pers. comm.).

Gournay and Ribemont have provided graphic evidence for the treatment of human bodies in a ritualised context but which may again depict contempt: the regional diversity of the weapons suggests that the bodies subjected to ritual treatment – if the erstwhile owners of these arms – were war-captives. The great 'ossuaries' or bone-houses at the corners of the sacred enclosure at Ribemont were constructed from the limb-bones of about 1000 robust men in the prime of their fighting lives. The deposition of the long-bones was the end-product of a complex procedure, including decapitation and removal of the heads from the sanctuary, and dismemberment of the bodies while the flesh was still on the bones; more sinisterly, there is evidence from cutmarks that the bodies were suspended (perhaps upside-down)

FIG. 13: *Stone frieze depicting alternating horses and severed human heads; from the Iron Age* oppidum *at Nages, southern France. 3rd-2nd century BC. © Paul Jenkins.*

prior to (or for purposes of) their butchery (Cadoux 1984; Smith 2000: 152-63; Brunaux 2000: 14-18). If this were true (and the interpretation has been challenged, notably by Chris Knüsel of the University of Bradford: pers. comm.), then such inversion accords with other evidence cited in this paper, namely the late Iron Age inverted body at South Cadbury in Britain, the Bronze Age rock-figures at Hamn in Sweden and prisoners in Peruvian Moche ceramic imagery. We should remember, too, the hung bodies at Fesques (see above).

The human bone-assemblages at Ribemont's sister-shrine of Gournay exhibit a rather different pattern of ritual behaviour, particularly in so far as here severed heads were retained and displayed hanging from the entrance to the sanctuary. As at Ribemont, the bodies were mainly those of young adult men but there is evidence that three women were also present (Brunaux 1996: 69-77; Brunaux *et al.* 1985), reminding us that prisoners-of-war consisted not only of combatant males. The inclusion of women and children in the prisoner-of-war record is well-illustrated by examination of 14th century North American Plains Indian warfare. Zimmerman (1997: 84) has drawn attention to osteological remains of the 'Crow Creek Massacre' in South Dakota, which included those of fertile-age women, clearly selected as prisoners for their reproductive capacity.

In suggesting that the human heads at Gournay were those of defeated battle-victims, such practice accords with a convincing body of evidence for the collection of enemy trophy-heads in Iron Age and Roman Europe within a ritual context (*see* Fig. 10). Many Graeco-Roman writers allude to head-taking among the Gauls (for example Diodorus Siculus V:

29, 4; Livy X: 26). Livy even describes how the Boii of Cisalpine Gaul cut off the head of the Roman general Postumus in 216 BC and bore it off, with other spoils of war, to 'the most hallowed of their temples' (Livy XXIII: 24). There is some specific archaeological testimony to support such practice, notably at southern Gaulish shrines, such as Roquepertuse and Entremont, where the battle-scarred heads of young men were fixed in niches. Sculpture from these *loca sancta* depicts severed heads (Fig. 13), sometimes hung from the saddles of war-horses (Benoit 1969: pl. XXII), the latter an image repeated in Celtiberian imagery (Fig. 14) (Green 2001: fig. 39, col. pl. 14). What is more, elsewhere in Gaul there are images of warriors brandishing severed heads: an Iron Age potsherd from Aulnat in the Auvergne bears such a depiction (Green 2001: 99, fig. 37), and Gallic coins – from Alesia (Fig. 15) (Rapin 1991: 322; Green 2001: 98, fig. 38), for instance – show similar representations. Trophy head-taking was by no means confined to Iron Age Europe: several panels on Trajan's Column depict Roman soldiers bearing aloft the decapitated heads of Dacian prisoners, including that of king Decebalus himself, as offerings to the emperor (Le Bohec 1994: figs. 51, 109; Settis *et al.* 1988: pl. 28). The act of taking enemy heads combines ritual, aggression, violence, triumph and contempt (Fig. 16).

The foregoing discussion reveals a persistent pattern of trophy-taking behaviour in barbarian Iron Age Europe, from Gaul to Denmark, in which the weapons of the defeated were either displayed as testimony to victory or were ceremonially destroyed in flamboyant performances of riddance or closure. The bodies of the captured, too, were sometimes

FIG. 14: *Late Iron Age Celtiberian copper-alloy fibula in the form of a horseman with a severed human head beneath the horse's mouth; Numancia, northern Spain. © Paul Jenkins.*

similarly treated and, in some communities, particular emphasis was placed on the significance of the human head. Such trophies were, perhaps, valued both for themselves, because of the perceived sacred symbolism of the head, and because they presented highly dramatic evidence of their owners' shame.

FIG. 15: *Late Iron Age coin depicting a Gallic warrior holding a severed human head in each hand; from Alesia, Burgundy, Central France. © Anne Leaver.*

Conclusion

This enquiry has focused upon patterns of ritualised behaviour associated with war, defeat and imprisonment as presented in ancient literature, iconography and material culture. Questions of high and low status, the abrupt transition from one to the other, issues of insult, contempt, fear and triumph have been explored, together with the imagery associated with foreignness, exclusion, dishonour, loss of identity, restraint, submission, ownership and sacrifice. In investigating the symbolism of submission within the context of conflict (Fig. 17), it has been useful to make comparisons between ancient and modern societies and, in antiquity, between communities in temperate Europe, Egypt and even as far away as Peru, for it is clearly possible to identify ways of presenting subjugation that recur over time and space, and familiarity with one society's 'currency' of meaning may well inform our understanding of another's practices. A good example is the gang-chain,

FIG. 16: *Skull of young adult man showing signs of trauma; from a grain storage pit at Danebury. © The Danebury Trust.*

FIG. 17: *Part of an engraved sword-scabbard, depicting a line of spear-bearing horsemen with a subjugated captive beneath the first horse's hooves. 4th century BC; from the cemetery at Hallstatt, Austria. © Paul Jenkins.*

a form of detention and restraint that is loaded with both practical and metaphoric significance; the wrought-iron gang-chains found in ritual contexts in late Iron Age Gaul and Britain gain resonance from an appreciation of the way such – essentially similar – chains were used within the 19th century French penal system.

In the same way, it is possible to make meaningful linkages between the contemptuous treatment of prisoners' hair in contexts as disparate as Roman Dacia, Pharaonic Egypt, the Peruvian Moche and the Taliban under American guard in 2002. Indeed, the ancient societies studied here present an uneasy similarity with the present. This paper has touched upon many uncomfortable aspects of conflict all too familiar to us: genocide, ethnic cleansing, mass deportation, retaliation, physical abuse, public and ritualised humiliation are all attested in the texts, archaeology and imagery relating to warfare, conquest and capture in the European Iron Age and beyond.

Acknowledgements

I should like to thank Dr Mike Parker Pearson for inviting me to give an address on ritual bondage at a conference held by the Department of Archaeology, University of Sheffield, a presentation that provided the germ of an idea for the present paper, and Andreas Hårde, of University of Aarhus for inviting me to speak on this subject at a day-conference on Warfare and Society in the Department of Prehistory at Moesgård. I also wish to express my gratitude to Professor Vandkilde for the invitation to contribute to the present volume and to the editors for their valuable and constructive comments on the first draft. I am grateful to my colleague Dr Josh Pollard for introducing me to Michel Foucault's work *Discipline and Punish*. Finally, I am greatly indebted to Dr Peter Liddle, Director of the World War II Experience in Leeds, who readily provided access to the Centre's impressive archive and gave me helpful guidance in an area of research new to me.

BIBLIOGRAPHY

- Alcock, L. 1972. *'By South Cadbury is that Camelot...'* Excavations at Cadbury Castle 1966-70. London: Thames and Hudson.
- Bates, S. 2000. 'Excavations at Quidney Farm, Saham Toney, Norfolk 1995'. *Britannia*, 31, pp. 201-38.
- Benoit, F. 1969. *L'Art primitif méditerranéen de la vallée du Rhône*. Aix-en-Provence: Publications des Annales de la Faculté des Lettres.
- Benoit, F. 1981. *Entremont*. Paris: Ophrys.
- Boardman, J., Griffin, J. and Murray, O. (eds.) 1988. *Greece and the Hellenistic World*. Oxford: Oxford University Press.
- Bradley, R. 1990. *The Passage of Arms*. Cambridge: Cambridge University Press.
- Bristow, P. 1998. *Attitudes to Disposal of the Dead in Southern Britain 3500 BC to AD 43*. (British Archaeological Reports British Series, 274[2]), Appendix 2 (Gazetteer of Sites Entry 1299). Oxford: Archaeopress.
- Brunaux, J.-L. 1991. 'The Celtic Sanctuary at Gournay-sur-Aronde'. In: S. Moscati, O.H. Frey, V. Kruta, B. Raftery, M. Szabo (eds.): *The Celts*. London: Thames and Hudson, pp. 364-65.
- Brunaux, J.-L. 1996. *Les Religions Gaulois. Rituels Celtiques de la Gaule Indépendente*. Paris: Errance.
- Brunaux, J.-L., Meniel, P. and Poplin, F. 1985. *Gournay I: Les Fouilles sur le sanctuaire et l'oppidum (1975-1984)*. Paris: Errance.

- Cadoux, J.L. 1984. 'L'ossuaire gaulois de Ribemont-sur-Ancre. Premières observations, premières questions'. *Gallia*, 42, pp. 53-78.
- Carman, J. (ed.) 1997. *Material Harm. Archaeological Studies of War and Violence*. Glasgow: Cruithne Press.
- Chapman, A. 2001. 'Excavation of an Iron Age settlement and a Middle Saxon cemetery at Great Houghton, Northampton, 1996'. *Northamptonshire Archaeology*, 29, pp. 1-42.
- Cliff, N. 1998. *Prisoners of the Samurai. Japanese Civilian Camps in China, 1941-1945*. Rainham (Essex): Courtyard Publishers.
- Coles, J. 1990. *Images of the Past*. Vitlicke: Hällristningsmuseet.
- Cooper, C. 2000. 'The log/autograph album of an R.A.F. Sergeant Wireless Operator and P.O.W.: Stanley Hope'. *Everyone's War. The Journal of the Second World War Experience Centre*, 1, pp. 42-48.
- Cooper, C. 2001. 'The diary of a far eastern prisoner of war'. *Everyone's War. The Journal of the Second World War Experience Centre*, 3, pp. 49-52.
- Creekmore, H. (trans.) 1963. *The Satires of Juvenal*. New York: The New American Library.
- Cunliffe, B. 1992. 'Pits, preconceptions and propitiation in the British Iron Age'. *Oxford Journal of Archaeology*, 11(1), pp. 69-83.
- Cunliffe, B. 2001. *Facing the Ocean. The Atlantic and its Peoples 8000 BC - AD 1500*. Oxford: Oxford University Press.
- Denison, S. 2002. 'All the emotions on display in Southwark Roman cemetery'. *British Archaeology*, 67, pp. 6.
- De Sélincourt, A. 1965. *Livy. The Early History of Rome*. Harmondsworth: Penguin.
- Ferris, I. 1994. 'Insignificant Others; images of barbarians on military art from Roman Britain'. In: S. Cottam, D. Dungworth, S. Scott and J. Taylor (eds.): *TRAC 94: Proceedings of the Fourth Annual Theoretical Roman Archaeology Conference. Durham 1994*. Oxford: Oxbow, pp. 24-31.
- Filer, J.M. 1997. 'Ancient Egypt and Nubia as a source of information for cranial injuries'. In: J. Carman (ed.): *Material Harm*. Glasgow: Cruithne Press, pp. 47-74.
- Foucault, M. 1977. *Discipline and Punish. The Birth of the Prison*. Alan Sheridan (trans.). Harmondsworth: Penguin.
- Freeman, C. 1996. *Egypt, Greece and Rome*. Oxford: Oxford University Press.
- Frere, S. and Fulford, M. 2001. 'The Roman invasion of AD 43'. *Britannia*, 32, pp. 45-56.
- Gebűhr, M. 2001. *Nydam and Thorsberg. Iron Age Places of Sacrifice*. Schleswig: Archäologisches Landesmuseum in der Stiftung Schleswig-Holsteinische Landesmuseum Schloß Gottorf.
- Glob, P.V. 1969. *The Bog People*. London: Faber and Faber.
- Grant, M. (trans.) 1956. *Tacitus. The Annals of Imperial Rome*. Harmondsworth: Penguin.
- Green, M.J. 1978. *Small Cult Objects from Military Areas of Roman Britain*. (British Archaeological Reports British Series, 52). Oxford: British Archaeological Reports.
- Green, M.J. Aldhouse 2001. *Dying for the Gods. Human Sacrifice in Iron Age and Roman Europe*. Stroud: Tempus.
- Green, M.J. Aldhouse 2002. 'Any old iron!' In: M.J. Aldhouse-Green and P. Webster (eds.): *Artefacts and Archaeology. Aspects of the Celtic and Roman World*. Cardiff: University of Wales Press.
- Hill, E. 2000. 'The embodied sacrifice'. *Cambridge Archaeological Journal*, 10(2), pp. 317-26.
- Hochschild, A. 1998. *King Leopold's Ghost. A Story of Greed, Terror and Heroism in Colonial Africa*. London: Macmillan.
- Homer. *The Iliad*. A.T. Murray 1963 (trans.). (Loeb Edition). London: Heinemann.
- Hooper, B. 1984. 'Anatomical considerations'. In: B. Cunliffe: *Danebury: An Iron Age Hillfort in Hampshire*. (Council for British Archaeology Research Report, 52). York: Council for British Archaeology, pp. 463-74.
- Hughes, D. 1991. *Human Sacrifice in Ancient Greece*. London: Routledge.
- Hunter, F. 2001. 'The Carnyx in Iron Age Europe'. *Antiquaries Journal*, 81, pp. 77-108.
- Jenkinson, S. (ed.) 2002. 'The witch of Northampton'. In: *Tales from the Grave*. (Channel 4 TV Documentary Series, 29.3.02).
- Keppie, L.J.F. and Arnold, B.J. 1984. *Corpus Signorum Imperii Romani. Corpus of Sculpture of the Roman World. Great Britain*. Vol. I, Fasc. IV: Scotland. London and Oxford: British Academy and Oxford University Press.
- Lambot, B. 1998. 'Les morts d'Acy Romance (Ardennes) à La Tène Finale. Pratiques funéraires, aspects religieuses et hiérarchie sociale'. In: *Etudes et Documents Fouillés 4. Les Celtes Rites Funéraires en Gaule du Nord entre le Vie et le 1er siècle avant Jésue-Christ*. Namur: Ministère de la Région Walionne, pp. 75-87.
- Lambot, B. 2000. 'Victimes, sacrificateurs et dieux'. In: V. Guichard and F. Perrin (eds.): *Les Druides*. (*L'Archéologue*, Hors Série No. 2). Paris: Errance, pp. 30-36.
- Le Bohec, Y. 1994. *The Imperial Roman Army*. Raphael Bate (trans.). London: Routledge.
- Liddle, P. and McKenzie, S.P. 2000. 'The experience of captivity: British and Commonwealth prisoners in Germany'. In: J. Bourne, P. Liddle and I. Whitehead: *The Great World War 1914-45*. Vol. I: *Lightning Strikes Twice*. London: Harper Collins, pp. 310-28.
- Macdonald, P. 1996. 'Llyn Cerrig Bach. An Iron Age votive assemblage'. In: S. Aldhouse-Green (ed.): *Art, Ritual and Death in Prehistory*. Cardiff: National Museum of Wales, pp. 32-33.
- Mattingly, H. 1948. *Tacitus on Britain and Germany*. Harmondsworth: Penguin.
- Murray, O. 1988. 'Life and society in Classical Greece'. In: J. Boardman, J. Griffin, O. Murray (eds.): *Greece and the Hellenistic World*. Oxford: Oxford University Press, pp. 198-227.
- Nikolaidou, M. and Kokkinidou, D. 1997. 'The symbolism of violence in Late Bronze Age Palatial societies of the Aegean: a gender approach'. In: J. Carman (ed.): *Material Harm*. Glasgow: Cruithne Press, pp. 174-97.

- Olmsted, G.S. 1979. *The Gundestrup Cauldron*. Brussels: Latomus.
- Parker Pearson, M. 2000. 'Great sites: Llyn Cerrig Bach'. *British Archaeology*, 53, pp. 8-11.
- Purves, L. 2002. 'Forget Cuba, what about our hell-holes?' *The Times*, 22.1.02.
- Randsborg, K. 1995. *Hjortspring: Warfare and Sacrifice in Early Europe*. Aarhus: Aarhus University Press.
- Rapin, A. 1991. 'Weaponry'. In: S. Moscati, O.H. Frey, V. Kruta, B. Raftery and M. Szabo (eds.): *The Celts*. London: Thames and Hudson, pp. 321-31.
- Rhoades, J. (trans.) 1957. *The Poems of Virgil*. London: Oxford University Press.
- Sanden, W. van der 1996. *Through Nature to Eternity. The Bog Bodies of Northwest Europe*. Amsterdam: Batavian Lion International.
- Scott-Kilvert, I. (trans.) 1960. *Plutarch. The Rise and Fall of Athens: Nine Greek Lives*. Harmondsworth: Penguin.
- Settis, S., La Regina, A., Agosti, G. and Farinella, V. 1988. *La Colonna Traiana*. Torino: Guilio Einaudi editore.
- Shinnie, P.L. 1967. *Meroe. A Civilization of the Sudan*. London: Thames and Hudson.
- Shinnie, P.L. 1996. *Ancient Nubia*. London: Kegan Paul.
- Smith, A. 2000. The Differential Use of Constructed Sacred Space in Southern Britain from the late Iron Age to the Fourth Century AD. Ph.D. thesis, University of Wales, Newport.
- Stampholidis, N.C. 1996. *Reprisals. Contribution to the Study of Customs of the Geometric-Archaic Period. Eleutherna, Sector III*, 3. Rethymnon: University of Crete.
- Stead, I.M., Bourke, J.B. and Brothwell, D. 1986. *Lindow Man. The Body in the Bog*. London: British Museum Publications.
- Szabó, M. 1991. 'Mercenary activity'. In: S. Moscati, O.H. Frey, V. Kruta, B. Raftery and M. Szabo (eds.): *The Celts*. London: Thames and Hudson, pp. 333-36.
- Taylor, T. 2002. *The Buried Soul. How Humans Invented Death*. London: Fourth Estate.
- Tierney, J.J. 1959-60. 'The Celtic ethnography of Posidonius'. *Proceedings of the Royal Irish Academy*, 60. Dublin: Royal Irish Academy, pp. 189-275.
- Walker, L. 1984. 'The deposition of the human remains'. In: Cunliffe, B.: *Danebury: An Iron Age Hillfort in Hampshire*. (Council for British Archaeology Research Report, 52). York: Council for British Archaeology, pp. 442-63.
- Warner, R. (trans.) 1958. *Plutarch. Fall of the Roman Republic*. Harmondsworth: Penguin.
- Wellesley, K. (trans.) 1964. *Tacitus. The Histories*. Harmondsworth: Penguin.
- Welsby, D.A. 1996. *The Kingdom of Kush. The Napatan and Meroitic Empires*. London: British Museum Press.
- Wiseman, A. and Wiseman, P. (trans.) 1980. *Julius Caesar. The Battle for Gaul*. London: Chatto and Windus.
- Zimmerman, L. 1997. 'The Crow Creek massacre, archaeology and prehistoric plains warfare in contemporary perspective'. In: J. Carman (ed.): *Material Harm*. Glasgow: Cruithne Press, pp. 75-94.

Rebellion, Combat, and Massacre: A Medieval Mass Grave at Sandbjerg near Næstved in Denmark

/21

PIA BENNIKE

Ever since the first fossil remains of australo-pithecines were discovered, we humans have speculated whether certain lesions on bones are evidence of aggression between individuals throughout the stages of our evolution. A few holes in the skull of an *Australopithecus africanus* from South Africa were initially interpreted as evidence of violence committed by other australopithecines. Some of our oldest ancestors were therefore seen as 'killer-apes'. Later on however, Bob Brain, a distinguished South African professor of anatomy, suggested that the holes matched puncture-holes made by the canines of a leopard. Our ancestors were therefore no longer considered to be aggressive killers, but vulnerable victims of the many roaming wild carnivores (Brain 1972).

Similarly, our view of aggression among the Neanderthals has gone through several stages of interpretation which mostly reflect the changing political and philosophical ways of thinking. At times, especially during periods of unrest and battle, the Neanderthals were seen and illustrated as aggressive creatures, whereas in other periods, i.e., the 1960s, they were mainly seen as peaceful, harmless 'hippie-like' individuals (Trinkaus and Shipman 1992). Our own perception of our 'natural instincts' changes: in war they are used as an excuse for man's cruelty to man. These fluctuations in perception run

parallel to the endless discussions on how nature and culture influence human behaviour and their levels of aggression. There are many ways of studying patterns of human aggression and violence and wars of the past. One way is to study the remains of bones; but once again the interpretation of lesions on bones is crucial when drawing decisive conclusions.

Human burials

Mass graves or individual graves situated outside a cemetery may indicate an unusual preceding event. If the remaining skeletal material is well preserved lesions, fractures and abnormalities of the bones may reveal the nature of the event. However, the discovery of a mass grave does not necessarily mean that a war was waged in the vicinity. It is well known that victims of various epidemics, such as plague, cholera etc. were buried in mass graves as well. Such graves contain a majority of the remains of children and old people, whereas young individuals, who supposedly have the strongest immune systems, had a better chance of recovery. The demographic pattern in a mass grave may therefore reflect that part of a population succumbed to a disease which does not necessarily leave any visible traces on the bones of a skeleton.

Mass graves are found all over the world, and can be traced to all periods in the history of mankind. Two Mesolithic skull-pits discovered at Ofnet in Germany contained the remains of c. 32 individuals: 5 males, 10 females and 17 sub-adults. Because of the unusual and rather strange arrangement of the skulls, the pits were named the 'skull-nests'. Many of the skull fragments exhibit a variety of lesions, and the find is thought to be related to a massacre (Frayer 1997). Collective megalith graves in Denmark have been dated to the Neolithic periods and represent a different kind of mass grave. In contrast to mass graves from later periods, collective megalith graves from the Middle and Late Neolithic were functional over a long period of time. They could contain over 100 individuals, and the number of injuries seen on the remaining bones is rarely unusually high (Bennike 1985a).

The remains of several humans have been discovered in several Danish bogs. The bones are mainly from the Early Neolithic and the Iron Age. It is still unknown whether these individuals ended their days in the bogs as votive offerings, as punishment for committed crimes or because they were enemies (Bennike 1985b). In a few cases, a number of injuries are visible on the skeletal remains (Sellevold *et al.* 1984).

A Viking Age cemetery with c. 132 inhumation graves was excavated at Trelleborg. One mass grave contained the remains of 12 individuals buried next to each other, suggesting that they were buried at the same time. The skeletons lay on their backs pointing East-West, and the majority was male. There were almost no traces of injuries on the bones, but that might be attributed to the fact that the latter were poorly preserved. The dental status was reported as 'very good' and might be related to the low average age of the individuals and/or favourable living conditions. The author of the report therefore suggested that the individuals may have been a group of elite soldiers (Christophersen 1941).

So far, no mass graves have been found in Denmark containing victims of the plague epidemics that swept the country in the mid-14th century and again later. However, 56 skeletons were found in a single burial plot connected to a building in Copenhagen used to house plague victims in 1711. Surprisingly, all the children (23%), males and females were buried in coffins, instead of having been pushed into the grave in a panic. Isolated bones

of several hundred victims of a cholera epidemic which hit Copenhagen during the 19th century (Bonderup 1994) have been excavated and are now stored at the Laboratory of Biological Anthropology in Copenhagen.

Considering the number of battles and wars that were fought in Europe throughout the past, only a few mass graves have been discovered and excavated. Many may still be intact, but it is also possible that victims were left on the battlefields to wild animals, weathering and post mortem decay, destroying the bone structure itself. This is what happened to bones discovered in Aljubarrota, Portugal which were dated to 1385 AD. Almost all that was left of the skeletal remains of at least 400 individuals was fragmented shafts of long bones (Cunha and Silva 1997). A recent study of 38 skeletons of victims of the battle of Towton in North Yorkshire in 1461 AD (Fiorato, Boylston and Knüsel 2000) suggests that some of the more than 28,000 men who are believed to have died during this particular combat suffered the same fate. The battle of Towton was part of the Wars of the Roses (1455-1487 AD), a civil uprising during the fight for the throne. The skeletal remains of the 38 individuals were found in a mass grave and the bones had a total of 113 skull injuries, an average of 3 per individual, of which 73 were sharp, 28 blunt and 12 puncture wounds. In addition, 43 postcranial injuries were found.

Until 2001, three mass graves (with the skeletal remains of c. 1,200 individuals) at the site of Visby in Gotland, Sweden, were the only ones to have been excavated in Scandinavia. During the battle of Visby in 1361, a large part of Gotland's male population succumbed to the Danish King Valdemar Atterdag's army. The skeletons of the 1,200 men were disposed of in three large mass graves, which are the largest medieval graves in Europe ever to be subjected to anthropological study – by Ingelmark in Bengt Thordemann's book published in 1939. The many skeletons lay in random positions except for in one of the graves, where the uppermost twenty skeletons lay parallel to one another with their heads to the west as prescribed by medieval tradition.

In 2001, an area of 7m² revealed the skeletal remains of c. 60 individuals divided into three pits with different categories of bone deposits 1) complete skeletons, 2) whole limbs and 3) disarticulated bones. The remains were probably gathered some time after the battle and buried in three different pits according to their stage of decomposition. The excavated area is believed to be only a smaller part of a much larger area containing one or more further mass graves which have not been excavated. The majority of the 60 skeletons were males, 20-30 years of age with an average height of c. 170 cm. Whereas the skulls and skull fragments had 86 cut lesions (an average of 1.6 cuts per skull) only a few (11) such lesions were found on the postcranial bones. The skeletal remains are believed to belong to the victims of 'The Battle of Good Friday' which took place in Uppsala, Sweden, in 1520 AD. According to a preliminary report presented as a poster (Kjellström 2002) the battle was fought between the Danish King Christian's troops and rebels loyal to Sten Sture, the Swedish national administrator.

The mass grave from Sandbjerget in Næstved

It seems only natural to compare the finds from previously excavated mass graves with a newly detected one in the so-called Sandbjerget ('Sand Hill') in the town of Næstved in Denmark. Three C-14 dates show that the grave is from around 1300 to 1350 AD. An anthropological study of the skeletal material found at this site is the subject of this article. Sandbjerget is a 41 m high ridge in Næstved, on Southern Zealand. After the town was established during the Iron Age (c. 1,500 years ago) the ridge functioned as a windbreak for the town, as grazing ground for cattle and as a source of sand for pottery, for floors and as foundations for cobble-stoned streets. After a preliminary excavation in summer 1994, it became clear that the hill also functioned as a burial site including a mass grave. Records from the old archives revealed that several graves had been observed in the 19th century. One grave contained a clay vessel which was dated to around 1,000 AD. Another grave contained three skeletons, one of which had its skull placed between the thighbones. This was customary when burying criminals. Written sources described that one grave was different from the others as it contained the skeletons of at least 30 individuals and that many of those had fatal injuries. It was reported that a spearhead was found among the many bones. Unfortunately, the remains of all the graves have been lost, including the bones (Hansen 1995; 1996).

The excavation of the new mass grave, an area of 12 m², took place during the winter 1994. An anthropologist (the author) and a museum conservator assisted the excavating archaeologists. With the little time granted for the excavation, an efficient collaboration between various specialists proved to be of great value in terms of paving the way for the subsequent anthropological studies of the skeletons. It became clear that every single bone had to be marked and removed separately. As there was no time to make drawings, the problem was solved by extensive use of the stereo-photogrammetry technique to document the exact position of the skeletons in relation to one another. Each bone was numbered on photographic enlargements and registered with accompanying remarks on its position and on the skeleton it belonged to. The mass grave appeared to be 1.5 m deep and four layers were produced using the computerised facilities. This method allowed the bones belonging to the individual skeletons to be identified and collected together in boxes later.

Correlating sections of the key diagram of all the skeletons to the number of excavated skulls revealed that the latter were more or less evenly distributed throughout the mass grave. As previously mentioned, there was no clear pattern with regard to the orientation of the bodies. Although some bodies lay more or less parallel to the edges of the grave, the heads were not placed in any particular direction. The preference for an orientation of the head to the west, north-west or south-west (22 cases) compared with an orientation to the east, north-east or south-east (16 cases), seems not to be significant.

It is impossible to reconstruct the manner in which the grave was filled, but the 18 m³ space available (assuming that the grave was 1.5 m deep) would have left very little room between the corpses. Had they been pushed in, one might have expected to find more of them lying on their back, their abdomen, or on their sides, parallel to the edges of the grave. If they were thrown into the grave by the arms and legs, they would certainly land in random positions. The survey diagram illustrates that the skeletons were found in all orientations and positions, with a few parallel to the edges of the grave suggesting that both scenarios are probably correct (Fig. 1).

The bodies were thrown or pushed into the c. 12 m² and 1.5 m deep grave without any regard for the customs usually adhered to in burial procedures during the Middle Ages. Their heads pointed in all directions; their arms were not positioned systemat-

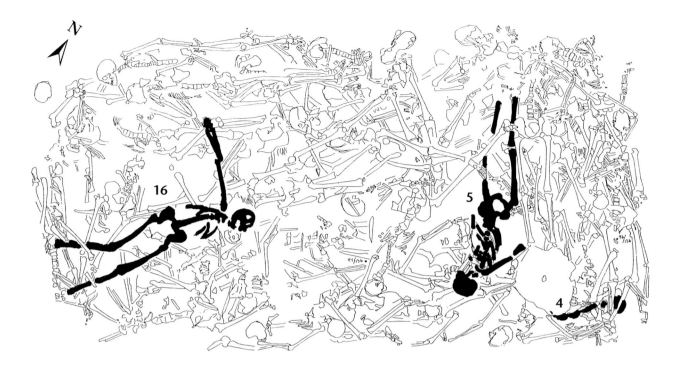

FIG. 1: *A. The layout of the skeletons in the uppermost of the four photographed grave-layers. Skeleton no. 16 (Fig. 2) was found with its mouth open and arm stretched out sideways. Skeleton no. 5 (Fig. 7) lay perpendicular to no. 16. Skeleton no. 4 (Fig. 5) is barely discernable. Drawing: Per Poulsen. B. Drawings compiled from photographs of the skeletons in each of the four layers of the mass grave using special photogrammetric technique. The characteristic medieval practice of laying out a corpse in a supine position with its head pointing west was clearly violated in this case. Drawing: The National Museum and Næstved Museum.*

ically according to tradition. The task of disposing of the naked, lifeless bodies, with numerous fatal skull injuries and faces disfigured beyond recognition by swords and axes (Figs. 2 and 3), must indeed have been an abhorrent one. Exceptionally few belongings were found among the bones: three small rivets, three buckles, two of iron and one of bronze. This suggests that the victims were stripped of everything before burial, coins in the pockets, belts and clothes with buttons. This was, however, the custom of war during the Middle Ages. All possessions and weaponry belonged to the victor. The skeletons in the Sandbjerg grave in Næstved may well be the remains of victims of a rebellion or battle (Hansen 1995; 1996).

During the Middle Ages, corpses were traditionally buried in consecrated soil lying on their backs with the head pointing west. Their arms were positioned in various ways, e.g. stretched alongside the body, or bent at different angles with the hands placed somewhere between the level of the hips and breastbone. The position of the arms was to a certain extent related to the different medieval periods. The hands

often rested high up on the chest during the late Middle Ages, while they were positioned lower down during the earliest periods (Kieffer-Olsen 1993). Gifts and possessions were rarely buried with the deceased.

Very few medieval skeletons have been found in other than the customary burial position, for example with their head pointing to the East. Such deviations may have been purely accidental or an expression of revenge or punishment for a social misdemeanour, so that when the deceased was resurrected he/she would not be facing the rising sun (equivalent to Christ). To be buried in unconsecrated soil, as was the case for five decapitated men and one woman (Bennike and Hansen 2001), who were found in the garden of Næstved Town Hall, was probably a more severe punishment. Numerous sharp cuts are seen on these skeletons as proof of decapitation. Two of the decapitated individuals were laid out with their heads pointing towards the west – either accidentally or as a gesture of respect – even though they were buried in unconsecrated soil. They were C-14 dated to 1300-1350 AD, and are more or less contempora-

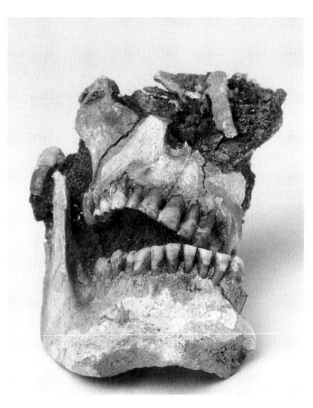

FIG. 2: *Skeleton no. 16 with numerous injuries to the back of the skull. The face with open jaws seems to reflect agony and despair. Photographer: Jens Olsen.*

FIG. 3: *Skeleton no. 31 with both sides of the chin cut off. Photographer: Jens Olsen.*

neous with the skeletons in the Sandbjerg mass grave in Næstved.

Deviations from normal burial habits, be they medieval or present day, are therefore an expression of extraordinary circumstances, e.g., a misdemeanour, a murder, a battle or massacre. Breach of tradition occurred primarily when one wanted to dispose of a body quickly, without any particular gesture of respect or visible evidence of ceremony. No attention was paid to positioning the body correctly. This is clearly illustrated in the large Sandbjerg mass grave in Næstved. However, deviations from normal medieval burial habits also occurred under other circumstances, for instance when many individuals died simultaneously, thus disrupting the social structure. Survivors would not have had the strength to carry out burials according to normal procedures. Moreover, owing to the dangers of contamination and since it was not known how viruses and bacteria spread at the time, the smell of death would be reason enough to bury the dead within twenty-four hours. Much later, in 1853, when Copenhagen was struck by a cholera epidemic (Bonderup 1994), the victims were buried in mass graves.

The number of victims

The anthropological examination of the single bones and bone fragments, some of which had more or less disintegrated, led to the conclusion that c. sixty individuals had been buried in the mass grave. This estimation was based on the number of bones listed in Table 1.

The sixty preserved right femurs and 60 right ulnae would seem to indicate that there were at least sixty individuals in the grave. However, one cannot exclude the fact that a few of these bones may accidentally have been registered twice due to their fragmentation. The registration of the sixty skulls/skull fragments was much more complicated, rendering it impossible to use the skulls as the sole indicator of the number of individuals; several separately registered skull fragments may have belonged to a single individual.

Table 1: Number of bones

	R	L	R+L
Femur	60	55	115
Tibia	55	56	111
Humerus	59	50	109
Ulna	60	49	109
Radius	56	48	104
Foot/parts of	57	45	102
Pelvis	51	51	102
Fibula	34	39	73
Hand/parts of	40	33	73
Patella	33	35	68
Clavicula	23	18	41
Spine/parts of			44
Ribs/parts of			42
Lower jaw			56
Upper jaw			53
Skull			58
Skull/lower jaw			56
Skull/skull fragments			66

Gender

Sex determination of a skeleton is based on a number of criteria, such as the larger, heavier bones and muscle attachments of the adult male. Moreover, the shape of several bones, for example the skull and pelvis, is also an indication of whether the skeleton is male or female. Unfortunately, many of the excavated pelvic bones and other fragile and spongious parts of the skeleton, which are included in gender-determination criteria, had disintegrated. But the remaining intact pelvic bones were those of adult males. Three had certain female characteristics, but other preserved parts of the respective skeletons exhibited prominent male characteristics. Gender-determination is rarely 100% accurate (5-10% uncertainty is not uncommon). This concurs with the fact that a few bones, including skull bones, among the Sandbjerg skeletons had female characteristics, even though the rest of the respective skeleton was distinctly that of an adult male.

There was no indication that any of the skeletons in the mass grave were female. The same conclusion was reached after the examination of the skeletons found in the three mass graves at Visby on Gotland (1361 AD).

Age

The age at death is determined according to a number of age-related characteristics on various parts of the skeleton. The extensive disintegration of many of the pertinent bones, for example the pelvis and ribs, precluded the use of some of the most dependable criteria for determining the age of the Sandbjerg skeletons. In most cases, the degree to which the sutures of the skull had ossified was definable. This criterion, however, only serves as a rough estimate of an individual's age: child or young, middle-aged and old adult. The estimated age is then compared to the average dental attrition of the first two molars, which was registered systematically (Smith 1991). Table 2 lists dental attrition values as the sum of the two molars' degree of attrition, whereas the age groups are derived from a number of other studies (Brothwell 1981, and others). This comparison gives rise to a slight discrepancy between the numbers of individuals in the various age groups based on the observed dental attrition, and those based on the degree of ossification of the cranial sutures; but in both cases the two youngest age groups (i.e., individuals under 30 years of age) add up to 57% of the total number of skeletons.

In a previous study of medieval male skeletons (Bennike 1985b) the age distribution based on the same criteria produced a similar slight discrepancy in the number of individuals; but 49% belonged to the two youngest age groups, while 41% and 10% were delegated to the older and oldest age groups, respectively. Thus, the age distribution of the male skeletons from Sandbjerg does not vary significantly from what is found among males in an ordinary medieval graveyard. Contrary to the age distribution in contemporary armies, and to what one might have expected in a medieval 'army', the victims in the Sandbjerg grave were not all young males.

When dental attrition is used as a criterion for age determination, it is essential to have some knowledge of the overall attrition pattern in the given population. The difference between attrition of the first molar (which erupts around the age of 6) and the second (which erupts around the age of 12) in very young individuals is part of the pattern. Dental attrition was much more severe in various age groups during the Middle Ages in Denmark as compared to the same age groups in the modern population. It was even more severe in prehistoric periods.

Table 2: Age distribution

Table 2: Age distribution

Age groups	Attrition Degree	Dental attrition		Cranial sutures	
Juvenis (< 18/20 yr.)	5 – 6	3	(6%)	6	(13%)
Adultus (18/20-25/30 yr.)	7 – 10	27	(51%)	20	(44%)
Ad/maturus (25/30-40/50 yr.)	11 – 14	17	(32%)	16	(36%)
Senilis (> 40/50 yr.)	15 – 16	6	(11%)	3	(7%)
Adult (unspec.) (> 18/20 yr.)		5		15	

Table 3: Height (men)

	Period	No.	Femur (cm)	Height (cm)
Sandbjerget, Næstved	1300	49	47.5	175.7
Tirup, Jutland*	1100 – 1300	82	47.0	174.6
St. Jørgen, Odense*	1250 – 1450	152	47.2	175.0
Council hall garden, Næstved	1300	4	46.0	172.3
St. Peder's graveyard, Næstved	1100 – 1800	13	44.5	168.8
St. Mikkel, Århus*	1000 – 1529	58	47.4	175.5

Calculated according to Trotter and Gleser 1952

*Boldsen 1993

The epiphyses (ends) of the long bones in three skeletons had not yet ossified, indicating that they were the remains of very young individuals. Judging from the size of the bones, these young persons were between 16 and 18 years old when they died. Several age-related changes on a skeleton are also among the criteria used to determine the age at death. These include osteoarthritis and degeneration of the jaws, both of which are rarely seen in young individuals. There were relatively few cases or traces of osteoarthritis of the hip, wrist or elbow among the skeletal remains. The scarcity of remains of very young individuals and the scarce evidence of osteoarthritis confirm the pattern of age distribution as established by the degree of dental attrition and ossification of the cranial sutures; there were very few individuals in the youngest and oldest age groups. The majority of the victims were between 18-20 and 50-60 years old when they died, most of them young adults.

Stature

It was impossible to measure the height of each individual *in situ* because of their position in the grave and the degree of preservation, fragmentation and disintegration. Instead, stature of adults was measured according to a method developed by Trotter and Gleser (1952) by which the length of the thighbone (femur) is used to calculate the total height of the individual. This method is not totally accurate as it does not take individual body proportions into account. The average length of the thighbone (47.5 cm) and the calculated average height of the victims in the Sandbjerg grave (175.7 cm) are listed in Table 3, together with the respective values derived from other studies of skeletons from various periods in Denmark.

In comparison, the average height of the victims from Visby on Gotland – calculated according to the same method as the Sandbjerg skeletons – was 170.4 cm. The Næstved victims were 5 cm taller. The difference in height may be explained by the fact that the majority of men from the Visby graves are presumed to represent local farmers, whereas the

Næstved skeletons may either represent a more professional group of 'soldiers' selected among the taller men, or individuals from a higher social class who grew up under more favorable conditions than their average contemporary farmer counterparts. However, a femur length of 47.5 cm (Table 3) is not entirely unusual in skeletal material from the Middle Ages, although they certainly were among the tallest. The distinct difference in height between the Visby and Næstved skeletons may represent geographical and even genetic differences expressed in two different populations.

Not all the Næstved skeletons were exceptionally tall. The plotted lengths of the femurs form a curve with a normal distribution. The longest femur measured 53 cm, which represents a height of 188.5 cm. The shortest femur measured 42.5 cm which corresponds to a total height of 164.1 cm.

Dentition

An examination of the dentition revealed that a few of the Sandbjerg victims suffered from a number of dental diseases which must have caused them considerable pain. Tooth-loss can either be calculated as the percentage of individuals who have lost teeth relative to the total number of individuals, or as the percentage of lost teeth relative to the total number of teeth per individual. Teeth are registered as 'lost' when the alveolae in the jaw have closed completely and been absorbed, which happens over time.

There was relatively little tooth-loss among the Næstved skeletons. Only 9% of the skulls had lost one tooth or more. By comparison, a general study of medieval populations in Denmark concluded that the loss of teeth per skull was 26% (Bennike 1985b). The percentage of tooth-loss in relation to the total number of teeth was 0.6% among the Sandbjerg skeletons, whereas in the 1985 study it was 2% (Table 4).

The number of cases of caries is calculated in much the same way as the number of lost teeth. Fifteen out of fifty-eight Næstved individuals had one or more teeth with caries – 26% as compared to 28% in 'other medieval populations' (Table 4). The frequency of cases of caries per tooth in relation to the total number of registered teeth is 2%, which is somewhat lower than the 4% found in 'other medieval populations'. However, the latter includes both male and female skeletons, except for tooth-loss per total number of teeth (Table 4). In the comparative group, the incidence of dental disease was higher among females than among males; this fact may distort the comparison between the two groups to a certain degree.

The relatively good dental condition among the Næstved victims could be linked to the fact that there were more young individuals than in the comparative group, although the difference is relatively small. The incidence of parodontal disease with exposed dental roots was low, which is confirmed by the low incidence of tooth-loss. There was little evidence of hypoplasia of the enamel, which is characteristic of crises such as diseases and/or malnutrition suffered during childhood and adolescence. This might be seen as an indication of relatively favorable conditions during the individuals' early years of life.

Lesions

One of the main objectives of the anthropological study was to determine the events that led to the presumably simultaneous death of so many men. As the mass grave is dated to the 1300s, the many

Table 4: Tooth-loss and caries pr. individual pr. total number of teeth

Tooth-loss	pr. individual		pr. total number of teeth	
Sandbjerget, Næstved	(5/58)	9 %	(8/1379)	1 %
Other medieval pop.*	(29/111 m/f)	26 %	(35/1541 m)	2 %
Caries				
Sandbjerget, Næstved	(15/58)	26 %	(28/1371)	2 %
Other medieval pop.*	(30/109)	28 %	(85/2283 m/f)	4 %

*Bennike 1985b

FIG. 4: *Skull (excavation no. 955) with a very deep, sharp-edged lesion on the left parietal bone. The blow came from above and was undoubtedly fatal. Photographer: Jens Olsen.*

Table 5: Cranial lesions

	R	L	R/L
Forehead	11	12	23
Nasal bone			2
Facial bones	5	8	13
Lower jaw	18	17	35
Temporal bone	3	9	12
Parietal bone	13	15	28
Occipital bone	3	6	9
Total	**53**	**67**	**122**

Front: (Forehead, nasal and facial bones, lower jaw) 56
Side: (Temporal and parietal bones, forehead, lower jaw) 50
Back: (Occipital and posterior part of parietal) 16

deaths might possibly be attributed to the so-called 'Black Death' plague epidemic, which broke out in the middle of the 14th century. However, this explanation can be rejected partly because there are no remains of women and children among the victims and partly because the skeletons bear evidence of innumerable and massive lethal injuries and lesions (Fig. 4).

Skeletons rarely reveal the cause of death; not even in the case of bubonic and lung plague. Most diseases affect the organs and soft tissues, which deteriorate soon after death and leave no visible traces on the skeleton.

Any evidence of healing processes of lesions on bones may indicate that the victim survived for more than 2-3 weeks. In contrast, lesions with sharp, unhealed edges without any cell reaction indicate that the victim either died immediately or very shortly after the wound was inflicted. The Sandbjerg skeletons provide innumerable examples of the latter. This raises a number of questions with regard to the number, shape and position of the lesions on the skeletons, to the weapons that were used to inflict the injuries, and to the context in which the weapons were used.

Unfortunately, most of the skulls were crushed by the pressure of the soil and should therefore have been examined minutely in situ. The lack of time

forced us to settle for the second best solution: the museum conservator encased the best-preserved skulls in plaster during the excavation. These specimens were later taken to the Laboratory of Biological Anthropology where they were 'unwrapped', cleaned, and studied by the author under more favourable conditions.

The disintegration of bone tissue and the pressure-induced crushing of the bones made it difficult to identify many of the lesions accurately. Moreover, lesions inflicted by blunt weapons, such as clubs, are more difficult to identify than the sharp cuts or slashes seen on many of the fragmented skulls. The same is true for lesions inflicted by arrows or spears. The only lesions that could be identified accurately and registered were those that left sharp and/or smooth edges on well preserved bones, e.g., wounds inflicted by swords and axes. There were no arrows, spearheads or parts of other weapons lodged in the bones or the soil.

Far from all the lesions on other parts of the skeletons could be accurately identified because of extensive damage to the bones. For example, the ribs and spines had disintegrated to a large extent, as had the pelvic bones and shoulder blades. As previously stated, the soft tissues deteriorate very soon after death, and lesions on them leave no traces on the bones.

In spite of the many limitations, it is quite remarkable that hundred and twenty-two lesions on the skulls could be identified accurately (Table 5); but it must be kept in mind that these only amount to a small minority of the actual number of injuries that must have been inflicted on the Sandbjerg victims. Most of the skulls, in fact over 90% of them, bear evidence of more than one lesion (Fig. 5). One skull has a total of nine lesions! The few skulls that do not seem to exhibit any visible traces of lesions may indeed have been injured as well; the evidence may have been lost through deterioration of the bone tissue around the affected areas. On average, each skull bore a little more than two lesions.

Figure 6 illustrates the position of the numerous lesions on a skull model. Only nine are situated on the occipital bone, with three on the right and nine on the left temporal bones, respectively. Roughly half the lesions are situated on the front of the skull, the facial bones and the mandible. They are almost evenly distributed on the left and right side in contrast to those on the right and left temporal bones.

Skulls excavated at Visby (Ingelmark 1939) and at Æbelholt Monastery (Møller-Christensen 1982) exhibit decidedly more injuries on the left side than on the right. It was therefore rather surprising not to find the same pattern on the skulls from Sandbjerg. The higher frequency of left-sided lesions on the Visby skulls, (69%) as compared to the right-sided lesions, and a majority of left-sided lesions on the Æbelholt Monastery skulls have been explained by the fact that, in face-to-face combat, a right-handed enemy would inflict most injuries on the left side of the adversary's skull. Right-handedness was as common as it is today, which is reflected in the fact that the vast majority of medieval skeletons have slightly heavier bones on the right side of the skeleton than on the left (Steele 2000).

The usually distinct left-/right-sided pattern is so vague on the Sandbjerg skeletons that it is questionable whether the combatants actually fought face-to-face. If the Sandbjerg victims were attacked by surprise, one would expect to find relatively more injuries to the back of the head. On the contrary, the number of injuries to the front of the skull/face is rather high, and would probably have been higher if the fragile facial bones had been as well preserved as the more robust cranial vault. It was also unusual to find that sixteen individuals (30%) had been struck in the teeth, which break in characteristic

FIG. 5: *Skeleton no. 4 with four almost parallel slashes on the cranial vault. The injuries were inflicted with a sharp weapon e.g. a sword or ax, and look like the result of a massacre. Photographer: Jens Olsen.*

patterns when struck with sharp weapons. In many cases the fracture continued into the adjacent part of the jaw (Fig. 7).

The postcranial bones exhibited sixty-three visible lesions (Fig. 8), which were more or less evenly distributed on the bones of the left and right sides. Most of them are seen on the larger, best-preserved bones, namely the bones of the arms and legs, although many of the absent, more fragile bones could have exhibited many lesions as well. There was no obvious difference in the number of injuries on the bones of the upper and lower extremities.

The distribution, gravity and number of injuries and lesions on the foreheads, faces and jawbones give rise to a number of questions. Did the warriors wear helmets to protect the back of their heads? Did they wear visors? Does the lack of a left-/right-sided pattern in the distribution of the injuries and the many lesions on the skulls mean that the warriors were also attacked after they had been fatally wounded and lay defenceless with their helmets removed? This still does not explain why there are almost twice as many lesions on the skulls than on the rest of bones, unless the body was protected by armour. Were the attackers on horse-back so that their blows mainly struck the men on the head? Is it conceivable that the victims were driven into the grave and attacked from above? There are endless possible

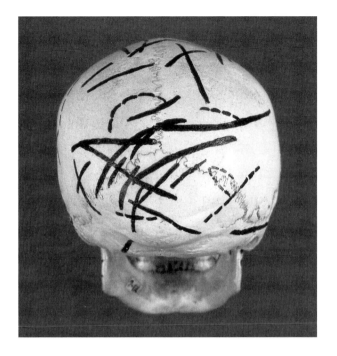

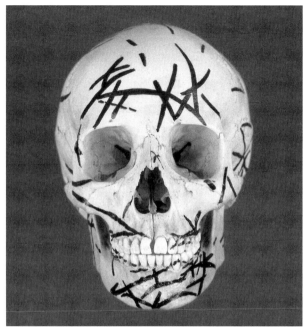

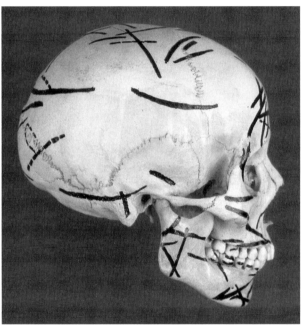

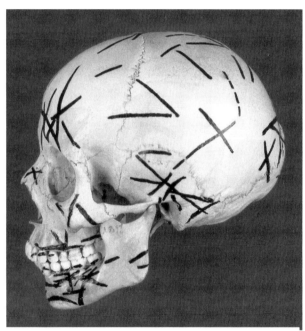

FIG. 6: *The injuries are more or less evenly distributed on the left and right sides of the skull model, but there are many on the facial bones. Drawing: Pia Bennike, photographer: K. Stub-Christensen.*

explanations, some of which would appear to be more credible than others. If these were professional soldiers they could well have carried armour including helmets, etc.

The patterns of injuries on the skeletal remains from the battles of Towton (Fiorata *et al.* 2000) and Uppsala (Kjellström 2002) seem to be rather similar to the distribution of lesions on the skulls from Sandbjerget. There are few differences between the number of injuries on the right and left sides, a large number of sharp injuries scattered on the cranial vault and fewer on the facial bones. In general, there seem to be fewer injuries on the postcranial bones than on the skull. The victims are believed to have been professional soldiers. The study of the skeletal material from Towton concludes that the battle must have been large, multifaceted and brutal and fought with very efficient weapons of war (Fiorato *et*

FIG. 7: *A violent blow with a sharp weapon displaced the right upper row of teeth inwards, and split two of the teeth on skeleton no. 5. Photographer: Jens Olsen.*

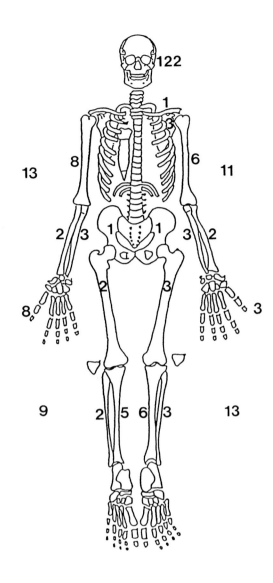

FIG. 8: *The distribution of the 63 lesions on the post-cranial bones. There is no significant difference between the number of injuries on the left and right sides or on the bones of the arms and legs. Drawing: Pia Bennike.*

al. 2000). The same seems to be true for the skeletal remains of the Good Friday battle at Uppsala in 1540 (Kjellström 2002). Our interpretation of the fate of the Sandbjerg skeletons suggests that the men buried in the mass grave were victims of professional warriors with efficient weapons. There is little to go by with regard to whether they were professional fighters themselves. The fact that they all had relatively few postcranial injuries may indicate that they wore some kind of protection (which the victims of the battle of Visby who were mainly local farmers had not). Furthermore, the Sandbjerg victims were rather healthy individuals as reflected by the few pathological bone changes and good dental conditions. Finally, the evidence of several hand injuries may indicate that they did fight before they were killed, which means that they may have been professional fighters like those from Towton and the Good Friday battle at Uppsala.

There have been discussions on why the skull injuries seem to be so numerous in the various skeletal samples from mass graves, because helmets were known and used for protection at the time. It has therefore been suggested that some of the skull lesions may have been inflicted after the enemy was deadly wounded and the helmet removed, but it cannot be proven. More detailed studies of the single lesions are necessary in order to elucidate this question.

Conclusion

The anthropological study of the Næstved skeletons revealed that around sixty male individuals were buried in the large mass grave at the Sandbjerg. A couple of skeletons were not yet fully developed and had only slightly worn teeth, indicating that they were the remains of very young men. There were also a few skeletons of individuals who were well over fifty years of age. Their teeth were very worn, and several had developed osteoarthritis of the joints. However, most of the victims were between the ages of 18 and 50; the majority was between 18 and 30 years old at the time of death. This might indicate that we are dealing with a group of professional soldiers.

There were undoubtedly many more than the hundred and twenty-two registered skull injuries. The number of lesions on each skull varied, and only a few skulls were devoid of any injury. One skull exhibited nine severe wounds, but on average each skull had two. There were sixty-three distinguishable lesions on the postcranial bones; each skeleton had an average of three lesions. The lesions were more or less equally distributed on the left and right sides of the skeletons and were most easily identified on the best-preserved bones, namely the bones of the arms and legs. The number of injuries on the bones of the extremities was evenly distributed between the arm and leg bones.

Generally speaking, the wounds were evenly distributed on the left and right sides of the cranial bones except for those inflicted on the temporal bones. Three were found on the right and nine on the left temporal bones, respectively. The overall pattern is rather inconsistent with what has been described in other studies of similar cranial lesions. Skulls from the battle of Visby on Gotland in 1361 and from the medieval Æbelholt Monastery have a distinct majority of injuries on the left side of the head, indicating that the blows were dealt by right-handed adversaries. The lack of such a pattern on the Næstved skulls could be indicative of the fact that this particular battle was not fought face-to-face, or that the victims wore protective helmets. Other forms of battle and/or protective amour should therefore be considered in the case of the Næstved skeletons. The pattern found at the Sandbjerg might, however, suggest that in this case swords were wielded in a cutting-fashion from right to left and back again.

The calculated average height of these men was 175.7 cm, which is considerably more than the average 170.4 cm for the skeletons found in the three mass graves at Visby. This might support the notion that the Sandbjerg victims were part of a small army of professional soldiers. However, the average height of the Sandbjerg victims correlates well with that of other skeletal finds from the Middle Ages in Denmark.

The dental condition among the Næstved skeletons was relatively good, with a lower frequency of tooth-loss and caries than that found among other Danish medieval skeletal remains. This seems to correlate well with the idea that those buried in the mass grave were relatively young healthy men employed in an army.

Three C-14 dates corrected for the reservoir effect show that the mass grave most probably dates back to 1300-1350 AD, during the civil war in Denmark which began after the death of King Valdemar Sejr in 1241 and lasted until Valdemar Atterdag re-established peace in the country around 1367.

Two historical sets of events took place around the time the mass grave was established and the 60 men died. The first seems to have taken place a little earlier than indicated by the C-14 dating. Three battles were fought between the Danish Royal army and groups of Danish or German noblemen. Wars were expensive and the country was poor. Money was borrowed from noblemen who gradually became more powerful; sometimes too powerful for the king's liking.

From 1288 to 1293, noblemen under the leadership of Marsk Stig Andersen Hvide and their Norwegian counterparts waged wars along the Danish coasts in revenge for having been outlawed for their involvement in the murder of a Danish king. Their war of revenge in 1289 is well documented in written sources. They terrorised and plundered many of the cities close to the west coast of Zealand. Næstved was one of the largest and wealthiest towns at the time (Hansen 1995). The town was definitely worth attacking, but whether it was worth dying for is another question. Outlaws were always buried in unconsecrated soil, which correlates well with the situation of the Sandbjerg mass grave.

The second historical event took place in 1344 when King Valdemar Atterdag besieged the Holsteiners' fortifications in Næstved and conquered the fort the following year. The mass grave may either have been established by the surviving Holsteiners after the siege in 1344 or by Valdemar Atterdag's people after the conquest in 1345. The victors may have cleaned up the battlefield, collected all military equipment and personal belongings before disposing of the corpses in the mass grave. Møller-Christensen (1955) excavated a leprosy cemetery with 700 individuals in Næstved, and studied the skeletal remains in great detail. It is interesting to note that none of the victims in the mass grave or in any of the other 'ordinary' cemeteries exhibited any evidence of having had the disease, in spite of the fact that the cemeteries are all from the same period.

A kinship-study would very likely establish whether the victims came from the local area or were foreigners. The skeletal material from several excavations at this particular site may provide a sufficient sample for such a future study. An additional study may include occupation and activity related bone changes. Such a study has recently shown interesting results based on the bones of the skeleton crew of King Henry VIII's Great Ship, the *Mary Rose* (Stirland 2000).

BIBLIOGRAPHY

- Bennike, P. 1985a. 'Stenalderbefolkningen på øerne syd for Fyn'. In: J. Skaarup (ed.): *Yngre Stenalder*. Rudkøbing: Langelands Museum, pp. 467-91
- Bennike, P. 1985b. *Palaeopathology of Danish Skeletons. A Comparative Study of Demography, Diseases and Injuries*. Copenhagen: Akademisk Forlag.
- Bennike, P. 1998. 'De faldne fra krigergraven'. *Liv og levn*, 12, pp. 14-21.
- Bennike, P. and Hansen, P. Birk 2001. '"Af med hovedet, damen først!" Næstveds middelalderlige rettersted i Amtmandshaven'. In: L. Esbjørn and L. Plith Lauritsen (eds.): *Død og Pine*. Nykøbing Falster: Middelaldercentret, pp. 18-24.
- Boldsen, J. 1993. 'Height variation in Denmark A.D. 1100-1998'. In: E. Iregren and R. Liljekvist (eds.): *Populations of the Nordic Countries. Human Population Biology from the Present to the Mesolithic*. Lund: Institute of Archaeology, University of Lund, pp. 52-60.
- Bonderup, G. 1994. *Cholera-morbro'er og Danmark*. Aarhus: Aarhus University Press.
- Brain, C.K. 1972. 'An attempt to reconstruct the behavior of australopithecines: the evidence for interpersonal violence'. *Zoologica Africana*, 7, pp. 379-401.
- Brothwell, D.R. 1981. *Digging up Bones*. Oxford: Oxford University Press.
- Christophersen, K.-M. 1941. 'Odontologiske Undersøgelser af Danmarks Befolkning. III. Om Tændernes Tilstand hos en Vikingetids Befolkning. Undersøgelser fra Gravpladsen ved Trelleborg, Sjælland'. *Tandlægebladet*, 1, pp. 1-20.
- Cunha, E. and Silva, A.M. 1997. 'War lesions from the famous Portuguese medieval battle of Aljubarrota'. *Int. J. Osteoarchaeology*, 7, pp. 595-99.
- Fiorato, V., Boylston, A. and Knüsel, C. (eds.) 2000. *Blood Red Roses. The Archaeology of a Mass Grave from the Battle of Towton AD 1461*. Oxford: Oxbow Books.
- Frayer, D.W. 1997. 'Ofnet: Evidence for a Mesolithic Massacre'. In: D.L. Martin and D.W. Frayer (eds.): *Troubled Times. Violence and Warfare in the Past*. (War and Society, 3). New York: Gordon and Breach Publishers, pp. 181-216.
- Hansen, P. Birk 1995 'Krigergraven på Sandbjerget'. *Liv og Levn*, 9, pp. 18-23.
- Hansen, P. Birk 1996. 'Nedkuling'. *Skalk*, 1, pp. 5-10.
- Ingelmark, B.E. 1939. 'The Skeletons'. In: B. Thordeman (ed.): *Armour from the Battle of Visby 1361*. Stockholm: Almquist och Wiksells Boktryckeri AB, ch. 4, pp. 149-209.
- Kieffer-Olsen, J. 1993. Grav og Gravskik i det middelalderlige Danmark. Ph.D. dissertation, University of Aarhus, Aarhus.
- Kjellström, A. 2002. A Sixteenth-Century Warrior Grave from Uppsala, Sweden – the Battle of Good Friday. Poster text presented at the XIVth Palaeopathology Association meeting, Coimbra, September 2002.
- Møller-Christensen, V. 1955. *Ten Lepers from Næstved in Denmark*. Copenhagen: Danish Science Press.
- Møller-Christensen, V. 1982. *Æbelholt Kloster*. (2nd edn.). Copenhagen: Nationalmuseet.
- Sellevold, B. Jansen, Hansen, U. Lund and Jørgensen, J. Balslev 1984. *Iron Age Man in Denmark*. Copenhagen: Kgl. Nordiske Oldskriftsselskab.
- Smith, B.H. 1991. 'Standards of human tooth formation and dental age assesment'. In: M. Kelley and C.S. Larsen (eds.): *Advances in Dental Anthropology*. New York: Wiley-Liss, pp. 143-68
- Steele, J. 2000. 'Skeletal indicators of handedness'. In: M. Cox and S. Mays (eds.): *Human Osteology*. London: Greenwich Medical Media, pp. 307-23.
- Stirland, A.J. 2000. *Raising the Dead. The Skeleton Crew of King Henry VIII's Great Ship, the Mary Rose*. Chichester: John Wiley and Sons.
- Trinkaus, E. and Shipman, P. 1992. *The Neandertals*. New York: Vintage Books.
- Trotter, M. and Gleser, G.C. 1952. 'Estimation of stature from long bones of American whites and Negroes'. *American Journal of Physical Anthropology*, 10, pp. 463-514.

Society and the Structure of Violence:
A Story Told by Middle Bronze Age
Human Remains from Central Norway

/22

HILDE FYLLINGEN

Introduction

In the Early Nordic European Bronze Age, almost 4000 years ago, people began, more scrupulously than before, to make artefacts specifically designed for warfare. These artefacts were weapons forged in metal, shaped for use in close combat. Swords, daggers, axes and spears became important personal possessions and depositing them in burials and hoards reinforced their symbolic meaning. Despite this fact, archaeologists have traditionally believed the Bronze Age to be a peaceful and prosperous period in our prehistory. One may ask why large investments of energy were put into the weapons if these weapons were not used at all. Did the Bronze Age people live in harmony only driven by the desire to secure agricultural wealth and religious prestige?

As mentioned above, 150 years of research has presented us with the view that the Early and Middle Bronze Age (Montelius' period II-III) was a time of prosperity and peace and weapons were made for symbolic use hence ignoring the potential of bronze weaponry in war. Or as Øystein K. Johansen put it in his latest work from 2000:

Researchers as a whole agree that the Bronze Age appears to have been a quiet and calm period. We cannot detect any major changes in people's living conditions during the Bronze Age. Neither violent events, nor migrations or anything similar can be detected in the material (Johansen 2000: 144, my translation)

I do not claim that this is altogether wrong. However, my point is that the skeletal material tells us stories about a less peaceful and easy life. Kristian Kristiansen's recent work supports such a more nuanced view. He argues that the inability to recognise warfare in prehistoric societies can be blamed on the academic traditions developed after World War II. Warfare in the past did not fit well with the idea of building a modern welfare society (post-war) and archaeologists were therefore more focused on rituals and religion (Kristiansen 1999b: 175). Additionally, his studies of wear patterns on swords should be taken as strong indications of violent encounters.

In connection with writing my thesis at the University of Bergen, I have analysed extensive Norwegian skeletal material dated to the Middle Bronze Age. The focus has been on an assemblage of skeletons with trauma at Sund, Inderøy in Nord-Trøndelag, comparing these with the results of an analysis of skeletons from burials, notably cairns, from the same geographical area – at Toldnes. The results of these examinations are presented in this paper and brought together with Kristiansen's analyses of swords. Ethnographic examples will serve as analytical tools in approaching Middle Bronze Age warfare in Norway.

This article will thus focus upon the organisation and perception of warfare in Bronze Age society in Central Norway (Nord-Trøndelag), in particular based on the evidence collected from human skeletal material.

Violence as structuring principle

Violence, or rather the prospect of violent acts, can clearly be seen as having a structuring effect on society as it is part of definition of 'us' versus 'them' and acts as 'moral glue' to unite a society. Schröder and Schmidt (2001) looks at violence as

not necessarily confined to situations of inter group conflict but as something related to individual subjectivity, something that structures people's everyday lives, even in the absence of an actual state of war. (Schröder and Schmidt 2001: 1)

Violence can, in some cases, even be perceived as a ritual activity in itself, and hence as a normal part of everyday life. The Apache tribe of North America exemplifies nicely the connection between violence/war and rituals. An ethnographic description of Western Apache society, during the nineteenth century, distinguishes between two kinds of warfare. On the one hand there were raiding parties, which had the explicit goal of obtaining material goods, notably food. On the other hand were the revenge parties that set out to wage war on a specific enemy group in response to the death of a relative. War was a ritual practice charged with religious meaning and an important element in upholding culturally defined moral values. War expeditions, under strict ritual regulations, formed part of the rites of passage of boys becoming men. These male rites were strongly connected to the puberty rites performed for, and by, the girls. Organised violence was considered just as important as resource procurement and various household activities reflected in the girls' rites of passage, as both were believed to be vital in securing the survival of society.

Thus, war would have both collective and individual sides to it. The collective side would be to take revenge, to demonstrate power, and to fulfil social responsibilities. The individual side would be to obtain personal gratification, material wealth and social prestige (Schröder 2001: 146-49).

The Cheyenne tribe, also in North America, is a classic example of the importance of alliances and warrior societies. The Cheyenne had a military organisation based on military/warrior societies, which cut across normal kinship relations. Each warrior society was named, had an internal organisation, and consisted of men of all ages, who joined voluntarily and became members for life. The formal organisation was only operational during the summer months, so for the rest of the year the warriors would live in scattered camps with their own kin.

The Cheyenne war chiefs were the officers of the military societies, each of which had two headmen and two 'servants'. The war chief was elected for life but would normally appoint a younger man as chief if he were no longer fit to lead himself. His 'membership' would still be valid – securing the personal status – but he would be free of any responsibilities. In cases where the war chief died, all the members of the warrior society would come together to elect a new chief. A war chief could not be tribal chief at the same time. This ensured that military and social power were never in the hands of a single person (Llewellyn and Adamson Hoebel 1942: 99-102).

In modern day society the evidence of violence as a structuring principle can be found on all levels – from the domestic to state. On a state level, the possibility of war helps defining 'us' as a group against 'others' or the 'enemy'. Hence the military, which would be redundant without the prospect of war, is a powerful marker in the creation of national or ethnic identity (Fyllingen 2002a: 58).

I would like to advocate the following view on violence and war: violent acts are hardly ever random but express some sort of relationship to the opposing part. Violence is never completely without meaning to the actor and is never an isolated act; it is part of a historical process and a part of defining a group's ideology, and can actually be seen as a meaningful action when associated with ritualised operations like sacrifice, hunting and war (Schröder and Schmidt 2001: 1-14). It becomes ritualised precisely because it crosses the boundaries between life and death (Blok 2000: 24-29).

War is a state of confrontation in which the possibility of violent encounters is always present. This means that people do not necessarily get physically hurt on an everyday basis, but the threat will always be present, and is not gender specific. The decision to go to war will not be unanimously reached, but will be made – even in the most egalitarian societies – by those holding power. Still, the entire society will profit from (the cultural meaning of) war as it contributes to legitimating power, and possibly create material goods for the victors (Schröder and Schmidt 2001: 1-14). *Individuals are creative forces in society*, with a 'free will', but still (self-evidently) structured by norms and laws in society. A person always has a choice and will make this choice based on personal and social needs (both culturally defined), in other words, what will benefit the ego both biologically and socially (Fyllingen 2002a: 59).

My attempt to connect violence and rituals, as described above, is based on Schröder and Schmidt 2001. Violence is considered not necessarily as an act but as an idea. Violence is meaningful action with a ritualised operation (Blok 2000: 24-29). Violence is not a ritual in itself, but the ritual lies in taking part in actions of violence, notably war. It is the concept of violence, manifested in warfare or raiding, which is structuring. This means that the act of violence, e.g. war, does not structure society but the *idea* of 'us' and 'them' (mentioned above), the fear of attack and collective memory (history) – all contributing to the decision of going to war – does. Within a social setting this would imply that personal status is obtained through participation in a ritualised action. A social setting will also contribute to social status on the individual and collective levels. The status of the individual could be described as that of a 'war hero' who carries both the physical evidence of war, i.e. physical disabilities, scars etc, as well as personal attributes like a sword or other insignia. On a collective level, personal status is substantiated through burial or, as we will see, rejected through the lack of such.

As I will show below, such 'war heroes' – or those of special status if you will – may be detected in skeletal material through the physical evidence of old and new violent trauma inflicted by metal weapons and in burials through the weapons themselves.

Archaeological evidence of violence and battle

Evidence of combat and violence in Middle Bronze Age may be found in three specific fields: I) Weapons were made for the first time specifically for hurting humans, as they had little or no value in hunting; II) Many rock art motifs from Norway and Sweden, for example, depict armed males – with axes or swords – in addition to scenes of 'fighting', i.e. several individuals with weapons and individuals in boats with arms (mostly axes); III) Skeletal material shows both healed and fresh evidence of violent trauma (Harding 2000: 271, 275; Fyllingen 2002a; 2002b; 2002c). K. Kristiansen (1983) has analysed sword types from the Early Bronze Age in Denmark in order to establish whether or not all swords were used in battle and if so, how they connected to specific groups within society. He found that the majority of swords had been damaged and had patterns of sharpening that indicated long-term use. During period II (1500-1300 BC), the position of the sharpening shows that stabbing must have played an important role (during battle). In period III (1300-1100 BC), by contrast, the entire blade (not only the point) tends to be heavily sharpened. This suggests that the swords were now also used for cutting. In both periods, the luxurious metal hilted swords show few traces of wear and damage compared to other types of swords, most particularly the flange-hilted swords, which quite evidently were made for use in battle. Kristiansen believes the differences between the metal hilted swords and the flange-hilted swords to be the result of social relations. The metal hilted sword belonged to the chief, and primarily embodied symbolic meanings and social functions; these swords actually occur in the wealthiest graves. The flange-hilted swords used in battles would, by comparison, have belonged to a class of warriors, even if the chiefs and the warriors belonged to the same social sphere. Kristiansen consider this to correspond to the difference between the political- ritual and the military kind of power. The two kinds of power helped to maintain a warrior tradition and to enhance the high esteem of leading families (Kristiansen 1983).

At Toldnes, one of the areas in question, two swords have been recovered – one metal hilted and one flange-hilted – both belonging to period II/III (Gaustad 1965: 12-14; Rygh 1906: illustrations).

The interpretation of combat scenes on rock carvings in Scandinavia will not be discussed in length here, since this subject clearly deserves thorough attention. I will limit myself to making the reader aware of the fact that many rock carving scenes, primarily from Southeast Norway and Bohuslän, depict people engaging in fighting. They seem to be male figures carrying swords, or wielding axes ready to strike, and often the figures appear as groups in boats. Also, boats seem to have played an important part in rock art, which may reflect their importance as a medium of transportation, for example in raiding or warfare. It should be mentioned that such human figures do not appear on rock carvings from Trøndelag, the area of study. Boat carvings, on the other hand, are rather abundant in the area.

As for the evidence of violence on human skeletons, this will be discussed below.

Archaeological evidence of violent trauma

Archaeological artifacts like weapons highlight the threat of violence, but not its outcome. Investigation of injury and mortality patterns may help assess the environmental and social influences on behaviour (Larsen 1997: 109-19) as the location of the injury on the body provides information on how the violence was inflicted. Ethnographic evidence suggests that in some instances the intention was only to maim the opponent while other sources, i.e. from the Middle Ages, shows that one struck to kill. The latter can be seen on skeletons where some of the (lethal) injuries were inflicted after the victim was on the ground or fleeing from the scene, for instance lesions on the right posterior of the cranium or to the back. A large number of cranial injuries (from the Stone to the Middle Ages) suggests face-to-face battle with lesions on the left side indicating a right-handed attacker, but could also be implying that helmets were not worn for protection. Additionally, lesions to the arms are most often defense injuries while injuries in the abdominal area indicate that the person was not using a shield at the time of impact (Larsen 1997: 157ff).

Pia Bennike (1985) has examined all skeletal material from Danish prehistoric and medieval periods. When it comes to violent trauma, she found that adult, often middle aged, men are the victims and that most attacks have been with the intention to kill, as most lesions are found in the chest area. A few people also seem to have been executed. When looking at the prehistoric skeletal material and the bog corpses together, it seems that the common methods of execution have been decapitation, strangulation and cutting of the throat – only the first being detectable in skeletons and seemingly quite

'common' in the Iron Age. In cases where the weapon is still present in the body, it is most often an arrow. Other injuries are consistent with warfare as the large number of trepanations also indicates (Bennike1985: 104-19).

Scenarios like the ones described above can assist the archaeologist on the road to 'discovering' social patterns in prehistory. In theory, injury patterns can provide us with clues to how an incident took place. They will tell us whether people engaged in face-to-face battle, distance battle – e.g. shot in the back – or if the person in question was executed. Old injury patterns, i.e. healed lesions, are indications of past conflict, and combined with fresh lesions may lead us on the track of an ongoing violent conflict. Also, the lesion in itself can provide information on which type of weapon inflicted the injury. In other words, were the flange-hilted swords in Kristiansen's study used on human flesh?

A case study from Norway

In 2000 I carried out osteological analysis of nine Early and Middle Bronze Age sites – eight cairns and one mass grave – which comprised between thirty-five and forty individuals. The osteological reference books consulted, and the methods of analysis used, are Brothwell 1981; Larsen 1997; Mays 1998; Ortner and Putschar 1981; Roberts and Manchester 1995; Sneppen, Bünger and Hvid 1998; Trotter and Gleser 1958; Trotter 1970 and Ubelaker 1989.

The main material was a mass grave/burial pit from Sund, Inderøy, and a large burial site at Toldnes, Sparbu, in the county of Nord-Trøndelag. As there is no reference material of Norwegian human bones, I decided to analyse five other cairns from Nordland County in the north to Østfold County in the southeast. All the skeletons are from inhumation[1] burials and have been dated to the period 1800-1100 BC based either on 14C analyses and/or burial goods or burial type.

Toldnes/Holan is an area between two farms with at least twenty-two burial cairns dated to the Bronze Age.[2] Excavations were carried out between 1879 and 1905 by the eminent archaeologist Karl Rygh. His descriptions are very thorough and tell us about twenty-two cairns containing some forty stone-cists, bronze burial goods (twenty-one artifacts) and at least fifteen preserved inhumations. By Norwegian standards Toldnes/Holan is our richest Bronze Age burial site. There are two swords from the site, one metal hilted from Toldnes (Cairn I, T.2204) and one flange-hilted from Holan (Cairn XIII, T.7501) – both period II-III. Neither was found with the presented skeletal remains.

Unfortunately, at the time of excavation, skeletons were not valued for their information and only the burials with well-preserved skulls were taken to the museum. Today, seven individuals – six adults and one child – can be identified as belonging to three different cairns, all from the area of the Toldnes farm. Of the six adults – three female and three male – pathological conditions could be detected in three.

Toldnes I – is a female (T.1267/A.I.3772)[3] aged 25-50 with *rickets* in the ulna (Fig. 1). The skeleton is from Cairn II (Rygh's number) and she was buried with a bronze dagger (T.1265) and a bronze celt (socketed axe) (T.1266).[4]

Toldnes II – Male (T.2408/A.I.3729) aged 35, 173 cm tall. He has *rickets* in the right tibia. This male was buried with a female, aged 25-35, and a child under the age of 5. There were no artifacts in the grave.

Toldnes III – Male aged 25-35, 185 cm tall. The tendon at the right radial tuberosity has been severed – possibly at a very early age – resulting in the tuberosity being turned inwards. Abnormally marked blood vessels on the lower extremities – especially the femur – are indications of a cardiovascular condition.

Generally the bones from Toldnes are well preserved. None of the individuals shows any degenerative changes to the skeleton or any evidence of violent trauma (Fyllingen 2002a; 2002b; 2002c; Rygh 1906).

Sund (T.18863/A.I.5186) is situated on a long, small strip of land 12 km southwest of Toldnes. In 1967 large concentrations of inhumed human and animal bones were found on gravel deposit in connection with commercial exploitation of the gravel. The bones lay on top of the gravel, right below the topsoil, covering an area of 10-15 m, and could be divided into six mixed concentrations. Northwest of the skeletons there was a coal pit, possibly a hearth, and a ditch was visible as a semi circle in the gravel. None of the structures could be directly connected to the skeletons. Radiocarbon dating pointed towards the middle part of the Bronze Age – 1500-1100 BC. The only recovered artifact was a bone needle belonging to the Late Neolithic/Middle Bronze Age.

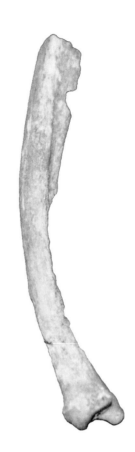

FIG. 1: *Toldnes I (A.I. 3772). Female, age 25-50, with rickets in the ulna. Notice the severe curving of the bone.*

Within the piles of human bones were also bones from rodents, a small ruminant, horse and mammals. Nothing in the find makes it stand out as an obvious burial. The excavators did not agree on how to interpret the find, and have suggested it being either a destroyed burial or remnants of a house. So far, no Bronze Age burial has been recorded in the vicinity (Farbregd, Marstrander and Torgersen 1974).

Due to the fragmentary nature of the material, identification of individuals was extremely difficult and sexing only possible in four cases. I operate with a minimum number of twenty-two, but the real number is considered to be higher. 50% of the individuals were children under the age of 15 – somewhat uncommon in Bronze Age burials. 54% of the adults show evidence of both healed and fresh trauma, which is what I will concentrate on here. In addition there were several cases of degenerative changes – arthritis and spondylosis (63%) – and evidence of osteochondritis dissecans, cribra orbitalia and enamel hypoplasia in the children (10%).

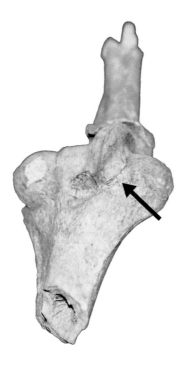

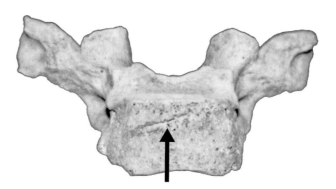

FIG. 2: *Sund II (A.I. 5186). Older adult with healed cut to the left humerus between the medial epicondyle and the throchlea. View of distal portion of the humerus and proximal portion of the ulna.*

FIG. 3: *Sund IV (A.I. 5186). Adult individual with a cut going across the anterior part of a thoracic vertebra. The cut shows no signs of healing and would have been fatal.*

Sund I – This is an individual between 17 and 25 years; sexing was not possible. There is a 2-3 cm long cut to the temporal bone. The cut has healed, but the edges have not fused. The cut is shallow and would not have been a serious injury. The lesion was probably inflicted with a metal blade.

Sund II – This is an older adult, probably over 35-40 years; sexing was not possible. One lumbar vertebra shows a small calcification of the body due to either a slipped disc or a small piece of broken bone. This must not be confused with Spondylosis. The left ulna has degenerative changes (lipping) proximally – at the elbow joint – a (secondary) result of the trauma to the humerus. The left humerus has a 1-1.5 cm long cut between the medial epicondyle and the trochlea. The cut is healed, but not fused, and must have severed several muscles, hence the arthritis (Fig. 2). This injury is a defense injury, inflicted by a metal blade, which would have occurred as the person protected him/herself from an attack by raising the arm towards the attacker. The reason why the damage was not more severe can be put down to the blow either not being very well carried out or

going through a shield, or other means of protection, before connecting with the arm.

Sund III – This is also an older adult; sexing not possible. The individual has a healed fracture of a metacarpal.

Sund IV – Adult individual over the age of 25; sexing was not possible. One thoracic vertebra with a 1,5 cm long, 1-2 mm deep cut across the anterior side of the vertebral body. The cut has one deep and one shallow end. There are no signs of healing (Fig. 3). The injury was inflicted by a metal blade, possibly a sword, which had been thrust through the abdomen and into the spinal column.

Sund V – This is an old adult between the ages of 45 and 60; sexing was not possible. There is evidence of a healed lesion, possibly a fracture, at the distal end of the right radius resulting in deformation of the styloid process and arthritis of the wrist.

Sund VI – An adult (over 25 years) female with three parallel cuts going across the shaft of the first foot phalanx. No signs of healing. The cuts were made with a sharp blade, and it appears as if someone tried to cut off the big toe.

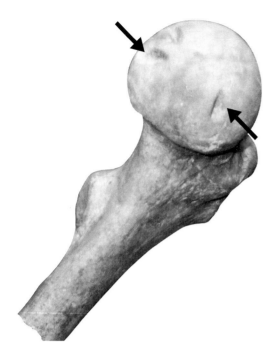

FIG. 4: *Sund VII (A.I. 5186). Adult male with deep impressions in the head and neck of the left femur. The lesions were probably inflicted by a triangular shaped metal blade and show no signs of healing. This is possibly a fatal wound as it probaly severed a major artery in the groin area.*

FIG. 5: *Sund VII (A.I. 5186). Adult male with healed crush injury to the os sacrum. Notice the calcified ligament. Posterior view.*

Sund VII – Adult (over 25 years) male. The neck of the left femur has a 1.5 cm long cut across the anterior surface that appears to be peri-mortem. At the fovea capitis there are two triangular lesions. The lesions show no healing and appear to have been made with a pointed object. The angle of the wound shows that the stab must have been inflicted from the front, through the lower abdomen/pelvic region. This wound is very likely to have been the cause of death (Fig. 4). The left ilium has a healed, but not fused, fracture at the auricular surface. The os sacrum has a fracture of the 2-3 first vertebrae posterior and seems to have been completely dislocated. The fracture is healed, but L5 has fused with sacrum and there is only one joint between L4 and L5 on the left side (Fig. 5). The supraspinal ligament has calcified, so the trauma would have taken place several years prior to death. This kind of severe fracture cannot be caused by falling, but is a consequence of the person being hit or kicked in the lower back. This sort of injury could very well be inflicted by a stone axe or club.

There was also a fragment of a thoracic vertebra with the left transverse process broken off almost completely. A small piece must still have been intact, but has broken off post-mortem. A healed, but not fused, fracture of the upper part of the left lamina, probably partially disconnected the spinous process. The superior articular process has a new facet formed laterally. The superior costal demifacet is pushed distally as a result of the rib not being in its place. The new facet is larger and more pronounced (Fyllingen 2002a).

Ritual and violence
– an alternative social structure

After going through the skeletal evidence and now comparing this to the archaeological finds and the theories presented I believe that there seem to have been two strata among the Bronze Age people – one of them being the Warriors. I will now explain why this conclusion is plausible.

Malnutrition, starvation, dental attrition, cribra orbitalia, rickets and 13C results all point towards a diet based on farm products and meat. This is surprising as these people lived by the sea and close to an abundant source of food. Could this be related to religious practice? It is possible that a food taboo existed among the elite preventing them from consuming fish. Another angle is that when agriculture replaced hunting/fishing as the main means of subsistence religious changes also took place. Agriculture is often accompanied by fertility rites in order to secure the crops and make the animals breed. Social innovations could have been dominant enough to replace the existing way of life over time. A historically known example from Iceland describes how knowledge gets lost over time due to the social implications of farming and strong links to the ancestor's (ideal) way of living:

K. Hastrup (1995) uses historical sources to describe events taking place between AD 1400 and 1800. On Iceland a clear social division existed between farmers and fishermen. Farming was controlled by strict laws and was also connected to the ways of life in the 'old land' – the homeland of the first settlers. After the Plague the population decreased by 40% and in order to get farming started again, the government passed laws designed to make fishing a less attractive occupation. The result was that during the next 200 years Iceland suffered from a collective loss of skills, and the failure to exploit the potential of fishing entailed increasing material poverty. The powerful self-image was found in the Sagas representing traditional values rendered the Icelandic people helpless when it came to creating their own history (Hastrup 1995: 108-18).

Kristiansen (1987) has shown how swords are only found in elite burials, with certain types of swords showing evidence of battle. A sword in itself is a huge investment and a strong symbol as we can see for example in the rock art. It also appears to have been a personal object as it followed the person in the grave and is not part of the (collective) hoards. Kristiansen has presented a different ritual symbolism represented by the metal hilted sword and the flange-hilted sword as traces of use can barely be detected on the metal hilted swords while the flange-hilted show extensive use. Both are found in chiefly burials, on occasion accompanied by 'ritual' war axes (i.e. axes not intended for practical use), and are thought to represent two strata among the elite.

In his paper from 1999 he develops this idea further. Through the burial goods, head gear, dress code and rock art he finds evidence of both a ritual chief and a warrior chief – twin rulers. These two chiefs

ruled side by side and were responsible for the political and ritual power in society. In some instances they might have been buried next to each other, evident in two cists as primary burials, but normally they were buried individually under a 'chiefly' barrow (Kristiansen 1999a: 540-49).

I would like to look more closely at the chiefly systems presented by Kristiansen (1983; 1987; 1999a; 1999b) and Harding (2000) as I find them very useful in understanding and interpreting Sund and its relationship to the burials at Toldnes.

The well-known term 'chiefdom', used by the archaeologists presented in this paper, derives from the systems of social organisation developed by E.R. Service based on ethnographic investigations and published in 1971. He divided social organisation into an evolutionary system starting with the Band, followed by the Tribe, the Chiefdom and finally the State. The chiefdom is a kinship based hierarchical system governed by a chief and with an economic system based on redistribution. As the chief is the person in power he (as most chiefdoms are believed to be patriarchal) is responsible for distributing food and material goods among his people. Only one man occupies the office and the position is inherited. Chiefdoms will always be riddled with conflict as the chief will have to fight other groups in order to maintain the territory of the chiefdom and secure the flow of goods and possibly also women (according to Service) as one tends to practice exogamy, marrying outside the clan, in order to keep up political relations. Having connections to other chiefdoms through marriage is a good way to maintain peace but also to secure a loyal alliance in the case of war.

Religious practice will include ancestors and has a function in establishing social statuses among the living as genealogies become important. The ritual leaders in the chiefdom will consist of a priesthood that governs society together with the chief. The religious and the secular leader will often descend from the same family, or may even be one and the same person, securing political and religious power within a single family/clan (Service 1971: 145-68).

I suggest that the people at both Toldnes and Sund belonged to elite groups. By elite I am referring to groups in society with a strong social position, both ritually and military, controlling large areas of land and resources – independent of access to bronze as this might not have been as vital in Norway as it was further south. As there are only 12 km between the two places they might even belong to the same clan. Settlements have not been located, but it is possible that the population in this area originated from the same settlement.

The skeletal material proves that 'everyone' shows signs of malnutrition continuously through life.[5] Hence there is no difference between each individual when it comes to distribution of food, or which types of food were consumed. A difference only becomes apparent when we look at the inflicted violent trauma, as no lesions have been detected on the bodies from the cairns.[6] This could be important for two reasons: 1) It may tell us who the Toldnes people were, and 2) It may tell us why the individuals at Sund were targeted.

These questions can be answered by comparing new and old injuries of a violent nature. Judging from the injuries on the Sund population and the manner in which the bodies are treated, I believe that the Sund people belonged to a warrior segment in the local area of Inderøy/Sparbu. As some of the injuries are old, at least 5-10 years, the occurrences of violence had been repeated. Of course we do not have enough bodies from Toldnes to exclude them totally as participants in warfare; the presence of both a metal and a flange-hilted sword in the burial cairns and the absence of trauma make the interpretation ambiguous.

Still, taking both Kristiansen's articles from 1987 and 1999 into consideration together with ethnographic description of both the Apache and the Cheyenne, I suggest a possible division between the ritual leader and the war leader in the Nord-Trøndelag Middle Bronze Age society. The evidence of violent attacks is present and as both sword types are also found there should, according to Kristiansen, be a specialisation of social responsibilities.

The result of trade and war

The theory of how bronze was obtained is based on a system of trade, redistribution and possibly reciprocity. This is dependent on an economic surplus, as one needs goods to trade. Long-term starvation is hardly consistent with an economic surplus, so maybe we need to look at alternatives for how this metal was obtained. I do not exclude trade as the source of the bronze found in Central Norway, but only want to suggest an alternative view based on the osteological analyses of health, nutrition and trauma.

It is obvious that bronze was a material of high value both economically and symbolically, so we can imagine that owning it and being able to deposit it in burials and hoards would add to the status of an individual or a family. With the bronze being such an important social communicator, people would have gone to great lengths to secure the supply.

I have suggested, based on Harding (2000) and Kristiansen (1983; 1987; 1999a; 1999b), that the elite had a 'professional' military system with its own warrior society. The reasons for going to war or raiding are many. It could be a strictly personal motive – a revenge party – or a raid setting out to obtain goods. During a raid it would be in the group's best interest to try to avoid violent encounters by getting in and out fast. On the other hand, violence would be important during war, as one attempts to hurt the opponent and return as victors. The warrior group would have been respected for its ability to secure a flow of bronze and protect society, and their statuses were probably different from, but just as important as, those of the religious leaders.

It is not possible to distinguish between warfare and raiding in the archaeological material (presented here), hence I tend to alternate between the two terms. The way I use the terms here is that raiding was used to obtain material goods, while warfare was more of a political act. Raiding would probably also mean that one set out with a smaller group of warriors than one would during war. Raiding and warfare could have been used to obtain bronze directly or to secure the flow of bronze by making sure alliances were maintained and trade routes safe.

The warrior society at Sund would have had a special position. The ability of a warrior to walk the fine line between life and death – and protecting society – would have resulted in achieved status according to the accomplishments of the individual warrior. Alliances would have existed between societies, through negotiation and possibly marriage, securing the military force of the local communities. The presence of violent conflict, either immediate or as a possibility, would structure society, as it kept people together and ensured the status and power of the elite. As seen in the ethnographic examples, the violent encounters could have had ritual implications that added to personal status and to the idea of an esoteric section in society. Old injuries would have been carried with pride, as symbols of noble deeds in the past. 'War heroes' carrying their sword, the strongest symbol of power, must have had a calming effect in the every day struggle for existence, as one knew crops, cattle and people would be protected. As a result the violent and deadly outcome of war also symbolised life and survival.

War and punishment
– the desecration of war heroes

In order to understand the actions behind the Sund society it might prove useful to look at historical sources.

In contrast to physical death, social death might not be irrevocable, but the dead person could be symbolically resurrected as a praised martyr. Therefore, in the exercise of political power, also the social-symbolical killing of the dead may become of utmost importance. (Aijmer 2000: 5)

The implication of this is that moral violation could be pursued into death by not allowing funerals to take place, or by desecration of the body. Postmortem beheading was used as an extra punishment for criminals or to ensure that the dead did not return. In Christianity, retrieval of all the body parts and burial in sacred ground was of utmost importance in order to secure the passing into Heaven. During the Crusades it was not unusual to cut up the corpse so it would be physically easier to bring the dead home. Also, amputated body parts like hands and feet were often kept and can be found next to the dead in burials or kept somewhere else as a relic. Hence the actual body was important for (later) rituals in connection with the social death of a person (Aijmer 2000; Bennike 1985; Kaliff 1997; Larsen 1997 among others).

When going through the photographs and drawings from the excavation at Sund it is possible to notice a pattern within the disarticulated remains. It seems that skulls were placed on top of the bone pile, which could be interpreted as a desecration of the dead, stripping them of their status. This kind of behavior is to be expected when new groups take over a territory as it sends a message to the remaining population – the family members of the desecrated dead.

If Bronze Age burials in cairns were symbols of personal status and group power, failing to recognise this status would be a display of contempt. In this, a double ritual meaning may be recognised. First, the warriors at Sund were not granted respect in accordance to the status they would have achieved through participation in ritual activity, i.e. war, during life. Second, the victors would themselves be taking part in ritual activities through this desecration, consequently enhancing their own personal status.

It is safe to say that something out of the ordinary happened at Sund around 1300 BC. Through the skeletal remains of a people long gone, we have been granted a peek at the realities people had to face. A population comprised of two to three families – possibly a farmstead – was massacred and left on display in the most disrespectful manner. Ongoing conflicts finally took their toll and, with this mass homicide, marked the end of a local era.

NOTES

1 At this time it has not been possible to retrieve or analyse cremations from the BA.

2 There is reason to believe this number to be too small. The cairns are constructed in the same manner as other cairns in Norway and South Scandinavia from the same period, with one or several dry wall cists at the bottom of the cairn which was built from boulders mixed with gravel. None of the cairns was covered by soil. Cremated burials were secondary deposits and thought to belong to the Late Bronze Age or Early Iron Age.

3 T= museum-number for the Museum of Trondheim, NTNU, Norway. A.I. = anatomical identification number given by the Anatomical Institute, U. of Oslo, Norway.

4 It is not common to find axes in Middle Bronze Age burials, but they are known from Late Bronze Age Russia (Mälardal type). Additionally, two other axes have been found in the burials at Toldnes and a fourth was found close to the cairns and described as a stray find.

5 This includes cairn burials from Nordland, Trøndelag and Vestfold. See Fyllingen 2002a.

6 This is evident for all cairn burials in the study (see note 5 above).

BIBLIOGRAPHY

- Aijmer, G. 2000. 'Introduction: The Idiom of Violence in Imagery and Discourse'. In: G. Aijmer and J. Abbink (eds.): *Meanings of Violence. A Cross Cultural Perspective*. Oxford: Berg Publishing, pp. 1-21.
- Bennike, P. 1985. *Palaeopathology of Danish Skeletons. A Comparative Study of Demography, Disease and Injury*. Copenhagen: Akademisk Forlag.
- Blok, A. 2000. 'The Enigma of Senseless Violence'. In: G. Aijmer and J. Abbink (eds.): *Meanings of Violence. A Cross Cultural Perspective*. Oxford: Berg Publishing, pp. 23-38.

- Brothwell, D.R. 1981. *Digging up Bones*. New York: Cornell University Press.
- Farbregd, O., Marstrander, S. and Torgersen, J. 1974. 'Bronsealderens skjelettfunn på Sund, Inderøy, Nord-Trøndelag på bakgrunn av andre bronsealders funn'. (Det Kgl. Norske Videnskapers Selskab Museet Rapport Arkeologisk Serie 3). Trondheim: Det Kgl. Norske Videnskapers Selskab, pp. 1-23.
- Fyllingen, H. 2002a. The Use of Human Osteology in the Analysis of Violence and Ritual During the Middle Bronze Age. A Case Study from Nord-Trøndelag. Unpublished Cand.phil. thesis, Departement of Archaeology, University of Bergen.
- Fyllingen, H. 2002b. 'Massakren Inderøy – vold og uro i bronsealderen'. *SPOR,* 2002(1), pp. 41-43.
- Fyllingen, H. 2003 'Society and Violence in the Early Bronze Age: Analysis of Human Skeletons from Nord-Trøndelag, Norway'. *Norwegian Archaeological Review,* 36(1), pp. 27-43.
- Gaustad, F. 1965. Tidlig metalltid i det nordenfjelske Norge. Unpublished Magistergrads-thesis, Nordic Archaeology, University of Oslo.
- Harding, A.F. 2000. *European Societies in the Bronze Age.* Cambridge: Cambridge University Press.
- Hastrup, K. 1995. *A Passage to Anthropology. Between Experience and Theory.* London: Routledge.
- Johansen, Ø.K. 2000. *Bronse og Makt. Bronsealderen i Norge.* Oslo: Andresen og Butenschøn.
- Kaliff, A. 1997. *Grav och kultplats. Eskatologiska föreställningar under yngre Bronsålder och äldre järnålder i Östergötland.* (AUN, 24). Uppsala: Uppsala University.
- Kristiansen, K. 1983. 'Kriger og høvding i Danmarks bronzealder. Et bidrag til bronzealdersværdets kulturhistorie'. In: B. Stjernquist (ed.): *Rapport från det tredje nordiska symposiet før bronsåldersforskning i Lund 23.-25. april 1982.* (Institute of Archaeology Report Series 17). Lund: Lund University, pp. 63-87.
- Kristiansen, K. 1987. 'From Stone to Bronze – the Evolution of Social Complexity in Northern Europe, 2300-1200 BC'. In: E.M. Brumfield and T.K. Earle (eds.): *Specialization, Exchange and Complex Societies.* Cambridge: Cambridge University Press, pp. 30-51.
- Kristiansen, K. 1999a. 'Symbolic structures and social institutions. The twin rulers in bronze age Europe'. In: A. Gustafsson and H. Karlsson (eds.): *Glyfer och Arkeologiska rum – Vänbok til Jarl Nordbladh.* Gothenburg: Gothenburg University, pp. 537-52.
- Kristiansen, K. 1999b. 'The Emergence of Warrior aristocracies in later European Prehistory and their long-term history'. In: J. Carman and A. Harding (eds.): *Ancient Warfare. Archaeological Perspectives.* Stroud: Sutton, pp. 175-90.
- Larsen, C.S. 1997. *Bioarchaeology. Interpreting Behavior from the Human Skeleton,* Cambridge: Cambridge University Press.
- Llewellyn, K.W. and Adamson Hoebel, E. 1942. *The Cheyenne Way. Conflict and Case Law in Primitive Jurisprudence.* Norman: University of Oklahoma Press.
- Mays, S. 1998. *The Archaeology of Human Bones.* London and New York: Routledge.
- Ortner, D.J. and Putschar, W.G.J. 1981. *Identification of Pathological Conditions in Human Skeletal Remains.* (Smithsonian Contributions to Anthropology 29). Washington D.C.: Smithsonian Institution.
- Roberts, C. and Manchester, K. 1995. *The Archaeology of Disease.* (2nd edn.) Stroud: Sutton.
- Rygh, K. 1906. 'En gravplads fra broncealderen'. *Det Kgl. Norske Videnskabers Selskabs Skrifter,* 1906(1), pp. 3-30.
- Schröder, I.W. and Schmidt, B.E. 2001. 'Introduction. Violent imaginaries and Violent Images'. In: B.E. Schmidt and I.W. Schröder (eds.): *Anthropology of Violence and Conflict.* London and New York: Routledge, pp. 1-24.
- Schröder, I.W. 2001. 'Violent events in the Western Apache past: Ethnohistory and ethno-ethnohistory'. In: B.E. Schmidt and I.W. Schröder (eds.): *Anthropology of Violence and Conflict.* London and New York: Routledge, pp. 143-57.
- Service, E.R. 1971. *Primitive Social Organization: An Evolutionary Perspective.* New York: Random House.
- Sneppen, O., Bünger, C. and Hvid, L. 1998. *Ortopædisk Kirurgi.* Copenhagen, Aarhus and Odense: Foreningen af Danske Lægestuderendes Forlag a/s.
- Trotter, M. and Gleser,G.C. 1958. 'A Re-Evaluation of Estimation of Stature based On Measurement of Stature taken during life and Long Bones after death'. *American Journal of Physical Anthropology,* 16(1), pp. 79-123.
- Ubelaker, D.H. 1989. *Human Skeletal Remains. Excavation, Analysis, Interpretation.* (Manuals on Archaeology 2). Washington D.C.: Smithsonian Institution.

The Dead of Tormarton:
Bronze Age Combat Victims?

As bronze may be much beautified
By lying in the dark damp soil,
So men who fade in dust of warfare fade
Fairer, and sorrow blooms their soul.
(Wilfred Owen)

/23

RICHARD OSGOOD

Postgraduate research by the author at Oxford University into Bronze Age warfare (published for the most part in Osgood 1998 and Osgood and Monks with Toms 2000) conducted a detailed examination of published palaeopathological reports from Late Bronze Age sites of Britain and North Europe in the hope of finding evidence for combat wounds. Perhaps unsurprisingly there were precious few examples; human remains from this period are quite scarce, and those displaying trauma as a result of fighting an even smaller sub-set. One site, however, intrigued me. Not only did it reveal the best evidence for warfare in the British Bronze Age, but there also lay the possibility that more information could be recovered. This site was Tormarton, a rural location in the west of England which had yielded the skeletons of two men with ancient weapons injuries. An excavation proposal was established and funding was raised through the University, BBC television, and the British Academy. The aim of this proposal was to establish a context for the human bodies and to try to provide a cogent argument as to the framework of combat in this period, and whether or not they did indeed relate to warfare.

The discovery of the site

In 1968 a gas pipeline was cut into the Jurassic Limestone in West Littleton Down, Tormarton, South Gloucestershire. What was uncovered remains one of the most intriguing Bronze Age discoveries in the British Isles.

Local farmer Dick Knight had been following the progress of the pipeline with his family, keeping a careful lookout for archaeological finds. His watchfulness was rewarded by the finding of a number of human remains in the disturbed soil east of Wallsend Lane (ST 76737667). Initially thought to be the bones of two individuals dumped without ceremony into a ditch or pit, later palaeopathological work revealed that at least three individuals, all young males, were represented. The remains were studied and then placed at Bristol City Museum where they are now on display.

What made this archaeological discovery so significant was the presence of dramatic weapons injuries suffered by the unfortunate victims. The oldest individual, a man in his mid-late 30s in age and c. 1.75-1.76 ms in height, had twice been speared from behind – a lozenge shaped hole perforated one side of the pelvis (Figs. 1 and 2).

Another of the men had suffered wounds that are shocking to anyone who sees them. He too had been

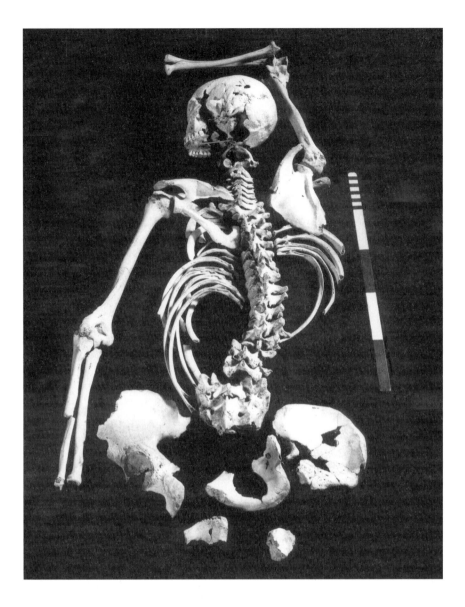

FIG. 1: *The Skeleton of the victim that suffered wounds to the head, spine and pelvis (photograph by Ian Cartwright).*

stabbed in the pelvis – the bronze spear had been thrust in and twisted so that it broke off, remaining within the bone (Fig. 3). He had been speared, again from behind, with force great enough to pierce the lumbar vertebrae and sever the spinal cord – an act that would immediately have rendered him paralysed. As with the wound to his pelvis, this spear also broke off (Fig. 4). A circular perforation in the left side of his skull was inflicted at the same time or soon after – perhaps representing the 'coup-de-grace' of the encounter (Fig. 5).

This latter spear seemed to be of a type found in the Middle Bronze Age and, according to Dr Peter Northover of Oxford University, of an alloy of metals that had perhaps originated in Austria or Switzerland (Northover, forthcoming) – with loops on the side of the spearhead that would have been used to haft it to the spear shaft. To back up this theory, a sample of bone from the leg of the person that had suffered the spine and head wound was sent for radiocarbon analysis, a date of 2970 + 30 BP (around c. 1315 -1045 BC) was obtained (Sample number OxA-13092). This fits well with a date at the end of the Middle Bronze Age in the British Isles (Fig. 6).

A brief report on the finds was made in the Transactions of the Bristol and Gloucester Archaeological Society. This report mentioned that not all of the skeletal material had been recovered (Knight *et al.* 1972: 14) and thus small-scale excavations were undertaken by the author from 1998-2000 to establish whether further remains were indeed present, and exactly what the circumstances of their deposition were.

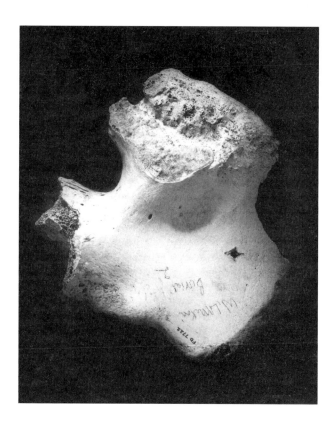

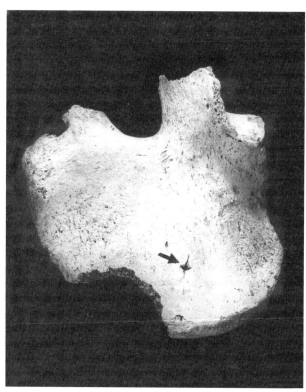

FIG. 2: *Both sides of the pelvis wound suffered by the oldest victim (photograph by Ian Cartwright).*

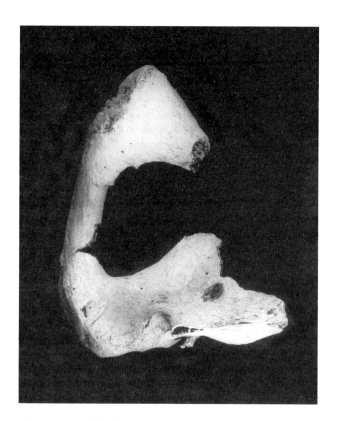

FIG. 3: *A Middle Bronze Age spearhead, still embedded in the pelvis of its victim (photograph by Ian Cartwright).*

FIG. 4: *A Middle Bronze Age side-looped spearhead pierced through the lumbar vertebrae (photograph by Ian Cartwright).*

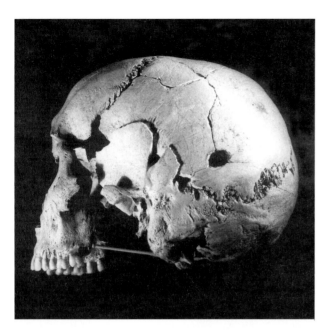

FIG. 5: *The 'coup de grace' wound to the left side of the skull (photograph by Ian Cartwright).*

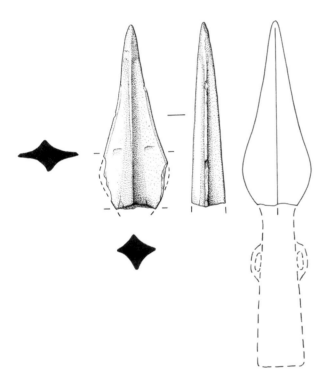

FIG. 6: *The bronze spear from the lumbar vertebrae (length 49 mm) (drawn by Ann Linge).*

Initial investigation

Preliminary work took place in 1998 with the permission of the farmer, Mr George Gent. Field-walking in the region only revealed a couple of flint flakes and quite a large quantity of clay pigeon was present. Other studies were thus undertaken - this included both a magnetometer and also a resistivity survey. The resistivity survey results were confused and revealed little whilst the magnetometer survey simply showed the line of the gas pipe. The signal from this pipeline was so strong that no other archaeological features were discernible. Some distance from the line of the gas pipe, further magnetometer work by Stratrascan revealed the presence of a large linear feature, ending, more or less, by the gas pipe itself.

Aerial photographs of the field in which the site is present clearly show the line of the gas pipe, whilst a series of prehistoric field systems are located in the vicinity along with several features possibly representing trackways.

The excavations

A week of excavation with a small team took place in August 1999, and a further week in 2000. Under the supervision of representatives of Transco (the firm responsible for gas pipelines in the UK), grid squares were mechanically stripped of topsoil by JCB to reveal the line of the pipe. Extensions were made in either direction (E and W) along the pipe to reveal any archaeological features. The initial clearing revealed the top of the linear feature present in the magnetometer survey, running, roughly, North-South. Further cuttings across this feature were made in 2000.

The context

The only archaeological feature in all of the trenches of 1999-2000, proved to be a large V-shaped ditch (Fig. 7). This was truncated by the gas pipeline and lay exactly on the grid reference of the previous discoveries. Quite steep-sided, c. 3.1 ms wide and c. 1.4 ms deep, this feature had, in part, been deliberately filled with limestone slabs. The cuttings were all of similar fill cycles either side of the pipeline. Three cuttings to the south of the ditch had no archaeological artefacts and were fully excavated by hand to define the profile and fill of the ditch.

The cutting immediately to the north of the

FIG. 7: *The V-shaped linear ditch at Tormarton into which the bodies were thrown (photograph by Ian Cartwright).*

pipeline (Cutting 2) was machine excavated to the level of the limestone rubble (Layer 6) and then the ditch was excavated by hand. The next layer (Layer 7) contained large quantities of human bone, much of which was in a fairly poor condition; not as well preserved as the bone recovered in 1968. Excavation proved difficult as the sticky soil matrix adhered to the bone making successful cleaning tricky. Fragments of human skull were present alongside vertebrae, jaws and limb-bones.

A trench across the northern terminal of the ditch, several metres from the finds of bone in 1968 and 1999, was cut in 2000. The terminal was cut through solid bedrock but contained no material whatsoever – which might suggest a lack of nearby settlement sites.

All of the human bone from the excavations was added to the material found in 1968 and sent to a palaeopathologist for analysis. Dr Joy Langston of Newcastle University concluded that there were now at least four and probably five individuals in the ditch (Langston, forthcoming). The identifiable remains were all male and their ages ranged from around 11 years of age to the late 30s. As we have seen, two of the males display savage wounds and it seems likely that, as this was a single-phase episode of violence,

the others in the ditch also suffered a similar fate though their end left no visible trace on their skeleton. The adult with the spear in his spine was in his early – mid 20s in age at the time of the attack, and was the tallest in the group at 178 cm (5ft 11″). Perhaps unsurprisingly, many of the bones showed traces of 'Schmorl's Nodes', indicating that these males had had a physical lifestyle – possibly farming.

We had thus established that there were probably five young male victims that had been killed with spears in one bloody encounter and that their bodies thrown into the bottom of a ditch.

The environment

Snails are very particular about their habitat, as a result of this the specialist can often determine the environment of an archaeological site at a particular time. Several samples was taken in the 1999 and 2000 excavation seasons for the recovery of molluscs and these were processed by hand at the University Museum, Oxford (Robinson, forthcoming). By analysing the molluscs, Dr Mark Robinson has discovered that the bodies at Tormarton were thrown into the ditch which was cut through recently cleared woodland; species such as *Acicula fusca* were recovered and this genus requires a relatively old, undisturbed woodland habitat. The ditch had not been left open for very long when it was filled in. Substantial numbers of carnivorous molluscs – *Oxychilus cellarius* – were also present – these had fed on the brains of the human victims that lay in the ditch, a phenomenon often occurring in bodies held in the chamber tombs of the Neolithic in the British Isles (Robinson, pers. comm.).

The British Middle Bronze Age was a time of much change in terms of agricultural practice. Spelt wheat was increasingly grown, in some regions replacing emmer wheat as the staple crop. Field systems too are increasingly common in the agricultural record, most famously with the 'reaves' system on Dartmoor (Fleming 1988), and those present on the aerial photographs of Tormarton may reflect this. Indeed, many such prehistoric field systems are located in close proximity to lengths of linear ditch – perhaps the latter demarcated territories, which enclosed parcels of land, probably illustrating a pressure for good agricultural land. Thus it might appear strange that no plant remains were found in any of the samples taken from the excavation, but this may be explainable if the ditch was a) only open for a short period of time or b) there was in fact no major agriculture close to the ditch. The fact that the molluscs reflect a mainly woodland fauna may suggest the second option to be correct.

The significance of the site

The finds made in 1968 and the excavations of 1999 and 2000 established a number of important facts about the site of Tormarton:

- At least four, and probably five, human individuals were killed with spears and then cast into a large V-shaped linear ditch.
- The bodies were thrown in without ceremony.
- The burials were covered in a single phase by casting in large limestone slabs – perhaps slighting defences in the process.
- The ditch had been cut in a landscape of recently cleared woodland.
- The condition of the bones in the ditch had deteriorated significantly in the last 30 years.
- This is the best skeletal evidence for Bronze Age combat in the British Isles.

As concluded in the report of 1972, the bodies were buried without ceremony. Dumped into a linear ditch, they were then covered with large limestone slabs which were, on excavation, seen to be surrounded by voids. These slabs perhaps represent the upcast material excavated initially to create the ditch, and used subsequently as an internal bank. One possibility is that the ditch is part of a demarcated tract of land, an element of the later Bronze Age divisions of landscape, which was attacked. The ditch was around 70 ms in length and thus perhaps not a huge physical obstacle in the landscape, especially by comparison with some of the major Bronze Age linear ditches of Wessex (see Bradley *et al.* 1994; Cunliffe and Poole 2000) notwithstanding the fact that some elements of the Bronze Age have disappeared. However, it was probably a fairly important statement of intent, of a claim for territory.

If the covering material over the bodies was part of a bank inside the ditch, then this might indicate that the victims were those that dug the ditch and laid claims to the land. As the feature was slighted in covering the bodies, this would be an act one probably wouldn't have undertaken if one had invested large amounts of effort and energy in digging it. One might also be witnessing a deliberate attempt to deprive the victims of an afterlife – their method of burial differing conspicuously from the norm. As a connoisseur of Bronze Age funerary rites, Mike Parker Pearson (Parker Pearson 1999), said to me – their deposition certainly doesn't smack of the actions of grieving relatives!

In some ways it is this linear ditch or boundary which is of central importance even in later periods. Increasingly, the study of these features is being deemed important with the work of Barry Cunliffe at Windy Dido in Hampshire (Cunliffe and Poole 2000) and Richard Bradley *et al.* (1994: 42, 60) on Salisbury Plain in Wiltshire. The former found some human remains of Iron Age date that had been inserted into one of the Windy Dido Bronze Age linear ditch segments, whilst Bradley also found Iron Age human skeletal material placed deliberately close to the terminal of a Bronze Age linear at Sidbury Hillfort in Wiltshire (op. cit.).

Pre-existing Bronze Age monuments were certainly important to later societies and the insertion of human remains within them as special deposits may have been vital to these groups to claim ancestors. As Richard Bradley states,

If we were to correlate this (deposition of bone) sequence with what is known about the linear ditches, it would suggest that they originally acted in rather the same way as earthworks around settlements and hillforts. Human identities may have been dispersed among the bone deposits associated with both kinds of boundary. (Bradley 2000: 150-51)

Perhaps the situation at Middle Farm in Dorset is a similar one – here human remains were placed close to the terminal of a (roughly) North-South running Middle Bronze Age linear ditch. At Middle Farm, the bodies may have been 'trussed' as they were deposited (Smith *et al.* 1997: 78, 157). The three (possibly four) human burials had not been put into a proper grave, but rather had been deliberately laid near the floor of the ditch and sealed in. Were they used to dedicate the founding of the ditch in much the same way as one might expect a local dignitary to lay the foundation stones of a community building in the current era?

The fact remains, however, that the Tormarton bodies were Middle Bronze Age, and were placed in a Bronze Age ditch. Tormarton and Middle Farm were contemporary, and in both cases the bodies were deposited without ceremony. At Tormarton, the boundary is central to events. It may well have been one element of the landscape clearance of the period – connected to pressure for land and resources and the emergence of new group identities. Intriguingly, the presence of a boundary continued to be important here. Tormarton lies on the current Wiltshire/-South Gloucestershire border and was also on the territorial edges of the Kingdoms of Wessex and Mercia in the Anglo-Saxon period. *Tormarton* itself means 'high point on the boundary', neighbouring Marshfield 'field on the Marches (borders)', and Rodmarton/Didmarton have similar poignancy. Although not suggesting that these names be connected to Bronze Age boundaries, there is nevertheless a rather neat indication of the continued importance of boundaries in the region's past.

This situation is not unique in Britain by any means. The recent work of the Oxford Archaeology Unit in Ashton Keynes on the current Wiltshire/-Gloucestershire border revealed an extensive alignment of Middle and Late Bronze Age pits with some special deposition of human skeletal material. These ran directly along the current county and parish boundaries, boundaries that were certainly already in place in the Domesday book of AD 1086. Rather than a genuine physical barrier, this pit alignment

was a symbolic border close to prehistoric field systems. As the Tormarton ditch was only c. 70 ms in length, and thus perhaps no real barrier, the parallels seem quite strong.

In addition (moving from micro to macro) the general location of the ditch in the landscape is also significant. It is one of a number of elements which are present along the scarp of the Cotswolds. There is quite a steep drop down into South Gloucestershire (as anyone who has driven along the M4 motorway to Bristol can testify) and a series of Iron Age 'hillforts' are sited across the length of the drop; Sodbury and Horton Camps for example. These are both within 7 km of Tormarton and thus the linear ditch with the war victims was sited in a region of rich agricultural land which also lay by a very real natural physical boundary.

A further possibility is that the cutting of the ditch was part of a ritual activity and the bodies represent victims of some type of votive practice, perhaps in the same style as the many bodies at Velim in Bohemia (Harding 1999: 158). Prehistoric linear ditches are often known to have special deposits close to or in their terminals so this might fit. The clear lack of ceremony in the manner of the deposition of the bodies speaks against this interpretation and there are no accompanying deposits, unless one counts the embedded spears themselves. Besides, many of the ritualised mass graves from the Bronze Age in east central Europe are likely to be war-related.

The bodies had been speared from behind whether on the ground or not. They may have been humiliated as part of their death as shown by the spearing of the buttocks of the oldest male. Whatever is the case, it is clear that they were killed in a savage encounter leading one to feel that there is at least some validity to the famous claim of the English philosopher, Thomas Hobbes, that early life was 'nasty, brutish and short' (Hobbes 1651: ch. 13). Some of the skeletal material appeared to be more disarticulated than others. Were some of the body parts exposed by the ditch as a warning to others – in much the same way that the heads of traitors were displayed in the Medieval period? Were parts scattered around by scavengers prior to the ditch being filled? These are questions that must remain unanswered at this point in time, especially given that there are no other known Middle Bronze Age elements within this landscape, apart, perhaps, from the field systems mentioned above (though an Early

Bronze Age round-barrow is situated on a slight rise to the north).

Site parallels

Within the Middle and Late Bronze Age of the British Isles skeletons are rare, individuals with weapons injuries rarer still. Only one other later Bronze Age skeleton with wounds exists in Britain, from Dorchester on Thames in Oxfordshire, and this is far from complete. This pelvis of this body had been pierced by, and retained, the blade of a spearhead – in this case a triangular-bladed basal-looped spearhead. An accelerator date was obtained for this body from the Oxford University AMS laboratory. The date was 2900 ± 40 bp (1260-990 BC) – firmly in the Late Bronze Age of the British Isles (Lab Reference: OxA-6883 –20.5).

There are examples of other bodies of Middle Bronze Age date that show evidence for combat trauma on the mainland of Europe. Perhaps most dramatic is the mass grave at Wassenaar in the Netherlands. Here some of the twelve bodies were seen to have suffered blade wounds and projectile injuries; in one case, a flint arrowhead remained lodged in the ribcage of a victim (Louwe Kooijmans 1993: 15-16). The mass grave at Sund in Norway is a contemporaneous example of a mass grave with several injured individuals. In this case the trauma seems also to be clearly war-related (Fyllingen chapter 22). The communal stone cist at Over Vindinge, near Præstø in Denmark, c. 1600-1500 BC, provides similar evidence even if the mode of burial accords better with normal funerary practice. One of the skeletons, a mature male, had been speared from behind. The tip of the spearhead, of Valsømagle type, was still embedded in the pelvis, and the anthropological examination showed that he had lived for a while with the wound (Bennike 1985; Vandkilde 2000).

Further notable Middle Bronze Age examples come from grave 122 at Hernádkak in Hungary where, in similar fashion to Dorchester and Tormarton, a spear (more or less complete) transfixed the pelvis of a male individual. This is particularly interesting as this came from an established cemetery site, which provided grave goods, including boar's tusks, for a number of the other individuals interred (Bóna 1975: 150). A further Tumulus Culture example is that of Klings in Germany, which also revealed a projectile wound; in this case a bronze arrowhead embedded in a vertebra (Feustel 1958: 8).

One should perhaps sound a note of caution at this point in relation to equating all bodies with spears embedded as necessarily being combat victims. The work of Ian Stead on the Iron Age cemeteries of East Yorkshire is important in emphasising that the completion of social death may sometimes have required post-death violence. The sites of Rudstone and Garton Slack both had several graves in which the interred body had been pierced by several spear-heads, Stead's interpretation was that this had happened whilst the body lay in the grave. One grave at Garton Slack (GS10) had fourteen spearheads, six having been driven into the corpse after death. GS7 had eleven spearheads, five transfixing the body (Stead 1991). Certainly there seems to have been an element of deliberate 'ritual' killing of an already deceased individual, perhaps comparable to the destruction of shields (South Cadbury – see Coles *et al*. 1999), swords (Flag Fen – see Pryor 1991: 114), and Phalerae (Melksham – see Osgood 1995) we find in Late Bronze Age Britain. The context in these particular cases clearly indicates something else than warfare as the reason for the injuries.

Conclusions

This was a paper on the site of Tormarton, which has been interpreted in terms of warfare in the Middle Bronze Age. The site has yielded the bodies of men killed in combat over 3000 years ago. It is my belief that combat of this period was undertaken by relatively small war bands, raiding sites and prestige goods en route to their destination (Osgood 1998). One possibility is that these men were the remnants of a defeated raiding party and thus in that sense a Bronze Age equivalent to the deposit of the Hjortspring boat (Randsborg 1995). However, as the linear ditch was filled after their demise this seems something that a victorious group that has defended the site was unlikely to do.

The finds at Tormarton also illustrate an important fact about the uses of bronze spears. It has sometimes been assumed that the smaller spears were used for throwing and larger variants for stabbing (in lance form). The fact that the Tormarton spears were both small and had been used for stabbing refutes such a claim. One would have used one's weapon in whatever mode was successful for killing

– if that meant using a small spear to stab at close range, so be it. I have looked at all the Bronze Age spears in the Ashmolean Museum in Oxford and the only fact one can really glean is that the longer spearheads tend to be the heavier ones!

Spears of all shapes and sizes would be thrown or thrust depending on the use that the encounter demanded, and the physical abilities of the user. Spears seem to have been the killing weapon of choice in the Middle Bronze – perforations both on human skeletons and also items of protection such as shields seemingly bearing this out. According to Dr Peter Northover of the Department of Materials of Oxford University, the smaller blades of Middle Bronze Age side-looped spearheads were not cold-hardened but cast in high-tin bronze to increase hardness (Osgood 1998: 116).

Defensive equipment would also have been available to the Middle Bronze Age Warrior in Britain; broadly contemporary leather and wooden shields have been recovered from Ireland and it is possible that helmets, greaves and breastplates were worn (see Osgood and Monks with Toms 2000: 25-30). No evidence for such protection was recovered from the bodies at Tormarton, though it is likely that accoutrements of the dead would have been removed before deposition; this was no ceremonial burial with grave goods. Furthermore, any organic elements would have rotted away.

The Tormarton victims remain the only tangible example of British Middle Bronze Age violence and are amongst the best examples for Bronze Age warfare in northern Europe. They represent an example of a skirmish with at least four (and probably five) males being killed. We will, of course never know exactly how many people were involved in the encounter but can postulate that the bodies represent a failed attempt to make a territorial claim by the digging of a large boundary ditch. This claim was refuted in the most final of manners; with the killing of those that had made it.

Acknowledgements

I would like to thank Mr George Gent and family of Tormarton for the kind permission to work on the site and for their enthusiasm. Transco gave advice and supervision on excavating safely round the gas pipeline. Mr Jim Kennedy undertook all the JCB work with great expertise. Dr Mark Robinson and Ms Ruth Pelling of the University Museum, Oxford have worked on the sampling, whilst Dr Tyler Bell was invaluable as Assistant Director of Excavations and Helen Simons of St Hugh's College, Oxford, had the unenviable task of cleaning and marking the bones. All excavation was undertaken with Julian Richards, Bill Locke and a BBC crew as part of a possible TV series; their exertions were also much appreciated. The British Academy provided a grant essential to the funding of the post excavation work. Dr Frances Healey pointed out the intriguing report of Middle Farm in Dorset to me.

Dick Knight, Charles Browne and their families made the initial discoveries and report and also helped with the excavations of 2000. Without their efforts we would have lost a most important piece of the understanding of life in the Bronze Age of Europe.

BIBLIOGRAPHY

- Bennike, P. 1985. *Palaeopathology of Danish Skeletons. A Comparative Study of Demography, Disease and Injury.* Copenhagen: Akademisk Forlag.
- Bóna, I. 1975. 'Die Mittlere Bronzezeit Ungarns und ihre südöstlichen Beziehungen'. *Archaeologia Hungarica*, 49.
- Bradley, R. 2000 *The Archaeology of Natural Places.* London: Routledge.
- Bradley, R., Entwistle, R. and Raymond, F. 1994. *Prehistoric Land Divisions on Salisbury Plain: The Work of the Wessex Linear Ditches Project.* (English Heritage Reports, 2). London: English Heritage Publications.
- Carman, J. and Harding, A.F. (eds.) 1999. *Ancient Warfare.* Stroud: Suttons.
- Coles, J.M., Leach, P., Minnitt, S.C., Tabor, R. and Wilson, A.S. 1999. 'A Later Bronze Age shield from South Cadbury, Somerset, England'. *Antiquity*, 73(279), pp. 33-48.
- Cunliffe, B.W. and Poole, C. 2000. *The Danebury Environs Programme: the Prehistory of a Wessex Landscape. Volume 2 – Part 7. Windy Dido, Cholderton Hants, 1995.* (English Heritage and Oxford University Committee for Archaeology, Monograph 49). Oxford: Institute of Archaeology.
- Feustel, R. 1958. *Bronzezeitliche Hügelgräberkultur in Gebiet von Schwarza.* Südthüringen Southern Thuringia: Herman Böhlaus Nachfolger.
- Fleming, A. 1988. *The Dartmoor Reaves.* London: Batsford.
- Harding, A. 1999. 'Warfare: a defining characteristic of Bronze Age Europe?' In: J. Carman and A. Harding (eds.): *Ancient Warfare.* Stroud: Suttons, pp. 157-73.
- Hobbes, T. 1651. *Leviathan.* London.
- Knight, R.W., Browne, C., and Grinsell, L.V. 1972. 'Prehistoric skeletons from Tormarton'. *Transactions of the Bristol and Gloucester Archaeological Society*, 91, pp. 14-17.
- Langston, J. (forthcoming). *The Human Skeletal Material*

from the Bronze Age Burial Site at Tormarton, Gloucs.

- Louwe Kooijmans, L.P. 1993. 'An Early/Middle Bronze Age multiple burial at Wassenaar, the Netherlands'. *Analectica Praehistorica Leidensia*, 26, pp. 1-20.
- Northover, P. (forthcoming). *Analysis and Metallography of a Side-Looped Spearhead (#R1613).*
- Osgood, R.H. 1995. 'Three bronze phalerae from the River Avon, near Melksham'. *Wiltshire Archaeologicla Magazine*, 88, pp. 50-59.
- Osgood, R.H. 1998. *Warfare in the Late Bronze Age of North Europe.* (British Archaeological Reports International Series, 694). Oxford: Archaeopress.
- Osgood, R.H. and Monks, S. with Toms, J. 2000. *Bronze Age Warfare.* Stroud: Suttons.
- Oxford Archaeology Unit 2000. Cotswold Community, Ashton Keynes, Wiltshire and Gloucestershire SKCC99: Archaeological Interim Report.
- Parker Pearson, M. 1999. *The Archaeology of Death and Burial.* Stroud: Suttons.
- Pryor, F. 1991. *The English Heritage Book of Flag Fen: Prehistoric Fenland Centre.* London: Batsford/English Heritage.
- Randsborg, K. 1995. *Hjortspring: Warfare and Sacrifice in Early Europe.* Aarhus: Aarhus University Press.
- Robinson, M. (forthcoming). *Land Snails from the Bronze Age Ditch and Burials at Tormarton.*
- Smith, R.C., Healy, F., Allen, M.J., Morris, E.L. and Woodward, P.J. 1997. *Excavations along the Route of the Dorchester By-Pass Dorset.* (Wessex Archaeology Reports, 11). Salisbury: Trust for Wessex Archaeology Limited.
- Stead, I.M. 1991. *Iron Age Cemeteries in East Yorkshire.* (English Heritage Archaeological Reports, 22). London: English Heritage.
- Vandkilde, H. 2000. 'Material culture and Scandinavian archaeology. A review of the concepts of form, function, and context'. In: D. Olausson and H. Vandkilde (eds.): *Form, Function, and Context. Material Culture Studies in Scandinavian Archaeology.* (Acta Arrchaeologica Lundensia series, 31). Lund: Wallin and Dalholm Boktryckeri, pp. 1-47.

Funerary Rituals and Warfare
in the Early Bronze Age Nitra Culture
of Slovakia and Moravia

ANDREAS HÅRDE

/24

The dawn of the Bronze Age in south-western Slovakia can be ascribed to the Nitra culture, a cultural group believed to have immigrated into the Nitra and Vah basins through the Moravian Gate from Little Poland and western Ukraine. During the end of the Eneolithic the inhabitants of the river valleys in south-western Slovakia comprised of peoples belonging to the Bell Beaker, Corded Ware and Nagyrev cultures. According to Anton Točik (1963) and Jozef Bátora (1991), the expansion of these cultural groups was now halted and pushed back by the newcomers who were thought to be more aggressive than their neighbours, thus explaining their quick spread in the area. The Nitra group was characterised by introducing not only a new burial custom but also a new copper industry, thus denominating a cultural sphere that would define the onset of the Bronze Age in the region and in neighbouring areas. The possibility cannot be dismissed that the conception of the Nitra culture is that of a culture intimidating its neighbours through continuous warfare, and in which affiliation to the culture was distinguished through social practice expressed especially in burial customs and the trade of certain prestige goods.

Although different in many aspects, the Nitra culture has to a large extent a similar material culture to the Eneolithic cultures in the area – especially the Bell Beaker culture – making it highly probable that the people in the Nitra river valley were to a great extent of the same population rather than members of an incursion. From an overall view the Nitra culture appears to be a unified cultural complex but when looking closer it is apparent that there are significant variations in the way this culture was expressed and how it incorporated traditions from neighbouring regions. These fluctuations vary throughout the different phases and geographical regions and can be considered as inherent phenomena deriving from contacts with other tribes in neighbouring geographical regions.

The Nitra culture is highly interesting when it comes to the study of warfare. Although it covers a small geographic area, extensive research has been conducted in recent decades that cast new light on the subject. The major burial sites are well documented and several osteological analyses have been made giving valuable insights in the nature of violence. Točik (1963; 1979) and Bátora (1991) have presented overviews of the culture giving detailed information on settlement structure, burial customs and social organisation. Recently, Bátora (1999a) also presented an outline of warfare in the Early Bronze Age in Slovakia mainly discussing the role of war in the Nitra and Únětice cultures. His article is the first attempt at placing warfare as an important feature in the development of Early Bronze Age cultures.

With these sources it is possible to give an outline of the cultural complex and most important of all to see how warfare was a defining character for this cultural group when it came to expansion – to acquire new resources – and retaining control over the flow of prestige goods.

Settlement structure

The Nitra culture was located within rather clear boundaries in Slovakia and Moravia. Although many new sites have been found in the last two decades (cp. Bátora 1991) the extension of the culture has not changed much since Točik (1963) presented his overview. Remains from the Nitra culture can be found between the Žitava river in south-western Slovakia and the Morava River in eastern Moravia. To the south the area is limited by the Little Danube and to the north by the beginning of the central Slovakian highlands near the cities of Topol'čany and Trenčin (see Fig. 1) (cp. Točik 1963; 1979; Bátora 1991; Furmánek et al. 1999). This region can in turn be divided into three major areas; (1) the Nitra Basin, (2) the Váh Basin in Slovakia, and (3) the Rusava and Olšava valleys in Moravia.

Of these areas, the Nitra Basin is the most important with many large burial sites such as Branč, Nitra-Čermáň, Jelšovce, Výčapy-Opatovce and Mýtna Nova Vés. Cemeteries from this region usually comprise of 300-500 graves. Most burial sites in this area lay on small terraces close to the river. Only a few were located on sand and loess dunes. Although traces of settlements are scarce, the fact that many contemporary burial sites lay close to each other suggest that settlements must have been close to the cemeteries.

The Nitra Basin is divided geographically into two regions with a southern lowland – the Lower Nitra Basin – and a northern hill-land – the Middle Nitra Basin. The border between these two regions can be set at the highest point of the town of Nitra where the lowland changes into a landscape of rolling hills. This division not only marks different geographical areas, but also social ones. Considering burial customs, there are distinct differences between burial sites from these regions concerning the sacrifice of animals at the occasion of a funeral. In the

northern part sheep and goats prevail, whereas cattle are found in the south (Bátora 1991).

The Váh Basin covers the western extension of the Nitra culture. Unlike the area round the Nitra River burial sites are rare, and most finds derive from scattered locations with few remains. In fact, many finds have been made at locations strongly influenced by Bell Beaker cultures. In some cases we are dealing with single burials from the Nitra culture at Bell Beaker cemeteries (e.g. Točik 1963). It should be questioned whether we are dealing with two different cultures at all in this area since they are so hard to tell apart. It is due to this that the western frontier of the Nitra culture is difficult to define. This is also the case when it comes to the Únětice culture in this area. How the Nitra and Únětice cultures interacted with each other is still much debated (cp. Novotná 1999; Vladár and Romsauer 1999).

Important burial sites are Abrahám and Vel'ký Grob – medium sized with less than 200 interred individuals at each site. As with the Nitra Basin there are few defined settlements in this area, but many finds and remains of small burial sites on loess and sand dunes show that settlements were scattered over a larger area and that they were fairly small. In the northern part of this region there are a few settlements on hill-tops, but these are exceptions to the normal distribution of settlements in the Nitra culture mainly in valleys.

In eastern Moravia the river Morava runs through two smaller river valleys in a highland area near Kroměříž where the third extension of the Nitra culture is located. Although there is only a small number of finds in this area the cemetery at Holešov is an important one with its several hundred graves and high frequency of weapons (Ondráček 1972; Ondráček and Šebela 1985). It was located on the flood plain, which is rather uncommon for this cultural group.

The location of settlements was by no means at random, but followed a very strict pattern. There are clear connections between topographic setting, soil type and the size of cemeteries (Bátora 1991). The largest burial sites are located on river or brook terraces with black soil, suggesting that prime agricultural land was sought after. Not surprisingly these sites were inhabited for long periods. Branč was estimated to have been in use for 120 years and Výčapy-Opatovce for 150 years. Sites situated on black soil are mainly located in the Middle Nitra Basin, but

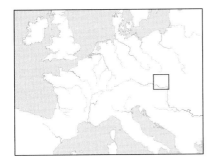

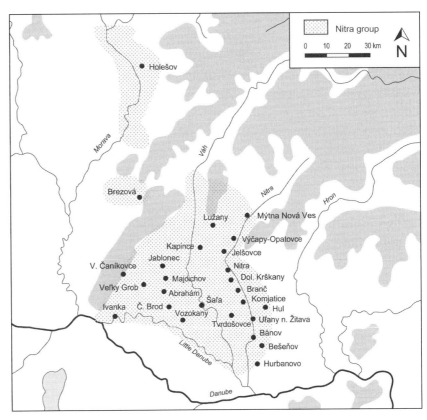

FIG. 1A: *Geographical distribution of the Nitra group in Moravia and Slovakia (After Točik 1963, Furmánek et al. 1999 and Stuchlík 2001. B. Comparative chronology of the Early Bronze Age in the study region (after Bátora 2000).*

Map legend: Nitra group — 0 10 20 30 km — N

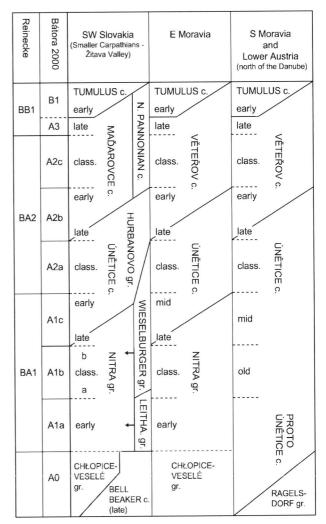

Reinecke	Bátora 2000	SW Slovakia (Smaller Carpathians - Žitava Valley)	E Moravia	S Moravia and Lower Austria (north of the Danube)
BB1	B1	TUMULUS c. early	TUMULUS c. early	TUMULUS c. early
	A3	late	late	late
BA2	A2c	class.	class.	class.
	A2b	early	early	early
		late	late	late
	A2a	class.	class.	class.
BA1	A1c	early	mid	mid
		late	late	
	A1b	b class.	class.	old
		a		
	A1a	early	early	
	A0	CHŁOPICE-VESELÉ gr.	CHŁOPICE-VESELÉ gr.	
		BELL BEAKER c. (late)		RAGELS-DORF gr.

Diagonal culture bands: MAĎAROVCE c., HURBANOVO gr., ÚNĚTICE c., WIESELBURGER gr., NITRA gr., LEITHA gr., N. PANNONIAN c., VÉTEŘOV c., ÚNĚTICE c., NITRA gr., PROTO ÚNĚTICE c.

can also be found in the Váh Basin (Vel'ký Grob, Abrahám), Moravia (Holešov) and the Lower Žitava Basin although they are few in numbers. Second in popularity to black soil was the alluvium found in the lower basins of the Nitra, Váh and Žitava rivers. The least preferred soil type was brown soil, found mainly in the middle and northern part of the Váh Basin region. Occasional settlements on brown soil are found also in the Middle Nitra Basin, which is rather surprising.

The people of the Nitra culture preferred to live on fertile black soil in river valleys. The reason that some of their settlements were found elsewhere must be considered in a chronological perspective. The Nitra culture can be divided into three chronological phases. (1) An *early phase* that is distinguished by being influenced by Chlopice-Veselé, Kosihy-Čaka groups and Corded Ware Culture. This is followed by (2) the *classic phase* in which the Nitra culture developed a distinct material culture. Later in this phase, the Únětician tradition from Moravia begins to influence Nitra material culture. (3) The *late phase*, or *Nitra-Únětice phase*, marks a transition into an Únětice-influenced society in which stone, bone and copper artefacts are entirely replaced by their counterparts in bronze (Točik and Vladár 1971).

The oldest settlements are located in the Middle Nitra Basin, which was inhabited by the Čaka group during the Eneolithic Period (Pavúk 1981). During the classic phase settlements spread into the Lower Nitra Basin, and it is here that we find the earliest Únětician settlements from the Hurbanovo group at the end of the younger phase (Dušek 1969; Vladár and Romsauer 1999). Contemporary with the earliest settlements along the river Nitra are also some settlements in the Váh Basin. Although these appear to derive from settlements of the Chlopice-Veselé group the Nitra Culture does not properly populate this area until at the end of the classic phase (Pavúk 1981; Vladár and Romsauer 1999). The same is valid for the Nitra settlements in Moravia that are dated to the classic and late phase (Ondráček and Šebela 1985). Bátora (1991) shows that settlements on terraces prevail in the early and classic phases and were gradually replaced by settlements on dunes that became predominant during the Nitra-Únětice period. Settlements on hill-tops appear with the emergence of the Únětice culture.

Economy

Pastoralism and arable farming were the basis of the economy. In the previous period of the Chlopice–Veselé group the main economic activity was stock breeding with small and mobile settlements (Pavúk 1981; Bátora 1991). In the Nitra culture the deployment on black soil enabled an extensive use of cultivated cereals, and although this can only be attested by a few finds of stone sickles, cereals must have been important in supporting a large population along the river Nitra. Breeding of cattle, pigs and sheep and goats was also essential. This is suggested by several burial sites where animal sacrifice formed part of the funeral ceremony. Ribs of cattle in graves are frequent at Branč (Vladár 1973) and at Holešov (Ondráček and Šebela 1985). At the latter site skulls of cattle accompanied the dead body. At Výčapy-Opatovce, Jelšovce and Mýtna Nová Ves sheep and goats prevailed as cattle were lacking (Bátora 1990; 1991; 1999b; 1999c). This difference in stock breeding indicates that cattle were preferred in the lowlands whilst sheep and goats belonged to the hill- and highland regions.

Alongside domesticated livestock the hunting of wild boar, roe deer and red deer was important. It is not clear to what extent they were hunted to provide meat and hide, as the domesticated animals must have met requirements in this respect. However, it is striking that many prestige goods in the Nitra culture were manufactured from raw material provided through the hunting of wild animals (cp. Točik 1963; Bátora 1991). Antler beads were vital in the manufacture of large and conspicuous necklaces. More than 25,000 beads of antler have been recovered in Slovakia, and they occur primarily in the cemeteries of the Nitra Basin and only rarely at sites outside this region. Even the shells from gathering shellfish were used in prestigious necklaces.

The tusks were taken from wild boars and worn by warriors as ornaments. Since the wild boar's is a fierce and dangerous opponent only the bravest hunters would have dared to hunt it. The sizes of the tusks, with a mean measure of 15 cm, indicate that the boars were of impressive size and no doubt a difficult foe to kill. Wearing wild boar's tusks was probably considered an insignia of bravery and courage.

Prestige goods were numerous in the Nitra culture and consisted mainly of female ornaments and, in later phases, of metal weapons (see Fig. 2). Apart from necklaces of antler and shell beads most prestige goods were made of copper. In particular, spiral rod necklaces, diadems, dress pins and ear- and arm-rings in many different shapes predominate (Točik 1963; Bátora 1991). Different kinds of rings clearly dominate. They are often willow-leaf shaped, but during the classic phase a double-wire shape becomes common. The copper ornament industry is largely conservative in its design, keeping the inventory intact throughout the period. The copper industry of the Nitra culture stands out as a major source of inspiration for the production of metalwork in the Early Bronze Age of Central Europe. In the early phase the copper industry is very simple. The willow-leaf industry does not have the conspicuous mid-rib and the wire industry is in its infancy with a simple design. During the classic phase a diverse metal industry is developed with a willow-leaf shape with mid-rib, copper sheet tubes and spirals in addition to double-wire trinkets. With the advent of the Únětice culture much of the metallurgy changes as bronze superseded copper. Old forms of prestige goods were no longer circulated and were replaced by Únětician objects, mainly dress pins, smaller rings and bronze axes.

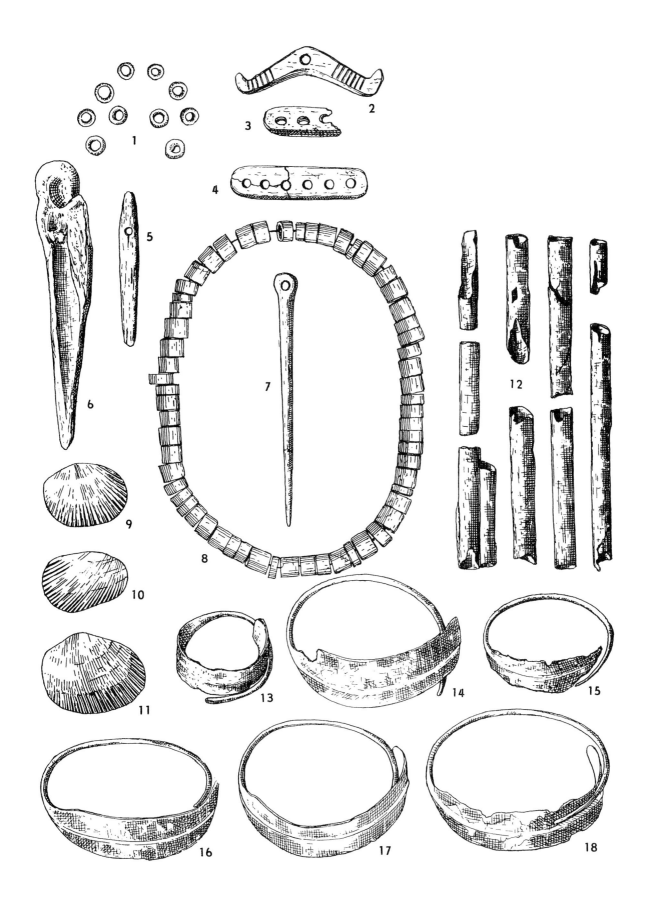

FIG. 2: *Rich female grave at Branč (grave 142) dating to the classic phase. 1. Faience beads. 2.-5. Necklace spacer-plates of bone. 6. Bone awl. 7. Bone needle or dress pin. 8. Antler bead necklace. 9.-11. Cardium shells. 12. Copper sheet tubes. 13.-18. Willow-leaf rings. Scale 1:1. (after Vladár 1973).*

The single most valuable resource the Nitra culture had at its disposal was copper. Although it alone cannot explain the development of this cultural group, access to the metal gave an advantage over neighbouring societies. The mountains of central Slovakia are rich in copper and many sites have been in use continuously from the Bronze Age until the medieval period, such as Spania Dolina at Banská Bystrica (Točik and Vladár 1971). Although the Nitra culture to a large extent occupied flat landscapes copper was not a scarce resource due to the proximity of the copper-rich highland massif of the Small Carpathians. This massif contains several deposits of copper pyrites at sites such as Baba Mountain, Pezinok-Cajla, Červen y Kameň and Selec only to mention a few. The greatest amounts of copper ore in south-western Slovakia are otherwise found in the Štiavnické Vrchy mountains (Bátora 1991). It is also relevant to mention that tin occurs in the Tribeč mountains, though this metal did not come into use until the Nitra-Únětice phase.

Burial customs and funerary rituals

Since settlements are absent, our knowledge of the social organisation of the Nitra culture comes entirely from studies of burial sites. In comparison with later Bronze Age cultures the Nitra culture exhibits diverse and complex funeral practices. The complexity of the burial customs has been discussed in detail by Bátora (1990; 1991; 1999b; c), and he has shown that in order to understand the social organisation it is not only of importance to study the grave goods, but also the arrangement of graves within the cemetery and their construction. In fact, the construction of the tomb sometimes reveal more about the social position than the accompanying grave goods.

Burial practice in the Nitra culture conformed to a standard structure in which the graves have a similar construction and the grave goods are determined by the sex and social position of the deceased. The graves are usually oblong with sharp or rounded corners and the deceased was placed in a contracted position on the side: males were placed on their right side and females on their left side, both facing south. All graves were inhumations and organised into flat cemeteries with an oblong plan extending usually in S-N/N-S direction. Three patterns in the arrangement of the graves can be outlined, although the internal organisation of the cemeteries is the same throughout the period. During the early phase the graves were positioned in small groups while during the classic phase the graves line up into settled rows. In the late phase the graves are placed in larger groups in a more scattered pattern similar to the custom in the Únětice culture (Bátora 1991; Točik 1979).

Some graves deviate from the ordinary burial practice, often regarding size and construction of the tomb. These particular tombs reflect social differentiation within society, inasmuch as they were reserved for warriors and the social elite. They differ from other graves in the following features to a greater or lesser extent:

1. Space – large burial chamber
2. Special position of the body
   a. skeleton extended on its back
   b. displacement of body
3. Construction of the grave pit
   a. wooden chamber or wood lining
   b. stone lining
   c. prepared walls
   d. red ochre
4. Construction above the grave pit
   a. death house – with either (1) post holes at the corners or (2) oblong ditch
   b. burial mound – with either (1) concentric ditch or (2) without ditch

Many features of the burial customs were new to the region and have more in common with funerary practices in eastern and Danubian central Europe. This may indicate that the people of the Nitra culture had recently moved into the region (Točik 1963). A major change to the funerary practice was the orientation of the body to a W-E, E-W direction with men and women placed on opposite sides – a gender differentiated position, which is uncommon within the Únětice culture but common in the Early Bronze Age and Bell Beaker periods of Danubian central Europe. The use of washed walls in grave pits, red ochre and wooden crypts originates in the east, but disappears during the late phase to give way to the more informal burial rites of the Únětice culture. Death houses, however, occur at sites in Austria, Germany, Moravia and Poland from the Únětice period through to the Urnfield period (Bátora 1999c).

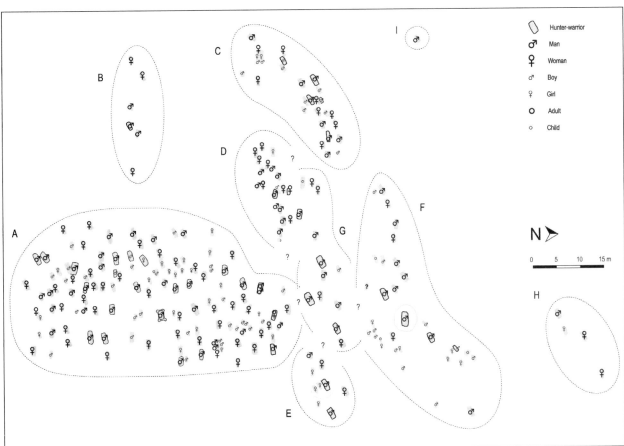

FIG. 3: Branč. Family groups (below) and warrior burials (above) from the Nitra culture.

Considering the spatial arrangement of the graves, their construction and wealth in grave goods, it is possible to distinguish three social categories, each including men, women and children, by elaborating on the discussion of Bátora (1991). At the top of the social ladder stood high-ranking warriors and their families with ostentatious tombs rich in grave goods. Below them were commoners who made up the majority of the population and whose graves were less elaborate, although not poor in grave goods. Women are frequent in this category, which also includes the remaining warriors. At the bottom of the society were those who had no grave goods or just a few items. Men and children are overrepresented in the buried population. The division between the last two categories does not necessarily indicate two separate social levels. Female grave goods should be seen as an investment in wealth displayed at the funeral. Male grave goods by contrast consisted mainly of raw material for manufacturing tools except if they were provided with ornaments such as copper rings and beads. In other words, while burial customs ordained grave goods for women, men were not likely to be buried with anything special – apart from their personal belongings. As a general rule women and girls were of higher rank within their social class. It is remarkable that girls often have larger tombs than boys, as seen at Branč (see Fig. 3), which adds to the impression that they were provided with more grave goods, and thus in some sense were more important. As I discuss below, women and children are often interred near a high-ranking warrior, which can be taken to indicate that they too had an important position in the social system.

Warriors and their social organisation

From the various burial sites it is clear that warriors conformed to a hierarchical structure, which in death entitled them to an extraordinary burial and specific grave goods. Burials from Branč and Mýtna Nova Ves indicate that we are dealing with two classes of warriors – high-ranking warriors, or chieftains, as opposed to common warriors (figs. 4-5). They are recognisable through differentiation in grave goods and the construction of the tomb. Common warriors are buried in ordinary graves with weapons and perhaps ornaments of bone or copper. High-ranking warriors often have several weapons and an array of ornaments and pots, including amphorae.

The tomb itself often possesses specific features, as mentioned above, such as a wooden lining, and was sometimes covered by a mound or a death house. The position of graves at burial sites is also an indication of social status. At Branč, warriors related to a chieftain have different war gear and grave constructions to other warriors, although both groups can be classified as belonging to the commoners.

Equipment
All warriors in the Nitra culture were equipped in a similar way, their paraphernalia consisting of an array of weapons, tools and ornaments (Figs. 6-9; Branč: Fig. 10). Amongst the arms, equipment for archery predominates with arrowheads, bracers and an occasional fragment of a bow. The warrior also carried other arms such as daggers, knives and axes. Archery was dominant in the early and classic phase, whereas daggers and knives were preferred during the transition to the Nitra-Únětice phase. Axes are few in number and belong to the late phase. Their appearance, together with bladed weapons, is a result of influences from the burial customs of the Únětice culture. Tools connected to warriors were whetstones used to sharpen daggers and knives. Stone raw material in male graves cannot be considered as warrior equipment since it is difficult to assess its application, although it might have been used to manufacture arrowheads. Ornaments belonging to the insignia of a warrior are boar's tusks, bone discs, ornamented bone tubes and belt fittings of copper. These ornaments belong to the early and classic phases, and can be found in contemporary Eneolithic cultures in the region and in the Mierzanowice and Koš'any cultures (Bátora 1999c). With the exception of copper belt fittings, these ornaments disappear during the late phase and do not seem to have been replaced by other artefacts.

Boar's tusks occur at many burial sites and their use spread well beyond south-western Slovakia. They often have drilled holes at the proximal end and in some cases also at the distal tip and their position in graves near the chest or head indicate that they were worn either as an ornament on the body or as a hair decoration. A warrior usually wore between one and three boar's tusks, but some had as many as eight (Branč, grave 104). Perhaps the number of tusks symbolised social rank amongst warriors.

The bone disc, with its many drilled holes, was usually found at the back of the head suggesting that it formed part of the warrior's hairstyle. These discs are mainly found at sites in the Nitra Basin (Mýtna Nová Ves, Branč and Nitra-Čermáň) and only on one occasion in the Váh Basin (Veľký Grob), and have their origin in the Mierzanowice and Košťany cultures, dating them to the early and classic phases of the Nitra culture (Bátora 1999c, cp. Chropovský 1960).

Ornamented bone tubes have been found at a number of sites (Alekšince, Branč, Černý Brod, Jelšovce, Mýtna Nová Ves, Nitra-Čermáň, Šaľa I and Tvrdošovce), and their use is debated. Bátora (1991; 1999c) suggests that they were some kind of whistle used by hunters, but they could also have been a case for an awl or a pin, or simply an ornament for dress or body. They are made of the *tibia* from a sheep or the *humerus* from large birds and are associated with high-ranking warriors. Similar bone tubes are frequent in Little Poland and eastern Slovakia at this time, but not elsewhere (Bátora 1999c).

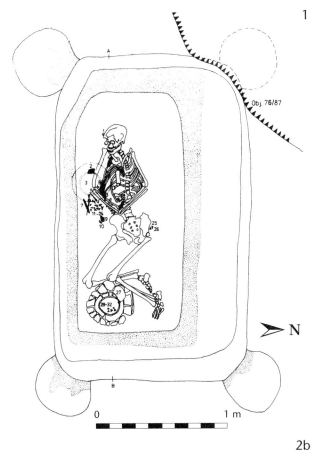

FIG. 4: *Death house with burial of high-ranking warrior (Mýtna Nová Ves grave 262). 1. Plan of grave. 2. Reconstructed grave: a. horizontal and b. vertical view. (after Bátora 1991).*

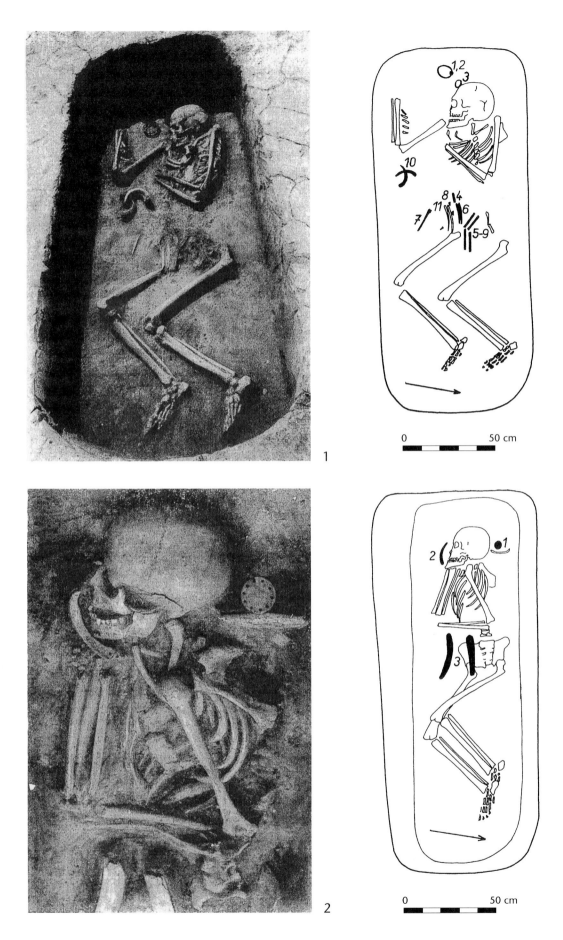

FIG. 5: *1. Warrior from high-ranking family. 2. Common warrior. (after Vladár 1973).*

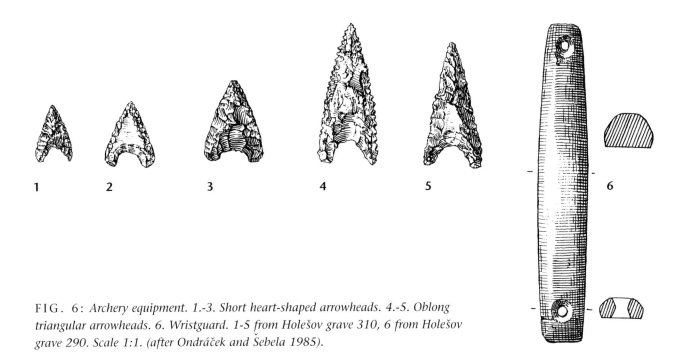

FIG. 6: *Archery equipment. 1.-3. Short heart-shaped arrowheads. 4.-5. Oblong triangular arrowheads. 6. Wristguard. 1-5 from Holešov grave 310, 6 from Holešov grave 290. Scale 1:1. (after Ondráček and Šebela 1985).*

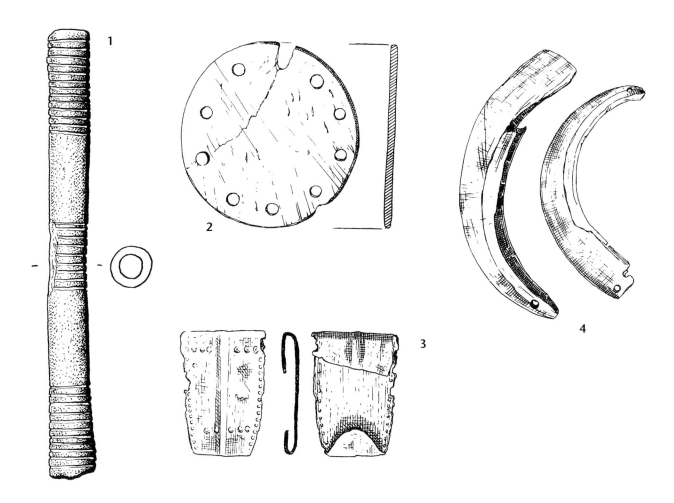

FIG. 7. *Material symbols of warriorhood. 1. Ornamented bone tube (Mýtna Nová Ves grave 262). 2. Bone disc (Branč grave 179). 3. Belt fittings (Branč grave 31). 4. Boar's tusks (Branč grave 74). Scale 1:1. (after Vladár 1973; Bátora 1991).*

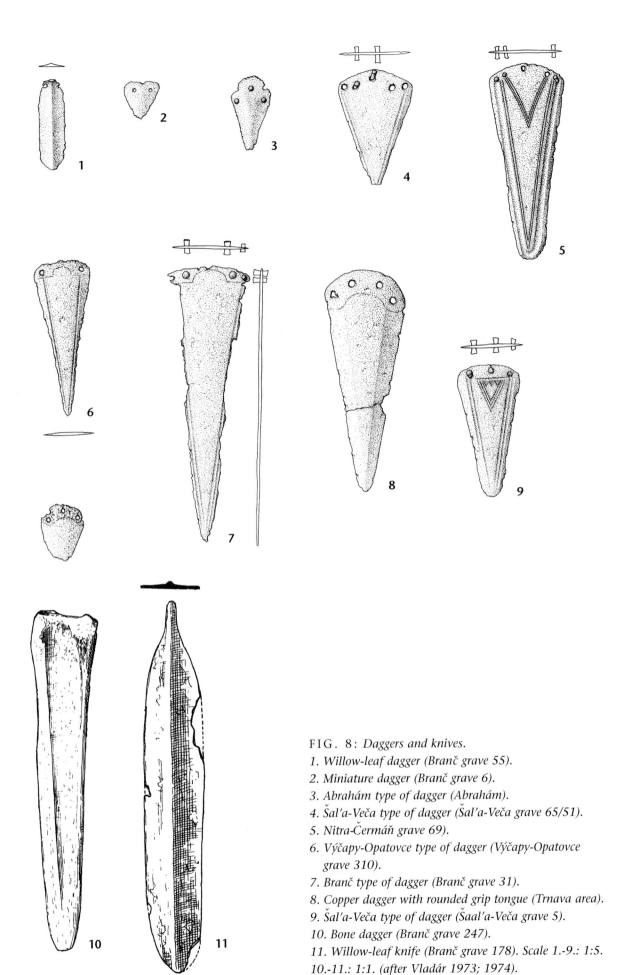

FIG. 8: *Daggers and knives.*

1. *Willow-leaf dagger (Branč grave 55).*
2. *Miniature dagger (Branč grave 6).*
3. *Abrahám type of dagger (Abrahám).*
4. *Šaľa-Veča type of dagger (Šaľa-Veča grave 65/51).*
5. *Nitra-Čermáň grave 69).*
6. *Výčapy-Opatovce type of dagger (Výčapy-Opatovce grave 310).*
7. *Branč type of dagger (Branč grave 31).*
8. *Copper dagger with rounded grip tongue (Trnava area).*
9. *Šaľa-Veča type of dagger (Šaaľa-Veča grave 5).*
10. *Bone dagger (Branč grave 247).*
11. *Willow-leaf knife (Branč grave 178). Scale 1.-9.: 1:5.*
10.-11.: 1:1. (after Vladár 1973; 1974).

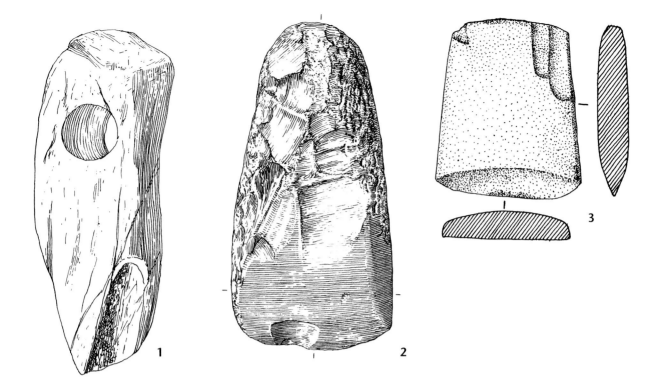

FIG. 9: *Axes. 1. Antler axe (Branč grave 85). 2. Polished stone axe (Holešov grave 68). 3. Polished stone axe (Branč grave 222). 4. Symmetrical hammer axe (Mýtna Nová Ves grave 262). Scale 1.-3.: 1:1, 4.: 1:1. (after Vladár 1973; Ondráček and Šebela 1985; Bátora 1991).*

Social structure and burial practice

In order to apprehend the social organisation of warriors and their importance within the society I have chosen the cemetery of Branč as a case study. Being one of the largest burial sites from the Nitra culture and located in the very heartland of the Nitra Basin, as well as being well documented (Hanulík 1970; Vladár 1973; Shennan 1975), makes it an ideal object of study. The cemetery at Branč was in use throughout the Early Bronze Age – from the classic Nitra phase to the Únětice-Mad'arovce horizon. It is difficult to draw any clear chronological boundaries amongst the graves.

I have divided the Nitra culture part of the cemetery into nine sections (Fig. 3) based on chronology and spatial relationships. These are only rough divisions since some groups overlap and have unclear boundaries: groups D, E, F, and G. The earliest graves at Branč are dated to the early classic phase. The cemetery seems to have been abandoned some time during the Nitra-Únětice phase, but was resurrected again further to the north during the Mad'arovce phase. Groups A, B C, D and E cover the classic phase,

whereas the late phase consists of groups F, G, H and L (The graves from the Mad'arovce phase are not considered in this discussion).

The cemetery extends in a south-north direction in both a chronological and spatial sense. Later graves respected existing ones, as no tombs cut into each other. This indicates that graves were marked above the ground and were visible for a very long time. When the later Mad'arovce cemetery was established it was located at some distance from the primary burial site indicating that the old graves were still marked. This phenomenon has also been observed at Výcapy-Opatovce and Jelšovce (Tocik 1979; Bátora 1991). In comparison it should be mentioned that at Mýtna Nová Ves the graves from the Únětice period are placed directly over the burial site from the Nitra culture, cutting into older tombs (Bátora 1991). There are reasons to believe that small mounds or house-like constructions, to mark the tombs of warriors, covered some of the graves at Branč (Vládar 1973) (Fig. 11).

Out of the 237 graves from the Nitra culture, 40 (17%) tombs belonged to warriors (Fig. 10). Including

Grave	Sex	Orientation	Arrow	Wristguard	Axe (stone)	Axe (bone)	Dagger (Cu)	Dagger (bone)	Knife (Cu)	Boar tusk	Bone disc	Ornamented bone	Beltfittings (Cu)	Whetstone
2	m	SW-NE	-	-	-	-	-	-	1	-	-	-	-	-
6	M	SW-NE	-	-	-	-	1	-	-	-	-	-	-	-
15	m	W-E	-	-	-	-	-	-	1	-	-	-	-	-
18	M	W-E	-	-	-	-	-	-	-	-	1	-	-	-
22	M	W-E	4	-	-	-	-	-	-	-	-	-	-	-
24	M	W-E	-	-	-	-	-	-	-	1	-	-	-	-
31	M	SW-NE	-	-	-	-	-	-	1	2	1	-	1	-
32	M	W-E	6	-	-	-	-	-	-	1	-	-	-	-
39	M	W-E	-	-	-	-	-	-	-	2	-	-	-	-
54	M	SW-NE	-	-	-	-	-	-	1	-	-	-	-	-
55	M	W-E	-	-	-	-	1	-	-	-	-	-	1	-
57	M	W-E	-	-	-	-	-	-	-	-	-	1	-	1
62	M	SW-WE	-	-	-	-	-	-	-	-	-	-	1	-
70	m?	SW-NE	-	-	-	-	-	-	-	2	-	-	-	-
72	M	SW-NE	-	-	-	-	-	-	-	1	-	-	-	-
74	M	SW-NE	1	-	-	-	-	-	-	3	-	-	-	-
83	M	SW-NE	-	-	-	-	-	-	1	-	-	-	-	-
85	M	W-E	-	-	-	1	-	-	-	-	-	-	-	-
87	M	SW-NE	-	-	-	-	-	-	1	-	-	-	-	-
88	M	SW-NE	-	-	-	-	1	-	-	-	1	-	-	-
96	M	W-E	-	-	-	-	-	-	1	-	-	-	-	-
104	M	SW-NE	7	-	-	-	-	-	-	8	-	-	-	-
121	c	?	-	-	-	-	-	-	1?	-	-	-	-	-
168	f	?	-	-	-	-	-	-	1	-	-	-	-	-
173	M	WSW-ESE	-	-	-	-	-	-	-	2	-	-	-	-
178	M	WSW-ESE	3	-	-	-	-	-	1	2	-	-	-	-
179	M	W-E	-	-	-	-	-	-	-	1	1	-	-	-
182	M	WSW-ESE	6	-	-	-	1	-	-	2	-	-	-	-
188	m	WSW-ESE	-	-	-	-	-	-	1	3	-	-	-	-
191	M	W-E	-	-	-	-	-	-	-	1	-	-	-	-
192	M	WSW-ESE	-	-	-	-	-	-	-	2	-	-	-	-
195	M	SW-NE	-	-	-	-	1	-	-	-	-	-	-	1
211	M	WSW-ESE	-	-	-	-	-	-	-	1	-	-	-	1
222	M	W-E	-	1	1	-	-	-	-	-	-	-	-	-
223	F	ENE-WSW	-	-	-	-	-	-	-	1	-	-	-	-
232	F	ENE-WSW	-	-	-	-	-	-	-	1	-	-	-	-
233	M / M	WSW-ESE	8	-	-	-	-	-	-	1	-	-	-	-
247	M	WSW-ESE	-	-	-	-	-	1	-	-	-	-	-	-
264	M	WSW-ESE	-	-	-	-	1	-	-	-	-	-	-	-
298	M	W-E	-	-	-	-	-	-	-	1	1	-	-	-

FIG. 10: *List of graves with warrior equipment at the Nitra culture cemetery of Branč.*

a double grave, the total number of warrior burials was 41. Not surprisingly, 81% of them were men, while children make up 15% and women 5%. Children and women with warrior's gear are thus not uncommon. Warrior equipment in children's graves at Branč comprises only knives, and this should indicate that these children were expected to become warriors when they reached adolescence. Why some women were buried as warriors is unclear. A female warrior is also known at Mýtna Nová Ves (grave 177), who has been buried with several arrowheads (Bátora 1991). At Branč two women received a boar's tusk each.

These may perhaps be interpreted as an honour given to the dead person as part of the funerary ritual, rather than an ornament the women wore in life. Women with warrior equipment are known from other contemporary cultures. In the Mierzanowice culture some women had arrowheads as grave goods and in Košťany-related cultures some were buried with daggers (Bátora 1991). Nevertheless, we should not be unfamiliar with the possibility that some women might have been warriors.

The organisation of warriors in the Nitra culture may follow some sort of kinship structure. Family groups can be claimed to exist at Branč. They are not clearly discernable in the early stage (group A), but more so in the later phases (cp. groups B – I). The possible family groups can be outlined with help of the warrior graves: each family's warriors were equipped in distinguished ways. Group A was apparently shared by two larger families, or clans, dividing the burial area into a western and an eastern section. The graves of the sections appear distinct, and the warriors' gear changes over time, but is confined to a specific pattern for its own group. This is evident in the western part where graves usually contain boar's tusks whereas these artefacts are conspicuously rare in the eastern part. Within the western part a further division can be made: burials in the south-western part have boar's tusks and weapons are almost absent, whereas burials in the western central part have knives and daggers, while the graves to the north-west have arrowheads together with boar's tusks. In the south-eastern part there are graves with boar's tusks, arrowheads, knives and daggers. A distinction between the two parts is also found in the manner in which the tombs were constructed (Fig. 11). The other groups were not divided by different families, as the warrior tombs

are situated in the centre of each group. Graves with boar's tusks predominate in the far western part of the cemetery (groups B, C, D) and knives and dagger marks graves to the east (groups E, F, G). It is interesting that axes are found only in groups C and D.

From the weaponry in the graves we can reconstruct three categories of warriors: (a) an archer, (b) an archer with a close-range weapon and (c) a warrior with a close-range weapon only. At Branč the archer type of warrior is dominant. The archer's attributes – arrowheads, boar's tusks and bracers – are found amongst those belonging to the common warriors buried at the western part of the cemetery. This warrior type is confined to these attributes until the transition to the Nitra-Únětice phase, only adding an occasional knife or axe to the outfit. The distinction between warriors of type (a) and (b) is mainly a chronological division, although both kinds are contemporary during the end of the classic phase.

Warriors armed only with daggers or knives are different from the archers. They are mainly found in the eastern part of the cemetery amongst high-ranking warriors, and it is remarkable that they change their outfit from a Nitra style one with bow and arrow, to a style closer to Únětice practice: the fact that these warriors distinguished themselves by carrying daggers earlier than the rest of the warriors indicates that they considered themselves as part of an elite, perhaps with connections to an Únětician elite outside the region. The Únětice culture influenced the Nitra culture more and more, and the Únětician connection may have been utilised by the local elite to obtain control of the flow of prestige goods. The foreign impulses in new rites and material things appear to have been used locally by the elite to promote and maintain rank and status.

Within each family group the warriors were often buried side by side in pairs or in groups of three in which the tombs often contain the same kinds of grave goods, whilst the construction of the tomb might vary. The warrior graves are located in the middle of the family groups surrounded by women and children. Girls were placed nearer the warrior's tomb than boys. Most striking are the two girls (graves 98, 94) buried together with a male warrior (grave 88) under the same mound in group E. At Branč, children's graves are particularly frequent in the eastern part of the cemetery near the so-called death house (grave 31) and the tomb with a concentric ditch (grave 62). The same pattern has been

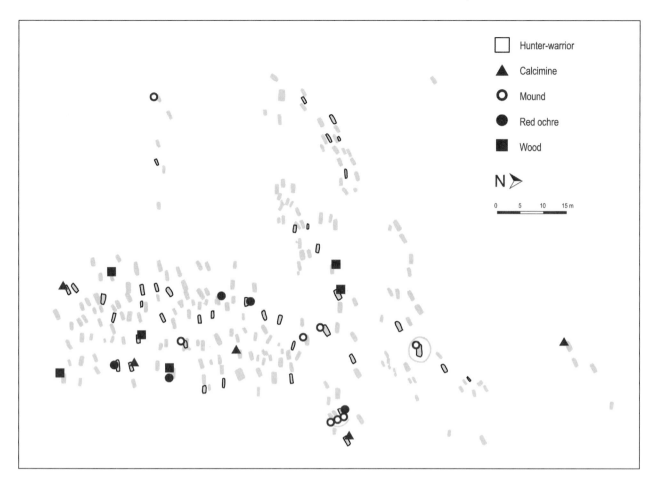

	Hunter-warrior
	Calcimine
	Mound
	Red ochre
	Wood

N

0   5   10   15 m

FIG. 11: *Map showing the location of special-constructed tombs from Branč in the Nitra culture period.*

observed at Mýtna Nová Ves (Bátora 1991). This practice marks the importance of these two groups in the society, considering they are strongly associated with high-ranking warriors. A chieftain's base of power may well have been conveyed through the number of women and children related to him. Although high-ranking male warriors appear to have been individuals of power the wealth of the community was mostly on the women's side. Possibly inheritance passed through a matrilineal system, requiring chieftains to acquire women in order to enforce and expand their authority.

The basic division between elite and common warriors can usually be determined by the position and construction of the tomb as well as the grave goods. At Branč there are three tombs (graves 31, 88, 62) belonging to elite warriors. They are all positioned in the eastern half of the cemetery and mark the centre of a family group. Most remarkable is grave 31, with a construction of a death house above the grave pit. This kind of construction is rare and occur

elsewhere only at Mýtna Nová Ves (graves 206, 262, 305, 509, 513) and Moravská Nová Ves (grave 19) (Bátora 1999c; see Fig. 4). In Jelšovce a similar construction has been observed (grave 444). Here an elite warrior was buried in a grave surrounded by an oblong slot with traces of timbers – suggesting a form of wooden construction above the ground (Bátora 1990). The tomb construction was the same at all these sites. A large pit was lined with wood, which sometimes also covered the opening, creating a wooden crypt for the dead person. Wooden posts were erected at the corners of the pit supporting a roof – or in the case of Jelšovce a construction of logs dovetailed at the corners. Inside the death house, and above the grave, a large amphora was placed as a funeral offering. Since the area in the immediate vicinity of the death houses lacked graves, these tombs must have been accessed regularly, perhaps to carry out certain rituals or offering gifts to the ancestor. Warriors in these graves were often given many grave goods. At Branč the grave contained a

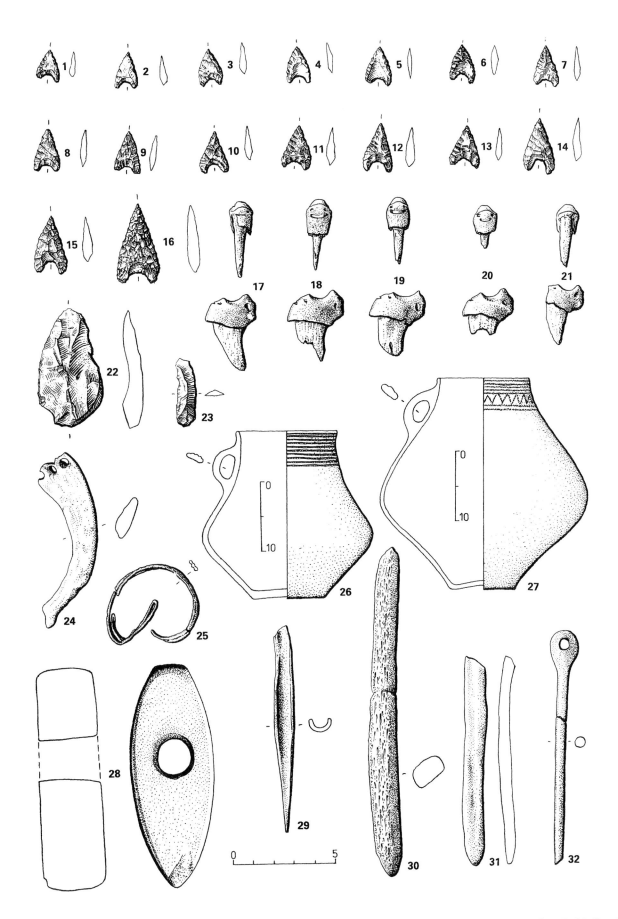

FIG. 12: *Grave goods from elite warrior's grave at Mýtna Nová Ves (grave 262): 1.-16. flint arrowheads. 17.-21. Bear's claws. 22.-23. Stone raw material. 24. Boar's tusk? 25. Double copper wire ring. 26.-27. Amphorae. 28. Symmetrical hammer axe. 29. Bone awl. 30.-31. Part of composite bow? 32. Bone needle or dress pin. (after Bátora 1991).*

knife, belt fittings, boar's tusks and a bone disc. The death houses at Mýtna Nová Ves contained similar grave goods. The most spectacular was grave 262 with its many arrowheads, a stone axe, a fragment of a bow, a boar's tusk, pots and bear claws from the funerary shroud (Fig. 12).

The death houses belong to the early and classic part of the Nitra culture and resemble chamber graves in Ukraine and Russia. The custom of building death houses was adopted by the Únětice culture from which it spread further west during the Middle Bronze Age (Bátora 1999c). In the Nitra culture grave mounds with a concentric ditch replaced them during the Nitra-Únětice phase. The mounds have less conspicuous constructions, but the tradition of placing an amphora or pots on top of the grave continued, and the tombs often exhibit rich grave goods, such as the dagger in grave 88 at Branč. The third chieftain's tomb (grave 62) had been plundered, but pieces of belt fittings indicate an originally rich outfit.

There are other tomb constructions worth mentioning when considering the warriors of the Nitra culture. These are graves with washed walls, wood-lining – without a construction above ground – or sprinkled with red ochre. These constructions were widespread throughout Nitra culture burial sites and have their origin in the Kost'any culture (Bátora 1990; 1991). At Branč these graves are mainly found among the warriors at the eastern side of the cemetery. Considering that these graves are located within a family group that distinguished themselves by hosting the cemetery's warrior elite it seems as if the practice of painting the grave sides with a wash, using red ochre and lining graves in wood was a privilege restricted to a high-ranking family and their kin. It is also striking that the use of these different constructional elements was evenly distributed over time, as if only a certain individual of each generation was entitled to be buried in this particular fashion.

It must be noted that it is unclear to what extent constructions above ground were used. The mounds at Branč were observed only because of ceramic vessels put on top of the grave inside the barrow. A closer look at the cemetery reveals that mounds must have been more frequent than originally acknowledged. It is striking how graves with wood lining, washed walls or red ochre are located in similar areas to mound–covered tombs: they are placed some

distance away from the nearest graves. This might indicate that barrows also covered these graves.

A further interesting phenomenon is the occurrence of paired tombs, such as graves 191, 192 and 182, 178. Why warriors were buried in pairs, or in groups of three, is difficult to answer unequivocally. One reason could be that warriors were customarily buried that way – probably covered by a common mound – to present partnerships in war in life and death. It should be mentioned that double graves with warriors are known from Branč (grave 233) and Mýtna Nová Ves (grave 29) (Vladár 1973; Bátora 1991).

Cenotaphs

To receive a proper burial was crucial for warriors, in particular for those of high rank. Proximity to the newly dead and the ancestors was evidently important to the community's ritual and social life: warrior graves clearly had a central role as being the hub of a larger family group. The social significance of warrior tombs is emphasised by the phenomenon of cenotaphs in the Nitra culture (Fig. 13-14). From the early and classic phases there are in total six cenotaphs equipped with the outfit of an elite warrior. They are all from Mýtna Nová Ves (graves 116, 117, 308, 458) and occur in the vicinity of the five death houses. Apart from grave 458 with a wooden burial chamber, these tombs might be classified as belonging to common warriors as they were of standard construction. However, the occurrence of pots, either in the grave or on top of it, shows that they belonged to elite warriors. Thus they were kinsmen to the chieftains in the death houses in the same way as those buried in the eastern part of Branč.

The same pattern is observable among cenotaphs dating to the Nitra-Únětice phase, but here two features become distinct. First, animal parts replace the absent human being. Second, founder's gear is placed in many graves. These new features may well derive from the Únětice culture in which cenotaphs are well known. Of special interest is grave 6 from Mýtna Nová Ves where the body of a sheep or goat was buried (Bátora 1999b). It lay on its right side with the face to the south as if it were a man. The grave goods were placed near the head and a copper dagger was stuck between the cervical vertebrae. A similar tomb was found at Branč (grave 4), dating to the beginning of the Mad'arovce culture, with pig bones and bronze rings (Vladár 1973).

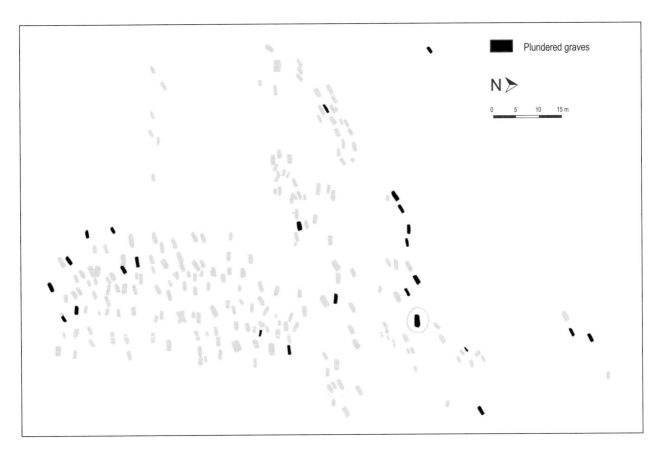

FIG. 13: *Plundered graves from the Nitra culture at Branč.*

Site	Grave	Position	Description	Reference
Early Phase				
Mýtna Nova Ves	116	E	Double Coned vessel, boar tusk, radiolarite arrowhead, 2 radiolarite flakes, bone awl	Bátora 1999b
Mýtna Nova Ves	117	E	Double Coned vessel, quartz arrowhead	Bátora 1999b
Mýtna Nova Ves	308	C	Willow-leaf copper knife, silex arrowhead, bracer, ceramic vessel	Bátora 1999b
Mýtna Nova Ves	458	C	Double Coned vessel, wooden burial chamber	Bátora 1999b
Tvrdošovce	23	?	Ceramic vessel, copper arm-ring, copper burled ring, 2 copper belt fittings, 13 small copper tubes, copper ring, copper knife, antler bead necklace	Točík 1979
Nitra-Únětice Phase				
Matúškovo	50	S	2 boar tusks, 4 end-pieces to bellows, 3 slabs of sandstone, ceramic cup, fragments of ceramic vessel	Točík 1979
Mýtna Nova Ves	6	W-E	Body of sheep (*Ovis ammon* f. *aries*) or goat (*Capra ibex* f. *hircus*), copper dagger, copper awl, Cyprus pin, copper ear-ring, Double Coned vessel, radiolarite arrowhead	Bátora 1999b
Mýtna Nova Ves	201	E	Ceramic vessel: Kegelhalsgefäss	Bátora 1999b
Velký Grob	50	NNW	Ceramic vessel, body of a calf (*Bos taurus*)	Chropovský 1960
Velký Grob	51	N	Ceramic vessel	Chropovský 1960
Výčapy-Opatovce	23	?	2 horse teeth (*Equus caballus*)	Točík 1979

FIG. 14: *List of cenotaphs from the Nitra (and Únětice) culture. E = eastern part, W = western part, C = central part, N = northern part, S = southern part.*

It is difficult to discuss Nitra culture cenotaphs in general terms, since the majority of them are from the cemetery of Mýtna Nová Ves. However, they do give some valuable information about the character of warfare related to this site. The fact that cenotaphs always belong to warriors suggests that they have their origin in armed conflict where the body of a fallen warrior companion was never retrieved. In this sense it is interesting that such a fate only befell elite warriors. Skeletal trauma from the same cemetery, as I will show below, suggests that elite warriors were subject to immense aggression. These warriors often had several severe cranial fractures, whereas the postcranial skeleton showed surprisingly few signs of injury. Many of the skull fractures must have left the victim mortally wounded, and the warrior was even hit repeatedly while on the ground. If not recovered by his companions the warrior would have fallen into the hands of the enemy and thus never returned to kin and community.

Grave robbing

Another feature linked to warfare is the massive occurrence of plundered graves. Grave robberies are known from the early phase of the Nitra culture at Branč, Mýtna Nová Ves and Tvrdošovce (Vladár 1973; Točik 1979; Bátora 1991), but the majority of robbed graves date to the late phase of the Nitra culture, and this increase in plundered graves may be due to influence from the Únětice culture. Whereas plundered graves are exceptions in the Nitra culture they are almost the norm in the Únětice culture, at least on the larger burial sites in Austria and Moravia (cp. Neugebauer 1991). The proportional variation in this practice is illustrated at Jelšovce where 64 out of 103 graves from the Únětice period were plundered, whereas only a few graves were touched during the Nitra period (Bátora 1991).

The practice of grave robbing follows the same pattern throughout the Early Bronze Age but the reasons for it, and the effect on society, probably differed. Grave robbing seems to have been widely accepted by people during the Únětice period as it does not affect the burial practice much. However, it is still unknown whether cemeteries were plundered during a single session, or if they were plundered over a longer period of time. The bones in plundered graves were usually scattered, both inside and outside the pit, and therefore the grave cannot have been opened until several months after the burial.

The main reason for plundering graves may have been to obtain metal objects, mainly daggers, but taking body parts, such as skulls, was also on the agenda.

The plundered graves at Branč are of special interest as they can be related to warfare. In this cemetery 11% of the graves have been subjected to looting, and they are mainly located at the northern rim of group F and at the southern fringes of group A (Fig. 13). A few also occur in the other groups, notably groups H and L. Tombs of both men and women were affected, and there are no indications that warrior graves were the main focus, even if it must be noted that some of the robbed male graves could have been warrior tombs. Warriors in group G and F merely received a dagger or a knife as an indication of status – if these objects were stolen from graves in the northern part the latter would escape categorisation as warrior graves. Grave 56 in group D probably housed a warrior – although no grave goods attest it – since this man had three healed cranial fractures showing that he had been in close combat several times and survived the encounters. In some warrior graves (graves 74, 173, 192) the looters did not care for arrowheads or boar's tusks, which enable us to identify the dead as warriors.

There are two indications that plundered graves in Branč were the result of troubled times. First, it should be remembered that people in the Nitra culture paid great respect to their dead. Looting a grave must have been an act of desecration – a method used to insult and bring harm to an enemy. Second, the cemetery fell into disuse during the Nitra-Únětice phase inasmuch as the youngest burial groups, H and I, only contain very few burials. The cemetery is then not in regular use until the Maďarovce period, when the new graves are established further to the north. With these two factors in mind the conclusion is that in the late phase the Branč community were involved in a conflict, which they lost. This resulted in a population decrease in the region, and the conquered group were subsequently humiliated by having their burial site plundered by the victors.

Weaponry

Weaponry in the Nitra culture is rather diverse in character and only copper daggers can be said to be clearly characteristic. Weapons have no or little unity in style, and most weapons are the same as those found in the preceding Eneolithic cultures in Central Europe, notably the Corded Ware and Bell Beaker cultures. This is especially valid for the axes, hammers or hammer-axes, which are found only occasionally. All weapons are from burials (ritual hoarding has not been recorded for the Nitra culture).

The weapon classes are represented as follows. *Long-range weapons* consisted of bow and arrow only. There are no indications of javelins, slingshots or throwing axes. *Close combat weapons* covered axes, hammers and axe-hammers. Pole arms, such as spears or lances, are absent on all sites with the exception Holešov (grave 174) – here the tip of a lance head was recovered (Ondráček and Šebela 1985). *Close quarter fighting* weapons consisted of daggers and knives.

Long-range projectiles and weapons for close quarter fighting dominate the weaponry. This contrasts markedly with the recorded trauma on skeletons, which show that close combat weapons caused the majority of wounds (see below). This apparent paradox is important in order to understand the symbolic meaning and display of weapons and their relationship to the deceased in death and in life.

The occurrence of weapons in graves varies significantly between different burial sites and thus does not follow a consistent pattern throughout the entire society. This does not mean that weaponry cannot be used for chronology, or as evidence of social structure and organisation. As for chronology, as in the case of Branč, different kinds of weapons were put into the graves at different times, but these preferences are not necessarily the same at a neighbouring burial site. Only daggers have proved to be useful in establishing a useful chronology as Vladár has shown (1974). Other weapons are not suitable, either because their form does not change, as for arrowheads and knives, or because they occur rarely and in small numbers, as notably the case with axes and hammers.

As social markers, weaponry separated not only warriors from other men and women, but also functioned as an internal marker within the warrior group in terms of their relationship to different families. This is, however, at its clearest when it comes to arrangement of the grave and its furnishing, as shown above. Various preferences in regard to the display of weapons in the graves probably reflect the ideological attitude of the deceased's family more than the kind of weaponry each warrior carried on war raids.

Arrowheads and archery equipment

Arrowheads are by far the most common kind of weaponry. In south western Slovakia 220 arrowheads have been recorded from the Nitra culture (Bátora 1999a). This number should be considered as low taking the entire culture area in account and half of the arrowheads have actually been located at Mýtna Nová Ves. Also, in Holešov, in eastern Moravia, 132 arrowheads were recorded from the graves. Fairly large cemeteries such as Branč had only 35 arrowheads (Fig. 10) and many of the others have still fewer or none. The significance of this will be discussed below.

The majority of the arrowheads are made of lithic material – with radiolarian rocks and linmoquartzite predominant, but a few were made of bone or antler (Bátora 1991; 1999a). The arrowheads made of stone can be divided into two groups – (1) short heart-shaped arrowheads and (2) oblong triangular (Fig. 6). Both these forms probably stem from the preceding Eneolithic cultures in the region; there are no indications that they are the invention of the older or later Nitra culture phase. Their shape is likely to be closely related to their function (cp. Vladár 1973): all projectile points were produced in a similar manner, with saw-teeth edges projecting into wings at the base to form barbs. They are of eminent craftsmanship and designed to be highly aerodynamic. When hitting the target the edges were intended to cause blood loss and the barbs to prevent the arrow falling out or, in the case of human enemies, to be difficult to pull out without causing yet more blood loss. Unlike the arrowheads of bronze and iron from the middle and later Bronze Age (cp. Eckhardt 1996) it is not possible to distinguish between projectile points designed for either hunting or war in the Nitra culture. Both were functional for either intention.

Other indications of archery are bracers and bows. Bracers were strapped onto the wrist of the hand holding the bow in order to prevent the bowstring from snapping back at the underarm. All Nitra bracers are made of fine-grained sandstone, although many might be expected to have been made of leather as was the case in later times. Bracers are scarce in Slovakia and only a few have been found (Branč, Mýtna Nová Ves) (Bátora 1991). Many are instead found in eastern Moravia and especially at Holešov (Ondráček and Šebela 1985). Together with the many arrowheads from this region the bracers suggest that archery was here highly favoured by warriors and was an important item in warfare strategies.

Bows may be attested by one fragment from Mýtna Nová Ves (grave 262). It consists of a thin antler rod measuring half a metre. It has rounded ends and a rectangular cross-section with one broad side arced, and similar items (termed 'spatulae') actually occur at contemporary sites in England, Germany and Poland. Bátora (1991) interprets them as a device for retouching stone tools, but they all have a similar shape and length to the antler strips used in composite bows (cp. Eckhardt 1996). I would therefore suggest that the antler object found with the high-ranking warrior in Mýtna Nová Ves formed part of such a bow.

Axes, hammers and hammer-axes

Among close combat weapons we find a series of blunt and edged weapons that can be labelled as axes, hammers and hammer-axes (Fig. 9). This group is not homogeneous either in form or depositional type. They occur only seldom in graves, and none can be regarded as specifically tied to the Nitra culture. The origin of their form can be traced to the Bell Beaker, Corded Ware and Chlopice-Veselé cultures (Bátora 1991). Axes, hammers and hammer-axes were almost exclusively made of stone, but there are exceptions such as axes of bone in Branč (grave 85) (Vladár 1973) and Holešov (graves 46, 141, 381) (Ondráček and Šebela 1985), and a bronze axe from the early Únětice culture in Velký Grob (grave 8) (Chropovský 1960).

Bátora has divided hammer-axes into three types: 1. small heart-shaped (Černy Brod, grave 65), 2. with a facetted butt (Brezová pod Bradlom, Komjatice) and 3. with an asymmetric body and a pointed butt (Mýtna Nová Ves, grave 262). Hammers are represented in Čachtice, grave 1. This is a grooved hammer of a type known to have been used for grinding copper ore. Hammer picks occur in Výčapy-Opatovce (Bátora 1991) and Holešov (grave 141) (Ondraček and Šebela 1985). Polished axes of different sorts occur at Branč (grave 222), Výčapy-Opatovce (grave 231), Mýtna Nová Ves (grave 260) (Bátora 1991), Holešov (grave 68, 261) (Ondraček and Šebela 1985) and Jelšovce (Bátora 1986).

The division of these close combat weapons into three groups might be somewhat deceptive inasmuch as it derives from the study of their form rather than their use. To be kept in mind is their function in battle: trauma on skeletons show that both the edge and neck of these weapons were used, hence they can all be considered as having been handled as hammer-axes where both the sharp edge and the blunt neck were used. This same point is valid for the bronze axe with pointed neck from Vel'ký Grob (grave 8) (cp. Chropovský 1960).

Hammers, axes and hammer axes are, as mentioned, also likely to be greatly under-represented at the cemeteries. It is certainly worth noting once more that although the majority of trauma has been caused by this kind of weapon, they were not associated with warriors to any great extent. Less than one percent of the graves at Branč and Mýtna Nová Ves displayed axes or hammer-axes hence showing that their symbolic association with war and warriorhood is rather weak. The axes from Branč do, however, show an interesting pattern. They all occur in the western part of the cemetery, each within a group of graves of their own – groups C and D respectively, which date to the classic/late phase. Whilst contemporary graves at the same site contain daggers and knives as the main weaponry for warriors, the warriors of groups C and D were equipped with axes and bows. This strongly suggests that the presentation of various kinds of weaponry in graves was a matter of social identity – expressed by different families through the burial custom.

Daggers

The dagger is a common weapon in the earliest Bronze Age (Fig. 8). In south-western Slovakia there are approximately 40 daggers from this period (Bátora 1999a), and all are made of copper with the exception of a bone dagger from Branč (grave 247)(Vladár 1973). Daggers made of bronze do not appear until the beginning of the Únětice culture.

No other weapon in the earliest Bronze Age shows such a variety of forms as the copper dagger. It was developed from the willow-leaf shaped knife (see below), and it is a matter of debate whether this so-called knife is in reality more a heavily worn dagger (cp. Vladár 1974). Nevertheless, one important difference between daggers and knives in the Early Bronze Age is the particular technique of joining hilt and blade. Daggers have a grip tongue, with the hilt fastened by rivets onto the blade, whereas knives have a tang.

The willow-leaf shaped dagger has its origin in the Chlopice-Veselé culture and occurs at Branč,

Nitra-Čermáň and Holešov (Vladár 1974). The triangular dagger was probably developed in the Nitra and the Kosť any cultures. The different types and subtypes tend to occur at particular burial sites, suggesting several local metal industries in which copper objects rarely left their region of origin. The chronology of daggers has been worked out by Vladár (1974): the early phase is characterised by two main types – 1. willow-leaf daggers and 2. triangular copper daggers. In the later phase only triangular copper daggers are present. Types and subtypes of daggers dating to the early phase are: willow-leaf dagger, miniature dagger, the Abrahám type, the Šaľa-Veča type, the Nitra type and the Výčapy-Opatovce type. In the late phase we have the Branč type, the copper dagger with rounded grip tongue, and the Šaľa-Veča type.

The willow-leaf dagger is characterised by an oblong or oval blade with a rounded tip. The dagger is thicker at the centre of the blade creating a smooth ridge laterally. Two rivet holes provided fastening for a handle. This type is only found at Branč (graves 55, 88) (Vladár 1974). None of these daggers have decoration. It should be mentioned that a similar blade exists from Tvrdošovce, which has been classified as a knife, although it has a grip tongue and is slightly larger than most knives from this period (Točik 1963).

The triangular copper daggers exist in many varieties depending on the cemetery in which they have been found. Daggers of Abrahám and Šaľa-Veča types lie typologically close to the willow-leaf form, but usually have three to five rivet holes on a slightly rounded grip tongue. The blade is flat and has a more triangular shape, and daggers of the two types are undecorated. They occur at Abrahám (grave 62, 64), Branč (grave 195), Černý Brod (42, 65), Tvrdošovce (grave 14/60) and Holešov (grave 50) (Vladár 1974; Ondráček and Šebela 1985). Some are even almost equilateral, either in miniature form, as at Branč (grave 6), Abrahám (grave 1) and Holešov (grave 46), or in a larger form, as at Šaľa-Veča (grave 51/65) (Vladár 1974; Ondráček and Šebela 1985). The Kosihy-Čaka and Bell Beaker cultures south of the Carpathians are the sources of influence for these early triangular daggers in the Nitra culture. It is not until the beginning of the Únětice culture that this kind of dagger becomes common in Central Europe.

In the early phase there are also blades with incised decoration, such as the Nitra type, where the decoration consists of grooves and/or thin incised lines alongside the cutting edges forming a triangle. Some daggers also have a smaller triangular pattern at the centre of the blade. The blade is flat and has a slightly rounded grip tongue with five rivet holes. This type has been found in Nitra-Čermáň (grave 69), Šaľa-Dusikáreň (grave 7), Výčapy-Opatovce (grave 187), Tvrdošovce (grave 48/60) and Šaľa-Veča (graves 29/65, 54/65) (Vladár 1974).

Daggers of the Výčapy-Opatovce type differ from other triangular blades by having an oval cross-section instead of being completely flat. It has also a rounded grip tongue with three rivet holes. It is found at Výčapy-Opatovce (grave 310) and Nitra-Cermáň (grave 52) (Vladár 1974).

Many of the daggers in the later phase are developments of earlier forms. Copper daggers with rounded grip tongue are similar to the Výčapy-Opatovce type, but have four rivet holes. One specimen has been found at Branč (grave 264) (Vladár 1974). Similar to the Nitra type are daggers belonging to the Šaľa-Veča type that are placed at the very end of the Nitra culture sequence. These blades have a rounded grip tongue with three rivet holes, and their decoration differs from the Nitra type by having an incised shaded V-pattern at the proximal end of the blade. They occur at Šaľa-Veča (graves 5, 6/64), Veľký Grob (grave 7), Abrahám (grave 142) and Holešov (grave 83) (Vladár 1974; Ondráček and Šebela (1985). Chropovský (1960) has interpreted the blade from Veľký Grob as a halberd, since the blade was inserted obliquely into the hilt. This would be one of the earliest finds of halberds in the Bronze Age. They are otherwise considered to belong to the Únětice culture.

From Branč (grave 182) there is a single blade classified as the Branč type 2 (Vladár 1974). With its rather straight grip tongue with four rivets it clearly does not resemble any other type of blade. The dagger in question was most likely an import from the region south of the Carpathians.

The role of daggers in warfare can be of course debated, but there can be no doubt that they were used. Many daggers, as well as knives, are heavily worn and show signs of having been sharpened by whetstones. Some even show damage at the blade's proximal end with broken tangs and torn rivet holes.

Knives

Before the appearance of daggers the copper blade industry of the Nitra culture was characterised by knives, which seem foreign to the region and bear no resemblance to Eneolithic copper blades such as the tanged copper daggers of the Bell Beaker culture (Točik 1963). The Nitra culture knives are all willow-leaf-shaped with a tang to fit into the organic hilt (Fig. 8). In the early phase knives are flat, but in the classic phase a mid-rib appears extending medially across the blade's dorsal side and giving the knife a triangular cross-section (Bátora 1991). As mentioned above, daggers were developed from knives, but while the latter grew more triangular the former maintained the willow-leaf shape throughout the period. Though daggers to a large extent replace knives as prestige goods the knife never really went out of use, as documented by finds at Branč (Vladár 1973). Knives have been found at Branč, Černý Brod, Lužany, Tvrdošovce, Jelšovce, Mýtna Nová Ves and Holešov (Točik 1963; 1979; Ondráček and Šebela 1985; Bátora 1991).

Spears and lances

Pole arms were not employed by warriors in the Nitra culture to judge from the evidence at various burial sites. However, the tip of a lance of copper was found at Holešov (grave 174). It was placed in the hands of the dead. Its origin is uncertain, but may belong to the early Únětice culture in Moravia (Ondráček and Šebela 1985).

Trauma[1]

Traumatic traces on skeletons from normal burials (Figs. 14-18), due to intentional violence, are common in the Nitra culture. The presence, frequency and style of the fractures indicate that clashes between groups were the grim reality of ongoing war. From the many osteological analyses conducted it is possible to reconstruct the nature of violent encounters and to outline the strategy behind the skirmish as well as fighting techniques. Numerous skeletons from many of the larger burial sites have been examined (cp. Furmánek 1997; Stloukal 1985; Bátora 1999a; Jakab 1999), and so far there are 22 cases of cranial fractures, 12 fractures on arms and 3 on legs (Bátora 1991) (Figs. 15-16). Skeletal trauma occurs predominantly on male skeletons (21) followed by women (10) and children (5). The majority of traumata were hack, slash and crush injuries to

the left side of the body, mainly on the skull. The nature of the wounds and their position on the body indicate that they were the result of a clash between armed persons. Fractures such as depressions on skulls, broken clavicles and arm bones might have been caused by accidents not relating to warfare, but they are fewer than those arguably resulting from the confrontation of close combat. It should be noted that some bodies show signs of post-mortem trauma (Fig. 17): these are mutilations where parts of the body have been butchered. The reason for this is unknown but is surely connected to rituals performed at the time of burial (and hardly as suggested by Bátora (1991) and Chropovský (1960) a result of vampire slaying).

Blows to the head encompass the widest variety in terms of battle wounds and battle techniques. There are basically three kinds of skull fractures from the Early Bronze Age: 1. hack and slash wounds. 2. crush injuries and 3. depressions (Fig. 18). Hack and slash wounds are characterised by a broken bone wall with a hole that takes its shape from the weapon. The impact of the weapon has usually flaked the bone, due to the stroke itself or due to the weapon being pulled out. Such wounds are often fatal inasmuch as the flaking of the bone, including the splints, have caused immense blood loss and/or severe brain damage. The man from grave 206 from Mýtna Nová Ves has a clear hack fracture on the left frontal bone with severe flaking on the inside of the skull (Jakab 1999). In contrast, injuries from crushing blows are characterised by broken and at the same time shattered bone, and the bone wall is not necessarily broken. Such fractures may not be fatal unless they are extensive. A large crush fracture shattered the zeugmatic and temporal bones of the head's left side on the male skull from grave 262 at Mýtna Nová Ves (Jakab 1999). The last category of cranial injuries consists of small rounded depressions. Unless these wounds are clearly visible they are easily overlooked. They are either results of minor crushing blows failing to break the bone or a result of well-healed fractures. In this latter case the fracture might only be visible as a small hollow in the skull and could easily be ignored, especially if the skeleton is not well preserved.

Judging from the appearance of cranial fractures they are either round- or quadrilateral, long and narrow with marked edges. The latter suggests that they were caused by blunt or heavy weapons with a

Site	Grave	Sex	Age	Description	Reference
Abrahám	95	M	Maturus I	Fracture on left os frontalis (1)	Bátora 1999a
Abrahám	100	?	?	Trepanation (1)	Bátora 1999a
Branč	55	M	?	Fracture over left orbita in os frontale (Healed) (1)	Bátora 1999a
Branč	56	M	Maturus	Three fractures on right os parietale (Healed) (1-3)	Vyhnánek and Hanulík 1971
				Two fractures on right os frontale (Healed) (1-2)	
Branč	86	M	?	Fracture on left os parietale (Healed) (1)	Bátora 1999a
Cermán	44	M	Adult	Trepanation on the left os parietale (Healed) (1)	Bátora 2002
Matúškovo	10	c	Infans III	Impression on the left os parietale (1)	Bátora 1999a
Mýtna Nová Ves	29 (A)	M	Adultus II	Round fractures on the left side of os parietale (1)	Bátora 1999a
	29 (B)	M	Maturus I	Round fractures on the left side of os parietale (1)	Bátora 1999a
Mýtna Nová Ves	39	c	Infans III	Sharp and oblong fracture on os parietale (1)	Bátora 1999a
Mýtna Nová Ves	132	M	Adultus I	Impression on the right side of os frontale (1)	Bátora 1999a
Mýtna Nová Ves	206	M	Adultus I	Round fractures on the left side of os parietale (1)	Jakab 1999
Mýtna Nová Ves	226	M	Senilis	Three irregular cranial fractures (1-3)	Bátora 1991
Mýtna Nová Ves	262	M	Adultus I	Fracture on proximal ulna (Healed) (1)	Jakab 1999
				Large oval opening on os frontale (1)	
				Small fracture on the left side of os frontale (1)	
				Two small concentric openings on left os parietale (1)	
				Fracture on left os temporale and os zygomaticum (1)	
Mýtna Nová Ves	305	M	Adultus I	Fractures on the left side of the mandible (1)	Jakab 1999
				Four broken costae – two on each side (1-2)	
				Fracture on radius and ulna (1)	
				Fractures on the right scapula (1-2)	
Mýtna Nová Ves	319	F	Maturus I	Fracture on right os parietale near the sutura sagittale (1)	Bátora 1990
				Trepanation on os parietale (-)	
Mýtna Nová Ves	509	M	Adultus I	Fracture on left os parietale (1)	Jakab 1999
				Fractures on os temporale and os zygomaticum (1)	
				Fracture on os occipitale (1)	
				Fracture on right os parietale (1)	
Mýtna Nová Ves	513	M	Adultus I	Fracture on distal humerus and proximal ulna (1)	Jakab 1999
Příkazy	?	M	Maturus	Round fracture on the leftside of os parietale (1)	Bátora 1999a
Příkazy	?	M	Senilis?	Three fractures from blunt weapon on the craniums left side (1-3)	Bátora 1999a
Senkvice	?	?	?	Fracture on the left side of the cranium (1)	Bátora 1991

Blunt and sharp force trauma. Nitra group.

FIG. 15: *List of hack, slash and crush fractures on skeletons in normal burials of the Nitra (and Únětice) culture (cases known to the author).*

Site	Grave	Sex	Age	Description	Reference
Veselé	12	F	?	Silex arrowhead in vertebrae	Budinský-Krička 1965
Jelšovce	436	M	Maturus I	Silex arrowhead in a rib on the chest's left side	Bátora 1999a
Mýtna Nova Ves	325	M	Maturus I	Silex arrowhead in cervical vertebra	Bátora 1999a

FIG. 16: *Trauma from stab-wounds noted on skeletons in normal burials of the Nitra (and Únětice) culture (cases known to the author).*

Site	Grave	Sex	Age	Description	Reference
Branč	82	F	?	Right leg chopped off	Vládar 1973
Velkký Grob	?	?	?	Mutilated	Chropovský 1960
Příkazy	?	M	?	Decapitulated	Bátora 1991

Mutilated bodies

FIG. 17: *Skeletons with evidence of mutilation from normal burials of the Nitra (and Únětice) culture (cases known to the author).*

crude edge, such as stone clubs or axes. This seems a paradox since these kinds of weapons are rare in graves of the Nitra culture. In fact, the popularity of knives, daggers and archers equipment stands in contrast to the absence of injuries from these weapons among the cranial fractures.

The second largest area of injuries is the arm, especially the left elbow and left lower arm. These fractures are similar to each other, with either a blow to the elbow that fractures the *distal humerus* and *proximal ulna* or a strike to the lower arm that breaks the *ulna* and *radius* into two. This kind of injury may well be connected to attempts to parry an incoming blow to the head. Although a broken arm would make the person an easy target, many such parry injuries have healed showing that the victim survived the incident. The same can be said for leg injuries, which are rather few in number and add little to the understanding of organised violence.

Although one could expect the opposite, injuries to the torso are almost absent. Such trauma consists of broken ribs and arrowheads shot into vertebrae and ribs. I expect that the poor preservation of skeletons and maybe in some cases the ignorance of the osteologist could be the main reason of the small numbers of this kind of trauma. It has often been mentioned that stabbing weapons, such as daggers and arrows, would not leave any marks in the skeleton as they hit the chest. We would most likely be able to observe traces on both ribs and vertebrae from stabbing weapons if a careful search was conducted. But admittedly, it remains difficult to separate cut wounds from scratches and dents deriving from taphonomic processes.

The distribution of trauma among the dead persons allows a division by sex and age. Cranial fractures occur mainly on mature male skeletons, except at the cemetery of Mýtna Nová Ves where young male adults suffered most skull injuries (see below). Males are, then, by far the group of individuals most frequently engaged in warfare. Broken arms are, however, rather frequent among women. It is interesting that trauma owing to violence often occurs on individuals we would expect not to be active in warfare; that is older men and especially women. The high frequency of skeletal trauma on these persons may well show that they were less able to defend themselves, either because they were unarmed, or simply because they were not agile enough to dodge an incoming attack.

It is important to point out that skeletal trauma occurs throughout the Nitra phases but is most frequent in the classic and late phases. According to Bátora (1991; 1999a), most of the traumatic evidence dates to the transition between these two phases, and to the transition to the Únětician period. When looking at the occurrence of trauma in the Nitra culture one must recognise that each site had a distinct traumatic history of its own although a general pattern in the traumatic history can also be outlined for the culture as a whole. These distinctions are important in understanding the role of violence in the community and its consequences. At Branč it is surprising that cranial fractures have actually healed in each case to occur. The man in grave 56 is a case in point with five healed fractures on the skull indicating that he has participated in warfare several times and survived the encounters. Healed fractures may show that the agent escaped his antagonist quickly. This is interesting since a blow to the head normally turns the victim into an easy kill as he staggers around or falls stunned to the ground. Either the enemy did not have time to finish his victim off, or it was not their aim to do so.

The situation at Mýtna Nová Ves is completely different. Contrary to other burial sites trauma is

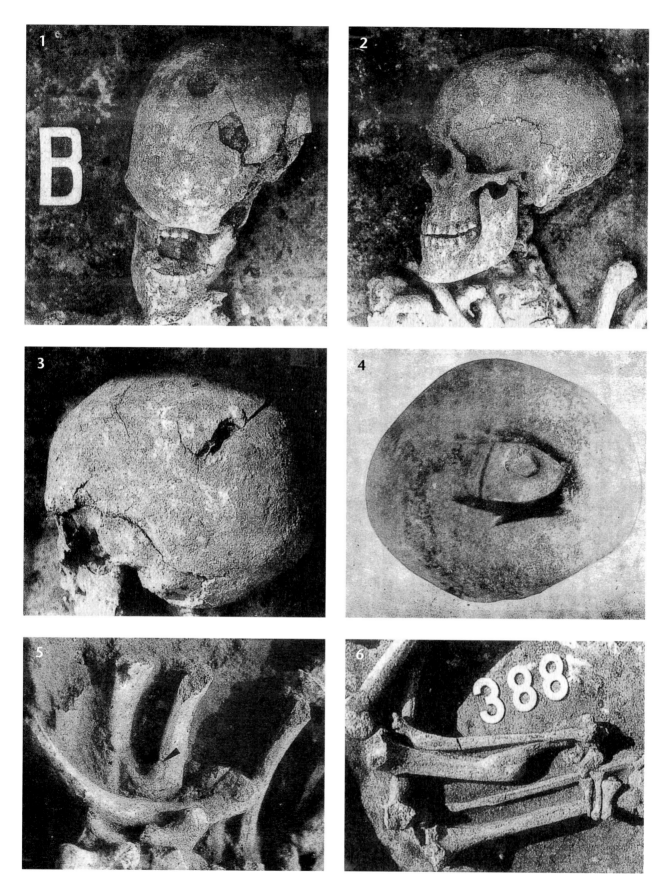

FIG. 18: *Skeletal trauma found in normal burials of the Nitra (and Únětice) culture. 1.-2. Round crush fracture on parietal bone (Mýtna Nová Ves grave 29/84a). 3. Hack trauma on parietal bone (Mýtna Nová Ves grave 39/84). 4. Depression on parietal bone with trepanation attempt (Mýtna Nová Ves grave 342/88). 5. Arrowhead stuck in rib bone (Jelšovce grave 436/85). 6. Fractured, but healed, tibia (Jelšovce grave 388) (after Bátora 1990; 1991).*

most common among younger men and women. Those being most prone to violence were the young chieftains buried in death houses, and it is worth noting that they also show traces of multiple fractures, which is otherwise a rare phenomena. While the men at Branč survived a fight, the combatants at Mýtna Nová Ves were victims of intense and fatal aggression. Multiple fractures on each skeleton show that the victim was severely mauled by the antagonist, even if the fallen individuals must have died instantly in some cases. Blows were often directed towards the head, smashing the skull several times while leaving the postcranial body alone (graves 226, 262, 509), but there is also one case in which the whole body was injured, resulting in a smashed shoulder blade and several broken ribs (grave 305).

That warriors of high rank were prime target for assailants is not only evident from the evidence of multiple traumas, but also by the occurrence of cenotaphs for the same warrior class. The fact that bodies were heavily molested on the battleground and others were carried away by the enemy indicates that it must have been essential to desecrate the body of the fallen warrior as far as possible. The death of an elite warrior must have been a severe blow to a community, altering the balance of power in the region, perhaps with consequences for rights to land and women. To kill a warrior of high rank was perhaps seen as especially courageous, granting the killer special status.

Extensive and detailed osteological research in skeletal trauma has been conducted on the remains from warriors buried in the conspicuous death houses at Mýtna Nová Ves. The fractures on these skeletons provide vital information on combat techniques used by warriors in the Nitra culture as well as giving hints on warfare strategy and its effects on the people. None of the wounds show signs of healing, except one *ulna* (grave 262). Only fractures on the head were directly lethal, while other fractures could have been healed. The traumatism of some of these elite warrior burials are described in some detail below:

Grave 206 – male, adultus I
On the left side of the forehead (*os frontale*) was an oval fracture measuring 28 x 41 mm on the outside. The fracture is cone-shaped with a triangular cut at its bottom. On the inside of the skull the fracture is larger, and a piece of the cranial vault has been broken off, aggravating the wound. Pieces from the triangular fracture were still present, showing an elliptical splitting of the bone. The fracture was a mortal wound due to the splitting effect of the bone on the inside of the cranium.

Given its heavy impact on the cranium the fracture must have been caused by a weapon with both a crushing and stabbing effect, being cylindrical in shape. Bátora (1999c) suggests a hammer-axe as the cause, whereas Jakab (1999) claims it to have been a spearhead. It seems most likely that an axe was used, since a spear would not have such a crushing effect on the cranium, unless it was very blunt from extensive use. The fact that spearheads were more or less unknown in this region at this time also tends to make this implausible. The fracture was either caused by the neck or with the pointed edge of the axe. A very similar fracture occurs on a man from Příkazy in Moravia, and on skeletons from a mass grave in Elben in Germany (Bátora 1999c).

Grave 262 – male, adultus I
This skeleton has the following fractures: 1. A healed fracture on the elbow (*proximal ulna*). 2. An oval opening measuring 39 x 74 mm between the left side *tuber frontale* and the sagittal bones. The fracture goes in an arc from the nasal root and goes alongside the sagittal bones and turns through the *tuber frontale* and down through the eye-socket. On the outside the edge of the fracture is sharp, but rough on the inside. This fracture, which smashed the face of the warrior, was caused by the violent force of an axe or club. Post-mortem change and taphonomy made the fracture wider. 3. In the medial direction from the fracture mentioned above there is a smaller fracture measuring 6 x 9 mm. It is larger on the inside of the cranium. 4-5. There are two openings on the left parietal bone (*os parietale*) near the *sutura sagittalis*. There might have been more, but the rest of the skull is missing. Each fracture is round in shape, measuring 12 x 13 mm, and the edges were heavily corroded, making it difficult to assess what kind of weapon that caused them. A sharp and pointed weapon must have caused the puncture fractures, since the fracture is not much larger on the inside of the skull. 6-7. There are splintered fractures on the left temporal bone (*os temporale*) near the *foramen magnum* and on the *processus frontalis* on the zygomatic bone (*os zygomaticum*).

FIG. 19A-B: *Mass grave from Nižná Myšľa in Slovakia (after Jakab, Olexa and Vladár 1999). Abnormal multiple burials of men, women, and children are frequent in east central and south east Europe during the latest Early Bronze Age, and osteological and forensic studies show that the skeletons often carry marks of violence (see Figs. 20-22).*

Grave 305 – male, adultus I

This skeleton has following fractures: 1. The mandible has a vertical fracture splitting the bone in half. The fracture is sharp on the dorsal side, but on the ventral side a splinter of bone has been broken off. On the mandible's left side the bone has been splintered at the line dividing *ramus* and *corpus*, showing a sharp fracture line. These fractures on the mandible show that a powerful blow had been struck on the left side, causing a direct fracture there and an indirect fracture running through the bone at the front teeth. A punch with a fist, or a club, caused the blow. 2. The chest has four fractures on the ribs (*costae*) – two on either side of the rib cage of the same pair of ribs. 3. A fracture on the left lower *radius* and *ulna* broke the bones into two. This kind of fracture derives from a reflex action in an attempt to parry an incoming blow. 4. The right shoulder blade (*scapula*) has two fractures. One fracture extends from the *margo lateralis* to the *spina scapulare* measuring 41 mm and the other fracture extends 15 mm down from the *margo cranialis*. A sharp and crushing weapon such as an axe or hammer-axe caused the fractures.

Grave 509 – male, adultus I

Many cranial fractures were noted for this man, but due to bad preservation parts of the right side and the back of the head was missing. 1-2. A large fracture (80 mm) on the left parietal bone (*os parietale*), where a piece of the bone was chipped away, was caused indirectly by two smaller direct fractures on the left temporal bone and left sphenoid bone. One of the blows hit the *sutura squamosa* and the other the mastoid process. A blunt object caused the fractures. 3. On the occipital bone a 47 mm long fracture running to the back of the cranium is visible. 4. A fracture on the right side of the skull is barely visible due to heavy fragmentation of the bone. Its appearance cannot be assessed.

Grave 513 – male, adultus I

This skeleton has a fracture on the distal *humerus* and *proximal ulna*. This is probably a parrying fracture like those on the skeletons in graves 262 and 305.

Warfare and social change

Whether warfare characterised the population of the Nitra culture from its earlier phase as Točik (1963)

and Bátora (1991) stated is difficult to say, but warfare certainly grew in importance during the period. Although several customs in the Nitra culture seem to have had an east European origin this is not enough to postulate a substantial migration into the area. Rather the river Nitra enabled people in this region to create a wealthy society. The river served as a route of communication facilitating the distribution of prestige goods and commodities between the Hungarian plain and Little Poland. The fertile soils of the river valleys and easy access to copper ore, moreover, made population increases possible and a larger accumulation of wealth than in the neighbouring areas.

The presence of elite warriorhood and warfare in society might explain the shift in settlement structure and economy that occurred during the Early Bronze Age. Bátora (1999a) has studied the frequency of weapons and the frequency of skeletal trauma and concluded that the bow and arrow together with axes gave way to daggers during the course of the Nitra culture. Trauma increased during the end of the classic period, but declined slightly with the advent of the Únětice culture.

It has been argued (Bátora 1991; 1999a) that the greatest threat to the Nitra cultural group came from neighbouring regions. The highest frequencies of weaponry are found in Ša'la I, Tvrdošovce, Holešov and Mýtna Nová Ves, which are all situated on the borders of the region occupied by the Nitra culture. This should be compared to Vel'ky Grob, Branč, Výčapy-Opatovce and Jelšovce, where weapons are relatively few. Little knowledge, however, exists concerning the people living in directly neighbouring regions at this time and we may also assume that war was a driving force within the Nitra culture itself. The frequency of warrior graves and weaponry surely indicates internal conflict between areas over the control of resources rather than defence from external forces. It must also be emphasised that the occurrence of weapons alone does not give a valid picture of the extent to which a population engaged in bellicose activities. At Branč some warrior graves did not contain weapons, but there is a high frequency of trauma, indicating that this population waged war or defended themselves against attacks. The high frequency of cenotaphs and grave robbery adds to this picture of local societies in a state of war. The political nature of warfare must surely also be considered, such as the organisation of warriorhood and the relation between warriors and other social identities in the society.

Clear evidence of war appears in the classic phase, and at the threshold to the Nitra-Únětice phase. During this period there is a settlement expansion from a core area in the middle Nitra basin to the regions in the south and into less fertile areas in the south-west and north-west. This expansion coincides with a marked increase in trauma on human skeletons, with a change in weaponry, and with an increase in the number of cenotaphs and plundered graves. It seems as if a shift in both the economy and social structure was connected to war-related violence.

A shift in power structure obviously happened when the elite from the Nitra culture began to ally with that of the Únětice culture. One reason for increased competition and hostilities may have been the change towards an Únětician way of living among the elite. This included control of bronze objects and their production and distribution. In the classic Nitra phase there were not yet bronze artefacts in the burials, but it is evident that the social elite changed their grave goods from the bow and arrow to the dagger, which is a fashion seen in the Únětician regions of, for instance, Moravia and Austria at this time. When bronze finally began to appear in south-western Slovakia it is amongst this elite that it is found.

Although the changes in power structures were connected to external contacts, the main dynamics of change are located within Nitra society itself. As outlined above, this society rested upon three foundations – as regards settlement organisation and economy: 1. Domestic production and control of arable land. 2. Control of copper production. 3. Control of the manufacture and distribution of prestige goods, notably in metal. The evidence from the cemeteries of Branč, Jelšovce and Mýtna Nová Ves suggests a ranked or stratified society in which certain family groups had obtained dominant positions entitling them to rich grave goods and elaborate tombs. These high-ranking families may have been in control over both land and the distribution of prestige goods. The settlement expansion described above may furthermore be interpreted as an attempt to enlarge the subsistence base through the incorporation of neighbouring villages. The strategy for this might of course have varied, but acquisition of women would have been an easy way of legitimating

the take-over. The fact that women and girls often had rich grave goods (Shennan 1975) and were often buried near high-ranking warriors suggests that society's wealth was often invested in females. Women might have inherited land, or be entitled to wealth and power, which could be taken over when kidnapped women were married into the hostile group. Abduction of women could simply be the means to enhance the preconditions of biological reproduction. However, to completely depopulate a conquered area was not desirable – many burial sites were in use for a very long time. The conquering party may well have forced the subjugated group to pay tribute to them, and this tribute could have been in the form of food, prestige goods and/or women. The relationship between the dominant and dominated groups may perhaps be visible in the presence or absence of an 'upper class', including warriors, on the burial sites. Such a relationship of dominance-subordination may be expressed in the northern part of the Nitra basin where Mýtna Nová Ves has many high-ranking graves and a high frequency of weapons, whereas Výčapy-Opatovce, a few kilometres away, has no elaborate tombs and very few weapons. A similar pattern is visible in the Vah basin, but here the Únětician burial custom with its absence of weaponry may be a source of error.

As pointed out above, the expansion of the Nitra culture occurred in a southern and a western direction towards the Únětician region of Moravia and Lower Austria. Whether there was any expansion eastward is difficult to say, but there are surprisingly few finds east of the river Nitra. Although it might be difficult to talk about major economic and political centres in the early Nitra phase, some sites – such as Nitra, Mýtna Nová Ves and Holešov – were settled early and grew rapidly in size. With the advent of the classic phase it is possible to discern sites that held an important position – to judge from the wealth invested in grave rituals, the number of warriors interred and the building of elaborate tombs for the elite. In the north, Holešov in Moravia and Mýtna Nová Ves in Slovakia are situated close to copper ore sources. In the middle Nitra basin there is Jelšovce, and further south Branč, where wealth may have been built on access to trade and arable land. Towards the end of the classic phase the structure of the economic and political core regions changes. Výčapy-Opatovce apparently fell under the hegemony of Mýtna Nová Ves, and it is possible that people from this site even exercised some authority over those living at Jelšovce. Although Jelšovce held an important position earlier on, its greatest days were in the Únětice and Mad'arovce periods. The burial sites at Nitra and Branč indicate that these sites lost authority to sites in the south and west such as Tvrdošovce and Šal'a. In Moravia, Holešov strengthened its power by expanding along the river Morava.

When it comes to war strategies, two radically different phenomena can be discerned from the burial sites of the Nitra culture. On the one hand there are sites with a high frequency of weaponry as well as an increasing number of individuals who had suffered deadly violence. On the other hand there are a number of sites where weapons are relatively few and where the frequency of trauma is high, but not of a deadly nature. Mýtna Nová Ves belongs to the former category and Branč to the latter. The evidence of trauma from Branč indicates that killing an enemy was not the primary aim in war. Rather, raiding and ambushing must have been the underlying strategy when cattle were stolen or women abducted. Skirmishes between warriors alone might also have been on the agenda, but the many injuries to women and older men suggest that settlements were attacked rather than a battle being fought between warriors on a battle field away from home. Since the bow and arrow were the most common weaponry it is easy to conclude that war was waged mainly at a distance and that close combat rarely took place, but as emphasised repeatedly above, the majority of all traumatic injuries are caused by weapons such as axes and clubs. These were used in order to close in on the enemy, but were never accepted as a proper weapon to be displayed in burial ritual. The warrior of the Nitra culture was presented as a hunter in death and not as a close combat fighter. The violent encounters might be understood as mostly sporadic incidents initiated by daring individuals during a fight rather than the united attack of a whole group. This might be the answer to the question of why so many individuals survived close combat. The attacker would be on unsafe ground while launching the close combat and given little time to finish the attack before the foe's friends came to the rescue. Although this war strategy does not appear as a lethal one overall, there were periods when people were driven away from their home area by force, as possibly apparent from the burials

at Branč. War was indeed a power struggle even if not always deadly.

At Holešov and Mýtna Nová Ves the many weapons indicate that warriors were here important to an extent not recognised elsewhere in this cultural group. The prominence of warriors and the many traces of deadly wounds are reason to believe that the reality of war was different in the northern part of the Nitra culture. Overall, war tactics in the north must have been similar to the ones in the southern part of the Nitra culture region. Raids and ambushes by archers were the norm even here, but it seems as if the struggles between high-ranking individuals were more violent and that rivalry over land and resources was very common. It is striking that high-ranking warriors have multiple peri-mortem wounds and that this trend coincides with the emergence of cenotaphs. As mentioned above, killing a high-ranking warrior may have enhanced the prestige of the successful warrior considerably. The northern part of the Nitra culture was, with its access to copper ore, an important area, also to non-residential groups, which may have been among the attackers. It is not surprising that the burial sites here are both large and wealthy and in general testify to the overriding political and economic significance of this region.

Epilogue: a forensic study of war-related ritual killings in the Early Bronze Age

In the Únětice phase all these power struggles seem to have been settled, and the boundaries were strengthened by the construction of fortified sites in both the middle and southern Nitra basin as well as in the highland regions between the Vah and Morava rivers (cp. Fig. 23). These fortified sites came to dominate the political climate until the Middle Bronze Age (Novotna 1999). At some of these sites – as well as at some of the very similar sites that were constructed all over eastern central Europe at this time – there are mass graves containing individuals of different age and sex (Figs. 20-22) with various degrees of trauma and mutilations (cp. Rittershofer 1997). This certainly testifies to ritual forms of killing and indirectly to a general climate of violence and war, which is also evident from other sources. This peculiar ritual custom is briefly described below through two forensic studies of skeletal remains from Kettlasbrunn in the north-eastern part of Lower

Age groups	AUSTRIA Lower Austria	CZECH REPUBLIC Bohemia	Moravia	SLOVAKIA
Children (0-15 yrs)	36,6%	48,3%	41,8%	52,9%
Juveniles and young adults (15-25)	14,6%	3,4%	11,1%	9,8%
Adults (20-60+)	48,8%	48,3%	47,1%	37,3%

FIG. 20: *Age distribution in a sample of mass graves or abnormal graves.*

Austria and Nižna Myšľa in eastern Slovakia. Both mass burials are dated to the later Early Bronze Age and occur under similar circumstances in storage pits on settlements.

These studies are central because they underline not only the pathological aspect of the skeletal remains, but also the ritual behaviour behind the peculiar find complexes and more indirectly their socio-political background in war and warfare. As such, they provide us with tools for approaching burials of the same unusual kind elsewhere in this region and period. These case studies are exceptional inasmuch as the use of forensic methodology during excavation and examination of the bodies managed to reveal vital information about the peri- and post-mortem changes to the buried bodies that is usually lost or ignored. These case studies serve as an indication that burials in settlement and storage pits must have had a special status, very different from the status and feelings presented in the ordinary burials described above and that these ritual killings in some way were connected to the practice of warfare.

Kettlasbrunn

The pit was located near some house constructions on an open settlement and can be dated to the Early Bronze Age – most likely the late Únětice or Věteřov culture. The storage pit contained one child (age and sex unknown) and three adults: one male (41-50 years) and two females (25-35 and 31-40 years). The individuals were placed in two pairs with a thick layer of soil in between. The child and the mature woman were near the bottom of the pit, whereas the mature man and younger woman were close to the pit surface. Apart from a small bone needle and a canine tooth on the child's body no artefacts were found together with the skeletons, although some pottery sherds were in the filling of the pit (Winkler and Schweder 1991: 79).

The bodies of the child and mature woman were in an almost complete state of disarticulation as only a few bones, notably cranial and torso parts, were in their anatomically correct position. Both skeletons showed traces of weathering, i.e. they had been exposed to the taphonomic effect of wind, sun and rain. This had bleached and made some of the bones brittle. Also, gnaw marks from rodents were numerous on the skeletons. Thus the disarticulation, traces of weathering and the gnaw marks show that the bodies had been exposed for an extended period before they were covered with soil. This indicates that the pit must have remained open until the next burial took place (ibid.: 89f).

Before the second burial occurred the remains of the first one was covered with a thick layer of soil. The mature man and the younger woman were then buried and their taphonomic history is similar to those earlier described. The man was apparently buried more quickly than the woman, as his body only showed a minor degree of disarticulation and no notable effect of weathering. Thus his body had been covered soon after his death. Yet his disarticulated ribs show that the body had been exposed, albeit for a shorter time.

The skeleton of the younger woman shows almost complete disarticulation and the bones were weathered and in poor condition. This is in complete contrast to the male whose bones were well preserved. Whether the woman had been buried at the same time as the man is difficult to say with certainty, and it can only be stated that her body was uncovered for a longer period and subjected to severe taphonomic factors (ibid.: 90).

So far it can be stated that the multiple burial at Kettlasbrunn is not an ordinary burial as the bodies had been buried in a storage pit at different moments and the different bodies had only been partially covered in the meanwhile. Thus, we are here observing a burial that apparently has little in common

Human skeletal remains at settlements and in 'cultural pits'

No.	Site	Id.	Object	Fr.	B.P.	Part.B	C.B.	Trauma	Spec.	Other remains	Date	Notes
					Human remains							

No.	Site	Id.	Object	Fr.	B.P.	Part.B	C.B.	Trauma	Spec.	Other remains	Date	Notes
Austria (Lower Austria)												
1	Bernhardstal (Mistelbach)	V/13	Pit	-	x	-	-	-	? (juv.)	T, P	Veterov	
2	Böheimkirchen (St. Pölten)	?	Layer	x	x	-	-	-	-	-	Veterov	
3	Fels am Wagram (Tulln)	Pit 39 (Leithen)	Pit	-	-	-	1	-	?	-	Unetice	
		Pit 10 (Kogel)	Pit	- -	-	-	1	-	?	-	Unetice	
		Pit 16 (Kogel)	Pit	-	-	-	2	-	Adult, child	O, P	Unetice	
		Pit 22 (Kogel)	Pit	-	-	-	1	-	Adult?	O	Unetice	
		Pit 24 (Kogel)	Pit	-	-	-	1	-	Adult?	-	Unetice	
		Pit 28 (Kogel)	Pit	-	-	-	1	-	Adult?	-	Unetice	
		Pit 39 (Kogel)	Pit	- -	-	-	1	-	Adult?	P, T	Unetice	
		Pit 44 (Kogel)	Pit	-	-	-	1	-	Adult?	-	Unetice	
		V 55	Pit	-	-	-	2	Yes	Adult, child	P, O	Unetice	Building
		(1936)	Pit	-	-	-	3	-	Adults (2), child (1)	-	Unetice	SMI
4	Franzhausen (St. Pölten)	V 76	Pit	-	-	1	-	Yes	Male (30 yrs.)	T	Unterwölbing	
5	Friebritz (Mistelbach)	V 93	Storage pit	-	-	-	1	-	Female (13-14 yrs.)	-	EBA	Building, illness: anemia
6	Gaindorf (Hollabrunn)	?	Pit	-	-	-	x	-	?	-	Unetice	
		?	Storage pit	-	-	-	2	-	Female (adult), child	-	Unetice	
7	Gaubitsch (Mistelbach)	Grave 2 (1986)	Pit	-	-	-	2	-	Males ((S1) 17-20, and (S2) 40-60 yrs.)	-	Unetice	Disturbance, burial in two phases, (S2): (R)
		(1989)	Pit	-	-	-	2	-	?	-	Unetice	
8	Großweikersdorf (Tulln)	Pit 6	Pit	-	-	-	1	-	?	-	Veterov	
		Pit 11	Pit	-	-	-	1	-	?	-	Veterov	
		Pit 14	Pit	-	-	-	1	-	Child (2-3 yrs.)	P	Veterov	Malnutrition
		(1928)	Pit	-	-	-	2	-	?	-	Unetice	
		(1927)	Pit	-	-	-	3	-	?	-	Unetice	SMI
9	Guttenbrunn (Mistelbach)	Pit 34	Pit	-	-	-	1	-	Child	-	Veterov	
		Pit 45	Pit	-	1	-	-	-	?	-	Veterov	
10	Herrnbaumgarten (Mistelbach)	?	Storage pit	-	-	-	1	-	Child	P	Unetice	
11	Hollabrunn	(1985)	Pit	-	-	-	2	-	Female (adult), male (adult)	SR	Unetice	Tied (xh)
12	Jetzelsdorf (Hollabrunn)	(1905)	Pit	-	-	-	3	-	?	-	Unetice	
		V.22	Storage pit	-	-	-	1	-	Child (3 yrs.)	SR	Unetice	
13	Kettlasbrunn (Mistelbach)	(1990)	Storage pit	-	-	-	4	Yes	Female (2) (25-35, and 31-40 yrs.), male (41-50 yrs.) (R), child (1) (1-3 yrs.)	SR	EBA	SMI, male: (R)
14	*Ladendorf (Mistelbach)	(1988)	Pit?	-	-	-	1	-	Female (25-35 yrs.)	P	Veterov	
15	Oberndorf a.d. Ebene (St. Pölten)	V 1	Pit	-	-	-	2	Yes	Male (50 yrs.), female (18-20 yrs.)	SR	Unetice	
16	Peigarten (Hollabrunn)	Pit 1	Pit	-	-	-	1	-	Child	-	Unetice	
		Obj. 3	Storage pit?	x	-	-	-	-	Child	-	EBA	
17	Poysbrunn (Mistelbach))	Pit 10	Storage pit	1	-	-	-	Possible	Female (25-35 yrs.)	A, P	Veterov	
18	Poysdorf (Mistelbach)	?	Pit	5	-	-	-	(Yes)	Adults? (3), male (41-60 yrs.)	P	Veterov	Male: trepanation
		(1979)	Pit	-	-	-	1	-	Female (25-35 yrs.)	-	Unetice	Malnutrition, illness
19	Prinzendorf (Gänserndorf)	?	Storage pit	-	-	1?	-	-	?	P	EBA	
20	Roggendorf (Horn)	Grave 11	Pit	-	-	-	2	-	?	-	Unetice	
		Grave 24 (?)	Pit	-	-	-	3	-	?	-	Unetice	Buried on top of each other, disturbance
		(1931, 1937, 1939)	Pit	-	-	-	3	-	?	-	Unetice	
21	Schleinbach (Mistelbach)	Pit 3	Storage pit	-	-	1	-	-	Child	A, P	EBA	
		Pit 34	Pit	1	-	-	-	-	?	P	EBA	
		Pit 71	Storage pit	-	1	-	-	-	?	A, P	EBA	
		(1927)	Grave?	-	-	-	2	-	Male (adult), female (adult)	-	Unetice	Tied
		(1931)	Pit	-	-	-	4	Possible	Children (3) (4-5, 10, and 12 yrs.), male (1) (35-40 yrs.)	A, P	Unetice	SMI, children: tied
		(1938)	Storage pit	1	-	-	-	-	?	A, P	Unetice	
22	Stillfried a.d. March (Gänserndorf)	(1939)	Pit	-	-	-	2	-	Adults?	A	Unetice	Fortified site, cranium of a dog (Canis familiaris)
		Pit 11 (1939)	Storage pit	-	-	-	1	-	?	A, P, O	Unetice	Fortified site
		Pit 13 (1939)	Storage pit	-	-	-	1	-	?	P	Unetice	Fortified site
		V 10 (1968/69)	Storage pit	-	-	-	1	-	?	P	Unetice	Fortified site
		V 19 (1968/69)	Storage pit	-	-	1	-	-	?	P, S	Unetice	Fortified site, disturbance
		V 20 (1968/69)	Pit	-	-	1	-	-	?	P	Unetice	Fortified site, disturbance
		V 26 (1968/69)	Pit	-	-	1	-	-	Child	O	Unetice	Fortified site
		V 32 (1968/69)	Pit	-	-	1+2	-	-	?	-	Unetice	Fortified site, overlapping pits
		V 34 (1968/69)	Pit	-	-	1	-	-	?	O	Unetice	Fortified site
		V 36 (1968/69)	Pit	-	-	-	2	-	?	-	Unetice	Fortified site
		1553/3 (1987)	Pit	-	-	-	2	(Yes)	Males ((S/1) 25-35, and (S/2) 19-22 yrs.)	P, S, T	Unetice	Fortified site, (S/1): trepanation, (S/2): (R)
		1551/2 (1989)	Pit	-	-	-	1	(Yes)	Female (S/3) (maturus)	-	Unetice	Fortified site, trepanation
		1552 (1989)	Pit	-	-	-	2	-	Children (S/4, S/5)	A, As, P, S, T	Unetice	Fortified site
23	Stoitzendorf (Horn)	?	Storage pit	-	-	-	4	-	Children (7, 8, 11, and 14-15 yrs.)	A, P, SR	EBA	SMI, disturbance
24	Stotzing (Eisenstadt-Umgebung Burger)	?	Pit	-	-	-	1	-	Female (adultus - senilis)	A, P, O	Wieselburg	
25	Unterhautzental (Korneuburg)	V 16	Storage pit	-	-	-	1	(Yes)	Male (35-50 yrs.)	A, O	Unetice	
		V 18	Storage pit	-	-	-	1	-	Adolescent (15 yrs.)	SR	Unetice	
		V 23	Storage pit	-	-	-	1	-	Male (30-50 yrs.)	A, P	Unetice	
		V27	Pit	-	-	-	2	-	Children (2-2.5, 6-7 yrs.)	-	Unetice	
		V 95	Rek. pit.	-	-	-	3	-	Female (adult), children (2)	P, O	Unetice	Grave?
26	Waidendorf (Gänserndorf)	S 7	Pit	x	-	-	-	-	?	SR?	Unetice	
		(1971)	Pit	-	3	-	-	-	Female (1) (17-20 yrs.), children (2) (3-4, and 4-5 yrs.)	A, P	Unetice	Malnutrition, illness
27	Zellerndorf (Hollabrunn)	(1911)	Pit	-	-	-	5+	-	?	-	Unetice	SMI
Bohemia												
28	Blazim (Louny)	?	Pit	-	-	-	1	Yes?	?	-	Unetice	
		?	Pit	-	-	-	1	Yes?	?	-	Unetice	
29	Bisany (Louny)	Pit 1/1950 (Gr. 26)	Grave?	-	-	-	1	-	Male (30-35 yrs.)	P	Unetice	
		Pit 2/1950 (Gr. 27)	Pit	-	-	-	1	-	Female? (7 yrs.)	A, P, S	Unetice	
30	Brezno (Louny)	House 54	Pit	x	-	-	-	-	?	-	Unetice	Building, cremation
		House 62; house 86	Ditch	-	-	-	2	-	Adults?	-	Unetice	Building
		House 64	Pit	-	-	-	1	-	Infant (infans I)	-	Unetice	Building
31	Cerhynky (Kolin)	?	Pit	-	-	-	2	-	Adults?	-	EBA	
32	*Dobromerice (Louny)	Obj. 22	Pit	-	-	-	1	-	?	-	EBA	
		Obj. 71	Pit	-	-	-	1	-	?	-	EBA	
		Obj. 104	Pit	-	-	-	1	-	?	-	EBA	
33	Hlizov (Kolin)	(1925)	Pit?	2	-	-	-	-	Adults?	P	EBA	
		(1925)	Pit	-	-	-	9	-	Females (2) (adults), male (maturus?), children (4) (5, 6, 10, and 14 yrs.)	P	Unetice	Pit house?
34	Kamenná Voda (Most)	Pit 63	Pit	-	-	-	1	-	Infant	-	EBA	
35	Klucov (Kolin)	Pit 2	Pit	-	-	-	2	-	Children (10, and 15 yrs.)	-	Unetice	
36	Kolin (Kolin)	?	Pit	-	-	-	6	-	Males (3) (adults?), females (3) (adults?)	-	Unetice	SMI
37	Kozly (Brandýs n Labem)	Hut II	Pit	-	-	-	1	-	Child (4 yrs.)	-	Unetice	Building
38	Kremýz (Teplice)	Pit 17	Pit	1	-	-	-	-	?	SR	Unetice	
39	Mcely (Nymburk)	Obj. I-1	Pit	-	-	1	-	Yes	Male (mat. I 40-45 yrs.)	As, P, SR	Unetice	Charred bones
40	Meclov-Brezi (Domazlice)	?	Pit	x	-	-	-	-	?	P	EBA	Cremation
41	Plotiste n. Labem (Hradec Králové)	(1957)	?	x	-	-	-	-	?	-	Unetice	
42	Praha	Bubenec	Pit	2	-	-	-	-	Children	-	EBA	
		Lysolaje	Pit	-	-	x	-	-	?	-	EBA	
		Kobylisy	Pit	-	1	-	-	-	Male (sen.: 60 yrs.)	A, P	EBA	
43	Praha-Zahradni Mesto	Pit 1	Pit	-	-	-	2	-	Adult?, child	A, P, S	Veterov	
44	Roztoky (Praha-západ)	?	Pit	x	-	-	-	-	?	-	EBA	
45	Tvrsice (Louny)	?	Pit	-	-	-	x	-	?	-	EBA	
46	Vedomice (Roudnice)	(1956)	Pit	-	-	-	1	-	Child	O, S	EBA	
47	Vraný (Kladno)	a	Pit	-	-	-	1	-	?	P, T	Unetice	
		b	Pit	-	-	-	3	-	?	P	Unetice	
		c	Pit	1	-	-	1	-	Adult?, child	-	Unetice	
Moravia												
48	Bánov (Uherské Hradiste)	?	?	2	-	-	-	-	'Adults'	A	EBA	
49	Bezmerov (Kromeriz)	Nr. 101	Grave?	-	-	1	-	-	Child (7 yrs.)	A, P, O	Veterov	Disturbance
50	Blucina "Cezavy" (Brno-Venkov)	Pit (1945)	Storage pit	-	-	-	1	-	Children	P	Veterov	Settlement refuse above the skeletons
		Pit 1/1950-51	Storage pit	3	-	-	-	-	?	P	Veterov	
		Pit 11/56 (Sh.5)	Ditch	1	-	-	-	-	?	SR	Unetice	
		Pit 2/57 (Sh. 4)	Storage pit	-	-	-	1	-	Female (20-30 yrs.)	A, P	Veterov	
		Pit 6/57 (Sh. 4)	Storage pit	-	-	-	1	-	Child (7-8 yrs.)	A, P, S	Unetice	Above burial: A, P, O, S
		Pit 1/58	Pit	-	-	-	1	-	Child	A, P, T, O	Unetice	
		Fl.II-3/60	Pit	-	-	-	1	-	Child (infans I)	P	Veterov	
		Fl.III-13/60	Storage pit	1	-	-	-	-	?	P, S	Veterov	
		Fl.VII-15/60	Storage pit	1	-	-	-	-	?	A, P, S	Veterov	
		Fl.VII-16/60	Storage pit	1	-	-	-	-	?	A, O (?)	Veterov	
		(1962)	Storage pit	-	-	-	6	-	Male (1) (adult), females (2) (adults), children (3)	-	Unetice	SMI
		Pit 5/85	Pit	17	-	-	-	Yes	Adults (6); Children (11) (2-6 yrs)	A, As, P, SR	Unetice - Veterov	Charred bones
51	Branisovice (Znojmo)	1196/46	Pit	-	-	-	1	-	Female	-	EBA	
52	Bratcice (Zidlochovice)	Obj. I	Pit	1	-	-	-	-	'Adult'	A, P	EBA	
53	Brno-Cerna Pole (Brno-mesto)	"Grave 1"	Grave	-	-	-	1	-	Female (30-40 yrs.)	As, O	Unetice	
		Pit 28	Storage pit	-	-	-	2	-	Female (40-45 yrs.); male (40-45 yrs.)	A, As, P, M	Unetice	Buried on top of each other

					Human remains							
No.	Site	Id.	Object	Fr.	B.P.	Part.B	C.B.	Trauma	Spec.	Other remains	Date	Notes
		Pit 31	Storage pit	-	-	-	2	-	Female (40-45 yrs.); male (40-45 yrs.)	A, P, M	Unetice	Buried on top of each other
		Pit 32	Storage pit	-	-	-	2	-	Children (8-9, and 12-15 yrs)	A, P, M	Unetice	Buried on top of each other
		Pit 35	Pit	-	-	-	1	-	'Adult'	A	Unetice	Head rested on cranium of a dog
		Pit 42	Pit	-	-	-	2	-	?	-	Unetice	
54	Brno-Zidenice (Brno-mesto)	Pit 2	Pit	-	-	-	1	(Yes)	Male (50 yrs.)	-	Unetice	
55	Budkovice (Brno-venkov)	Obj. I	?	x	-	-	-	-	?	-	Veterov	Building
		Pit 3	Pit	x	-	-	-	-	?	-	Veterov	Charred bones
		Pit 29	Storage pit?	-	-	-	1	-	Male (40-50 yrs.)	As	Veterov	
56	Bystrocice-Zeruvky (Olomouc)	?	Storage pit	-	-	-	3	-	Female (40 yrs.), male (40-50 yrs.), male (18-20 yrs.)	-	EBA	Buried on top of each other, massburial
57	Dobsice (Znojmo)	Pit 19	Pit	-	-	-	1	-	Adolescent (juv.)	A, P	Unetice	
58	Hodonice (Znojmo)	Pit P1	Storage pit	1	-	-	-	-	infant (inf. I)	A, P	Veterov	
		Pit 1	Pit	1	-	-	-	Possible	Child (7-8 yrs.)	P	Veterov	
		Pit 2	Pit	-	-	1	-	-	Adolescent (14-16 yrs.)	-	Veterov	
		Pit 21	Pit	-	-	1	-	-	Adolescent (13-14 yrs.)	-	Veterov	
		Pit 38	Pit	-	-	-	1	-	Child (9-10 yrs.)	A, P	Veterov	
59	Hradisko (Kromeriz)	WI/1950-51	Pit	-	-	1	-	-	Adolescent?	P	Veterov	Fortified site
		Gr. 1/55	Grave?	-	-	1	-	-	Male (18 yrs.)	P	EBA	Fortified site, amphora burial
		Gr. 2/55	Pit	-	-	-	1	Yes	Adult (20-40 yrs.)	-	EBA	Fortified site
		Gr. 3-4/55	Pit	-	-	-	2	-	Female (18-22); child (6 yrs.)	O	EBA	Fortified site
		Gr. 5/55	Pit	-	-	-	1	Yes	Female? (20-30 yrs.)	P, O	EBA	Fortified site
60	Hrusky (Vyskov)	?	Pit	-	-	-	1	-	?	-	Unetice	
61	Hulin	H I	Storage pit	-	-	-	3	-	Male (1) (25-30 yrs.) (R), children (2) (3, and 5 yrs.)		Veterov	Child (3 yrs.) was tied, male and child (5 yrs.): anemia
62	Klucov (Cesky Brod)	Pit 2	Storage pit	-	-	-	2	-	Female? (15 yrs.), child (10 yrs.)	SR	Unetice	.
63	Knezdub-Sumarnik (Hodonin)	?	Layer	-	-	-	1	-	Female (7 yrs.)	P	Veterov	
64	Lovcice (Hodonin)	?	Pit	x	-	-	-	-	?	-	EBA	
		?	Pit	-	-	-	1	-	?	O, T	EBA	Grave?
		?	Pit	1	-	-	-	-	?	-	EBA	
65	Lovcicky (Vyskov)	Pit 4	Pit	-	-	-	1	-	?	P	Unetice	
		Pit 88	Pit	-	-	-	1	-	Child	A, P	Unetice	
		Pit 149/67	Pit	-	-	-	1	-	?	A, P, O	Unetice	
		Pit 180 (= 5/67)	Pit	-	-	-	3	-	?	A, P	Unetice	
66	Marefy	(1927)	Pit	3	-	-	1	-	Adults	P, O	Unetice	Disturbance
67	Nemcany (Vyskov)	1961/60	Pit	-	-	-	1	-	Female	O	EBA	Grave?
68	Nosislav (Breclav)	Pit 2/73	Pit	1	-	-	-	-	?	A, P	Unetice	
69	Pavlov (Breclav)	Obj. 40	Pit	-	-	-	1	-	Female (30 yrs.)	-	Unetice	
		Pit 378	Pit	-	-	-	1	-	Male (40-60 yrs.)	-	Unetice	
		Pit 388	Pit	-	-	1	-	-	Adolescent	-	Unetice	
		Pit 822	Pit	-	1	-	-	-	Female? (16-20 yrs.)	-	Unetice	
		Pit 859	Pit	-	2	-	-	-	Adolescent (juvenilis), child (4 yrs.)	-	Unetice	
		Pit 860	Pit	-	-	1	-	-	Child (2-3 yrs.)	P	Unetice	
70	Pohorelice (Breclav)		Pit	-	-	-	1	-	Male	-	Unetice	
71	Prasklice (Kromeriz)	(1959)	Storage pit	-	-	-	3	-	Adults (2), child (1)	As	Unetice	Massburial
72	Pribice (Breclav)	?	Pit	-	-	-	1	-	Female (30-40 yrs.)	-	Unetice	
73	Prerov	(1968)	Storage pit	-	-	-	3	-	Females (2) (20-40, and 14-20 yrs.), child (5 yrs.)	O, S	Veterov	Bed of shells
		(1985)	Pit	x	-	-	2	-	Male (1) (15-17 yrs.), child (1) (9-10 yrs.), adult? (1)	SR	Veterov	Massburial
74	Pribice (Breclav)	(1969)	Storage pit	-	-	-	1	-	Female (30-40 yrs.)	P, S	Unetice	
75	Rajhrad (Brno-venkov)	(1672)	Storage pit	-	-	-	5	Yes	Male (1) (adult), female (1) (adult), children (3)	As	Unetice	SMI, Piglet (Sus scrofa)
		(1957)	Pit	-	-	-	1	-	?	A, S, T	Unetice	
76	Rataje (Kromeriz)	(1954)	Pit	16	-	-	-	?	Female (1) (adult), males (5) (adults), adolescents (2), children (8)	-	Veterov	
77	Svatoborice (Hodonin)	?	Pit	-	-	-	1	-	?	-	EBA	
78	Satov (Znojmo)	?	Pit	-	-	-	1	-	?	-	Unetice	
		Obj. I-7/1962	Pit	-	-	-	1	-	?	-	Unetice	
		Obj. III-44/1962	Pit	-	-	1	-	-	?	-	Veterov	
79	Slapanice (Brno-venkov)	(1934)	Pit	-	-	-	1	-	?	P	Unetice	
		Pit 1/74	Pit	-	-	-	1	-	Adult	-	Unetice	
		Pit 5/74	Pit	-	-	-	1	-	Child	-	Unetice	
		Pit 9/74	Pit	-	-	1	-	-	?	-	Unetice	
		Pit 10/74	Pit	-	-	-	1	-	Adult	-	Unetice	
		Pit 5/75	Pit	1	-	-	-	-	?	-	Unetice	
80	Telnice (Brno-venkov)	Obj. 5/1959	Pit	-	-	1	-	-	?	S, SR	Unetice	
81	Tesetice (Olomouc)	?	Pit	-	-	-	1	-	?	-	Unetice	
82	Tesetice (Znojmo)	Obj. 7	Pit	-	-	-	1	-	Child (10-15 yrs.)	SR	Unetice	
83	Tesetice-Kyjovice	(1991)	Storage pit	-	-	1	3	-	Females (2) (adult, and juvenile), infant (1), male (juvenilis)	O, P	Veterov	Disturbance, dog (Canis familiaris), pig (Sus scrofa)
84	Trbousany (Brno-venkov)	(1960)	Pit	-	-	-	1	-	?	SR	Veterov	
85	Tvrdonice (Breclav)	Pit 1	Storage pit	-	-	1	-	-	?	P, S	Unetice	Disturbance
86	Újezd u Brna (Brno-venkov)	(1969)	Storage pit	-	-	-	1	-	Female (20-25 yrs.)	SR	Veterov	
		(1971)	Storage pit	-	-	-	1	-	Male? (15-16 yrs.)	A, P	Veterov	
87	Velesovice (Vyskov)	Obj. 16	Pit	-	-	1	-	-	Male	-	EBA	
		Obj. 49	Pit	-	-	-	1	-	Female (20-30 yrs.)	-	EBA	
88	Velké Pavlovice (Breclav)	Pit 10/81	Pit	-	-	-	1	(Possible)	Adolescent (16-18 yrs.)	A, P, S, T	Veterov	
		Pit 6	Pit	-	-	-	1	-	?	-	Veterov	
		Pit 49	Pit	1	-	-	-	-	?	-	Veterov	
		(1981)	Storage pit	-	-	-	8	-	Male (1) (40 yrs.), female (1) (30-40 yrs.), children (6) (3, 5, 6, 7, 8, 9)	-	Veterov	SMI
89	Veterov (Hodonin)	(1928)	Pit	-	-	-	3	-	?	-	Veterov	
		Obj. 1/56	Pit	-	-	-	1	Yes	Male (35-45 yrs.)	-	Veterov	
		Obj. 7/56	Pit	-	-	1	-	Yes	Child	A	Veterov	Disturbance
		Obj. 3	Pit	-	-	-	1	-	Male (30-40 yrs.)	-	Veterov	
		Obj. 4	Pit	-	-	-	1	Possible	Female (30-40 yrs.)	SR	EBA	
90	Visnové (Znojmo)	?	Pit	-	-	-	-	-	?	-	Unetice	
91	Vyskov	?	Pit	-	-	1	-	Possible	Male (maturus)	A, P, O, T	Unetice	Disturbance (older grave?), remains of three piglets (Sus scrofa)
92	Znojmo	?	Pit	-	-	-	4	-	Female (adult) (1), male (adult) (1), children (2)	-	Unetice	SMI
93	Gánovce (Poprad)		Well	-	x	6	-	-	Adults (2), children (4)	O, SR	Otomani	Fortified site, the arm still carried two bronce rings
94	Hoste (Galanta)	Pit 1	Pit	-	-	-	2	-	?	-	Madarovce	Fortified site, animal burial on site
		Pit 11	Pit	1	-	-	-	-	Child	-	Madarovce	Fortified site
		Pit 52/84	Pit	-	-	-	1	-	?	-	Madarovce	Fortified site
95	Ivanovce (Trencin)	S. A-L/66	Pit	-	1	-	-	Yes	Female (juvenilis?)	-	Madarovce	Fortified site
96	Jelsovce (Nitra)	?	?	-	-	-	9	-		-	EBA	
		?	?	-	-	-	6	-		-	EBA	
		?	?	-	-	-	5	-		-	EBA	
		?	?	-	-	-	5	-		-	EBA	
97	Kosice-Barca (Kosice)	?	?	x	x	x	x	-	?	-	Otomani	Fortified site
98	Malé Kosihy (Nové Zámky)	Pit 37	Storage pit	-	-	-	1	-	?	P, SR	Madarovce	Fortified site
		Pit 57	Storage pit	x	-	-	1	-	?	A, SR	Madarovce	Fortified site, disturbance
99	Nitra	?	Layer	x	-	-	-	-	Child	-	Madarovce	Fortified site
100	Nitransky Hrádok (Nové Zámky)	Pit 3	Pit	3?	3	-	-	-	?	?	Madarovce	Fortified site
		Pit 27	Pit	-	x	-	-	-	?	A	Madarovce	Fortifiued site, high amount of animal bones
		Pit 28	Pit	?	?	?	?	?	?	?	Madarovce	Fortified site
		Pit 105	Pit	?	?	?	?	?	?	?	Madarovce	Fortified site
		Pit 116	Pit	?	?	?	?	?	?	?	Madarovce	Fortified site
		Pit 120	Pit	?	?	?	?	?	?	?	Madarovce	Fortified site
		Pit 134	Pit	-	-	x	-	-	?	A	Madarovce	Fortified site, high amount of animal bones
		Pit 216	Pit	?	?	?	?	?	?	?	Madarovce	Fortified site
		Pit 220	Pit	?	?	?	?	?	?	?	Madarovce	Fortified site
		Pit 228	Pit	?	?	?	?	?	?	?	Madarovce	Fortified site
		Pit 230	Pit	?	?	?	?	?	?	?	Madarovce	Fortified site
		Pit 237	Pit	?	?	?	?	?	?	?	Madarovce	Fortified site
		Pit 242	Pit	-	-	-	1	-	Child	P, Sa	Madarovce	Fortified site
		Pit 243	Pit	?	?	?	?	?	?	?	Madarovce	Fortified site
		Pit 252	Pit	?	?	?	?	?	?	?	Madarovce	Fortified site
		Pit 254	Pit	?	?	?	?	?	?	?	Madarovce	Fortified site
		Pit 266	Pit	?	?	?	?	?	?	?	Madarovce	Fortified site
		Pit 268	Pit	?	?	?	?	?	?	?	Madarovce	Fortified site
		Pit 285	Pit	?	?	?	?	?	?	?	Madarovce	Fortified site
		Pit 296	Pit	?	?	?	?	?	?	?	Madarovce	Fortified site
		Pit 297	Pit	?	?	?	?	?	?	?	Madarovce	Fortified site
		Pit 298	Pit	x	-	-	-	-	?	As	Madarovce	Fortified site
		Pit 300	Pit	x	-	-	-	-	?	-	Madarovce	Fortified site
		Pit 306	Pit	?	?	?	?	?	?	?	Madarovce	Fortified site
		Pit 325	Pit	?	?	?	?	?	?	?	Madarovce	Fortified site
		G/12	Layer	-	-	-	?	-	?	?	Madarovce	Fortified site
		G/18	Pit	-	-	-	x	-	Child	?	Madarovce	Fortified site, amphora burial
		H/22	Pit	-	-	-	x	-	Child	?	Madarovce	Fortified site, amphora burial
		N/22	Pit	-	-	-	x	-	Child	?	Madarovce	Fortified site, amphora burial

No.	Site	Id.	Object	Human remains Fr.	B.P.	Part.B	C.B.	Trauma	Spec.	Other remains	Date	Notes
101	Nizná Mysľa (Kosice)	Obj. 308	Storage pit	-	-	-	5	Yes	Male (1) (14-18 yrs.), females (2) (19-24, and 30-40 yrs.), child (2) (3-5, and 9-13 y	SR	Otomani	Fortified site, SMI
102	Prasník (Trnava)	S. VI/76	?	1	-	-	-	Yes	Adult	-	Madarovce	Fortified site
103	Spišský Stvrtok (Spisská Nová Ves)	Obj. 40/74	Storage pit?	-	-	2	7	Yes	Male (1) (25 yrs.), females (3) (adults), children (5)	P	Otomani	Fortified site, buried in two groups
104	Unín (Senica)	?	Pit	-	-	-	1	-	Child	-	Veterov	Fortified site
		?	Layer	x	-	-	-	-	.	-	Veterov	Fortified site
105	Včelince (Rimavská Sobota)	Obj. 9	Pit	-	-	-	1	-	Infant	A	EBA	
		Obj. 13	Pit	-	-	-	1	-	Infant	A	EBA	
		Obj. 23	Pit	1	-	-	-	Yes	Child (8-9 yrs.)	A	EBA	
		Obj. 24	Pit	1	-	-	-	Yes	Child (6-7 yrs.)	A	EBA	
		Obj. 28	Pit	1	-	-	-	Yes	Male (20-30 yrs.)	A	EBA	
		Obj. 29	Pit	1	-	-	-	Yes	Child (6-15 yrs.)	A	EBA	
		S.II-C/6	Layer	1 -	-	-	-	Yes	Adolescent	A	EBA	
		S.II-E/4	Layer	1	-	-	-	Yes	Adolescent?	A	EBA	
106	Veselé-"Hradisko" (Trnava)	SZ 24, Pit 5	Pit	1	-	-	-	-	?	A, As, P, SR, T	Veterov	Fortified site
		Pit 46S	Storage pit	1	-	-	-	-	?	A, As, P	Veterov	Fortified site, charred remains
		Pit 103S	Storage pit	1	-	-	-	-	?	A, P	Veterov	Fortified site
		SZ 20, Pit 11	Storage pit	1 -	-	-	-	-	?	A, As, P, T	Veterov	Fortified site
		SZ 29, Pit 8	Storage pit	1	-	-	-	-	?	A, As, P, T	Veterov	Fortified site
		Pit 7 ZS	Storage pit	1	-	-	-	-	?	A, P	Veterov	Fortified site
		?	?	2+	-	-	-	-	Females (adult)		Madarovce	Fortified site, Cult mask made of human face

No.: Refers to numbers in the text and on the maps.
Site: Find location
Id.:
Object Object where the burial/deposit was located.
  Gr.=grave
Human numbers represents the numbers of individuals present
  Fra. = (Fragments) fragmented and heavily damaged bones; B.P. = (Body Parts) dislocated body parts
  Part.B. = (Partial Body) body parts in correct anatomical position where not more than half of the complete body is present
  C.B. = (Complete Body) complete human body with little or no dislocation of its parts.
  Trauma = notes on peri-mortem violence.
  Spec = (Specification): Notes on sex, age and other significant features: (R) = heavily built.
  Other : A = animal bones; As = ash; P = pottery; T = tools; O = ornaments; S = shells; Sa = sand; SR = settlement refuse (smaller fragments of pottery, adobe, bones etc.)
Dates:
Notes: SMI (special massinhumation)

FIG. 21: *List of abnormal burials from Early Bronze Age settlements (cases known to the author).*

with the ordinary mortuary practice of the Early Bronze Age. Thus, it is important to take a closer look at the circumstances around the burial in Kettlasbrunn in order to place it within a relevant mortuary paradigm. Investigation of the causes of death for the interred individuals revealed that at least the mature man and woman had died under violent circumstances. The cause of death for the child and younger woman have not been possible to determine as their bodies were too fragmented.

It is of note that both the mature man and woman display similar injuries: puncture fractures on the upper part of the torso and sharp-force fractures to the base of the neck (ibid.: 94). The puncture fractures on the left shoulder blade of the female show that she has been stabbed twice through her heart with a sharp implement. This had been done with such force that the ribs behind the shoulder blade had been splintered. Also the male had had a stake hammered through his chest. It had entered from the front and punctured his heart and even damaged the vertebrae behind it, fracturing the left shoulder blade as well. The puncture fractures present on both individuals imply that a sharpened bone tool was used. Sharp-force trauma to the base of the neck, where the axis and *foramen magnum* connect, implies that a sharp tool, presumably a knife, had been thrust into the brain. For the woman there are no further indications that she had been subject to peri-mortem violence. The peri-mortem trauma his-tory for the man on the other hand reveals some interesting notions. To begin with, the hands were tied behind his back at the time of his death. He had fractures on the right shoulder blade and vertebral column deriving from at least two slashes with a sharp instrument, probably a knife. The position of these injuries together with the trauma to the chest shows that his hands were behind his back at these moments (ibid.: 94).

Furthermore, the man had been engaged in violent action shortly before his death. The lower jaw was fractured and splintered as a result of blunt-force trauma. A sharp-force trauma to his left thigh near the pelvis and cut marks on the ribs near the breast-bone were also observed. None of these wounds show signs of healing, which means that they were caused shortly before his death.

It is clear that the perpetrators were particularly aggressive towards the mature male in comparison to their other victims. The apparent different treatment of the man, with hints of warfare, raises the question of the social status these victims had in life. Are we dealing with war captives, social outcasts, slaves or perhaps criminals?

The osteological analysis implies that both the older male and the female belonged to the upper echelons of society since their skeletons show little pathological change for their age. It would be expected that the skeletons of both the man and woman should show some indications of stress relating to

· WARFARE, RITUALS, AND MASS GRAVES

No.	Site	Id.	Object	Ind.	Trauma	Date	Notes
Austria (Lower Austria)							
3	Feis am Wagram (Tulln)	V 55	Pit	Adult	Cranial fractures	Unetice	Building
				Child	Cranial fractures		
4	Franzhausen (St. Pölten)	V 76	Pit	Male (30 yrs.)	Painful death, soft tissue trauma?	Unterwölbling	
13	Kettlasbrunn (Mistelbach)	(1990)	Storage pit	Female (31-40 yrs.),	Fracture: os zygomaticum (H)	EBA	
					Puncture fracture: scapula, sin.		
					Blunt fracture: scapula, med. sin.		
					Blunt fracture: costae (flakeing near scapula)		
				Male (41-50 yrs.)	Fracture?: fibula, dist. sin. (H)		
					Sharp fracture: ossa nasalia (H)		
					Fracture: ulna and ossa metacarpalia (H)		
					Blunt fracture: mandibula		
					Sharp fracture: condylus occipitalis, dx. + sin.		
					Puncture fracture: scapula, dx.		
					Puncture fracture: scapula, sin.		
					Sharp fracture: femur, prox. sin.		
					Sharp fracture: vertebra thoracicae (2nd), cau.		
					Sharp fracture: vertebra thoracicae (8th)		
					Fracture: costae, dist.		
15	Oberndorf a d. Ebene (St. Pölten)	V.1	Pit	Male (50 yrs.)	16 severe cranial fractures	Unetice	
				Female (18-20 yrs.)	? severe cranial fractures		
17	Poysbrunn (Mistelbach))	Pit 10	Storage pit	Female (25-35 yrs.)	Possible fracture on the mandibula	Veterov	
18	Poysdorf (Mistelbach)	?	Pit	Male (41-60 yrs.)	Trepanation (H)	Veterov	
21	Schleinbach (Mistelbach)	(1931)	Pit	Children (3) (4-5, 10, and 12 yrs.), male (1) (35-40 yrs.)	Violent death?	Unetice	SMI
22	Stillfried a.d. March (Gänserndorf)	V 19 (1968/69)	Storage pit	?	Violent death?	Unetice	Fortified site, disturbance
		V 20 (1968/69)	Pit	?	Violent death?	Unetice	Fortified site, disturbance
		1553/3 (1987)	Pit	Male (S/2) (19-22)	Trepanation (H), (S): os parietale, dx.	Unetice	Fortified site, trepanation
		1551/2 (1989)	Pit	Female (S/3) (maturus)	Trepanation	Unetice	Fortified site, trepanation
25	Untehautzental (Korneuburg)	V 16	Storage pit	Male (35-50 yrs.)	Impression fracture (H): os parietale, sin.	Unetice	
Bohemia							
28	Blazim (Louny)	?	Pit	?	?	Unetice	
		?	Pit	?	?	Unetice	
39	Mcely (Nymburk)	Obj. I-1	Pit	Male (mat. I: 40-45 yrs.)	Fracture: butchering	Unetice	Charred bones
Moravia							
50	Blucina "Cezavy" (Brno-Venkov)	Pit 5/85	Pit	Adults (6), Children (11) (2-6 yrs)	Fractures: butchering	Unetice - Veterov	Charred bones
54	Brno-Zidenice (Brno-mesto)	Pit 2	Pit	Male (50 yrs.)	Blunt fracture: os parietale, sin. (H)	Unetice	
					Blunt fracture: os parietale, dx. (H)		
58	Hodonice (Znojmo)	Pit 1	Pit	Child (7-8 yrs.)	Violent death? Cranial fractures?	Veterov	
59	Hradisko (Kromeriz)	Gr. 2/55	Pit	Adult (20-40 yrs.)	Cranial fractures	EBA	Fortified site
		Gr. 5/55	Pit	Female? (20-30 yrs.)	Painful death, soft tissue trauma?	EBA	Fortified site
75	Rajhrad (Brno-venkov)	(1872)	Storage pit	Male (adult)	Decapitated	Veterov	
				Adolescent (16-18 yrs.)	Soft tissue trauma?		
88	Velké Pavlovice (Breclav)	Pit 10/81	Pit	Male (35-45 yrs.)	Soft tissue trauma?	Veterov	
89	Veterov (Hodonin)	Obj. 1/56	Pit	Child	Fracture: os parietale, os temporale and os occipitale, sin.	Veterov	
		Obj. 7/56	Pit		Fracture: butchering?	Veterov	Disturbance
					Fracture: os temporale, sin.		
		Obj. 4	Pit	Female (30-40 yrs.)	Soft tissue trauma?	Veterov	
91	Vyskov	?	Pit	Male (maturus)	Cranial fractures?	Unetice	Disturbance
Slovakia							
95	Ivanovce (Trencín)	S. A-L/66	Pit	Female (juvenilis?)	Sharp fracture: os frontale, dx.	Madarovce	Fortified site
101	Nizná Mysla (Kosice)	Obj. 308	Storage pit	Male (14-18 yrs.),	Soft tissue trauma	Otomani	Fortified site, SMI
					Blunt fracture: os frontale (H)		
				Females (19-24 yrs.)	Soft tissue trauma		
					Fracture: mandibula		
				Female (30-40 yrs.)	Soft tissue trauma		
				Child (3-5 yrs.)	Thermal trauma (boiled)		
					Cranial fractures?		
				Child (9-13 yrs.)	Soft tissue trauma		
102	Prasnik (Trnava)	S. VI/76	?	Adult	Fracture: butchering	Madarovce	Fortified site
103	Spissky Stvrtok (Spisská Nová Ves)	Obj. 40/74 [A]	Storage pit?	Male (25 yrs.) No.1	8 cranial fractures	Otomani	Fortified site
				No. 2	Cranial fractures		
				No. 4	Cranial fractures		
				No. 8	Cranial fracture		
105	Vceilnce (Rimavská Sobota)	Obj. 23	Pit	Child (8-9 yrs.)	Fractures: butchering	EBA	
		Obj. 24	Pit	Child (6-7 yrs.)	Fractures: butchering	EBA	
		Obj.28	Pit	Male (20-30 yrs.)	Fractures: butchering	EBA	
		Obj. 29	Pit	Child (6-15 yrs.)	Fractures: butchering	EBA	
		S.II-C/6	Layer	Adolescent	Fractures: butchering	EBA	
		S.II-E/4	Layer	Adolescent?	Fractures: butchering	EBA	

No.   Refers to numbers in the text and on the maps (see Tab. 1).
Trauma: Evidence of intentional violent interference. (H) = healed wound, (PM) = perimortem, (S) = symbolic
         cau. = caudale, dx. = dexter, lat. = laterale, med. = mediale, sin. = sinister

FIG. 22: *List of trauma and pathological change in skeletons from Early Bronze Age settlements (cases known to the author).*

malnutrition, infections and physical strain as these were common in this period (cp. Teschler-Nicola and Gerold 2001). However, the slender female skeleton bears no witness of hard labour and her teeth indicate good health with no signs of malnutrition. Apparently she had led a healthy life from childhood onwards. Her joints and vertebras were hardly worn (Winkler and Schweder 1991: 94).

The male has seemingly a similar background within the upper parts of society. His large muscular attachments suggest a heavy build and although he had suffered from a mild form of arthritis since his youth his skeleton show little indication of being subject to stress. Thus, this was not someone who had partaken in heavy labour, although his build implies he had been involved in significant amounts of physical activity. Winkler and Schweder (ibid.: 95) suggest that this man was either a warrior or at least someone who had led an adventurous life. Healed fractures imply that he had been in violent conflicts several times earlier in his life. Fractures on both his arms and hands are similar to those of a boxer and the healed lesion on his nasal bones shows that he was once struck in the face with a sharp object.

In summary, the significant feature in Kettlasbrunn is the repetitive behaviour in the burials, killings

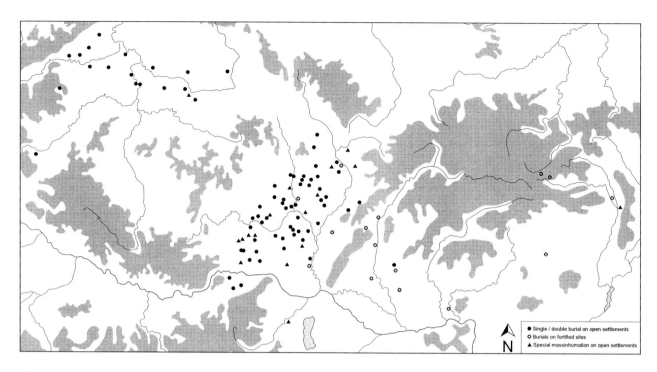

FIG. 23: *Geographical distribution of special burials from settlement sites in Slovakia, Moravia and Bohemia.*

and choice of victims, which can be interpreted as some kind of ritual practice. The similar pattern of peri-mortem fractures for the mature man and woman implies killings in a ritualistic manner – a conduct governed by certain regulations and reproduced over time. In addition, the implications of the social rank of the deceased and his assumed warrior identity point to a ritual practice embedded in an overall socio-political structure.

Nižna Myšľa

To further strengthen the discussion of the violent and ritual aspects of burials in storage pits I would like to draw attention to a multiple burial in Nižna Myšľa (Fig. 19) in eastern Slovakia from the early Otomani culture (Jakab *et al.* 1999).

Five individuals were recorded in the burial: one adolescent male (14-18 years old), two adult females (19-24 and 30-40 years old) and one child (sex and age unknown). They were located in the centre of the pit within distinct layers indicating subsequent interments over a period of time.

The burial was located on a fortified site on the outskirts of a settlement area and close to an older cemetery (ibid.: 91). The hourglass shape of the pit suggests that it had been used as a storage pit prior to its use as a burial ground. It had been filled with

soil and settlement refuse in subsequent phases, and there were at least nine distinct layers. The composition of the layers suggests three major phases: a phase of primary use with skeletal remains, a closing ritual phase, and a waste disposal phase. All human remains were located in the five successive layers in the lower half of the pit and each body was in a separate layer. The pit was then 'sealed' with a thick layer of soil without any vestiges of cultural activities. The upper part of the pit was afterwards used for waste disposal – slowly filled over an extended period. This final major phase appears not to be connected to the two preceding phases.

Within the burial phases animal bones, pottery sherds, clamshells, polished stones and clay weights accompanied the human bodies. The arrangement of these items appears to be intentional, in comparison to the other layers without human remains where they are intermixed. It is noteworthy that in the layers with the female bodies these items lie near the sides of the pit whereas in the layer with the adolescent man they have been heaped on top of him. Also, it seems as if the composition of animal bones is different in each layer. The small child near the bottom of the pit lies close to some horse bones and there are bones of bird and cow near the female bodies (ibid.: 113-177).

Above the accumulation of the human bodies there was a thick layer of loess soil. It contained no remains of settlement refuse, and therefore it appears as if this soil layer was intended to cover and seal off the burials (ibid.: 94). Thus this can be regarded as a closing phase for the funerary activities connected with the pit. After the burials were sealed the remainder of the pit was apparently used as a waste dump as the succeeding two layers were filled with settlement refuse and charcoal in a haphazard manner. There are also indications of re-cutting in the top layers, an activity not recorded for the lower levels of the pit.

The five individuals were buried in a fashion deviating from conventional burials in cemeteries. Forensic analysis suggests that four of them had been subject to a ritual killing and the fifth had apparently been boiled prior to the burial. It could also be concluded that they were buried during at least four events, and that the bodies had decomposed between these events. From these four events, together with the major three phases of the filling of the pit, it is possible to draw an outline of a ritual cycle.

Near the bottom of the pit close to the northern side the right parietal bone from a smaller child was found. No other skeletal remains from this individual was located in this layer and the preservation of this bone and the rest of the human remains in the pit indicate that this cranial fragment had been buried in isolation. Odd as it may seem, the rest of this child was actually discovered with the topmost burial. The bone surfaces of the cranial fragments from this child are different from the skeletal remains of the other individuals and resembles bone material subject to thermal changes. It has therefore been suggested that the skull of the child had been boiled before it was buried (ibid.: 106).

Close to this child, the body of a younger woman was found. She was lying on her left side with slightly bend legs close to the north-eastern wall. Whereas her legs and pelvis rested on their sides the upper parts of the body had been turned to the right towards the base of the pit. The chest was lying against the floor level with the bent left arm beneath it with the palm pressed against the chest. The right arm projected slightly from the body with the hand near the skull. The skull had been turned sideways to the left so that the face looked over the left shoulder and the mandible touched the cervical vertebrae. This position of the body is by no means a natural one. Judging especially by the way the skull had been twisted it is apparent that the upper part of the woman's body had been deliberately turned to the ground by forcing the head to the left. The slight projection of the right shoulder blade and arm towards the head and the left shoulder towards the feet can also be taken as indicators that the body had been turned peri-mortem and not post-mortem. A smaller fracture on the left cheek appears to have been caused during the peri-mortem treatment (ibid.: 111).

In the succeeding layer above the young woman an adult woman was located near the centre of the pit. She was placed on her right side and like the younger woman the torso and skull had been turned towards the pit base. Her arms were bent and pressed to the chest and the legs were flexed and crossed.

Directly above the adult woman was the body of an older child. It was placed on the stomach in a stretched position. The right leg was outstretched along the axis of the body whereas the left leg was heavily bent beneath it. Similarly the left arm was pressed alongside the side of the body and the right one bent and projected out from the body. Oddly enough, the palm of the right hand was turned towards the top of the pit instead of towards the base as would be expected. The mandible was to the left side of the body, whereas the first cervical vertebra and the calvarium were missing. These parts were, however, located some distance away and were apparently disturbed soon after burial. The skull rested on the right side. The position of the legs, arms and skull leads to the conclusion that the child had its skull forced to the side just like the rest of the individuals in the pit.

Concerning the disturbance of the head, no cut marks are visible on either the skull or axis. Considering that the rest of the body is in an anatomically correct position, this suggests that the movement of the axis and calvarium can be explained as a natural consequence of the taphonomic processes (ibid.: 97f, 107). The nature of the ritual killing might explain this disarticulation.

In the last burial layer there was the burial of an adolescent man placed crouching in a kneeling position at the centre of the pit, and the remaining elements of the boiled child skull, part of which was found at the bottom of the pit some 80-90 centimetres lower down. The scattered cranial parts of the child were located at the western edge of the pit. All bones were disarticulated and no postcranial body parts of this individual were found. The well-preserved skeleton of the man indicates that if the rest of the body of the child had been buried it would have been found. Thus there are strong indications that this was not the case. Concerning the adolescent man, his limbs had been pressed tightly beneath his torso as if he had been bound (ibid.: 113). He rested on the pit fill on his knees and hands, and his head had been turned to the right. It should be mentioned that the adolescent man has a healed depressed fracture on the brow (ibid.: 106).

Similarly to the multiple burial in Kettlasbrunn there are several interesting ritual aspects, both concurrent and coeval: especially the selection of the buried individuals, their peri- and post-mortem treatment and the apparent demarcation of the beginning and end of a ritual cycle. With the exception of the small child, who is only represented by cranial fragments, all the individuals have apparently been disposed of in a similar fashion, and they had obviously been alive when placed in the pit. The forcefully wrung heads suggest that they were killed by having their throats slit (ibid.: 113-17, 223). Both women had apparently been placed sitting in the pit judging by the position of their legs. They were then forced down on one side so that their chest touched the ground. The perpetrator must have pressed his or her body against the victim and by twisting the head further to the side, so that the body became forced around its own axis, the victims body was locked in the grip of the perpetrator. This may also be the case with the child who was placed on the stomach. As I see it, the perpetrator, standing to the left, turned the child's body towards the ground and blocked the movement of the child's left arm and leg with his or her body. As a response to the forced movement, or perhaps as an attempt to free itself, the child's free limbs stretched out. Jakab *et al.* suggest that the child's left arm and leg were tied together, which might also be possible (ibid.: 115). It can also be assumed that the adolescent man was killed after he was placed in the pit. Not only was his head turned sideways like the others, his palms were pressed to the ground as if he was stabilising his kneeling body or trying to push away from the ground (ibid.: 113).

Estimating the time span between the first and last burial is difficult. Considering that the bodies are located in different layers it can at least be assumed that some time had passed between each killing. Since there is no evidence of weathering, as at Kettlasbrunn, I assume that each body was covered with soil soon after each instance of killing. It is of note that the older child had apparently been placed on a thicker bed of organic matter, such as hay or shrubbery, as its body has sunk and collapsed during decomposition, leading to the disarticulation of the skull, right foot and left hand (ibid.: 97ff, 113ff). Also, the surface beneath the older child was trodden down, indicating that the bodies from the earlier interments had decomposed. The situation is similar for the adolescent man. When he was placed in the pit the body of the older child must have been more or less decomposed as there are no signs that the man's body had been displaced due to decomposition of the layers beneath him.

Females and children seem to be preferred when it comes to the killings at Nižna Myšľa. To determine a possible social status in life for these individuals has proven to be difficult and is not discussed by Jakab *et al.* (1999). There are no notable pathological changes on the skeletons that would indicate a distinct social class as was the case with the individuals in Kettlasbrunn. The smaller child might be an exception, as it is recorded that its skull is abnormally large (ibid.: 123). Concerning the ritual treatment of their disposal and objects accompanying them the deceased can be divided into three distinct groups. The first group consists of the small child. As I have mentioned, its thermal treatment and burial are clearly distinct from the rest of the bodies. It is also of note that it was lying close to some horse bones (ibid.: 115). The second group is made up of the two women and the older child – who might be a girl (ibid.: 122). These are all disposed of in a similar manner, as mentioned above, and the accompanying animal bones, clamshells and pot sherds are placed near the sides of the pit. The third, and last, group is that of the man who had been tightly tied and was partially covered in animal bones. In sum, it appears as if the ritual conduct and the stage in which each individual was killed were influenced by the sex of the victim.

If a body had time to decompose between each burial the pit might have been in use for some months, or perhaps a year. It is important to bear in mind the notion that the beginning and end of the ritual activities in the pit was marked through the burial of the fragments of the smaller child's boiled skull at the bottom of the pit and later in conjunction with the last burial of the adolescent man. As there are no signs of disturbance between the layers it must be assumed that these skull fragments were buried on purpose. Thus, those who carried out the ritual had kept the remaining skull pieces and were waiting for the cycle of ritual killings to be completed before the last fragments could be interred. When this was done the lower levels of the pit were sealed, preventing later access.

The significance and meaning of all these details of ritual action of course need a wider analysis than can be given here. Evidence from the rituals of normal burials needs to be incorporated too, and the tie between the abnormal burials of ritual killings and the fortified settlements needs to be explored further. The widespread appearance of hill-top settlements with defence systems around this time cannot be a coincidence. The interpretation must therefore take account of the generally war-like and socially disruptive context of the latest Early Bronze Age at the transition to the Middle Bronze Age in eastern and south eastern central Europe c. 1700-1500 BC.

NOTE

1   I have not been able to examine the skeletons myself, but have relied on already conducted research. These are not as detailed as could have been wished for – Jakab's detailed analysis (1999) being an exception in order to ascertain the exact nature of a fracture, such as the relationship between multiple fractures being the result of one or several blows, or the effects of taphonomic processes after burial. Due to this, the exact number of fractures resulting from different encounters cannot be established and I have therefore chosen to indicate both the lowest and highest possible number of fractures. However, this state of affairs merely has a minor negative effect on the understanding of the nature of injuries. Moreover, it should be noted that Bátora's compilations (1991 and 1999a) contain somewhat inconsistent information about the trauma found on Nitra culture skeletons. This is of course unfortunate, and I have not been able to consult unpublished osteological reports. The published material, however, verifies a rather high frequency of trauma.

BIBLIOGRAPHY

- Bátora, J. 1986. 'Výskum pohrebiska zo staršej doby bronzovej v Jelšoviach – Die Ausgrabung des Gräberfeldes aus der älteren Bronzezeit in Jelšovce'. *Archeologické Rozhledy*, 38, pp. 263-74.
- Bátora, J. 1990. 'The latest knowledge on the burial rite of the people of the Nitra group'. *Anthropologie*, 28(2-3), pp. 169-74.
- Bátora, J. 1991. 'The reflection of economy and social structure in the cemeteries of the Chlopice–Veselé and Nitra Cultures'. *Slovenská Archeológia*, 39(1-2), pp. 91-142.
- Bátora, J. 1999a. 'Gräber mit Totenhäusern auf frühbronzezeitlichen Gräberfeldern in der Slowakei'. (beitrag zu Kulturverbindungen zwischen Mittel-, West- und Osteuropa). *Praehistorische Zeitschrift*, 74, pp. 1-57.
- Bátora, J. 1999b. 'Waffen und Belege von Kamptreffen während der Frühbronzezeit in der Südwestslowakei'. In: J. Bátora and J. Peška (eds.): *Aktuelle Probleme der Erforschung der Frühbronzezeit in Böhmen und Mähren und in der Slowakei*. Nitra: Archaeologica Slovaca Monographiae.
- Bátora, J. 1999c. 'Symbolische(?) Gräber in der älteren Bronzezeit in der Slowakei'. In: J. Bátora and J. Peška (eds.): *Aktuelle Probleme der Erforschung der Frühbronzezeit in Böhmen und Mähren und in der Slowakei*. Nitra: Archaeologica Slovaca Monographiae.
- Bátora, J. 2000. *Das Gräberfeld von Jelšovce / Slowakei. Ein Beitrag zur Frühbronzezeit im nordwestlichen Karpatenbecken.* (Prähistorische Archäologie in Südosteuropa, 16) Kiel: Verlag Oetker/Voges.
- Bátora, J. and Peška, J. (eds.) 1999. *Aktuelle Probleme der Erforschung der Frühbronzezeit in Böhmen und Mähren und in der Slowakei*. Nitra: Archaeologica Slovaca Monographiae.
- Budinský-Krička, V. 1965. 'Gräberfeld der späten Schnurkeramischen Kultur in Veselé'. *Slovenská Archeológia*, 13(1), pp. 51-106.
- Chropovský, B. 1960. 'Gräberfeld aus der älteren Bronzezeit in Vel'ký Grob'. In: B. Chropovský, M. Dušek and B. Polla (eds.): *Pohrebiská zo staršej doby bronzovej na Slovensku I – Gräberfelder aus der älteren Bronzezeit in der Slowakei*. (Archaeologica Slovaca Fontes, 3). Bratislava: Vydavatel'stvo Slovenskej akadémie vied Bratislava.
- Dušek, M. 1969. *Bronzezeitliche Gräberfelder in der Südwestslowakei*. Bratislava.
- Eckhardt, H. 1996. *Pfeil und Bogen. Eine archäologisch-technologische Untersuchung zu urnenfelder- und hallstattzeitlichen Befunden.* (Internationale Archäologie, 21). Espelkamp: Verlag Marie Leidorf.
- Furmánek, V., Veliačik, L. and Vladár, J. 1999. *Die Bronzezeit im slowakischen Raum.* (Prähistorische Archäologie in Südosteuropa, 15). Rahden: Verlag Marie Leidorf.
- Furmánek, V. 1997. 'Stand der demographischen Erforschung der Bronzezeit in der Slowakei'. In: K.-F. Rittershofer (ed.): *Demographie der Bronzezeit. Palädemographie - Möglichkeiten und Grenzen*. Espelkamp: Verlag Marie Leidorf.

- Jakab, J. 1999. 'Anthropologische Analyse der Gräber mit Totenhäusern des frühbronzezeitlichen Gräberfeldes in Mýtna Nová Ves'. *Praehistorische Zeitschrift*, 74, pp. 58-67.
- Jakab, J, Olexa L. and Vladár J. 1999. 'Ein Kultobjekt der Otomani-Kultur in Nižná Myšla.' *Slovenská Archeológia* 47.1, pp. 91-127.
- Neugebauer, J.-W. 1991. *Die Nekropole F von Gemeinlebarn, Niederösterreich. Untersuchungen zu den Bestattungssitten und zum Grabraub in der ausgehenden Frühbronzezeit in Niederösterreich südlich der Donau zwischen Enns und Wienerwald.* Mainz am Rhein: Verlag Philipp von Zabern.
- Novotná, M. 1999. 'Die Aunjetizer Kultur in der Slowakei. Stand der Erforschung'. In: J. Bátora and J. Peska (eds.): *Aktuelle Probierne der Erforschung der Frühbronzezeit in Böhmen und Mähren und in der Slowakei.* Nitra: Archaeologica Slovaca Monographiae.
- Ondráček, J. 1972. 'Pohřebiště nitranské skupiny Holešově – Das Gräberfeld der Nitra-Gruppe in Holešov'. *Archeologické Rozhledy*, 24, pp. 168-72.
- Ondráček, J. and Sebela, L. 1985. 'Pohřebiště nitranské skupiny v Holešově (katalog nálesů)'. *Studie Muzea Kroměřížska*, 85, pp. 2-130.
- Pavúk, J. 1981. 'Die ersten Siedlungsfunde der Gruppe Chlopice-Veselé aus der Slowakei'. *Slovenská Archeológia*, 29(1), pp. 163-175.
- Rittershofer, K.-F. (ed.) 1997. *Special burials in the Bronze Age of Eastern Central Europe. Proceedings of the Conference of the Bronze Age Study-Group at Pottenstein 1990.* Espelkamp: verlag Marie Ledorf. Internationale Archäologie.
- Stloukal, M. 1985. 'Antropologický rozbor koster z Pohřebiště v Holešov'. *Studie Muzea Kroměřížska*, 85, pp. 131-69.
- Strouhal, E. 1978. 'Demography of the Early Bronze Age Cemetery at Výčapy – Opatovce (Southwest Slovakia)'. *Anthropologie*, 16(2), pp. 131-35.
- Stuchlík, S. 2001. 'Die Besiedlung Ostmährens durch die Aunjetitzer Kultur und den epischnurkeramischen Komplex zu Beginn der Bronzezeit – The settlement of Eastern Moravia by the Únětice culture and the Epi-corded Ware complex at the beginning of the Bronze Age'. In: A. Lippert, M. Schultz, S. Shennan and M. Teschler-Nicola (eds.): *Mensch und Umwelt während des Neolithikums und der Frühbronzezeit in Mitteleuropa. Ergebnisse interdisziplinärer Zusammenarbeit zwischen Archäologie, Klimatologie, Biologie und Medizin. Rahden/Westf.*, Verlag Marie Leidorf GmbH. Internationale Archäologie: Arbeitsgemeinschaft, Symposium, Tagung, Kongress 2.
- Teschler-Nicola, M. and Gerold, F. 2001. 'Ergebnisse intraserieller und interserieller paläodemographischer und paläoepidemiologischer Analysen. Zur Erfassung von Lebensraum-Konstituenten in der frühen Bronzezeit am Beispiel Traisentalserien'. In: Lippert, A. Schultz, M., Shennan S.J. and Teschler-Nicola, M. (eds.): *People and their Environment during the Neolithic and the Early Bronze Age in Central Europe. Results of interdisciplinary cooperation between archaeology, climatology, biology and medicine. International workshop November 9-12, 1995 Institute for Pre- and Protohistory – University of Vienna.* Vienna: Verlag Maria Leidorf GmbH, pp. 245-56.
- Točik, A. 1963. 'Die Nitra-Gruppe'. *Archeologické Rozhledy*, 15(6), pp. 716-74.
- Točik, A. 1979. *Výčapy-Opatovce – a d'alšie pohrebiská zo staršej dobz bronyovej na juhozápadnom Slovensku und weitere altbronzezeitliche Gräberfelder in der Südwestslowakei.* Materialia Archaeologica Slovaca. Nitra: Archeologický ústav Slovenskej akadémie vied.
- Točik, A. and J. Vladár 1971. 'Prehľad bádania v problematike vývoja slovenska dobe bronzovej – Übersicht der Forschung in der Problematik der bronzezeitlichen Entwicklung der Slowakei'. *Slovenská Archeológia*, 19(2), pp. 365-422.
- Vladár, J. 1964. 'Vplyvy kultúry zvoncovitých pohárov v náplni nitrianskej skupiny – Einflüsse der Glockenbecherkultur in der Kulturfüllung der Nitra-gruppe'. *Študijné zvesti*, 13, pp. 111-26.
- Vladár, J. 1973. *Pohrebiská zo staršej doby bronzovej v Branči – Gräberfelder aus der älteren Bronzezeit in Branč.* (Archaeologica Slovaca Fontes, 12). Bratislava: Vydavateľstvo Slovenskej akadémie vied Bratislava.
- Vladár, J. 1974. *Die Dolche in der Slowakei.* (Prähistorische Bronzefunde, VI:3). München: C.H. Beck'sche Verlagsbuchhandlung.
- Vladár, J. 1981. 'Zur Problematik osteuropäischer und südöstlicher Einflüsse in der Kulturentwicklung der älteren Bronzezeit im Gebiet der Slowakei'. *Slovenská Archeológia*, 29(1), pp. 217-33.
- Vladár, J. and Romsauer, P. 1999. 'Zur Problematik der Aunjetitzer Kultur in der Slowakei'. In: J. Bátora and J. Peška (eds.): *Aktuelle Probleme der Erforschung der Frühbronzezeit in Böhmen und Mähren und in der Slowakei.* Nitra: Archaeologica Slovaca Monographiae.
- Winkler, E.-M and Schweder, I.M. 1991. 'Die Skelette aus der fruhbronzezeitlichen Siedlungsgrube von Kettlasbrunn in Niederosterreich'. *Archaeologia Austriaca*, 75, pp. 79-105.
- Vyhnánek, L. and Hanulík M. 1971. 'Multiple cutting wounds on the calva'. *Anthropologia. Acta Facultatis Rerum Naturalium Universitatis Commenianae*, 17, pp. 199-203.

Warfare, Discourse, and Identity /25-30

Warfare, Discourse, and Identity: An Introduction

TON OTTO

/25

Warfare involves the use of violence to achieve one's goals by forcing other people to submit; it is 'politics by other means' (von Clausewitz 1989). Violence is thus a resource for controlling people. However, both violence and warfare are greatly enmeshed with meaning, just as all other forms of human action. To enter into war normally requires a cause and motivation, but apart from this basic premise variety abounds. The meaning of war can be tied to religious conviction or to a struggle for resources; it can be presented as morally justified and unavoidable or as a way to obtain glory and honour (cp. Warburton chapter 4). In one way or other the goals and motivations for warfare are embedded in a discourse that has currency at the time of action. I use the notion of discourse somewhat loosely – well aware of its complex intellectual history – to refer to all forms of meaning-giving activity as well as the products of that activity (in the form of documents and material culture). Subscribing to a Foucauldian perspective (cp. Dreyfus and Rabinow 1982), I see these meaning-giving activities as embedded in power relations while simultaneously engendering the subjectivities that define these relations. Warfare creates a context of acute contest for power that affects people's subjectivities and identities while coping with the situation. By identities I mean more or less stable ascriptions of social position, which are established in negotiations and sometimes contestations between those who subscribe to the identity and others who ascribe it to them (cp. Jenkins 1996; Otto and Driessen 2000; Vandkilde chapter 26).

Warfare can generate war-specific identities in many different ways. In societies that do not have specialists for conducting war, men (as is mostly the case) may prepare themselves for war through special rituals, thus becoming temporary warriors able to kill. Harrison (1993) describes how this happened in a Melanesian society living on the banks of the Sepik River in Papua New Guinea. The Avatip did not conceive of humans as inherently violent and their men had to undergo

special rituals and war magic in order to dissociate them from their normal identities, which would lead them to extend sociality to most people. Harrison describes
warfare as a central resource for creating identities. Because Avatip people could
in principle extend their peaceful exchange relationships without any limit and
were even morally obliged to do so, warfare helped them to define and create more
bounded units. The violence used to kill people was thus a necessary ingredient
for establishing local communities with claims to land and other resources. In
the following quotation Harrison sums up this point of view pointedly:

In Melanesia it is not so much groups that make war, but war that makes groups. That is to say,
war is part of the way in which groups having claims or interests in resources assert their existence and identities in the first place. It is through conflict that these groups separate themselves out from each other and constitute themselves as distinct entities capable of competing
for resources. (ibid: 18)

Another scholar who has emphasised the role of violence and warfare in the creation of identities is Anton Blok (2000). According to Blok, lack of difference
between existing groups threatens their identity and thus, in a way, their existence. This may lead to explosive situations and the use sometimes of extreme
violence, as was the case in Rwanda and the former Yugoslavia. Blok finds theoretical inspiration in Freud's notion of the 'narcissism of minor differences'
and especially in Bourdieu's thesis that social identity is based on difference and
that difference is asserted against those who are closest. Clearly, the violence in
the former Yugoslavia and in Rwanda has done much to emphasise ethnic distinctions between groups that are very similar and therefore violence can be
seen as an important resource for marking identities. Whether Blok is right in
positing a structural relationship between levels of difference and levels of violence, is much more debatable however.

Two of the contributions in this section deal with the situation of former
Yugoslavia and therefore there will be an opportunity to return to this region in
more detail. All contributions address some different aspect of the relation
between discourse and identity, namely the role of material culture, of narratives, of agency, of the practice of everyday-life, and of researchers. I will deal
with them in the order of appearance.

Material discourse and the discovery of warrior identities

As an archaeologist Helle Vandkilde (chapter 26) is primarily interested in how
a discourse about warriorhood can be inferred from the material finds from past
societies. Concretely she considers these remains as part of a material discourse
that reflects central aspects of the past society, which can be read and interpreted
by the archaeologist. Before starting her reading Vandkilde develops a model
that should help her to interpret the material signs. Her main focus of interest
is on the development of warrior identities and warrior institutions in Copper
Age Europe. Warriorhood is one among many possible identities and the question is when it can be clearly identified in the archaeological record. Aspects of
this identity will be expressed somehow in objects that are related to certain
people; in particular, weapon finds in graves. Vandkilde argues that warrior
identities imply warfare but not necessarily warrior organisations. The ethnographic record of Amazonia and Highlands New Guinea provide many examples

of societies in which warfare is prevalent but where warrior organisations are absent. All men are potential warriors in the case of war and return to other roles and identities when the war is over.

Warrior organisations probably have their basis in male clubs or brother-hoods, which can be found in many societies. So-called warrior bands specialise in violence and war, which provide them with legitimacy, identity and sustenance (through war spoils and booty). They are often tightly knit together because of the long-term reciprocity relations between their members, especially the war leaders and their followers. Access to warrior bands can be regulated in different ways, namely through age, through personal qualities and status, or through rank. Once warrior roles are organised in special institutions, they carry a certain weight and can therefore have an impact on the way a society develops and changes. This appears to happen especially under conditions of external pressure and crisis, which give warrior organisations a clear function and related power within society.

Looking at the material discourse that has been uncovered by archaeologists, Vandkilde concludes that war was unmistakably an aspect of northern and central European societies from the 6th to the 4th millennium BC. Some skeletons reveal war-related traumata and some settlements were fortified. It is, however, not possible to identify institutionalised warrior identities on the basis of the objects found in burial sites; often tools that also functioned as weapons. The situation changes substantially with the Corded Ware culture that began around the beginning of the 3rd millennium BC and spread over large parts of central and northern Europe around 2800 BC. Taking Bohemia as a case study Vandkilde demonstrates that these societies were organised along the cross-cutting principles of gender and rank. Men and women of high rank can be identified by the number and nature of objects found in their graves, while age appears to be of little importance. High ranking men carry battle-axes or mace-heads, which most likely signify their warrior status. This development continues in the Bell beaker culture (from c. 2500 BC), which shows greater internal variation among the warriors. Warriorhood had clearly established itself in Europe as an organising aspect of society.

Narratives and the persistence of warrior identities

Sanimir Resic (chapter 27), in contrast to Vandkilde, does not look at material discourse but instead at the rich historical record of warrior tales. From the four thousand year old Sumerian epic *Gilgamesh*, via the Homerian *Iliad*, the medieval French *Chansons de Geste* up to present day Hollywood films like the *Terminator*, warrior values have been sung and celebrated across time and space. Resic recognises warriorhood as a universal value of manhood, at least in recorded history, and participation in war and battle as a common *rite de passage* marking the transition from boy to (real) man. Warrior values have wide appeal across cultures and are often identified with manliness: honour, loyalty, duty, obedience, endurance, strength, sexual potency, courage and camaraderie. These values are clearly an enforcing ideological basis for military institutions such as warrior bands and armies and their maintenance and fostering may well be connected with warrior organisations.

Resic recognises that there are temporary depressions in the popularity of warrior values, such as after the wars of religion in the 16th and 17th century, but that these values have had an enormous resilience throughout history.

Following the French revolution, the transition from professional armies to citizen armies with a strong commitment to nation and cause had a strengthening effect on the celebration of warriorhood. Also the 20th century provides many examples of the exaltation of military virtues and manhood, with a sinister climax in fascism in its various forms. Resic observes that warrior values played an important role in the Balkan wars of the 1990s. More generally, the modern soldier has become more of a mechanical cyborg with all the new war technology, but in spite of the increasing number of women soldiers, warrior identities are still predominantly masculine.

Resic's long list of examples certainly makes the point that warrior values have been a pervasive part of Western history, but I think that we should investigate the fluctuation and variation of these values more consistently. Warfare, the celebrated 'baptism of fire' making true men, may also – and often does – lead to disillusionment and a reassessment of values. It is necessary to find out which forces promote the maintenance of a war-celebrating discourse. Hereby it is certainly not enough to refer to the universal cravings of young men for adventure, thrill, fame and power, as Resic does. It is probably correct that young men are most susceptible to the myths of warriorhood, but it requires more than groups of adventure seeking youngsters to sustain and reproduce these myths. Therefore a focus on discourse is required, that takes into account the multiple interests and power groups that gain from military discourse and the subjectivities it produces.

Discourse and agency

Stef Jansen (chapter 28) focuses on the situation in five villages in the Krajina, which are now part of the Republic of Croatia. Before the war four of the five villages were predominantly Serbian and the fifth predominantly Croatian. At first the Croatian inhabitants were driven away or killed; then, later on, all Serbs had to flee and their houses were looted and burned. At present a number of the Serbs have returned to be among a population consisting otherwise of returned, relocated and refugee Croats. Jansen investigates the dominant discourse among the Croatian population, that of exclusive nationalism. Croatia is considered as the – in principle, exclusive – national homeland of all Croats, quite unlike the pre-war multi-ethnic situation. The most current explanations for the rise of exclusive nationalism refer to two causes, often in combination. The acerbity of the ethnic nationalism is linked to the suppressed traumas of World War II massacres that had occurred in the region. In addition, political propaganda and media manipulation is seen as causing the spreading and embracement of nationalist ideas and sentiments. Jansen is not quite satisfied with these explanations, which he finds too deterministic as they pay insufficient attention to the agency of the people involved.

Looking for agency does not mean that the situation has to be understood from a perspective of resistance. Jansen discovers very little resistance in the sense of Scott (1990). Most people appear to accept and evoke the dominant discourse, which they reproduce in the form of catchwords and phrases – 'story lines' as Hajer (1995) calls them. A focus on agency thus replaces the question of determinism with the question of conformism. Why do most people underwrite the dominant discourse, even though in a reduced, imprecise and ambiguous way? According to Jansen we have to understand this against the background of the overwhelming experience of powerlessness vis-à-vis the powers-that-be, called *politika* or politics. Ordinary people experienced a deep cleavage between their everyday lives and the places where things were decided for them. By evoking the authoritative discourse of nationalism, they attempted to reassert some autonomy for their personal narratives and to exert some level of (discursive) control over their own lives and experiences. As the authoritative discourse was only partly embraced through reduced story lines, and weakened by telling silences, vagueness, and sarcastic resignation, people distanced themselves from responsibility for nationalist ideas at the same time as they tried to be empowered by them. This conformism thus provided some comfort in difficult and uncomfortable times, without taking into account the discomfort it caused for others, namely the Serbian returnees.

Jansen's analysis gives us insight into the complex ways a dominant discourse is reproduced and sustained by actors, who have ambivalent feelings and ambiguous experiences. Torsten Kolind (chapter 29) also looks at this relationship but among another group in the former Yugoslavia and with a somewhat different conclusion.

Discourse and the practice of everyday life

Stolac is a little town in Bosnia Herzegovina, which was predominantly Muslim before the war. During the war the Muslim inhabitants were expelled by Croat militia and all their monuments, mosques and public buildings destroyed. After the war the Muslim population, supported by the international community, has started to return to what is, in fact, a community that is sharply divided along ethnic lines. Therefore at first sight the case of Stolac confirms Blok's and others' (Malkki 1998) argument that violence creates or reinforces identities between people, who in other ways are very similar. Torsten Kolind (chapter 29) has, however, two fundamental objections to this argument based on his interpretation of events in Bosnia Herzegovina. In the first place the argument appears to confound cause and effect. A sharp ethnic division is unmistakably an effect of the war and violence used, but it is too farfetched to see diminishing differences as the cause of war. Kolind seeks the explanation rather in power-seeking nationalistic politicians, who developed ethnic cleansing as their explicit aim and strategy. I agree that power politics should be a central aspect of an explanation, but on the other hand the ideas of ethnic nationalism and separation obviously struck a cord among a wider group of the population; otherwise it would have been impossible to motivate so many to go into war. Once violence is used, it forces all the others to make a choice, also those who do not wish to do so, because war has the tendency to create two over-determining identities: a person is either an ally or an enemy.

Kolind's second argument against Blok's thesis is based on his detailed investigations into forms of identification among Bosnian Muslims in Stolac. Contrary to expectation, ethnic identification did not play a major role in Muslim everyday life; rather, it was the case that ethnic and nationalist references were mostly avoided. In their explanations of events and depictions of their own situation people would refer to feelings of local patriotism, to inter-ethnic tolerance, or to the opposition between Europe and the Balkans. They would sometimes also adopt the identity of victims, with clear ethnic undertones, but Kolind argues that, on the whole, local identifications did not support the dominant discourse of ethnic nationalism. He interprets this phenomenon as a kind of counterdiscourse against the dominant nationalist discourse, and as rooted in the necessities of everyday life. After the destruction or 'unmaking' of their life-world, caused by the war, people had to build up a new existence. The Muslims who had returned to Stolac could simply not afford to harbour strong feelings of ethnic hatred, because they had to live with the consequences of such antagonistic feelings. In order to remake their world, which had to be based on some form of coexistence with the Croats, they countered the dominant nationalist discourse by drawing on other discourses that downplayed or ignored ethnic differences and emphasised tolerance.

Local discourse and the role of anthropologists and archaeologists

The last contribution in this section turns our attention to the often unintended role that researchers play in the local politics of identity. The scene is Guatemala after the protracted civil war from the early 1960s to the end of 1996. Staffan Löfving (chapter 30) argues that the violence of the war has politicised and essentialised cultural differences. Guatemalan national identity became established at the expense of the oppression of a dispossessed majority, the Maya Indians. As an effect of the civil war the Pan-Mayan movement has a strong need to establish a distinct, vital and continuous Mayan identity. In this process anthropologists and archaeologists are both resource persons and potential enemies. Anthropologists who criticise the essentialising use of culture by indigenous movements, are seen as the true colonialists of the era of globalisation, because they counter local attempts to construct an identity based on continuous (even if invented) traditions (cp. Kolig 2005). The same applies to archaeologists who, rather than confirming a view of a peaceful Mayan past, focus on internal violence, sacrifice, blood symbolism, and the ultimate collapse of Mayan society. Where the assertion of cultural continuity is a political resource for Mayan activists today, the investigation of historical collapse and breaking points could possibly weaken such identity claims.

According to Löfving two different notions of the past are to be found in contemporary Guatemalan praxis. In the first place there is the past as lived experience, ritually commemorated in everyday practice and marked by the presence of the dead in the local community and the presence of geographical reminders of past events in the local landscape. This notion of the past is disrupted by the widespread forced migration that has removed people from their historical reminders and the graves of their ancestors. The other notion is the new ideology of continuous Mayan culture, which is partly defined by outsiders and the Maya movement. This is an invented tradition – to a large extent based on the work of foreign and local researchers – which is perceived as lost and in need of revitalisation. Löfving concludes that political disruption and cultural continuity do not exclude each other, but often go hand in hand. Foreign researchers have a duty to be sensitive to local uses of their knowledge and to engage actively in repatriating their research results.

Conclusion

The chapters in this section all demonstrate how violence and warfare intersect with local identities, discourses, and institutional arrangements. Violence, which is the main instrument of warfare, inscribes itself sharply in personal and communal experience. Therefore it functions as a clear sign or marker of difference. Theoreticians like Harrison and Blok have underlined the potential of warfare to draw lines and create or reinforce separate identities. But after a war people have to live with these social boundaries, which are often in the way of other possible identities and social interaction. Kolind and Jansen demonstrate how people in their daily lives negotiate a dominant discourse created in a war situation. They may evoke the discourse while simultaneously distancing themselves from it, as Jansen observes among the Croats of the Krajina. Or, as shown by Kolind concerning the Muslims of Stolac, they may obstruct the dominant discourse by referring to alternative ways of understanding the world and marking identities.

If warfare is an important part of the life of communities, warrior identities may be institutionalised in the form of specialised roles and organisations. Vandkilde demonstrates that this happened in Europe in the later Neolithic and the Early Bronze Age from around the 3rd millennium BC. From the 2nd millennium onwards Resic shows the existence of a continuous theme of celebration of warrior identities in narratives of the Western world. Even though there are fluctuations, this continuing theme demonstrates the overwhelming militarisation of our societies, as military organisations together with their providers and users (weapon producers, governments) are the main bearers of warrior values and myths. Researchers have an important role to contextualise these values and myths, just as their knowledge is an indispensable resource for the construction of identities by local political movements, as Löfving shows. Therefore, to analyse the power relations based in and expressed by discourses and identities of warfare, unavoidably implies reflections on the position of the researcher.

REFERENCES

- Blok, A. 2000. 'Relatives and rivals: The narcissism of minor differences'. In: H. Driessen and T. Otto (eds.): *Perplexities of Identification: Anthropological Studies in Cultural Differentiation and the Use of Resources*. Aarhus: Aarhus University Press, pp. 27-55.
- Clausewitz, C. von 1989. *On War*. M. Howard and P. Paret (eds. and trans.) Princeton: Princeton University Press.
- Dreyfus, H. and Rabinow, P. 1982. *Michel Foucault: Beyond Structuralism and Hermeneutics*. Chicago: University of Chicago Press.
- Hajer, M. 1995. *The Politics of Ecological Discourse*. Cambridge: Clarendon Press.
- Harrison, S. 1993. *The Mask of War: Violence, Ritual and the Self in Melanesia*. Manchester and New York: Manchester University Press.
- Jenkins, R. 1996. *Social Identity*. London and New York: Routledge.
- Kolig, E. 2005. 'The politics of indigenous – or ingenious – tradition: Some thoughts on the Australian and New Zealand situation'. In: T. Otto and P. Pedersen (eds): *Tradition and Agency: Tracing Cultural Continuity and Invention*. Aarhus: Aarhus University Press, pp. 293-326.
- Malkki, L.H. 1998. *Purity and Exile: Violence, Memory, and National Cosmology among Hutu Refugees in Tanzania*. Chicago: University of Chicago Press.
- Otto, T. and Driessen, H. 2000. 'Protean Perplexities: An Introduction'. In: H. Driessen and T. Otto (eds.): *Perplexities of Identification: Anthropological Studies in Cultural Differentiation and the Use of Resources*. Aarhus: Aarhus University Press, pp. 9-26.
- Scott, J.C. 1990. *Domination and the Arts of Resistance: Hidden Transcripts*. New Haven: Yale University Press.

Warriors and Warrior Institutions in Copper Age Europe

HELLE VANDKILDE

/26

This study[1] assesses the importance of warriorhood in the Copper Age of central Europe, particularly focusing on the Corded Ware culture of the early 3rd millennium BC. A central question is whether warrior institutions existed, and if they did, what their social significance and position was in societies of the past. While we cannot assume that warriors have always been present in all societies, the ethno-historical evidence allows us to say that – when they are present – warriors organise in a limited number of ways. The point of departure for this archaeological study is therefore a non-comprehensive sociological analysis of warriors that highlights the issues of identification and institutionalisation, an approach that is necessary in order to provide a tool that can enter into a dialogue with archaeological data.

Introductory notes

The warrior is a symbolic figure of power and a specialised user of physical violence. Warfare can be understood as violent social action and war as a situation of recurring warfare: in a study of warriors the phenomena of warfare and war are not out of focus; on the contrary warriors and warfare are interdependent phenomena. It is on the other hand quite clear that the presence or absence of warrior representations in the archaeological record cannot be used as a direct measure of war or peace in the past (cp. Robb 1997).

Why warriors?

Yet why spend time and energy on warriors in prehistory? There are several interconnected reasons for this: First of all, warriorhood is worth studying in any context because of its association with power, dominance, coercion, violence and bloodshed. When warriorhood becomes institutionalised it may even become a dominant field of power in society and hence a factor of oppression. In addition, when institutionalised, warriorhood might have a certain potential for producing, or contributing to, social change. It is also an identity interwoven with ideology, thus attracting opposite meanings of beauty and ugliness, gallantry and brutality, and bravery and cruelty. Furthermore, warriorhood needs an archaeological review because it has played, and still plays, a part in the discourse of large-scale cultural change within the discipline, having evoked scenes of fiercely armed male warriors on horseback conquering new land and of warrior aristocrats divided into war leaders and retinues of chiefly warriors. The question whether such a stereotypical picture can be sustained underlies much of the discussion in this article. The final reason is directly connected to our late-modern world: warriors, warfare and war, whether we like it or not, brutally interfere with human societies almost everywhere, and archaeology has an obligation to participate in the current debate within this field. There is no doubt that warriors require further study, more generally and in specific prehistoric settings.

Warriors and the archaeological sources[2]

The fundamental question arises whether and how warriors and their institutions can be identified in a fragmented archaeological setting with a silent discourse. The answer is in the affirmative – especially when the archaeological sources are optimal. Even associations of warriors are arguably identifiable, and this may be because social institutions are in general durable social structures (Giddens 1984) and therefore materially distinct.

Warriorhood is manifested in a variety of ways in archaeological 'texts', sometimes distinctly, sometimes vaguely, sometimes only in one domain, and sometimes in several domains. This inconsistency is not straightforward to explain since it is influenced by the variable state of the sources as well as embedded in past social practice – i.e., a particular relationship between action infused by ideology and stricter norms of 'how things should be done'. Complete or partial absence of the presentation of warriors in the funerary domain does not necessarily imply that warriors did not exist or were of no great concern. Instead of being placed with the bodies of the deceased, weapons were sometimes offered as gifts to the gods in sacred places, and such weapon offerings in hoards are arguably statements relating to warfare, either metonymically or metaphorically. Hence, they hint at the existence of warriors even if these are not fully displayed in the burial domain. Conversely, sometimes the importance of warriors might have been exaggerated in burials, but in my opinion displaying the identity of the deceased as a warrior, especially if on a massive scale, is bound to have a distinct bearing on the practices of the living society.

In an interpretive enterprise it is certainly best to use as many different sources as possible, but the sources can be one-sided and narrow in their scope. The Corded Ware culture is a case in point inasmuch as cemeteries constitute more or less the only access to knowledge of social practices. The Urnfield culture can be listed as a deviating case with a narrow

source situation: rich weaponry was placed ritually in sacred places, cremation burials only vaguely state the social identities of the deceased, and only the very top of the elite sometimes received monumental burials with presentations of (symbolic) warriorhood. The material culture of the Late Bronze Age nevertheless altogether indicates that warriors existed and quite likely organised formally in specific institutions. The burial sphere needs an additional comment because it is a central source of historical data on warriorhood.

Funerals and funerary monuments are indeed suitable means to question, confirm or legitimise existing relations of social power. Weapons in burials are a material extension of the dead body and are thus linked to the collective social identities of the deceased and probably also to his or her personhood. Personal equipment in burials can be understood in at least two different ways. First, as a direct metonymic statement of the lived identity of the deceased; the warrior may not be interred with full equipment but instead only those parts of the equipment that have a certain symbolic meaning. Second, as a symbolic metaphorical statement with a less direct bearing on the lived identities (cp. Whitley 1995). If, for example, a small, exclusive group of mature males are interred with rich weaponry, this is likely to reflect 'symbolic warriorhood' in the sense of one or more of the following possibilities: a former warrior identity, heroic status or ambition, political authority or high social rank. Likewise, weapons accompanying small children and young adolescents are unlikely to be actual signs of practiced warriorhood, but should rather be understood as metaphors originating in warrior values. Combinations of the two types of statements can of course occur.

The presence of specific weapons with certain males in Corded Ware and Bell Beaker burials certainly suggests that warriorhood formed an important part of social practice over wide areas of Europe during the 3rd millennium BC. Maceheads and battle-axes of the Corded Ware culture can only be interpreted as weapons for war, directly or on the symbolic level. The equipment and flint and copper daggers of the Bell Beaker and Early Bronze Age archer might have had uses outside the domain of war, notably hunting. But since these weapons are highly specialised and furthermore replace each other chronologically, their connection to warfare seems unmistakable. Prior to the 3rd millennium BC evidence of warriors is more sporadic and also difficult to interpret, especially because the shape of potential weapons allows several functions. As early as the middle of the 6th millennium BC, concurrently with the beginnings of agriculture, there is nevertheless evidence of war encounters: here and there cases of skeletal trauma, fortifications, and implements with a potential for war. This makes it likely that some form of warrior identity existed even then, but signs of institutionalisation are scarce. A possible conclusion is that warrior institutions emerged massively only after c. 2800 BC along with the spread of Corded Ware culture, and that this event had a lasting impact on subsequent social practices in the Bronze Age and the early Iron Age, where the warrior continued to be a key figure.

Theorising warriors

A ready-made theory of warriors, which can be directly adopted, does not exist, but can be provided by making use of current theories about action, structure, discourse, and especially identity. In addition, ethnographic descriptions are helpful in providing clues to a classification of warrior institutions. Warriorhood is, in short, a potentially powerful collective identity founded in war-related action and communicated through material, spoken and written discourse. Warriors must arguably be studied – not in isolation and not as a purely male habitat – but in relation to other social identities of age, gender, hierarchy, and even profession and ethnicity. Warriorhood is thus relational and interactive by nature; it is dependent on a variety of overlapping and disparate social identities and their interactions. Warriors, like any other kind of social actor, affect their context by participating in it, and they are themselves altered in the process. Warriors then contribute to the reproduction and transformation of society. The following sections will take a close look at the sociology of warriorhood in order to enable a better understanding of the presentation of warriors in various prehistoric settings.

The examination of social identity in archaeology is a road paved with danger because it is about classifying people and because it is easy to mistake our categories for theirs. It is nevertheless almost unavoidable to study social identity since social identification is basic to human social life in the three domains of day-to-day activities, the life cycle of individuals,

and the durée of institutions (cp. Giddens 1984). And inasmuch as the analytical-theoretical tools we employ are ours, not theirs, it is hardly realistic to completely escape our world and its pre-understandings. One step towards obtaining an insider view of culture and society is doubtless to avoid absolutist and static understandings of the other.

Warriorhood as identification

Social identities – often negotiable and intermingled – are quite central to the societal project, and warriors indeed find their place as agents and identity within particular social contexts. Warriorhood as a means of social identification is the theme of this section.

Fredrik Barth argues for a relational and changeable understanding of social identity in his classic introduction to *Ethnic Groups and Boundaries* (1969) and Richard Jenkins does so more explicitly in his book *Social Identity* (1996), which owes much to Barth's dynamic approach to ethnicity in anthropology. Barth defines the nature of social identity – and here I am merely translating his notions about ethnic identity into social identity in general – as inherently double because it simultaneously concerns people's identification of themselves and how other people classify them. He also emphasises that the need to mark identity increases with growing intercultural interaction, that material culture marks identity in the sense that social agents select what they believe is significant, and, finally, that in order to wholly understand social identity we must study the processes and practices that create it (Barth 1969: 3ff).

Social identity is in other words produced and reproduced by human actions and discourse, and the human body is the medium through which identification is enacted and signified. Even if changeability is inherent to all social identities, this is more valid for some identities than others. Among those most resistant to change are the primary identities of selfhood, humanness and gender, and in some cases ethnicity (Jenkins 1996: 21). Social identity is, in short, the systematic presentation and constitution of relations of differences and similarities between individuals, between collectives and between individuals and collectives (ibid.). Social identity is a process, something you are or become (ibid.). Social identity exists within and cuts across categories of gender, age, status, rank, profession and ethnicity.

Drawing on Marx, a distinction can be made between a collective identity, which is merely recognised internally – as a group *for* itself (*für sich*), and a collective identity, which is also identified externally by other people – as a group in itself (*an sich*) (cp. Jenkins 1996: 22f).

Being a warrior qualifies as an identity created and recreated through the duration of activities that altogether make up warfare. Being a warrior is then founded in warlike actions against the other, but it is also an identity which relies on non-coercive interaction among a particular group of people. Signification through various modes of discourse is logically included in the construction of warriorhood, not least the material mode. In general, warfare and related social activities constitute a framework within which social identities are formed, among these warrior identities. All agents take their social identity from certain networks of interactions. This is likewise true of warriors who position themselves in the power games of a field, be it a marginal or a dominant one. The warrior identity is also more or less enabled and constrained by social structure including the spectrum of other social identities; being a warrior is – like any other social identity – individually felt and collectively shared. The distinction between warriors as a group for itself and a group in itself is archaeologically important seeing that the first model offers a possible explanation in archaeological cases where traces of warriorhood are inconsistent.

The collective trait is essential since one cannot conduct war alone. Individualism is, following Clastres (1994: 186ff), simultaneously obligatory to the warrior since the warrior's desire is to increase personal glory. Being a soldier is similarly an identity with collective and individual traits, but the former is clearly more dominant than the latter. The warrior is a double being who has apparently opposite qualities such as chivalry and brutality, high esteem and cruelty. Related to this doubleness is the fact that the warrior is a boundary crosser who bestows violence and war upon outsiders and through these very acts seeks glory and fame from society. Internally, we can envision warriors as players distributed across a playing field with co-players and counter-players comprising other warriors and other social identities. Positioning in the internal game, however, depends on the outcome of external affairs. Externally, the warrior operates individually within the collective frame of companions in a flow of actions and interactions ending up with violent confrontations with the other.

Warriorhood is relational and interactive by nature in that it is dependent on negotiation with a variety of overlapping and disparate social identities and primary parameters in identification. More precisely, warriorhood is a secondary identity which combines other social identities, especially those based on the parameters of age and gender, but it may interfere, overlap or coincide with ethnic or cultural identity and with identities of class, typically elitism. It may, moreover, be perceived as a profession. Internally, a continual process of distinction also takes place – if it is not already formalised – separating mainstream warriors from successful and renowned warriors, and ordinary warriors from war chieftains. Warriorhood may – expressed in terms of practice and time – exist in three social dimensions of action (Giddens 1984) that can coincide: Firstly, warriorhood may be acted out in the daily routine activities of certain individuals (durée; reversible time). Secondly, warriorhood may be a phase in the life cycle of certain individuals (irreversible time). Thirdly, warriorhood can be organised as an institution (reversible time). Warrior institutions, which are particularly durative in character, will receive particular attention in the next section.

In conclusion, warriorhood is a negotiated difference that needs to be confirmed and maintained, and a dynamic process of self-definition and definition by others typically occurs. The dynamic component is rooted in the fact that warrior identities can be tightly or loosely knit, and the recruitment of warriors can be open or closed. It is also a relational difference, since warrior identities can crosscut, feed on, and interact with other identities, depending on the specific social context. Warriors gain their identities from the action, interaction and discourses in specific power networks, but only when integral to a dominant social field is warriorhood likely to be converted into a coercive power aimed at controlling other sections of society. The institutional aspect will receive further attention below.

'Männerbünde' and institutionalised warriorhood

On the field of battle it is a disgrace to the chief to be surpassed in valour by his companions, to the companions not to come up to the valour of their chief. As for leaving a battle alive after your chief has fallen, that means lifelong infamy and shame. To defend and protect him, to put down one's own acts of heroism to his credit – that is what they really mean by 'allegience'. The chiefs fight for victory, the companions for their chiefs. Many noble youths, if the land of their birth is stagnating in a protracted peace, deliberately seek out other tribes, where some war is afoot. The Germans have no taste for peace; renown is easier won among perils, and you cannot maintain a large body of companions except by violence and war. The companions are prodigal in their demands on the generocity of their chiefs.....Such open-handedness must have war and plunder to feed it. (Tacitus 1948: 112-13)

This quote from Tacitus' 'Germania', AD 97-98, in many ways captures the essence of warrior organisations: they nourish themselves on violence and war and the relationship between the companions of warriors is one of interdependency. Probably all warrior institutions contain elements of 'Gefolgschaft', defined as a long-term relationship of reciprocity between the members, especially the war leader and the followers (Bazelman 1999). The followers are loyal to their chief as long as he behaves honourably in battle and is generous when the spoils of plunders are distributed. Such a relationship has often been interpreted in purely economic terms, but should obviously also include an entire cosmos of morals and ethics, as made vivid by Tacitus' description of the phenomenon, by Homer's epics, and not least by the Anglo-Saxon poem of Beowulf (ibid.; Heaney 1999). A special tone of camaraderie often characterises the discourse in a warrior band, but can obviously become more an ideal of equality than a reflection of the real conditions of action. This is, for instance, the case in the aristocratic warrior bands of Homer's *Iliad*. Specific eating and drinking rituals – especially the ritual consumption of alcohol – are quite often reported on in historical and ethnographic sources as the glue that unites a warrior group by symbolising solidarity and particular paths of loyalty.

However, it should be stated from the onset that warriorhood as a specific individualistic social identity is not necessarily institutionalised. Non-institutionalised warriorhood exists, for example, in some bellicose tribal societies in Amazonia and Highland New Guinea. Warriorhood here occurs inseparably from being male and is not framed by institutions of war. War bands are organised *ad hoc* without formalised rules of behaviour, membership and leadership. In other words, war bands only exist for short periods of time – for instance, when necessitated

by a raid against a neighbouring group (Redmond 1994: 3ff).

Warriorhood can, by contrast, in a number of empirical cases, and in accordance with Tacitus, be classified as what Richard Jenkins labels an organisation: organisations are task-oriented collectivities 'constituted as networks of differentiated membership positions which bestow specifically individual identities upon their incumbents' (Jenkins 1996: 25). In addition to certain norms and rules of behaviour, such organisations include procedures for recruiting their members (ibid.). Likewise, the duality of structure and action inherent to an institution will occur in the three intersecting domains of domination, legitimation and signification (cp. Giddens 1984: 29ff): what is going on inside an institutionalised field will, one way or the other, necessarily be categorised as loaded with power; it will have to be legitimised ritually and socially, and it will be signified through different forms of discourse. Organisations are durative social constructs, as emphasised by Giddens' phrase 'longue durée of institutions', and therefore likely to be visible archaeologically. Institutions simply make the social identities they create and maintain much more durable and thus materially visible.

The organisational aspect of warriorhood will be inspected more closely in this section. The focus is upon the warrior institution – in many ways a potential unit of power: a club that excludes other community members but with obligations towards them and society as such. Such associations are always exclusionist in character in that membership is conditioned typically by parameters such as gender, age, status, and sometimes class. Warrior bands belong historically among a larger group of male clubs, so-called *Männerbünde*. Their formation relies on success in establishing a group of followers through personal qualities, manipulation and social indebtedness, whereas the internal structure and mode of reproduction can vary quite a lot.

Jenkins' description of organisations fits the male clubs found throughout history. The term 'fraternal interest group' is often encountered in the ethnographic literature, where it is used synonymously with male clubs, male societies, fraternities, male associations, secret societies, male segregation practices, sex-segregated fraternities, and so forth. A fraternal interest group is sometimes understood literally as tightly knit groups of male kin, 'brothers by blood' (see Paige and Paige 1981: 55; Harrison

1996). Mostly, however, male clubs refer to a symbolic kind of brotherhood that can cut across kinship ties, and herein lies their importance from the perspective of social change. They typically override the importance of descent groups and segmentary and clan structures of tribes, while emphasising the significance of solidarity and alliance among a group of unrelated men (cp. Bohle 1990: 288; Lipp 1990: 31). A male club forms a radically different type of organisation inside a segmentary society organised according to kinship relations and in which it often has some well-defined obligations, notably in the ritual domain. Langness (1974) has emphasised four purposes of male clubs: male bonding through secrecy and cult, power over biological reproduction and women, warfare, and the maintenance of political dominance. This is a generalising and absolutist view. Male clubs, for example, are not quite analogous to warrior clubs, and the women may not be, or feel, dominated. Warrior clubs are institutionalised war bands, which are interest groups with a warlike aim. In ethnohistorical settings, warrior clubs are mostly a male affair, but some have been known to include women. A few cases of female war clubs are, in fact, known, hence reminding us that there can be no prescribed logic attributing war and warriors solely to males (cp. Browne 1995). Institutions comprising only males sometimes have their female counterparts: female clubs, if less noticeable, are likewise widespread in time and space.

More or less exclusive men's clubs, encompassed by the German term *Männerbünde*, nevertheless exist in many societies, sometimes with a strong martial strain (Mallory 1989: 110f; Ehrenreich 1997: 117ff). Male clubs occur all over the world in a wide variety of societies, including our own world, and they are also common in societies known from historical sources. According to Bruce Knauft (1991: 403), so-called middle-range or complex prestate societies are more disposed to have fraternal interest groups than simpler societies. At the other end of the scale, state societies – including dissolving national states – have them, as recently demonstrated in Lebanon, the Balkans and elsewhere in our turbulent world. Even cultural subgroups in Western Europe are apt to form them. Other authors have emphasised that male associations found in ethnographic studies tend to concur with patrilocal residence, patrilineal descent, and stable valuable economic resources (Paige and Paige 1981). An origin in matrilineal societies has also been suggested (Lipp 1990), the idea being to unite the males to form a qualified opposition to the inherently strong position of women in these societies.

Male clubs cannot be considered as synonymous with military purposes. A warrior band can usually be classified as a male club, but by no means all *Männerbünde* are of a warlike or otherwise violent nature. In a cross-cultural and comparative perspective there is no direct correspondence between the occurrence of male interest groups on the one hand and warfare and aggressive behaviour on the other. However, they can indeed be described as associated factors (Schweizer 1990). A multivariate factor analysis processing selected data from 186 pre-industrial societies around the world ('World Cultures Database') demonstrates that male social groupings occur separately from, but adjacent to, such phenomena as high level of prestige based on warriorhood, warfare and plunder, and manly ideals of aggressiveness (ibid: table 1). It is likely that a similar overlap characterises male organisations in modern Western societies where, for instance, groups of soccer supporters possess an undercurrent of violence fuelled by ideals of aggressiveness.

Warrior bands may have several purposes, but their central objective is military. The difference between soldier and warrior has been explained above in terms of the relationship between individuality and collectivism. Among warriors individualistic behaviour is allowed and even encouraged, which is true of soldiers to a lesser degree. Warriors organise in warrior bands, while soldiers are organised in armies. The difference between warrior band and army is evident in the contrast between the individuality of the Homeric aristocratic warrior bands and the later hoplite army, which has a unified expression – quite in harmony with the collective ideology of the early city-state (Runciman 1999: 731ff). There may also be a difference of scale, of course.

In his book *The Evolution of War*, Keith Otterbein (1970: 19ff) distinguishes between military organisations with professional or non-professional agents. Among fifty ethnographically studied societies only four were completely without any kind of military organisation. The remaining forty-six societies all had military organisations divided almost equally between those with and without professionals. Twenty-four societies were without professionals. Here all healthy adult males fought as the situation required, organising war parties or defence *ad hoc*. Warfare was not considered a vocation as such, but could on the other hand be an important means of enhancing prestige. By comparison, twenty-two societies had professionals such as warrior associations, standing armies or mercenaries, but they frequently included non-professional warriors alongside professionals. Otterbein states that a coinciding relationship tends to exist between a high level of political centralisation and the presence of a professional military organisation (op. cit.: 22, 75), thus agreeing with Bruce Knauft (1991), who connects fraternal interest groups with complex societies. Warrior clubs are nevertheless quite common among those tribal societies that do not normally count as 'complex' in their social organisation – such as the tribes of the Great Plains. Mercenaries and standing armies are ignored in the following, where the focus is upon warrior institutions.

A reading of ethnohistorical sources (notably Otterbein 1970; Larick 1986; Völger and v. Welck 1990; Redmond 1994; Sanders 1999) suggests that warrior institutions can be roughly classified on the basis of their organising principles. In a basic sense warrior institutions are a means to regulate access to warriorhood generally or to supreme forms of warriorhood. Their main purpose is to wage war in the sense of a continued flow of social actions ending up with violence, but they usually carry out other functions besides strictly war within the social and religious life of society. It is possible to separate three different means of regulated access, all of them usually combined with gender: Age-based, status-based and rank-based access to the warrior band.

1. Warrior institutions in which access is regulated through age:
Warriorhood is a phase in the life cycle of all male individuals, who during their youth are organised in particular warrior institutions. In other words, all young males become warriors when they pass from boyhood to adulthood. Warrior bands comprise groups of young unmarried men who have undergone initiation together, and often divide into junior and senior members. Age is thus also important in structuring the interior of the warrior institution, which however also comprises achieved, and therefore unstable, positions of, for instance, leadership.

This form of warrior institution is well known, notably among some North American tribes and in East Africa among pastoral tribes. In the latter setting three different age grades can usually be distinguished, namely, boys, warriors and elders with different responsibilities, roles and privileges in society. Females also hold age-based statuses, which often correspond to those carried by males. A male then moves through the warrior age grade as a member of a specific cohort together with male companions born within a period of c. fifteen years. Warriorhood marks the proudest period in a man's life and is actively created by and reflected in their weaponry, hairstyle and dress. Age-based war bands are competitive and carry great esteem, but they are not an elite themselves; nor are they attached to any elite. Political influence in age graded societies is usually in the hands of older men (gerontocracy) whose interest it is to keep the warrior bands occupied with war raids and cattle herding and in this manner keep them away from the social fields of settlement and women. The material culture of the warriors is used to create and manipulate social position inside a cohort, between cohorts, and towards the outside world of other ethnic groups against which war is waged (cp. Larick 1986).

2. Warrior institutions in which access is regulated through personal qualities:
Warriorhood is also in this case a phase in the life cycle of most males, but membership in particular warrior bands, as well as the position within them, depends on personal qualities such as, notably, ferocity and prowess in war. Successful warrior bands with charismatic war leaders especially attract capable young males to become members. The classic spirit of mutual indebtedness, obligations and companionship is strikingly described by Tacitus in the opening quote, even if the social setting of Germania Libera in this period had grown quite hierarchical with an elite of warriors. The best examples, however, come from ethnographically described non-hierarchical societies in addition to modern warrior gangs operating in dissolving nation-states in, for example, ex-Jugoslavia.

Most military societies of the Great North American Plains belong in this category (Fig. 1). Almost all of the approximately fifty tribes had warrior bands with similar names, paraphernalia, obligations, and organisational principles: dog society, fox society, and so forth. The number of warrior bands among the Cheyenne is known to have varied

between four and six, and new bands were continuously formed on the basis of old ones. All males of a tribe usually belonged to a warrior association at some point in their life, and it was wholly possible to move from one association to another: warriors were usually invited to become a member of a particular warrior band, which was divided into war leader, ordinary members, and so-called officers. Positions as war leader and officers were not based on a right to command but on severe moral obligations to manage better than ordinary warriors. They were thus not officers in the normal sense of the word, but were obliged to stay put in a fight, if necessary to die as the ultimate expression of bravery. There was a high degree of competition inside and between warrior bands, which were rarely formally ranked though their reputation varied (cp. Wilderotter 1990). They were not elitist by nature, nor were they the instrument of an elite inasmuch as these societies were not hierarchically structured.

This status-regulated type of warrior band corresponds roughly to the early form of *Gefolgschaft* among Germanic tribes described by Heiko Steuer (1982: 58), Lotte Hedeager (1990: 184f), and Anne N. Jørgensen (1999: 194ff). This type of companionship has as a characteristic trait a high degree of mobility on behalf of the members, who could choose to leave the band if it was not successful enough. A high degree of movement across geographic space is quite clear from what Tacitus says (see opening quote). It also lacked the firm and formalised relationship between lord and retinue of later medieval times in which the retinue was, more or less, a small private army owned by the war lord. The Germanic warrior institution in its early form, however, deviates from the warrior bands of the Plain Indians in one important respect: even though war leaders had to be elected by the people's assembly, only males from the aristocracy had access to warrior institutions, which leads directly to a different type of warrior institution.

3. Warrior institutions in which access is regulated through distinctions of rank:

Active warriorhood is in this case a phase in the life cycle of certain males who belong to the elite or who seek to gain admittance to the elite through their occupation as professional warriors. Under circumstances of institutionalised social hierarchy, warrior institutions recruit their members mainly from the aristocracy. The military organisation of society is thus monopolised by an elite who dominate precisely through military sources of power (warrior elite), using the military to back up other sources of power (cp. Mann 1986). High-ranking males will tend to be warriors by birth even if this identity for the youngest and eldest is on a symbolic level. The warrior identity then tends to be formally inherited rather than achieved, and the leadership of a warrior institution is similarly a question of position in the social hierarchy. Each war leader has at his command a retinue of personal retainers who are dependent on their leader for work, payment and weaponry. The spoils of war are distributed according to certain established rules and may for instance follow positions in the hierarchy. This is the case notably in Homer's *Iliad*. On a more moral level, however, ideals of a balance between giving and taking may very

FIG. 2: *Members of the Abipón aristocratic warrior institution, Grand Chaco, South America. They could be easily distinguished by their feather headdresses, their specific dress, weaponry and drinking horns (after Lacroix 1990).*

well subsist and thus influence the social relationship between war lord and followers. War leader and ruler can be one and the same person, or the war leader and associated companions can represent the ruler.

Chiefdoms with paramount chiefs, subchiefs and chiefly followers, and aristocratic states with kings and their knights all organise their warrior institutions according to principles of rank; at the very least access to warrior clubs depends on relations of class. This is the dependent companionship with oathbound loyalty towards the war leader described by, for instance, Heiko Steuer (1982) in the hierarchical setting of the late Germanic and the later medieval periods. Nevertheless, the classic spirit of reciprocity, and hence of equality, between war leader and companions may still pertain as an ideal to follow. This is the sort of companionship described in Homer's epic with a permanent hierarchical command structure containing a complex system of dependencies and fellowships, but accompanied by ideals of reciprocity and equality among aristocrats, as reflected, for example, in the often quoted suffix of King Agamemnon as 'first among equals'. A similar structure is communicated in the Linear B texts of Late Bronze Age Greece with '*wanax*' (king), '*lawagetas*' (war leader), and the so-called '*hequetai*' (followers) (cp. Kilian 1988) and made material in the spatial planning and adornment of the two great halls in the Palace of Nestor at Pylos in Messenia (Davis and Bennet 1999; see Vandkilde chapter 34). The early Anglo-Saxon epic about the Swedish king Beowulf in the 7th century AD reflects a very similar organisation of war lords and warrior retinues (Bazelman 1999; cp. also Heaney 1999). Several mature Bronze Age communities in central and northern Europe could be understood within a similar frame. Similar retinues are held by chiefs in chiefdoms such as those of the Maori, Fiji, Hawaii, Grand Chaco, Cauca Valley, etc. and by kings in warrior states, such as the Zulu and Dahomian kingdoms in Africa and the Samurai-organised ancient state of Japan. The Abipón in the Grand Chaco of South America (Fig. 2) had specific warrior institutions solely consisting of horse-mounted aristocratic warriors, who however avoided hierarchical divisions among themselves and who used drinking rituals to symbolise and strengthen the unity of the warrior institution (Lacroix 1990). The Aztec sun warriors were similarly recruited among the aristocracy and organised in warrior institutions with at least two different status classes with different privileges and obligations

(Dyckerhoff 1990). Leadership and loyalty paid to a specific leader are, however, better observed in other cases such as chiefdoms and states in West Africa. Here the traditional age graded warrior institutions often became the tools of ambitious and successful war leaders, who were even given the right to mobilise certain age groups as followers (Tymowski 1981).

The presence of institutionalised warriorhood will automatically strengthen the position of warfare and warriorhood in society, which depends on organised violence, or the threat of such, for its maintenance internally and expansion externally. This is most particularly true within hierarchical settings. The need to legitimise and signify interactions and identities of war and rank will increase and they might therefore be carried – more symbolically – into other fields of social practice. In consequence, the material world will probably show increased visibility of weaponry in one or more social domains. This might arguably be what we see in later European prehistory where Neolithic and Bronze Age societies choose to present, or partly misrepresent, warrior identities and war in contexts of interaction not directly related to violent confrontation with the other: weaponry in the funerary domain and/or in the sacrificial domain could be understood as indirect evidence of warrior institutions being legitimised and signified materially in rituals. Warrior representations in burials tally well with this idea as do the conspicuous sacrificial offerings of weapons in wetlands during the European Bronze Age, notably emphasising the possibility that some of them may be the end points of actions directly related to warfare. It should at the same time be taken into account that bellicose rhetoric can misrepresent a social reality that also includes peace. The degree of actual war fluctuates quite a lot even in societies commonly categorised as being in a permanent state of war (Helbling 1996). This means that the disaster and horror of war are interrupted by situations and periods of peace.

What makes warrior clubs especially fascinating to study is their often-presumed association with social change. The idea is not new. Early in the 20th century Richard Lowie (1962[1927]) presented the hypothesis that military associations of the Plain Indians were central to the origin of the state due to their high standards of order and organisation, an idea already implicit in the concept of 'militärische Demokratie' launched earlier by Friedrich Engels (1977[1891]). The law and order of these warrior institutions (Fig. 3) certainly stood in some contrast to the otherwise decentralised societies of which they formed a part. One might say that originally these military clubs constituted well-defined but rather marginally placed social fields, and certainly not in themselves power fields with strong leadership and central functions in society.

However, an altered situation with increasing power in the hands of the warrior associations could be observed when the Plain tribes came under increasing pressure from the colonial state apparatus in terms of war, conquest, genocide, disease and socio-economic crisis. These indigenous societies certainly underwent social change in the direction of hierarchy, but never became state societies of their own. Rather they became incorporated into an existent state administration and more or less annihilated. A similar process of internal change triggered by external pressure can be outlined in northern Europe under the expansive politics of the Roman Empire. But here the outcome was entirely different in that warrior institutions – alongside attempts to emulate the Roman state administration – certainly played a leading role in state formation processes among indigenous Germanic tribes. This is suggested among others by Heiko Steuer (chapter 16). The formation of an expansive Zulu state is an appropriate analogue. The great warrior king Shaka organised his state and army on the basis of age grade warrior institutions, which became regiments with their own individual names and leaders, in this case also with colonial state administrations lurking in the background (Wilderotter 1990). A very similar development can be observed in the Grand Chaco, where decentralised and rather egalitarian matrilocal societies in the 16th and 17th centuries transformed into a hierarchical warrior society with chiefly families, aristocracy, commoners and slaves (Lacroix 1990). Adoption of horses and fierce resistance against the Spanish colonisation were key points in this transformation (ibid.), but it is evident that during the process warrior institutions became central sources of social power in contrast to the marginal position they had previously. The above examples seem to show that the potential for change inherent to warrior organisations can (only?) be activated under conditions of external pressure and crisis. Warrior institutions can then come within the reach of a dominant field or themselves become a power field; power understood as domination and coercion

FIG. 3: *A warrior band of hostile Chiricahua Apaches in Mexico, c. 1886 (after Barrett 1974). War and warrior institutions had then become the very essence of Apache society. In the phrasing of Pierre Clastres, the general warlike situation of 'primitive society' had transformed into actual permanent war (1994: 186). The Chiricahua Apache tribe under the war leader Geronimo (to the right) exemplifies such a chaotic situation of societal change with warfare and warriors as key variables.*

can consequently be activated and aimed at controlling the activities of fellow agents in society.

In summary, this section has described warrior institutions from the point of view of sociology and on the basis of data derived from ethnohistory. Warrior institutions belong historically among a larger group of male clubs. Warfare is not an unequivocal feature of the huge number of male clubs known from around the world, but relevant data show that it is a linked feature. Warrior institutions are in fact clubs with a warlike purpose mostly with male members, and there seems to be support for the hypothesis that warrior institutions in themselves carry a potential for social change, which can be activated under circumstances of external pressure and crisis. Warrior institutions divide roughly into three related categories on the basis of whether the main criterion of access is age, status or rank, all of them typically integrating elements of *Gefolgschaft* in the sense of a long-term-relationship of reciprocity between leaders and followers in terms of economy and ethics. Such a relationship of parity can, however, in hierarchical settings become more an ideal to strive for rather than strict social reality. This is notably evident in the *primus inter pares* ideology in Homer's epic, which stands out against the real world of a strictly hierarchical society. Sociologically, warrior institutions are durable social fields and therefore likely to be discernable archaeologically, especially in the three dimensions of power enactment (social organisation), legitimation (in rituals) and signification (material discourse). It can be argued that warriorhood and war, particularly in contexts of hierarchy and inequality, will tend to expand on a symbolic level into other sections of society.

Searching for early warriors

Warriorhood is an evasive identity in northern and central Europe in the 6th to the 4th millennium BC. Burial customs are not usually sex-differentiated in the clear manner of later burials (Häusler 1994), and this may have some bearing on how gender was perceived of in the living societies. Many of the tools have warlike potentials, such as transverse arrowheads and polished flat axe blades of flint or stone, but specific weapons, made unequivocally with the purpose of waging war, are harder to distinguish. Stone shaft hole axes of the latest Linear Pottery culture and succeeding Middle Neolithic cultures, including the Funnel-necked Beaker culture, could have been used in war or may have symbolised warriorhood. Sometimes these axes are, in fact, elegantly shaped and made of spectacular varieties of stone; hence recalling later battle-axes of the 3rd millennium Corded Ware culture (Zapotocký 1991: 68ff). In John Chapman's vocabulary (1999: 107-10), most axes of this period are tool-weapons and some are weapon-tools, thus designating their double use potential with emphasis on the first or the second function. However, it is a difficult matter to associate these finely polished stone axes with specific persons, or groups of persons, especially in the earlier period.

The presence of such axes, and a range of other tools, with a potential in the waging of war should nevertheless be seen on the background of evidence of war-related traumata and fortifications of settlements (e.g., Thorpe 2000 and this volume chapter 10; Wahl and König 1987; Vencl 1999; also Chapman 1999; Makkay 2000; Jensen 2001: 225, 230, 232f, 446, 448f). Taken together, this suggests that in the 6th to the 4th millennia BC warfare did indeed form part of social practice and thus played a role in the reproduction of society. How can this picture be interpreted in terms of warriors?

Warriorhood is a social identity which is created through the social processes of war and which is nourished on warfare as social action. The evidence, diffuse as it is, might then support the conclusion that warriorhood existed in the 6th to the 4th millennia, but generally not as a specific social identity; a group for itself and in itself. The best analogies for such a non-institutionalised warriorhood might be found among tribes in Amazonia and Highland New Guinea, where warriorhood – as mentioned above – is an inseparable part of being male and is not framed by institutions of war. War bands in the 6th to the 4th millennia could thus have been organised *ad hoc* without formalised rules of behaviour, membership and leadership. They therefore mainly existed to defend the settlement and to prepare and conduct war raids against neighbouring groups, ceasing to exist the moment these activities were concluded.

Whether warriorhood in this period was generally as closely associated with the presentation of masculinity as in the ethnographic cases is difficult to say, but at least it is not unlikely. In the later 5th and the 4th millennia, especially in eastern central Europe and in the Balkans, the specific association between male gender and weaponry emerges more consistently. There are even cases, prior to and around 4000 BC, which suggest that some form of institutionalised warriorhood may have existed here and there (if not everywhere). The social patterns that appear seem to point forward in time to the 3rd millennium BC and therefore deserve attention.

Age grades at Tiszapolgar-Basatanya

John Chapman (1999: 124ff) has convincingly argued for a state of increased warfare in the mature Copper Age societies of eastern and central Europe, c. 4600-3700 BC, focusing on a range of new potential weapons and exotic materials, first and foremost copper. Heavy shaft hole axes and adzes in copper, antler and polished stone, pressure-flaked bifacial arrowheads of flint or obsidian, and long dagger blades of copper, flint or boar's tusk should in particular be mentioned since their potential for war is rather apparent; hence the name weapon-tools. Their massive presence suggests that war was a recognised form of social action in these societies even if they do not tell of the frequency of war encounters. Settlement organisation is not of much help here since little is known.

In the mature Copper Age cemeteries of the Balkans, as well as in associated cultures of eastern central Europe, there is generally a clear association between male gender and these weapon-tools (cp. Zapotocký 1991). In some cases the association between male gender, weaponry and high-rank/wealth is obvious, as at the cemetery of Varna I, which is strategically situated where the Danube joins the Black Sea in Bulgaria (Chapman 1999: 126f with further references). In particular at Varna, high-ranking maleness seems to draw on prowess in warfare and the acquisition of exotic prestige goods. The unusually large number of cenotaphs with weaponry at this site suggests that a considerable part of the males did not return from war raids and/or travels into foreign territory. The material thus generally hints at the existence of warrior identities, but were they institutionalised?

In the Tiszapolgar and Bodrogkeresztur periods on the Hungarian Plain the burial custom is differentiated along gender lines as regards the precise orientation of the crouched position of the deceased as well as the accompanying objects. Particularly striking is the fairly strictly sex-differentiated burial custom, usually with males crouched to the right and females crouched to the left. Distinctly sex-differentiated burial customs are otherwise rare this early, appearing massively only in the 3rd millennium BC with the Corded Ware culture (Häusler 1994). This seems to reflect gender-specific attitudes to social life, and it is noteworthy that weapon-tools are only found in male burials.

The cemetery of Tiszapolgar-Basatanya (Bognár-Kutzián 1963) on the Hungarian Plain provides evidence of association between weaponry and specific persons. Basatanya is one of the best-known burial sites of the Copper Age Tiszapolgar-Bodrogkeresztur culture on the Hungarian Plain, and useful details have been made available about the gender, age and patterns of disease and trauma of the skeletons (ibid.). Fifty-seven skeletons have been anthropologically determined in the earlier Tiszpolgar phase and seventy-nine skeletons in the later Bodrogkeresztur phase. At Basatanya age and gender are in fact clearly mediated through the number and types of copper ornaments, but even vertical social status is communicated in the burials. Details in dress and equipment mediate a gendered life course with significant age groups among both sexes. Joanna Sofaer-Derevenski (2000: 392-95) states that at c. five years of age the children at Basatanya became gendered. In the early stage of the cemetery females underwent significant life-course changes at the age of twelve and forty, whereas males underwent similar changes at fifteen and again at eighteen-twenty years of age, and at c. twenty-five, thirty-five and fifty years of age. In the later phase of the cemetery twenty years of age is a significant stage for both sexes. Moreover, for males thirty years of age is a significant turning point (ibid.).

I have especially examined the distribution of weaponry and trauma at the cemetery (Fig. 4A-B). Trauma is a rather common phenomenon, which must be war-related, and mostly male skeletons show signs of violence, either healed wounds or possibly in some cases the cause of death. Likewise, the position of flint points on the skeleton, for instance among the ribs, is taken to indicate a violent death (cp. Bognár-Kutzián 1963: 392ff). Small pieces of flakes are common in these burials, but flint knives longer than seven centimetres are normally restricted to males. Whilst females in rare cases carry a flint knife, they are almost never accompanied by other types of weapon-tools. We can thus assume that it was males who first and foremost engaged in affairs of war, and warriors must be searched for in this group.

A sample of burials containing all adult male burials, in addition to male child burials with weaponry, was therefore selected for further inquiries, in total sixty-four burials (Fig. 4A-B). The age-regulated pattern found by Joanna Sofeer-Derevenski (ibid.) finds support in the distribution of weapon-tools. Long flint knives signify adult male-hood from c. sixteen years of age; only two boys carry such knives, and they are somewhat unusual in that they both seem to have died as victims of violence. It is characteristic that flint-knives tend to become longer when the man gets older: most knives longer than ten centimetres occur among older men.

In the early phase of the cemetery males do not, however, become genuine weapon carriers until they are around twenty-five years of age, usually signified by a boar's tusk knife, in rarer cases by a shaft hole axe. But it is especially mature and wealthy males who carry shaft hole axes and arrowheads (Fig. 5). This pattern changes somewhat in the later phase, but an age-regulated social structure is maintained. The age of thirty-five is the stage in a man's life which is signified by weaponry such as copper daggers, shaft hole axes and arrowheads. Again, a group of mature males around forty to sixty years of age is particularly spectacular as regards wealth and weaponry. Traumata cannot be argued to relate to any particular group of males; all ages seem inflicted and the trauma seemingly concurring with mature individuals might well have been acquired at a young age. Wealth, notably in copper, increases with age, but with twenty-five to thirty-five years of age as an approximate turning point when some of the weapon-carrying males emerge as the wealthiest in the male community.

Grave no.	Date	Age	Trauma	Hammer axes, maceheads, or axe blades	Arrowheads of stone or bone	Copper dagger	Flint/obsidian blade > 7 cms	Boar's tusk knife	Copper	Remarks
077	ECA	07	Lesions on skull, pelvis				X		X	buried like an adult man
040	ECA	09-10	Flint blade cause of death?				X (under ribs)			
065	ECA	16-18								
079	ECA	16-18								
050	ECA	18-19						X		
042	ECA	18-19					X			
061	ECA	18-20								dog
010	ECA	18-22					XX			
080	ECA	20-22								
023	ECA	24-29	Healed fracture on right femur				XXXX	XX	X	
052	ECA	25	Death by blade across cervical vertebrae?	X			X	X		dog
036	ECA	25		X						
056	ECA	25-30	Death by lesion on the skull?							
030	ECA	25-30								
053	ECA	25-30	Death by lesions on skull and arm				X			dog
035	ECA	30-35	Death by spine fracture, first lumbar vertebrae?				XXX	X		+ 2 children
038	ECA	35					XX			
013	ECA	35						X	XX	+ child
028	ECA	35-40			X				X	
045	ECA/MCA	35-40		X			X			
060	ECA	50-55		X			X	X		
039	ECA	50-55					XX		X	
012	ECA	55-60		XX	XX		X	X	XX	
068	ECA	55-60	Death by violence?					X		
067	ECA	60	Death by violence?	X	X		X			dog
014	ECA	60		X				X		
005	ECA	65-70					X			

F I G . 4 A : *The Copper Age cemetery of Tiszapolgar-Basatanya in Hungary, Tiszapolgar phase. Table showing burials with individuals who had received weaponry; mostly grown-up males but also two boys (data after Bognár-Kutzián 1963).*

Were all males at Basatanya warriors? No doubt warriorhood was a potentiality for all adult males in this community in both phases, as indicated by the presence of long flint knives with individuals of sixteen to sixty-five years of age. The fact that all other kinds of weaponry cluster with males above twenty-five years of age in the first phase and thirty-five in the second phase might be directly interpreted to indicate that these were the practicing warriors. This solution, however, becomes problematic as regards the wealthy and weapon-carrying elderly males.

Socially and materially they relate to other identities of age and gender presented at Basatanya, but they obviously share some features of identification in terms of their weaponry, wealth and mature age. The weapon-tools are likely to have had a bearing on the lived identities of these men. Their age, however, suggests that even though they were arguably presented as warriors by the surviving relatives, this identity must mainly have been on a symbolic level; hence the weapon-tools were metaphors of warriorhood and signs of wealth and power rather than a

Grave no.	Date	Age	Trauma	Hammer axes, maceheads, or axe blades	Arrowheads of stone or bone	Copper dagger	Flint/obsidian blade > 7 cms	Boar's tusk knife	Copper	Remarks
074	MCA	18-20					X	X		
146	MCA	18-20					X			
091	MCA	18-20								
102	MCA	18-20					X			
133	MCA	18-20						X		
122	MCA	19-20	Death by violence?				X			
125	MCA	20-23								
139	MCA	22-24								
156	MCA	25					X			
106	MCA	25-40					X			
128	MCA	28-30	Skull lesion				X			
143	MCA	28-30								+ woman
119	MCA	30-35								
149	MCA	30-35					X			
105	MCA	35				X	X	XX	XX	
098	MCA	35					X		X	
044	MCA	35				X	X		X	
135	MCA	35-40								
151	MCA	35-40					X			
137	MCA	35-40								
037	MCA	40		X	XXXX		X		X	
099	MCA	40-45					X		X	
152	MCA	40-45					X			
114	MCA	40-45								
071	MCA	45-50	Only the skull was buried		X		X		X	
140	MCA	45-50	Healed skull lesion							
145	MCA	50	Skull lesion				X			
101	MCA	50-55	Healed skull lesion				X			woman and child
072	MCA	50-55								
092	MCA	50-55	Skull lesion (lump)				X		X	
083	MCA	50-55	Lesion on first vertebrae below skull		X		X			
001	MCA	50-55					X		X	
003	MCA	54-59								
129	MCA	55-60		X			X		X	
117	MCA	60-65			X		X		X	
132	MCA	Full-grown					X			
141	MCA	Full-grown			X		X			

FIG. 4B: *The Copper Age cemetery of Tiszapolgar-Basatanya in Hungary, Bodrogkeresztur phase. Table showing burials with individuals who had received weaponry; all grown-up males (data after Bognár-Kutzián 1963).*

direct metonymic statement of the identity of the deceased in the period immediately preceding death. Earlier in life these men may have been warriors, but at the time of death they were hardly warriors *sensu strictu*; more like headmen, men of worth and influence.

Impressive copper axes and adze-axes formed an important part of this social setting of the mature Copper Age. Copper shaft hole axes are sometimes deposited in burials, always those of men: like their stone counterparts they are usually placed in front of the face with the shaft placed in the hands of the deceased man, underlining fierceness. The extremely wealthy and tall man buried nearby at Tiszawalk-Kenderföld with his copper axe and various other

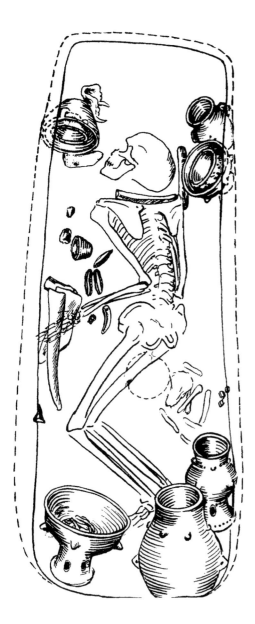

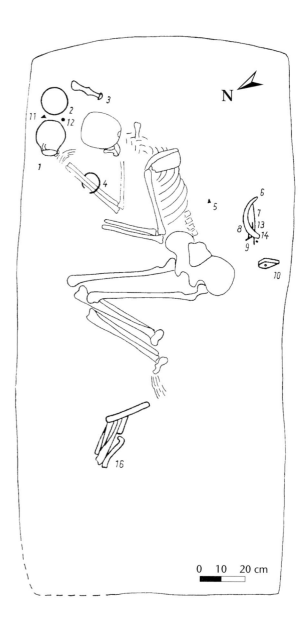

FIG. 5: *Grave 52 at Tiszapolgar-Bastanya. Young man, twenty-five years of age with weaponry; one of the youngest to have obtained a hammer axe (after Bognár-Kutzián 1963).*

FIG. 6: *Rich male burial with copper weapon-axe from the Copper Age cemetery of Tiszavalk-Kenderföld in Hungary (after Patay 1978).*

weapon-tools (Fig. 6) even lacked his left hand at the time of burial (Patay 1978: 21) and could thus hardly count as a practicing warrior. It seems though that such large copper weapons were also deposited in sacrificial hoards (cp. Todorova 1981; Patay 1984), an alternative ritual sphere. Weapon-tools were thus not only consumed in the funerary domain, but also offered as gifts to the gods, perhaps as the end point of a flow of war-like actions.

Of course, it is possible to interpret the scenario at Basatanya as indicating that warriorhood was

not tied to specific groups of persons, hence not institutionalised. Alternatively, the recurring signs of warfare all over eastern and central Europe in the Climax Copper Age, the specific social pattern of age cycles and the weapon-tools as tokens of power carried by mature males (gerontocracy?) at Basatanya, in addition to the custom of consuming valuable weaponry in a ritual domain other than burials, might better fit the hypothesis that age-regulated warrior institutions existed. Perhaps warrior bands consisted of young unmarried males between the age

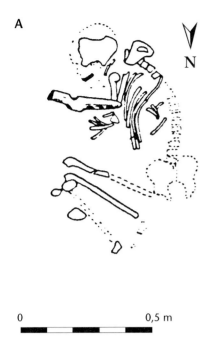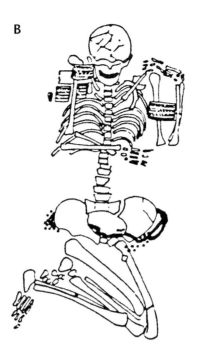

FIG. 7: *Male right-hocker burial with T-shaped antler axe placed in front of the face (A) and female left-hocker burial with ornaments at Brezsch Kujawski in Poland (B)(after Bogucki 1996).*

0   0,5 m

of sixteen/eighteen and twenty-five/thirty-five who were initiated together: this would be the age when they obtained the right to carry a long flint knife. Among males, mature persons doubtless attempted to control wealth, even if they were hardly in a position to control the war bands. They had gained the right to carry symbols of war as a result of their earlier life as warriors, probably reinforced by their personal qualities as community leaders and entrepreneurs. The latter would be in accordance with ethnographically analogous communities with age graded warrior institutions and with the strict gender-differentiated and age-regulated social structure presented at Basatanya, as well as with the fact that weapons accompanied males, not females.

Warriors at Brezsch Kujawski-Oslonki?

From the point of view of warrior identities, the contemporaneous Lengyel culture in the region of Brezsch Kujawski in central Poland is also worthy of note. The archaeological material is however less clearly readable inasmuch as the age pattern is unknown and information of trauma is not available. Burials as well as settlements are known from a number of investigated sites with well-preserved archaeological remains dating to the period 4500-4000 BC. Oslonki is a recently excavated site with thirty trapezoidal longhouses and more than eighty graves (Bogucki 1996; Bogucki and Grygiel 1997). At some point in time a ditched and palisaded enclo-

sure was established to fortify the peninsula on which the houses were located. This suggests that warfare formed part of social practice. In general, houses are organised in single farmsteads and small hamlets, each with a limited number of households (ibid.) and each with a small group of attached burials.

The burial custom at Brezsch Kujawski and Oslonki was strictly sex-differentiated – unusual for the time as noted above: males were buried crouched to the right, whereas women were buried crouched to the left. Approximately twenty percent of the males carried t-shaped antler axes (Zapotocký 1991), always in the same position, held in the hand and placed in front of face and body (Fig. 7). Bone points are also found with males: one grave contained an archer with five bone arrow points in a quiver worn at his back. Flint knives occur only with males. A corresponding group of females had dress fineries in exotic materials, notably belts of shell beads. Copper trinkets occur with both sexes, but especially with females (cp. ibid.). Of particular note is a female grave with an extraordinary amount of copper, among the earliest metal in northern central Europe, including a copper diadem. All burials at Oslonki were gender-differentiated in the orientation of the body, but less well-equipped burials than those mentioned above commonly occur. Like at Tiszapolagar-Basatanya gender was distinctly signified in burials, but vertical distinctions also seem to have existed singling out a particular group of agents with the males carrying specific antler axes and the females dressed in their copper fineries. The deposited copper trinkets and the gender-differentiated burial custom at Brezsch Kujawski-Oslonki suggest a tie with the Chalcolithic cultures of the Carpathian Basin; networks of exchange probably linked the two places.

However, were the males with antler axes at Brezsch Kujawski-Oslonki warriors? This central question cannot be answered at present, but the social pattern nevertheless recalls the situation described above for Tiszapolagar-Basatanya and also shows similarities to later Corded Ware practices. The antler axes could have had other uses besides war, and details such as the age distribution of the skeletons are not available. On the other hand, the number of axe-carrying men in the dead population is fairly large. They may well have to be identified as a group, and were in all likelihood also recognised as such; some sort of male bonding among a peer group is a possibility. Elitist warrior conduct and stereotyped gender differentiation were not to arrive massively on the European scene until more than a thousand years later with the Corded Ware cultures, then coinciding with a metallurgical revival. The right-crouched males with their antler axes in front of their face at Brezsch Kujawski are strikingly similar to a peer group of male burials of the Bohemian Corded Ware culture.

Corded Ware warriorhood

The first really convincing warriors in Europe north of the Alps were presented in a strikingly stereotyped manner in the funerary rituals of many communities at the beginning of the 3rd millennium BC. In death the warrior was accompanied by a slender battle-axe, or macehead, of stone; evidently as much a symbol of warriorhood as a weapon in itself. Deborah Olausson (1998) has made the important point that the battle-axes should not be understood as prestige goods, since they are not specialist products ordered by a patron and used in exchange activities. Rather they are 'do-it-yourself' objects, which did not require a great deal of expertise. Furthermore, they either occur in burials or are deposited singly on special locations (usually wetlands) as a ritual offering, and this indicates that they are very personal objects (ibid.). In Bohemia the battle-axes are tied intimately to specific males (see below), and this pattern may perhaps have a more general relevance. The key

importance of this weapon for the social identity of the deceased is emphasised by its position in the grave. The head of the axe is usually placed in front of the face of the male, who holds the end of the haft in his hands. The war-axe often goes along with a cord-decorated drinking beaker of pottery and in some cases an amphora, suggesting that warriors participated in communal drinking rituals. This idea of a relationship between male bonding, consumption of alcohol, and ornamented beakers in the Later Neolithic and the Copper Age has notably been launched by Andrew Sherratt in a series of articles (e.g., 1981; 1984; 1994a; 1994b).

Corded Ware funerary customs as evidenced over much of central Europe were extremely standardised (Fig. 8, lower). Differences of gender were in particular emphasised through the orientation of the body and the side on which it was placed and through particular gendered objects. Deviations from a stereotyped basic pattern occur, but rarely. Each deceased member of the community was usually inhumed in a single grave, which was often covered by a small earthen mound. Double burials of two persons, male and female, sometimes occur the bodies lying in gender-specific positions with the feet touching. The position of the body is always bipolar as well as side-specific with gender as the structuring principle. Males were usually buried crouched to the right with their head facing the west, whilst females lie crouched to the left with their head facing the east. Common to both sexes are the flexed position of 'hocker' and the face looking towards the south (cp. Häusler 1994; Wiermann 1998; Turek and Cerný 2001). Part of the material equipment is also gender specific. Jugs, shells, teeth and copper spirals tend to occur only in left-hocker burials, whereas battle-axes, maceheads, flat axeheads, bone daggers, chisels, bone dress pins and boar tusks are found mostly in right-hocker burials. It certainly looks like binary gender oppositions, but there are more dimensions to it.

The amazing spread of Corded Ware culture around 2800 BC over large parts of central and northern Europe has traditionally been thought to contain such social and material novelties as warriors, male dominance, and horse riding in addition to the overt presentation of individual potency, the ritual consumption of alcohol, innovations in copper metallurgy, and pastoralism (cp. Treherne 1995; Sherratt 1994b). One problem with the persuasive trilogy of horse, masculinity and battle-axe is that horses are

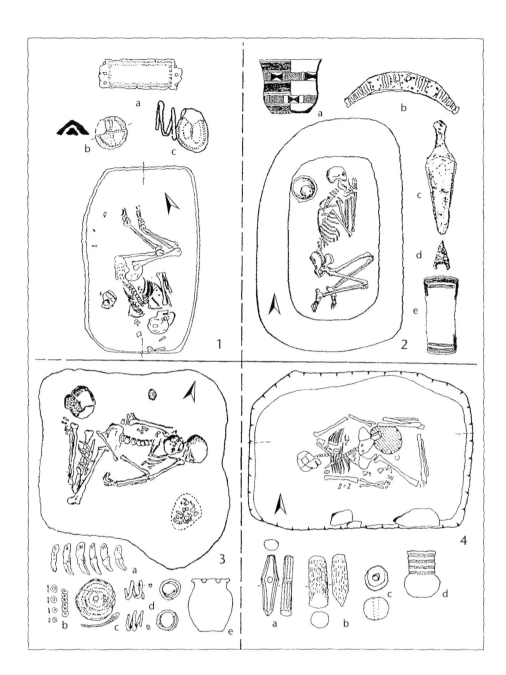

FIG. 8: *Corded Ware and Bell Beaker burial rites. Above: Bell Beaker. Below: Corded Ware. Right: male. Left: female (after Turek and Cerný 2001).*

still not well documented in central and northern Europe at this time. Nor does the archaeological evidence support a state of male dominance in Corded Ware society, this idea being linked to our own world. Warriors nevertheless did exist and in all likelihood occupied important positions in society even if the traditional picture has to be nuanced and the evidence substantiated. Social divisions along gender lines definitely became of central importance over vast areas, and at least in some regions warriors and high-ranking females came to form a sort of elite, as we shall see. That the overt presentation of individual potency and wealth now became possible for the first time, as argued by Paul Treherne (1995), is probably too simple a view since collective identities were as important as ever; they merely became structured differently than previously. In some regions pollen analyses suggest that everyday life became temporarily more mobile and the importance of cattle seems to have increased (e.g., Ethelberg *et al.* 2000), but evidence of pastoralism is generally scanty. It is true though that copper metallurgy was again integrated in society after a conspicuous absence in the preceding centuries, and the idea that beakers were used for the ritual consumption of alcohol gains at least some validity

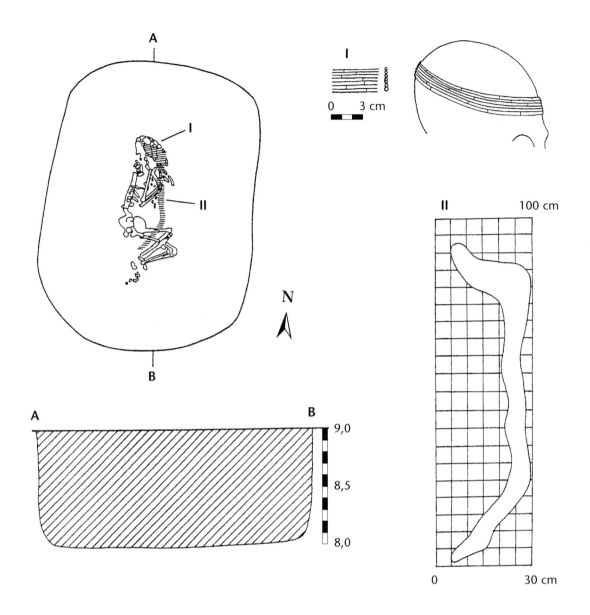

FIG. 9: *Corded Ware left-hocker burial with composite bow and copper diadem. Female?*
(after Czebreszuk and Szmyt 1998).

through the remains of a sort of honey-sweetened beer at the bottom of the bell-shaped beaker from Ashgrove, Fife in Scotland (Dickson 1978; Andrew Sherratt, pers. comm.).

The social and material uniformity often claimed to have existed in Corded Ware Europe should probably be understood as a macro-regional fashion movement containing material and social novelties which received a differential reception regionally and locally – not unlike modern global culture. I will not enter the discussion of whether it was primarily internal cultural construction or intruding ethnic groups, but merely state that a dialogue between local practices and foreign novelties must have played a vital part and that European societies were transformed in the process, some more radically than others. The new material similarities over vast parts of northern and central Europe in particular adhere to a series of particular objects buried with the dead. Digging into deeper levels of material culture reveals local and regional peculiarities and deviations from the overall pattern. It has, furthermore, proven impossible, or at least complicated, to pinpoint the place of origin of the Corded Ware culture.

Interpretations of the Corded Ware culture often have to rely widely on the evidence of cemeteries, since settlements from this period are still not well known even if the situation in this respect is improving (e.g., Simonson 1986; Nielsen 2000; Turek and Peska 2001). The populations generally seem to have organised their settlements in a scattered pattern of small hamlets and single farms. Likewise, pollen diagrams – especially from southern Scandinavia – provide evidence of extensive land clearings (Jensen 2001: 458ff). A wider representation of sources for Corded Ware society would be preferable, but the particularly rigorous funerary customs must be embedded in the social practices of the living community, if still on an ideal level. The deceased persons were for example hardly interred with full equipment and with all personal belongings, but merely with certain objects selected for their specific symbolic meaning; thus my claim is that these objects and their attached meanings were linked to the lived identities of the deceased and to the way they were categorised by their family and especially the wider community. The groupings that appear when examining and comparing the interments must thus have a direct bearing on Corded Ware social structure. When warriors went to war they may well have carried a range of other war tools besides the battle-axe. It is for example possible that wars were waged with bow and arrow as the primary weapons (Fig. 9). Moreover, the women might have worn dress fineries only on special occa-sions. This is difficult to gain any precise knowledge about, but what we can say is that certain social identities existed, which were communicated in the funerary custom. Some of these identities were probably institutionalised inasmuch as they reappear over time-space.

Case study
Bohemian Corded Ware burials are particularly well examined and have therefore been chosen as a case study. Corded Ware people buried their dead in small cemeteries probably situated close to their settlements on the slopes of the terraces of major rivers and their tributaries (Turek and Peska 2001). In the following examination of warriorhood, rank and gender in the Bohemian Corded Ware culture I make extensive use of statistics provided by Roland Wiermann (1998; 2002), a data collection of around two hundred burials from six cemeteries presented in diagram form (Fig. 10). These data can be used to reconstruct Corded Ware social structure, or more precisely, they provide a picture of the dead population. Wiermann has interpreted the social patterns underlying the diagram in terms of a big-man society of Melanesian type. Moreover, he sees traces of age grades with analogues among east African pastoralists. My interpretation, as presented below, differs from Wiermann's. What is immediately apparent to my eye is that gender and rank were the two main

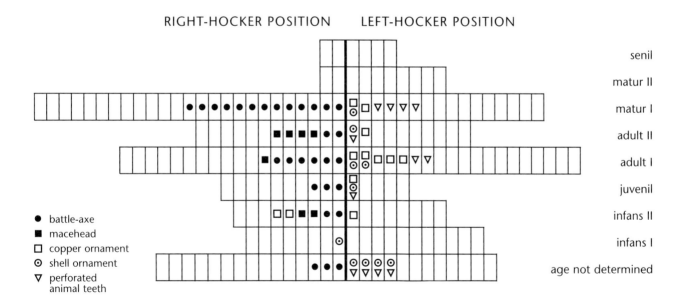

FIG. 10: *Burial statistics for the Bohemian Corded Ware culture sorted according to age, gender and wealth (after Wiermann 2002).*

principles along which this Corded Ware society was organised. It is especially striking that gender, and to a lesser degree rank, was interpreted and mediated very rigidly by these people. Similarly, the stereotype of mediated identities calls for particular attention since it hints at rather inflexible understandings of some forms of social identity.

Fortunately, sixty burials had well-preserved skeletons allowing for very reliable determinations of biological age and sex (ibid.). They could therefore be used as a control group. In addition, they demonstrated with clarity that males were buried in right-hocker position whilst females were buried in left-hocker position. Few deviations from this pattern exist. It is sometimes assumed that Corded Ware burials represented only the upper echelon of the population. This is perhaps possible, but the diagram sampled by Wiermann could be interpreted differently. According to the physical anthropologist Elizabeth Iregren (pers. comm.), the mortality pattern is quite 'normal', suggesting that the sample may represent some sort of complete society without missing groups. It should be mentioned, though, that the youngest individuals are underrepresented due to the poor preservation of these skeletons. There is, as one might expect, increased mortality among young women in the childbearing age and also among young men (adultus I), probably owing to war-inflicted deaths. Data on trauma has not been included in the analysis. However, it should be noted that five males and one not-sexed individual at Vikletice had healed cranium fractures (Buchvaldek and Koutecký 1970: 276). We can then assume that warfare formed part of the actions that created warrior identities.

With very few exceptions, biological sex corresponds wholly to social gender as reflected by burial position and types of grave goods. Gender is thus underlined in the position of the body as well as in the objects applied to the body. The gendered difference is probably the least negotiable in this society, and males and females each constitute groups that were clearly recognised as separate (in itself, 'an sich'). Importantly, the sex-differentiated burial position is maintained from earliest childhood to very old age. Personal appearance becomes gendered around the age of seven (infans II), or a little earlier, as suggested by very young boys, six months to six years old, buried with battle-axes or mace heads (Turek and Cerny 2001: 609). Bodily appearance may

have become partly un-gendered late in life. The few old persons present in the diagram are without particular gender-specific objects, but this could possibly be due to a problem with source representativity, or merely mean that these old people belonged to the less privileged part of the population (see below).

The strict gender differentiation is further reflected in the physical appearance of the skeletons. Males and females of the Corded Ware are very different in stature as demonstrated recently by Jan Turek and Viktor Cerný (2001), whereas such sexual dimorphism is less marked in the subsequent Bell Beaker population. Such a marked discrepancy between male and female skeletons is likely to originate in gender-specific divisions of labour, but to infer patriarchy and male dominance on this basis is too farfetched. Personal equipment in these burials communicates gender, but clearly also social status.

Social rank is another important social parameter in Bohemian Corded Ware communities. The diagram distribution (cp. Fig. 10) suggests the existence of two rank groups that ignore differences of gender: 1. High-status armed males and high-status more peaceful looking females with dress fineries. 2. Low-status unarmed males versus low-status females. The first group has other objects besides the characteristic cord-decorated pottery, especially in the personal field of body decoration – i.e., dress and weaponry – whereas the second group mostly has pottery. It is thus significant that males and females on each side of the diagram divide into two groups independently of age: males with and without weapons and females with and without dress accessories like copper trinkets. The balance in wealth between males and females within the high-ranking group was established early in life, since the number of grave goods is the same (cp. Turek and Cerny 2001: 608). Rank thus very clearly crosscuts gender as an identity-structuring factor. The social identities on the vertical level also appear quite distinct materially; hence we must assume that they were recognised as a group by other agents. Society thus has two almost equal-sized groups: a high-ranking group with particular objects and a low-ranking group without. There is more interaction within each rank group than between them. Some sort of elite apparently existed. This point is further illustrated in the cemetery of Vikletice where individuals with particular objects were buried close to each other within a particular area, hence emphasising their common identity

(Wiermann 2002: Abb. 2). Variability in the life cycles of individuals is, by contrast, not distinctly mediated in material culture – very unlike the situation at Tiszapolgar-Basatanya. Social age is not communicated through the number and type of personal objects associated with the body, even if the forms and size of pottery present in individual burials seems more sensitive to the age of the deceased.

Rank might well have been inherited rather than being achieved through particular actions inasmuch as the rank difference occurs from early childhood almost to very old age. Formal rules of endogamy may have structured this practice. It is equally possible to understand Corded Ware elite identity as having been transferred across generations as a particular kind of habitus, and endogamy may in this case have been the result even if not formally required. In any case the boundary between elite and non-elite is marked by some objects and interrupted by others. Vivid interaction seems to have taken place: cord-decorated pottery unites the two groups, thus showing that the whole community shared identity on some cultural level. It seems likely that the particular corded pottery style was associated with a broader kind of group identity, which we might call cultural identity or ethnicity.

Finally, it is noteworthy that inside each social group of gender and rank there is little variation. The number and classes of personal objects vary only slightly from individual to individual within the elite group. Again it is the number of pots that varies the most, and it is possible that the quantity of deposited corded pots relates to the social rank of the deceased (see, for instance, Buchvaldek and Koutecký 1970: grave 1963/110-111). We may nevertheless conclude that with regard to the internal structure of the elite there were, at least ideally, few differences of rank. Communal identities of rank, gender and culture have been emphasised whereas individualistic traits have been almost concealed. The stereotype within each of these groups is absolutely conspicuous, and beyond the ideology which was surely an active factor in creating this typecast, this may reveal to us a social structure which was more restrictive than enabling when it became established. Internal processes of rivalry and distinction, which usually characterises the various social fields of a society, seem in this case regulated by strict social norms (also Turek and Cerny 2001). The same rigorousness is valid for warriorhood.

High-ranking males carry battle-axes or mace-heads, sometimes a flat axehead as well: the position is classic with the haft in the hands and the axehead or blade in front of the face (Fig. 11). The elegantly shaped shaft hole axes and maceheads obviously did not merely symbolise maleness as opposed to femaleness since less than half the males carried them. They surely signified a superior kind of maleness, which included an identity as warrior. Hunting is less likely inasmuch as the selected weaponry clearly is war-related: it makes no sense to hunt with a macehead or a battle-axe! These males might be considered born members of a warrior sodality with common drinking rituals. The drinking beaker with cord impression or herring-bone pattern actually occurs predominantly in male burials (cp. Mallory 1989: 244; cp. Turek and Cerný 2001: 605f).

It follows from the above description of gender and rank in Bohemian Corded Ware burials that access to warriorhood was regulated through distinctions of rank (and male gender). The third group of the three aforementioned sociological types of war bands thus fits best. The rank-based structure in the Bohemian Corded Ware culture evidently contrasts with the age-regulated structure at the much earlier, copper age site of Tizsapolgar-Basatanya. All high-ranking males were by definition warriors even if this identity for the youngest and eldest must have been on a symbolic level; the war-axe was in these cases merely a metaphor for past or future warriorhood. In other words, I suggest that the warrior band as an active unit of war consisted of adult males from the elitist group. The individualistic traits often apparent among the members of such warrior clubs are not presented, or did not exist, among the warrior bands of the Bohemian Corded Ware culture. The Abipón in the Grand Chaco of South America offers a good analogy in this respect: warrior institutions of the Abipón social elite avoided hierarchical divisions among themselves and used drinking rituals to symbolise equality and to strengthen solidarity (Lacroix 1990). It is common even in strictly hierarchical societies that the companionship in warrior bands maintains an ideology of equality, but these associations of warriors were nevertheless a power source that could be used to back-up other kinds of power sources. In accordance with the rigid character of the burial rites, ideology may well have formed a central power source in Bohemian Corded Ware society and

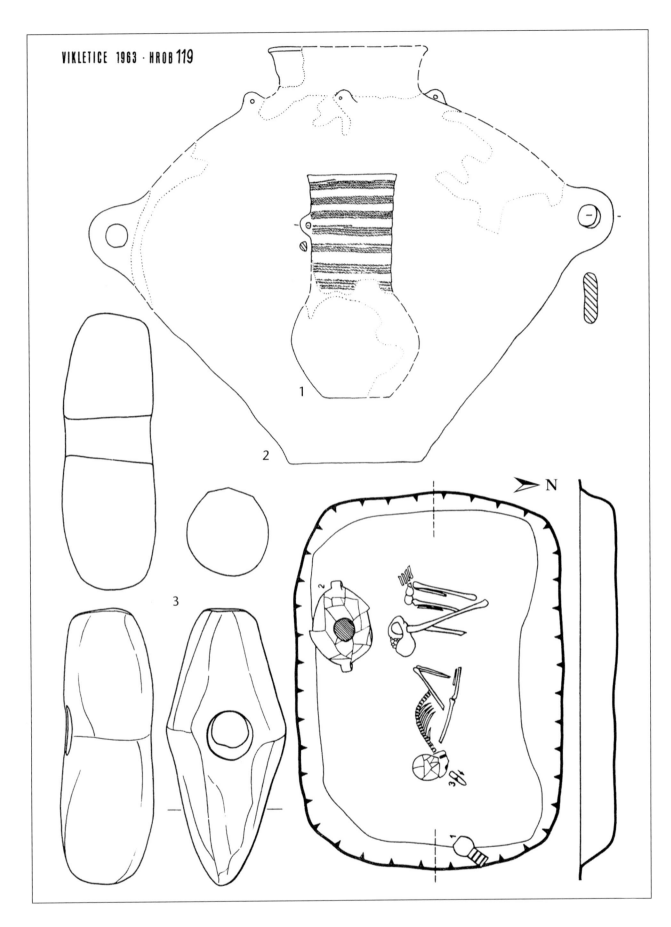

VIKLETICE 1963 · HROB 119

FIG. 11: *Corded Ware warrior burial from the cemetery of Vikletice in Bohemia. Note the battle-axe in front of his face and the corded beaker nearby (after Buchvaldek and Koutecký 1970).*

could if necessary be enforced through violence and coercion.

High-ranking females are buried with various trinkets of amber, shell, bone or copper. Likewise, age-regulated life cycles are not easily recognised, but obviously womanhood must have had a symbolic character among girls and women over forty to fifty years of age. Quite possibly high-ranking females also organised in social clubs. It is noticeable that the group of high-ranking females is almost of the same size as the group of high-ranking males. It certainly suggests some sort of common social ground, and it would be difficult on this background to argue for male dominance in Bohemian Corded Ware society. There are no signs that women participated actively in war raids but this does not mean that they did not interfere in matters of war.

Gender is a central feature of the Bohemian Corded Ware culture and elsewhere in Corded Ware Europe. The material equipment of high-ranking males is strikingly similar over wide distances in Europe – for example, between such remote regions as southern Sweden and Bohemia (Knutsson 1995). It is somewhat less striking as regards the personal appearance of high-ranking females in that dress accessories vary more. Even the interaction across geographical space into foreign territory thus seems gendered to a considerable degree, in addition to being highly sensitive to social rank, as already argued above. The description of the warrior as a boundary-crosser may contribute to explaining the macro-regional dispersal of material culture and social conduct.

In summary, it is fairly clear that specific objects like copper spirals, shells and weaponry belonged to a presumably privileged subgroup within society consisting of males and females of all ages from early childhood to a mature age. Privileged males, irrespective of age, carried a battle-axe or a macehead symbolising an identity as warrior, but high-ranking males were fully matched by high-ranking females in terms of particular objects. The specific corded drinking beaker occurs mostly in burials of males, probably implying associated drinking rituals. It seems reasonable to assume that Corded Ware males of high rank organised in warrior clubs with restricted access, the exclusiveness symbolised by specific weapons and most likely a particular social conduct. The burial custom is heavily ritualised, suggesting that the society that performed these uniform rituals

was strictly structured, specifically along the lines of gender and rank, and hence restricted rather than enabled social action.

The social pattern described above seems to carry some relevance for Corded Ware communities in other regions such as Switzerland, Austria, along the upper Danube, Saxo-Thuringia, the Netherlands and southern Scandinavia. However, skeletons are not necessarily well preserved, and it is often difficult to assess the social structure in any detail. Deviations from the social pattern encountered in Bohemia are probably not uncommon, but the cemetery at Schafstädt, ldkr. Merseburg-Querfurt in the classic Corded Ware area of Saxo-Thuringia, deviates substantially (Hummel 2000). A rather fluent gender categorisation was found among sixty-three burials. Here age seems to have been a decisive parameter in the acquisition of social status. A closer look at Schafstädt, however, also reveals that battle-axes or maceheads are absent. Likewise, the number of status-related trinkets is mediocre. Together this suggests the absence of elite warriors and their female counterparts at Schafstädt.

Conclusion: Bell Beakers and beyond

The search for early warriors and their associations ends here, but could obviously have ventured further into the Copper Age and later settings in Europe using the same or a similar method with elements from different disciplines besides archaeology. By way of conclusion, what follows is a few observations, and especially hypotheses, about warriorhood in the succeeding Bell Beaker communities of central Europe. This highlights the possibilities inherent to a comparative procedure since there are similarities as well as differences between practices in the Corded Ware and the Bell Beaker communities, which surely constitute guidelines for the interpretation. Like the Corded Ware community in Bohemia, the subsequent Bell Beaker communities in central Europe seem to have been ranked societies, and warrior institutions arguably existed which recruited their members mainly from an elite group.

The Bell Beaker culture is another macro-regional culture that appeared in large parts of Europe c. 2500 BC. Bell Beaker communities are also mostly known from their burial rites, but the number of settlements has been increasing over recent years. The funerary practice is highly uniform though less so

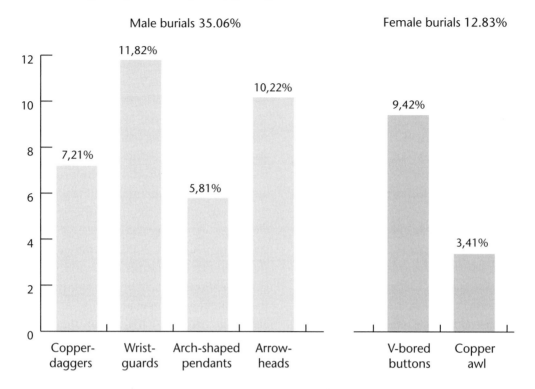

Percentage of specific non-pottery gravegoods in relation to total number of burials

Male burials 35.06%   Female burials 12.83%

FIG. 12: *Statistics showing the frequency of male and female Bell Beaker burials with particular objects (adapted from Müller 2001).*

than in the Corded Ware culture. The dead body was usually inhumed in a single grave and placed in a small flat-grave cemetery. The body position is likewise bipolar and side specific with gender as the structuring principle (Fig. 8, upper). Males were buried crouched to the left with their head facing north, whilst females were resting crouched to the right with their head facing south. It is remarkable that the orientation and position of the body are the opposite of what was practiced in the Corded Ware, but nevertheless strictly sex-differentiated. Common to both sexes are the flexed position of hocker and the face looking towards the east. Part of the material equipment is also gender specific. Males may be buried with archers' gear of pressure-flaked arrow-heads (in a quiver), a bracer (wristguard) to protect the hand from the backlash of the composite bow, boar's tusk, and sometimes a model of a bow, in addition to a flat tanged copper dagger and a zone-decorated bell-shaped beaker, probably for the ritual consumption of alcohol. Females may possess a few ornaments for hair, body or dress such as, notably, v-perforated buttons of bone or amber for the fas-

tening of woollen costumes, a copper awl, and sometimes a spiral-shaped trinket of copper, gold, or electron. Gender distinctions are maintained from earliest childhood to very old age. The Bell Beaker gender pattern corresponds almost exactly to biological gender as ascertained through anthropological examinations (Müller 2001). However, far from all individuals were buried with weapons and dress fineries (Fig. 12).

The Bell Beaker material is not as sociologically well investigated as that of the Corded Ware culture, but it seems clear that social rank was decided at birth rather than achieved later on. In the Bell Beaker group of East Central Europe, it is possible to outline a privileged group of males and females, with weaponry and dress accessories, respectively, covering all ages and hence recalling the social pattern of the Corded Ware culture. Thirty-five percent of male burials contained weaponry, whereas only thirteen percent of the female burials can be characterised as wealthy with v-bored amber buttons and copper awls. This seems to suggest that some sort of imbalance existed between males and females in the elite group; or perhaps a more distinct border between warriors and ordinary males existed than between privileged and ordinary females. Interestingly, the male group of weapon bearers, whose members were probably recruited to warrior associations, was more differentiated internally with varying numbers and classes of weapons. This feature of material variation internal to the social identities of the Bell Beaker culture is in contrast to the rigid stereotype of gender and rank in the Corded Ware culture; it could possibly relate to age-cycles, but this is a less well known aspect. Alternatively, it could be internal ranking and ongoing rivalry for power in an elite group, precisely, it would seem, as in some Bronze Age communities (Vandkilde 1999). This less rigid pattern of actions in the ritual domain of funerals must surely have a bearing on Bell Beaker social structure and practice in general.

The warrior has, in conclusion, appeared as a specific social identity who stands in a variable relationship to the parameters of age, rank, gender and culture – depending on the social context: warriorhood hardly remained the same over time, as indicated by the Copper Age case studies. Variability as well as certain recurring patterns take place over time and space, and this stands in contrast to the stereotypical ideologically loaded presentations of the warrior figure in the archaeology of the 20th century (Vandkilde 2003 and this volume chapter 5). Likewise, warrior identity receives its sustenance from war-related actions and interactions of animosity, rivalry and companionship. This should be remembered when dealing with the presentation of warrior ideals in the past and in the present.

NOTES

1 The birth of this study took place in a Dublin bookshop in the early Spring of 1997. I was looking for new archaeological books, and more vaguely for a new research topic. When the bookshelf dealing with prehistory did not contain anything of immediate interest, I moved on to the shelf on sociology and social anthropology. The first book I picked from the shelf was a collection of articles by Pierre Clastres: *Archeology of Violence* (translated into English from French 1994). My first thought was, rather naively, that the book had been misplaced! Slowly it dawned on me that the author was the late French social anthropologist of structuralist persuasion, and that 'archeology' was used in the sociological sense; to uncover layer by layer of a social phenomenon; in this case the elementary warlike structures of primitive society. The book, by closer inspection, proved to discuss warring tribes in the Americas: Tupi-Guarani, Yanomami, Blackfoot, Apache, Algonquin, Iroquois, *etc.* According to Clastres, warfare should simply be understood as the dominant structure, indeed the essence, of tribal society: warfare is omnipresent among tribes, and the very factor that prevents their transformation to state. What Clastres does is to reverse Thomas Hobbes (1991[1651]) famous dictum about the primitive being '*bellum omnium contra omnes*' by arguing that if the state is categorised as the pacifier of that being, then war in primitive society is war against the state; thus the phrase *society against the state* (Clastres 1977; 1994; Bestard and Bidon-Chanal 1979: 225). My study of warriors is in many ways indebted to Clastres work in that it became the genesis of a desire to uncover the more bellicose aspects of human life in a distant European past.

2 See also more generally my introduction to the section: warfare, weaponry and material culture.

BIBLIOGRAPHY

- Barrett, S.M. 1974. *Geronimo. His own story*. London: Ballantine Books.
- Barth, F. 1969. 'Introduction. Ethnic groups and boundaries'. In: F. Barth (ed.): *The Social Organisation of Culture Difference*. Oslo: Universitetsforlaget, pp. 9-38.
- Bazelman, J. 1999. *By Weapons Made Worthy. Lords, Retainers and their Relationship in Beowulf*. Amsterdam: Amsterdam University Press.
- Bestard, J. and Bidon-Chanal, C. 1979. 'Power and war in primitive societies: the work of Pierre Clastres'. *Critique of Anthropology*, 4, pp. 221-27.
- Bognár-Kutzián, I. 1963. *The Copper Age Cemetery of Tiszapolgár-Basatanya*. Budapest: Akadémiai Kiadó.
- Bogucki, P. 1996. 'Sustainable and unsustainable adaptations by early farming communities of Northern Poland'. *Journal of Anthropological Archaeology*, 15, pp. 289-311.
- Bohle, B. 1990. 'Ritualisierte Homosexualität – Krieg – Misogynie. Beziehungen im und um den Männerbund: Beispiele aus Neuguinea'. In: G. Völger and K. von Welck (eds.): *Männerbande-Männerbünde*. (Ethnologica, Neue Folge, 15). Cologne: Joest-Museum der Stadt Köln, pp. 285-96.
- Browne, N. 1995. 'Women and the war show: T.V.'s gendered construction of the homefront'. In: C.R. Sutton (ed.): *Feminism, Nationalism and Militarism*. Arlington (Va.): American Anthropological Association, pp. 57-60.
- Buchvaldek, M. and Koutecký, D. 1970. *Vikletice. Ein Schnurkeramsiches Gräberfeld*. (Præhistorica III). Prague: Universita Karlova Praha.
- Chapman, J. 1999. 'The origins of warfare in the prehistory of Central and Eastern Europe'. In: J. Carman and A. Harding (eds.): *Ancient Warfare. Archaeological Perspectives*. Stroud: Sutton Publishing, pp. 101-42.

- Clastres, P. 1977. *Society against the State*. Oxford: Blackwell.
- Clastres, P. 1994. *Archeology of Violence*. (Reserches d'anthropologie politique 1980). New York: Semiotext(e).
- Czebreszuk, J. and Szmyt, M. 1998. 'Der Epochenumbruch vom Neolithikum zur Bronzezeit im Polnischen Tiefland am Beispiel Kujawiens'. *Prähistorische Zeitschrift*, 73(2), pp. 167-233.
- Davis, J.L. and Bennet, J. 1999. 'Making Mycenaeans: warfare, territorial expansion, and representations of the Other in the Pylian Kingdom'. In: R. Laffineur (ed.): *Polemos. Le contexte guerrier en Égée à l'age du bronze*. (Actes de la 7e Rencontres egéenne internationale Université de Liège, 14-17 avril 1998). Liege: Université de Liège and University of Texas at Austin, pp. 105-18.
- Dickson, J.A. 1978. 'Bronze Age mead'. *Antiquity*, 52, pp. 108-15.
- Dyckerhoff, U. 1990. 'Adler, Jaguar, Otomí – Kriegerbünde bei den Azteken'. In: G. Völger and K. von Welck (eds.): *Männerbande-Männerbünde*. (Ethnologica, Neue Folge, 15). Cologne: Joest-Museum der Stadt Köln, pp. 251-57.
- Ehrenreich, B. 1997. *Blood Rites. Origins and History of the Passions of War*. New York: Metropolitan Books.
- Engels, F. 1977(1891). *Familiens, privatejendommens og statens oprindelse*, (Der Ursprung der Familie, des Privateigentums und des Staats). Copenhagen: Politisk revy.
- Ethelberg, P., Jørgensen, E., Meier, D. and Robinson D. 2000. *Det Sønderjyske Landbrugs Historie. Sten- og Bronzealder*. Haderslev: Haderslev Museum og Historisk Samfund for Sønderjylland.
- Giddens, A. 1984. *The Constitution of Society. Outline of the Theory of Structuration*. Cambridge: Polity Press.
- Grygiel, R. and Bogucki, P. 1997. 'Early farmers in North-Central Europe: 1989-1994 Excavations at Oslonki, Poland'. *Journal of Field Archaeology*, 24, pp. 161-78.
- Harrison, S. 1996. 'War, warfare'. In: A. Barnard and J. Spencer (eds.): *Encyclopedia of Social and Cultural Anthropology*. London and New York: Routledge.
- Häusler, A. 1994. 'Grab- und Bestattungssitten des Neolithikums und der frühen Bronzezeit in Mitteleuropa'. *Zeitschrift für Archäologie*, 28, pp. 23-61.
- Heaney, S. 1999. *Beowulf. A New Translation*. London: Faber and Faber.
- Hedeager, L. 1990. *Danmarks jernalder. Mellem stamme og stat*. Aarhus: Aarhus University Press.
- Helbling, J. 1996. 'Weshalb bekriegen sich die Yanomami? Versuch einer spieltheoreretischen Erklärung'. In: P. Bräunlein and A. Lauser (eds.): *Krieg und Frieden*. Bremen: Keo.
- Hobbes, T. 1991(1651). *Leviathan*. Richard Tuck (ed.). (Cambridge Texts in the History of Political Thought). Cambridge: Cambridge University Press.
- Hummel, J. 2000. 'Das schnurkeramische Gräberfeld von Schafsstädt, Ldkr. Merseburg-Querfurt: ein Beitrag zu der Bestattungsweise'. *Jahresschrift für mitteldeusche Vorgeschichte*, 83, pp. 25-51.
- Jenkins, R. 1996. *Social Identity*. London and New York: Routledge.
- Jensen, J. 2001. *Danmarks Oldtid. Stenalderen 13.000-2.000 f.Kr*. Copenhagen: Gyldendal.
- Jørgensen, A.N. 1999. *Waffen und Gräber. Typologische und chronologische Studien zu skandinavischen Waffengräbern 520/530 bis 900 n.Chr*. Copenhagen: Det kongelige nordiske Oldskrifselskab.
- Kilian, K. 1988. 'The emergence of the Wanax ideology in the Mycenaean palaces'. *Oxford Journal of Archaeology*, 7, pp. 291-302.
- Knauft, B.M. 1991. 'Violence and sociality in human evolution'. *Current Anthropology*, 32(4), pp. 391-428.
- Knutsson, H. 1995. *Slutvandrat? Aspekter på övergången från rörlig till bofast tillvaro*. ('Done roaming? Aspects of the transition from an itinerant to a settled life'). (AUN, 20). Uppsala: Societas Archaeologica Upsaliensis.
- Lacroix, M. 1990. '"Volle Becher sind ihnen lieber als leere Worte". Die Kriegerbünde im Gran Chaco'. In: G. Völger and K. von Welck (eds.): *Männerbande-Männerbünde*. (Ethnologica, Neue Folge, 15). Cologne: Joest-Museum der Stadt Köln, pp. 80-271.
- Langness, L.L. 1974. 'Ritual power and male dominance'. *Ethnos*, 2, pp. 190-212.
- Larick, R. 1986. 'Age grading and ethnicity in the style of Loikop (Samburu) spears'. *World Archaeology*, 18(1), pp. 269-83.
- Lipp, W. 1990. 'Männerbunde, Frauen und Charisma. Geschlechterdrama im Kulturprocess'. In: G. Völger and K. von Welck (eds.): *Männerbande-Männerbünde*. (Ethnologica, Neue Folge, 15). Cologne: Joest-Museum der Stadt Köln, pp. 31-40.
- Lowie, R.H. 1962(1927). *The Origin of the State*. New York: Russell and Russell Publishers.
- Makkay, J. 2000. *An Early War. The Late Neolithic Mass Grave from Esztetrgályhorváti*. (Tractata Minuscula, 19). Budapest. (Published by author).
- Mallory, J. 1989. *In Search of the Indo-Europeans. Language, Archaeology and Myth*. London: Thames and Hodson.
- Mann, M. 1986. *The Social Sources of Power*. Vol. I: *A History of Power from the Beginning to A.D. 1760*. Cambridge: Cambridge University Press.
- Müller, A. 2001. 'Gender differentiation in burial rites and grave goods in the Eastern or Moravian group of the Bell Beaker culture'. In: F. Nicolis (ed.): *Bell Beakers Today: Pottery, People, Culture and Symbols in Prehistoric Europe*. (Procceedings of the International Colloquium, Riva del Garda. Trent, Italy). Trent: Servizio Beni Culturali, Ufficio Beni Archeologici, pp. 11-16
- Nielsen, P.O. 2000. 'Limensgård and Grødbygård. Settlements with house remains from the early, Middle and Late Neolithic on Bornhom'. In: C. Fabech and J. Ringtved (eds.): *Settlement and Landscape. Proceedings of a Conference in Århus, Denmark. May 4-7 1998*. Aarhus: Jutland Archaeological Society, pp. 149-65.
- Olausson, D. 1998. 'Battle-axes: Home-made, Made to Order or Factory Products?' In: L. Holm and K. Knutsson: *Preceedings from the Third Flint Alternatives Conference at*

Uppsala, Sweden, October 18-20, 1996. (Occasional Papers in Archaeology, 16). Uppsala: Department of Archaeology and Ancient History, Uppsala University, pp. 125-40.

- Otterbein, K.F. 1970. *The Evolution of War. A Cross-cultural Study.* New Haven: HRAF Press.
- Paige, K.E. and Paige, J.M. 1981. *The Politics of Reproductive Ritual.* Berkeley, Los Angeles and London: University of California Press.
- Patay, P. 1978. *Das kupferzeitliche Gräberfeld von Tiszavalk-Kenderföld.* Budapest: Akadémiai Kiadó.
- Patay, P. 1984. *Die kupferzeitliche Meissel, Beile und Äxte in Ungarn.* Munich: C.H. Bech.
- Redmond, E. 1994. *Tribal and Chiefly Warfare in South America.* Ann Arbor: University of Michigan.
- Robb, J. 1997. 'Violence and gender in early Italy'. In: D.L. Martin and D.W. Frayer (eds.): *Troubled Times: Violence and Warfare in the Past.* Langhorne: Gordon and Breach Publishers, pp. 111-44.
- Runciman, W.G. 1999. 'Greek hoplites, warrior culture, and indirect bias'. *The Journal of the Royal Anthropological Institue. (N.S.),* 4, pp. 731-51.
- Sanders, A. 1999. 'Warriors, Anthropology of'. In: *Encyclopedia of Violence, Peace and Conflict.* Vol. III. London: Academic Press, pp. 773-84.
- Schweizer, T. 1990. 'Männerbünde und ihr kultureller Kontext in weltweiten interkulturellen Vergleich'. In: G. Völger and K. von Welck (eds.): *Männerbande-Männerbünde.* (Ethnologica, Neue Folge, 15). Cologne: Joest-Museum der Stadt Köln, pp. 23-30.
- Sherratt, A. 1981. 'Plough and pastoralism: aspects of the secondary products revolution'. In: I. Hodder, G. Isaak and N. Hammond (eds.): *Patterns of the Past: Studies in Honour of David Clarke.* London: Cambridge University Press, pp. 261-304
- Sherratt, A. 1984. 'Social evolution: Europe in the later Neolithic and Copper Ages'. In: J. Bintliff (ed): *European Social Evolution. Archaeological Perspectives.* Sussex: Department of Archaeology, University of Bradford, pp. 123-34.
- Sherratt, A. 1994a. 'Core, periphery and margin: perspectives on the bronze Age'. In: S. Stoddart and C. Mathew (eds.): *Developments and Decline in the Mediterranean Bronze Age.* Sheffield: Sheffield Academic Press, pp. 335-46.
- Sherratt, A. 1994b. 'What would a Bronze Age world system look like? Relations between temperate Europe and the Mediterranean in later prehistory'. *Journal of European Archaeology,* 1(2), pp. 1-58.
- Simonsen J. 1987. 'Settlements from the Single Grave culture in North-West Jutland. A preliminary survey'. *Journal of Danish Archaeology,* 5, 1986, pp. 135-51.
- Steuer, H. 1982. *Frühgeschichtlichen Sozialstrukturen in Mitteleuropa. Eine Analyse der Auswertungsmethoden des archäologischen Quellenmaterials.* (Abhandlungen der Akademie der Wissenschaften in Göttingen, Phil.-Hist. Kl., 3. Folge, 128). Göttingen.

- Sofaer-Derevenski, J. 2000. 'Rings of life: the role of early metalwork inmediating gendered life course'. *World Archaeology,* 31(3), pp. 389-406.
- Tacitus, P.C. 1948 (AD 97-98). *On Britain and Germany.* (A translation of the 'Agricola' and the 'Germania' by H. Mattingly). Harmondsworth: Penguin.
- Thorpe, I.J.N. 2000. 'Origins of war. Mesolithic conflict in Europe'. *British Archaeology,* (52), pp. 9-13.
- Torodova, H. 1981. *Die kupferzeitlichen Äxte und Beile in Bulgarien.* Munich: C.H. Bech.
- Treherne, P. 1995. 'The warrior's beauty: The masculine body and self-identity in Bronze-Age Europe'. *Journal of European Archaeology,* 3(1), pp. 105-44.
- Turek, J. and Cerný, V. 2001. 'Society. Gender and sexual dimorphism of the Corded Ware and the Bell Beaker populations'. In: F. Nicolis (ed.): *Bell Beakers Today: Pottery, People, Culture and Symbols in Prehistoric Europe.* (Procceedings of the international Colloquium, Riva del Garda. Trent, Italy, 11-16 May 1998). Trent: Servizio Beni Culturali, Ufficio Beni Archeologici, pp. 601-12.
- Turek, J. and J. Peska 2001. 'Bell Beaker settlement pattern in Bohemia and Moravia'. In: F. Nicolis (ed.): *Bell Beakers Today: Pottery, People, Culture and Symbols in Prehistoric Europe.* (Procceedings of the international Colloquium, Riva del Garda. Trent, Italy, 11-16 May 1998). Trent: Servizio Beni Culturali, Ufficio Beni Archeologici, pp. 411-28.
- Tymowski, M. 1981. 'The army and the formation of the states of west Africa in the nineteenth century: The cases of Kenedugu and the Samori State'. In: H.J.M. Claessen and P. Skalnik (eds.): *The Study of the State.* (New Babylon Studies in the Social Sciences, 35). The Hague: Mouton, pp. 427-42.
- Vandkilde, H. 1999. 'Social distinction and ethnic reconstruction in the earliest Danish Bronze Age'. In: C. Clausing and M. Egg (eds.): *Eliten in der Bronzezeit. Ergebnisse zweier Kolloquien in Mainz und Athen.* (Monographien des Römisch-Germanischen Zentralmuseums, 43). Mainz: Verlag des Römisch-Germanischen Zentralmuseums, pp. 245-76.
- Vandkilde, H. 2003. 'Commemorative tales: archaeological responses to modern myth, politics, and war'. *World Archaeology,* 35(1), pp. 126-44.
- Vencl, S. 1999. 'Stone Age warfare'. In: J. Carman and A. Harding (eds.): *Ancient Warfare.* Stroud: Sutton Publishing, pp. 57-72.
- Völger, G. and Welck, K. von (eds.) 1990. *Männerbande-Männerbünde.* (Ethnologica, Neue Folge, 15). Cologne: Joest-Museum der Stadt Köln.
- Wahl, J. and König, H. 1987. 'Antropologisch-traumalogische Untersuchung der menschlichen Skeletreste aus dem bandkeramischen Massengrab bei Talheim, Kreis Heilbronn'. *Fundberichte aus Baden-Württemberg,* 12, pp. 65-193.
- Whitley, J. 1995. 'Tomb cult and hero cult. The uses of the past in Archaic Greece'. In: N. Spencer (ed.): *Time, Tradition and Society in Greek Archaeology. Bridging the 'Great Divide'.* London and New York: Routledge.

- Wiermann, R.R. 1998. 'An anthropological approach to burial customs of the Corded Ware culture in Bohemia'. In: M. Benz and S. van Willingen (eds.): *Some Approaches to the the Bell Beaker 'Phenomenon'. Lost Paradise…? Proceedings of the 2nd Meeting of the ' Association Archéologie et Gobelets' Feldberg (Germany), 18th – 20th April 1997.* (BAR International Series 690), pp. 129-40.
- Wiermann, R.R. 2002. 'Zur Sozialstruktur der Kultur mit Schnurkeramik in Böhmen'. In: J. Müller (ed.): *Vom Endneolithikum zur Frühbronzezeit: Muster sozialen Wandels? (Tagung Bamberg 14.-16. Juni 2001).* (Universitätsforschungen zur prähistorischen Archäologie, 90). Bonn: Dr. Rudolf Habelt GmbH, pp. 115-31.
- Wilderotter, H. 1990. 'Prairiefüchse, Schwarze Münder, Hunde-Soldaten und irre Hunde; Militär-Gesellschaften der Plains und Prairie-Indianer'. In: G. Völger and K. von Welck (eds.): *Männerbande-Männerbünde.* (Ethnologica, Neue Folge, 15). Cologne: Joest-Museum der Stadt Köln, pp. 237-42.
- Zapotocký, M. 1991. 'Frühe Streitaxtkulturen in Mitteleuropäischen Äneolitikum'. In: J. Lichardus (ed.): *Die Kupferzeit als historische Epoche, Teil 1-2.* (Saarbrücher Beiträge zur Altertumskunde, Vol. 55). Bonn: Dr. Rudolf Habelt, pp. 68ff.

From Gilgamesh to Terminator:
The Warrior as Masculine Ideal
– Historical and Contemporary Perspectives

We few, we happy few, we band of brothers; For he today that sheds his blood with me
Shall be my brother; be he ne'er so vile. This day shall gentle his condition: And gentle-
men of England, now abed shall think themselves accursed they were not here, And hold
their manhoods cheap whiles any speaks That fought with us upon Saint Crispin's day.
King Henry V, Shakespeare (In: Ellis 1990: 7)

There's one thing you men can say when it's all over and you are home once more.
You can thank God that twenty years from now when you're sitting by the fireside with
your grandson on your knee, and he asks you what you did in the war, you won't have
to shift him to the other knee, cough and say, 'I shovelled crap in Louisiana'.
General Patton in 1944 (In: Ellis 1990: 7)

/27

SANIMIR RESIC

These two powerful and seemingly timeless quota- tions by Shakespeare and General Patton, separated by more than three centuries, truly illustrate the eternal summons to men, to participate when real men are moulded. Thus, the warrior mentality is deeply rooted in manhood. Moreover, the indoctri- nating message in these quotations, which histori- cally are very common both geographically and cul- turally, and in all shapes and forms, constitute a strong reminder to young men that surely nothing in the world will make them pass the test of man- hood like battle. To pass the so-called baptism of fire, has, in other words, been equal to passing the ultimate male *rite de passage*. Similarly, even to die bravely in combat has historically in many cultures promised the warrior-man eternal after-life, a place in *Elysium*, the happy hunting grounds or *Valhalla*.

Indeed, as the American Professor of War litera- ture, Samuel Hynes, correctly underlines, even if mes- sages such as the initial quotations by Shakespeare and General Patton are less explicit in late 20th century war narratives, the test of courage and man- liness, especially for young men, has continued to loom romantically 'beyond everything else that life is likely to offer them' (Hynes 1997: 137). It is hard not to agree with the English Professor of gender- studies, Jonathan Mangan's conclusion that the

Warrior as male hero has been a central and contin- uous icon in human history (Mangan 1999: 1).

It is similarly hard not to agree with the American historian Donald D.J. Mrozek when he argues that the need for defence has exaggerated behavioural differences between men and women, and thus given military institutions a special importance in preserving the distinctive sphere of male virtues. The military and soldiering have traditionally been linked to the ancient roles of hunter and defender, and have also created a relatively easy space in which to identify manliness (Mrozek 1987: 220). This is, to the best of our knowledge, in the end a gender construction.

Thus, American 20th century recruiting slogans like 'The Marine Corps Builds Men' and 'Join the Army and Feel Like a Man' are just two examples of military organisations taking advantage of this tendency, or rather this powerful construction. And, as the American Professor of Vietnam War literature, Milton J. Bates, argues, these slogans in a way reveal much of the masculine construction. After all, if the military promise to 'build men', it also confirms that 'one is not born, but rather, becomes a man.' To become a soldier is consequently to become a man, and the pre-military man has not yet acquired true masculinity (Bates 1996: 140-41). Overall, this chapter aims to explore, in a rather summary mode, the longevity in the human value of warriorhood despite the horrendous consequences of every war.

Celebrations of warriorhood in prehistory and ethnography

Looking back on human history and civilisation, it is literally filled with warrior-epics which corroborate the previous quotations by Shakespeare and General Patton. Already four thousand years ago, the first written warrior epic, the Sumerian *Gilgamesh* (George 1999: passim) story set a standard for future warrior stories and epics to follow. Homer's *Iliad* and *Odyssey*, written in the 8th century BC, had most likely been inspired by *Gilgamesh*, and the Homeric epics have made an impact on ideal warriorhood which has been hard to challenge (Homer 1997: lv). In the 11th century AD, in the oldest of the French epic *Chansons de Geste* (songs of deeds), the *Chanson de Roland* illustrates, through the Frankish Achilles, Roland, the model of Christian chivalry. This model was subsequently refined by Sir Thomas Malory in *Le Morte d´Arthur*, written in the mid-15th century. The striking thing is how Malory's opus, which elevates the warrior ideals and the chivalric codex to a climax, was written by a failed and drunken knight while he was in jail for *inter alia* rioting, looting and pillaging a monastery, and raping women. In other words, these stories of ideal warriorhood and manliness not infrequently hide the more common derelictions performed by the warrior-man.

Considering the various categories of warrior values we detect an overwhelming historical, cultural, and geographical universality in praising these features: honour, loyalty, duty, obedience, endurance, strength, sexual potency, courage, and camaraderie, the last best considered as the nucleus of the typical *Männerbund*. Of greatest importance is that all of these warrior values are traditionally also synonymous to manliness. Furthermore, in quite a few cultures, sexual performance has been, and most often still is, linked to some of the mentioned categories. Among the Maori warriors of New Zealand the word for bravery, 'toa', was synonymous with being sexually aroused. For many so-called 'tribal peoples' and cultures of hunters and gatherers, an erect penis symbolises aggressiveness and courage. Even in many current societies we find links between genitals, manliness, aggressiveness and courage, which

is adequately verified in the common expression to *have balls* which means to be courageous. The lack of them is synonymous with cowardice. Also, metaphorically speaking, to kill an enemy is to 'fuck' him (Duerr 1998: 192-94).

At least since the ancient Egyptians, cowardice in battle by warriors or their enemies has earned them epithets of female genitalia to indicate that they are 'weak' men. This tendency, the synonymity between the words *woman* and *coward*, has also diligently been used by drill instructors in modern armies while turning young men into soldiers. To this very day in Sweden, the land of gender equality, schoolteachers grapple with the problem of young boys disparagingly referring to weak boys and girls using words for female genitalia or homosexuality.

Long-term trends in the commemoration of warriorhood

By and large, what we see here is a very slow-to-change mentality and a long-term trend. Or what the French *Annales*-historian Fernand Braudel called a *longue durée*. The warrior mentality has in other words a rather timeless character. Even following periods of seeming unpopularity, as after the brutal wars of religion in the 16th and 17th centuries and during the Enlightenment, warrior values have had an extraordinary tendency to be resuscitated and experience a renaissance.

Returning to warrior values, it is hard to exaggerate their long-term popularity. And as I previously mentioned, even after periods of lesser importance, or rather an unpopularity of warriorhood, they show a cyclical tendency to be celebrated as manly ideals and virtues. Perhaps this is also a consequence of the popularity that these values generally have among human beings. For instance, both courage and honour have an intimate relationship to human sensibilities and virtues. Thus, in the Catholic faith courage is still considered one of the cardinal virtues. According to Plato, courage, together with justice, wisdom and moderation, is one of the main virtues. But, to Plato, the act of true courage is a wise courage, it is an act which follows rational deliberation. It is when a value, such as courage is abused, brutalised, and vulgarised that it becomes a problem. After all, atrocities committed by soldiers in the name of warrior values are more common than not, which Sir Thomas Malory perfectly exemplified.

Masculinities of the 18th-19th centuries

To understand warriorhood and masculinity in our own age we must pay some attention to the development in warfare and the emergence of what the historian George L. Mosse called the 'Myth of the War Experience', in the late 18th and early 19th century. Mosse turns to the French Revolution and asks why men unprecedentedly and eagerly rushed off *en masse* to face death in battle? In France in the decades prior to the Revolution, the martial spirit of the army, containing a mix of career officers, mercenaries, and militiamen, was declining. So what made young men, who would not 'brave danger and pain' before the first citizen army was created in 1792 willing do so afterward (Mosse 1991: 4-16)?

Around the mid-19th century, we also see a close to universal bond between warrior values and conventional notions of masculinity being revived. This revival was a renewal of medievalism augmented not least by Romanticism and Social Darwinist ideas, which largely re-established the status of the warrior in Western civilisation. The somewhat devalued image of warriorhood and martial values of the 18th century lost ground and was replaced by the age of legends and heroes, Teutonic warrior cult, Arthurian knighthood fantasies, warrior romances of Alexandre Dumas and Walter Scott, and Wagnerian operas. This is a paradoxical development in European history between 1815 and 1914: a relatively peaceful century saw the creation of the probably strongest warrior society the world had ever known. On the eve of World War I, almost every able bodied European man of military age had a soldier's identity card and was prepared for military duty and general mobilisation at any time (Resic 1999: 20).

Mosse suggests how a Myth of the War Experience was built up during the Wars of the French Revolution (1792-1799) and the German Wars of Liberation against Napoleon (1813-1814) in order to entice young men to war. There was within the myth a re-emergence and widespread praise of war, warrior values, and the warrior-man. Previous wars had mainly been fought by mercenary armies with little interest in the cause for which they fought. The citizen-armies, initially composed of large numbers of volunteers, fought with a commitment to their cause and nation. This new breed of soldiers was fed with the myth, which created a nationalistic glorification and romanticisation of war and a celebration of the warrior-man that culminated

with the generation of 1914. But the myth gradually also worked as a way to make the realities of war and battle easier to bear. The general evolution in warfare in the 19th century had, through the industrial revolution and total war, given new dimensions to death in war. Battlefields had become deadlier places than ever, so it called for much greater effort in masking and transcending deaths in war.

The World Wars and warriorhood

World War I, according to Mosse, further strengthened the emphasis and focus of militant masculinity, which also reached a climax in the search for a 'new man' that started before the war. By the early 20th century throughout the Western world, the ideal was physical, aesthetic and moral. Physical strength and courage was combined in the harmonious proportions of the body and the purity of soul. Many men belonging to the 1914 generation were motivated to volunteer because of the ideal of manliness. The soldier *per se* was praised as the true representative of the people, and admired for his physical and moral strength, common sense, and matter-of-fact courage (Mosse 1991: 222-23).

During World War II the idealised manly qualities of World War I were again demanded of young men. However, the fascist era had taken the concept even further, where the man of the future, the 'real man', besides being of fascist ideology, was truly soldierly, courageous and full of willpower. This was strongly connected to clean-cut appearance, hardiness, and self-discipline. Even for men lacking the extremes of the warrior ideal, this definition was imperative. In both the Italian and German concept of the 'new man', he was to emerge from the war and always be linked to the war experience. Clearly the history of militant masculinity reached a climax with fascism, which itself was a product of the Great War, and to all fascist regimes manliness was a vital national symbol. The 19th century ideal of masculinity was somewhat altered by the fascists to a glorification and vulgarised image of the World War I experience. Service and sacrifice for the nation was the true manly goal. To the fascists, there was no difference between a real man and a warrior, a warrior in love with war and combat.

In subsequent wars of the 20th century, according to Mosse and the majority of the scholars of the field, much of the powerful Myth of the War

Experience and warrior ideals were largely rejected and devalued. Supposedly the post-World War II era has not added anything to encourage the Myth of the War Experience. The vision of Armageddon conjured up in 1945, symbolically speaking after *Auschwitz* and *Hiroshima*, has led to a 'more realistic' attitude, one without national sentiments and stories of war as glorious. Martial enthusiasm and militant masculinity has *not* been resurrected (Mosse 1991). A highly esteemed historian like John Keegan even maintains that in the United States by the 1960s warrior values were extremely hard to impress upon Americans (Keegan 1993: 49-50). After the two world wars, Samuel Hynes argues, the romantic images of war 'fell out of fashion' (Hynes 1997: 41).

Vietnam and the recreation of war mythology

As we approach the Vietnam War, this was clearly not the case in the United States, as I have showed in my book *American Warriors in Vietnam: Warrior Values and the Myth of the War Experience During the Vietnam War, 1965-73* (1999). Studying combat soldiers' letters and diaries, written during the troops' active duty in Vietnam, which I subsequently compared with veterans' memoirs and interviews, it appeared that this generation of American soldiers had no lack of romantic masculine images of war and warriorhood. 1945 had hardly been that complete turning point in the United States. Rather, the American combat soldiers' conception of warrior values confirmed the long-term celebration of warriorhood during the Vietnam War. Primarily the early war phase volunteers showed us that the Myth of the War Experience was hardly a thing of the past in 1965. Indeed, the Vietnam War was no 1861, 1917 or 1941 and few of the 1960s' generation as a whole harboured a sense of crusade with regard to the war, but we still find that extolling warrior values was an important mechanism in this conflict.

Obviously those who actually did volunteer and who deliberately sought the male confirmation in the fight, and not the draftees and those avoiding military service and battle, are crucial to our understanding of the enchantment of warriorhood. The volunteers from 1965-1968 rather wanted to experience 'their' *Gettysburg* and 'their' *Normandy*. They wanted to repeat their fathers' deeds of the so-called *Good War* in the 1940s.

Throughout my study I could see differences in conception between those who had volunteered and those who had been drafted, especially as the war dragged on. This observation is not surprising: on the contrary, it is almost expected. Indeed, volunteers, in contrast to draftees, have traditionally been the group of men most susceptible to the Myth of the War Experience and in turn most inclined to choose to walk the warrior's path. It's always the young men who prove themselves susceptible to the glory, thrills and values of warriorhood who confirm the long-term trend of the worth of the warrior, war and combat experience. Ultimately these young men perpetuate the existence of the warrior in human society.

Many a young man has initially entered war believing in the worth of the warrior as a vital part in the construction of ideal masculinity, while subsequently with experience realising the inconsistency between ideal and the reality in war and battle. The American combat soldiers who volunteered for the Vietnam War were no exception to this. But as they saw their pre-experience expectations of warriorhood being crushed in Vietnam and in the United States, they now made an American exception. As the war progressed and the national project was shattered, much of the praising of warrior values lost its power. Admittedly the complete Vietnam experience has contributed greatly to further revealing the Myth of the War Experience, and has caused large cracks in the ideal construction of manliness in the guise of the warrior. But already during the war, even among peace-activists, the Vietnamese adversary's warrior abilities were honoured, which again proves the slow-to-change trend in praising the warrior. The peace movement often turned out be an against-the-war in Vietnam movement and anti-American involvement rather than anti-war as a general principle.

The American warrior had in the end lost the war to the opposite of his stereotype of manliness, which clearly was a blow to the traditional dominant image of masculinity. American men in general seemed to have experienced emasculation in large part because of the Vietnam War. After all, it became an experience of national trauma, which even some active anti-war men of the generation expressed in the aftermath. They had missed their opportunity to partake in the traditional and mythical male American experience, to have fought and won a war.

In any case, the indications that American manhood craved 'remasculinisation' during the post-Vietnam era appear to find overwhelming corroboration throughout American culture. Perhaps the most obvious testimony to this is found in the Rambo movies. In many veterans' accounts, their emphasis indicate that regardless of the outcome of the war, they had actually at the time believed that they were doing the manly thing by fighting. The veterans continuously played an important role in the project of remasculinisation. From being scapegoats and carriers of national shame and guilt, they slowly gained a profound role as 'spokespersons' for the remasculinisation of American culture and society.

As with the veterans of other conflicts, in many ways the Vietnam veterans, especially those who had volunteered early on, highlighted in their post-war accounts that despite the legacy of the war, they had done the manly thing. The tendency is comparable to the World War I experience, and is seemingly a universal trend among many veterans of war; a trend that as time goes by gains further momentum and strength. Perhaps the Vietnam veterans are particularly interesting in the light of this phenomenon, not least when juxtaposing the alternative masculinities that emerged during and since the war with the previously so dominant manly ideal in the form of the warrior. After all, the construction of ideal manliness in the shape of the warrior had successfully been challenged during the Vietnam War. Therefore in the final analysis, the Vietnam veterans' accounts are truly significant messages. Many veterans endeavoured to separate the bad Vietnam experience *per se* from the worth of the masculine warrior values. Through this effort of salvaging, they contribute to the reconfirmation of the long-term belief in warring as a vital male enterprise.

Contemporary trends

An interesting tendency, which becomes apparent when looking at the last two centuries, is how the worth of warriorhood and warrior values are strengthened during times when men either feel or display a sense of crisis in manhood. This phenomenon is often and partly explained by women's stronger position in society and a society in transformation. We find such a situation in the late 19th and early 20th century in both Western Europe and the United States, as well as in post-Vietnam America. In other words, when men seem to see their positions threatened, and experience a crisis in male identification, there is an increase of longing for the warrior. I believe that we currently have a similar situation.

Despite the obvious weakening of classical militarism in today's Western countries, we hardly lack military values and trends both from a domestic and a global perspective. Perhaps the so-called 'post-military society' has emerged from a continuing civilising process. But as numerous commentators have underlined, new forms of warriors, wars and a still fairly strong militarisation of the world, combined with an extreme excitement and fascination with the powers of military technology, render a far more complicated, multiple, and versatile reality. We see a proliferation of weapons of mass destruction in the hands of autocratic regimes, traditional guerrillas in Mexico or Kosovo applying the internet as a weapon, the creation of the 'post-modern guerrilla hero', and moves towards the creation of the cyborg warrior. The latter would be a human integrated with a machine, not unlike Arnold Schwarzenegger in the Hollywood blockbuster *Terminator*.

Even in the strongly masculine business/corporate world, not least in Japan and America, warrior values both in language, metaphors and company strategies has been transferred and adopted. Many companies are structured along military lines, mimicking army hierarchy and authority. In general there is a gradual and continual evolution of what could be called techno-scientific post-modern war. Correspondingly the horrible images broadcast to us on TV from conflicts around the world offer few romantic messages of war, but we get used to the images and grow numb: the horrors and the familiar patterns no longer shock us.

Admittedly, the warrior has in many ways gone underground, but it is still easy to spot him in his contemporary form. Take, for instance, middle class professionals spending weekends in woods playing war games with paint guns, family fathers re-enacting famous historic battles, and youth and city gangs organised as paramilitary units based on crude military values.

In 1991, there were an estimated 100,000 gang members organised in 1,000 different gangs in Los Angeles County alone, modelled as warrior societies in which violence and killings are rites to manhood.

Gangs committed a third of all the homicides in Los Angeles County (i.e., over 700 murders) that same year (Gibson 1995: 305; Moore and Gillette 1991: 5). One scholar even explains much of the daily violence inflicted by young men in Western societies as simply another way to prove their manliness, and street violence then serves as a surrogate for war (Gerzon 1992: 174). On the whole, this is a rapidly growing phenomenon in European cities as well, not least among marginalised young men, often with an immigrant background in large urban centres. This is a development which not even 'peaceful' Sweden has escaped. The 1999 Governmental Committee for the Evaluation of the State of Contemporary Swedish Democracy identified a troubling tendency among young immigrant boys – hostility towards Swedes in general, and the Swedish establishment in particular. These young men see themselves as 'warriors' fighting against a society which has not been able to integrate them into civil society.

Throughout the Western world today we also see Fascist movements re-emerging and mesmerising in particular unemployed young men with messages of violence. In the United States, paramilitary right-wing extremists bombed the Federal Building in Oklahoma City in 1995 with 168 innocent individuals being killed. The responsible and convicted bomber, Timothy McVeigh, had been decorated for his service in the Army during the war in Kuwait and dreamed of becoming a Green Beret, the ultimate American warrior. There was also the recent sinister tendency among armed schoolboys in America to take out their anger by shooting fellow students in pure massacres. In the shooting at Thurston High School in Springfield, Oregon, in 1998, the teenage killer was even equipped and dressed for war.

The American sociologist James W. Gibson points to the disturbing emergence of a powerful paramilitary culture of war and warriors in the 1980s, which he calls a 'New War culture'. In the decades following the Vietnam War millions of American men have purchased combat weapons and begun to train for wars. Most of these men missed out on the Vietnam War and wish to prove their masculinity and rightful place in the long tradition and mythology of aggressive American males. New indoor pistol-shooting ranges have expanded, and men dressed in camouflage clothes all over the United States are playing survival games and paintball wars. The new warrior hero as presented and portrayed by the New War culture is a lone fighter or part of a small group of elite warriors, and the warrior is presented as an ideal identity for all men disregarding occupation. Within the phenomenon of New War a number of racist paramilitary groups appeared, which in the wake of the failure of the Vietnam War consider the white man's world in grave jeopardy. These groups of white supremacist New War warriors believe they are morally permitted to do battle even beyond the law (Gibson 1995: 8-9, 195-96, 268-69, 284, 294).

Looking at Hollywood, we have seen an increase in the production of warrior nostalgia movies. *The 13th Warrior, Gettysburg, Gladiator, Braveheart, We Were Soldiers Once*, even *Saving Private Ryan*, are only a few examples. Interestingly enough, even if some of these new films present an anti-war message, many young men still manage to interpret and find the promise of an exclusively male adventure offered by warriorhood in these films. After seeing *Saving Private Ryan*, while leaving the cinema in Malmö Sweden, I listened to the reaction of young men after this cinematographically powerful depiction of the senseless carnage of modern battle. Not surprisingly, most comments of the young men carried the message that this film had the coolest and best combat-scenes they had ever seen in a movie. Similar comments followed the very powerful and extremely tragic *Gladiator*.

Perhaps it is impossible to produce a really off-putting war-movie, which truly manages to induce anti-war feelings. After all, the eternal message remains obvious, i.e., despite the horrors, disasters and carnage of war and battle, the camaraderie among brothers in arms is best found in war and has no other true competition. This is a message to which many young men are very susceptible.

Revival of warriorhood and violent masculinities in the Balkans

Looking at the Balkans, warrior values linked to manliness are hard to exaggerate. In fact, I am convinced that in order to understand the wars in the 1990s the construction of ideal masculinity in the Yugoslav successor states is absolutely necessary. There is a most powerful machismo among many men in this part of Europe. Moreover, in the national historiography, legends and myths of Southeast Europe male heroes and warriors are found in abundance, just as the status of war-veterans and heroes of World War II during socialist Yugoslavia is hard to exaggerate. By and large, the complete Yugoslav system was based on a profound celebration of military and warrior values. Likewise, the region and beyond the old *Vojna Krajina* or *Militärgrenze* between the Habsburg and Ottoman empires, where much fighting in the 1990s took place, has harboured a warrior masculine ideal for centuries. It is a long-term trend and culture that had its importance when turning the ignition-key for war in 1991. In his research the Croatian historian Ivo Žanić has convincingly established the important historical and mythical *Hajduk* tradition linked by *guslars* (bards) from the days of resistance *vis-à-vis* the Ottomans up to the war in the 1990s (Žanić 1998).

Even in Sweden, the UN missions have been, and still are, a perfect excuse for young Swedish men to seek the male warrior adventure. Many young Swedish UN peace-keepers repeatedly apply for duties in the Balkans. Some UN veterans have done three or four tours of duty in Bosnia and Herzegovina and in Kosovo. This does not include the Swedish mercenaries who fought in the war in the 90s. The infamous neo-nazi and war criminal Jackie Arklöf set a perfect and frightening example. After returning from his mercenary 'combat-tour' fighting Muslims in Bosnia and Herzegovina, he gunned down and murdered two Swedish police officers. He is currently serving a life-sentence in prison where he laments the fact of not being able to live the life of the warrior (Hildebrandt 2000: 74-83).

There are also Croatian-Swedish war-veteran clubs, where much of the war culture is recycled. A Croatian war-veteran and the organiser of one such veteran club, whom I interviewed in Sweden last year, stressed that he had not volunteered for military duty and combat, claiming to have been unwillingly shanghaied from his town by the Croatian Army. But while he was organising the veteran club in Sweden, he made a distinction between those who had been 'ranjeni' (wounded) and those who had not been in combat. In other words, even this forced draftee placed a specific honour on the fact that he had been bloodied and 'baptised' in combat. As we talked, while listening to the lamenting voice of musician and war-veteran Thompson, the veteran I interviewed pointed at another veteran and said: 'You know, he was just a cook, he never had to prove his manhood.'

Similarly, according to the interviews carried out by the German scholar Natalija Bašić, many Serbian veterans consider themselves 'better men' after experiencing combat and due to their war-experience. To most of these veterans, Bašić emphasises, the combat experience in Croatia and BiH is regarded as their male *rite de passage* (Bašić 1999). The Serbian ethnologist Ivan Čolović has also showed how, for instance, football hooligan organisations in Belgrade were, before the outbreak of the war, in fact a training ground for vulgar and macho paramilitary groups, which were later to be unleashed in Croatia and Bosnia (Čolović 2000: 371-96).

It seems logical to link much of this behaviour to early education and schooling in Serbia. According to the Serbian Sociologist Vesna Pešić, throughout their eight years of elementary school, children are methodically socialised into worshiping warriorhood and male heroes. As the students advance through eight years of a belligerent syllabus, by the age of 14 they know perfectly well who the enemies are, and that the brave die only once while the coward dies a thousand times (Pešić 1994: 55-75).

Sport, violence, and manliness

Warrior values as part of manliness are also very much alive in the world of sports and so-called martial arts. Perhaps nowhere are these values as conspicuously expressed as among athletes in contemporary society, and in popular imagination sports still reflect what have been regarded as manly virtues. Physical exercise in general has had a historic importance to the construction of modern masculinity. The Australian scholar R.W. Connell calls attention to the development of team sports at the end of the 19th century across the English-speaking world as a heavily convention-bound arena. As a test of masculinity, sports were even in some cases developed as a deliberate political strategy. It was a conscious construction and had a part in turning young schoolboys into tough men. Sports today, as Chris H. Gray maintains, are in many ways closer to ritual war than war. Mark Gerzon even considers sports a rite of passage to manliness in a Western world in the absence of war (Gray 1997: 106; Connell 1996: 30-37; Gerzon 1992: 173-74).

Likewise, most spectators still express nationalism, symbolically using facial war paint in the national colors while demanding that the athletes deliver victories in a warrior-like fashion. Similarly there is an enormous increase in adventure sports, where young men, and women, dare mountains, rivers, climates etc. in order to test their courage. This has even turned into a worldwide billion-dollar tourist industry.

Female fierceness

Recently we have seen the advent of new structures and attitudes, not least with the current debate about the wisdom of creating 'gender-blind' combat units. This phenomenon is interestingly enough a part of the debate for 'gender equality' in the United States and most Western countries, and is focused on women in the Armed Forces as being frustrated by not being allowed in combat units. Already in 1988 the U.S. Marine Corps opened its door to the 'Basic Warrior's Training Course' for women. By the late 1990s, 12% of both the U.S. and British Army were women. Women soldiers cannot be shrugged off simply as auxiliaries to fighting men in an age when centuries-long military tradition and assumptions about the roles of women and men are in a state of change.

This so-called feminisation of the military and militarisation of the feminine have implications for soldiers as well as the civilian world. In other words, can soldiering be the ultimate seal of manhood if women do it too? Will women still be routinely considered as the second sex if they can face the mental and physical challenge of battle and, with so-called male courage, sacrifice themselves for their country? What will this ultimately mean for the predominant image of manhood, since, as R.W. Connell correctly maintains, masculinity is, no matter what, defined as 'not-femininity' (Connell 1996: 70).

Indeed, in the 1991 Gulf War, the military gender-structure displayed a somewhat new face. Over 34,000 women fought in this conflict, and eleven were sent home in body-bags. We could see mothers leaving husband and children for service in the war, or soldiers getting pregnant and being sent home from the War, or women troops being captured and killed. One American female soldier in Saudi Arabia maintained 'There aren't any men or women here, just soldiers' (Gray 1997: 175). However, the matter is even more complex. Although the post-modern soldier/warrior has become closer to the cyborg and gender identity more blurred, military cyborgs are still rather masculine (Gray 1997: 42-43, 175).

Concluding remarks

The high-tech nature of contemporary warring and soldiering, with its masculine predominance, has transferred much of the traditional male emphasis of physical force to the mechanical cyborg soldier giving it a masculine high-tech dominance. This 'new male' version of the warrior is easier for women to adopt. We have seen the advent of a soldier type constructed as basically male, vaguely female and vaguely mechanical, but as a whole a masculine cyborg postmodern warrior. A terminator of whom, unfortunately, we have not yet seen the last.

R.W. Connell stresses that it should be 'abundantly clear' that the history of masculinity is not linear. (Connell 1996: 198) Accordingly, as we have seen from the present study, not least in comparison and in corroboration to other similar surveys, the praising of warriorhood, its values and its ideals has a historical cyclic tendency among men as a vital part of the construction of masculinity. At times war romantics and warriorhood are presented and appear almost as being part of a human or male destiny. Perhaps there is a true need for romantic images in life, and perhaps we need to know, or at least believe, that the world has some adventure to offer, especially when we are young. Indeed the call to war and warriorhood has always been powerful among young men. During the Middle Ages the evidence for this phenomenon is found in profusion as the youths formed the spearhead of feudal aggression as knights. It was a matter of honour and reward that could be gained, but also a lust for adventure. From the violent and aggressive knightly society to the 'New War Culture' in the USA, we still see a predominance of young men. It goes thus without saying that historically and traditionally young men seem to have craved adventure, thrill, fame, power and respect, and it has most easily been provided through warriorhood. Ultimately then, no matter what reasons, constructions or factors, cultural, social, or biological for that matter, young men have been and continue to be those most susceptible to the myths of war and warriorhood.

After the setbacks of the masculine warrior ideals due to the Vietnam experience, warriorhood has reappeared and taken many new guises in contemporary society. Indeed as we could see above, there is scarcely a lack of warrior nostalgia, aiming at re-accommodating the spirit of the warrior. Even at the beginning of a new millennium, the path of the warrior in many ways appears as a way to escape boredom and the humdrum of everyday life, and the world has hardly turned into a peaceful place with no threatening conflicts on the horizon. Since the 1980s we have also seen an enormous industry skyrocketing in the United States of pulp fiction literature, novels, warrior magazines, comic books, military toys, and films which in one way or the other praise the warrior. Likewise, the sale of military weapons continues to grow. Hollywood, with its movies reaching a worldwide audience, is earning billions through its vulgar celebration of the warrior. Furthermore, we have even seen lately a growing tendency to let women into the world of the warrior. As I write both men and women in the guise of the warrior are getting ready to see the Elephant, i.e., meet their baptism of fire, in Iraq, and here we go again.

BIBLIOGRAPHY

- Bašić, N. 1999. 'Krieg ist nun mal Krieg'. *Mittelweg*, 36, pp. 3-19.
- Bates, M.J. 1996. *The Wars We Took to Vietnam: Cultural Conflicts and Storytelling.* Berkeley and Los Angeles: University of California Press.
- Čolović, I. 2000. 'Football, hooligans and war'. In: P. Nebojša (ed.): *The Road to War in Serbia.* Budapest: Central European University Press, pp. 373-98.
- Connell, R.W. 1996. *Masculinities.* Cambridge: Polity Press.
- Duerr, H.P. 1998. *Obscenitet och Våld: Myten om Civilisationsprocessen.* Vol. III. Stockholm: Brutus Östlings Bokförlag Symposion.
- Ellis, J. 1990. *World War II: The Sharp End.* London: Windrow and Green.
- George, A. (trans.) 1999. *The Epic of Gilgamesh.* London: Penguin Books.
- Gerzon, M. 1992. *A Choice of Heroes: Changing Face of American Manhood.* New York: Houghton and Mifflin.
- Gibson, J.W. 1995. *Warrior Dreams: Paramilitary Culture in Post-Vietnam America.* New York: Hill and Wang.
- Gray, C.H. 1997. *Postmodern War: The New Politics of Conflict.* London: Routledge.
- Hildebrandt, J. 2000. 'Jackies Krig'. *Magazine Café*, 9. Stockholm. September 2000, pp. 74-83.
- Homer 1997. *The Iliad.* S. Lombardo (trans.). Cambridge and Indianapolis: Hackett Publ. Co.
- Hynes, S. 1997. *The Soldiers' Tale: Bearing Witness to Modern War.* New York: Allan Lane, Penguin Press.
- Jeffords, S. 1989. *The Remasculinization of America: Gender and the Vietnam War.* Bloomington: Indiana University Press.
- Keegan, J. 1993. *A History of Warfare.* New York: Alfred A. Knopf.
- Mangan, J.A. 1999. *Shaping the Superman: Fascist Body as Political Icon.* London: Frank Cass.
- Moore, R.L. and Gillette, D. 1991. *King, Warrior, Magician, Lover: Rediscovering the Archetypes of the Mature Masculine.* San Francisco: Harper Collins.
- Mosse, G.L. 1991. *Fallen Soldiers: Reshaping the Memory of the World Wars.* Oxford: Oxford University Press.
- Mrozek, D.D.J. 1987. 'The habit of victory: the American military and the cult of manliness'. In: J.A. Mangan and J. Walwin (eds.): *Manliness and Morality: Middle-Class Masculinity in Britain and America 1800-1940.* Oxford: Oxford University Press.
- Pešić, V. 1994. 'Ratničke vrline u čitankama za školu', ('Warrior virtues in school books'). In: R. Rosandić and V. Pešić (eds.): *Ratništvo Patriotizam Patrijahalnost,* ('Warfare, Patriotism, Patriarchy'). Belgrade: Centar za antiratnu akciju, pp. 55-76.
- Resic, S. 1999. *American Warriors in Vietnam: Warrior Values and the Myth of the War Experience During the Vietnam War, 1965-73.* Lund: Lund University Press.
- Žanić, I. 1998. *Prevarena povijest,* ('Deceptive History'). Zagreb: Derieux.

The (Dis)Comfort of Conformism: Post-War Nationalism and Coping with Powerlessness in Croatian Villages

/28

STEF JANSEN

Village snapshots: failed rendezvous after violent displacement

Depending upon one's perspective, this article refers to villages in 'Krajina' or villages in the 'Formerly Occupied Territories of the Republic of Croatia'. What is certain is that they are located in an area between a main Croatian transit road and the new border with Bosnia Herzegovina. As an activist working with a dialogue project, I had access to people in a set of five such villages within one municipality of Croatia.[1] Walking into the villages in the late 1990s, visitors would first of all be struck by the contrast between Plavo, consisting entirely of newly built houses, and Bijelo and its surrounding villages, where the visible remains of material destruction were still shocking. In the latter inhabitants had only just begun to repair the ruins using the UNHCR plastic sheeting common to all post-war settings in the region. Not much economic activity was taking place apart from a timber mill and some subsistence agriculture, which was greatly impeded by several minefields. Further landmarks included a police station and a bar across the road from it mainly frequented by the numerous police officers, some remnants of destroyed Partisan monuments and an enormous Croatian flag on the central crossroads.

The relatively few inhabitants of these villages in the late 1990s were on the whole elderly and female. One thing that could *not* strike the visitor upon arrival would be signs of the national composition of the population: differences in this domain were neither visible nor audible. Diametrically opposed narratives of the past claimed either a historical Serbian or Croatian majority; but attempting to avoid the terror of national mathematics, I would argue that the area had been nationally mixed for centuries, with smaller villages often including large majorities of one or the other nationality.[2] Unsurprisingly, the region's recent history was subject to an intense struggle of representation. During WWII, a key moment in all versions of local history, the region was the scene of horrific violence, which pitted Croatian fascist Ustaše against multi-ethnic (but in this area mainly Serbian) communist-led Partisans. Massacres and starvation left few, if any, families intact, an enormous demographic and political legacy that later determined a good part of the power balance in Yugoslavia. Of the villages in question, Bijelo had been the main centre with a mixed but majority-Serbian population. Reflecting participation in the Partisan army, there had been a high degree of Party membership, with a similar pattern as in the smaller and predominantly

Serbian-inhabited villages of Sivo, Zeleno and Crno. Plavo, mainly Croatian-inhabited, had been known as a hard-core 'Ustaše village' and was therefore relatively deprived of state privileges.[3]

The villages in this study were at the heart of the post-Yugoslav conflict during the final decade of the twentieth century. In 1990, the Serbian nationalist revival ignited by Slobodan Milošević was countered in Croatia, where the first post-communist elections were won by the nationalist HDZ, led by Franjo Tudjman. Local and 'imported' Serbian hardliners engaged in provocations with the support of the locally based Yugoslav Army (JNA) division, and in previous 'black sheep' villages, such as Plavo, a wave of Croatian national euphoria gave way to a climate of revenge. The situation became extremely polarised and paramilitary groups carried out acts of violence against civilians on both sides. After the 1991 referendum, facing Milošević-supported Serbian rebellions against an alleged revival of Ustaša fascism, Croatia declared its independence. In response, 'Serbian Krajina' seceded from Croatia and almost all of its Croatian inhabitants were expelled in a collaborative operation by militant local Serbs, the JNA and volunteer militias from Serbia. Plavo was completely destroyed and the few elderly Croats who stayed put were killed. In the other villages, most Serbs remained in place during the four-year 'war republic' of Krajina and were joined by displaced Serbs from throughout the rest of Croatia. The *Oluja* offensive of August 1995 integrated the area into the Republic of Croatia – this time all Serbian inhabitants fled and their abandoned houses were looted and burned.

Hence, all villagers were displaced at some point during the 1990s. The scale of material destruction was enormous. A number of displaced Croats began returning in 1996, while a slow trickle of refugee-return commenced on the Serbian side in 1997, albeit consisting almost exclusively of elderly ladies, sometimes accompanied by their sick or disabled husbands. Many pre-war inhabitants simply never returned. Thus, when the fieldwork for this article was carried out, the national composition had changed dramatically as a result of war, refuge, relocation and ethnic engineering. A 'Yugoslav' identity was no longer viable and the former predominantly Serbian-inhabited villages were now housing a mixture of Serbian returnees, Croatian refugees from Bosnia, relocated Croats from other areas, a few

'mixed' couples and some others. The destruction of Yugoslav landmarks, exclusive economic policies and state assistance, excessive symbolism and a strong, aggressive, police presence left no doubt that this was now Croatian territory.

Living conditions were harsh, particularly for Serbian returnees, since their houses had not been repaired; many lived off subsistence agriculture, sometimes complemented by humanitarian aid. Employment opportunities, which were scarce even for Croats, were non-existent for Serbs. After *Oluja*, most Croatian Plavo households, whose houses Serbian forces had destroyed in 1991 and who had been displaced to other parts of Croatia or abroad, were granted a newly built house by the Croatian state. Most of them had at least one member employed or on a state pension. Only one of the many Serbs that used to live in Plavo had returned – he was married to a Croatian woman.

After a war that could be seen as a process of programmed national un-mixing (Duijzings 2000: 37-64), communication between people of different nationalities, most of whom had spent all their lives as neighbours,[4] was sparse, particularly in public. Where contact did exist, verbal harassment and abuse of Serbian returnees was common, particularly by the police, and there were a few cases of arson and rape. Those returnees, mostly elderly people, lived in fear and poverty and complained of isolation. Most of them emphatically distanced themselves from the militants who had proclaimed Krajina a separate Serbian republic in 1991. They saw their return as sufficient proof of their desire to coexist with Croats.

Most Croats refused to communicate with yesterday's enemies, and they were particularly angry about what they saw as the Serbian refusal to acknowledge what had happened. Pressure from the travelling Catholic priest, from local authorities and the police, from the mass media and from neighbours rendered any dialogue undesirable. In the dominant nationalist discourse of the day, Croatia was the exclusive national homeland of Croats. All others, it was argued, should know their place – Serbs, in particular, should not make any claims. The pattern of non-communication was only rarely broken by a few Croats, who said they understood the universal human need to return to one's birth place. A small minority even transcended greeting formalities, by helping out Serbian returnees with practical matters and sometimes socialising with them. For example, Nela, a young Croatian refugee from Bosnia with two small children, regularly had coffee with her Serbian neighbour in Bijelo, for which other Croats often criticised her. When I asked her about this she defensively snapped that as far as she was concerned it was the most normal thing in the world to have coffee with one's neighbours. She defiantly added that those who had a problem with that could 'go and fuck themselves'.

Another example was Davor, who had returned from Germany in 1991 to join the Croatian war effort. By some twist of fate, although he was from another part of Croatia, he ended up working in the Zeleno timber mill and found a friend in Nikola, an elderly Serb who lived nearby. It should be clear that communicating and certainly socialising with Serbs was the result of a conscious decision to break with the collectively sanctioned pattern of segregation. This may be one of the reasons why it was an easier step for relative outsiders to take.[5] Moreover, it was possibly less risky for Davor to socialise with Serbs because his war volunteer experience could counter suspicions of his being soft on Serbs and gave him the authority to be critical of the discourse of national liberation that he, after all, had embodied at the front.

Despite the presence of such exceptions, the picture arising from the above sketch is a bleak one, particularly in conjunction with the absolute political dominance of the nationalist HDZ in this region. Of course it should be noted that, in the early 1990s, this had been a solid and radical part of Serbian Krajina, cleansed violently of its Croatian inhabitants. For obvious reasons, during the research period in 1997-1998, Serbian returnees opted for a low profile and hard-liners either did not return or kept their heads down. This explains this study's emphasis on nationalism amongst Croatian villagers – a pattern that I consider strictly temporal and circumstantial, *not* cultural.

Explaining post-Yugoslav nationalisms: WWII trauma and media?

More than a decade after the start of the 1991-1995 post-Yugoslav wars, we are still in the process of attempting to understand the roots of that conflict, and, in particular, the appeal of the various nationalist discourses amongst broad segments of the population. For the sake of argument, I ignore the widespread racist-cum-culturalist approaches, which lay the blame for nationalism's success with atavistic Balkan hatreds, both as pseudo-explanations and as straw men for more critical analyses. Surely it is time to redirect our attention to a range of less essentialist explanations that have been put forward. Many of those tend to focus on suppressed traumas of WWII massacres, particularly when addressing the situation in the previously disputed areas of Croatia. My research on post-Yugoslav anti-nationalism[6] pointed out that local dissidents considered such abundant reference to WWII traumas highly problematic. Firstly, 'trauma-centred' explanations for the appeal of post-Yugoslav nationalisms were seen as unwelcome because they reproduce nationalist propaganda, since reformulations of these memories of terror also played a central role in the nationalist discourses that were instrumental in the build-up to the war and in its continuation. Secondly, such explanations prevent contextualisation of the actual importance of those memories with regard to recent events (see Jansen 2002).

A set of alternative explanations for the popular support of nationalism, put forward by local and foreign critics favour what we might call a constructivist perspective. They tend to attribute more explanatory power to political propaganda and media manipulation. Memories of WWII suffering, it is argued here, were first and foremost instruments in the hands of nationalist politicians and, when assessing them, it is hard to draw the line between indoctrination and trauma. Many valuable analyses have combined those two perspectives, at times emphasising the role of WWII legacies (Bowman 1994; Denich 1994; 2000; Hayden 1994), while at other times highlighting the importance of media manipulation by the nationalist regimes (Glenny 1992; Silber and Little 1995; Thompson 1994).

Strikingly, most approaches, whether journalistic, political or academic, converge on seeking causes for ordinary people's adherence to nationalism in collective, structural factors. I aim to draw the attention to a major problem arising from such a rather one-dimensional emphasis on collective patterns of thinking and/or behaviour: the lack of attention to the agency of the people involved. Let me make clear straightaway that I do not wish to underestimate the importance of WWII traumas, based on the very real horrors of that time. Similarly, the pernicious role of the mass media in the preparation and perpetration of the post-Yugoslav violence is beyond doubt. However, it seems all too easy to take their determining influence simply as a given. While we cannot dispute the existence of collective traumas of WWII massacres, we have very little evidence of their *direct* impact on events half a century later, nor can we assume that this impact is uniform in nature (see Jansen 1999; 2002). A similar argument could be developed concerning media manipulation. Certainly, some nationalist propaganda was extremely successful in mobilising some people into committing violence – but *which* messages, and *which* people are we talking about here?

With regard to both WWII trauma and political propaganda, we should take care to avoid the pitfalls of determinism. Without accounting for the mechanisms with which individuals in the post-Yugoslav context related to the dominant nationalist discourses communicated to them, we run the risk of reducing them to helpless victims, toyed with by structural factors and stripped of any form of agency. The empirical material in this article allows us to put these issues into critical perspective.

La vita é not so *bella*: agency and pessimism

But wait a minute. Of course we are dealing here with a situation in which many people *did* feel exactly that kind of powerlessness. For an outsider as well, at first sight, evidence of human agency certainly *did* seem rare. Therefore, overestimating either the role of WWII traumas or the importance of propaganda represented attractive options, given that they were reflected in widespread local representations of all-powerful regimes (whether good or bad) and helpless ordinary people. However, I believe that, ethnographically, such reductionist explanations, even though firmly entrenched in popular use, are only partially adequate at best. Moreover, ethically-politically, through their disregard for individual agency, they preclude questions of responsibility to an uncomfortable extent and further marginalise existing alternative narratives of past and present as well as dissident routes of action, which had been silenced in recent times.

So far, so anthropological: am I cruising towards yet another conventional exercise in uncovering agency and resistance in a context of apparent homogeneity? I believe there is another twist to my story. I would like to refrain from optimistically infusing 'resistance' into a situation that I myself considered depressing and hopeless, although I have much respect for others who have done this to great anthropological effect in other settings. The work of Scott, in particular, is characterised by this tendency to identify strains of oppositional behaviour in contexts where one would not expect them (1985; 1990). Scott argues against Gramscian approaches and claims that, in fact, subalterns are capable of seeing through hegemonic projects. What's more, while they feign compliance, they rely on 'hidden transcripts', collective alternative worldviews, which underlie mundane acts of covert resistance.

I had gone into this research finding Scott's ideas very inspirational but, sadly, my activist work in these Croatian post-war villages did not increase my hopes for critical grassroots action. I found Scott's model simply too optimistic: while he might be right that people are not simply passive recipients of hegemonic nationalist discourses, this did not automatically exclude their continued, enthusiastic adherence to it. His followers could easily argue that I did not look closely enough for examples of covert acts of resistance, but I believe that the overwhelming nationalist homogeneity in words and deeds was more important to people's lives, and more in need of analysis, than possible examples of hidden resistance to it. No thanks, then, no references to films like Robert Benigni's 'La vita é bella', please. Rather than fuelling an interest in heterogeneity for heterogeneity's sake, the research sharpened my awareness of the ways in which individuals *actively* engaged with 'structures' on the everyday level. It made me wonder in which ways villagers coped with, digested, used and even embraced trauma and propaganda. Crucially, rather than searching for anti-nationalist 'hidden transcripts', I became interested in how they were involved in the (re)production of nationalist homogeneity.

A key question in this text is: can such a largely pessimistic conceptualisation of agency offer valuable material in order to bring to light individual coping strategies in a context of relative powerlessness? Particularly, I analyse the role of strategic essentialising in people's positionings in relation to dominant nationalist discourses (see Berdahl 1999: 208; Herzfeld 1996). Compatibility, or at least the minimisation of incompatibility, between personal and 'large' narratives then becomes an important issue. If people are seen as at least partly capable of constructing everyday life formats which are not-all-too-incompatible with the discourses that are *de rigueur* at that moment, maybe this explains, to a certain extent, the impressions of internal homo-geneity and consensus that await many students of post-Yugoslav nationalisms. People's narratives con-structed around a set of catchwords and phrases can then be seen as mechanisms by which they position themselves, consciously and unconsciously, in rela-tion to dominant discourses in confusing times.

In what follows, I focus on some patterns perme-ating the lives of the villagers. First I look at how they constructed and reproduced a virtually absolute dividing line between their everyday experiences and the 'politics' of the moment, protecting themselves against the dangers of the latter. Then, I analyse some recurrent coping patterns involving subse-quent linkages of personal narratives with authori-tative discourses.

Powerlessness, silence, and self-protection

'Big politics' and 'small people'

Let us start from this observation: even though it was hard not to be shocked by the extreme character of national exclusivism in these war-affected villages, my strongest impression was not one of militant nationalist hatred. The situation, it seemed to me, was much more characterised by powerlessness, conformism and confusion. The experience of war, displacement and the political shifts on the state level had given rise to lives constructed around a defining break. The resulting confusion was rein-forced by feelings of extreme powerlessness, since most villagers experienced configurations of state, war, nation and territory – in short: *'politika'* ('poli-tics') – as largely objectified and out of individual control by ordinary people such as themselves. Let us admit immediately that this was a realistic atti-tude, both amongst 'winners' (here: Croats) and 'losers' (here: Serbs): in the past decade, many had lost their homes, their loved ones, their property, their jobs, and so on. No amount of academic insights into the dialectic relationship between agency and structure could change these people's experience that their lives had been eaten up by 'higher powers' beyond their control.

One of the most common interjections used in conversations in the villages, regardless of the

nationality of the interlocutors, was an expression of resignation: *'e, šta češ...'* ('what can you do...' or 'well, what are you going to do about it...'). The large majority of villagers, many of whom had per-sonally survived horrific experiences, settled for a rather phlegmatic approach. This testified to their perseverance in hard times, but it also reflected res-ignation. Experiences in the past served as an impor-tant counterpoint, as people took shelter in under-statements such as 'it could have been a lot worse', often referring to memories of *that* war (WWII), which relativised the horrors of *this* war, because, 'back then, things were much harder than now'. Thus, resignation functioned as a coping pattern.

Amongst Croats, evoking the authoritative dis-course of the blessing of simply having one's own state, was often enough to imply that no action was needed on the side of 'small people'. With regard to difficult living conditions and other problems, the standard attitude amongst Croatian villagers was one of widespread declared trust in the powers-that-be. 'The state will take care of all that,' they argued, 'but we can't expect results overnight'. This phrase, reproduced by many Croatian villagers, lit-erally reflected regime statements.[7] The deafening silences that surrounded it on all sides reflected a cornerstone of the dominant nationalism: non-engagement. Resignation helped people to get a grip on the situation and enabled an avoidance of indi-vidual responsibility and action. Serbian villagers, of course, could not rely on an equivalent discourse of trust, but their scarce references to life during the previous Krajina period reflected a similar emphasis on the gap between 'ordinary people' and 'politics'. Hence, underlying the differences between Croatian and Serbian villagers, there was a sense of resigna-tion to the absolute control of the powers-that-be, whether expressed through declared trust in the authorities (by Croats) or through the acceptance of powerlessness (by Serbs).

These feelings of powerlessness were coupled with a nearly complete absence of collective action. Despite dire living circumstances, there was no sign of protest and I witnessed no attempts to improve the village situation in any way, except on the pri-vate level. In fact, some Croatian villagers com-plained of the lack of mutual help, pointing out how people restricted their activities to their own family and refused to engage in collective efforts. In terms of communication between people of different nationalities, there was virtually none. Hence, indifference was much more prevalent than militancy. Obviously, we have to contrast this passivity with the situation in the early 1990s when these villagers had been in the frontline of their respective national revivals. Then, people in his region had taken the initiative and engaged in different forms of, sometimes violent, collective action. Now, the perception that ruled amongst Croats was one of benevolent, all-powerful state authorities and atomised – but not conflicting – village families who were awaiting the fruits of their sacrifices in the war. Their houses had been rebuilt already and they waited in the certainty that the rest would follow. Serbian returnees expected nothing of the kind and simply survived in silence. This dichotomous picture, suggesting two monolithic national patterns, was only rarely undermined. For example, there was significant resentment amongst Croatian villagers towards Croatian refugees from Bosnia, who were considered 'primitive' and said to engage in mafia-like practices. However, this did not express itself in any practical way and never led to any rapprochement between people of different nationalities.

Telling silences: the abdication of responsibility
In many ways, life in the villages was more striking with regard to what it systematically ignored than with regard to its actual content. The above-mentioned dominant refusal to communicate or otherwise engage with national 'Others' was a case in point, as was the reluctance to take social or political action on any level. In narrative terms, telling silences were crucial in the villagers' stories. Elsewhere I have analysed how war stories, both of WWII and of 1991-1995, usually concentrated on one period and one event only, without even mentioning the rest of the conflict (Jansen 2002). The key for attributing selective silences was almost always self-victimisation. In this discursive conflict, two versions of war history were mutually exclusive, whereas, sadly, there was plenty of evidence to support both of them. Such silence and the wider resistance against contextualisation were accompanied by a pervasive vagueness. Interestingly, people with dissident practices and views were much less vague, but almost everybody avoided going into details. It was virtually impossible to collect a concrete, chronological account of *any* event. Vagueness set the scene for sweeping accusations and served as an instrument of self-protection, particularly in relation to more powerful people of one's own nationality. Throughout the post-Yugoslav states, the period after 1991 was often summarised as '*sve ovo*' ('all this') or '*ovo sranje*' ('this shit'). This had to do with simplification and abbreviation, but it also reflected a wider reluctance to specify.

These patterns of silence, vagueness and resignation allowed villagers to carve out a niche for themselves as part of an anonymous victimised mass, void of responsibility for their current predicament. Underlying the popular phrase, 'there are reasons for all this' ('*sve ima to svoje*'), which was never further explained, there was the idea that, 'we don't know the reasons and it is better not to ask'. In a dangerous context of confusion, loss and despair, digging deeper was considered a job for politicians and, by not going into these issues, people also avoided being entangled in them. Nevertheless, resignation to 'higher powers' was often mirrored by scepticism, initially less obvious to outsiders. Sarcasm about the lack of control over one's own fate was a popular theme in pitch-black humour in many parts of former Yugoslavia, and people liked telling anecdotes in which they themselves figured as *schlemiel*-like losers (see Jansen 2000a; 2001).

Such sarcastic resignation supported ideas of self-victimisation and the almost ontological dividing line between everyday lives and 'politics' also allowed abdication of personal responsibility. As John Malkovich's character argued emphatically in the film *Dangerous Liaisons*: it was 'Beyond Their Control'. Hence, by postulating the existence of their everyday lives as at least theoretically independent from 'higher powers' many villagers also aimed to protect themselves against the overwhelming influence of the latter. It is this aspect that I turn to now.

Linking subjectivity to 'politics'

Evoking authoritative discourses
If most villagers considered 'politics' to be out of their control, and reinforced this perception in their everyday (in)action, did this mean they were resigned to fatalism? Not completely, I would argue. My research material indicates that, if people wished to protect themselves against what they perceived as the danger of 'politics', they also felt that some of its aspects and some of its uses were not quite as undesirable. They evoked authoritative discourses

and created a picture whereby their everyday lives were perhaps not reflective of 'politics', but at least compatible or not-too-badly-out-of-tune with them. In that way, they deployed large, complex and powerful discursive practices in an attempt to assert control over the present. Hence, through a twist of strategy, it was precisely by postulating the separateness of their everyday lives from 'politics' that villagers opened up the possibility of linking their subjectivity to authoritative discourses of state and nation, *on their own initiative*. We have to look, then, at the ways in which people related to the dominant nationalisms and to alternative discourses and how they (re)structured their own practices for private and/or public use.

If we consider the villagers' attitudes as a set of coping patterns with violence, loss, poverty, a narrative break in the life story, powerlessness and confusion, we are dealing here with contested constructions of ontological security (Giddens 1991: 35-69; see also Gillis 1994: 3). Making sense of experience, they relied at least partly on pre-existing discursive material, often of the more powerful and authoritative kind (see Herzfeld 1985: 21). Moreover, given the extreme context, they were continually expected to position themselves in relation to 'politics' through a process of ideological interpellation (Finlayson 1996). Usually they attempted to locate themselves favourably in relation to the dominant nationalist discourse – favourably, of course, in the eye of the beholder.

This does not mean that they were merely inscribing themselves into powerful discourses, although their choice was often extremely limited. Rather, I suggest, it was a question of practical sanity; if your everyday life was completely out of tune with *all* 'politics', you would not be able to function in a public environment and you'd probably be considered mad or dangerous. Obviously, it is a question whether this drive for compatibility was only for public use, or whether there were benefits of incompatibility, but it should be clear that in the post-Yugoslav context the need to reposition oneself in relation to powerful discourses was acute. But how did these repositionings take place? How did people link their everyday lives to the 'politics' available? And how did they attempt to position themselves favourably in relation to powerful forces?

In a period of intense turmoil, narrative can become a common tool to comprehend processes of

change, or to try to keep them in check.[8] Moreover, we have to take into account that the stories of the villagers were to an extreme extent performative utterances. The discourses enforced by powerful institutions in the post-Yugoslav context were tied together by a nationalist prism, but, importantly, they were polysemic. Given this context, narrating events, or choosing not to, particularly in terms of nationality, was one of the few political acts accessible to most people. In the villages at the heart of this study, certain phrases referring to, or better, evoking authoritative discourses, and sometimes literally taken from those discourses, were continually reproduced in everyday life. Particularly when confronted with outsiders, such as state officials, journalists, or NGO workers, many narratives resorted to such evocation (see McKenna 1996: 231-32).

Following Hajer, I use the concept of 'story lines' to clarify one way in which people can connect their everyday life experience to the authoritative discourses of 'politics'. The term 'story lines' refers to a 'generative sort of narrative that allows actors to draw upon various discursive categories to give meaning to specific [...] phenomena' (1995: 56). Crucially, story lines are characterised by a high degree of multi-interpretability. This, suggests Hajer, makes some sense of order in discursive praxis possible, because when an actor uses a certain story line, it is automatically expected that the addressee will respond within a similar framework. However, as a result of the multi-interpretability this does not mean consensus. Story lines, then, offer actors the opportunity to talk and think about a topic without having to grasp the whole problematic. By calling on a story line, complexity and conditionality are reduced and a certain implicit common ground is presupposed – in the process, the authoritative discourse and the different kinds of power that are associated with it are evoked, while a large degree of vagueness is retained.

'We have Croatia!'

Let us consider an example of a story line. A central axis of Croatian nationalism was the sanctity of the national state and this authoritative discourse was continually reinforced through symbolism, policies, propaganda, and so on. Nevertheless, the discourse of nationalism was never (and could never be) spelled out completely. Rather, it was condensed in phrases such as '*Imamo Hrvatsku!*' ('We have Croatia!'), that were literally reproduced in many of the villagers' accounts. In this way, it could be argued that powerful discourses of 'politics' controlled local everyday lives. However, my study indicates that people also actively used those lines in order to evoke the authority of large and complex 'politics' and thereby to assert control themselves. This could be in order to justify certain behaviour or situations, to avoid reflection about certain issues, to deny responsibility, or simply to survive and stay somehow practically sane (sanity in the eye of the beholder, again).

A very straightforward goal of evoking powerful discourses is to invest the speaker with authority. It also sets the rules for conversation on a supposedly 'generally accepted' level without having to specify. 'We have Croatia!' was one of the most widely used lines in this way. Sometimes, as in the case of thirty year old Robert, former officer in the Bosnian Croatian army, and now living in a Serbian-owned house in Bijelo, this was broadened to a more general alignment with the nationalist tenet of 'one nation, one state':

Everything will fall in place now that the Croatian people have their own state and the Serbs have gone. The Serbs have their own country. And the Muslims should be off to Turkey.

Again this provided justification for maintaining control over the house Robert occupied[9] and for a refusal to return to his native Bosnia. However, more frequently, the intrinsic superiority of having one's own state was expressed less specifically. The fact that we deal with a very old population plays an important role here. Many villagers were approaching death; they reflected on their lives, assessed their achievements and constructed evidence of continuity. For many in those post-war conditions, successful lives of their children or material achievements were inaccessible as symptoms of this, but nationalist discourses provided such evidence by using concepts of birth, life and eternity. Croatian nationalism, through the story line 'We have Croatia!', offered a morbidly enlarged version of everything the Croatian villagers lacked. Instead of saying 'we have nothing', it said 'we have Croatia' (which is everything we need and everything we always wanted). Instead of saying 'we'll die soon', it said 'Croatia will live' (and therefore we will too). And instead of saying 'we are lonely and abandoned', it said 'we are together at last' (with our own people).

National disambiguation

One of the most striking patterns in the evocations of authoritative discourses through the use of story lines was the way in which they rewrote the past as a straightforward preface to the current situation. Elsewhere, I have analysed this in detail (Jansen 2002) and some of the examples above illustrate this process of retrospective disambiguation. Here, I focus on a similar process of disambiguation of the present,[10] on patterns framed by the nationalist idiom or in reaction to it, which were at the time particularly prevalent amongst Croatian villagers. National disambiguation was the key to the construction of a social reality consisting only of discrete national groups: *us* (all of us) and *them* (all of them).

Let us look at some examples. Related to the historical cause of national liberation, Croatian villagers often evoked the powerful discourse of the preferability and superiority of national homogeneity to explain their reluctance to engage with Serbian returnees. They argued that everybody '*hoće biti svoj na svome*', a common phrase that says that it is only normal that a nation 'wants to be free in its own land'.[11] This was accompanied by the idea that everybody always feels better amongst 'his/her people'. Nada and Jozo, an elderly couple from Bosnia who fled to Croatia during the war and resettled in a Serbian-owned house in Bijelo, stated:

Here things are good. We always felt that nostalgia for our own state, for our own Croatia. We are glad to be amongst our own people. It is better to be with one's own.

Note that Nada and Jozo had lived in a highly mixed area in Central Bosnia for the previous sixty years of their lives – they had never spent a considerable period of time in Croatia before. They dramatically reformulated their narrative of 'home' and brought it in tune with the dominant nationalism. Thus the terms 'them' and 'us' now referred to a whole new family, village and state history and this allowed justification of the fact that they refused to leave the house they occupied and return to Bosnia.

On many occasions, Croatian villagers discussed contemporary events in other places – but only those that confirmed their perspective on the local situation. For example, they referred to heavily mediatised incidents in a distant Bosnian town in order to evoke the idea that all Serbs create problems, and therefore to reassert the impossibility of co-existence with local Serbs. Josip, a sixty-odd year old Croat in Plavo, explained his unwillingness to engage with his Serbian neighbours in these terms:

Living together with them? Phooo, look what they are doing to us in Derventa!

Again, in one narrative movement, the Serbs in faraway Derventa and the local Serbs were equated as an unambiguous 'them' – to be avoided at all cost.

More generally, there was a striking leaking of diplomatic language into the everyday discourses of many villagers, of whom the older ones were often only semi-literate. Peace treaties and political declarations provided useful story lines evoking the authoritative world of international geopolitics and people deployed them to retrospectively reclaim control over their everyday lives. In this way, a family from Gradić in Bosnia stated that they had moved to (a Serbian house in) Bijelo, because, 'we didn't want to be a minority'. In doing so, they invoked a reason of a diplomatic nature, rather than referring to the fear, uncertainty and lack of opportunity that probably lay at the root of their move (which took place *after* the war). Similarly, when justifying their reluctance to return to Bosnia, they argued:

We don't want our children to go to school in Gradić, where they can't study in their own language.

Both parents had always lived in Gradić, previously a mixed town in Bosnia, and they certainly had never had any communication problems with their Bosniac or Serb neighbours, who spoke the same local dialect. However, they retrospectively applied the current doctrine of discrete Bosniac, Serbian and Croatian languages on a previously ambiguous situation. Times changed more dramatically than language, and the evocation of the authoritative discourse of language rights allowed them to resist being subjugated by another, possibly threatening discourse, that of rights of property and return.

The persistence of ambiguity:
coping patterns of the marginalised
Given the centrality of strategies of disambiguation, the coping patterns of villagers in ambiguous positions deserve particular attention. We have already seen some snippets of this, when I explained that a very small minority of villagers consciously broke with the collectively sanctioned pattern of segregation. Such dissident practices were legitimised in different ways, but, interestingly, they also often relied on evocations of authority. One case in which one mighty discourse was introduced in order to avoid control by an alternative form of 'politics' was the introduction of an extra 'Other'. Some villagers, Serbs and Croats, did occasionally engage with national 'Others' and thus resisted the dominant denunciation of the standard 'Others'. When explaining this, they then often introduced the presumed danger of fundamentalism amongst 'Muslims', 'Turks' or *Mujahedini*, as the real problematic 'Other'. In addition to this minor phenomenon, dissident practice was usually justified with reference to non-national logic. I shall quickly distinguish three such alternative approaches, often deployed in combination with each other.

First of all, outsiders often strongly emphasised individual responsibility and refused to make generalisations about groups of people. Zoran, a Bosnian Serb married to a local Croatian woman, insisted on seeing 'a person as a person': he systematically used the first person singular, rather than plural, and often employed his own name when recounting past events. Moreover, he almost always qualified national labels: 'some clever/crazy/stupid Croats', or 'some mad/open-minded/aggressive Serbs', thereby providing an explanation for his practice to engage with some persons while steering clear from others, regardless of nationality.

A second alternative was the idea that the stakes of the post-Yugoslav conflict were not national but civilisational: a struggle between civilisation and primitivism. This was one of the underlying tenets of a large part of the critique of nationalism in all post-Yugoslav states. Rada, a fifty-year-old Serb who, because of her political stance, was shunned by Croats and Serbs alike, bewailed her victimisation by what she saw as an essentially primitive reflex of other villagers.

If they have something to tell me, if they want to discuss something with me, let them tell me! Let us sit down at a table and talk about it. I am always ready for that! But on the basis of arguments! Not on the basis of the fact that I belong to the Serbian nation! What kind of primitivism is that!

This attachment to values of 'civilisation' was often related to a strongly developed belief in education.

When condemning violence and hatred, Rada, one of the few highly educated villagers, often referred to others as 'illiterates', and explained how educated people would never do such things. When I pointed out that many of the politicians who had brought the country to war were highly educated, she reformulated that as a sharp accusation: an educated person should know better, and therefore s/he should *act* better. An educated person has no excuses for not respecting the rules of civilisation, which 'primitives' may not recognise.

A third widespread form of doubt in national disambiguation was a more empirical one, related to the above-mentioned process of sceptic resignation. Many villagers (of those present, obviously more Serbs than Croats) foresaw that they would live together again. 'It is normal', people would say, 'After "that war" we also lived together again.' Others, like Zoran, evoked the more general discourse of multi-ethnicity as the rule, rather than as an exception.

All that business about dividing people on national grounds is nonsense. There is no such thing as an ethnically clean village or town, and certainly not state. Multi-ethnicity is inevitable and completely normal.

Zoran, as I have mentioned before, was married to a Croat and he was the only Serb to have returned to Plavo. After half a decade of displacement, which had forced his family in and out of different places in Croatia and Bosnia, he now eked out a living of subsistence agriculture. Throughout the 1970s-1980s Zoran had been the village teacher, a Party member who had perceived himself as a Yugoslav. After the war, even amongst the now dominant hard-line Croats he was still credited for his consistent fairness in nationality issues in Yugoslav times. This, however, had not prevented his family from being the first one to be expelled, nor did it lead to any social interaction or assistance upon his return, except from his wife's (Croatian) family.

Nationalism, structure, and agency: making room for pessimism

Positioning and legitimising violence

Let us now return to the issues of WWII trauma and media manipulation. The material presented in this study on the role of nationalism in people's coping strategies encourages us to question certain central assumptions of prevalent approaches to the post-Yugoslav nationalisms. I have argued above that emphasising the role of individual agency in people's positionings in relation to dominant discourses allows for a less deterministic sketch of the situation in post-war Croatia than an approach which privileges WWII traumas or media manipulation as almost independent variables. In that sense, conformism, rather than determinism, becomes a central notion. And, to take this one step further: attributing some level of personal engagement to those who did conform – and they were a large majority – also allows us to begin to account for those who did not (see Jansen 2000b).

In a brief text written during the post-Yugoslav wars, Belgrade anthropologist Ivan Čolović addresses the issue of why war propaganda was so effective (1994: 57-62). He suggests that the key lies in the authority of the media, derived from their assumed identity as the voice of the regime. And, in effect, in a conversation with a villager in Bijelo, this was illustrated literally, when the man pointed at the television set every time he mentioned 'the state', 'our leaders' or 'Tudjman'. In an unintended twist of irony, he did not know that there was a colourful children's game show on. Čolović argues that many people, whenever asked for their opinions by someone from outside, tend to give the answer which they assume is the nearest to the line of the current centres of power of interest to them. People do not believe that these outsiders want to know their opinions at all – and usually they are probably right. Why would they be asked? 'This means', says Čolović,

that many people see television and other media not as a source of truthful information and convincing messages, but as a bulletin board on which daily orders are shown. Or as some kind of political traffic lights that tell you when you have to turn right or left, or if you should go straight on or simply wait until further notice. Otherwise, you are in danger [...] of being punished, excluded from traffic, or simply run over. (1994: 61)

Critics could argue that surely such an approach fails to take into account the existence of hidden transcripts, providing alternative visions of reality. While such resistant interpretations might have existed, and I hope they can play a significant role in the future, in my view they did not affect collective life in the villages in any meaningful way during my stay. Not only were the dominant nationalist discourses communicated through the media reproduced widely, but sometimes media messages were actively incorporated into personal narratives in order to legitimise certain acts. An extreme example was provided by Jozo and Nada, the elderly Bosnian Croat couple in Bijelo, for whom the equivalence between the Serbs who destroyed Vukovar televised on the news and the local Serbs had provided justification for ethnic cleansing.

Back in Bosnia, in the very beginning, we watched television… You must have seen it as well, how the Serbs were destroying Vukovar.[12] My God, it was horrible. They were burning and looting and killing. So we arranged with the Muslims to chase out our Serbs. Later, the *Mujahedini* turned against us. They wanted a fundamentalist state. And they drove us out.

Note the curious synchronic equivalence through disambiguation: in 1991 all of them, including *our* Serbs, were criminals, so they had to be removed preventively. The effect of propaganda through the media, then, did not simply take the form of crude manipulation; sometimes it functioned as a 'traffic light' and sometimes it provided a legitimising background for acts of violence. This is how my analytical focus has shifted from structure-based determinism to agency-oriented conformism.

The (dis)comfort of conformism

Nationalism was omnipresent in the post-Yugoslav context, amongst intellectual and political elites, but also in the daily lives of most other people, certainly in the war-affected areas. Exclusivist acts were rife and often considered acceptable and normal. Still, I would argue that a lot of villagers had not really reflected on many of the issues addressed by nationalism. Why should they? For most of them, there were more immediate worries. However, when the matter arose in conversation, they tapped into the always available, polysemic, contradictory discourse of nationalism. It was the most authoritative discourse of the moment and it served as a perfect *passe-partout* without really taking issue with one's own biography. More generally, in contrast to its alternatives, nationalism provided strikingly straightforward stories, an attractive attribute in the post-war confusion. Most of the people I encountered in these post-war villages did not seem to be fanatical

believers in the tenets of nationalism, but conformism with this dominant discourse provided them with comfort in uncomfortable times.[13] Often, the discomfort this caused for others was simply not taken into account.[14]

Again, I hasten to add that the material for this text was collected in a specific period after the war, when Croatian villagers were the ones for whom nationalism *worked*. It does not allow generalisations about the outbreak of the conflict in the early 1990s, when Serbian villagers demonstrated a similar enthusiasm for *their* nationalism. The background and logic of that outburst should be analysed in a different contextual light. But by the end of the decade, it was not a consolation to me that the frequent discriminatory acts and talk amongst Croatian villagers seemed to rely more on indifference and conformism than on hatred. And it was frustrating to be confronted with the same blanket explanations for a variety of phenomena. They were tired, of course. After years of war, displacement and loss, they did not want to reflect on events, on reasons or on guilt. Luckily for them, they did not have to – in fact, they were encouraged not to – quoting some story lines was sufficient. So a question by an outsider became more often than not an occasion to throw in some story lines from the nationalist discourse. Why? Because that's what you did. And because you assumed that that's what others did. And they did.

Reconquering everyday life

So: conformism. But not only conformism, of course. The consequences of the nationalist outbursts, the violence, the loss and the poverty were real (see Povrzanović 1997). Even if the existence of a 'national question' was a matter of debate in 1989, it was certainly reality now (Buden 1996: 171) and maybe it would be too painful to give up the enormous importance attached to nationality now, after all that had happened in the name of it. We should not forget that in the past decade many villagers had gone through experiences that had dramatically affected their everyday lives. This text attempts to take that into account by conceptualising their current practices as coping patterns, mainly consisting of 'favourable' repositionings in a compatible relation to the authoritative discourse of nationalism. McKenna's study of rank-and-file engagement with Muslim separatism in the Philippines puts forward a similar argument (1996). He points out that 'ordinary people' do not blindly reproduce the dominant narratives, but that their support relies on practical compliance on the basis of their own collateral goals.

However, just as I would argue that acts and statements which strike us as radically nationalist do not necessarily mean that the villagers in question were simply militant national believers, I think it also does not necessarily mean that they were always just hoping to get the most out of nationalism by aligning themselves with it. Certainly, this factor played a role; by positioning themselves as near as possible to the heart of the dominant nationalism, they sided with the strong and picked the fruits of crushing the weak. This process was alternately accessible to Serbs and Croats during the 1990s.

However, at least after the war, for many of the villagers nationalism seemed to fulfil another function as well. Repositioning strategies were not just about getting near the heart of power. They were also about keeping a distance from those centres, about reconquering the everyday life experience, about fighting the colonisation of 'ordinary' lives by 'politics'. Non-engagement, non-communication, vagueness, simplification, selective amnesia, sceptical resignation and various story lines allowed people to construct their everyday experiences in tune with the authority of nationalism, without coming too near to its risks. Thus they lived without demanding introspection, without posing nasty questions and without requiring an eye for complicated nuances. They evoked nationalism, without really going into its 'ins and outs', which meant that they did not come too near the power associated with the discourse. It also meant that that power did not come too near them. What else is resignation other than leaving it all to 'higher powers'? What else is vagueness other than keeping it all out of my house or my head?

World-wide activist experience in war areas teaches us that the most extreme crimes, the most radical forms of hate speech and the most violent attitudes are often not to be found amongst those people who have been victimised most by violence. Sociological research in Croatia confirmed this (Hodžić 1998). Maybe the idea of nationalism as a coping pattern which navigates between proximity and distance and which asserts non-responsibility *and* control

over everyday life can help us understand this. For some, the stakes are simply too high – they have to cling to a distance, they cannot afford to come that near to powerful discourses of hatred and violence. People whose stakes are not that high are in an easier position to align themselves freely with more dangerous 'politics'. In Bijelo, Serbian returnees experienced most provocations by the police, on- or off-duty. One officer informed me that the best solution would be 'to mine all Serbian houses' and prided himself that the area had always been a 'hard Ustaša region'. The man had never even been there before the war.

How does all this relate to the pessimistic conceptualisation of agency that I mentioned earlier? While acknowledging depressing levels of homogeneity around nationally exclusivist behaviour and positionings, this article undermines the argument of structural determination by WWII trauma or media manipulation – at least for the post-war period. Impressions of monolithic consensus, I would argue, do not necessarily rule out agency on the part of those expressing it (even if this sometimes takes place through non-action). I pointed out how virtually all villagers experienced a radical separation of their everyday lives from 'politics': they saw things as decided for them by powerful others. Through the same process, in a drive for self-protection, they postulated at least relative autonomy for their personal narratives. Villagers then attempted to exert control over their own lives by evoking authoritative discourses in their practice. Some life experiences were both retroactively and strategically brought in tune with exclusivist nationalism; in this way, paradoxically, they were reformulated as if they belonged to the individual's everyday life experience, rather than having been imposed by an uncontrollable force. These repositioning strategies allowed proximity *and* distance, innocence *and* merit, lack of responsibility *and* control. They allowed one to draw on 'politics', materially and psychologically, but simultaneously to keep the latter at a distance from one's personal everyday life. Crucially, this was compatibly at a distance. In this way these people exerted power in a situation of extreme powerlessness; it allowed them comfort in uncomfortable times, regardless of the discomfort it caused to others.

NOTES

1 Research carried out in 1997-1998. All persons are referred to by pseudonyms, as well as the villages, which I have called Bijelo, Plavo, Crno, Sivo and Zeleno. The NGO which carried out the project was local, Zagreb-based, with me being the only non-Croatian citizen. Many thanks are due to colleague-activists, particularly to Sanda Malbaša for support and constructive criticism. I also wish to thank Jody Barrett, Andy Dawson, Caroline Oliver, Ivana Spasić, Mark Johnson, Nerys Roberts and the participants of the *War and Society* Seminar at Aarhus University, 2001.

2 It should be emphasised that there were no phenotypic, clothing or dietary differences between Serbs and Croats either. All villagers spoke an identical local variant of what was previously called Serbo-Croatian or Croato-Serbian. Religion was relatively more important amongst Croats (Catholicism) than amongst Serbs, who were more likely to have a Partisan background. Still, in former Yugoslavia, these villages were different from a lot of other nationally mixed areas, and they presented a rather extreme situation. Unlike, for example, most larger places, the nationality of all inhabitants was known to all others.

3 The story of a 'privileged' Serbian-dominated 'Communist' village next to a 'deprived' Croatian-dominated 'Ustaša' village was a common one in this part of the Yugoslav Socialist Republic Croatia.

4 In my view, the term 'neighbours' occupies a problematic position in debates about the post-Yugoslav conflict. Both in media and in academic coverage, it is often inappropriately and uncritically employed when referring to Yugoslav times, resulting in an unquestioned, idealised representation of that past as co-operative and harmonious. In this text I straightforwardly use the term (ex-)neighbours for people who live(d) in the same neighbourhood, regardless of the warmth of their co-existence. See Jansen 2002 for a discussion of contested local memories with regard to previous relations between Serbs and Croats; see Jansen, forthcoming, for a similar study in a Bosnian context.

5 Davor and Nela had come from other regions. Note that, like many others, both of them occupied houses owned by pre-war Serbian inhabitants.

6 1996-1998 research for my Ph.D. dissertation in Zagreb and Belgrade (Jansen 2000b).

7 In many ways, this situation is only a reformulated continuation of prevalent patterns in Titoist Yugoslavia. Sociological research in the Former Yugoslavia always uncovered a strong adherence to authoritarian values (Golubović 1995; Biro 1994: 13-38; Hodžić 1998).

8 See Ricoeur 1990: 167-193; 1991: 32-33; Ganguly 1992: 29-30; Rapport 1997; Jansen 1998; 2000b.

9 Like many others, Robert kept a videotape he had made when first arriving here. It showed the heavily damaged house as he had found it. Since then he had made several improvements and kept the tape as proof of this. This is a clear indication that, despite his refusal to acknowledge the issue of property rights on an explicit level, implicitly he was aware of the possibility that they might apply to him.

10 Duijzings employs the useful term 'ethnic unmixing' to refer to the more material aspects of this process (2000). Bauman convincingly argues how the extermination of ambiguity lies at the basis of nationalist discourse (1992).

11 Literally, this story line means that everyone 'wants to be his own on his own', in other words, 'wants to own himself in a place that is his'. This refers simultaneously to two levels: nations and individual members of these nations.

12 They talk about 1991. It is not a coincidence that Nada and Jozo should mention Vukovar, a town that occupies a central position in Croatian nationalism as an icon of Serbian aggression and Croatian suffering and sacrifice. Note that Vukovar is situated in Croatia, about 200 km away from their Bosnian village, which in 1991 was not involved in the war.

13 See Bolčić 1995: 480-81.

14 In fact, many seemed to have come to a point where they excluded reflexivity on the issue (see Iveković 1994: 198).

BIBLIOGRAPHY

- Bauman, Z. 1992. 'Soil, blood and identity'. *Sociological Review*, 40(4), pp. 675-701.
- Berdahl, D. 1999. *Where the World Ended: Re-unification and Identity in the German Borderland*. Berkeley: University of California Press.
- Biro, M. 1994. *Psihologija postkomunizma*. Belgrade: Beogradski Krug.
- Bolšić, S. 1995. 'The features of a "nationalised" society'. *Sociologija*, 37(4), pp. 473-83.
- Bowman, G. 1994. 'Xenophobia, fantasy and the nation: the logic of ethnic violence in former Yugoslavia'. *Balkan Forum*, 2(2), pp. 135-64.
- Buden, B. 1996. *Barikade*. Zagreb: Biblioteka Bastard/Arkzin.
- Ćolović, I. 1994. *Pucanje od zdravlja*. Belgrade: Beogradski Krug.
- Denich, B. 1994. 'Dismembering Yugoslavia: nationalist ideologies and the symbolic revival of genocide'. *American Ethnologist*, 21(2), pp. 367-90.
- Denich, B. 2000. 'Unmaking multiethnicity in Yugoslavia: media and metamorphosis'. In: J.M. Halpern and D.A. Kideckel (eds.): *Neighbours at War: Anthropological Perspectives on Yugoslav Ethnicity, Culture, and History*. University Park (Penn.): Pennsylvania State University Press, pp. 39-55.
- Duijzings, G. 2000. *Religion and the Politics of Identity in Kosovo*. London: C Hurst.
- Finlayson, A. 1996. 'Nationalism as ideological interpellation: the case of Ulster Loyalism'. *Ethnic and Racial Studies*, 19(1), pp. 88-111.
- Ganguly, K. 1992. 'Migrant identities: personal memories and the construction of selfhood'. *Cultural Studies*, 6(1), pp. 27-50.
- Giddens, A. 1991. *Modernity and Self-identity: Self and*

Society in the Late Modern Age. Cambridge: Polity.

- Gillis, J. 1994. 'Memory and identity: the history of a relationship'. In: Gillis, J. (ed.): *Commemorations: The Politics of National Identity*. Princeton: Princeton University Press, pp. 3-26.
- Glenny, M. 1992. *The Fall of Yugoslavia*. Hammondsworth: Penguin.
- Golubović, Z. 1995. 'Social change in the 1990s and social character: the case of Yugoslavia'. *Sociologija*, 37(4), pp. 441-53.
- Hajer, M. 1995. *The Politics of Ecological Discourse*. Cambridge: Clarendon Press.
- Hayden, R.M. 1994. 'Recounting the dead: the rediscovery and redefinition of wartime massacres in late and post-communist Yugoslavia'. In: R.S. Watson (ed.): *Memory, History, and Opposition under State Socialism*. Santa Fe: School of American Research Press, pp. 167-84.
- Herzfeld, M. 1985. *The Poetics of Manhood: Contest and Identity in a Cretan Mountain Village*. Princeton: Princeton University Press.
- Herzfeld, M. 1996. *Cultural Intimacy: Social Poetics in the Nation-State*. London: Routledge.
- Hodžić, A. 1998. 'Intervju'. *Feral Tribune*, 660, 11/05/98, 40-41.
- Iveković, R. 1994. 'La Défait de la pensée'. *Kritika Centrizma/The Critique of Centrism* (*Beogradski Krug/Belgrade Circle*, 0), pp. 195-99.
- Jansen, S. 1998. 'Homeless at home: narrations of post-Yugoslav identities'. In: A. Dawson and N. Rapport (eds.): *Migrants of Identity: Perceptions of Home in a World of Movement*. Oxford and New York: Berg, pp. 85-109.
- Jansen, S. 1999. 'Identities, memories and ideologies'. *Social Anthropology*, 7(3), pp. 327-32.
- Jansen, S. 2000a. 'Victims, underdogs, rebels: discursive practices of resistance in Serbian protest'. *Critique of Anthropology*, 20(4), pp. 393-420.
- Jansen, S. 2000b. Anti-nationalism: Post-Yugoslav

Resistance and Narratives of Self and Society. Unpublished Ph.D. dissertation, University of Hull, Hull.
- Jansen, S. 2001. 'The streets of Beograd: urban space and protest in Serbia'. *Political Geography*, 20(1), pp. 35-55.
- Jansen, S. 2002. 'The violence of memories: local narratives of the past after ethnic cleansing in Croatia'. *Rethinking History*, 6(1), pp. 77-93.
- Jansen, S. (forthcoming). 'Remembering with a difference: clashing memories of conflict in Bosnian everyday life'. In: X. Bougarel and G. Duijzings (eds): *Bosnia: Picking Up the Pieces. Social Identities, Collective Memories and Moral Postures in Post-War Bosnia*. London: Hurst and Co.
- McKenna, T.M. 1996. 'Fighting for the homeland: national ideas and rank-and-file experience in the Muslim separatist movement in the Philippines'. *Critique of Anthropology*, 16(3), pp. 229-55.
- Povrzanović, M. 1997. 'Identities in war: embodiments of violence and places of belonging'. *Ethnologia Europaea*, 27, pp. 153-62.
- Rapport, N. 1997. 'Individual narratives: "writing" as a mode of thought which gives meaning to experience'. In: N. Rapport: *The Transcendent Individual: Towards a Literary and Liberal Anthropology*. London: Routledge, pp. 43-63.
- Ricoeur, P. 1990. *Soi-même comme un autre*. Paris: Editions du Soleil.
- Ricoeur, P. 1991. 'Life in quest of narrative'. In: D. Wood (ed.): *On Paul Ricoeur: Narrative and Interpretation*. London: Routledge, pp. 20-33.
- Scott, J.C. 1985. *Weapons of the Weak: Everyday Forms of Peasant Resistance*. New Haven: Yale University Press.
- Scott, J.C. 1990. *Domination and the Arts of Resistance: Hidden Transcripts*. New Haven: Yale University Press.
- Silber, L. and Little, A. 1995. *The Death of Yugoslavia*. London: Penguin/BBC Books.
- Thompson, M. 1994. *Forging the War: Media in Serbia, Croatia and Bosnia-Herzegovina*. London: Article 19.

Violence and Identification in a Bosnian Town: An Empirical Critique of Structural Theories of Violence

TORSTEN KOLIND

/29

The aim of this article is to account for some of the consequences of the recent war in Bosnia Herzegovina on matters of identification in everyday life among the Muslims of Stolac.[1] Or, to put it a little polemically, the lack of effect the war had.

On the level of experience it has had devastating effects. As ethnographic analyses of war experiences show, war often radically destroys the everyday taken-for-granted world of civilians, which, following the work of Scarry (1985), has been termed the 'unmaking of the world' (Das *et al.* 2001; Nordstom 1997; Povrzanovic 1997; Zur 1998; Maček 2000; Jackson 2002). Or, as one of my informants once said:

Nicolas [TK], you can't understand what the war has done to us. At first sight everything may look normal, but it's not. Nothing is normal. The war has changed everything.

But on the level of identification the changes have not been that absolute and radical. This came as a surprise to me during my fieldwork, and it also stands in some opposition to structurally inspired anthropological analyses of war and war-related violence.

Such analyses have primarily focused upon the inherent potential of violence and war to create identities. In a condensed form the line of reasoning goes like this: Identity is built on difference, and when differences become too small identity is at risk; violence then recreates or reinforces difference. This is, for instance, Blok's (2000) argument. He calls it, following Freud, the 'narcissism of minor differences' when violent practices are aimed at destroying resemblance and thereby creating 'the other'. As he writes in relation to the eruption of war in former Yugoslavia:

Once more we see the working of the narcissism of minor differences: the erosion and loss of distinctions and differences result in violence. (ibid.: 41)

Violence as a technique to create others is also present in Olujic's (1998) analysis of violence in Bosnia and in Malkki's (1998) study of Hutu narratives of Tutsi violence. As she concludes one of her chapters:

Through violence, bodies of individual persons become metamorphosed into *specimens* of the ethnic category for which they are supposed to stand. (ibid.: 88, original italics)

Violence creates the structural division on which identity is built: we are us because we fight against them and vice versa. This kind of argument is also present in Knudsen's (1989a) study of the vendetta on Corsica. Due to the FBD[2] marriage pattern the endogamous group grows and group distinctions become harder to make. The role of vendettas is therefore to recreate distinctions. Or consider Harrison's (1993) claim that violence in Melanesia has a structural function, that is, groups do not create war, war creates groups. As he sees it, both gift giving and violence (-giving) create social relations, which is contrary to the view of Mauss, who saw violence as the failure of the gift (Corbey 2000). Theorising on violence, Bowman (2001) suggests in his re-reading of Clastres that violence does not even have to be carried out physically to construct identity. Violence is a force that creates boundaries and may operate conceptually prior to manifesting itself in action. It is the imagining of violence that '…serves to create the integrities and identities which are in turn subjected to those forms of violence which seek victims' (ibid.: 27) and it is the imagining of violence against the other that is the medium through which (embedded) societies are represented to themselves.[3]

On quite a different scale but using the same kind of reasoning, Appadurai (1999) tries to understand contemporary ethnic atrocities worldwide in relation to a general culture of modernity. Modernity and globalisation, he claims, have disembedded social relations and created uncertain and alienated identities. In a grotesque way violence thus uses the body to recreate certainty and intimacy.[4]

I have two objections against structural approaches to violence, the second being the most essential in the present context. First, even if we accept the idea that violence creates unambiguous identities, such consequences should not be confused with explanations of why violence occurs in the first place. Taking Bosnia as an example, the war did not come from the bottom up, owing to social relations that had become disembedded or people being alienated (Appadurai's argument), nor was it the result of differences that had been erased. Rather, the war had its origin in a struggle for power on the political level (Ramet 1992; Cohen 1993; Malcom 1994; Naughton 1994; Bennett 1995; Woodward 1995; Glenny 1996; Hayden 1996; Gallagher 1997; Sofus 1999; Oberschall 2000). The war then to some extent generated its own dynamic, as for instance shown by Bax (1997; 2000), and war-related violence reinforced ethnic identities (Sorabji 1995; Olujic 1998; van de Port 1998). However, this is not the same as accounting

for the eruption of war, as the structural approach implicitly does, operating with a kind of thermo-dynamic argument: at a certain point differences become too small and therefore violence erupts.

Second, claiming that violence creates unambiguous identities only accounts for part of the process relevant for understanding the relationship between violence and identification. It is probably fair to say that violence plays a part in constructing a general polarised atmosphere of 'us and them', but this does not say anything about how people react or relate to such a dichotomised space of identity. As I see it, it is not violence that is creative, but rather it is people's reactions to violence that constitute the creative element.[5]

As regards Stolac, violence has unmistakably created potentials for unambiguous identities (Kolind 2002b) by politically ethnifying all aspects of every-day life. And national identities have on the public and political level been promoted as the only salient ones. Furthermore, today Stolac is totally separated ethnically on the local level of public organisation. But when analysing everyday identifications of the Muslims of Stolac another picture emerges, one that is more complicated and less clear cut. I have previously looked at how the Muslims of Stolac refrain from using ethnic stereotypes when explaining the war and finding out where the blame lies (Kolind 2002a).[6] Here I shall examine how the Muslims of Stolac, when identifying themselves as Muslims, often refrain from exclusive antagonistic identifications, but instead highlight coexistence, tolerance and inter-ethnic respect, all of which constitute patterns of identifications with clear ties to pre-war inter-ethnic social life.

I shall take a look at different kinds of identification of relevance to the Muslims, each responding to a different level: a local-patriotic identification on the local level, an ideal of tolerance on the national level, and identification with the Balkans and Europe on the global level. The general picture that emerges through these three identifications is that people use and mould already existing categories of identity to fit their new reality. Only on one front have I encountered what can be regarded as an exclusive identification that both sets the Muslims explicitly apart and stereotypes the ethnic other: when they identify with the role of the victim. The other identifications stress commonality, coexistence and tolerance.

This pattern of identification among the Muslims of Stolac is part of what I label a *counterdiscourse* (Kolind 2004), which is not necessarily a conscious or outspoken kind of resistance. But taking into account a) the creation of exclusive national identities on the public level among leading Muslims in Bosnia prior to the war, b) the crystallisation of ethnic identities due to the war-related violence, and c) the present division of power along ethno-religious lines in Stolac, I term the mere non-use of ethnic and/or religious identification, as well as the insistence on clinging to already existing ideals and categories of tolerance and coexistence, as a kind of resistance or counterdiscourse.

Before analysing the four different kinds of identification in present-day Stolac, I shall first a) outline some general features of the *public* creation of an explicit and non-ambiguous Muslim national identity throughout the war, and b) account for the *everyday pre-war* status of ethnicity. We are dealing with two different scales here, both of which are relevant to understanding the counterdiscourse among my informants as well as the anchoring of this counterdiscourse in pre-war everyday identification.

Muslim national identity in Bosnia Herzegovina

At the end of the 1980s and the beginning of the 1990s, the Yugoslavian society eroded and the country experienced religious revitalisation, party pluralism, a centralisation of power and most of all a surge of nationalism (Friedman 1996: 143-77, Ramet 1992: 176-86, Höpken 1994: 231-39, Malcom 1994: 193-212).

From the 1950s up until the 1980s, the Bosnian Muslims' interests were primarily secular, and one did find many of Tito's stern followers among this group of secularised Muslims. Religious Muslims only played a secondary role in the political life, and attempts at merging religion and national interest were harshly suppressed by the Communist regime. Therefore, a Muslim by faith and a Muslim by nationality were not the same. Just before the coming of the war and certainly throughout the war these two aspects became rather suddenly conjoined in public. Muslim national identity became increasingly defined on the background of faith, that is, Islam and religious institutions came to play a central role in the nationalist mobilisation of the Muslim community.

In 1990, inter-ethnic relations among the people of Bosnia Herzegovina were becoming increasingly tense, and with the coming of the multi-party election, Bosnia Herzegovina saw the emergence of distinct political parties representing each of the republic's three ethnic communities, which was a rather crucial change in the formation of Muslim identity. The Muslim Party of Democratic Action (SDA – *Stranka Demokatske Akcije*) was formed in 1990, mainly by devout religious members directly involved in Islamic religious activities, and in only a couple of months it managed to mobilise a huge part of the Muslim population of Bosnia Herzegovina (Bougarel, 1996: 96-97, Cohen 1998: 58-61, Malcom 1994: 218-22). Inter-ethnic tensions grew throughout Bosnia Herzegovina, but also became an important condition for the existence of nationalist parties, a factor making the nationalist parties cooperate in grotesque manners at times (Bougarel 1996: 98).

Muslim religious nationalism became further strengthened throughout the war. As the ethnic logic (the logic of carving up the territory of Yugoslavia and Bosnia Herzegovina along ethnic lines) gained more and more predominance – heavily supported by diplomatic activities of the international community – and as the war progressed with immense assaults on Muslims by both Croats and Serbs, leading Muslims became increasingly sectional, religious and nationalistic, and the space left for imagining a secular multiethnic society became more and more limited.

Religious Muslim nationalism spread throughout the society and its institutions – for instance, the Bosnian army (*Armija BiH*), which was multi-ethnic at the beginning of the war, soon became almost dominated by Muslims and controlled by the SDA (Cohen 1998: 69-70; Mojzes 1998: 95). In the educational system, religious Muslim nationalism increased with the introduction of Islam into most schools, and especially the subjects of 'History' and 'Language and Literature' changed in accordance with the ongoing process of formation of Bosnian Muslim identity (Maček 2000: 173). Also language gained national importance, which is not surprising, since language is traditionally an important signifier in the creation and imagination of national identities. Leading Muslims renamed their language Bosnian and furthermore tried to introduce 'h' into words formerly without it (because the 'h' sound was associated with Turkism) or to use as many Turkish synonyms as possible. Greetings were also affected: on an everyday level, the choice of greeting became a political act and religious greetings became more frequent. Street names were changed into old or new Muslim names, the colour green (associated with Muslim religious identity) became increasingly important in symbolic practice, religious holidays became more and more popular, and TV and radio broadcasting became oriented towards promoting a Muslim religious national identity, and so forth (Maček 2000: 172-86).

A last example of Bosnian Muslim nation building and the politicisation of Muslims' ethnic identity is the decision taken by the Bosniac Assembly (*Bošnjački Sabor*) in September 1993 to replace the national name 'Muslim' with the new name 'Bosniac'. This can be seen as an attempt to resolve the issue of national and territorial claims on behalf of the Muslims (Bringa 1995: 33-36; Bougarel 1996: 109). 'With the new name the politicians stressed the transformation of the Bosnian Muslim community into a political and sovereign nation, closely linked to the territory of Bosnia Herzegovina' (Bougarel 2001: 8).

Thus, despite contradictions and ambivalence, on a general level Bosnia Herzegovina saw the rise of a Bosnian Muslim nationalist party and of nationalist politics in a relatively short period of time. And though secular opposition existed throughout the war both outside and partly inside the ruling SDA (the Muslim nationalist party), a religious, antagonistic Muslim nationalism dominated in many public and political settings, a situation not many people would have imagined only a few years before. But the situation was special. A war was going on, the Muslims were under attack from both the Croats and the Serbs, and the international community (aside from the Islamic world) had abandoned them. Bosnian Muslim nationalism grew, therefore, not because it had been lying dormant throughout the years, but because the logic of the war created it. In fact, Muslim nationalism had always been weak, among other things because attempts at expressing a solid and unifying Muslim identity had traditionally been centred on confession and/or cultural traditions or pan-Islamism, aspects hindering the growth of a strong national identity (Bougarel 1996; 1997).

Bringa (1995: 36) sums up the tragic creation of Muslim nationalism in this way:

The Bosnians have apparently been organized into tidy, culturally and ethnically homogeneous categories, and the Muslims seem finally to have become a neat ethno-national category its neighbours and the international community can deal with and understand. They have been forced by the war and the logic of the creation of nation-states to search for their origins and establish a 'legitimate' and continuous national history.

Bosnian Muslim identity in everyday practice

While Bosnian Muslim nationalism only seriously began developing in the late 1980s,[7] this does not mean that a strong sense of ethnic belonging was absent among the Bosnian Muslims, particularly as regards the countryside. But it is a sense of belonging one should not confuse with the Muslim nationalism of the late 1980s.

Several writers have argued that Bosnia Herzegovina has never been truly politically modernised, that communitarianism (the prevalence of ethnic identities in social relations) has continued to be the most prevalent characteristic of the society, that the state project never succeeded, and that over-communal identifications never managed to surpass local ethnic identifications (Simić 1991; Bougarel 1996; Sunic 1998). The central element in these ethnic identifications has been religion: Serbs have been Orthodox, Croats Catholic, and also for the Muslims a strong relationship between religion, faith and ethnic identity existed (Lockwood 1975; Bringa 1995).

To clarify the characteristics of this local identification in which religion plays a constitutive part, Bringa (1995) explores the native concept of *nacije*. *Nacije* is a combination of religious, cultural and social identities, something to which people felt strong emotional attachment, something into which one was born and socialised, something that was normally 'inherited' from parents, something unquestionable. National identity in Yugoslavia was, on the contrary, something one could choose from the options given by the state, and it often lacked the essential feeling attached to *nacije* in everyday identification. Bringa translates *nacije* as 'ethnoreligious identity' (ibid.: 22) and concludes that the Muslims of Bosnia Herzegovina had a strong sense of ethno-religious identity but a weak sense of national identity. They referred to their collective identity not in a idiom 'of shared blood and a myth of common origin', but 'in an idiom which de-emphasized descent and focused instead on a shared environment, cultural practises, a shared sentiment, and common experience' (ibid.: 30). The difference between ethno-religious identity and nationality was then exploited in the nationalist projects of the late 1980s in Yugoslavia; in Maček's words:

...[T]he new national political elites could mobilise the ethno-religious notions of belonging into the new national projects of constitution of sovereign states for Muslims, Serbs and Croats respectively, [by] filling the new national identities with the old ethno-religious feeling of essential belonging. (2000: 157)

In what follows, I shall highlight four related characteristics of pre-war embedded, local and everyday ethno-religious identification.

First, though Islam was an important identity marker for Muslims in Bosnia Herzegovina it should not be understood simply as sets of clearly definable rules of practices, but rather, as Sorabji (1996: 54) suggests, as 'a domain of loose moral imperatives'. She mentions hospitality, cleanliness, generosity, honesty, kindness, courtesy, industry and the like. Though these values are central to Muslims they are not exclusively Muslim virtues, but rather part of a general moral codex in which, for instance, there are overlaps with the ideology of Communism that says 'work hard, don't cheat your neighbours, redistribute your wealth, and so on' (ibid: 55). Sorabji seldom found her informants evaluating each other's actions in religious terms. It was thus, as she reports, not 'haram' (Arabic: forbidden by God) to slander someone, but rather 'ne valja' (Serbo-Croatian: no good), and it was not 'sunset' (Arabic: recommended by the Prophet and pleasing to God) to wash your hands before meals, but rather 'fino' (Serbo-Croatian: good) (ibid.: 55). Lockwood, who did fieldwork approximately fifteen years before Sorabj and Bringa, came to similar conclusions. Though Islam was much more visible and pronounced in 'his' village, he emphasised that: '[r]elatively little stress is placed upon religious doctrine; much more important are outward signs and symbols' (Lockwood 1975: 48).

The second characteristic is the contrasting aspect of Muslim identity. It is not so much that the Catholics and Muslims are different *per se*; rather they continually accentuate differences in establishing identity. The different ethno-religious groups needed each other (or each other's otherness) to construct identity, and this otherness – though using, for instance, religious practices – was embedded in a local setting and everyday practices in contrast to the national identities constructed and represented in the public and national spheres. If modernisation means an increasing disembeddedness of social relations, then greater parts of Bosnia Herzegovina can be considered not very modern. Identity in Bosnia Herzegovina was to a great extent premised upon actual local face-to-face interaction. So religion was not that important in itself but rather due to the difference it was able to make in everyday local life.

Third, though pre-war ethno-religious identity was fundamental in everyday life, it was one among many types of identity. Work relations, type of education, class, gender, degree of culturedness were all identities that in certain situations could be more important than ethnicity (Allcock 2000: 170-211). In other words, relatively 'ethnic-free' zones of social interaction existed in which identification depended on the context. Ethno-religious identification should furthermore be contrasted with higher levels of identification. In *Dolina*, Bringa (1995: 65-73) reports, ethno-religious identity was connected to the household, the family and sometimes the neighbours, but at a village level a unifying localistic supra-ethnic identification held sway, defined by an ethos of hospitality and neighbourliness (*komšiluk*). The practice of *komšiluk* was, as Bougarel defines it, a 'permanent guarantee of the pacific nature of relations between the communities, and thus a security of each of them' (1996: 98). At an even higher level of identification was the Yugoslav state, which with its various institutions penetrated village life and enforced unification.

The fourth and last element relates to the capacity to live with difference, an ability which many Bosnians have developed throughout the years and one that should not be romanticised. Mono-ethnic communities existed all over Bosnia, and as Lockwood (1975) shows, inter-ethnic interaction in, for instance, a weekly market does not necessarily lead to further ethnic integration. Furthermore, we also have examples of villages and areas where inter-ethnic coexistence has been marked by continual and cyclic outbursts of violence and inter-ethnic (blood) strifes (Bax 1995; 1997; 2000; Boehm 1984); researchers have also observed how traumatic memories from the Second World War have been passed on to the younger generations (Denich 1994). Some even talk about a myth constructed by Western intellectuals in which the Bosnian mentality is presumedly tolerant. The reality is rather – it is argued – one of inter-ethnic distrust (Simić 2000). I shall not revitalise this often rather intense debate that has been running in the discipline of anthropology especially in and after the recent war (see Brandt (2002) for a thorough discussion). My point is rather that throughout their socialisation, the people of Bosnia Herzegovina had come into contact with members of different ethno-religious groups in different areas, both in the public sphere and often also in the private sphere, and one should not forget that the young and the middle-aged had lived in peace for almost fifty years when the war broke out in 1992. Ethno-religious identity could then be a significant factor shaping the content and atmosphere of inter-ethnic interaction, but it was seldom a hindrance thereof, and the interaction between the different ethno-religious groups was often characterised by respect. A central point here is that differences were not downplayed; equality as sameness was not the ideal, as can be seen, for instance, in Scandinavian contexts (Gullestad 1992). Instead, differences were persistently nurtured, but in a way that Georgieva has described as 'familiarisation of differences' in a different context (Georgieva 1999),[8] that is, the net of interconnected lines compromises all levels of everyday life, eliminates otherness, and changes it into familiar difference. The 'others' are therefore 'perceived not as a menace, but rather as an inseparable part of the complex world of everyday life' (ibid.: 68).

The war and the war-related nationalism in many ways attempted to destroy these aspects of the Bosnian Muslims' everyday practice of identification. Religion became much more closely connected to a more doctrinaire reading of Islam. Ethnic identity became increasingly disembedded and was represented in new and more rigid and politicised categories detached from people's everyday experiences. Ethnic identity surpassed all other identities, colonising every aspect of life, and was increasingly presented as a hindrance for interaction.

So far I have presented two rather opposing approaches to ethnic identity: a religiously based public nationalist identity highlighting difference and arguing for ethnic segregation, and a local embedded practice of ethno-religious identity stressing coexistence. Following the logic of the structural approach to violence one should expect the former to prevail after the war. In the following I shall analyse four different types of identifications currently at work among the Muslims of Stolac, which in sum depict a world characterised by complexity, ambiguity and contradictions rather than clarity, certainty or simplicity; a world resting more upon pre-war everyday inter-ethnic experiences and cultural categories than upon the sudden emergence of a public discourse of nationalist exclusion. Inertia seems to be a leitmotiv.

Muslim identifications

Considering the logic of the structural approach and that the war was presented as a struggle based upon religion and nationality, one might presume that Islam and Bosnia are the two most important signifiers in creating a distinct and unambiguous Muslim identity. It is my impression that they are not.

Bosnian nationalism?

Bosnia Herzegovina as a nation-state is not a central element in the identification of the Muslims of Stolac. Neither are the typical Herderian aspects through which a nation is often imagined relevant: language, history, blood, flag, race, the dead soldier, and so forth (Mosse 1990; Anderson 1991; Knudsen 1989b; Foster 1991).

For instance, when talking about actually fighting the war, people seldom say they fought for Bosnia, territory, cultural traditions, and so on; instead they claim to have fought for survival, for their homes, property, families, and because they had nowhere to run. People are not proud of Bosnia, but they are proud of having survived. One of my informants once said that 'everyone who survived the war is a success'. For the Muslims, survival was not at all taken for granted during the war.

Emir fought in the war.[9] Once during an interview he showed me some pictures of himself and a friend in uniform, smiling and holding machine guns in the air. Emir's mother felt embarrassed and did not want me to see the pictures; Emir, on the other hand, was proud of what he and his comrades did in the war, but his pride was not related to a Bosnian nation:

Emir: We did not fight for any ideas at all. We did not try to create some kind of state or country; we just did it in order to survive. I fought at the front with my weapon just to prevent them from coming and taking my mother. So that they wouldn't kill her or the children who were with her. There were no ideas in my head at that time, because I was hungry and thirsty and without anything. There were people who only had 10 bullets in their guns, and they fought against tanks in that way. And in that situation you don't think about creating a state.

During the war, the Muslims of Herzegovina were pressed from two sides and had no possibilities of escape. Many of my informants see this as a major tactical mistake on behalf of the Croats and Serbs. They would have escaped if only they had been given the opportunity. People therefore feel they became Bosnians not by choice, but by fate.

Even today, *Bosnia* does not serve as a meta-narrative that can bestow meaning upon one's sufferings, traumas and material losses. People are simply not proud of their country and do not identify with it. They are fed up with the corrupt and worthless politicians; the educational system does not work or qualify people for jobs – only private connections do; the unemployment rate is disastrous; people do not get their pay or pension in due course; the division between rich and poor has increased strikingly; there is no future for the kids, and so on and so forth. Similar attitudes are revealed in a survey conducted in Bosnia Herzegovina in 2000, which reveals that sixty-two percent of all youth in Bosnia Herzegovina would leave the country if given the opportunity, and among those wanting to stay only ten percent gave patriotism as a reason (UNDP 2000).

People also feel that Bosnia Herzegovina has lost prestige internationally. As Amra said, making a sad joke, 'today for us Albania is the West'. Albania is reckoned as by far the most backward country in Europe.

Below is a quotation that, even though it offers a rather hopeless perspective, expresses what many people feel at least part of the time.

Lamira: He [her husband] was a soldier; I was a nurse and worked all the time at the hospital with wounded soldiers. People came in to the hospital without legs, arms, totally destroyed people with psychic traumas. I gave birth to a child, and then another. We have yet another boy, but he is not here now. He is three years old. Everything is a big trauma. We

stayed here all the time as *domoljubi* [those who love their country]. We hoped things would be better. Now our country does not make sense. No law. Anarchy, anarchy...criminals. It's like the things we only saw in the movies before.

Religion

In the war, religion became politicised and served as a vehicle for expressing a separate Muslim identity. My findings suggest that today religiosity has found a level more or less equal to the pre-war situation and does not play a major role in people's everyday practices of identification (see also Bougarel 2001). To be a good Muslim means to behave decently and to be morally upright and does not relate to religious practices. For some, Islam is totally unimportant. People generally have a very pragmatic relation to their faith.

During the war, people were more religious than today, which they find normal due to the immense psychic pressure they endured, and consequently there is nothing strange about the decline of religious observance when the war stopped. Osman mourns, however, the disappearance of religious solidarity, feeling that it vanished too soon and that people forget the things that happened too quickly. Often people lament the loss of sense of community and social interaction, but the issue is not religion but *drustvo* (approximately: social intermingling), and I also suppose that Osman mourns the lack of solidarity in general more than religious solidarity.

Following the Koran one should not drink alcohol, but the majority of my informants had a relaxed attitude to alcohol. I do not recall being in a Muslim home where I was not offered *loza* (grape brandy), and often it was homemade. Homemade wine was also highly estimated. Not that people drank a lot, but it was just not an issue. Only few people regularly attended the Mosque (or what was left of it) – the same number of people as before the war, I was told. At the Friday prayer there were about five to seven young people, the remaining twenty to twenty-five persons being older than sixty. And never did I observe people praying during the day/evening. Possibly the use of religious greetings has increased, but only rarely did I hear them; instead people said *dobra dan, šta ima, kako si/ste, sta radi* to each other, all of which are secular and religiously neutral greetings.

Before the war, children who attended *mekteb* (Koran school) would sometimes learn to recite the Koran in Arabic. Mensur's wife knows some of the

Koran in Arabic, but she does not get much credit for that. As Mensur said to her once: '...well it doesn't matter – God understands all languages. Why then should we pray in Arabic?'

The last example refers to the Ramadan of November and December 2000, when I did fieldwork. As a codified ritual in Islam, Ramadan is an individual statement of one's personal commitment, and it could easily be used as a way to identify and collectively emphasise Muslim identity. But not many people in Stolac observed the fasting. Some did it some of the days; others ate and drank less during this period. Often only one or two in a family fasted or partly fasted. In fact, it was not a big issue. Several times I heard people make jokes about those who fasted. Once when Sefer's mother talked about waking up before sunrise to have something to drink, her husband laughingly replied, 'this is surely not for me', and later when she told me that smoking is not allowed during the Ramadan, as did her daughter-in-law, her husband said, 'well then Ramadan is good for something'.

Moreover, those who fasted or partly fasted focused mainly on the practical aspects: how it is easier to fast in the wintertime as the days are shorter, how to wake up at four o'clock to have a morning coffee, how drinking a little water is allowed if one needs to take pills, and so on. Or they said that they did it because it was a nice tradition.

These examples are not meant to illustrate that Islam does not have a place in people's lives. Many people see themselves as believers and respect their God. And often people said that if the Croats and Serbs really believed they would behave like the Muslims who also believed. In other words, to be a believer is about being a decent human being, behaving properly, taking care of your family and fellow man, and so on. It is not about knowing the Koran, praying, fasting, attending the Mosque, greeting the right way, and so on. In fact, I often heard people dissociate themselves from more dogmatic religious manifestations. Instead they want to stick to what they regard as the typical Bosnian way of practising religion. Anvere is what one might regard as a normal believer, to the extent that such a concept makes sense; she believes a bit, fasted a bit during the Ramadan, and she knows the Koran a bit. She said:

The Muslims here in Herzegovina were not particularly religious. Maybe it is different north of Sarajevo, in Zenica. The Serbs weren't either, but the Croats go to church every Sunday... During the war we were more religious – we were scared...Before the war I never saw anyone with a veil, but during the war there were quite a few donations from Saudi Arabia and Iraq, and they brought it with them. But it is not autochthonous [real] Bosnian. The Muslims here are different... It seems as though everybody today is afraid of Islam [everybody in the Western world]...but our Islam is different than in the Arabic countries; we are not particularly religious.

Throughout the war and afterwards, religious symbols have been manipulated, religious differences reinforced, and many people have been killed or had to flee in the name of religion. Today many of the Muslims in Stolac are fed up with politicians' nationalistic religious rhetoric. People do not identify much with their country, neither during the war when they mainly fought for survival and not for a state, nor today when the positive symbolic value of Bosnia Herzegovina is slight. Religion is more important; I only met a few people who declared themselves atheists. But their religious identity has not colonised more of their identity as Muslims than it possessed before the war; on the contrary, it seems as though people distance themselves from a radical and dogmatic form of Islam.

In the remainder of the article I shall analyse how the Muslims of Stolac define themselves as Muslims, considering the fact that national and religious identities do not suffice.

Localistic identification

Local patriotism in Stolac is strong. Not everybody is necessarily content with living in Stolac, especially the young find the city boring; nevertheless, most acknowledge that the city is (or at least was) special, with a unique atmosphere, beauty and spirit. This local patriotism focuses upon Stolac as a centre of open-mindedness, a meeting point for different cultures, and states that Stolac does not exist without its three ethnic groups. But at times it also serves as a critique of the Croats, who are considered blameworthy.

Coexistence

The ideal of coexistence is of central importance to the Muslims of Stolac and is often associated with a typical Stolac mentality: people from Stolac are seen as simply predisposed to co-exist. To be a real citizen of Stolac is to accept and enjoy multiculturalism; in fact, it runs in the cultural genes of the people. They see Stolac as a crossroads for different cultures and stress that all newcomers have been welcomed. People mention, for instance, the Jewish grave just outside town commemorating the Rabbi Mosha Danon from Sarajevo, who died on his way to Jerusalem and was buried here in 1830. Or they refer to ornamentation on one of the world-famous Stećaks at the *Necropolis Radimlja*, two kilometres outside town. This stone is decorated with the figure of an upright man whose hand is lifted upward at the elbow and whose palm is open, greeting the spectators. This hand is read as a gesture saying that all strangers are welcome.

Nusret expressed his idea of local inter-ethnic coexistence in Stolac in the following way:

Our wish is that people return. We want the Stolac spirit to return. And that spirit is the citizens who lived here before the war – *bosnjaki, i serbi, i hrvati.*[10] And as for those coming from outside [*strani*], the best thing would be if they left.

Another element of the Stolac spirit relates to the feeling of oneness with the beauty of the city. Several times I was reminded that before the war Stolac was a candidate for UNESCO's list of cities worthy of preservation. People tell about the four picturesque Mosques, the big city market built in old Ottoman style; *Begovina*, the once so beautiful housing complex owned by the Rizvanbegović family; the old watermills; the Hotel Bregava, a newly constructed but architectonically well-integrated building; all the cafes, the shops, and so forth. Today everything is mined or burned. They tell about the lovely Mediterranean climate – 'It is November and I haven't even fired yet, this city keeps me young', as Senad said – about the lovely Bregava River running through the town, where one can swim and fish and which provides water for the gardens situated alongside; about all the fruits: 'in the summer we have kiwi, peaches, apricots, figs, apples, grapes, pomegranate, plums. In the summer you don't have to be hungry – fruits are so plentiful', as Sefer's mother said. And they tell about all the industries and factories in town, which employed a lot of people and exported to the whole of Yugoslavia and abroad. In short, they are/were very proud of their city.

The reason why I write 'are/were' is that the people of Stolac often have ambivalent feelings for their town. Visual memories from before the war, pictures from pre-war Stolac, present images, dreams – they all merge together. Some say they cannot remember what the town looked like; others say they cannot drive through the town without seeing the 'real' Stolac in their mind's eye. Everybody thinks the city has become ugly, but nonetheless I was asked – often in the expectant way one normally asks visitors to one's town – how I liked the city, and it was obvious that the appropriate and polite answer would be 'it is very nice', but I just could not say it, because Stolac is ruined, dirty and ugly. And when I said that, people became quiet for a while and then said, 'yes, that's right'. Despite such contradictory feelings, Stolac is generally constructed in people's minds as a beauty, as a pearl on earth with a fantastic climate. When people are in a realistic mood their Stolac does not exist anymore; instead they see an ugly city dominated by Croats, but when talking about Stolac in a more insubstantial but still physical manner, the old and real Stolac seems to rise from the dust and people show great pride for their town and consider it the best place on earth. It is as though the beauty of the city coalesced with the spirit of coexistence, and together they functioned as a symbolic or cultural space or an idea with which one could identify.

However, identification with pre-war Stolac is also a critique of the Croats who demolished the old Ottoman-inspired town and who today dominate it and are trying to create a new Catholic and modern Stolac. So just talking about the past and pre-war Stolac is a strongly political act. Furthermore, people say that the Croats do not love the city or do not have any real attachment to it. If they did they would not have treated it the way they did, and at least they would have cleaned up.[11] As Nermin said, 'They want a dump rather than cultural monuments'. Implicitly they are saying that the Croats do not deserve the city.

The last element in local patriotism has to do with the value attached to everything homemade – *domaci* – in particular, food. People would proudly let me know that the bread, the cheese, the yoghurt, the juice, the *loza* (grape brandy), and so on, which they served were *domaci*. The homemade food was regarded as superior because the food products were produced in the Stolac area without chemicals or fertilizers, which in turn made the *domaci* clean (*čisto*). The positive connotations of everything homemade affected other kinds of classifications. For instance, people would say that they did not have a problem with Croats originating from Stolac, and they labelled them *nasi domaci hrvati*: our home-made Croats.

Struggle for local identity

There are not many visible Muslim markers of identity in Stolac; the Croats dominate the public space and the Muslims do only little to change this. Many are probably afraid of the reactions that visible demonstrations could provoke (Mahmutcehalic 2001), but more importantly this attitude is in line with the Muslims' self-perception. As I demonstrate below, the Muslims identify themselves *by not using* the same methods and rhetoric as the nationalists (the Croats and Serbs). Waving a flag or making nationalistic graffiti simply would not give respect. Let me nevertheless turn to some of the few physical markers of identity I did encounter, which I relate to a 'Muslimness' and just as importantly to the Stolac spirit.

Before the war, a modest local historical pamphlet was published in Stolac with illustrated articles on the history, archaeology and architecture of the area. After the war and the total destruction of the Stolac heritage, these booklets suddenly gained new importance in 1997 when some prominent people originating from Stolac initiated a reprint of these booklets as well as of a section comparing photographs of Stolac then and now in a book format (Dizdar *et al.* 1997). The book became evidence of a time and a city which no longer existed. The articles, which nearly nobody read before the war, now became an important source of identity for people living in Stolac, since the book helped to mentally (re)construct the picture of beautiful pre-war Stolac that the Croats had destroyed and now deny ever existed. Without these publications the ethnic and cultural cleansing of Stolac would have been total.[12]

Another source of local identity is the Bregava River running through town. People often talked about the river. 'You should be here in the summertime, Nicolas – it is so beautiful then. We sit down at the river, we barbeque, play football, and when it's to hot we go swimming', or, 'the river is so clean that you can drink from it'. Sometimes people would just say, 'at least we have *Bregava*'. I think that *Bregava* functions as a central component in the Muslims' identification with Stolac and the Stolac spirit. Here is a small excerpt from my notebook written after a conversation with Anvere:

...[S]he was laughing a little at the name Sanpero, the name of the restaurant where the cultural centre was before. She also said that before [the war] there was a gallery next to café Galleria, where artists from Stolac exhibited. She then tells me that it is so strange to realise that all this is destroyed and gone. The town really had soul before, a spirit: 'now they have destroyed everything, the Mosques, Begovina...it is only Bregava which they haven't destroyed'. When she said this I realised why people talk so much about Bregava; maybe they did before, but now it is of extra importance, because it is the only thing people have left. It is like a symbol saying 'we have not given up'. Bregava is what is left of the real Stolac.

Or, as Alen answered when I asked him if the *Bregava* River is the spirit of Stolac: 'Yes, it is the only thing left that is ours'. One of the Muslim cafés in Stolac is called *Café Benat*. The name refers to the place in town where the river is most scenic. And the name is not chosen arbitrarily; rather it is part of the seizing of the symbolic/physical space of Stolac.

In Stolac is another Muslim café, *Café Galleria*. It marks a more contested struggle for regaining a territorial marker – not Muslim territory but local territory, so that the Muslims can say, 'we are *stolčanici*' or 'the Stolac spirit still exists'. The café, located in the centre of town opposite three Croatian cafés, is popular. The owners of the place spend a lot of energy making it cosy and pleasant. It has big panorama windows so that one can look out over the centre of town, outdoor serving, flowers on the tables, and on the walls hang beautiful black and white drawings of pre-war Stolac – drawings being part of the mental (re)construction of the pre-war Stolac. The café has been destroyed twice by Croats. Accepting the café would be tantamount to accepting the presence of the Muslims of Stolac. It should also be taken into account that Muslim houses were also continuously mined or burned (especially in 1996-1998) in order to frighten the Muslims from returning. On both occasions the café has been reopened, which is of great importance to many Muslims, not in an antagonistic way, but rather as a symbol of the integration of the city and the possible coexistence of the ethnic groups in Stolac. Several people said that Croats also went to the café; however, I never met any. I suppose that people wanted them to use it and thereby legitimate it and play their part in constructing the Stolac spirit. And though the owner was one of the more antagonistically minded people in Stolac, he as well as many others hoped the café could play a part in reintegrating people in Stolac, and they saw it as a fight for the city and the Stolac spirit of beauty and coexistence.

These three examples are some of the few visible attempts at redefining the physical/symbolic space of Stolac that I encountered. In addition, there are all the minor daily acts, which are primarily about creating an everyday world but which nonetheless also relate to the creation of the Stolac spirit. It is difficult to establish a dividing line between the everyday construction of small life-worlds and the more visible conquests of physical/symbolic space – for instance, families who keep the road outside their house and garden extremely clean by sweeping it every day, even though the house next to them lies in ruins with a huge pile of litter in front. Are they just cleanly people who in an ugly and ruined city try at least to keep some dignity? Or is it possible to see such acts as efforts to clean up (parts of) Stolac to create a contrast to the total decay in which the Croats have left the town? My guess is that both alternatives are true. And what about returning to a demolished house, reconstructing buildings, insisting on being positive towards the future if for nothing else then for the children, the opening of Muslims' shops, baking cakes for a birthday party, tidying op the street, planting flowers, tilling the soil, going on visits, and so on – all these are actions that attempt to recreate and normalise life, which also relates to creating the Stolac spirit. The Stolac spirit is about the creation of a physical/symbolic space – that is, a space existing in reality: for instance, *Bregava, Galleria*, the Mosques – and a cultural space, which is more a (re)construction of ideals and the aesthetic of the town with which one can identify. The creation of a cultural space is simultaneously a struggle for physical space, and the struggle for physical space is also an attempt to reconstruct Stolac as a cultural place and thereby as a source of identification.

The Stolac spirit (the ideal of coexistence, seeing Stolac as a beautiful city and the importance of things homemade) probably existed before the war but has gained new significance today. The Muslims returned to a city from which they were expelled, their houses were destroyed and/or robbed, everything Muslim was destroyed, and the Croats now control all public administration and buildings, so to be a *stolčanici* is not that certain anymore. Hence, just to talk about the beauty of the city, the excellent

climate or the autochthonic and tolerant soul of Stolac, to open a café, stroll through the streets, use the Bregava River, produce and praise homemade food, identify *nasi domaci hrvati*, and so forth, is now more or less politically significant. The Stolac spirit is a source of identification but it is also a sort of political manifestation, saying we (also) have the right to live here. But it is important to distinguish such a local identification from a nationalistic or religious one. Evoking the Stolac spirit is not about being Muslim, but about being a Muslim citizen of Stolac.

Ideal of tolerance and coexistence

The Muslims of Stolac hold an ideal of tolerance and wish for coexistence with the other ethnic groups, but they are also rather conscious of this tolerant attitude when it is used to identify Muslimness in a self-reflexive manner. People's talk about the past and their steady wish for coexistence are linked together, for when they talk about the past they focus upon how everybody lived peacefully together and how ethnicity did not matter. This fusion reflects people's actual experience, their pre-war habitus. But talk about the past is not just nostalgic, it is also a political commentary on the present, and it is a source of identification.

Differences were an advantage

Talking about the past, people often stressed that cultural diversity enriched everyday life. People learned from each other, had more festivals in which to participate, and their social network was large. Ethnicity mattered, but the different cultural practices were not related to a larger political (nationalistic) framework. They were just differences articulated in relation to everyday habitual practices of identification, as also argued above in the section about the identity of everyday Muslims before the war.[14] Today people mourn the loss of these differences, or rather the respectful and habitual handling of them.

The war has highlighted and created ethnic differences, but it has also changed the positive connotations related to these and instead tried to promote an ideology where ethnic dissimilarity is seen as a hindrance for interaction and a threat against cultural survival. Therefore, talking about ethnic differences and evaluating them in a positive manner is a means to resist the rhetoric of ethnic separation that nationalistic politics communicates today in

Stolac. And for the Muslims it is a way to identify themselves by saying 'we have no problems interacting with the different ethnic groups; we are a tolerant people'.

Take, for instance, Osman, who tells how the Muslims' tolerant attitude during the war was turned against them.

In school I had Serbian and Croat friends. And for instance at Bajram, I invited my Serbian and Croat friends home, and at Christmas, the 25th [of December] and the 7th of January, respectively, the Muslims were invited home to them to roll eggs, but you do not do that today; it is unthinkable. Before the war the other cultures were an advantage for us; if there are only Muslims everybody knows each other, but with other religious communities we learned from each other – it was enrichment. Differences did exist, but it was a difference that enriched us. For instance, one participated in a Serbian wedding and saw how they did it, had some different food. But with the war the differences became our disadvantages. We were conscious about being different before the war, but it was a difference that enriched us. With the war this difference was turned against us [They were targeted because of their different ethnic identity].

So in this perspective, difference, respect and tolerance are three inseparable issues.

Nationality did not matter
But people also said that before the war there were no differences; everybody lived together and had a good time, and no one cared about ethnicity. At first glance this would seem to contradict the aforementioned views, but in fact they are rather similar. When my informants say they were not aware of ethnic affiliation before, it is probably not true, but this is not the point. They knew which religious community people came from. What they are expressing is that they did not care. This is in keeping with the official ideology of the Yugoslav regime (Brotherhood and Unity), which tried to make ethnicity a thing of the past, and to many of the Bosnian generations from after the Second World War a pan-ethnic Yugoslav identity was important.

Nijaz, a man of about thirty-five, depicts the days before the war as joyful, a time when ethnicity was unimportant. I think this memory is important to him, enabling him to imagine a better future and reminding him of what he is, of his values, inasmuch as he now lives in a totally different context.

TK: Do you remember how life was before the war?
Nijaz: Let me tell you one thing: everybody had their life. We had work – I began building my house before the war, I also started with my business and I was about to open a coffee bar. Things were going well for me. I had a good group of friends. You are born here in Stolac, grow up together and have your friends until you get married, and I married before the war. We had a circle of friends. And that's enough…[…]…You have a job, family, some everyday obligations; it was a normal life. You knew where to go in the evenings, and that you would meet your friends at that place. Then came the war and turned everything upside down. Now I can find some of my friends all the way from Canada to Australia. And not only Bosnians; also Serbs and Croats. I had a big circle of friends; we were about ten to twenty in that circle. It was my generation who was in it. Before the war it did not matter if you were Serb, Muslim or Croat. You did not know what people were, you were not weighed by it …[…]…When I was married, my best man was Croat, and at my boy's first birthday his godfather was Serb. Now they live in Canada. That's how it was.

The contradiction Nijaz expresses – that 'you did not know what people were' and at the same time, in line with the tradition of respectful interaction between the different ethnic groups, his friendship with and deliberate choice of a Croat and a Serb to be best man and godfather, respectively – does make sense. Firstly, claiming to be unaware of the ethnic identity of one's friends simply expresses indifference. Secondly, the habitual pre-war pattern of tolerant and respectful inter-ethnic interactions is now objectified and constructed by many Muslims as conscious knowledge, relevant in identifying the present Muslim identity. This should become more evident in the following examples.

People sometimes told me they have Croat friends, mingle with Croats and use Croatian cafés and shops and vice versa. Sometimes this is true, but often this is not quite correct. It was as if people wanted to assure me and themselves that they still uphold the ideal of tolerance and inter-ethnic interaction. Sefer, for instance, told me when I first met him that he had several Croat friends and acquaintances. But some months later when I asked him if he knew some Croats I could interview, he said that he did not really know any. And once when I invited Mensur, who had told me that he frequented the Croatian cafés, for a cup of coffee as we were standing outside a Croatian coffee bar, he said 'yes, but then we go to Galleria', which was the nearest

Muslim café, even though he was in a bit of a hurry. The point is that it is important for people to state that they are tolerant and to identify with the ideal of coexistence, but in practice this is rather difficult.

Critique of the Croats – identification of the Muslims

In talking about past inter-ethnic interaction or present tolerance, sometimes the Muslims blame the other ethnic groups – especially the Croats – for abandoning these ideals. This blaming and the implicit identification of oneself as tolerant is part of the same logic.

Below is an excerpt from an interview with a schoolteacher and one of her colleagues. She lives outside Stolac, in an area with many Croats. She wishes for a better future but is rather despondent.

TK: Did you know who was Muslim or Serb or Croat before the war?
Teacher: Yes, I did, but there were no differences; it did not matter. On the street in Stolac where I lived, there lived Serbs, Croats and Muslims. You received the flats from the factory where you worked and nobody asked about who the neighbours were – if you where Serb, Croat or Muslim. There were also mixed marriages. And nobody felt it mattered. We were all as one. My first neighbour was Croat, and my sisters and I raised their children. We were close. Not a day went by without visiting each other. We talked about everything. We knew each other's problems. Today she [the Croat neighbour] doesn't even bother to visit my parents, or even to talk to them. Our window is placed next to their terrace; you can jump from our apartment to their flat. It is senseless...when you realize all these things...you have to hope...but it is difficult.

Other examples more explicitly touch upon the issue of identification. People said, for instance, that 'Muslims read Serbian or Croatian literature but they do not read ours', or 'we read Cyrillic letters but they do not learn our alphabet', or 'we listen to their music, but they do not listen to ours'. Comparing in this way means that 'our' tolerance in fact exists at the expense of 'their' intolerance and ignorance. Whether the intention is to criticise them (Croats and Serbs and the whole nationalistic project) or if it is to approach what it is to be Muslim is hard to tell. It is probably both.

The war has forced people to reflect upon pre-war practices in order to make some sense of it all. Ethnic difference, including the way this difference was nurtured, articulated and created before, has become objectified and part of the Muslims' conscious identification. Practices of interaction (cp. the 'familiarisation of difference') have become ideals of identification. But persistently highlighting the trouble-free and joyful interaction among ethnic groups in the past, and stressing the possibility of resuming interaction in the future is also a way of rejecting the nationalistic rhetoric and practice of ethnic exclusiveness.

The Balkans – Europe

Maria Todorova's (1997) analysis of the Western image of the Balkans has shown how the Balkans as a distinct geographical area with specific cultural traits was only discovered/invented at the end of the 18th century, mainly through European travel literature. The Balkans then became associated with the traditional, non-civilised, tribal (politically undeveloped), rural and unenlightened. During the Balkan Wars and the First World War, violence was added to the picture and has remained a leitmotiv of the Balkans ever since. After the Second World War, a new demon, Communism, was grafted onto the image (see also Bakić-Hayden and Hayden 1992: 3-4), and during the recent war the image of the Balkans has come full circle, Western observers explaining the war as, if not unavoidable, then proof of the typical Balkan mentality (Kaplan 1993). Europe/Balkans has thus been constructed as a hierarchical symbolic geography in which West and East are valorised in terms of religion; the tendency seems to be the more eastern the more primitive and conservative. (Bakić-Hayden and Hayden 1992; Bakić-Hayden 1995; Ritman-Auguštin 1995; Zarkov 1995).

Europe/Balkans is a discursive field in which my informants were often caught, but rather than maintaining solid and unambiguous positions their evaluations and identifications were placed in the middle of an existing predicament. Sometimes Europe was seen as decadent and unsympathetic; at other times as the future and as civilised. Sometimes the Balkan-like was authentic and honest; at other times it was violent and crude. And at times the Balkan-like was depicted as an unavoidable destructive force which people could not refrain from acting out.

Balkan

In everyday life, when Balkan is used to evaluate behaviour and attitudes, it often bears negative connotations such as laziness, callousness, disrespect and violence, and it is associated with a lack of structure and aesthetics, trouble, complexity and corruption. It is, for instance, possible to say that a person not behaving properly is displaying a typical Balkan mentality.

But Balkan is also used in a more fundamental way to explain and comprehend the war and the post-war violent and complicated situation. Balkan then becomes associated with violence, madness and death. As Senad once said:

This war could never have occurred in Europe; it is a typical Balkan war. It's insanity and savagery. We have to try to get rid of it, this Balkanism, but it is a part of us.

The Balkan metaphor is employed to make sense of the madness. People in the Balkans are subsequently represented as being used to recurrent insane violence; they simply know it by heart. For instance, I often heard people say that every twenty-five or thirty years there is war, or 'we build up everything – houses, family, economy – and then we destroy it all'. One might anticipate that the Muslims of Stolac would associate the other ethnic groups with this Balkan mentality, as an ethnic critique in disguise. But they seldom do. The war, not the other ethnic groups, was Balkan-like. People perceive the Balkan mentality as a force operating in the world, an unavoidable aspect of living in the Balkans, or something that is capable of explaining without having to be explained itself. People sometimes describe themselves as crazy Balkan people, but do not conceive of themselves as acting subjects; rather it is the Balkan mentality and its irreversibility that are acting through them. As one of my informants said in reference to this mentality, 'maybe there is something in the air, or maybe in the cigarettes – maybe they put marihuana in the cigarettes here in Bosnia'. The wild and chaotic Balkan mentality is in this way created as an explanatory yet also isolated factor, thereby helping to uphold a normal and ordered world.

Other times Balkan has positive connotations. Sometimes when compared with the European mentality it represents the authentic, warm, real, impulsive and unpretentious, whereas Europe represents the decadent, commercial, false and cold. I have heard people say that it was only after the war, with the presence of the international community, that drugs and prostitution entered Bosnia for real; that Westerners are often very patronising and conde-

scending; that the kind of democracy that, for example, America defends only pretends to be democracy (this was after the election in USA, when the few votes in Florida determined the outcome), and so on. And many people fancy employing the well-known stereotypes of cold (northern) Europe versus the warm and hospitable Balkans, emphasising that they are more impulsive than Europeans, less stressed and more emotional.

Balkan identity is complex and contradictory. It is both a violent and destructive force, and an authentic, real and creative attitude. And sometimes people embrace the contradictions and adopt the wild and chaotic madness because it also represents something real, that is, non-European. Here is an excerpt from my diary:

He [Nihad] says that 'people in the West look at us and think we are crazy; they all think we are crazy', and he says something like 'our mentality is different'. He is not explicitly referring to a Balkan mentality, but something like it. This thing where you at one and the same time hate your culture, the Balkan-like – which is chaos, craziness, violence, war, the incomprehensible – and if not exactly love it then also identify with it. It is like saying, 'I am crazy, insane, and I am proud of it'. Like when we hear this powerful and crazy yell from down the road and I think some kind of catastrophe has happened; I ask what it is, and Nihad replies: 'Nothing, it is normal here'. And then they laugh a strange laughter, but also one that is meant to tell me [as a European], 'this is how we are: we are wild and we are proud of it'. It is this embracing of the otherness that they are perfectly aware Westerners see in them. As when Fahrudin says, 'people from the West look at us and say we are crazy, abnormal, but here everything is abnormal; that's why I feel normal'.

Balkan, then, is negative and positive, and also unavoidable. It is the opposite of the European mentality – sometimes better, sometimes worse. It is destructive and creative. It is used to understand the war, violence and destruction, and it is a central though contradictory element in the process of identification in which the Muslims of Stolac engage.

Europe

Opposite Balkans is Europe. The Muslims of Stolac have three ways of comparing and identifying themselves with Europe: as lagging behind, as being alike, and/or as being better – not by being Balkan but by being more European than Europeans.

Many associate Europe with peace, prosperity and orderliness and believe it offers a better life. 'Towards Europe', as an election poster stated in the general election in November 2000. Europe is a dream, and for many an unattainable one. Bosnia Herzegovina is seen as lagging behind materially and technologically in regard to their legal systems and societal development. I remember once Nusret showed me some pictures, one of which was of his family during the war. Its quality is bad, the background is a wall with graffiti on it, his children, who look sad, are wearing ragged clothes, and everything seems to be run down. Another picture is taken inside an apartment, in front of an empty peeling wall; one of the boys is sitting on an old mattress full of holes. That is all. Right next to the pictures in the album is one taken by his sister, who lives in Germany. The home is pretty and the quality of the picture is good. The children are well dressed and well groomed, pictures are hanging on the wall, and there is nice furniture and china on the table. Everybody looks healthy and lively. Nusret's comment was, 'it is the same, right? Totally the same; can't you see it? [He laughs a little]'. I know that his dream is to be able to give his children and himself such a future. In fact, he emigrated with his family to Sweden two months later.

On other occasions when people compare the Balkans with the pregnant symbol of Europe, they do not depict themselves as lagging behind, but rather as being no different. They stress that it is people in [Western] Europe who deem them underdeveloped and abnormal. As Mensur once said, 'The Bosnian Muslims are the most intelligent and most educated Muslims in the world, probably because we are from Europe'. Sometimes I was confronted with the view – contrary to the Balkan explanation – that the war could have occurred anywhere, that Europe also has a bloody history, or that it all was due to international as well as national politics, a view with which people resist attempts at linking up the war with a typical Balkan mentality.

The last aspect in comparisons between the Balkans and Europe is when people state that Bosnians are in fact better than Europeans, not because they are more authentic, emotional and hospitable, as was the case when the Balkan discourse was embraced and the otherness internalised, but because they are more European. The main arguments employed to support this identification relate to education and general knowledge.

Nihad's father, for instance, clearly places Bosnia Herzegovina at the top of Europe. He is a well-read man and was part of an intellectual group in Stolac which does not exist anymore because, as he says, the majority of the intellectuals have left the country. He often talked about the high intellectual level of former Yugoslavia: '…we had one of the best school systems in Europe – philosophy, architecture, mathematics…Plato, Aristotle, everything', and he tells me that the West does not reckon the people in the Balkans as capable of anything. One day he told me about a Spanish SFOR soldier who used to come and visit him: '…I asked if he knew [he mentioned a Spanish writer whose name I did not catch]; he didn't, so *I* had to teach *him* Spanish literature'. He also talked about civilisation, about the idea that a culture has to settle. He said that civilisation has existed in Bosnia and in Yugoslavia for many years, which may be regarded as a tree with close rings, while American culture has the same quality as a tree growing very fast. He speaks Russian and taught himself German from a tape. One day he told me about the superiority of the letter *c* in the Bosnian language; in German you need three letters to create the same *sch* sound.

In their attempt to define their identity as Muslims, my informants draw on the discursive construction Balkans/Europe, which in the war became even more relevant due to extensive use of this construction by ex-Yugoslav politicians trying to legitimate succession, for example, Western politicians, journalists and academics trying to Balkanise the war and thereby legitimate non-interference; and civilians trying to grasp the madness by encapsulating the violence as the result of some kind of unavoidable Balkan mentality.

The Muslims of Stolac employ the opposition in different and rather incongruous ways. Often the same persons made different statements depending on the situation or the topic discussed. The Muslims' dilemma of being (n)either Balkan (n)or European is a central aspect in everyday processes of identification and it is an identification which avoids ethnic connotations. To be a Muslim of Stolac means to continually place oneself somewhere in the discursive tension of Balkans-Europe.

The role of the victim
This last section relates to the Muslims' identification of themselves as victims, and in contrast to the previous identifications it serves as an explicit critique

of the Croat/Serbian other. The role of the victim is the only identification I have encountered that explicitly focuses upon the Muslims as an isolated group, that sets the Muslims apart, and that in a way reinforces the nationalistic rhetoric of exclusiveness and non-ethnic interaction. The role of the victim works, therefore, in opposition to the other three types of identification, but it is also the only antagonistic identification I have encountered among the Muslims.

The victim's role is linked to the national political rhetoric. Though not well documented it seems as though Muslims as a group were being promoted as victims in Bosnian politics throughout the war in order to gain sympathy internationally and to legitimatise the preservation of an ethnically heterogeneous Bosnia Herzegovina (Velikonja 2001: 282, Bougarel 1996: 108-109). For instance, looking at the many terrible rapes committed during the war it appears that the raped women only became important politically when they disappeared as individuals and were transformed into a symbol of the raped nation (Meznaric 1994; Morokvasic 1998, Olujic, 1998). As Olujic (1998: 45) writes, '... the rapes of individual women were microcosms of the larger invasions of territory'. In other words, the raped women symbolised the raped Bosnian nation.

In Stolac, Muslims recurrently talked about themselves as victims in the war, and how they had not committed any atrocities. Often it was a kind of mutual confirmation of their own fate – the situation, the humiliation and fear they had lived through and their communality as Muslims. The discourse of the victim's role was a way of representing identity. One of the aspects of this role concerns the Muslims as victims of a conspiracy.

Almost everybody I talked to recalled how the war came as a surprise, and they explain this by saying that it was all a conspiracy and that they were betrayed. People believed in Yugoslavia, in Brotherhood and Unity (*bratstvo i jedinststvo*), in their army JNA (the people's army), and later when the war had started they believed that they and the Croats were facing the same enemy – the Serbs. But when the war started the Muslims realised that the Serbs had had it planned out for a long time, and the Croats had intended to go against the Muslims when the two groups were still allied in 1992/1993. Nermin often talked about the deceit he, his family and the Muslims had experienced.

The Croats had planned it all far in advance. Before the war, in 1991-1992 they had already begun printing their own birth certificates. It was all planned.

The Serbs and Croats had a deal. The Serbs should have the territory to Neretva [the great river running through Hercegovina and also Mostar], and the Croats the territory from Neretva. What about us then? We were supposed to end up in the river.

Anvere (his wife): Nermin, he didn't think there would be a war; he just worked all the time. He thought that if one just works nothing would happen. At that time, in 1991, he helped a friend build a house in Dubrave. Then a Croatian friend of ours came by and asked, 'why are you working, Nermin; the war will still come and destroy everything'.

Nermin: I was standing there by the road working, and he passed and asked, 'Nermin why are you working? The war will soon come and everything will be destroyed'. I didn't take any notice of him; he has always been a little nervous. It was all planned. The Croats, they stabbed us in the back.

People also feel that the West let them down by not having intervened soon enough, by not having allowed the Muslims to arm themselves (due to the weapons embargo), and by having pretended not to know what was going on in Bosnia Herzegovina (especially the existence of the Croat prison camps). They feel as though they were part of an experiment. It was seldom explained what the experiment was about, but it relates to being part of a political game that they were unable to see through: old alliances, territory, military access to the sea, international politics, arms deals, and so on. On the other hand, people stress that it was the West that finally ended the war and ensured peace. People are therefore grateful, yet they ask themselves why the West did not intervene long before and thereby save thousands of lives.

To be a Bosnian Muslim, then, is to be betrayed by everyone and only have oneself to turn to, but at the same time people keep expecting the international community to sort out their present problems. It is as though they feel that the West 'owes them' for having let them down in the war.

Another aspect of the victim's role is that, when talking about the betrayal, people perceive themselves as credulous and naive. They ask themselves why they did not see things coming, why they believed the Croats and why they did not prepare themselves militarily for war. But at the same time this naiveté is also read as evidence of the Muslims' honesty, trustworthiness and decency.

Consider what Armin's mother once said:

In the beginning it was the Serbs [who started] and then they persuaded the Croats to do the same. Tudjman and Milošević were in Karadjordjevo, where they planned how to divide up the country. Then Milošević said to Tudjman: 'You can have your country to Neretva, you can have half of Bosnia Herzegovina and then I want half of Bosnia Herzegovina, then we'll make a deal and we will kill all the Muslims'. We Muslims did not know anything about the deal at that moment. We are decent people and want a good relationship with everybody; we are not bad people. There is one word very characteristic of us Muslims and that is *merhametli* [helpful; afterwards, while listening to the tape, one of my interpreters said that 'some think we were too *merhametli* during the war and that's why all this happened; we were too tolerant']; we are not bad people. We cannot all be brothers and sisters, but we have to be good friends, we have to understand each other, take care of each other and be together, communicate. And if I do not have anything today and you do, you should help me, and tomorrow if you do not have anything I'll help you.

The old lady clearly communicates the feeling of deceit, and the sense that the Muslims were victims, but with this fate as her starting point she also moves towards a definition of Muslimness: a deceived, naive, but decent people. The Muslims feel they showed their true nature during the war by not 'speaking' the same language as the other ethnic groups – that of destruction, looting and killing. Once I asked Emir why the Croats behaved so cruelly during the war, especially in the prison camps. He answered, 'I don't know; if I knew I would be like them'. The central strategy is to turn the wrongdoers' actions back against them. The misdeeds affected the Muslims, but they talk about their executers. It is also a way for the Muslims to identify the Muslim mentality. For example, people refer to the horrible atrocities against civilians committed by Serbs and Croats, but not by Muslims. They became innocent and are astonished by the brutal violence committed in the war. As Aida once said, 'we showed our nature by not being inhuman'.

A third element of people's talk about the betrayal is their lack of trust in the Croats. They can live together but will not trust each other again. And many feel an obligation to tell their children the truth, so the war will not repeat itself. When we discussed the curriculum in the school, Ljiljana, a schoolteacher, made the following remarks:

I cannot teach my child to forget it all. My mother never taught me about the things that happened before [the Second World War], and that's why these things happened to us. If we were a little stronger and if we didn't believe in everything they say... The worst thing is that we believed in them. We paid the highest price because we believed in them. We shall never forget. If we forget, it will happen again. They raise their children to become the greatest Croats or the greatest Serbs, and things will remain the same until they change their education.

The different elements in the role of the victim (we were deceived by everybody; we were naive but decent; we are not like them, and we will never trust them fully again; this will never happen again) are all parts of Muslim identification. But it is also a critique of the ethnic other and a legitimisation of the Muslims' return to Stolac. The Muslims have a dual way of legitimising their return, which at first glance can appear contradictory: legitimising by downplaying the relevance of ethnic categories (as seen in the *ideal of tolerance*), and legitimising by criticising the Serbian and Croatian aggressions and attempts at exterminating the Muslims. This duality was clearly expressed by Nusret during an interview:

There are no Bosnjaks without Bosnia Herzegovina, but there is no Bosnia Herzegovina without Serbs or Croats. Three different people live here; that is our fate.

The first part of the citation refers to people's criticism of the other ethnic groups and their attempt to exterminate the Muslims, and it is inside this reasoning that the victim's role is played out. The other part refers to the downplaying of antagonisms between the ethnic groups, simply because 'we have to live together'. The Muslims at one and the same time are saying 'we want to live here' and 'we have to live here together'. Besides being part of an ideal of tolerance this is also due to the fact that the Muslims, contrary to the other ethnic groups, do not have any adopted country to go to. They *have* to live in Bosnia Herzegovina.

By writing about how the Muslims of Stolac construct themselves as victims in order to identify Muslimness and not about how the Muslims in fact *were* victims, I feel I fail to do justice to their actual experiences. However, this tension also exists in the Muslim community of Stolac and relates to generational differences. Many of the young simply feel that the persistent talk about how Muslims are victims is a hindrance to creating a future. For them there is no status in dwelling upon the role of the victim. They sometimes entered the discourse of victimhood, but mostly they preferred to talk about football, cars and music, and they were often tired of the older generation clinging to the past.

I shall conclude with an example of this kind of conflict acted out between a mother and her son. One day, talking to Amra about my trip back home to Denmark, I said that in Sarajevo taxis are very expensive but are the only way to get from the hotel to the airport. Then Amra said, 'Why don't you just take the bus to X [I do not remember the name] and from there it's only a 250-300-meter walk. Everything is destroyed there, but just follow the road, because there are mines, so don't go off the road'. Her son, who obviously had overheard our conversation, interrupted in a somewhat resigned manner, 'there *are* no mines, mother'. And she replied, 'Yes there are. A lot of things were destroyed during the war; the frontline was there, and there are mines'. And her son said in a slightly unfriendly way, 'Mother there is no war anymore, the war is over!'

Discussion

Though not originating in ethnic and religious differences, the war in Bosnia surely exploited, enforced and to some extent constructed them. And while it is true that Bosnian Muslim religious nationalism came late and was not as articulate and extreme as Serbian and Croat nationalism, it did exist and it did preach and practice ethnic exclusion despite being somewhat contradictory. Considering the Bosnian Muslims' nationalism, the structural theories of violence, and finally the fact that today Stolac is totally ethnically divided and marked by nationalistic politics, one might expect the Muslims' identification to be unambiguous, exclusive, based on religious and national values, and strongly focused upon 'Muslimness'. My results show on the contrary that local Muslim identifications in Stolac are much more complex, ambiguous, and not at all as clear cut as structural theories of violence would have it.

The new Bosnian nation-state neither serves as a central imagined community for people nor has positive connotations. Religion is important, but in the sense of values attached to an everyday morality and not a doctrinaire reading of Islam.

I have pointed out four different identifications central to people's lives in trying to ascertain what it means to be a post-war Muslim resident of Stolac/- Bosnia. Three of these were not concerned with defining Muslimness as an exclusive and detached category; rather they centred on what it means to be a Stolčanici, a good neighbour living with difference and finally a Bosnian (Balkanian) in Europe.

The first one – local patriotism – was about (re)defining Stolac as a place for coexistence partly by constructing an endemic Stolac aesthetic. Stolac was/is a pearl on earth; it is beautiful and characterised by its unique broad-minded mentality. The second one referred to an objectified habitual pre-war practice of local inter-ethnic interaction characterised by tolerance ('familiarisation of difference' and the value of *komsjiluk*), an ideal of tolerance that partly was associated with a Muslim mentality. The third one placed the people of Stolac in relation to a more global discourse of the connections/relationship between Europe and Balkan. Different and contradictory positions exist here, but all are characterised by not focusing upon ethnicity. It is not the position toward the other ethnic groups which is central, but the position *vis-à-vis* Europe. Only the fourth identification, the role of the victim, was antagonistic, exclusive and focused explicitly on the Muslims as a separate group. And it was furthermore an unmistakable criticism of the ethnic others, predominantly the Croats.

In sum, the war has definitely changed the way the Muslims of Stolac identify themselves. But the changes have not been as radical as one would expect. Bosnian/Muslim nationalism is not very important, religion has returned to a more or less pre-war level, and pre-war values such as tolerance, coexistence, solidarity, neighbourliness and civilisation (Balkans/Europe), though somewhat changed, are still central.

Practices of violence certainly have the potential to consolidate, accentuate or even create identities. However, this is not the same as explaining why violence or war arises, which some structural theories of violence more or less explicitly attempt to do (see references in the introduction). Such a functional argument in which violence is supposed to erupt when differences become too small should – at least in the Bosnian case – be replaced by a focus upon political strategies. It was the war and post-war politics which initiated the ethnic division of Stolac, because ethnic cleansing was the explicit aim and method of power-seeking nationalistic politicians. War did not occur due to an intrinsic potential of violence; rather the violence in Stolac manifested and augmented the politically created division (Kolind 2004).

Secondly, the potential of violence for consolidating differences should be related to moves in the opposite direction. That is, everyday attempts at making social relations ambiguous and indistinct. In this paper I have analysed such an opposite move: resistance to nationalism. This counterdiscourse – as I call it – exists through the many ways the Muslims of Stolac reject the nationalistic paradigm and insist on pre-war values of tolerance, coexistence, neighbourliness and – as I have shown elsewhere (Kolind 2002a; 2004) – through their refusal to employ ethnic categories when accounting for the war and the erosion of their everyday world. However, this counterdiscourse is often flawed. Contradictions exist between the role of the victim and the other identifications, and 'inside' the different types of identification lurk potentials for ethnic hatred. But the ordinary people of Stolac, contrary to nationalistic politicians, cannot afford the luxury of ethnic hatred, because they have to live with the consequences of such antagonistic feelings (see also Jansen in this volume). What we see instead, to paraphrase Das and Kleinman (2001: 16), are not necessarily grand narratives of forgiveness and redemption, but small local stories in which people are experimenting with ways of inhabiting the world together. So when people stress ideals of tolerance and coexistence, but tend to attach such values to a Muslimness; when people identify themselves as belonging to Stolac, but implicitly question (some of) the Croats' love for the town, and when people project themselves as victims and identify the others as dealers in atrocities, we have examples of contradictions, but also attempts at reconciliation and the creation of possibilities for future coexistence. As I see the situation in Stolac, potentials for peace and (pragmatic) co-existence come from the bottom up – a situation contradicting the Hobbesian dictum that in the absence of inter-social regulating structures the world would be all against all. What we see in Stolac is an example of another 'human nature', one characterised by a will to build enduring social relations, not in a romantic or unproblematic way, but pragmatically, contradictorily and unavoidably.

NOTES

1 Stolac is situated in southern Hercegovina, with approximately 5,000 inhabitants. During the war, the Croats violently expelled the entire male population and placed them in horrendous prison camps, whereas the women and children were expelled to Muslim controlled areas. The whole of Stolac was then razed. The Croats destroyed all the buildings, objects of art, monuments, Mosques, and so on, bearing witness to the former Ottoman occupation, and they robbed all the Muslim houses. In 1996-1997, the Muslim population started to return supported by the international community. In the initial phase of their return, the Muslims were persistently met with violence and threats and their houses were mined. But gradually such incidents stopped, and during my fieldwork in September 2000 to April 2001, it seemed as though the Croats had accepted the idea that the Muslims would stay; however, the town remains totally ethnically divided.

2 Father's brother's daughter.

3 Cp. Allcock, inspired by Durkheim, who suggests that we view atrocity as sacralisation, that is, a symbolic focus for solidarity (2000: 401).

4 Analyses on such analytical levels are often rather speculative. One could argue otherwise and claim that violence itself has become a disembedded practice. As Bauman's (1994) study of the Holocaust shows, violence in modern bureaucratic systems has in fact been lifted out of concrete social relations and developed into a rational bureaucratic tool. In a similar vein, Pick (1993) argues that the conduct of violence and slaughter (both of animals and men) in Western Europe in the recent 150 years has followed the overall pattern of systematic industrial rationality.

5 This point comes from Ivana Maček in personal correspondence.

6 See also Jansen (2000) for similar findings from Belgrade and Zagreb.

7 Attempts were made, but they had been persistently politically marginalised (Bougarel 1997; Babuna 1999).

8 Interaction between Muslims and Christians in Bulgaria.

9 All the informants' names in this article are pseudonyms.

10 The reason why I write this in the vernacular is that I often heard exactly this expression: 'Bosnjaks, and Serbs and Croats'. It seemed to be a deeply felt and engraved memory device.

11 But not all Croats are blamed. Mostly people feel that the Croats from other parts of Bosnia who moved to Stolac during the war are the guilty ones. They do not come from the town and are therefore not seen as having any deep feelings for it; that is, they do not share the Stolac spirit.

12 Several times leading Croats have denied the existence of Mosques in Stolac, all remnants of them having been removed. For these Croats pre-war Ottoman Stolac has been repressed.

13 Remember that during and after the war many housewives did not bake cakes, as cakes symbolise joy and celebration.

14 Fatima's story below is an example of a story about respectful and tolerant interaction among the three ethnic groups: Our whole family was mixed. It didn't matter if one was Serb, Croat or Muslim; everybody was good enough. When my oldest brother was going to be married, he came with his bride, and her name was Olga, and I knew her from beforehand. We used to go down to videvo polje, and there she saw me with my son. 'Well you know who I am'. 'Yes, you are going to marry my brother'. When my brother came, he said to me: 'go to our father and tell him that I will be coming with my bride. How is he going to react when he hears that?' I went home and told him, and my father said: 'why are you asking me? If he loves her, so do I. If he can sleep with her, I can dine with her'. My sister-in-law says that she can never think of my parents without starting to cry. She also says that they received her well and weren't too religious, and that it did not bother them at all that she was Serb. And my uncle, he married a Croat woman. We took good care of each other. And one of my uncle's daughters married a Serb and they lived in Serbia. And another one married one from Montenegro... Now I can't remember them all. Everything was mixed in our family'.

BIBLIOGRAPHY

- Allcock, J.B. 2000. *Explaining Yugoslavia*. London: Hurst and Company.
- Anderson, B. 1991. *Imagined Communities. Reflections on the Origin and Spread of Nationalism*. London: Verso.
- Appadurai, A. 1999. 'Dead certainty: Ethnic violence in the era of globalization'. In: B. Meyer and P. Geschiere (eds.): *Globalization and Identity. Dialectics of Flow and Closure*. London: Blackwell, pp. 305-24.
- Babuna, A. 1999. 'Nationalism and the Bosnian Muslims'. *East European Quarterly*, 33, pp. 195-218.
- Bakic-Hayden, M. 1995. 'Nesting Orientalism. The case of Former Yugoslavia'. *Slavic Review*, 54, pp. 917-31.
- Bakic-Hayden, M., and Hayden R.M. 1992. 'Orientalist variations on the theme "Balkans": Symbolic geography in recent Yugoslav cultural politics'. *Slavic Review*, 51, pp. 1-15.
- Bauman, Z. 1994. *Modernitet og Holocaust*. Copenhagen: Hans Reitzels Forlag A/S.
- Bax, M. 1995. *Medjugore: Religion, Politics, and Violence in Rural Bosnia*. Amsterdam: VU Uitgeverij.
- Bax, M. 1997. 'Mass graves, stagnating identification and violence: A case study in the local sources of "the War" in Bosnia Hercegovina'. *Anthropological Quarterly*, 70, pp. 11-20.
- Bax, M. 2000. 'Planned policy or primitive Balkanism? A local contribution to the ethnography of the war in Bosnia-Herzegovina'. *Ethnos*, 65, pp. 317-40.
- Bennett, C. 1995. *Yugoslavia's Bloody Collapse. Causes, Course and Consequences*. New York: New York University Press.
- Blok, A. 2000. 'Relatives and rivals: The narcissism of minor differences'. In: H. Driessen and T. Otto (eds.): *Perplexities of Identification. Anthropological Studies in Cultural Differentiation and the Use of Resources*. Aarhus: Aarhus University Press, pp. 27-55.
- Boehm, C. 1984. *Blood Revenge. The Anthropology of Feuding in Montenegro and Other Tribal Societies*. Kansas: University Press of Kansas.
- Bougarel, X. 1996. 'Bosnia and Hercegovina - state and communitarianism'. In: D. A. Dyker and I. Vejvoda (eds.): *Yugoslavia and After. A Study in Fragmentation, Despair and Rebirth*. London: Longman, pp. 87-115.
- Bougarel, X. 1997. 'From Young Muslims to Party of Democratic Action: The emergence of a Pan Islamist trend in Bosnia-Hercegovina'. *Islamic Studies*, 36, pp. 532-49.
- Bougarel, X. 2001. Islam and Politics in the Post-Communist Balkans (1990-2000). Unpublished manuscript.
- Bowman, G. 2001. 'The violence in identity'. In: B. Schmidt and I. Schroder (eds.): *Anthropology of Violence and Conflict*. London: Routledge, pp. 25-47.
- Brandt, E. 2002. *On War and Anthropology. A History of Debates Concerning the New Guinea Highlands and the Balkans*. Amsterdam: Rozenberg.
- Bringa, T. 1995. *Being Muslim the Bosnian way. Identity and Community in a Central Bosnian Village*. Princeton: Princeton University Press.
- Cohen, L. 1993. *Broken Bonds. The Disintegration of Yugoslavia*. Boulder: Westview Press.
- Cohen, L. 1998. 'Bosnia's "Tribal Gods": The role of religion in nationalist politics'. In: P. Mojzes (ed.): *Religion and the War in Bosnia*. Atlanta: Scholars Press.
- Corbey, R. 2000. 'On becoming human: Mauss, the gift, and social origins'. In: A. Vandevelde (ed.): *Gifts and Interests*. Leuven: Peeters, pp. 157-74.
- Das, V., Kleinman, A., Lock, M. Ramphela, M. and Reynolds, P. (eds.) 2001. *Remaking a World. Violence, Social Suffering, and Recovery*. London: University of California Press.
- Das, V., and Kleinman, A. 2001. 'Introduction'. In: V. Das, A. Kleinman, M. Lock, M. Ramphela, and P. Reynolds (eds.): *Remaking a World. Violence, Social Suffering, and Recovery*. London: University of California Press, pp. 1-31.
- Denich, B. 1994. 'Dismembering Yugoslavia: Nationalist ideologies and the symbolic revival of genocide'. *American Ethnologist*, 21, pp. 367-90.
- Dizdar, M., S. Mulać, A. Pirie, F. Rizvanbegović, and M. Sator. (eds.) 1997. *Slovo Gorčina*. Mostar: Stamparija Islamskog centra Mostar.
- Foster, R.J. 1991. 'Making cultures in the global ecumene'. *Annual Review of Anthropology*, 20, pp. 235-60.
- Friedman, F. 1996. *The Bosnian Muslims. Denial of a Nation*. Boulder: Westwiev Press.
- Gallagher, T. 1997. 'My Neighbour, My Enemy: The manipulation of ethnic identity and the origins and conduct of war in Yugoslavia'. In: D. Turton (ed.): *War and Ethnicity. Global Connections and Local Violence*. Rochester: University of Rochester Press, pp. 47-75.

- Georgieva, C. 1999. 'Coexistence as a system in the everyday life of Christians and Muslims in Bulgaria'. *Ethnologia Balkanica*, 3, pp. 59-84.
- Glenny, M. 1996. *The Fall of Yugoslavia*. London: Penguin Books.
- Gullestad, M. 1992. *The Art of Social Relations: Essays on Culture, Social Action and Everyday Life in Modern Norway*. Oslo: Scandinavian University Press.
- Harrison, S. 1993. *The Mask of War. Violence, Ritual and the Self in Melanesia*. Manchester: Manchester University Press.
- Hayden, R. 1996. 'Imagined communities and real victims: Self-determination and ethnic cleansing in Yugoslavia'. *American Ethnologist*, 23, pp. 783-801.
- Höpken, W. 1994. 'Yugoslavia's communists and the Bosnian Muslims'. In: A. Kappler, G. Simon, and G. Brunner (eds.): *Muslim Communities Reemerge. Historical Perspectives on Nationality, Politics and Oppositions in the Former Soviet Union and Yugoslavia*. Durham: Duke University Press, pp. 214-47.
- Jackson, M. 2002. *The Politics of Storytelling. Violence, Transgression, and Intersubjectivity*. Copenhagen: Museum Tusculanum Press.
- Jansen, S. 2000. Anti-nationalism: Post-Yugoslav Resistance and Narratives of Self and Society. Unpublished Ph.D. dissertation, University of Hull, Hull.
- Kaplan, R. 1993. *Balkan Ghost. A Journey Through History*. New York: St. Martin's Press.
- Knudsen, A. 1989a. *En Ø i Historien. Korsika. Historisk Antropologi 1730-1914*. Copenhagen: Basilisk.
- Knudsen, A. 1989b. *Identiteter i Europa*. Copenhagen: Christian Ejlers Forlag.
- Kolind, T. 2002a. 'Non-ethnic condemnation in post-war Stolac. An ethnographic case-study from Bosnia-Herzegovina'. In: S. Resic and B. Törnquist-Plewa (eds.): *The Balkans in Focus. Cultural Boundaries in Europe*. Lund: Nordic Academic Press, pp. 121-37.
- Kolind, T. 2002b. 'Vold, identitet og modstand i Bosnien-Herzegovina'. *Jordens Folk*, 27, pp. 51-57.
- Kolind, T. 2004. Post-War Identifications. Counterdiscursive Practices in a Bosnian Town. Ph.D. thesis, University of Aarhus.
- Lockwood, W. 1975. *European Moslems. Economy and Ethnicity in Western Bosnia*. New York: Academic Press.
- Maček, I. 2000. *War Within. Everyday Life in Sarajevo under Siege*. (Uppsala Studies in Cultural Anthropology, 29). Uppsala: Acta Universitatis Upsaliensis.
- Mahmutcehalic, R. 2001. *The Agony of Stolac*. (The Bosnian Institute Series, 21/22). London: The Bosnian Institute.
- Malcom, N. 1994. *Bosnia. A Short History*. New York: New York University Press.
- Malkki, L.H. 1998. *Purity and Exile. Violence, Memory, and National Cosmology among Hutu Refugees in Tanzania*. Chicago: University of Chicago Press.
- Meznaric, S. 1994. 'Gender as an ethno-marker: Rape, war, and identity politics in the former Yugoslavia'. In: V. Moghadam (ed.): *Identity Politics and Women. Cultural Reassertions and Feminisms in International Perspective*. Boulder: Westview Press, pp. 76-97.
- Mojzes, P. 1998. 'The Camouflaged Role of Religion'. In: P. Mojzes (ed.): *Religion and the War in Bosnia*. Atlanta: Scholars Press, pp. 74-99.
- Morokvasic, M. 1998. 'The logic of exclusion: nationalism, sexism and the Yugoslav war'. In: N. Charles and H. Hintjens (eds.): *Gender, Ethnicity and Political Ideologies*. London: Routledge, pp. 65-90.
- Mosse, G.L. 1990. *Fallen Soldiers. Reshaping the Memory of the World Wars*. Oxford: Oxford University Press.
- Naughton, A. (ed.) 1994. *Forging war: The Media in Serbia, Croatia and Bosnia-Hercegovina*. London: Article 19, International Centre against Censorship.
- Nordstrom, C. 1997. *A Different Kind of War Story*. Philadelphia: University of Pennsylvania Press.
- Oberschall, A. 2000. 'The manipulation of ethnicity: from ethnic cooperation to violence and war in Yugoslavia'. *Ethnic and Racial Studies*, 23, pp. 982-1001.
- Olujic, M. 1998. 'Embodiment of terror: Gendered violence in peacetime and wartime in Croatia and Bosnia-Herzegovina'. *Medical Anthropology Quarterly*, 12, pp. 31-50.
- Pick, D. 1993. *War Machine. The Rationalisation of Slaughter in the Modern Age*. New Haven: Yale University Pres.
- Port, M. van de 1998. *Gypsies, War and Other Instances of the Wild. Civilisation and its Discontents in a Serbian Town*. Amsterdam: Amsterdam University Press.
- Povrzanovic, M. 1997. 'Identities in War. Embodiment of Violence and Places of Belonging'. *Ethnologia Europaea*, 27, pp. 153-62.
- Ramet, S.P. 1992. *Nationalism and Federalism in Yugoslavia, 1962-1991*. Bloomington: Indiana University Press.
- Ritman-Auguštin, D. 1995. 'Victims and heroes. Between ethnic values and construction of identity'. *Ethnologia Europaea*, 25, pp. 61-67.
- Scarry, E. 1985. *The Body in Pain: The Making and Unmaking of the World*. New York: Oxford University Press.
- Simic, A. 1991. 'Obstacles to the development of a Yugoslav national consciousness: Ethnic identity and the folk cultures in the Balkans'. *Journal of Mediterranean Studies*, 1, pp. 18-36.
- Simic, A. 2000. 'Nationalism as a folk ideology. The case of former Yugoslavia'. In: J.M. Halpern and D.A. Kideckel (eds.): *Neighbours at War. Anthropological Perspectives on Yugoslav Ethnicity, Culture, and History*. University Park: Pennsylvania State University Press, pp. 103-15.
- Sofus, S.A. 1999. 'Culture, media and the politics of disintegration and ethnic division in former Yugoslavia'. In: T. Allen and J. Seaton (eds.): *The Media of Conflict. War Reporting and Representations of Ethnic Violence*. London: Zed Books, pp. 162-75.
- Sorabji, C. 1995. 'A very modern war: Terror and territory in Bosnia-Hercegovina'. In: R.W. Hinde and H.E. Watson (eds.): *War: A Cruel Necessity. The Bases of Institutionalized Violence*. London: Tauris Academic Studies, pp. 80-99.

- Sorabji, C. 1996. 'Islam and Bosnia's Muslim nation'. In: F.W. Carter and H.T. Norris (eds.): *The Changing Shape of the Balkans*. London: University College Press, pp. 51-62.
- Sunic, T. 1998. 'From communal and communist bonds to fragile statehood: The drama of ex-post-Yugoslavia'. *The Journal of Social, Political, and Economical Studies*, 23, pp. 465-75.
- Todorova, M. 1994. 'The Balkans: From discovery to invention'. *Slavic Review*, 53, pp. 453-82.
- UNDP 2000. *Human Development Report Bosnian and Herzegovina 2000 Youth*. Sarajevo: Independent Bureau for Humanitarian Issues (IBHI).
- Woodward, S.L. 1995. *Balkan tragedy. Chaos and Dissolution after the Cold War*. Washington DC: Brookings.
- Velikonja, M. 2001. 'Bosnian religious mosaics. Religion and national mythologies in the history of Bosnia Herzegovina'. Unpublished manuscript. Ljubljana.
- Zarkov, D. 1995. 'Gender, Orientalism and the history of ethnic hatred in the former Yugoslavia'. In: L. Helman, A. Phoenix and N. Yuval-Davis (eds.): *Crossfires. Nationalism, Racism and Gender in Europe*. London: Pluto Press, pp. 105-20.
- Zur, J. 1998. *Violent Memories. Mayan war Widows in Guatemala*. Boulder: Westview Press.

War as Field and Site: Anthropologists, Archaeologists, and the Violence of Maya Cultural Continuities

11 Ahau was when the mighty men arrived from the east. They were the ones who first brought disease here to our land, the land of us who are Maya, in the year 1513.

Book of Chilam Balam of Chumayel (Roys 1967: 138)

Control of the Maya past is equal to the control of our power in the present, for history is the basis for demanding respect for our political and cultural rights as a People.

Maya intellectual Avexnim Cojtí Ren (2002)

/30

STAFFAN LÖFVING

Post-war Guatemala is marked by unprecedented levels of non-political violence. Organised crime, hijackings, and lynching of suspected outlaws by local mobs in rural Maya-populated areas threaten the yet to be fully implemented peace agreement of December 1996 (Moser and McIlwaine 2001).

Another marker of contemporary Guatemala is the growing presence on the national political arena of the Pan-Maya movement – an anti-racist initiative for the support of the Maya people and of its threatened values. In its most central documents it claims that the pre-Columbian Maya past was one of peace, and that violence – like disease and every kind of human evil – was brought by 'the mighty men' of the 16th century conquest (COMG 1995 [see the Chilam Balam quote above]). It is hardly surprising that a cultural movement, resembling a nationalist project in every aspect but the claim to their own state, makes such an effort to connect the plans for future harmony to the notion of a harmonious history. What is remarkable, however, is the absence of similar statements among those rural Mayas whom I encountered during the different phases of my anthropological fieldwork in the 1990s.

My assumption in this chapter is that culture in war is subject to processes of politicisation and essentialisation to which violence is crucial. Violence here is not a consequence of cultural difference but, rather, the means by which cultural difference is politicised (Harrison 1993; Warren 1993). In order to prove that assumption right I explore the *location* of knowledge about the Maya past and about what that location implies in terms of different notions of the very *content* of contemporary Maya culture.

Both anthropology and archaeology have a stake in the identity politics of post-war Guatemala. The role of foreign anthropologists in the production of knowledge is nowadays a highly contested issue among, in particular, the activists of the Pan-Maya movement (see Warren 1992; 1998; Watanabe 1995; Montejo 1999). Many of them perceive of anthropologists as the true colonialists of the era of globalisation. When American anthropologist Kay Warren discussed this with Maya intellectuals she was told that 'the appropriate role for North American Anthropologists should be one of helping to identify continuities in Maya culture, the timeless characteristics that make Mayas Maya' (Warren 1998:74). Post-modern critiques of cultural essences were thus seen as colonialist attempts at stripping the Mayas of their very last possessions – their culture and their pride.

The role of archaeology, and of 'foreign' archaeologists in the production of knowledge, is equally contested. Maya archaeology has developed considerably during the last twenty years, and this has had a particular effect on the Maya movement. Prior to the deciphering of 'the Maya code' (see Coe 1992) the mystery of Classic Maya civilisation (AD 300–900) inspired not only the New-Age-movement, but also the Maya revivalists themselves. Practically anything that served the interests of Maya identity politics could then be written on the *tabula rasa* of the Maya past. The problem now lies in the discoveries made by linguists and archaeologists of the violence from within that contributed in bringing about the collapse of Maya civilisation (see Sharer 1994); and of the role of sacrifice and the symbolism of blood in Maya religion and culture (see Schele and Miller 1992). The idea that the Spaniards brought violence to the Americas, and that pre-Columbian culture was one of peace and harmony is, to the regret of a certain strand of Maya activism, now difficult to sustain.

In a self-evaluating conference on the role of archaeology in the perpetuation of discriminatory Guatemalan politics, archaeologists from the University of British Columbia, Canada, admitted the impact of excavations and research on Maya communities. The practices of archaeological research have created new incentives for tourist economies that sometimes, but far from always, benefit the community in which the excavations take place (Castañeda 2001). At the same meeting Canadian archaeologist Marvin Cohodas (2001) argued that: 'archaeologists have appropriated ancient Maya history to exemplify what it means for a society to fail.' This equation of collapse with failure has, according to Cohodas, damaged Maya interests since it separates the past from the present, and, in the end, because the alleged break permeates the portrayal of contemporary Mayas as inferior to their ancient forebears. Such an inferiority legitimises contemporary discrimination. Maya intellectual Avexnim Cojtí Ren writes:

[The notion of] [c]ollapse denies our continuity as the original people of countries in which we reside and our possible claims to that historical continuity. The result of that discontinuation of our history is that Maya are denied a true identity; we are regarded still by the general population and even ourselves as just 'Indios,' with no history nor culture but customs and traditions, no land title, with dialects, and so forth. (Cojtí Ren 2002)

The object of this criticism is thus not the revelation of Maya violence, but the overemphasis on it, and the widely spread association of ancient Maya civilisation with the notion of abrupt political transformation.

It is important to incorporate a discussion on Guatemalan racism, and the satisfaction with which a non-Indian anti-Indian discourse now concludes that it is not only the present of the backward Indian that breathes chaos and violence, but also the past. Victor Montejo, exiled Maya and trained anthropologist suggests that the turn away from archaeologist Eric Thomson's 'peaceful Maya', to a new archaeological interest in blood and internal violence, coincided with the mayhem of the Guatemalan highlands of the early 1980s for political reasons (Montejo 1999; see Wilk 1985). In his paper for the British Columbia Conference, Montejo claims to be representing an 'insider's perspective' and argues that:

archaeology has played an endemic role in perpetuating the contemporary indigenous people in a timeless past or allochronic time. The continuous aim of some archaeologists in excavating tombs and sacred sites is to perpetuate their power (intellectual colonialism) over those whose remains and ancestors are being desecrated. In this process of free enterprise, working at the margins of ethical and human rights issues, the new and old archaeologists need to be reminded of their accountability to indigenous people as they transform their speculations into science and 'the Truth' about Mayans. (Montejo 2001)

But the attack on archaeology is countered by archaeologists in the field. One of them, Arthur Demarest, claimed in a lecture recently that Mayas are no less capable of objectively understanding the decline of Classic Maya civilisation than are Europeans the fall of the Roman Empire. He went on to say that it is a misguided and paternalistic North American attack on Maya archaeology that is to blame for the idea today that contemporary Mayas cannot accept new archaeological discoveries. The target in Demarest's critique is thus a growing reflexivity among western archaeologists (see, in particular, Pyburn 1998; 1999, and the British Columbia Conference 'Towards a more Ethical Mayanist Archaeology' at www.ethical.arts.ubc.ca). What interests me here is *the location* of the contested knowledge. On the one hand, Montejo is a born Kanhobal Maya. His critique of North American archaeology is admittedly from within a Maya position. But Montejo is also a U.S. trained anthropologist, a fact that Demarest implicitly chose to emphasise since Montejo's critique, by him, is referred to as North American paternalism. This connects to the violent break between culture as lived experience and culture as ideology in Maya nationalism. Is it possible to identify the roles, not just of violence, but also of anthropology and archaeology in that break?

Experiencing and establishing continuities

During fieldwork in the mid 1990s when I once had to leave Guatemala for a couple of days, I decided to embark on an excursion to the Maya Copán ruins located in Honduras just south of the Guatemalan border. I was lucky to travel in the company of one of the intellectuals of the Pan-Maya movement – then newspaper columnist Miguel Angel Velasco. To me the trip represented a short vacation, but for Miguel it was a pilgrimage to what once was the cultural and artistic centre of the Maya world. The Hieroglyphic Stairway of Copán (see Sharer 1994: 300-10) took our breath away and we engaged in long discussions and pondered on the possible significance of such a magnificent architecture. My friend is not an archaeologist by training, but what struck me as so fascinating was the eagerness and wit by which he interpreted the signs of the past in a culture and language of the present. He was certain that the patterns of the stone walls mimicked the patterns of the *huipiles* – the women's blouses back in Guatemala – and he repeated, without any hesitation and despite the fact that we were exploring the territory of Honduras, that he felt like 'coming home'.

I have also encountered more fundamentalist claims to Maya particularities. Another friend of mine, a young woman whose wedding was about to take place near a Maya mountain shrine in the Kakchikel speaking community of Chimaltenango invited me to partake in the ceremony. It turned out to be a beautiful event and the fact that I was the only *canche* (blond person) in this domestic, yet politically significant, celebration made me enjoy it even more. That is, until the Maya priest switched from blessing the wedding couple to a political agitation against the evil of foreign influence. The priest turned his own, and the others', attention to

me and my association with what he termed a 'European evil'. I was then advised to leave.

Theoreticians associated with the Latin American left (e.g., Friedlander 1975; Hawkins 1984) have argued that the very process of state formation in Latin America needed a stereotyped 'Other' and that indigenous identity was formed by the oppression of a dispossessed majority. The experience of being ascribed an identity as the 'Other' of the state thus resulted in a 'being'. The disciplines of both archaeology and anthropology have certainly proved that argument overly simplistic, but it is my position here that the anti-Western sentiments of the Maya priest in Chimaltenango, and also the doctrines of the Pan-Maya movement, cannot be understood without taking into account the racial foundations of the Guatemalan nation-state.

Demitrio Cojtí Cuxil, both analyst and ideologue of the Pan-Maya perspective in Guatemala, explores the arrogance of the state in relation to its Maya majority (1994). He argues that the Maya People is the rightful owner of the archaeological sites currently and 'legally' in the hands of either the state or private owners – neither of which has been Maya by any standard. In a sinister way, this way of exploiting Maya property claims an ownership of a Maya past, i.e., in administering and marketing the archaeological centres as treasures of the nation, the Guatemalan state competes with the Pan-Maya movement over the political and economic benefits of the positive symbolism of the Maya past. Cojtí Cuxil writes:

With regard to the access of the Mayas to the archaeological centres of their forefathers, one could say that today, the Mayas not only have to pay the entrance fee to see the 'ruins' of their forefathers (Tikal, Iximche, Saqulew, etc.), but they are also treated as third class citizens by the employees who work in these sites. (1994: 70, my translation)

The possible discovery of ancient wisdom, and/or of remnants of an older knowledge is, contrary to what is claimed by Maya critics, what inspires a great number of academics to pursue research on the present day cultures of Guatemala and southern Mexico (Watanabe 1992). In the last chapter of the epic 1993 volume on 'threads' connecting the ancient Maya with their contemporary inheritors (Freidel, Schele and Parker 1993), the late American art historian Linda Schele reflexively narrated an encounter that made her aware of the fact that the ancient Maya ballgame was, in a way, alive and well in parts of highland Guatemala. In 1989, in the colonial Guatemalan capital Antigua, Schele was hosting a workshop with forty Maya participants representing no less than eleven contemporary Maya language groups. After having explained the process of deciphering the glyphs of a Maya tablet to her Maya students, Schele states:

I looked around the room to see if my Maya listeners understood this complicated argument, especially since it had been delivered in my personal brand of Spanish [...]. To my right I spotted a handsome young Q'anhob'al named Ruperto Montejo, who has taken the Maya name Saq Ch'en. He sat with a beatific grin on his face, his hand palm up, moving as if he were bouncing an imaginary ball. I stopped talking in mid-sentence as it dawned on me what he was indicating.

'You know this word?' I asked in excitement. 'Do you have it in Q'anhob'al?'

His grin deepened as he explained that *pitz* referred to a children's game played with a grass ball.

'Jugador llamamos "pitzlawom,"' he said with equal excitement.

'Qué me dices?' I asked swiftly, not sure I had understood the Spanish correctly.
'*Pitzlawom* is the word for "ballplayer,"' he said in Spanish that I understood clearly. A feeling of elation expanded in my chest. (1993: 339)

Archaeology's part in the identity construction of the present is far from an exclusively Guatemalan phenomenon, and archaeologists have discussed and critically reflected on the role of the discipline in relation to nationalism and ethnic war worldwide (e.g., Meskell 1998; Kohl and Fawcett 1995). Regions like the Balkans and the Middle East are particularly relevant since many 'overlapping pasts' exist within a common framework. But just as in the case of the Schele quote above, it is not a focus on warfare and violence, but rather the establishment of a certain kind of continuity for a certain kind of people that politicises the past. When the writings of Schele are now criticised from a Maya position, it is an irony that the strong emphasis on continuity is ignored in favour of the issue of the violence within.

If Linda Schele managed to bridge millennia of socio-cultural transformations, anthropologists have successfully identified Maya cultural continuities through the more recent history of colonialism and

modern war. Barbara Tedlock's *Time and the Highland Maya* (1982) demonstrates how ancient Maya conceptions of qualitative time organise relations and ritual in Quiché-speaking Momostenango, and Robert Hill and Jonathan Monaghan dedicate their book *Continuities in Highland Maya Social Organization* (1987) to presenting 'evidence' that the closed corporate community – coined by Eric Wolf in 1957 as a socio-cultural survival mechanism emerging in the encounter with colonial oppression – actually stretches beyond pre-colonial times into the past and also into the post-colonial, post-genocidal present.

War...

The Guatemalan war officially extended from the first insurgent activities in the early 1960s to the signing of the peace agreement on December 29, 1996, making it then the longest ongoing armed conflict in Latin America. The total death toll was estimated to 200,000, of which the national army directly and indirectly bore a responsibility for 99 percent and the insurgents for the remaining one (CEH 1999).

The politico-military usage of the concept of war, and its roots in the writings of Prussian military strategist Carl von Clausewitz (1976[1832]), conceals a number of experiential and qualitative dimensions of political violence recently explored in a growing body of anthropological studies of political violence (see, e.g., Daniel 1996; Nordstrom 1997; Aretxaga 1997; Mamdani 2001). This literature argues for the need to explore how violence becomes rooted in local communities, resulting in silence and diminishing spheres of loyalty and trust; how structures of inequality and poverty are dialectically linked to processes of political and ethnic mobilisation; and, how contemporary warfare resists spatial containment through its intrinsic connections to the global network of markets, to forced migration and the cultural production of diasporas and exiled nationalism.

My research has been inspired by this new anthropology, and I have examined the Guatemalan war in a very specific location in time and space. In August of 1996 I went to a guerrilla occupied 'pocket' in the western highlands, inhabited by an ethnically mixed group of displaced Maya peasants of some 10,000 individuals. From the ethnography of that society (Löfving 2002) I will focus on two interrelated examples and arguments. The first is an alternative perspective on cultural continuity, an analysis of how cosmology orders the powers of the world, both the divine and the political, and how cosmology thereby serves as a sense-making structure that admits radical political changes without loosing its own explanatory power. The second is the example of a group of organised Maya diviners and the changing meaning of culture and history when the trauma of war and displacement is expressed in the language of cultural revitalisation and indigenous nationalism.

…and cosmology

One anthropological way of approaching the past is emic, that is, ethnographers tend to explore not the factual history, but how the past as a notion enters the life worlds of people in the present. It has been argued that the writers of the Popol Vhu, the Maya Quiché book of creation, mixed the known facts and noble lineages of then-recent years (16th century) with poetic narration about pre-historic events. This phenomenon is said to constitute a culture-specific view of the past (Tedlock 1985). About the people who are the subject of my studies, Colby and van den Berghe write: 'The Ixil have a strong consciousness of their past as shown in their ancestor worship and especially in their oral traditions where the historical and the mythical are intermixed' (1969: 146).

Don Nicolas Toma, a Catholic former politician in his early sixties, lived on the periphery of guerrilla controlled domains, and the power of his age and experience made him an unusually communicative informant. He told me how he had been persecuted in his hometown Cotzal, caught and tortured at the army barracks in the neighbouring town of Nebaj. He wanted me to understand that he had been forced by army repression to take refuge with the guerrillas and emphasised not having had knowledge of the activities of the rebels before he was accused of terrorism. But in the eyes of the local and army connected land-owning elite in Cotzal, Nicolas Toma was not without guilt. His position within a radicalised Catholic Church, in the peasant league and trade union of *Finca* San Fransisco – the local coffee-producing and non-indigenous grand estate – and in the then politically radical Christian Democratic Party made him an ideal, albeit presumptive, guerrilla. Don Nicolas said:

When they [The Christian Democratic Party] presented this Rios Montt as candidate for the presidency, and myself in Cotzal as mayor [in the elections of 1974], the people were very satisfied. I triumphed in politics, won the candidacy, but they robbed my votes. The name of the mister who robbed me was Gaspar Perez, the one of the MLN [*Movimiento para la Liberación Nacional*, a right-wing political party in Guatemala with close ties to hard-line officers of the army]. As a result of this he came to fear me a great deal. But I didn't want to do anything to him. You know, I am a worker as opposed to him, who lived by his party. […] Eventually he took office as mayor. But he couldn't let go of the thought that I would cause him trouble. So he accused me of being a guerrilla. He informed the army, even though I did not have any connections to the guerrilla. Even before this, they [adherents of the MLN] had depicted us [the Christian Democrats] as guerrillas, but the governor of Quiché didn't believe them. They went to inform on me in Quiché. 'The guerrilla is in Cotzal,' they told him. 'What's his name?' 'His name is this and that,' and they gave my name to the governor. 'And what about this person?' he asked them. 'He is a businessman. He is with the political party.' The governor asked whether I had a woman. 'Yes he has a woman.' And if I possessed lands and house. 'Yes he does.' 'Then listen Gaspar,' the governor said. 'You are accusing this person of nothing else than the problems you have with his political party. Don't do that, because a guerrilla doesn't have a woman, he doesn't have a house or any other possessions. He is roaming the mountains. So you should not go spreading such lies.' Ha! Gaspar returned sad and empty-handed to Cotzal. Next time he went all the way to Guate [the capital] to meet with the president.

In the above Nicolas takes the liberty of adding to his story the exact dialogue of events that occurred more than twenty years before in a place a day's journey from where he was at the time. This technique becomes more frequently employed as the story unfolds. Let us follow local mayor Gaspar's journey to the capital – which proved to be his last – in the words of his rival Nicolas:

So he went to the palace to obtain an audience with President Laugerud. There he announced that I was a guerrilla. But what he had been told by the governor of Quiché, he was told by the president as well. 'Don't get yourself mixed up in these things, hey, Gaspar! You might get yourself killed,' the president said. 'Ah,' said Gaspar, 'is that so?' So he took leave of the president and went to sit down in the Central Park. He was sad where he sat, because the government hadn't accepted his plea. But all of a sudden… there came the owner of *Finca* San

Fransisco – Osmundo Brol! 'What are you doing here, hey Gaspar [*voz Gaspar* – indicating a friendship between them]?' he said. 'I am announcing that the guerrillas are fucking things up in Cotzal, but this government is also guerrilla since they do not accept my complaints.' 'Ah, don't worry Gaspar. Be patient. If you'd like you could stay here a couple of days, and then I can help you with the audience,' Brol told him. 'Okay, I'll wait a couple of days,' he replied, and then they both went to see the president. The thing was that they [the guerrillas] had killed the brother of Osmundo – Jorge Brol – before, when I was younger. So now they told the president and he said: 'Is that so, is that so, ah. Then you must listen, Gaspar. The army will come from the city to your village. But be prepared, it will come like lightning or like a river and it will be very dangerous. If like a river it will destroy everything in its way. If like lightning it will burn everything. It won't even spare you. But okay, you took heart to tell it all and that is good. You will see what will happen.' And Gaspar took leave of the president once again, and once again he was sad. He took to drinking and very drunk he returned to Cotzal. He kept tippling for two weeks, and a couple of months later they killed him. When I was held hostage [by the army] they told me that the guerrillas had killed him. The lieutenant of the army said: 'Listen, you [*Ustedes* – plural] killed Gaspar. You are powerful indeed.' 'What do I know?' I said to him. So due to [my role within] the political party in Cotzal they accused me of being a guerrilla.

Note the president's verdict on the infidels. With the powers of a god he sentenced the people of the highlands to extinction. His words and metaphors echo methods usually associated with gods: lightning would strike the villages; a flood would sweep everything in its way. In the existing body of ethnography, I have found a proof-like linkage between Nicolas' telling of war memories and pre-war Ixil mythology. Anthropologists Colby and van den Berghe write:

According to current oral tradition [Nebaj in the 1960s], there was a time when the high god, Q'esla Kub'al, was more directly connected with earthly events. He became angry with the inhabitants because they were too close to him and spied on him. He sought to destroy them with flood. By building boats and tall houses the people survived. He then made plans to destroy them with fire. Word got around that the fire would penetrate the earth by two staff-lengths. Many people accordingly buried themselves in large urns just below the two-staff line. Others hid in rivers, caves, and mountains. When the fire came it penetrated not two, but three staff-lengths and killed all the people hiding in the urns. However, those who had taken refuge in the caves and rivers survived. (1969: 96-97)

The flood and the lightning are thus not metaphorical inventions of Nicolas Toma. Instead they are symbols of destruction and change. In the Ixil myth of creation documented by Colby and van den Berghe the high god later forgave the survivors and assigned them the duty of watching the next inhabitants of the earth. These were the humans. The survivors thus turned into intermediary deities whose power was founded on their connection to a high god: Q'esla Kub'al. If the humans failed in pleasing the deities in rituals and offerings, Q'esla Kub'al would send sickness as punishment upon the humans. 'The actions of the intermediaries thus depend entirely on the behaviour of humans' (Colby and van den Berghe 1969: 97). The location of Osmundo Brol as the one mediating the contact between Gaspar (the People) and the president (the God) is one avenue to ponder in this juxtaposition of political and cosmological orders, but so is the position of the rebels. Their stated goal, of overthrowing and substituting the earthly power with another, means that they are playing the subversive, but also powerful, role in the cosmology of Nicolas' narration.

The issue of culpability seems inevitable if we elaborate further on the connections between the myth of creation and the memory of war in northern Quiché. It is one of Colby and van den Berghe's main points that disaster is believed to be brought on a people that has itself to blame. The effects of guerrilla presence among the Ixils and on their sense and conceptualisation of guilt escapes simplistic reasoning in terms of cause and effect. It seems as though we are facing two alternative interpretations. On the one hand, there is the personal portrayal of the enemy – the name of the president, the identity of the close political rivals back home. This intimacy is pronounced by the tale-like representation of their dialogue. Accordingly, it could be claimed that the effort on behalf of the guerrilla project to change the direction of guilt out towards external agents has been successful; blame seems to work as long as you know its object.

On the other hand, if the president of the republic is equated with a god and the *finca* owners with intermediary deities superintending the people, then the guerrillas could be said to have failed completely. It follows that the threatening 'other' is impersonal and that the fatalism associated with a politically passive life of the oppressed prevailed. I noted the frequency of expressions like 'they perse-cuted us', and 'they exploited us' in the stories I heard and documented during fieldwork. The unwilling-ness to name the enemy is thus not a measure of security, nor a direct reference to chaotic percep-tions. Instead we could position 'they' – meaning anything from the neighbouring Ixil-speaking Chajules to the Spaniards during colonial times and the Pentagon – at the same level of abstraction as god. And ultimate power, then, is repositioned (or maintained) beyond the community.

An essentialised perspective on this cosmology would maintain that the identity of the God could (read *should*) not be changed. The quest for estab-lishing continuities (see Cojtí's exhortation to Warren above) tends to put the emphasis in the wrong places. Essentialism in this case would imply an application of a monotheist notion to a much more diversified theological context. The equation of God and President in the war memory of this (Catholic) Maya politician does not mean that the political power of the world is perpetual, or that it is beyond questioning. His own commitment to the revolution, and to the 'disobedient' guerrillas, retold in a triumphant narrative in the context of a still vibrant resistance before the peace agreement of December 1996, indicates that culture/cosmology mediates political views and identities, rather than determines them. It also questions the alleged con-servative essence of contemporary Maya politics (e.g., Stoll 1993). In line with the constructivism I cherish I would argue that Don Nicolas masters the complicated situation in which history and oppres-sion locates him, he is a victim neither of the Guatemalan army, nor of the essences of his culture and cosmology.

… and ethnic resurgence

Ever since the displaced communities in the Ixil area were guerrilla-organised during the war and until the disarmament of the rebels in 1997, *Majawil Q'ij* – the guerrilla-initiated and politically radical branch of the Pan-Maya movement – was the people's self-evident representative among the Maya revivalist institutions in the capital. This contact required messengers between organisations in Guatemala City and the rural areas, several of whom were recruited from the displaced people. These messengers did not lack agency of their own, however. One of my

key informants, a relatively young Mayan priest and former guerrilla hard-liner, defected from the *Majawil* in protest during the first half of my field-work because he was dissatisfied with the lack of individual freedom in his work, and critical of rebel involvement. The guerrillas responded aggressively and forced him to return to his politico-religious duties. He obeyed, and others who had followed his escape from guerrilla influence turned from being his friends to what seemed implacable enemies. My key informant was later rewarded for his loyalty and given opportunities to move up in the organi-sation towards a position in the nationwide indige-nous coalition COPMAGUA (*Co-ordinadora de las Organizaciones del Pueblo Maya de Guatemala*).

If we follow the messengers who trudge the path between the local and the national spheres of Pan-Mayanism into the local field, we find that the image of Maya culture is as contested on the village level as it is among political organisations in the capital. To both Catholics and Protestants in the villages drinking represented the dark force that threatened the physical and spiritual well-being of the community, while the drinking customs were defended by the Maya traditionalists, indeed pro-moted by them. If the Catholics and the Protestants saw a common denominator of evil in the practice of the locally produced liquor, *kuxa*, the traditional-ists had a new rhetoric imported from the *Majawil Q'ij* that condemned the two churches as represent-ing a common evil, that is, the connection to, and historical legitimisation of, colonialism and genoci-dal political practices. Members of all three religious groups thus acted in replacing the more common divide in other parts of Guatemala between Catholics and traditionalists on the one hand and Protestants on the other, with one that isolated the self-pro-claimed 'Mayas' against the other two.

New socio-cultural phenomena emerged in this contested field; in the interface between the *Majawil Q'ij* and its local partner – the organisation of the traditionalists. There was an altar with a wooden cross that was placed at the gable opposite the entrance of the traditionalist church in the village where I spent most of my time. About a hundred similar but smaller crosses with names of the dead hung on the walls of the log-house. This was their final destination. The crosses had once been made by relatives of the dead and used in ceremonies and marches to commemorate dead loved ones. The house was a place with dense symbolism, frequently exploited by photographers and organisations that had been in the area. The picture depicting a grieving man or woman with such a cross is, I believe, the most common visual representation of the displaced. And members of their communities are (without access to statistical evidence) the most common visual representation of contemporary Mayan war victims. This Maya church could, in other worlds, be said to occupy a central location in the political imagery of indigenous martyrdom in Guatemala. Such symbolism, when consciously used by local and guerrilla politicians, was directed inwards as well as outwards. The martyr, and the survivor's relationship to the martyr, had come to shape the identity of people in resistance. Here, however, I got the impression that the crosses and the very church or temple of the traditionalists were mainly for show; its symbolism was directed solely outwards. The house was always empty, except when a delegation visited, or when a meeting with the traditionalists was centrally organised. Nobody remembered any church of this kind in their places of origin before the war. There are no references in the ethnography of the Ixils and the Quichés of a locale for worship in the shape of a public house. Instead, it seems as if this one, referred to as 'the *church* of the Mayan *priests*' constituted a political attempt at equalising the importance of what was here perceived as three religious branches: hence the word '*sacerdote*' (priest) instead of 'shaman' (as used by ethnographer Tedlock 1982), 'diviner' (as used by ethnographers Colby and Colby 1981), or 'father before god' (the meaning of the local term in the Ixil language), and hence the word 'church' and the construction of a spatial centre for their organisation.

What appears here is a contradiction between the alleged continuity in the religious aspects of Maya culture on the one hand, and the practices and ideologies of new 'inventions of tradition' (Hobsbawm and Ranger 1983) on the other.

The difference between ongoing practices and new dictates from the Maya movement emerged as small conflicts now and then. When old Andrés decided to organise a ceremony in his new house, with the stated purpose of thanking the angels for their generosity, he picked 'the day of the diviners'. He bought firecrackers in the neighbouring town and drums of *kuxa* from his relative. He contracted the marimba players and spent the small fortune of

150 *Quetzales* on the event. But Juan, his literate son, checked the calendar produced and distributed by the *Majawil Q'ij*, and found out that Andrés had picked the wrong day. Juan announced the fallacy in public. The ceremony was celebrated as planned, but uncertainty as to the power of the day was a fly in the ointment.

One result of this conflict appeared in chats and interviews with older men, who repeatedly said that they felt confused over what it was that their culture was supposed to entail, what this 'new project' of politics and identity was all about. Many confessed that they had indeed lost the knowledge of their forefathers, but that they were struggling to regain it through studying. In ways I found ironic, I was consulted in this matter. One of the young Basque volunteers in the area invited me to a meeting with the traditionalists in which they wanted me to explain the meaning of the concepts of 'spirituality' (Paragraph III: C of the peace accord), and the very content of 'customary law' (Paragraph IV: E). The volunteer thought that an anthropologist should know. I told what I knew but emphasised the right of the Mayas themselves to define the content of their culture, which of course did not help them at all. The concepts were made topical in the authorities' attempts to explain the details of the peace agreement. The awkward request that was made to me was thus instrumental. People needed to know, not what their culture was, but rather how to make good use of the accord in order to gain something in the ongoing negotiations with the government and former landowners. The dilemmas posed by war, displacement and the new political framework within which a resettlement was about to begin indicates that the meaning of the concepts of *culture* and *history* now in use are defined by outside powers, such as the national discourse of the peace accords, or the Maya movement.

That is not to say that history had been unimportant before the war. The perception of past events in Maya rural areas was and is marked by the presence of the dead and the presence of geographical reminders of past events – indeed a geography that received the inscriptions of prayers and offerings. Dennis Tedlock compares the mountain shrines of Quiché-speaking Momostenango with books, and the prayer makers with writers who ritually document events among the living (1985). But forced migration interrupts the rhythm of this documentation as it separates the dead from the living through the separation of the living from the places of worship. The absence of relations and of a previously known landscape that served as ordering mechanisms for Self and Community, creates, under the influence of violence, a void that new discourses on power, morality, and order seem eager to fill (Scarry 1985). Ideologies with clear-cut messages of resurrection and revenge are likely to be the most attractive. What we have are thus two different notions of the past that the displaced traditionalists tried to understand. Rather than negotiating between them, between the past as lived experience ritually commemorated in everyday practice and the past as part of a new ideology of culture, they tried to understand the former through the latter.

Some of the *Balbastixhes* – 'fathers before God' – knew how to interpret dreams, while some did not. Some dream interpreters were not *Balbastixhes* but instead dedicated Catholics or Protestants. Some knew how to cure with 'magic' while others were in the process of learning what they needed to know to prescribe the right herbs. Some knew the names of the days, and hence their meaning, but most did not. In order to find a common denominator among them, we need not explore further their professional expertise, but, rather, their positions within social hierarchies. In certain conceptions that could perhaps be labelled Mayan, God is inseparable from the cosmological order, which in turn is inseparable from the social order (see the story of Nicolas above). 'Father before God' would accordingly mean a person worthy of fatherly responsibilities before the collective; for example the extended family and patched-up household, the guerrilla unit, or the resistance community. The politicised version of culture distorts those responsibilities as it models God after Christian ideas. Maya culture is no longer related to relations, nor to experiences, but rather to an invented tradition that is of necessity perceived as lost.

This sense of loss was repeatedly expressed, and many elders complained about the ignorance among the young, the Catholics, and the Protestants. The *Balbastixhes* also lamented their own lack of knowledge, and their own 'lives in oblivion'. This connects to the distribution of authority and influence among the *Balbastixes* according to a 'knowledge scale': the greater the knowledge, the more important the authority. Along this scale, or in this hier-

archy, people move at different speeds. For a man like my key informant, who acted as a messenger between the revitalisation movement in the capital and the group of traditionalists among the displaced, the effort of moving upwards was notable. Others were studying the pamphlets and calendars of *Majawil Q'ij*, unhappy about their present condition but struggling ahead. Yet others seemed to be quite content with what they knew and what they were able to accomplish with what they knew. They belonged to a category of people who did not preach, who could not read and hence not study in a western sense, but who practiced the knowledge of previous generations. Their category lacked official leadership status, however. Thus the scale also had a dimension leading from cultural knowledge as practice to cultural knowledge as textually objectified systems of thought, which by the definition I propose is the essentialisation of culture. Without violence, no such essentialisation occurs.

Conclusion

The voice of the dispossessed is no longer that of the rebellious, which, according to the interpreters of Marx, was bound to transform oppression into universal equality. Rather, it is that of the uprooted, deprived of both territory and language (Bhabha 1994). Maya criticism of foreign expertise exemplifies the epistemological relocation of the concepts of culture and identity, from the narrative of those describing to the narrative of the described. That is the still understudied link between the content and the location of knowledge.

In the neo-liberal world, the issues of 'labour' and 'state' are not viewed as the sources of power and identity they once were (Hale 1997). Since 'identity' – through the cultural essentialising processes of war – and 'history' – through the replacement of 'race' with 'origin' in discourses on political legitimacy and territorial rights – have taken their place, anthropologists, the explorers of identities, and archaeologists, the explorers of the past, find themselves in the loophole of anti-neo-colonialist artillery.

What then, if anything, is wrong with this angry criticism? It is certainly not incomprehensible, but from whatever location we approach the violence of the past we are bound to accept the fact that violence is, and always has been, a socio-culturally informed ingredient of every human society. So much for the peaceful Maya. But our next step would be to explore what it was that the Spaniards after all did bring five centuries ago. If it was not violence *per se*, it most certainly was a transformation of power and a violent dislocation of political agency. Anthony Giddens' model (1987) of the characteristics of the modern nation-state, in terms of the degrees of penetration of everyday life on the part of the state and through its monopolisation of legitimate violence, adheres to an analytical toolkit that would enable a combined grasp of political disruption and cultural continuity.

My second concluding emphasis would be a reflexive one. I have been urged by Maya friends in Guatemala to work for the translation of my own studies so that my *foreign* arguments can be made transparent, criticised, or, possibly, come to be of use to the people who are the subject matter of my research. That quest represents one of the locations where an abstract reasoning about the production of knowledge becomes concrete. Anthropologists, like Kay Warren, make it difficult for others to come up with excuses. In Guatemala, the Maya publishing house Cholsamaj is printing a Spanish edition of her 1998 book on the Maya movement, in which she writes: 'Repatriating research is particularly important because the traditionalist archive, the specialised knower or *k'amöl b'ey*, has disappeared in many communities, and *costumbre* is being revitalised by various groups with their own political agendas' (Warren 1998: 77). If a scent of heroism can be detected in such an effort, I think a final quote from Linda Schele, from a context where she praises the linguists who work with, rather than on Maya communities in Guatemala, reveals what is at stake for contemporary scholarship:

In our experience, other disciplines [i.e., other than linguistics], like archaeology, have not concerned themselves with returning knowledge and experience to the Maya community. We believe the world has changed in the last decade and that in the future academics of the developed world must consider the needs and goals of the people whom they study. The cultural scientists who should be in the forefront of this kind of interaction, in fact, are often the most unaware. We counsel our colleagues to try it. They will learn far more than they teach, and the returns are beyond value. (In Freidel, Schele and Parker 1993: 139)

BIBLIOGRAPHY

- Aretxaga, B. 1997. *Shattering Silence: Women, Nationalism, and Political Subjectivity in Northern Ireland.* Princeton: Princeton University Press.
- Bhabha, H.K. 1994. *The Location of Culture.* London and New York: Routledge.
- Castañeda, Q. 2001. Ethics and Intervention in Yucatec Maya Archaeology, Past and Present. Paper presented at the Conference 'Towards a More Ethical Mayanist Archaeology'. University of British Columbia, Canada.
- CEH 1999. *Guatemala: Memoria del Silencio. Tomo I-XII. Informe de la Comisión para el Esclarecimiento Histórico.* Guatemala: UNOPS.
- Clausewitz, C. von 1976(1832). *On War.* Princeton: Princeton University Press.
- Coe, M.D. 1992. *Breaking the Maya Code.* London: Thames and Hudson.
- Cohodas, M. 2001. Politicizing the 'Maya Collapse'. Paper presented at the Conference 'Towards a More Ethical Mayanist Archaeology'. University of British Columbia, Canada.
- Cojtí Cuxil, D. 1994. *Politicas para la reivindicación de los Mayas de hoy: Fundamento de los Derechos Específicos del Pueblo Maya.* Guatemala: SPEM y Editorial Cholsamaj.
- Cojtí Ren, A. 2002. Maya Archaeology and the Political and Cultural Identity of Contemporary Maya People. Paper presented at the Conference 'Towards a More Ethical Mayanist Archaeology'. University of British Columbia, Canada.
- Colby, B. and Berghe, P. van den 1969. *Ixil Country: A Plural Society in Highland Guatemala.* Berkeley: University of California Press.
- Colby, B. and Colby, L. 1981. *The Daykeeper: Life and Discourse of an Ixil Diviner.* Austin: Texas University Press.
- COMG 1995. *Construyendo un futuro para nuestro pasado: derechos del Pueblo Maya y el Proceso de Paz.* Guatemala: COMG/Editorial Cholsamaj.
- Daniel, V.E. 1996. *Charred Lullabies: Chapters in an Anthropography of Violence.* Princeton (New Jersey): Princeton University Press.
- Demarest, A. and Garcia, D. 2002. Archaeology, Sacred Site Development, and Community Identity: The View from Ongoing Multicultural Collaborations in Post-war Guatemala. Paper presented at the 2002 American Anthropological association Meeting in New Orleans, Louisiana.
- Freidel, D., Schele, L. and Parker, J. 1993. *Maya Cosmos: Three Thousand Years on the Shaman's Path.* New York: William Morrow and Co.
- Friedlander, J. 1975. *Being Indian in Hueyapan: a Study of Forced Identity in Contemporary Mexico.* New York: St Martin's Press.
- Giddens, A. 1987. *The Nation-State and Violence.* Berkeley: University of California Press.
- Hale, C.R. 1997. 'Cultural politics of identity in Latin America'. *Annual Review of Anthropology,* 26, pp. 567-90.

- Harrison, S. 1993. *The Mask of War*. Manchester: Manchester University Press.
- Hawkins, J. 1984. *Inverse Images: The Meaning of Culture, Ethnicity, and Family in Post-colonial Guatemala*. Albuquerque: University of New Mexico Press.
- Hill, R.M. and Monaghan, J. 1987. *Continuities in Highland Maya Social Organization: Ethnohistory in Sacapulas, Guatemala*. Philadelphia: University of Pennsylvania Press.
- Hobsbawm, E. and Ranger, T. (eds.) 1983. *The Invention of Tradition*. Cambridge: Cambridge University Press.
- Kohl, P.L. and Fawcett, C. (eds.) 1995. *Nationalism, Politics and the Practice of Archaeology*. Cambridge: Cambridge University Press.
- Löfving, S. 2002. *An Unpredictable Past: Guerrillas, Mayas, and the Location of Oblivion in War-Torn Guatemala*. Dissertation, Uppsala University.
- Mamdani, M. 2001. *When Victims Become Killers: Colonialism, Nativism and the Genocide in Rwanda*. London: James Currey.
- Meskell, L. (ed.) 1998. *Archaeology Under Fire: Nationalism, Politics and Heritage in the Eastern Mediterranean and Middle East*. London: Routledge.
- Montejo, V. 1999. *Voices from Exile: Violence and Survival in Modern Maya History*. Norman: University of Oklahoma Press.
- Montejo, V. 2001. Speculations on Archaeology's Turn in the Critique of Mayan Studies. Paper presented at the Conference 'Towards a More Ethical Mayanist Archaeology'. University of British Columbia, Canada.
- Moser, C. and McIlwaine, C. 2001. *Violence in a Post-Conflict Context: Urban Poor Perceptions from Guatemala*. Washington D.C.: The World Bank.
- Nordstrom, C. 1997. *A Different Kind of War Story*. Philadelphia: University of Pennsylvania Press.
- Pyburn, K.A. 1998. 'Opening the door to Xibalba: the construction of Maya history'. *IJHL Indiana Journal of Hispanic Litteratures*, 13, pp. 125-30.
- Pyburn, K.A. 1999. 'Native American religion versus archaeological science: a pernicious dichotomy revisited'. *Science and Engineering Ethics*, 5, pp. 355-66.
- Roys, R.L. 1967. *The Book of Chilam Balam of Chumayel*. Norman: University of Oklahoma Press.
- Scarry, E. 1985. *The Body in Pain: The Making and Unmaking of the World*. Oxford: Oxford University Press.
- Schele, L. and Miller, M.E. 1992. *The Blood of Kings: Dynasty and Ritual in Maya Art*. London: Thames and Hudson.
- Sharer, R.J. 1994. *The Ancient Maya*. Stanford: Stanford University Press.
- Stoll, D. 1993. *Between Two Armies in the Ixil Towns of Guatemala*. New York: Columbia University Press.
- Tedlock, B. 1982. *Time and the Highland Maya*. Albuquerque: University of New Mexico Press.
- Tedlock, D. (trans.) 1985. *Popol Vuh: The Definitive Edition of the Maya Book of the Dawn of Life and the Glories of Gods and Kings*. New York: Simon and Schuster.
- Warren, K.B. 1992. 'Transforming memories and histories: the meaning of ethnic resurgence for Mayan Indians'. In: A. Stepan (ed.): *Americas: New Interpretive Essays*. Oxford: Oxford University Press, pp. 189-219.
- Warren, K.B. (ed.) 1993. *The Violence Within: Cultural and Political Opposition in Divided Nations*. Boulder: Westview Press.
- Warren, K.B. 1998. *Indigenous Movements and their Critics: Pan-Maya Activism in Guatemala*. Princeton: Princeton University Press.
- Watanabe, J.M. 1992. *Maya Saints and Souls in a Changing World*. Austin: University of Texas Press.
- Watanabe, J.M. 1995. 'Unimagining the Maya: anthropologists, others, and the inescapable hubris of authorship'. *Bulletin of Latin American Research*, 14(1), pp. 25-45.
- Wilk, R.R. 1985. 'The ancient Maya and the political present'. *Journal of Anthropological Research*, 41(3), pp. 307-26.
- Wolf, E. 1957. 'Closed corporate peasant communities in Mesoamerica and central Java'. *Southwestern Journal of Anthropology*, 13(1), pp. 1-18.

Warfare, Weaponry, and Material Culture /31-34

Warfare, Weaponry, and Material Culture: An Introduction

So the possession of costly bronze daggers, swords, and rapiers consolidated
the positions of war-chiefs and conquering aristocracies as did the knights's
armour in the Middle Ages.
(Childe 1941: 'War in prehistoric societies'. *The Sociological Review* 33, p. 133)

/31

HELLE VANDKILDE

The three articles in this section all deal with warfare in societies of the Bronze
Age. In addition, they share a material perspective. In their articles Anthony
Harding (chapter 33) and Henrik Thrane (chapter 32) assess the reliability of
material culture as a source of data about Bronze Age warfare asking the essential question of how far the material evidence can bring us in understanding
affairs of war in the Bronze Age. My own contribution reflects on the social role
of weaponry, in strategies of power taking place in the ideal society described so
persuasively by Homer.

Archaeology and material culture

Material culture has more than one application in archaeological research. First,
it is the main source of empirical data about the past: only material remains
have been preserved for long periods of time in the past, and these are the sole
means of gaining access to prehistoric society. Archaeological sources for prehistory are, basically, material fragments of past human actions carried out by
situated actors. On the basis of archaeological data past human actions can be
reconstructed, and their structuration – their repetition and change – in
time/space can be figured out through analysis. More precisely, archaeology
may be described as a discipline that studies, mainly through material means,
the reproduction of and changes in human action, which is in fact embodied
social structure across time and space. Second, material culture is inevitably a
third party in the relationship between human actors and society, and for that
reason it is central to any interpretative enterprise. In archaeology material
culture therefore holds a twofold position as data material and as a principal
agent that has structured the lives of prehistoric beings.

The nature of material culture, identification, and warfare

Material culture functions intentionally as a silent form of discourse, especially in the domain of the human body, but also as a surrounding material setting, which furthermore constrains as well as enables action. Personal appearance can be emphasised as basic to organising and maintaining all sorts of social identities and therefore deserves particular attention. Objects certainly inspire social identification in various domains. Material culture is indeed the mediator and creator of various distinctions in society: age, gender, social status, profession, and so forth. And warriorhood combines such distinctions in different ways.

Weapons make warriors, in a manner of speaking, and weapons are tools of war. The material link makes it realistic to identify and study warriors and warfare in prehistory. Specific dress and weaponry are notably important in creating and recreating the warrior as identity and ideal. It should be repeated, however, that function is largely independent of material form. A stone axe, or a sword, does not automatically – and certainly not outside its context – allow interpretation in terms of warfare and warriorhood as emphasised in the succeeding articles. The function of objects sometimes falls well beyond the obvious, and since function and meaning are constructed through the social context, the latter needs to be studied carefully. Structuration theory maintains that discourse always forms part of the duality of structure and agency as 'signification', and this is important in archaeology, in which material culture as a silent form of discourse is intimately related to traces of action/structure. Signs, whether material or spoken, have no existence beyond the meaning-creating and communicative processes of interaction; they are, in Giddens' phrasing, the medium and outcome of these (cp. Giddens 1984: 31f).

It may nevertheless be added that material culture is in itself very persuasive and durable and creates social relationships in an interactive and recursive process. Material culture thus holds an unintentional stain in the sense that it continues to influence people and their actions and thoughts even when the memory of its original functions and meanings has been lost. Things, cultural landscapes and material settings are in themselves potent and persuasive due to their materiality. Exotic objects and inherently attractive materials like copper, gold and amber readily serve as valuables with potential uses as cultural, symbolic and economic capital in the quest for wealth and social position. Material culture is in sum a substantial and meaningful ingredient in any society, past or present, and therefore also for the understanding of war and warriors. The archaeological sources for warriors and warfare can be summarised in four groups:

Weaponry

Weaponry divides arbitrarily into implements with a potential for war and weapons intended for offensive and/or defensive purposes. Bows and arrows, points, axe blades, knives, daggers, and so on made of various organic or non-organic materials belong to the first category (cp. Capelle 1982; Vencl 1984; Chapman 1999). The second category comprises mainly swords, spearheads, shields, and body armour, but also maceheads and battle-axes.

This rough division indicates that most so-called weapons have potential uses outside warfare, notably hunting. Bones from Neolithic and Bronze Age settlements normally show little inclusion of wild animals, and I therefore presume that hunting in the Neolithic, and most certainly in the Bronze Age, was more a matter of prestige than economic necessity. The line between prestige-hunter and warrior can be blurred, as illustrated by the lion-hunt dagger from Shaft Grave IV at Mycenae, showing warfare against the lion, which is likely to be a metaphor for the elite warrior.

The multi-functionality, or ambiguity, of material culture in general implies that weapons cannot be reduced merely to tools of war. A series of other potential functions and meanings presents itself. Notably, weapons often maintain identities of gender, age, rank, or various other types of group identity like kinship, ethnicity and profession. In European prehistory the cultural biographies of weaponry typically ended with a ritual deposition either in burials or in votive offerings. Such depositions may comprise one item or a combination of several items depending on cultural and social factors. The archaeologist thus obtains intimate knowledge mostly of the destination, the ritual death of things, which may have had several very active lives prior to the final deposition. The destination of weaponry, however, may still be able to tell stories about the living society. Let us first look briefly at the deposition of weapons in burials, weapons in sacrificial deposits, weapon combinations and finally weapon technology:

Weaponry in burials

Objects can be deposited in burials for a variety of overlapping reasons related to culture and society such as emotion, social ambition, social rivalry and social identity, cultural values and norms, and religion. Burials can be carried out routinely as the way things should be done or more deliberately as political announcements, although also the former of the two modes may be carried out from the perspective of social strategy and power. Funerals and funerary monuments are suitable means to question, confirm or legitimise existing relations of social power.

Weapons in burials are often a material extension of the dead body and thus somehow relate to the collective social identities of the deceased, and perhaps in some cases also to his or her personhood. Using James Whitley's classification (1995: 5ff), a metonymic and metaphoric statement can be distinguished in the funerary domain: 1. Weapons and dress accessories may be a specific (metonymic) statement of the roles played by the dead individual in life. If, for instance, a sword accompanies a young man in one burial, and this relationship is repeated in other burials, the conclusion would be that among the identities young males carried in that society the warrior identity was an important one. 2. Weapons and dress accessories may alternatively be used to make more general (metaphoric) statements and have a less direct bearing on the lived identities of the deceased. If a small, exclusive group of mature males are interred with rich weaponry, this is likely to reflect symbolic warriorhood in the sense of an earlier warrior identity, heroic status or ambition, political authority or high social rank. Likewise, weapons accompanying small male children and young adolescents are likely to be metaphors originating in warrior values rather than actual signs of warriorhood.

Weaponry in votive deposits

A large number of weapons are known from sacrificial deposits all over northern and central Europe, and many of them have been retrieved from wetlands (Bradley 1990). Behind this specific practice, which is in general thought to involve a larger number of people than funerary rituals (Vandkilde 1996; 1999), are several social as well as religious motives such as commemorating the past, gifts to the gods, rites of passage, and potlatches to promote the importance of particular persons and social institutions and to confirm, question or legitimise power relations. These motives need not exclude each other.

In the Neolithic Period we are dealing mainly with implements with the potential for war, while in the Bronze Age the association with war often becomes unequivocal, due to the distinct warlike character of the objects. War booty offerings are a generally accepted find category in Iron Age Europe, probably because it is a practice sustained by written sources. Klavs Randsborg (1995) has suggested that Neolithic, and especially Bronze Age, depositions of weaponry have a similar background. A proportion of Neolithic and Bronze Age ritual depositions of weaponry might well be considered the culmination of actions relating directly to warfare, which even today is intertwined with rituals. Some weapon hoards may alternatively have been ritually deposited as part of 'fighting with property' – that is, the potlatch as a sort of surrogate warfare (Bradley 1990: 139f; Paige and Paige 1981: 52). Among the Northwest Coast Indians, however, potlatch rituals only became truly elaborate when the Canadian government prohibited warfare (ibid.), and therefore this analogy may not be fully appropriate to draw on in the interpretation of prehistoric ritual depositions. Warfare cannot in general be understood as a ritual, but it nearly always contains aspects of ritual (e.g., Ehrenreich 1997; Otterbein 2000). Warfare is in reality a flow of social actions that may be initiated and concluded by rituals. Ritual depositions of weapons in watery places can thus arguably be considered from the perspective of war, particularly if such an interpretation is supported by the specific social context. A weapon offering may then be seen as a metonymic statement of concluded warlike actions in terms of victory, peace-making, and alliance maintenance, or perhaps as an offering made to ensure luck in an impending war. Weapon offerings may also be metaphoric statements, which more distantly relate to the waging of war. It can be argued that the increasing institutionalisation of warriors and war generally implies that these two phenomena will be carried more symbolically into other social fields as part of their signification and legitimation. In either case, the war hypothesis works well with the series of motives mentioned above.

Weapon combinations
Weapon combinations can, used with caution, tell us about the equipment and fighting methods of the warrior, notably a preference for close or distant combat. It is, however, also necessary to note – and this tallies with what has already been mentioned above – that the warrior may not be interred with their full equipment but rather only those parts of the equipment that have a certain symbolic meaning. Weapons may be very personal objects (swords usually are) and thus invested with personalised names and meanings and at the same time function as insignia of gender or rank. Alternatively, weapons may have whole biographies and stories attached to them, which could have been invoked on special occasions (cp. Vandkilde chapter 34: the famed boar's tusk helmet which Odysseus puts on for a nightly spying expedition). Such particular weapons are perhaps more likely to be inalienable possessions that remain in the family as tokens of memory and inheritance than ones ritually exchanged with the dead or the gods.

Weapon technology

Weapon technology is not unimportant for the outcome of war. Improvements in weaponry have been known to escalate the degree and viciousness of warfare in numerous ethnographic cases. It is thus possible that innovations in the technology of war could have increased the speed of social change in prehistory. A comparison of weapons over time can give clues to significant changes in weapon technology, understood as the quality of weapons and styles of fighting. The emergence of the sword around 1600 BC in central and northern Europe is likely to have been such an important innovation with considerable social and symbolic effects in warlike as well as in more peaceful interactions inside society and between societies. Likewise, the introduction of horses would have changed conventional warfare as well as the whole social habitat, as documented, for instance, among the Grand Chaco Abipón in South America during the eighteenth century (Lacroix 1990).

Fortifications

In central Europe fortified settlements occur from early on and progressively more frequently, while in southern Scandinavia they remain more or less absent throughout prehistory and earliest history. This is certainly not because Scandinavians are more peaceful than central Europeans during these periods. Clearly the presence or absence of fortified settlements cannot be used as evidence of war and peace, but might be rooted in different cultural practices. Cultural tradition, degree of openness and topography of the landscape, settlement organisation, and social and inter-societal structure are relevant factors that might explain why settlements are sometimes fortified in central Europe and rarely so in southern Scandinavia.

Fortifications are as multi-functional as other kinds of material culture. Fortified settlements are sometimes urban or proto-urban centres for a limited region or even nodal points for super-regional trade and crafts. At other times they are marginal phenomena placed at the boundaries of settled areas to scare off the enemy and to protect against intruders. In the latter case, they may not be permanently settled, or perhaps only with certain specialised personnel. Fortifications and territorial marking go well together, but distinct territoriality can exist without fortifications.

Skeletal trauma

Traumata on skeletons are caused by various forms of interpersonal violence like warfare, homicide, gang aggression, intra-family fights, and forceful kinds of sport. The boundary between these forms of violence can be quite subtle, and ethnographic examples suggest that a high occurrence of inter-personal violence other than warfare quite often coincides with situations of war (e.g., Chagnon 1968). Skeletal traumas are, in fact, relatively frequent in European prehistory when it is taken into account that in some areas skeletons are not well preserved, that they are not routinely examined for marks of violence, and that much physical violence does not leave visible traces on the skeleton.

Abnormal mass graves with several injured individuals as well as normal single interments with an injured individual occur throughout the Neolithic and the Bronze Age all over Europe. The injuries vary from projectile wounds to marks and cuts from weapons. Some wounds are healed up, while other wounds were the cause of death. Skeletal trauma is a valuable source of additional information when it comes to detecting warrior identities, but it should also remind the archaeologist that war is always abominable. Prehistoric humans actually suffered from war through the loss of loved ones and through the injuries they received as combatants, supporters or victims.

Iconographic presentations in art and rituals

The Scandinavian Bronze Age rock carvings with – amongst other themes – representations of warfare and warriors are particularly relevant to this introduction. A similar reasoning may, however, pertain to warlike presentations in the field of wall painting and bellicose imagery on, for instance, grave stelae and vases. Without discounting the vast amount of literature interpreting the southern Scandinavian rock carvings, I would like to propose that the images of war – and perhaps in general the more elaborate of the themes – on the carvings are narratives of a highly ideal nature, but possibly rooted in historical events. The same realist root arguably adheres to the scenes of warfare carved on rock in southern Africa and attributable to groups of San Bushmen (Campbell 1986: 265). I suggest that the images carved in rock in Scandinavia were tales about privileged groups in society that probably created them as part of their rich oral tradition as well as part of their tradition of imagery. The carvings are analogous to Homer's epics in the sense that that there are certain recurring and therefore easily recognisable traits, the actions and myths of 'gods' and mortals occur intertwined, and the narrative, and certainly the interpretation of it, continues forever. Both are arguably narratives of an ideal nature maintaining the interests and ideology of a particular social group.

Jarl Nordbladh (1989) has analysed patterns of fighting at the rock art site of Kville in Bohuslän. Single combats predominate, the combatants are of equal size, and weapons never touch. When fighting occurs on ships, only a few crew members, made larger than the rest, carry weapons, and some ships have more than 135 crew members. Nordbladh finds this number much overrated, and therefore hints at the idealised character of the narrative. He nevertheless suggests that the pictures of combat at Kville provide a fairly correct picture of how fighting was actually carried out in Bronze Age society. Richard Osgood presents a similar view (1998; Osgood *et al.* 2000: 34). Fighting supposedly took place as heavily ritualised action or performances of a sport-like character (ibid.) – that is, as a rule with a non-deadly outcome.

War scenes in rock carvings undoubtedly reduce fighting to the demonstration of potency among high-ranking warrior heroes, who fight as equals and according to certain aristocratic rules and ethics. A rather similar pattern – combining sport-like duelling with glimpses of cruelty and death – is encountered in the Homeric epics, but here there are traces of other kinds of war, much more vicious and much less heroic, like raids on settlements to obtain slaves, women and portable wealth. It is highly likely that the rock carvings – like the poems of heroes and war in early history – overemphasise the ideals of war combat and aristrocratic companionship, and consequently underrate another, much more violent face of war. Skeletal trauma (e.g. Kjær 1912; Vandkilde 2000; Louwe Kooijmans 1993; Fyllingen 2002; Fyllingen chapter 22) and damages and sharpening traces on swords (Kristiansen 1984; Bridgford 1997) suggest as much.

A great potential for studies of war and peace

The obstacles of obtaining true insight into prehistoric warfare and warriors are surely many, but let us not become totally overwhelmed. Harding's contribution (chapter 33) is a sharply source critical analysis tending to uncover cracks in the evidence rather than attempting to link together existent evidence in an effort to understand. Such moderate pessimism does serve the purpose of reminding fellow archaeologists that our tales of war and peace in prehistory are utterly dependent on the archaeological sources. I must also agree that the evidence of the earlier Bronze Age is generally weaker than the later Bronze Age of the Urnfield Period. Still, I believe that context-based analyses of selected areas such as the Nordic Older Bronze Age with its numerous grave mounds, ritual deposits and settlements provide a very great potential for examining actions and identities of war and peace across a considerable period of time. Likewise, the Danubian-Carpathian Early Bronze Age with its geopolitical hotspots, extensive inhumation cemeteries, rich hoards, and defended hill-top settlements provides excellent possibilities.

Thrane (chapter 32) presents an optimistic, if still source-critical, view highlighting the rich Scandinavian evidence of weaponry in various ritual contexts and concludes that what we see in Bronze Age Scandinavia may well have been similar to the patterns of warfare and warriorhood described in the Homeric poems. Used with caution, Homer may indeed serve as an appropriate analogy to Bronze Age society north of the Alps. This is in tune with my own contribution, which examines the social and material world of Homer's epics with the underlying assumption that a comparative analogical enterprise could bring forth new insight on the subjects of warfare, gender and materiality in the illiterate Bronze Age of northern Europe. However, a precondition for advances in this respect might well be that the four classes of material evidence considered above are combined and thoroughly brought into dialogue with adequate theoretical platforms.

BIBLIOGRAPHY

- Bradley, R. 1990. *The Passage of Arms. An Archaeological Analysis of Prehistoric Hoards and Votive Deposits*. Cambridge: Cambridge University Press.
- Bridgford, S.D. 1997. 'Mightier than the Pen? An Edgewise Look at Irish Bronze Age Swords'. In: J. Carman (ed.): *Material Harm. Archaeological Studies of War and Violence*. Glasgow: Cruithne Press, pp. 95-115.
- Campbell, C. 1986. 'Images of war: a problem in San rock art research'. *World Archaeology*, 18(1), pp. 255-68.
- Capelle, T. 1982. 'Erkenntnismöglichkeiten ur- und frühgeschichtlicher Bewaffnungsformen. Zum Problem von Waffen aus organischem Material'. *Bonner Jahrbücher*, 182, pp. 265-88.
- Chagnon, N.A. 1968. *Yanomamö. The Fierce People*. New York, London and Chicago: Holt, Rinehart and Winston.
- Chapman, J. 1999. 'The origins of warfare in the prehistory of Central and Eastern Europe'. In: J. Carman and A. Harding (eds.): *Ancient Warfare. Archaeological Perspectives*. Stroud: Sutton Publishing, pp. 101-42
- Childe, V.G. 1941. 'War in prehistoric societies'. *The Sociological Review*, 33, pp. 126-39.
- Ehrenreich, B. 1997. *Blood Rites. Origins and History of the Passions of War*. New York: Metropolitan Books.
- Fyllingen, H. 2002. *The Use of Human Osteology in the Analysis of Ritual and Violence during the Early Bronze Age. A Case Study from Nord-Trøndelag*. Cand.Philol. thesis. Department of Archaeology, University of Bergen, Bergen.

- Giddens, A. 1984. *The Constitution of Society. Outline of the Theory of Structuration.* Cambridge: Polity Press.

- Kjær, H. 1912. 'Et mærkeligt arkæologisk-antropologisk fund fra stenalderen'. *Aarbøger for Nordisk Oldkyndighed og Historie.* Copenhagen: Det Kongelige Nordiske Oldskriftselskab, pp. 58-72.

- Kristiansen, K. 1984. 'Krieger und Häuptling in der Bronzezeit Dänemarks: Ein Beitragzur Geschichte des bronzezeitliches Schwertes'. *Jahrbuch des Römisch-Germanisches Zentralmuseums Mainz* (31), pp. 187-208.

- Lacroix, M. 1990. '"Volle Becher sind ihnen lieber als leere Worte". Die Kriegerbünde im Gran Chaco'. In: G. Völger and K. von Welck (eds.): *Männerbande-Männerbünde.* (Ethnologica, Neue Folge, 15). Cologne: Joest Museum der Stadt Köln, pp. 80-271.

- Louwe Kooijmans, L.P. 1993. 'An Early/Middle Bronze Age multiple burial at Wassenaar, the Netherlands'. *Analecta Praehistorica Leidensia,* 26, pp. 1-20.

- Nordbladh, J. 1989. 'Armour and fighting in the south Scandinavian Bronze Age, especially in view of rock art representations'. In: T.B. Larsson and H. Lundmark (eds.): *Approaches to Swedish Prehistory: A Spectrum of Problems and Perspectives in Contemporary Research.* (British Archaeological Reports International Series). Oxford: British Archaeological Reports, pp. 323-33.

- Osgood, R.H. 1998. *Warfare in the Late Bronze Age of North Europe.* (British Archaeological Reports International Series, 694). Oxford: Archaeopress.

- Osgood, R.H. and Monks, S. with Toms, J. 2000. *Bronze Age Warfare.* Stroud: Sutton Publishing Limited.

- Otterbein, K.F. 2000. 'A History of Recearch on Warfare in Anthropology'. *American Anthropologist,* 101(4), pp. 794-805.

- Paige, K.E. and Paige, J.M. 1981. *The Politics of Reproductive Ritual.* Berkeley, Los Angeles and London: University of California Press.

- Randsborg, K. 1995. *Hjortspring. Warfare and Sacrifice in Early Europe.* Aarhus: Aarhus University Press.

- Vandkilde, H. 1996. *From Stone to Bronze. The Metalwork of the Late Neolithic and Earliest Bronze Age in Denmark.* Aarhus: Jutland Archaeological Society and Aarhus University Press.

- Vandkilde, H. 1999. 'Social distinction and ethnic reconstruction in the earliest Danish Bronze Age'. In: C. Clausing and M. Egg (eds.): *Eliten in der Bronzezeit. Ergebnisse zweier Kolloquien in Mainz und Athen.* (Monographien des Römisch-Germanischen Zentralmuseums, 43). Bonn: Habelt, pp. 245-76.

- Vandkilde, H. 2000. 'Material culture and Scandinavian archaeology: A review of the concepts of form, function, and context'. In: D. Olausson and H. Vandkilde (eds.): *Form, Function and Context. Material culture studies in Scandinavian archaeology.* (Acta Archaeologica Lundensia, Series in 8, no. 31). Stockholm: Almqvist and Wiksell International, pp. 3-49.

- Vencl, S. 1984. 'War and warfare in archaeology'. *Journal of Anthropological Archaeology,* 3, pp. 116-32.

- Whitley, J. 1995. 'Tomb cult and hero cult. The uses of the past in archaic Greece'. In: N. Spencer (ed.): *Time, Tradition and Society in Greek Archaeology. Bridging the 'Great Divide'.* London and New York: Routledge, pp. 119-26.

Swords and Other Weapons in the Nordic Bronze Age: Technology, Treatment, and Contexts

HENRIK THRANE

/32

Warriors were mentioned early on in the study of the Nordic Bronze Age (Worsaae 1843; Müller 1897) and so was warfare, but not at any length. Any implication that violence of any kind was an important part of life was not manifested. Warfare and warriors were mentioned but not prominently and weapons were treated just like ornaments and pots. It didn't take long for the myth of the peaceful, happy and sunny Bronze Age as the first Golden Age to become the truth (Brøndsted 1939). In his book on the Bronze Age, Gordon Childe only mentions war when he reaches the Late Bronze Age Urnfield migration, but then he does so extensively (1930: 43, 192).

Curiously enough, C.J. Thomsen's three-age system was a reaction to the Latin poetical division of the past, in which the golden period was introduced as a term. In the 1840s people were not aware that they lived in the Danish Golden Age, and they did not use the expression, but by the end of the century and through the twentieth century that was how the Bronze Age came to be viewed, consciously or unconsciously. Of course, in a Golden Age peace and prosperity ruled, so there was no place for warfare – that only came with the crude and cold Iron Age.

Retrospectively we may find this curious, considering how many swords were known already by the time Thomsen formulated his model or by the time J.J.A. Worsaae, who knew several hundreds, divided the Bronze Age into early and late phases in 1859 (Worsaae 1843: 24). Swords, spearheads and axes (palstaves) were prominent among the early finds that filled the showcases of Thomsen's museum because they were big and solid and therefore observed and noted when farmers (or archaeologists) broke into the burial mounds. Thus the dichotomy between the material evidence and its interpretation existed right from the beginning of serious research on the Bronze Age. With the 'War and Society' project it became imperative to examine the material base for information on warfare in order to present a state of the arts and formulate a theory of the role of war in the Bronze Age. A couple of studies had already been made (Kristiansen 1984; Nordbladh 1987; Randsborg 1995) but no comprehensive study was available. We have tried to examine the data that may indicate the presence of war.

The first problem is that we only have sources which, at their best, indicate violence. How this violence was organised and whether it to a degree deserves the name War (Steuer 2000; Steuer chapter 16) is an open question. A general scale from group to individual violence can be made, however, and we can examine how the archaeological sources fit into this scale before moving on to an interpretation of the existence of warfare. Social groups normally find the will and means to defend themselves in societies where warfare is a normal phenomenon. Here it is unnecessary to speculate upon when this stage was reached. Suffice it to say that by the Bronze Age, European societies must have reached a stage where there was enough wealth around to tempt aggression and where social organisation had reached a level that made organised violence at a level of proper warfare possible.

Defence

An immediate effect of warfare in archaeological material would be in the settlement organisation. If it had not been organised into larger units already – as a means of providing the basis for the levying of forces for warfare – one effect of war could be to move settlements together in larger and better defensible units.

We know of a large number of Bronze Age settlements by now, probably more than 1000, covering the period from c. 2200 – 500 BC, albeit with some problems in certain periods, and so far there is no clear evidence of large agglomerated settlements of a kind that we may term village, whether fortified or open, as they are known from Central Europe from the Early Bronze Age (Jankuhn *et al.* 1977; Jockenhövel 1990). It is only in Central Sweden (Apalle and Hallunda) and perhaps at Voldtofte on Funen (Thrane 1995) that more than a couple production units – farms – form such settlements, which were also more permanent than normal. Yet even these deserve reservations. Only by c. 300 BC do villages bounded by palisades, which are thus clearly definable spatially, turn up in Jutland (Becker 1982; Rindel 2003) and Holland (Waterbolk 1977).

The tradition from the Late Neolithic went on through the Bronze Age with single farms or at the most two–three farms, each for a single family, forming somewhat dispersed settlements (Nielsen 1999). None has yielded any form of defence in the shape of palisades or ditches. The position on elevated parts of the landscape may, at best, be regarded as a strategic solution, enabling the inhabitants to escape should an enemy appear (as on Bellona: Kuschel 1988). Destruction levels are also absent. There is no clear evidence of war at any scale to be gleaned from the settlements (cp. the survey by Olausson 1993).

This negative evidence is all the more remarkable since substantial defensive earthworks at elevated settlements or using lake situations (Biskupin and Sobijuchie) form integrated elements of the Late Bronze Age Lausitz and Urnfield cultures in Poland (Bukowski 1962; Harding and Ostoja-Zagorski 1993) and as far west as Brandenburg (for the latest version of the distribution see Köpke 2002: map 5) and Central Europe (with the strange Early Bronze Age group in Slovakia and neighbouring countries [Jockenhövel 1990]).

The still ambiguous and certainly atypical settlement of Vistad in Östergötland (Larsson 1993) rather confuses the picture and could perhaps best be seen as an intrusive element, not representing the Nordic tradition. At the level above the settlements, communal defences are not absent from Scandinavia but none can be dated earlier than the early centuries AD (Jørgensen 2001). So, again we cannot adduce any positive archaeological evidence for organised group aggression or defence, and war may not be the motivation behind the suggested use of natural features as territorial borders (Thrane n.d.).

Traumata

At an individual level the direct evidence of violence is, of course, the traumata on skeletons – with the Norwegian case as the exception (Fyllingen chapter 22). Here the archaeological (de)formation processes have restricted our position severely. In spite of the hundreds of inhumation burials from the Early – Middle Bronze Age, c. 2200 – 1100 BC, very few have left more than soft traces of bones, normally just of the enamel of the teeth. Even graves with more or less complete skeletons have yielded little information – even if the bones have been kept. From Denmark (Brøste *et al.* 1956) and Scania (Håkansson 1985) we have just a single unambiguous instance of violence – the well-known broken off spearhead tip from period I at Over Vindinge (Aner and Kersten: no.1291 I; Bennike 1985: 109f). It is interesting that this observation fits remarkably well with the skeletons from Tormarton (Osgood chapter 23) and Dorchester (Osgood 1998: 19; Osgood chapter 23; Thorpe chapter 10) all of which have spearheads embedded in the pelvic region, curiously enough entering from the back.

The only partly preserved find from the bog of Granhammar in Central Sweden (Lindström 1999) and the cape from Gerumsberg (with its stabbed holes from a sword blade – von Post *et al.* 1925) indicate that the burials may not be as reliable or representative sources as we would like to think. The unexpected find from Norway (Fyllingen chapter 22) adds weight to this notion. It should be expected that 14 C-dates may bring more skeletal evidence into the Bronze Age when more stray skeletons are dated – cp. the Neolithic date of Sigersdal (Bennike *et al.* 1987).

If we know of skeletons in a condition where traumata may be recognised from only few of the graves with inhumation burials, and one of these gives undeniable evidence of violence by a proper weapon, how does this represent the Nordic Bronze Age skeleton population? The Over Vindinge grave. It could be argued that traumata are over-represented rather than under-represented in the preserved sources. Admittedly, we only have a fragile base for further inference, but it could also be argued that such violence was more common than would appear at first sight. Still, we are not yet in contact with proper war. One trauma does not make a war.

Weaponry

The trauma evidence leads to the dominant potential source for warfare, the weapon arsenal. While lethal damage may be inflicted by any casual means such as muscle power, simple stones and a big enough bone, or by common tools like clubs, hammers and axes, or by hunting implements like harpoons, arrows or spears (Kouwenhoven 1997), proper tools for violence made exclusively as weapons seem to be a metal age phenomenon. Developing from the small copper daggers of the Bell Beaker culture and the Early Bronze Age ubiquitous daggers (Thorpe chapter 10), the sword came to be the most impressive weapon from the Bronze Age until the breakthrough of firearms millennia later. In the Nordic Bronze Age we find the same types of weaponry as elsewhere in Europe although with a different emphasis. Defensive weapons are nearly absent, or are at least much less prominent than in the other European Bronze Ages (Harding 2000). Up here we find offensive weapons like swords, daggers, axes, spears and arrows but no evidence of slings or maces.

These five main weapon types vary according to time and place as expressed in the archaeological record, and even a brief examination demands a look at such aspects of the formation processes as production, treatment, deposition, context and retrieval (or availability to us).

Axes

Leaving aside the earliest (flint) daggers and the later (Early Iron Age) spear tips of bone (Randsborg 1995), the only types that were made of non-metal materials to any extent were axes (Baudou 1960; Horst 1986) and arrow heads. Shaft hole axes continued to be made of stone well into the Late Bronze Age with specific local or regional types indicating a widespread use (mainly outside the central regions of the Nordic Bronze Age). They continued the Neolithic tradition of 'battle-axes'. Whether this term is accurate is another matter and it may be argued that the stone axes have more in common with the bronze 'cult axes' than with the ordinary tree felling tools, which dwindled during the Bronze Age to the rather diminutive socketed axes of period VI (Baudou 1960).

Stone axes and metal 'cult axes' do not occur in graves after the Middle Bronze Age (periods II-III). Interesting though they are, I prefer to leave them out of this paper. The axe had a long tradition in the North, so it is no wonder that it was the first tool to be made locally in the new material in the Early Bronze Age (Vandkilde 1996). Soon the opportunities for a better connection of blade and helve offered by the bronze led to a series of experiments via the palstave to the socketed axe. Even if this was a broad European trend (Struve 1979: Taf. 53), we are able to observe the stages of a local Nordic development during the Early – Middle Bronze Age (Montelius I-III).

The socketed axe was obviously the solution that suited the Nordic metallurgists who produced them for 700 years in ever more economic versions. Even if one can hurt an opponent with a small socketed axe it hardly makes an ideal weapon. No emphasis was made on individual decoration – unlike the early flanged axes or the finer so-called weapon palstaves of period II. The exception is the axes with very high flanges and over-dimensioned ornate chapes (Asingh 1988; Vandkilde 1996: 114ff). Their weapon quality seems rather doubtful. Presumably the axe was no longer used as a weapon by the end of period II. That left the stage to the double-edged weapons. There are hardly any axes in certified contexts with period III swords.

Spearheads

The separate spearhead was an invention of the metal age. From the small but nearly always expressively decorated Bagterp type (Vandkilde 1996) to such monstrosities as the 45 cm long and richly decorated Valsømagle (ibid.) or the up to 50 cm long, stereotypic Baltic spearheads of period V, a long and highly varied development can be traced (Jacob-Friesen 1967; Thrane 1975 and below). Similar lengths are known from Minoan and Mycenaean (Höckmann 1980 and 1982) and Irish spearheads (Ramsey 1995). Their lethal qualities are well documented (Osgood chapter 23; Thorpe chapter 10). Whether these heavy points were thrown or thrust will remain a question, and quite likely both methods were used (Harding chapter 33).

Bow and arrow

Arrowheads of flint are not common in dated contexts but they do occur occasionally in graves during the Bronze Age – normally, alas, unassociated with other objects (Aner and Kersten: passim; Baudou 1960; Oldeberg 1974: nos. 169 I, 527, 694, 702, 715; Strömberg 1975: grave 33, normally just a single arrowhead per grave).

Of these few graves hardly any contain swords (Aner and Kersten: nos. 72, 4690 full hilt; 1835 plate, 4296, 4648, 4711A and 5029A flange hilt). This Nordic usage contrasts with other neighbour groups such as Mecklenburg (Schubart 1972) and the Lüneburg Early Bronze Age graves where several arrows, obviously placed in a quiver, like the Hallstatt burial in Hohmichele at the Heuneburg (Clausing 1998) – and presumably accompanied by the bow – belonged to the male equipment (Laux 1971: 90f). Other European groups like Wessex (Thorpe chapter 10) or Brittany, or the central European Urnfield, Hallstatt and Lausitz cultures did the same (Eckardt 1996).

Metal and bone were used in these cultures along with flint. In Scandinavia the cremation grave from Stora Vikers on Gotland is the only comparable find (Rydh 1968) with its eight long, lethal bone points which seem rather too numerous to kill a man (cp. Thorpe chapter 10). They had been burnt with the corpse, which the bronzes in the burial had not. So, like single calcinated flint arrowhead in some Late Bronze Age cremation graves it may well be that they sat in the corpse during cremation (cp.

Harding chapter 33; Osgood chapter 23; Thorpe chapter 10).

In an inhumation grave from Himmerland without weapons a male body had a flint arrowhead in the right chest (Aner and Kersten: XIV ms; cp. Härke 1992: 211), and another from Zealand with a dagger had 11 flint arrowheads, presumably in a quiver, and an additional one at the throat (Anon. 1989: no. 3). It was not common to place the equipment for the dead person on the funeral pyre in the Nordic region – in contrast to the Lausitz and Urnfield cultures (Thrane 1984: 134ff). Thus these rare cases could, like the Over Vindinge grave, represent a continuation of violence. Still, this is no more an indication of war than the Neolithic Porsmose man killed by a bone arrow in the Neolithic (Bennike 1985: 110ff). These cases may just as feasibly be the result of ambush or individual revenge as war incidents.

Sword production

Nordic metallurgy concentrated on the cire perdu technique, although solid moulds were used for tools such as palstaves, socketed axes, knives and sickle blades. Larger and more complicated objects were cast in clay moulds, which we only know from the Late Bronze Age. While moulds for sword blades or hilts have rarely been recognised (Neergaard 1908; Oldeberg 1960; Thrane 1995; Aner and Kersten: XI no. 5273), a whole set of spearhead moulds is known from Galgedil (Jensen 1995) and other fragments have been recognised (Nielsen 1956; Oldeberg 1960; 1974: no.1575; Thrane 1971: fig. 15; Kaliff 1995: 69). Our knowledge of weapon production is therefore nearly exclusively based upon the finished objects.

Apart from special objects like lurs (Thrane 1995), the production of large objects like swords and spearheads seems to have been widespread. Perhaps it was not as ubiquitous as smaller and simpler objects like rings or buttons, but it was concentrated in regionally working workshops where the necessary skill and artistic experience were available. This is what Ottenjahn suggested for the Middle Bronze Age sword production of bronze-hilted swords (1961; 1969). The use of 'once-casting' clay moulds (à cire perdue) resulted in a high degree of individuality, which explains Ottenjahn's difficulties in establishing clearly defined chorological groups (cp. Preben Rønne's style studies [1987]). So far no such workshop has been identified, but settlements (of periods II-IV) with good – i.e. find productive – rubbish dumps are hardly known either.

We therefore have to infer from the objects themselves. The swords and spearheads show great variation not only over the centuries but also regionally within the archaeological periods (Baudou 1960; Ottenjahn 1961; Jacob-Friesen 1967; Thrane 1975). There is a striking difference between the early multitype generations in periods II-III through period IV, with its restriction of the number of sword types but a broad variety of spearheads continuing to the final stage in period V, with its uniform sword blades and the nearly mono-typic Baltic spearheads. This development towards standardisation must reflect the role of weapons in contemporary society, but the question is how. A move from individual to more regulated – i.e., communal or controlled – production, and use, may be an option.

Period V is also the time when Nordic sword production really deviated from the otherwise dominant trend in Central Europe. Instead of adopting the tendency towards cut and thrust swords with willow shaped blades (Harding 2000), the final Nordic bronze blades were more rapier like, pointed with a strong tip with flanking grooves (Baudou 1960: type 2). This is a thrusting weapon. Even if there was an intermediate period (IV) with blades that look well suited for cut and thrust fighting, that type of sword never dominated in the North.

An extensive import of swords from Central Europe can be observed during the Middle Bronze Age, comprising fully hilted and flange-hilted swords. The scale was such that the Nordic region showed a high proportion of finds compared to Central Europe. This took some explaining (Sprockhoff 1931; 1934; Holste 1952), but it was only later that more intensive studies produced the right explanation, namely, that the numerical preponderance of octagonal hilted swords in the north is caused by the large number of local copies, combined with a stronger emphasis here on burying swords with the dead (Hachmann 1957; Quillfeldt 1995). A similar study of flange-hilted swords has not been made, but I am quite confident that they followed a similar trend (Cowen 1955; Thrane 1964: 156ff). The remarkable similarity of the flange-hilted swords over large tracts of Europe caused names such as the common European sword to be coined (Cowen 1955). It is tempting to see this main type as part of a cross-cultural set up – an international warrior segment.

The whole innovative or renewal process that can be followed in the armament presents some pertinent questions. Was it simply prompted by metal workers bringing the new fashion to foreign patrons or did warriors travel, displaying the latest improvements of their panoply? If the latter option is considered, a mercenary set up or travelling warriors joining the local leader's troop may be a possibility. Obviously, applying much later models and texts to a wholly Dark Age period is thin ice to walk on. Still, it may be worthwhile to ponder this kind of interaction, which could explain the preference for foreign types of male equipment, mainly weapons that we can follow through the Bronze Age. Clearly, above all arms were imported and adopted (Thrane 1975: 259). More comparative research is needed, however.

The situation is radically different in the Late Bronze Age. Great numbers of one type are no longer found. Only single specimens are known, graves no longer contain swords (cp. below), and the local sword production proceeded largely uninfluenced by the foreign sword types. Miniature swords take their place in the burials, curiously enough indicating the width of the import better than the proper swords. The change could reflect a change in the attitude towards weapons and their masters – the warriors – restricting the deposit of weapons in the graves (a grave from Löderup contains a fragment of a sword blade exactly like the bits found in period IV hoards [Strömberg 1975]). It is one of the changes which distinguish the Late Bronze Age from the preceding centuries.

Treatment

Any suggestion of how swords were used must be based upon the morphology of the blades and hilts (Fig. 1) and, more recently, upon observations of the wear and tear that only the best preserved swords permit. Kristiansen (1984) suggested that blades were subject to a secondary surface treatment in order to keep the edges sharp. This treatment could be so coarse that the whole profile of the blade was altered, reducing the edges by several millimetres. The only good reason for this would be such deep cuts in the edges that these could not be repaired simply by hammering and filing the original cutting edge. Here we have a contradiction. Such damages cannot result simply from stabbing at the opponent.

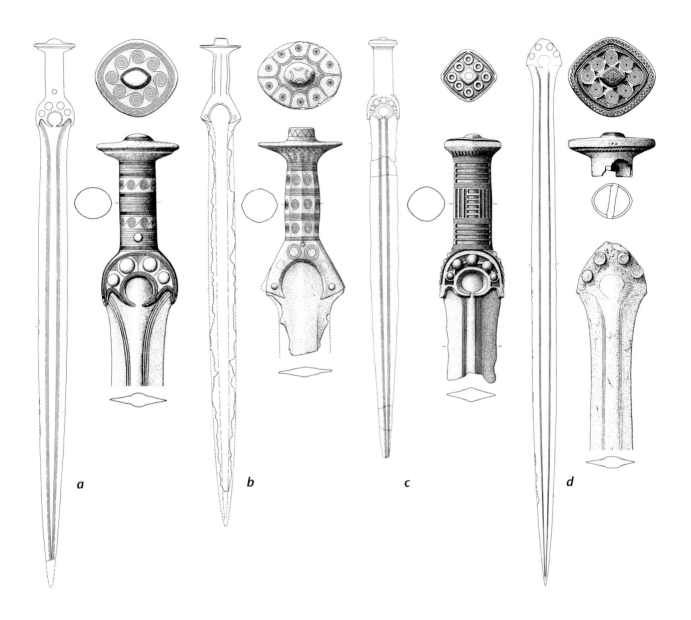

FIG. 1: *The main types of swords from periods II-III in Denmark: a. metal-hilted period II; b. hexagonal metal-hilted period II; c metal-hilted period III; d. plate-hilted period II; e. plate-hilted period III; f. tanged-hilted period III; g. flange-hilted per II; h. flange-hilted per III (adapted from Aner and Kersten 1973ff).*

They must mean that the damaged blades had cut into the opponent's sword (hardly a spearhead) with considerable force. This may serve to demonstrate that real life transcends the stiff rules that we may suggest (for other conclusions see Harding chapter 33; Thorpe chapter 10). The pre-depositional damages show that these swords were actually used against other sharp-edged objects (weapons), swords being the most likely candidates. Even if the edge damages are clearly observable only on a minority of the swords – due to post-depositional damage or corrosion – it seems reasonable to infer that the damage

happened to most of the swords – indeed, that it was part of their purpose. It is curious that a different pattern with fewer damages emerges in Southern Germany (Quillfeldt 1995: 5ff, 21), where at least the Late Bronze Age swords look so much more functional.

Kristiansen claimed a significant difference in the degree of re-sharpening of bronze-hilted and flange-hilted swords (1984: 195ff), the bronze-hilted swords more often having mint blades. Even granting that the sample is just 260 out of c. 2160 swords (only Aner and Kersten: volumes I-III), the trend is clear.

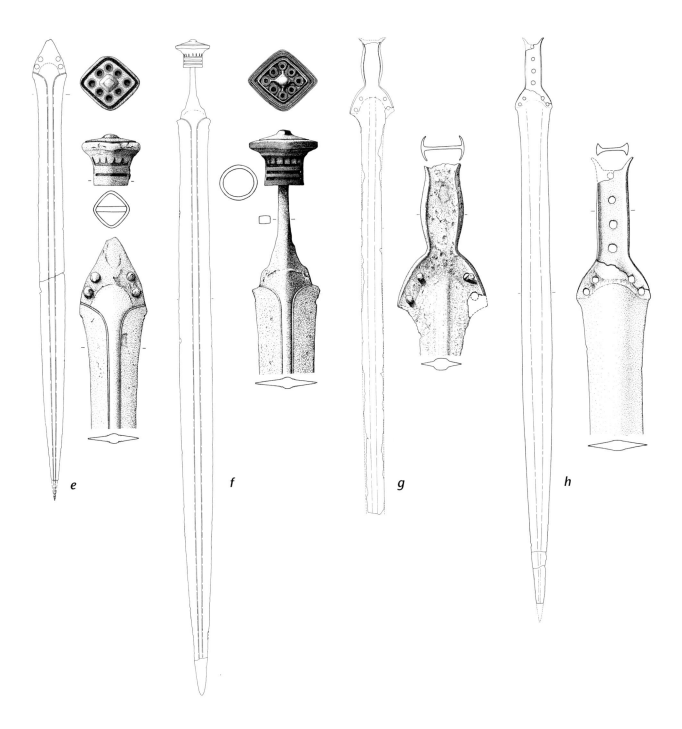

e   *f*   *g*   *h*

We look forward to the full documentation recently promised by Kristiansen. The metal hilted swords – 94 of 168 observed period II swords and 28 of 92 period III swords in Kristiansen's sample as well as 21 period II and 29 period III flange-hilted swords – display nearly identical percentages of wear in both periods. The blades generally show many more signs of wear and re-sharpening in period III, which matches the trend of less metal and longer use than he advocated earlier (1978). The swords with organic hilt plus metal pommel function like the flange-hilted swords in both periods.

Kristiansen maintains that there are more fully hilted swords with worn hilts but fresh blades than swords with worn hilts and blades.

The suggestion that the bronze-hilted swords, as the most precious 'schwersten und kostbarsten', were taken out of circulation faster than the other sword types does not carry much weight. If the other types were used more frequently for fighting, the risk of damage and diminished value would presumably increase and lead to a shorter circulation period. Wear is an effect of the intensity of use as well as of the length of use.

Table 1. Regional samples of metal-hilted bronze swords; cp. Fig. 1

South Germany. *Bronze Age including Urnfield Period (Quillfeldt 1995; in addition c. 300 flange-hilted and grip plate swords are known cp. Schauer 1971).*

	No.	Percent
Deposition in graves	66	22%
Deposition in hoards	23	7%
Deposition on dry land	48	16%
Deposition in wetlands	120	40%
Deposition on settlements	5	2%
Unknown	38	13%
Total	300	100%

Poland. *Bronze Age and Hallstatt Period (Fogel 1988).*

	No.	Percent
Deposition in graves	9	4%
Deposition in hoards	83	41%
Single finds	116	57%
Total	204	100%

Mecklenburg. *Early – Middle Bronze Age periods I - III (Schubart 1972).*

	No.	Percent
Deposition in graves	74	63%
Other kinds of deposition	44	37%
Total	118	100%

East Germany. *Bronze Age – Hallstatt Period (Wüstemann 2004).*

	No.	Percent
Deposition in graves	c. 230	29%
Deposition in hoards	111	15%
Single finds	c. 250	31%
Unknown	c. 200	25%
Total	c. 800	100%

Denmark – Scania – Schleswig. *Early – Middle Bronze Age periods I-III (Aner and Kersten 1973ff: volumes I-XI, XVII and manuscript to volumes XII –XIV; Oldeberg 1974).*

	No.	Percent
Deposition in graves	2103	83
Deposition in hoards	44	2
Single finds	376	15
Total	2562	100%

Kristiansen concludes that the bronze-hilted swords were only rarely used in combat. He further concludes that the use of these swords for fighting was secondary to their symbolic use, and that they therefore indicate chieftains while the swords with loose pommels or flanged hilts were wielded by warriors (see below). This interprets two technically rather different solutions as equally effective weapons – which seems doubtful.

The analysis of contexts shows different equipment for the two groups, apart from the octagonal hilted swords, which resemble the two simpler types in their contexts. Unfortunately, the octagonal hilted swords are included with the other bronze-hilted swords in Kristiansen's re-sharpening analysis. It will be interesting to see how different their patterns are. The octagonal hilted swords complement or substitute the Nordic metal hilted swords chronologically on Zealand (Rønne 1987: figs. 52-53). Their position in the hierarchy of the graves is still not clear (cp. above) but rather interesting because of the many copies that were made in the Nordic Bronze Age (Quillfeldt 1995). Generally they have few associated objects (Table 2).

Only 21 of the 168 swords from period II used by Kristiansen (1984: 205ff) and 13 of the 92 from period III belong to well documented contexts and are therefore reliable according to my definition of closed finds.

Deposition

The deposition manners which enabled us to retrieve the weapons were not uniform, neither at a European nor a Scandinavian level. A sample of some available regional counts will show something of the variation, which depends partly – and mainly – on different deposition rules in the Bronze Age (Table 1). From the end of period I to the transition from period III to IV, nearly all swords and a large portion of the spearheads and axes in the Nordic West Baltic area come from graves. This allows us to study the contexts with other objects and with the construction of graves and tumuli – including energy expenditure.

The change during period IV may partly be seen as a result of the cremation rite which demanded less room in the grave. After a transitional stage with full-length graves with swords, especially in North Zealand, the deposition ceased. That is hardly the

Table 2: Associations in Danish sword burials from periods II-III. Only finds documented by professionals, amateurs or observant other persons are included (Aner and Kersten 1973ff: volumes I-XI and manuscript to volumes XV-XVI). Association with gold is marked by *.

Type	Number of associated metal and amber objects 1	2	>2	Σ	Percent
Metal-hilted swords Period II	27 *4	7 *1	15 *7	49 *12	24 %
Hexagonal-hilted swords Period II	11 *2	3 *1	2	16 *3	19 %
Metal-hilted swords Period III	8 *5	6 *2	3 *1	17 *8	47 %
Plate-hilted swords Period II	21 *1	14 *1	17 *8	53 *10	19 %
Plate-hilted Swords Period III	2	5	6 *2	13 *2	15 %
Tanged-hilted swords Period II	2	4 *3	4 *3	12 *6	50 %
Tanged-hilted swords Period III	27 *9	29 *9	28 *8	84 *26	31 %
Flanged-hilted swords Period II	13 * 3	5 * 1	4 * 2	16 *4	25 %
Flanged-hilted swords Period III	64 * 26	32 *10	25 * 8	121 *42	35 %
Sword type unknown Period II	117 *4	10 *3	7 *2	134 *9	7 %
Sword type unknown Period III	4 *1	6 *3	4 *4	14 *8	57 %
Total	300 *46	124 *47	116 *94	529 *136	

whole explanation, as indicated by the occasional (exceptional) grave with sword even during period V (Thrane 1984: 142). A change in attitude is also witnessed by the introduction of miniature swords in period IV (Thrane 1969). These never approached the Middle Bronze Age swords in number, but compared to the genuine Late Bronze Age swords the mini swords are not unimportant numerically. They are also relevant insofar as they copy a wider range of Urnfield culture types than we know as proper swords – indicating a problem in the survival of the sources. We may infer that only some of the imported swords were buried and that our picture is thus eschewed.

Swords were, above all, placed in graves during periods II and III. This is fortunate because the position and the accompanying objects inform us how the swords were carried and which other weapons were worn with them. Needless to say, this applies only to well-documented inhumation burials. I shall return to the complexes of objects and the position of the sword graves in the mounds below.

The swords are interesting enough by themselves. Under good conditions we see details of swords' hilts and scabbards and shoulder straps and belts in Middle Bronze Age contexts (Aner and Kersten: nos. 2242C, 2529, 2663A, 2667, 2669, 3817A; Oldeberg 1974: nos. 247 and 540). The sword – scabbard relationship is interesting. Considering the small number of preserved scabbards of wood or wood and leather, it is surprising how many give the impres-

sion of not being in an original combination. There are scabbards which are much too long and have quite a different mouth from the sword hilt so that sword and scabbard do not match (Aner and Kersten: no. 2726E). The classic case is the young man from Borum Eshøj (Boye 1896: pl. X) with a complete scabbard holding but a dagger blade. This indicates some manipulation before interment, perhaps related to the burial situation. This lack of coherence between sword and scabbard may indicate that the swords had complicated histories and suggest manipulation at burial for various reasons.

Another fact apparent from the scabbards is a change from the handsomely carved broad wooden ones of period II to narrower and simpler leather scabbards with metal chapes in period III. This would seem to reflect an increase in the importance of functionality to the detriment of the significance of display. Unfortunately, we do not have anything comparable from the Late Bronze Age because of the cremation rite: swords deposited outside burials never have preserved scabbards. Only a single scabbard is known from the Late Bronze Age (Femhøje, see Broholm 1946: no. 394a).

Hoards

Weapons were deposited in the open countryside even before period I (Vandkilde 1996), and this rite continued to the end of the Bronze Age. The dominance of simple axe blades changed to the broad variety of weapons and weapon-effective tools through periods II-VI and hundreds of single or multiple depositions on dry or wet land took place.

Randsborg tried to interpret the mixed Smørum-ovre hoard and the Thorsted spearhead hoard retrospectively from the interpretation of the Iron Age weapon depositions as equipment of armed units (1995). This may be questionable methodologically and certainly has little support in the Bronze Age material. Depositions of weapons normally consist of a single weapon, sometimes of two. More than two (six swords at most [Aner and Kersten: nos. 831, 984, 2114 (3), 2075; Kristiansen 1984: 194]) carefully bundled and placed in lakes or bogs (or on dry land like the eight period I swords from Dystrupgård – Anon. 1993: 232f; Boas and Rasmussen pers. comm.) are rare. What do they reflect? The property of a local group or a larger entity? Was the deposition motivated by the same ideas as the other depositions of non-belligerent bronzes? Or were they genuine precursors of the Iron Age weapon deposits made by Celtic and Germanic tribes, as inferred by Randsborg (1995)?

If the latter is accepted then the forces involved must have been very different from the Iron Age ones. A comparison with the Iron Age officers' precious belts and so forth may be more to the point, indicating a 'state of warre' involving a few warriors fighting (duels?) with their swords. Did the buried swords perhaps belong to the victors (and not to the losers) who abandoned them to the powers of the earth or the water in gratitude? On the other hand, the composition of hoards need not be a faithful reproduction of the conditions of the living culture, but instead may serve to the purpose of communicating with the higher powers.

Hoards of weapons mixed with ornaments may have been deposited by a group comprising men and women as opposed to the exclusive weapon hoards (which are rare, apart from one-piece deposits) or ornament hoards which may be individual or group based. Even a hoard like Sørup with an imported sword and a big spearhead does not look like a single warrior's equipment because of the additional sword blade (Thrane 1969).

The interpretation of hoards has always been problematic, the alternatives being too numerous. The elements for any interpretation are rarely well documented. Even when they are, the conclusions are seldom safe.

I have suggested that single weapons – spears and swords alike – were placed in wet areas, functioning as a no-man's-land between the settlement areas, but so far this is a hypothesis (Thrane n.d.). It just indicates that weapons were used politically but does not inform us about the level of aggression, although a group level seems most likely. So we are still not at war but almost.

Contexts

Graves provides the best documented and most varied Bronze Age contexts. A source-critical analysis of them is a prerequisite for any use of the contexts and normally reduces the number available for further interpretations considerably. Most graves were not observed or documented by professionals or well-instructed amateurs and some of the most promising contexts have to be left out of consideration because of insufficient documentation.

It is important to know whether a sword was the only metal object in a burial. Equally interesting from our point of view, and more enlightening, is the selection of objects associated with the swords (Table 2). The list of combinations with ornaments, personal toilet items, other weapons, containers for drink and food, furniture and unique objects interpreted as cultic, and from the simplest materials – bone, wood or flint – to the most expensive, like bronze or gold and the occasional amber, glass or jet is long.

The contexts are crucial for Kristiansen's assessment (1984: 19ff; 1999) of a dichotomy between chiefs and warriors, each with their own favourite sword type and each characterised by different contexts. The chiefs' graves contained Nordic bronze-hilted swords and a wide (and richer) range of extras, sometimes very special, while the warriors' graves had functional swords and little else – apart from a razor and that sort of personal tool. This hypothesis should be controllable since, assuming that chiefs had easier access to wealth, the fully hilted swords ought to be found in richer associations than the flange-hilted swords. The evidence does not support a clear-cut division.

Kristiansen's key find is a double grave from period II at Norby in Schleswig, which does not have the best documentation (Mestorf 1890) but seems to consist of one skeleton with two spearheads, a bronze clad staff, an awl and a pair of tweezers, and a palstave with a fully hilted broadsword inhuma-tion and another with just a flange-hilted thrust sword (Aner and Kersten: no. 2538). To Kristiansen (1999: 544), this is the 'chiefly high priest' and his '"twin" ruler, the warrior chief', which creates a partnership otherwise quite unknown in a Bronze Age society. The person with no fewer than four weapons is the peaceful chief, and the unique staff is his 'ruling staff', while the person with just the sword is a warrior chief. I fail to see the logic of the inference. The set up is without doubt special, but other interpretations that do not equate the social and political status of two such different burials may be equally plausible. Obviously, a retrospection of much later texts on Germanic Iron Age tribes has been made (Thompson 1965; see Steuer chapter 16).

Taking Denmark as a whole, if one looks at the combinations of the four main types, a rather dif-ferentiated picture emerges of marked change from period II to period III. The dominance of the full hilt and plate swords in II gives way to the dominance of the tanged and flanged types in period III. Not only the frequency but also the richness of the asso-ciated objects shows the same trend.

Further details are interesting. The period II flange-hilted swords with richer associated finds are nearly all from oak coffins in wet core tumuli and one of four contains gold. Gold rings, gold wire and sheet gold are found with all four types, but in period II they prefer full hilt swords (Table 2). Gold is more frequent in the richer graves, whatever the sword type may be – tanged, flanged or fully hilted swords – from forty to twenty-five percent.

In period III the move from fully hilted and plate to tanged and flanged is not only numerical but accompanied by a similar rise in rich combina-tions for the latter types. It should be noted, how-ever, that grip-plate (Griffplatten) swords occur in equally as modest contexts, and there are exceptions to all the 'rules'. The richest period III grave with its gold arm ring and bronze wheeled cauldron con-tained a flange-hilted sword (Aner and Kersten: II: no.1269), thus flatly contradicting his thesis. Who was buried in this thirteenth-century grave in Trudshøj? Was he a big man, a chief, a ritual leader, a warrior chief, or a warrior with a special cultic responsibility and power? The answer will depend upon how we interpret the structure of Bronze Age society and how far we are prepared to go on the path of inference.

There are other graves with rich associations doc-umenting the social rank of persons buried with flange-hilted swords in both period II and III – e.g., bronze cups, even stools which are considered high status (symbolic) furniture (Werner 1994). Gold wire rings and solid gold arm rings cross the boundaries between the sword types. They may make a smaller percentage of the flange-hilted sword graves but they are by no means restricted to fully hilted or plate swords in either period II or III.

Whatever reservations we may have about a social-rank interpretation of grave finds, when a culture expresses itself so blatantly in the burial equipment, we can use the evidence for relative ranking within this culture.

The swords that we are able to determine by types form a qualified minority. There is a marked distribution and change from period II to period III, with the fully hilted sword as the leading type in period II in the Danish provinces (except volume 9 and Randers, where grip plated or flange-hilted swords, respectively, are slightly more frequent). It is a close race between flanged and tanged swords, with, in Jutland, a heavy preponderance of flange-hilted swords (Aner and Kersten: volumes VII-XI) – i.e., the two more effective technical solutions came to dominate. The desire for a posh hilt was frequent-ly met in period III by adding a series of bronze discs to the tanged hilt so that it came to look very much like the (much rarer) fully hilted swords.

Whether the change is a reaction to a shortage of fresh metal (as suggested by Kristiansen 1978) or had other causes, we cannot know. The connection of technically better editions with aesthetically pleasing solutions does not need to have a purely economic background.

Conclusion

This somewhat cursory review of the arms situation in the Nordic Bronze Age reveals one salient feature: variation, which is not restricted to the armoury. However, one may find so much variation that pat-terns seem to be non-existent or one may hesitate to attribute much importance to the patterns observed.

Although noted before, it seems worth repeating that we have to calculate with a range of hostile actions that will be indistinguishable from full-scale war in an archaeological record without oral or his-toric sources. I am referring to feuds, piracy and raids.

The victims of such actions would presumably be cared for as individuals, much like ordinarily deceased persons (cp. the graves with flint arrows mentioned above). Mass graves are only to be expected where the magnitude of the disaster prevented proper care by the relatives or where the kinsmen were no longer around to attend to a proper burial. Ritual may be another cause.

What is impressive is the massive existence of a broad range of proper offensive weapons combined with a nearly complete absence of defensive means. It is also striking that the weapons are all for close (to medium) range fight, not for long range fight. Finds indicating the existence of troops exceeding a handful are rare. The emphasis was consistently on individual equipment whether just a single weapon – most often the sword – or a combination of different types of weapons. The maximal but rare combination is sword, spear and axe.

The change from period III to IV, gradual though it looks, nonetheless seems significant and not just conditioned by the complete and exclusive dominance of the cremation rite. The period III cremation graves are closer to the period II inhumation graves than to the later period IV cremation graves when it comes to types and the multitude of objects buried with the bones. The Urnfield culture shows us that cremation was no hindrance to rich and diverse equipment, whether the bronzes were cremated with the deceased or placed in the grave unscathed by fire (Harding 2000). However, the distance from poor to average funerary equipment grew in the Late Bronze Age both in the Urnfield culture and later in the Hallstatt culture and in the North.

Weapons, again above all the sword, were predominant among the types signifying a rich grave. Thus a common attitude, maybe better expressed in the Late Bronze Age, is observable in those parts of Europe where burials were used to mark differences between individuals, especially males. To associate swords (and spears) with warriors does not seem to require a great mental leap. Nor does an interpretation of rich graves with weapons as those of leading persons in society (members of leading kinship groups). The fact that the rich and richest graves nearly always contain at least one weapon whereas not all graves with weapons are equally rich does not mean that the male sphere was split into three – chiefly warrior, non-chiefly warrior and non-warrior. Why assume that there were no local differ-

ences in affluence which were reflected in the burials of the leading men? Already a cursory review shows significant differences in the use of gold with or without weapons (Randsborg 1974; Hartmann 1982). It would be much more remarkable if the rich graves were all at a similar level, no matter what the local subsistence and exchange conditions allowed in the way of accumulated wealth.

The Nordic Bronze Age certainly had its own rules and norms, and the wealth of material may blind us to the fact that it was not an isolated, self-sufficient phenomenon but always part of a wider European network or system, whether centre-periphery or otherwise. Thus it is highly likely that norms and ideals were transmitted – and transformed – from the more complicated and in some aspects more advanced Bronze Age societies further south and southeast.

The existence of a Homeric model or code for the Warriors in Europe's Bronze Age seems supported by several kinds of material evidence which compare extraordinarily well with the oral tradition of Homer – e.g., burial rites like the Caka mounds or Lusehøj – and somewhat less so with the contemporary archaeological evidence from the Aegean. This observation or interpretation does not mean that every aspect of the Homeric warrior ethos was copied and was present in every nook of Europe whither the Bronze Age had reached. But I do think that some elements were adopted from the source that is now best known from the *Iliad*. Therefore, it is a proper reference for our understanding beyond the purely material interpretations that we would be left with if we were to study the Nordic Bronze Age solely on its own premises.

BIBLIOGRAPHY

- Aner, E. and Kersten, K. 1972-2002. *Die Funde der älteren nordischen Bronzezeit in Dänemark, Schleswig-Holstein und Niedersachsen*. Vol. I-XI. Neumünster: Wachholtz Verlag.
- Anon. (eds.) 1984-. *Arkæologiske Udgravninger i Danmark*. Copenhagen: Det Arkæologiske Nævn.
- Asingh, P. 1988. 'The excavation of a Complex Burial Mound and a Neolithic Settlement at Diverhøj'. *Journal of Danish Archaeology*, 6, pp. 130-54.
- Baudou, E. 1960. *Regionale und chronologische Einteilung der jüngeren Bronzezeit im nordischen Kreis*. Stockholm: Almqvist och Wiksell.
- Becker, C.J. 1982. 'Siedlungen der Bronzezeit und der vorrömischen Eisenzeit in Dänemark'. *Offa*, 39, pp. 53-71.

- Bennike, P. 1985. *Palaeopathology of Danish Skeletons: A Comparative Study of Demography, Disease and Injury.* Copenhagen: Akademisk Forlag.
- Bennike, P. and Ebbesen, K. 1987. 'The Bog Find from Sigersdal'. *Journal of Danish Archaeology*, 5, pp. 85-115.
- Boye, V. 1896. *Fund af egekister fra bronzealderen i Danmark.* København: Høst og Søn.
- Broholm, H.C. 1946. *Danmarks Bronzealder.* Vol. II. København: Nyt Nordisk Forlag.
- Brøndsted, J. 1939. *Danmarks Oldtid.* Vol. II. *Bronzealderen.* Copenhagen: Gyldendal.
- Brøste, K., Balslev Jørgensen, J. and Becker, C.J. 1956. *Prehistoric Man in Denmark.* Copenhagen: Munksgaard.
- Bukowski, Z. 1962. 'Fortified Settlements of Lusatian Culture in Great Poland and Kujawy in the Light of Research Carried out in the Years 1945-60'. *Archaeologia Polona*, 4, pp. 165-80.
- Childe, V.G. 1930. *The Bronze Age.* Cambridge: Cambridge University Press.
- Clausing, C. 1998. 'Zu Köchern der Urnenfelderzeit'. *Archäol. Korrespondensblatt*, 28, pp. 379-90.
- Cowen, J.D. 1955. 'Eine Einführung in die Geschichte der bronzenen Griffzungenschwerter in Süddeutschland und den angrenzenden Gebieten'. *BRGK*, 36, pp. 52-155.
- Eckardt, H. 1996. *Pfeil und Bogen.* Espelkamp: Verlag Marie Leidorf Gmbh.
- Fogel, J. 1979. *Studia nad uzbrojeniem ludnosci kultury luzyckiej w Dorzeczu Odry i Wisly.* Poznan: Uniwersytetu im. A. Mickiewicza.
- Fogel, J. 1988. *Militaria kultury luzyckiej z derzecznu Odry i Wisly.* Poznan: Uniwersytetu im. A. Mickiewicza.
- Hachmann, R. 1957. *Die frühe Bronzezeit im westlichen Ostseegebiet und ihre mittel- und südeuropäischen Beziehungen.* Hamburg: Flemmings.
- Harding, A. 2000. *European Societies in the Bronze Age.* Cambridge: Cambridge University Press.
- Harding, A. and Ostaja-Zagorsky 1993. 'The Lausitz Culture and the Beginning and End of Bronze Age Fortifications.' In: Chapman, J. and Doloukhanov, P. (eds.): *Cultural Transformation and Interactions in Eastern Europe.* Aldershot: Avebury, pp. 163-77.
- Hartmann, A. 1982. *Prähistorische Goldfunde aus Europa.* Vol. II. (Studien zu Anfänger der Metallurgie, 3). Berlin: Mann.
- Holste, F. 1952. *Die bronzezeitliche Vollgriffschwerter Bayerns.* München: C.H. Beck.
- Horst, F. 1986. 'Die jungbronzezeitlichen Stenäxte mit Nackenknauf aus dem Elb-Oder-Raum'. *Bodendenkmalpflege in Mecklenburg, Jahrbuch*, 1985, pp. 99-123.
- Härke, H. 1992. *Angelsachsische Waffengräber des 5. bis 7. Jarhrhunderts.* Köln: Rheinland-Verlag.
- Höckmann, O. 1980. 'Lanze und Speer'. In: Buchholz, H.G. and Wiesner, J. (eds.): *Kriegswesen 2. Angriffswaffen.* (Archaeologica Homerica, I.E:2). Göttingen: Vandenhoeck und Ruprecht, pp. 275-319.
- Höckmann, O. 1982. 'Lanze und Speer im spätminoischen und mykenischen Griechenland'. *Jahrbuch RGZM*, 27, pp. 13-158.
- Håkansson, I. 1985. *Äldre bronsålders gravfynd från Skåne.* Lund: Liber.
- Jacob-Friesen, G. 1967. *Die bronzezeitliche Lanzenspitzen Norddeutschlands und Skandinaviens.* Hildesheim: Lax.
- Jankuhn, H., Nehlsen, H. and Roth, H. (eds.) 1977. *Zum Grabfrevel in vor- und frühgeschichtlicher Zeit.* Göttingen: Vandenhoeck und Ruprecht.
- Jankuhn, H., Schützeichel, R. and Schwind, F. (eds.) 1971. *Das Dorf der Eisenzeit und des frühen Mittelalters.* Göttingen: Vandenhoeck und Ruprecht.
- Jensen, N.M. 1995. 'Galgedil ved Otterup'. *Fynske Minder*, 1995, pp. 54-58.
- Jockenhövel, A. 1990. 'Bronzezeitlicher Burgenbau in Mitteleuropa'. In: Schauer, P. (eds.): *Orientalisch-ägäische Einflüsse in der europäischen Bronzezeit.* (RGZM Monographs, 15). Oxford: Oxbow, pp. 209-28.
- Jørgensen, A.N. 2001. 'Sea defence in the Roman Iron Age'. In: Storgaard, B. (eds.): *Military Aspects of the Aristocracy in Barbaricum in the Roman and Early Migration Periods.* Copenhagen: Nationalmuseet, pp. 67-82.
- Kaliff, A. (eds.) 1995. *Skenet från det förflutna.* Linköping: Riksantikvarieämbetet.
- Keegan, J. 1993. *A History of Warfare.* London: Hutchinson.
- Kouwenhoven, A.P. 1997. 'World's oldest spears'. *Archaeology*, 50(3), pp. 25.
- Kristiansen, K. 1978. 'The consumption of wealth in Bronze Age Denmark'. In: Kristiansen, K. and Paludan-Müller, C. (eds.): *New Directions in Scandinavian Archaeology.* Copenhagen: Nationalmuseet, pp. 158-90.
- Kristiansen, K. 1984. 'Krieger und Häuptlinge in der Bronzezeit'. *Jahrbuch des Römisch-Germanisches Zentralmuseums Mainz*, 31, pp. 187-208.
- Kristiansen, K. 1999. 'Symbolic structures and social institutions'. In: Gustafsson, A. and Karlsson, H. (eds.): *Glyfer och Arkeologisk rum – en Vänbok till Jarl Nordbladh.* (Gotarc Ser. A, 3). Gothenburg: Department of Archaeology, Göteborg University, pp. 537-52.
- Kuschel, R. 1988. *Vengeance is Their Reply.* Copenhagen: Dansk Psykologisk Forlag.
- Köpke, H. 2002. 'Der Burgwall von Zützen'. *Veröff. des Brandenburgischen Landesmuseum für Ur. -u. Frühgeschichte*, 30, pp. 41-120.
- Larsson, T.B. 1993. *Vistad.* Umeå: Arkeologiska Institutionen, Umeå University.
- Laux, F. 1971. *Die Bronzezeit in der Lüneburger Heide.* Hildesheim: Lax.
- Lindström, J. 1999. 'Ett mord kastar ljus över bronsåldern'. *Bygd och Kultur*, 1999, pp. 4-8.
- Mestorf, J. 1890. 'Ausgrabungen des Professor Pansch II. Grabhügel der Bronzezeit, nenannt Moritzenberg, bei Norby, Ksp. Rieseby'. *Mitteilungen des Anthropologischen Vereins in Schleswig-Holstein*, 3, pp. 17-24.

- Müller, S. 1897. *Vor Oldtid*. Copenhagen: Det Nordiske Forlag.
- Neergaard, C. 1908. 'Haag-Fundet. En Affaldsdynge fra en Metalstøbers Hytte'. *Årbøger for Nordisk Oldkyndighed og Historie*, 1908, pp. 273-352.
- Nielsen, J.V. 1956. 'Vindblæsfundet'. *Kuml*, 1956, pp. 41-49.
- Nielsen, S. 1999. *The Domestic Mode of Production – and Beyond*. Copenhagen: Kongelige Nordiske Oldskriftselskab.
- Nordbladh, J. 1989. 'Armour and Fighting in the South Scandinavian Bronze Age'. In: Larsson, T.B. and Lundmark, H. (eds.): *Approaches to Swedish Prehistory*. (British Archaeological Reports International Series, 500). Oxford: British Archaeological Reports, pp. 323-33.
- Olausson, M. 1993. 'Predikstolen. A Bronze Age Hillfort in Eastern Central Sweden'. In: Arwidsson, G., Hanson, A.-M., Olausson, L.H., Johansson, B.M., Klockhoff, M., Lidén, K. and Nordström, H.-Å. (eds.): *Sources and Resources. Studies in Honour of Birgit Arrhenius*. (Pact, 38). Court St.Etienne: FLTR, Université Catholique de Louvain, pp. 53-64.
- Oldeberg, A. 1960. *Skälbyfyndet*. (Antikvarisk Arkiv, 15). Stockholm: Kungl. Vitterhets Historie och Antikvitets Akademien.
- Oldeberg, A. 1974. *Die ältere Metallzeit in Schweden* I. (Kungl. Vitterhets Historie och Antikvitets Akademiens Monografiserie, 53). Stockholm: Kungl. Vitterhets Historie och Antikvitets Akademien.
- Osgood, R. 1998. *Warfare in the Late Bronze Age of North Europe*. (British Archaeological Reports International Series, 694). Oxford: Archaeopress.
- Ottenjahn, H. (eds.) 1961. *Werkstätten nordischer Vollgriffschwerter der älteren Bronzezeit*. Berlin: Mann.
- Ottenjahn, H. 1969. *Die nordischen Vollgriffschwerter der älteren Bronzezeit*. (RGF, 30). Berlin: Walter de Gruyter.
- Post, L.v., Walterstorff, E.v. and Lindquist, S. 1925. *Bronsåldersmanteln från Gerumsberget, Västergötland*. (Kungl. Vitterhets Historie och Antikvitets Akademiens Monografiserie, 15). Stockholm: Kungl. Vitterhets Historie och Antikvitets Akademien.
- Quillfeldt, I. von 1995. *Die Vollgriffschwerter in Süddeutschland*. (Prähistorische Bronzefunde, IV, 11). Stuttgart: Franz Steiner.
- Ramsey, G. 1995. 'Middle Bronze Age Metalwork: Are Artefact Studies Dead and Buried?' In: Waddell, J. and Twohig, E.S. (eds.): *Ireland in the Bronze Age*. Dublin: The Stationary Office, pp. 49-62.
- Randsborg, K. 1995. *Hjortspring. Warfare and Sacrifice in Early Europe*. Aarhus: Aarhus University Press.
- Rindel, P.O. 2003. 'Eine befestigte Siedlung der jüngeren vorrömischen Eisezeit bei Lyngsmose'. *Archäol. Korrespondensblatt*, 33, pp. 123-43.
- Rydh, S. 1968. 'Ett gotlandskt fynd av benpilspetsar från bronsåldern'. *Fornvännen*, 1968, pp. 153-65.
- Rønne, P. 1987. 'Stilvariationer i ældre bronzealder'. *Aarbog*, 1986, pp. 71-124.
- Schauer, P. 1971. *Die Schwerter in Süddeutschland, Österreich und der Schweiz* I. (Prähistorische Bronzefunde, IV, 2). München: C.H. Beck.
- Schubart, H. 1972. *Die Funde der älteren Bronzezeit in Mecklenburg*. Neumünster: Wachholtz.
- Sprockhoff, E. 1931. *Die Germanischen Griffzungenschwerter*. Berlin: Walter de Gruyter.
- Sprockhoff, E. 1934. *Die Germanischen Vollgriffschwerter*. Berlin: Walter de Gruyter.
- Steuer, H. 2000. 'Kriegswesen'. In: (eds.): *Hoops Reallexikon der germanische Vorgeschichte*. Band 17. Berlin and New York: Walter de Gruyter, pp. 347-73.
- Struve, K.W. 1979. 'Die jüngere Bronzezeit'. In: Struve, K.W., Hingst, H. and Jankuhn, H. (eds.): *Von der Bronzezeit bis zur Völkerwanderungszeit*. (Geschichte Schleswig-Holsteins, 2). Neumünster: Wachholz.
- Strömberg, M. 1975. *Untersuchungen zu einem Gräberfeld von Löderup*. Lund: C.W.K. Gleerup.
- Thompson, E.A. 1965. *The Early Germans*. Oxford: Clarendon Press.
- Thrane, H. 1964. 'The Earliest Bronze Vessels in Denmark's Bronze Age'. *Acta Archaeologica*, XXXIII, pp. 109-43.
- Thrane, H. 1969. 'Eingeführte Bronzeschwerter aus Dänemarks jüngerer Bronzezeit'. *Acta Archaeologica*, XXXIX, pp. 143-218.
- Thrane, H. 1971. 'En broncealderboplads ved Jyderup Skov i Odsherred'. *Nationalmuseets Arbejdsmark*, 1971, pp. 141-64.
- Thrane, H. 1975. *Europæiske Forbindelser*. Copenhagen: Nationalmuseet.
- Thrane, H. 1984. *Lusehøj ved Voldtofte*. Odense: Odense Bys Museer.
- Thrane, H. 1995. 'Stand und Aufgaben der Bronzezeitforschung im westlichen Ostseegebiet während der Per. III-V'. In: Erbach, M.z. and Schauer, P. (eds.): *Beiträge zur Urnenfelderzeit nördlich und südlich der Alpen*. (Monographien des Römisch-Germanisches Zentralmuseums). Mainz and Bonn: Rudolph Habelt, pp. 429-52.
- Thrane, H. n.d. 'Aggression, territory and boundary – and the Nordic Bronze Age'. In: Olausson, M. and Olausson, L.H. (eds.): Conference report forthcoming Stockholm.
- Vandkilde, H. 1996. *From Stone to Bronze*. Højbjerg: Jutland Archaeological Society Publications and Aarhus University Press.
- Waterbolk, H.T. 1977. 'Walled Enclosures of the Iron Age in the North of the Netherlands'. *Palaeohistoria*, XIX, pp. 97-172.
- Werner, M. 1994. 'Faltstuhl'. In: (eds.): *Hoops Reallexikon der germanische Vorgeschichte*. Vol. VIII. Berlin and New York: Walter de Gruyter, pp. 176-84.
- Worsaae, J.J.A. 1843. *Danmarks Oldtid*. København: Selskabet for Trykkefrihedens rette Brug.
- Wüstemann, H. 2004. *Die Schwerter in Ostdeutschland*. (Prähistorische Bronzefunde, IV, 16). Stuttgart: Franz Steiner.

What Does the Context of Deposition and Frequency of Bronze Age Weaponry Tell Us about the Function of Weapons?

ANTHONY HARDING

/33

In considering the nature and significance of warfare in the Bronze Age, we are inevitably drawn to a discussion of the topic in the light of the archaeological evidence as we find it. True, analogies drawn from ethnography, and comparison with what Homer tells us of warfare practices in the heroic age of Greece, are tempting and possibly enlightening – if we could only be sure that they are apt. One of the many things that recent debates on warfare have taught us is the range of possible material that one could bring in to any discussion of prehistoric warfare and its significance for ancient society. But they also show how uncertain one must remain about the relevance of this material. If we are to use analogy in the study of warfare, we need to find a methodology that reassures us that our analogies are appropriate ones.

The various categories of evidence that bear on Bronze Age warfare have been discussed many times before. Apart from the weaponry itself, there are the very familiar Scandinavian rock art depictions that appear to show people brandishing various items of weaponry – axes, swords, spears – sometimes placed in pairs as if duelling is taking place. The significance of sword wear has often been stressed, apparently eloquent testimony to the way in which these weapons were intensively used, at least in some places at some times (Kristiansen 1984). Others have devoted studies to other categories of weaponry, for instance the Early Bronze Age daggers of copper and their flint imitations (Lomborg 1973; Vandkilde 1996). Here I propose to concentrate on aspects of the material that have not received the attention they deserve: the context of deposition of the weaponry, and what it can tell us about the function of the weapons, in warfare or other aspects of Bronze Age life.

Assuming that we have correctly identified the purpose of the items in question, we are dealing with the following offensive weapons: bows and arrows; daggers and related items; swords and rapiers; spears and lances; battle-axes; and perhaps slings or other equipment designed to hurl projectiles, such as catapults. In terms of defensive weaponry, helmets, corslets, greaves, and shields are involved. There is a lot of such material in the Bronze Age, though much more offensive than defensive weaponry. But, and this is an important but, it is not evenly distributed in time and space. Some places at some periods have much more of it than others, so whatever it was for, its role was not constant, and people either

had differential access to it, or used it in different ways. In other words, there is no one story to be told about Bronze Age warfare and warriors, at least on the evidence of weaponry.

In studying all this material, we have at our disposal a rich resource in the form of the *Prähistorische Bronzefunde* series, while many individual studies of weapon types have appeared in other places. Not all classes of object are equally well served: swords and daggers have occupied pride of place, while spearheads and arrowheads are relatively poorly provided for. I want to take as my starting-point, therefore, the information to be derived from a study of selected classes in particular areas. I intend to make use of the term 'warrior' to indicate the person who used the weaponry, without embarking on an analysis of the appropriateness of the term. I am conscious, however, that the production, availability and use of weapons need not necessarily lead one directly to the conclusion that they were inevitably and invariably associated with a warrior caste or a mode of action that presupposes the existence of warrior elites. Other modes are possible (associations with hunting; ritual; emblemic use) if less likely, but for present purposes the designation 'warrior' will continue to be used. Furthermore, the assumption that the warrior was male will also be made, though there are interesting discussions which consider the possible role of women in weapon use and fighting, and it is certain that not all instances of weapon burials were those associated with men. This female role is notable, but beyond my present scope.

Early Bronze Age: bow-and-arrow, dagger, halberd

The Early Bronze Age warrior possessed two main forms of offensive weapon: the arrow and the dagger. Another utilised form was the halberd, which bears some formal similarity to the dagger but was hafted and used in a quite different manner. In the latter part of the period, the spearhead made its appearance. Deposition is usually in burials, though hoarding started to become common in the period, and a number of well-known hoards from the latter part contain daggers (though not normally arrows). There are also a number of well-known depictions of bow- and dagger-bearing warriors, for instance on the stelae at Petit Chasseur, Sion (see below).

I do not attempt here to quantify the distribution

of daggers by deposition type on a European scale, but if one takes as an example von Brunn's study of Early Bronze Age hoards in Saxony and Thuringia (1959), out of 94 hoards only ten contained daggers (or parts thereof), seven contained halberds, and two spearheads. The great majority of the content of these hoards consisted of rings, flanged axes and ring ingots; weapons clearly occupied a lesser position (which does not mean they were unimportant in hoard deposition, however). Conversely, if one looks at a corpus of Early Bronze Age daggers such as Gallay's for France (1981), it is clear that the great majority are from graves. Only with material of the developed Early Bronze Age do hoards and wet finds start to make an appearance (where find circumstances are known). Interestingly, in Italy the situation is somewhat different, with a sizeable number of Early Bronze Age daggers coming from settlements and hoards, and only a few from graves (Bianco Peroni 1994: 181; four out of 80 are in tombs); the settlements are all in the north and are mainly terremare and pile sites.

For Britain, Gerloff's (1975) corpus shows that of 352 listed daggers (including knife-daggers) only a handful – mostly of the (late) Arreton series – came from hoards, and while there were many single finds and some river pieces, the great majority of those with known provenance came from burials. Other corpora of daggers illustrate much the same point. In eastern Germany, according to Wüstemann (1995: 39), 98 Early and Middle Bronze Age daggers are from graves, and 13 from hoards, while 82 are single finds and 53 are of unknown context.

For the student of warfare, what is uncertain is the extent to which these things were used in combat with humans, as opposed to hunting implements, and here the context of deposition has little to say. There are of course a number of instances where arrowheads are embedded in the skeletons of people, for instance a young male of the Beaker period in the ditch at Stonehenge (Evans et al. 1983), and these people presumably came by their death because of those arrow shots. But especially in such a context, it is far from certain that this act of interpersonal violence was an act of war; it could equally well have been some kind of ritual killing. Ötzi the Iceman had a bow and quiverful of arrows, as well as a flint knife-dagger (Egg 1992), and until recently it was thought unlikely that he was a warrior, though the recent discovery of an arrowhead in his shoulder has reopened this question (though not, in my view, to the extent that a warrior role can seriously be considered for him). Like most people, he had these things as part of his equipment for obtaining food through shooting game, and only on occasion for shooting at other people. Such a use continued, no doubt, throughout prehistory and indeed most of history. The bow and arrow was used in warfare into Medieval times and later, until the invention of gunpowder and the musket rendered the bow obsolete, and it certainly seems to have been used in the Iron Age (e.g., Chochorowski 1974; Dušek and Dušek 1984); there is good evidence that it was also much used in the Neolithic, as attacks on causewayed enclosures vividly demonstrate (Mercer 1999).[1]

Daggers, too, must surely have started life as functional implements for use in hunting, or even as knives. A use in warfare could only mean that hand-to-hand combat was practised, and while this may have been the case, it is hard to be sure that it was, at least as a normal aspect of life in the Bronze Age. In fact both the arrow and the dagger must have changed their meaning over time. Even if hunting was their primary role, that itself must have taken on different meanings at different times, since history and mythology show us that hunting was frequently a prestige activity that had little to do with mere subsistence. Nimrod, 'a mighty hunter before the Lord' (Genesis 10:9), was presumably not so highly esteemed because he was a good food producer, even if in origin that was the function of the activity. Thus the great stelae from Petit Chasseur at Sion (Sauter 1976; Bocksberger 1978) and other sites (Ambrosi 1988) certainly do show us stylised versions of prestige manipulators of the bow(s) and dagger(s), but we cannot directly tell if they are being depicted as warriors rather than as top-rank hunters. Probably both are involved, as hunter prestige may have been part of a broader status-related male identity.

Now Kristiansen (2000) and others have made a strong case for viewing Early Bronze Age dagger owners as warriors, part of a warrior-based society in which possession of elite weaponry marked individuals off as members of special groups, and there are good reasons for believing this to be true. What it does not tell us, however, is how those weapons were used, and one may be sceptical that the weaponry of the Early Bronze Age has much to tell us about warfare practices, however much they tell us about the status that followed from successful manipulation of these items. But it is certainly true that during the course of the period, the dagger started to assume a different role, as production of the form in metal became standard. Thus its presence in hoards, its hafting in metal as well as in organic materials, the variety of types that come into being, and above all the different contexts in which the dagger appears, show us how its function was becoming flexible. One of those functions was that of an item of prestige, carried and used as a badge. Such a badge may not necessarily have been that of a warrior, even if skill in combat was one desirable attribute. It was as much the attribution of huntsman/warrior status that may have mattered as anything to do with the ability to kill animals or humans. This transition from hunting role to warrior role, as seen in the carrying of dagger and bow, recalls the transition charted by Chapman (1999) for tools to weapons in the Late Neolithic, Eneolithic and Copper Age of southeast Europe.

Halberds clearly played an important role at certain times and in certain places in the Early Bronze Age, though for a relatively short period. While we know from the finds of metal-hilted halberds how they were hafted and therefore what they looked like (they are also depicted on rock art), it is much less certain how they might have been used. Apparently they were wielded so as to bring a blow down onto an opponent from above or perhaps the side, and thus to have served something like a dagger held at a distance (though not for thrusting or stabbing blows – that was the role of the spearhead when it arrived). If two warriors were both using halberds to fight each other then the contest might be regarded as an equal one (depending on skill and strength). At least they may have served to prevent an opponent from getting in close where he could make use of his dagger for thrusting.

In contextual terms, the situation is markedly ambiguous. In Ireland, where around 40% of all known examples occur, contextual information is in short supply. 'If a dead warrior of the Early Bronze Age brought with him a personal weapon to the grave, it was a dagger, but never a halberd which was chosen' (Harbison 1969: 35). According to Harbison, no halberd has ever been found in a grave in Ireland, while only 12 out of 150 are associated finds and all are from hoards. In eastern Germany, most are from hoards (such as the famous Dieskau

finds), though a certain number occur in high-status burials – such as Leubingen, or, further east in Poland, Leki Male – so that there was a 'personal' usage as well (Wüstemann 1995: 87). Hoards that include metal-hafted halberds are, however, preferentially located on moors or in bogs, clearly indicating a special means of selection and a special role for halberds in deposition (ibid.: 34). This tells us something about hoards but little about the function of halberds as weapons – though it can be argued that their occurrence in hoards but only rarely in graves – and then in special graves – is important in itself.

Middle and Late Bronze Age: sword and spear

With the increase in length of the dagger, first to the relatively modest proportions of the rapier, and later to that of the sword, we enter a new arena. Similarly with the spear, which appears in later contexts of the Early Bronze Age, we are no longer looking mainly at hunting, useful though spears can be to the huntsman. Now we enter a phase when items were being produced that can have had little or no purpose other than inflicting damage on people, actually or symbolically. But even here caution is necessary, as the depictions on Aegean seals and signets show us: both swords and spears are shown in use against animals, as well as humans (Kilian-Dirlmeier 1993). So no doubt they too had their origins in hunting. But several types of evidence illustrate to us graphically that a use in fighting became the norm: rock art, where people wave weapons at other people, not at animals; defensive armour in leather and sheet bronze (however impractical it may have been in actual use); and trauma on individuals.

When we turn to the finds themselves, there are intriguing regularities to explore. The obvious difference between swords hilted in bronze and those hilted with organic materials has often been thought to signify some special aspect of warrior practice, apart from certain technical aspects which meant that bronze-hilted swords were more difficult to produce than organic-hilted ones. It has been demonstrated, for instance, that solid-hilted swords (*Vollgriffschwerter*) tend to be less heavily used than flange-hilted swords (*Griffzungenschwerter*), indeed they were quite often not used at all, to judge from the lack of edge damage. Kristiansen (1984) has provided figures which appear to show that in Periods

II and III of the Nordic Bronze Age 65-70% of *Vollgriffschwerter* had not been sharpened at all and only 10% or less heavily sharpened; sharpening being taken as an indication of heavy use. What is also of interest is that their relative numbers vary considerably. Usually, taking Europe as a whole, *Griffzungenschwerter* are commoner than *Vollgriff-schwerter*, but the Nordic area is a notable exception to that: the numbers are approximately equal. In central Europe and Hungary there is a modest bias towards *Griffzungenschwerter* (1.37:1 and 1.12:1); in Italy, Yugoslavia and Romania the bias is more marked (2.57:1, 3.59:1 and 3.41:1); and in Britain and Ireland the bias is extraordinary (33.7:1 in Britain, while in Ireland there is no relationship because there are no *Vollgriffschwerter* at all). Extraordinary, that is, if you think that *Vollgriffschwerter* were important for people in Bronze Age Britain and Ireland – but they clearly were not. They used other things for whatever purpose *Vollgriffschwerter* were intended to fulfil, or they did not do those things at all. If it is true that *Vollgriffschwerter* were largely for parade and display, then perhaps more fighting and less display occurred in these far western areas. On the other hand, Bridgford (1997) was able to demonstrate that different classes of Irish Bronze Age sword had different wear and use patterns, with Ewart Park swords having significantly more edge damage than early Hallstatt swords, where the numbers of damaged and undamaged pieces was not greatly different.

Equally, if we look at density, there are striking differences (Table 1). Some areas, notably Italy, Yugoslavia, Romania and Britain, were relatively poorly provided with swords. Other areas, such as the south German/Austrian/Swiss area or Hungary, were well provided for. But compared with Denmark, north Germany, or – above all – Ireland, these figures are modest. In Ireland, indeed, where there is on average a sword roughly every 130 km², the number and density of swords is truly remarkable.

Now this may be telling us something about how swords were viewed, and potentially how they were used, in different parts of the Bronze Age world. But there is a further important aspect to be examined before we can consider such a hypothesis, and that is context. Swords are deposited in different contexts, most obviously some in burials, some in hoards, and some in wet places. (Most of the rest are isolated finds, which might belong to any of these three categories, though many are from gravel deposits that were almost certainly wet in prehistory.) I have not conducted a full examination of this matter, though this deserves to be done. In some areas, however, the figures are readily available and speak for themselves. In Britain, the 769 swords are divided as seen in Table 2.

This picture, while not unexpected to anyone who knows the British material, throws an entirely different light on the absolute figures mentioned above. In particular, the tiny number of swords known or thought to have been found with burials is remarkable, if these really were the fighting equipment of warriors, owned and used by them.

However, there are further aspects to this matter buried within the crude figures: there is a big difference in context between swords of the main part of the Late Bronze Age, down to and including the Wilburton period, and those of the latest part of the period, characterised by Ewart Park swords. Material deposited in wet places is distributed pretty evenly across all periods, but that found in hoards is predominantly from the Ewart Park phase (about three-quarters of the total). This means that even in so prolific a metalworking phase as Wilburton, with its numerous hoards, the majority of the weapon finds occur in wet places. So the dominant context of deposition changed even between two adjacent phases in this instance. In interpreting this pattern, much will depend on one's thoughts on hoard deposition, i.e. whether or not they were intended for recovery. If they were not, then of course wet deposition and hoards become rather similar in nature, or at least in end result. Since much of the material in hoards is broken or scrap metal, however, things are more complicated, and it would be foolish to maintain that scrap metal hoards represent the same intention as collections of whole and perfect objects, like those from Vyšný Šliac or the Elbe at Velké Žernoseky (Novotná 1970: 123, pl. 24; Plesl 1961: 155, pl. 54). In fact weapons are integral to wetland deposition, and incidental to hoards. So the explanation must lie elsewhere – and here one recalls the various ideas that have been advanced to account for the extraordinary number of hoards of the Ewart Park phase, for instance the dumping of bronze in response to the arrival of iron (Burgess 1979). If that is the case, then the appearance of swords in hoards is not particularly significant, certainly not by comparison with their presence in wetlands and their absence in burials. On the other hand, there are regularities to

Table 1: Numbers of metal and organic-hilted swords in various countries of Europe

	Organic	Metal	Area (1000km²)	Density (swords per 1000 km²)
Switzerland, Austria and South Germany[1]	672	489	275	4.22
Italy[2]	167	65	301	0.77
Romania[3]	273	80	238	1.48
Hungary[4]	226	202	93	4.60
Former Yugoslavia[5]	183	51	256	0.91
Denmark and North Germany[6]	604	641	181	6.88
Britain[7]	641	19	230	2.87
Ireland[8]	624	0	82	7.61

References: 1. Schauer 1971; Krämer 1985; Quillfeldt 1995. 2. Bianco Peroni 1970. 3. Bader 1991. 4. Kemenczei 1988; 1991. 5. Harding 1995. 6. Sprockhoff 1931; 1934; Ottenjann 1969. 7. Burgess and Colquhoun 1988. 8. Eogan 1965.

be observed in the state of swords deposited in wet places. In Ireland, Bridgford's analysis (1997: 110) has shown that a higher than expected proportion of swords from wet places had little or no damage – over 70% of those from rivers, for instance. Clearly selection was at work, so that undamaged pieces were selectively deposited in such contexts.

Before we explore this further, let us look for comparison at south Germany, Austria and Switzerland, where the picture emerges as shown in Table 3.

Even allowing for a number of structural differences in the figures presented here, it is immediately obvious that the picture is quite different from that in Britain: while the proportion found in water is very similar, the situation with burials and hoards is completely reversed. Admittedly a sizeable number of the burials belong to the Middle Bronze Age, and the British figures do not include dirks and rapiers which belong to that phase, but in Central Europe there are plenty of burials in the Late Bronze Age and Urnfield period as well. There may be more hoards with sword fragments to include than are listed here, but even so the imbalance is remarkable.

Such a scenario may well be repeated in other parts of Europe. We know, for instance, that in the Nordic area many swords emanate from graves, especially in Periods II and III, and though no figures have been calculated here they will assuredly repay study. What the two case studies illustrate is that swords were deposited in different ways, so they had different life-histories ('biographies') that may reflect different uses. It is natural to think of a sword found in a grave as the personal weapon of the deceased,

Table 2: British swords by context (after Burgess and Colquhoun 1988)

Context	No.	%
Water	213	27.7
Hoard	260	33.8
Single	167	21.7
Burials	3	0.4
Cave	2	0.3
Settlement	1	0.1
Unknown	123	16.0
Total	769	100.0

but of course the converse is not necessarily true: swords found in other contexts may well also have 'belonged' to someone. The important thing is that in the British context they became divorced from their ownership at the end of their life, whereas in central Europe their specific biographies meant that they continued to maintain that aspect.

In the former Yugoslavia, there is another striking pattern: in parts of the country that are mountainous and harder of access (parts of Serbia, Bosnia and Macedonia), swords are usually found in burials and hoards are rather infrequent (Harding 1995). Along the great river valleys (Danube, Sava, Morava near its confluence with the Danube), by contrast, almost all sword finds are from hoards, with a fair sprinkling from rivers. This is a complex issue that cannot be dealt with in detail here, but one may suggest that the presence of swords in burials in mountainous areas is connected with a need to express warrior identity in ways that were not appropriate where

Table 3: Context of sword deposition in South Germany, Austria and Switzerland (after Schauer 1971; Krämer 1985; Quillfeldt 1995)

Context	No.	%
Water	286	26.7
Hoards	67	6.3
Single	222	20.7
Burials	305	28.5
Settlements*	42	3.9
Unknown	149	13.9
Total	1071	100.0

* The settlement finds may in fact represent 'water' depositions, as all are from Swiss lake sites, where some authors have considered the deposition of bronzes to be non-functional (Müller 1993).

hoard deposition represented the dominant ideology.

Does this mean that swords in central Europe were considered more personal objects than those in Britain? Were swords in the mountainous regions of the Balkans objects more to be prized than those in the Danube-Sava corridor? This is certainly one possibility. The further removed from production centres they travelled, the more they might have gained in value, and thus the greater the prestige they conferred on their 'owners'. In the case of the Mycenaean swords of Macedonia and Kosovo, no doubt this was the case, but elsewhere it is harder to sustain this argument. Many fine swords could have been considered prestige objects in this way, but they did not end up in graves.

I have referred to the biographies of swords, and this does seem to be an apt expression to account for their life histories. Weapons that were consigned to the ground, intact or damaged, had served their purpose. They had been held, admired, handled, waved around in mock (or real) battles, used to threaten or frighten, and in the end, placed where custom and usage dictated they should go, in the ground or the water. Across Europe, there was a body of opinion which held that this was the proper thing to do. Only in some areas was it permissible for swords to retain their owner's identity and be buried with him or her.

Something of the same sort may also apply to spears, though unfortunately there is too little published information for comparable figures to be produced. I pass over here the whole question of the method of use of spears as opposed to lances, i.e. whether or not spears were thrown or held and thrust.

Probably both were done. The question is, in what circumstances? One study that gives the necessary detailed information on context is that by Říhovský (1996) on the Moravian spears and arrows (Table 4).

In other words, in this area at least spears were placed predominantly in hoards, while arrowheads occur in them infrequently, appearing instead with burials and on settlements – the latter potentially in their role as attacking weapons. It is notable that there seems to be no evidence in this area for deposition in wet places – though this is information that is not readily available. Spears occur in burials only in certain instances of very high status burials, such as at Velatice where they were found along with a sword and other rich grave goods (Říhovský 1958).

But can we take these figures at face value? In particular, what are we to make of the fact that in most Urnfield cemeteries the amount of bronzework deposited with burials is trivial, compared with what we know from hoard finds was around at the time? Thus a cemetery with many inhumations such as Przeczyce in Upper Silesia (Szydtowska 1968-1972) has graves that might be furnished with trinkets like buttons, but only a tiny minority have tools and only one or two have weapons (daggers, not swords). Does this mean that for most people acquisition of a sword was an impossibility? Or does it suggest that deposition practices dictated that in that cemetery, at least, swords were not to be deposited with the dead? If the former, then large parts of Europe were warrior-less, which, in view of the British-Irish situation, is hard to believe. Graves like that from Wollmesheim (Schauer 1971: 168f), or the Seddin 'Königsgrab' (Kiekebusch 1928), were surely those of elite warriors; here, it was acceptable for this warrior identity to be preserved in death, whereas at Przeczyce and other sites like it this was not the done thing.

How does this information relate to the function and role of weaponry? Arrowheads were used for attacks on sites, especially fortified sites, but also as part of the archer's personal possessions. Spearheads, like swords in the latest phases of Bronze Age Britain, were taken out of the personal sphere and placed in the realm of the inaccessible.

Does this relate to other things that were happening at the time? From various lines of reasoning, it is possible to conclude that the dominant form of warfare in Late Bronze Age Europe was raiding, coupled with the rise of personal display and combat

Table 4: Spear and arrowheads in Moravia according to context (after Říhovský 1996)

	Spearheads	%	Arrowheads	%
Single/unknown	86	36.0	39	13.8
Hoards	122	51.0	10	3.5
Cemeteries	17	7.1	100	35.3
Settlements	14	5.9	121	42.8
Caves etc	0	0	13	4.6
Total	239	100	283	100

or duelling (Osgood 1998; Harding 2000). The quasi-territorial nature of fort spacing across landscapes can hardly be called state formation except in Greece and Italy, but it was a process of coalescing of small groups into larger groups, centred on strongholds. In terms of elite manifestations, there is little or no evidence of special marking out within settlements; high status individuals were only marked in certain ways. In some areas, those individuals were able to utilise the need for common strategies to assure defence (and where appropriate, attack) to bolster their personal position in terms of enhanced visibility. There are areas of high visibility of personal status, and areas of low or no visibility, and at present we cannot discern what rules applied to determine which would be dominant.

There is abundant evidence that fighting was intimately connected with the ritual sphere, for instance parade armour that was not intended for actual use, the deposition of weaponry in wet places, the curious positioning of some forts in 'special' places that were not chosen for defence. Does the pattern of weapon use and deposition then become intelligible if seen as predominantly a part of the ritual sphere? Deposition consists of a series of prescribed and apparently repeated actions. Weaponry was thus preferentially removed from the personal sphere and put into the ritual one, out of reach of everyday use. So the answer to that question is probably yes; but it brings us back to the variability of the record.

At the end of the day, weapons tell us what was possible and what was probable in terms of warfare practices. They also shed unexpected light on the way they were used in terms of deposition. We can also use the weapons to give the lie, finally, to the notion that the past was ever a 'pacified' place. The manufactured controversy that Keeley (1996) introduced was never appropriate, at least not in a Bronze Age context, since we have always known from the volume and variety of weapons that it was a far from peaceful time. We may not know exactly how, when and where the weapons were used, but no serious Bronze Age scholar would maintain they were not used at all. The Bronze Age world was a dangerous place to live in, however exciting its technological, political and ideological advances.

NOTE

1 I thank Henrik Thrane for the observation that the bow and arrow are well attested in Danish weapon offerings of the Roman period.

BIBLIOGRAPHY

- Ambrosi, A.C. 1988. *Statue Stele Lunigianesi. Il Museo nel Castello del Piagnaro*. Genova: Sagep Editrice.
- Bader, T. 1991. *Die Schwerter in Rumänien*. (Prähistorische Bronzefunde, Abteilung IV, 8). Stuttgart: Franz Steiner.
- Bianco Peroni, V. 1970. *Die Schwerter in Italien /Le spade nell'Italia continentale*. (Prähistorische Bronzefunde, Abteilung IV, 1). Munich: C.H. Beck.
- Bianco Peroni, V. 1994. *I pugnali nell'Italia continentale*. (Prähistorische Bronzefunde, Abteilung VI, 10). Stuttgart: Franz Steiner.
- Bocksberger, O.-J. 1978. *Le Site préhistorique du Petit-Chasseur (Sion, Valais) 4, Horizon supérieur, Secteur occidental et tombes Bronze ancien*. (Cahiers d'Archéologie Romande, 14). Lausanne: Bibliothèque Historique Vaudoise.
- Bridgford, S. 1997. 'Mightier than the pen? An edgewise look at Irish Bronze Age swords'. In: J. Carman (ed.): Material Harm: *Archaeological Studies of War and Violence*. Glasgow: Cruithne Press, pp. 95-115
- Brunn, W.A. von 1959. *Bronzezeitliche Hortfunde, Teil 1. Die Hortfunde der frühen Bronzezeit aus Sachsen-Anhalt, Sachsen, Thüringen*. Berlin: Akademie-Verlag.
- Burgess, C. 1979. 'A find from Boyton, Suffolk, and the end of the Bronze Age in Britain and Ireland'. In: C. Burgess and D. Coombs (eds.): *Bronze Age Hoards, some Finds Old and New*. (British Archaeological Reports, 67). Oxford: British Archaeological Reports, pp. 269-83.
- Burgess, C. and Colquhoun, I. 1988. *The Swords of Britain*. (Prähistorische Bronzefunde, Abteilung IV, 5). Munich: C.H. Beck.
- Chapman, J. 1999. 'The origins of warfare in central and eastern Europe'. In: J. Carman and A. Harding (eds.): *Ancient Warfare: Archaeological Perspectives*. Stroud: Sutton, pp. 101-42.
- Chochorowski, J. 1974. 'Bemerkungen über die Chronologie der Pfeilspitzen skythischen Typs im

Nordteil von Mitteleuropa'. *Studien zur Lausitzer Kultur* (*Prace Archeologiczne*, 18), pp. 161-82.

- Dušek, M. and Dušek, S. 1984. *Smolenice-Molpír. Befestigter Fürstensitz der Hallstattzeit, I*. (Materialia Archaeologica Slovaca, 6). Nitra: Archeologický Ústav SAV.
- Egg, M. 1992. 'Zur Ausrüstung des Toten vom Hauslabjoch, Gem. Schnals (Südtirol)'. In: *Der Mann im Eis, Band 1. Bericht über das Internationales Symposium 1992 in Innsbruck*. (Veröffentlichungen der Universität Innsbruck, 187). Innsbruck: Universität Innsbruck, pp. 254-72.
- Eogan, G. 1965. *Catalogue of Irish Bronze Swords*. Dublin: National Museum of Ireland.
- Evans, J.G., Atkinson, R.J.C., O'Connor, T. and Green, H.S. 1983. 'Stonehenge - the environment in the late Neolithic and Early Bronze Age and a Beaker Age burial'. *Wilts Archaeol. Magazine*, 78, pp.7-30.
- Gallay, G. 1981. *Die kupfer- und altbronzezeitlichen Dolche und Stabdolche in Frankreich*. (Prähistorische Bronzefunde, Abteilung VI, 5). Munich: C.H. Beck.
- Gerloff, S. 1975. *The Early Bronze Age Daggers in Great Britain, and a Reconsideration of the Wessex Culture*. (Prähistorische Bronzefunde, Abteilung VI, 2). Munich: C.H. Beck.
- Harbison, P. 1969. *The Daggers and the Halberds of the Early Bronze Age in Ireland*. (Prähistorische Bronzefunde, Abteilung VI, 1). Munich: C.H. Beck.
- Harding, A.F. 1995. *Die Schwerter im ehemaligen Jugoslawien*. (Prähistorische Bronzefunde, Abteilung IV, 14). Stuttgart: Franz Steiner.
- Harding, A.F. 2000. 'Warfare: a defining characteristic of Bronze Age Europe?' In: J. Carman and A. Harding (eds.): *Ancient Warfare: Archaeological Perspectives*. Stroud: Sutton, pp. 157-73.
- Keeley, L.H. 1996. *War before Civilization*. New York and Oxford: Oxford University Press.
- Kemenczei, T. 1988. *Die Schwerter in Ungarn I (Griffplatten-, Griffangel- und Griffzungenschwerter)*. (Prähistorische Bronzefunde, Abteilung IV, 6). Munich: C.H. Beck.
- Kemenczei, T. 1991. *Die Schwerter in Ungarn II (Vollgriffschwerter)*. (Prähistorische Bronzefunde, Abteilung IV, 9). Stuttgart: Franz Steiner.
- Kiekebusch, A. 1928. *Das Königsgrab von Seddin*. (Führer zur Urgeschichte, 1). Augsburg: Benno Filser.
- Kilian-Dirlmeier, I. 1993. *Die Schwerter in Griechenland (ausserhalb der Peloponnes), Bulgarien und Albanien*. (Prähistorische Bronzefunde Abteilung IV, 12). Stuttgart: Franz Steiner.
- Krämer, W. 1985. *Die Vollgriffschwerter in Österreich und der Schweiz*. (Prähistorische Bronzefunde, Abteilung IV, 10). Munich: C.H. Beck.
- Kristiansen, K. 1984. 'Krieger und Häuptlinge in der Bronzezeit Dänemarks. Ein Beitrag zur Geschichte des bronzezeitlichen Schwertes'. *Jahrbuch des Römisch-Germanischen Zentralmuseums*, 31, pp. 187-208.
- Kristiansen, K. 2000. 'The emergence of warrior aristocracies in later European prehistory and their long-term history'. In: J. Carman and A. Harding (eds.): *Ancient Warfare: Archaeological Perspectives*. Stroud: Sutton Publishing, pp. 175-89.
- Lomborg, E. 1973. *Die Flintdolche Dänemarks: Studien über Chronologie und Kulturbeziehungen des südskandinavischen Spätneolithikums*. (Nordiske Fortidsminder, Serie B, 1). Copenhagen: H.J. Lynge og Søn.
- Mercer, R.J. 1999. 'The origins of warfare in the British Isles'. In: J. Carman and A. Harding (eds.): *Ancient Warfare: Archaeological Perspectives*. Stroud: Sutton Publishing, pp. 143-56.
- Müller, F. 1993. 'Argumente zu einer Deutung von "Pfahlbaubronzen"'. *Jahrbuch der schweizerischen Gesellschaft für Ur- und Frühgeschichte*, 76, pp. 71-92.
- Novotná, M. 1970. *Die Bronzehortfunde in der Slowakei. Spätbronzezeit*. Bratislava: Vydavateľstvo Slovenskej Akadémie Vied.
- Osgood, R. 1998. *Warfare in the Late Bronze Age of North Europe*. (British Archaeological Reports International Series 694). Oxford: Archaeopress.
- Ottenjann, H. 1969. *Die nordischen Vollgriffschwerter der älteren und mittleren Bronzezeit*. (Römisch-Germanische Forschungen, 30). Berlin: de Gruyter.
- Plesl, E. 1961. *Lužická kultura v severozápadních Čechách*. Prague: Czech Academy of Sciences.
- Quillfeldt, I. von 1995. *Die Vollgriffschwerter in Österreich und der Schweiz*. (Prähistorische Bronzefunde, Abteilung IV, 11). Stuttgart: Franz Steiner.
- Říhovský, J. 1958. 'Žárový hrob z Velatic I a jeho postavení ve vývoji velatické kultury'. *Památky Archeologické*, 49(1), pp. 67-118.
- Říhovský, J. 1996. *Die Lanzen-, Speer- und Pfeilspitzen in Mähren*. (Prähistorische Bronzefunde, Abteilung V, 2). Stuttgart: Franz Steiner.
- Sauter, M.-R. 1976. *Switzerland from earliest times to the Roman conquest*. London: Thames and Hudson.
- Schauer, P. 1971. *Die Schwerter in Süddeutschland, Österreich und der Schweiz I (Griffplatten-, Griffangel- und Griffzungenschwerter)*. (Prähistorische Bronzefunde Abteilung IV, 2). Munich: C.H. Beck.
- Sprockhoff, E. 1931. *Die germanischen Griffzungenschwerter*. (Römisch-Germanische Forschungen, 5). Berlin and Leipzig: de Gruyter.
- Sprockhoff, E. 1934. *Die germanischen Vollgriffschwerter der jüngeren Bronzezeit*. (Römisch-Germanische Forschungen, 9). Berlin and Leipzig: de Gruyter.
- Szydtowska, E. 1968-1972. *Cmentarzysko kultury lużyckiej w Przeczycach, pow. Zawiercie*. (Archeologia, 5, 8, 9). Bytom: Rocznik Muzeum Górnośląskiego w Bytomiu.
- Vandkilde, H. 1996. *From Stone to Bronze. The Metalwork of the Late Neolithic and Earliest Bronze Age in Denmark*. (Jutland Archaeological Society Publications, 32). Aarhus: Aarhus University Press.
- Wüstemann, H. 1995. *Die Dolche und Stabdolche in*

Ostdeutschland. (Prähistorische Bronzefunde, VI, 8). Stuttgart: Franz Steiner.

Warfare and Gender According to Homer:
An Archaeology of an Aristocratic Warrior Culture

HELLE VANDKILDE

/34

The constitution of Homeric society is the main focus of this article, which in particular highlights aspect of gender, warfare, and materiality. The underlying expectation is that such a study may ultimately lead to a better understanding of the social world of the illiterate Bronze Age north of the Alps. A social study of Homer[1] can, it may be argued, form the basis of a contextually based comparison with Bronze Age societies in temperate Europe, using the principle of relational analogy (Wylie 1985; Ravn 1993). However, such a comparative enterprise is not the immediate objective of the present study, which merely aims at calling attention to the existence of such a potential. The issue of ideology, which is important in Homer's epics as well as in current archaeology, will receive a few theoretical comments at the end of the article.

The Research Council project 'War and Society. Archaeological and Social-Anthropological Perspectives' has put me on the track of past war heroes, martial ideologies and, not least, military societies or warrior bands. Evidently the warrior role has a strong influence on our understanding of European prehistory (Vandkilde and Bertelsen 2000; Vandkilde 2003). When we speak of war and warriors we are not only speaking of social organisation, ideology, prestige and power, but also to a great degree of gender.

More or less exclusive men's clubs, encompassed by the German term *Männerbünde*, exist in many societies, often with a strong martial strain (Mallory 1989: 110f; Ehrenreich 1997: 117ff; Vandkilde chapter 26). To be a warrior is therefore a demonstration of a specific male identity, which, of course, cannot be assessed without including other gender identities: all in all, the specific social context cannot be ignored.

It is central to the discussion undertaken that the warrior identity can only be understood in its social context against the background of other social identities and confronted with the current warrior ideal. The Dutch scholar Hans van Wees (1992; 1997) has thoroughly studied the *Iliad* and the *Odyssey* in a social perspective, primarily the relationship between status rivalry, war and social hierarchy. Van Wees has indeed been a source of inspiration, particularly in the sense that the epics are considered a meaningful entity. I have, however, extended van Wees' main theme and added new themes and approaches especially as regards material culture, gender identities, and the warrior retinue. In particular, material culture cannot be ignored, forming as it does the foremost source material in archaeology in addition to being a powerful silent discourse in any society, past or present. Inspiration has also come from studies by Moses Finley (1972), Ian Morris (1987), and Otto Steen Due (1999a; 1999b). It may be added that the approach to Homer in the present study is archaeological in the sense that it focuses upon relevant themes in the discipline's discourse and in that it attempts to uncover 'layer by layer' levels of social practice in Homeric society (cf. Foucault 1985).

In the following attention is upon the Homeric epics, especially the *Iliad*, as a structured entity, but I do think that the archaeological value of the two interconnected stories has been widely underestimated. To use the Homeric epics as a supplement to the archaeological sources has long been *faut passé*, although there are brave exceptions (Frankenstein and Rowlands 1978; Rowlands 1980). The rejection can – in the case of Northern and Central Europe – be blamed on the geographical distance but also on the fact that the *Iliad* and the *Odyssey* are the written final products of a century-long oral epic tradition, in the same way as for example, the Beowulf-epic of the Early Medieval Period. In a chronological-historical sense, the epics do not therefore comprise an entity, as they incorporate elements and situations from different eras. It can nevertheless be argued that the epics are relevant to the study of the European Bronze Age. This is partly because the historical focal point of the epics is around 1200-900 BC, the so-called 'Dark Ages', although with distinct Mycenaean elements (see Page 1959; Finley 1972; 1973: 29; also the discussion in Morris 1987: 44ff; Morris and Powell 1997), partly because the epics are a logically reasoned, thoroughly elaborated whole, describing a specific society over a period of 20-30 years.

The epics may then, to a certain extent, be source material for the illiterate European Bronze Age, but they mainly emerge as a meaningful and complete whole, describing ideal and real features of a specific aristocratic society (cf. van Wees 1992). The epics are valuable as a relevant analogy, which can lead us to reflect on the social dimension in mute archaeological material.

Homer as an analogous social context

Opinions have differed within Greek Bronze Age and early Iron-Age archaeology, but after a long period of rejection, Homer has again been accepted and is used as a supplementary text to archaeology (see Morris 1987: 22ff; Morris and Powell 1997).

The epics contain antiquated features in language

and content, which directly refer to the characteristic Late Bronze Age material culture and political structure of the Aegean area. Certain types of weapons, especially boar's tusk helmets (Fig. 1.) and the tower or figure-eight-shaped shields, as tall as a man (The *Iliad* X: 261-271; IV: 404-405), for example, clearly belong to the Late Bronze Ages (Lorimer 1950: 133ff, 212ff; Page 1959: 218ff; Snodgrass 1964; Bloedow 1999). The same goes for the landscape of power, with the well-known cities and palaces like Pylos, Knossos, Tiryns, Mycenae and Orchomenos (Page 1959: 218ff; Chadwick 1976: 180ff; van Wees 1992: 262; Bennet 1997).

The grouping of warriors around the most outstanding heroes, the individualising hero worship of the warrior company, its asymmetrical construction and its internal rivalry belong to the Late Bronze Age and/or early Iron Age, since this form of military organisation differs distinctively from warfare in later Greek periods. In the Archaic and Classical periods, the military hoplite-system was a regular army with anonymous soldiers and a communal expression, which is in harmony with the collective ideology of the city-states (Runciman 1999: 732f).

However, the attitude towards death, the description of burial rituals and the social organisation in the epics is not really in keeping with the evidence from the Linear B texts and the archaeology the Late Bronze Age (Dickinson 1994: 81). It is more in accordance with the post-Mycenaean Period's simpler hierarchy and varying burial rituals, which shift between inhumation and cremation in ceramic containers (see Finley 1972: 51ff; Morris 1987: 18ff, 46, 53, 178; van Wees 1992: 1ff, 262).

Despite discussions about the historical reference point of the individual parts of the epics, and despite uncertainties about the exact time of writing,[2] it is often forgotten that the Homeric epics, with their maximum dispersion between 1700 BC and 500 BC (cf. Bennet 1997: 531), are the only writings from the European Bronze Age seen as a geographic whole[3]. Precisely that fact makes the epics suitable to be brought into interaction with archaeological sources, but caution obviously should increase with geographical distance. However, the epics may be considered – in terms of time, place and subject – a more obvious choice than the ethnographic observations and analogies often used in archaeology.

The Homeric epics can therefore – on a completely general level, and with the above reservations in

F I G . 1: *Head of a warrior wearing a boar's tusk helmet; lid of small ivory box (pyxis) dating to c. 1400 BC. Mycenae, chamber tomb 27 (after Borchhardt 1972).*

Basically, the epics, especially the *Iliad*, are about men and war (Due 1999b); 'the world' according to a male warrior élite. This imbalance or distortion may especially cause problems of interpretation if you want to use Homer as a source for specific historical conditions. This report, on the other hand, views the Homeric epics as an analysable context with a value in themselves; as a closed social context with a complexity of real and ideal ingredients. An analysis of these various ingredients – at times full of contrasts – can give new insight, although the narrowly elitist, male-dominated point of view that the narrator holds must naturally be taken into consideration and kept in mind in the evaluation of the interpretations. Despite these reservations as to sources, I think that it is possible to form a fairly exact impression of social practice, including class relations and ideals, identities and gender dominance. In other words, the Homeric epics can be used as an archaeological 'manual' on heroes, war and society, precisely because they comprise complete social contexts with many layers of meaning: something which archaeology does not exactly lack, but which is more difficult, and especially more demanding, to reconstruct. This is in keeping with recent Homer research (van Wees 1992: 262; Morris 1987: 22ff, 44ff, 53, 90f, 196ff with references) which agrees that the epics deliver a coherent, meaningful picture of an aristocratic society. Even the structural contrasts that also characterise the epics make sense when they are seen on the basis of the social contexts in which they are found.

The *Iliad* and the *Odyssey* can be understood against the background of three different contexts. The first is the epics in themselves, i.e., the aristocratic society that the poet describes, and which constitute a frame for the actors' thoughts, stories and actions. That is the context on which the following analysis is based. The second is that of the society which created, recreated and developed the epics through verbal recital, mainly in the period between the 13th and 9th century BC,[4] the Late Mycenaean and post-Mycenaean periods. The third is that of the society which had the epics 'frozen' by writing them down, probably around 700 BC or later, i.e. the period when the Greek city-state took form.

The two epics can, with their explicit ideals regarding, for example, gender, status, political leadership, war, trade and gifts, be explained with reference to their social potential in the user-societies, which are

mind – be regarded as an alternative way in to the illiterate Bronze Age societies north of the Alps. Far more important, however, is the fact that the epics have an internal logic that makes them useful for comparative purposes. Mind you, this logical structure reflects on the societies that created them. In our source-criticism, we must recognise that the epics are an ideological construction made by and for a male-dominated social élite. In the epics there are social groupings whose points of view are oppressed or disregarded. Above all, the conditions of the lower classes are out of focus, and they have no voice. Women are only rarely the centre of attention, and it can be argued that the point of view is primarily male. This special perspective of the epics can, at least partly, explain their more recent popularity; from Antiquity to the Middle Ages and up to the present, the two epics have been diligently used in reproductive strategies of the social élite, in the cultivation of a violent masculine identity as a war hero, and to legitimate men's authority over women.

the second and third contexts. The societies which used the epics were aware of their historical background, and they used history actively and strategically. In the post-Mycenaean Period, the epics represented respectable aristocratic ideals from a glorious but not so-distant past – perhaps in contrast to a chaotic present. In the early Greek city-state, the epics offered a common identity through glorification of a heroic past, thus making them suitable to legitimise a new type of state society, the *polis* (Morris 1987; van Wees 1992). The glorification of a heroic past, but with a material expression, is also substantiated in Archaic times by the hero cult, which took place in front of the monumental graves of the Bronze Age (Whitley 1995). However, the epics also legitimise asymmetry and action on other levels, especially men's authority over women, and war as a means of attaining prestige in society. The following analysis concentrates on the first of these three contexts, i.e., the epics as a logically structured whole.

Ideology and social organisation in Homeric society

The ideals of the epics do not always correspond to reality; there are both large and small deviations in a number of areas. Put simply, unrealistic elements inspired by the ideology of society are introduced into the story (van Wees 1992: 153). The epics are contradictory in places, especially the way that ideals of equality are contradicted by a hierarchical reality.

Homeric social stratification as a distinctly bipartite pyramid, with a ruling class of aristocrats at the top, is well documented[5] (Finley 1972: 61ff). Thus there are two distinct classes: the people and the aristocracy (*Iliad* II: 365f), added to which were slaves, often prisoners of war. The boundaries cannot be overstepped, since class affiliation is determined by birth (Morris 1987: 94ff). The relationship between the lower and upper classes is not ideologically coloured, probably because the allocation of power between those who dominate and those who subject themselves is undisputed. The ideology has free play, but only in the upper part of the pyramid.

The poet uses an inordinate amount of space in describing the personal qualities of the individual heroes and the advantages and disadvantages of their activities, and in so doing indicating their social position in the group of heroes. The aristo-

cratic warriors seem almost obsessed with their need for performance. Thus they constantly compete for social status on- and off the battlefield; therefore the nickname 'status warriors' (van Wees 1992; also Finley 1972: 132ff). In this way, the epics create an impression of an egalitarian state amongst the aristocrats, in which social status and authority depend on personal qualities, strength, influence, and success in war (van Wees 1992). This focus on individual performance is, however, misleading, because fame and success have no real influence on the social order.

If we look beneath the surface, it is evident that the aristocracy itself is subordinate to a strict hierarchical structure, where kinship, inheritance, political power and wealth are determining factors. The aristocracy divides into a ruling élite of kings and princes, and a non-ruling élite. Power and social position are inherited in certain families according to the principle of male primogeniture. Each aristocrat takes his position in the social hierarchy of aristocrats, defined in relation to the local dynasty, and the kings themselves are placed in a superior hierarchy in relation to the Dynasty of Pelopides of Mycenae (see below) and in relation to measurable property such as wealth, slaves and the size and geographical position of the kingdom. Thus the aristocrats form their own hierarchy, in which allegiance and contracts of service, for example of a military nature between ruler and subordinates (vassals) on several levels, seal power relations, as in a feudal society (see Finley 1972: 109ff).

Personal ability is good to have, but not a requirement, and there is a limit as to how far individuals can advance through the system purely on the basis of determination, ability and bravery on the battlefield. First-class war heroes exist both among kings and their aristocratic companions, but this status actually seems unimportant. Although competence in war is appreciated, it does not change the essential point, which is that hierarchy is predetermined. For example, Agamemnon has a supreme hereditary status in the hierarchy of princes, with given privileges and obligations which are not dependent on personal capability (Finley 1972: 87), and actually, his personal accomplishments on the battlefield are quite moderate. Besides, he is described as 'first among equals', which, all things considered, is an idealistic description, since he single-handedly and arbitrarily distributes honours and the spoils of war.

However, Agamemnon can assume an egalitarian, paternalistic, image precisely because he is the king of the most powerful state, namely Mycenae, with 'many isles and all of Argos' (*Iliad* II: 108). It is without doubt a strategic choice of tone and expression.

The *Iliad* bears witness to the existence of a centralised structure, in which the ruler of Mycenae – although to an unknown extent – exercises authority over other political units (see Page 1959: Maps I-III p. 121ff). This authority is sealed through gifts and services (*Iliad* II: 254f; IX: 129ff; XXIII: 296f). And Homer does describe Agamemnon with the suffix *wanax*, which, in contrast to the more common word for king, *basileus*, has the significance of paramount ruler – a post which is symbolised by a special sceptre, and sanctioned by the gods (see *Iliad* II: 100-108). Thus Agamemnon must necessarily be 'king over men' while the other princes bear more neutral labels. For example, Achilles is 'the fast runner', Odysseus is 'clever' and 'brave', Diomedes is 'good at war cries' and Nestor 'the old Gerenian charioteer'. This is a reflection of the actual power relations, which are clearly revealed when Achilles, for example, is thoroughly put in his place by Odysseus; 'Though thou be valiant, and a goddess bare thee, yet he (Agamemnon) is the mightier, seeing he is king of more' (*Iliad* I: 280-81).

Ideology comes in as a manipulative element amongst the aristocrats where, as mentioned, an ideal of equality flourishes. The background for this contrast between an ideal of equality and an actual hierarchical structure amongst the aristocracy can of course be debated, but two possibilities, which do not exclude each other, must be mentioned. Power is never permanent and unchangeable. It may therefore be the ruler of Mycenae who forces a 'false consciousness' on the elite in order to retain power. On the other hand, the aristocrats could have assumed the ideal of equality as a part of their self-image as a class and as a part of a strategy in which they see power relations as partly unclear. Basically, it gives them more freedom of action; an incentive to compete on and off the battlefield. The use of the princely appellation 'first amongst equals' can therefore be read either with emphasis on the 'first' or on the 'equals', depending on the person's position in the hierarchy of aristocrats and depending on the strategy the individual aristocratic warrior adopts.

Material culture and society in Homer

In Homer, material culture reflects to a greater degree the social hierarchy than the ideology of equality. Palaces, castles, spectacular gifts and drinking equipment, magnificent weapons and stately burial rituals are reserved for the aristocracy, and luxurious presentation of material goods increases concurrently with the position in the hierarchy of society. As mentioned, the hierarchy also saturates the upper classes of aristocrats, and also here the material culture follows along. In the so-called catalogue of ships 'the well built citadel of Mycenae' provides by far the strongest force, and can present the best and wealthiest war material:

… of these was the son of Atreus, lord Agamemnon, captain of an hundred ships. With him followed most warriors by far and goodliest; and among them he himself did on his gleaming bronze, a king all-glorious, and was pre-eminent among all the warriors, for that he was noblest, and led warriors far the most in number. (*Iliad* II: 575ff: also XI: 15ff)

It is a huge demonstration of power and social status. It is also a clear demonstration of violent potential – not only to the Trojan enemy but to a great degree also to the allied princes of the Achaean army. In exactly the same style, Agamemnon offers Achilles great fortunes and privileges to forget his anger and return to the ranks of the fighters (*Iliad* IX: 120ff). No other king could afford to be that generous.

Burial rituals and personal equipment follow the hierarchy to a large extent, whereby the ruling élite presents itself in the best equipment and receives the stateliest funerals. For example, in the *Odyssey* (XXIV: 32) we are told that the whole Achaean army would have co-operated in the construction of a grave mound for Agamemnon, which forms a contrast to the more moderate funerals of common aristocrats in the epics. It is, however, important to notice that the great hero Patroclus receives a funeral worthy of a king (*Iliad* XIII). But Patroclus is not a king, but a common aristocrat, and besides foster brother to the king of the Myrmidons, Achilles. Through a stately funeral, ending with the construction of a monumental mound, and followed by grandiose burial games in Patroclus' honour, a far higher social status is reflected than his position in the social hierarchy and his fame as a war hero can justify. The ideology of equality and free competition is expressed here when the material world –

especially weaponry and burial rituals – supports the idea of personal achievement and fame as a source of power and influence. Weapons and burial rituals are actively used in the constant status rivalry amongst the heroes; but the reality is that both high and low social positions are controlled by other, less easily manipulated mechanisms than the ability to wage war. Neither Patroclus' heroic deeds on the battlefield or the, to say the least, conspicuous burial perform-ance changes his social position in the hierarchy; roughly speaking, society is ideologically reproduced.

But there is more to it: when Achilles arranges a state burial for Patroclus, he also questions the exist-ing power relations. This material display may well be in accordance with the ideology or, if you like, with the latitude for social rivalry allowed by the social system. But the whole séance has a more pro-found meaning, which can be readily understood by everyone in Homeric society: even power is nego-tiable; power is never unchangeable! Social identi-ties, existing – or, as here, under construction – are communicated through material culture. Material culture has obviously tangible practical and social functions as well as more abstract symbolic meanings attached to the social organisation, ideology and history of society (Vandkilde 2000: 21ff).

Certain material things can have legends and sto-ries attached. This is illustrated by the story about the boar's tusk helmet (Fig. 1), which Odysseus puts on before he and Diomedes leave for a nightly spying expedition:

And Meriones gave to Odysseus a bow and a quiver and a sword, and about his head he set a helm wrought of hide, and with many a tight-stretched thong was it made stiff within, while on the outside the white teeth of a boar of gleaming tusks were set thick, to and fro, well and cunningly, and on the inside was fixed with a lining of felt. This helmet Autolycos on a time stole out of Eleon when he had broken into the stout-built house of Amyntor, son of Ormenus; and he gave it to Amphidamas of Cythera to take to Scadeia, and Amphidamas gave it to Molos as a guest-gift, but he gave it to his own son Meriones to wear; and now, being set thereon, it covered the head of Odysseus. (*Iliad* X: 260-71)

It is truly a helmet with soul and personality that Odysseus puts on his head. The expedition into the enemy camp means that a new story can be added to the helmet's cultural identity, which is therefore constantly changing. The story of the helmet shows how things can be bearers of history and memory, which are thus transferred in time and space. The helmet and its symbolic goods become a part of Odysseus' personality and social identity as a war hero in that situation.

The Homeric world is a highly hierarchical society, which becomes a natural state through material means. As a whole, the material culture follows the steps of the hierarchy but reflects to a certain degree the ideology of the upper classes. The war heroes and other people in the poems consciously or uncon-sciously use material culture in strategies and actions which maintain and create social identity. In addi-tion, we can see how the material culture so to speak influences the actors recursively. Material visibility, not least bodily appearance, is effective in itself, although within already designated limits.

War and aristocratic warriors in Homeric Society

In the epics, war and violence are closely connected to status rivalry (van Wees 1992). Moses Finley (1972: 131) expresses it this way: 'Warrior and hero are synonyms, and the main theme of a warrior culture is constructed on two notes – prowess and honour. The one is the hero's essential attribute, the other his essential aim.' War is officially waged to defend insulted honour, to gain the respect of friends and enemies, and to achieve fame. But, really, war is instead waged in order to confirm power and as a means of enrichment and expansion of territory. Wealth is a goal, which ideally is achieved through gift-giving, trade and exchange, but actually more often reached through war and raiding expeditions (van Wees 1992: 101).

The significance of the status rivalry expressed in duels between the heroes on the battlefield is no doubt highly exaggerated in the poems, again as a consequence of ideological distortion. The Homeric society is a warrior culture, in which war and violence are important structuring elements. It is the fine heroic duels that attract attention, because through them the warrior ideal can be reproduced almost infinitely, and the ugly face of war, which is also a part of it all, tends to be hidden. Attacks, piracy, and pillaging are clearly a very substantial part of the war actions, and at the same time an economic necessity both before and after Troy. For example, Odysseus reports on a predatory attack in this way:

From Ilium the wind bore me and brought me to the Cicones, to Ismarus. There I sacked the city and slew the men; and from the city we took their wives and much treasure, and divided it among us, that so far as lay in me no man might go defrauded of an equal share. (*Odyssey* IX: 39-42)

Homer gives the classic institution of war heroes a prominent place in society; in war and peace. A warrior band can be defined as a more or less exclusive club of specialised warriors. Ethno-historically, the institutionalised warrior band divides itself into three categories, with different recruiting, leadership and internal structure (Vandkilde chapter 26): first, warrior institutions in which access is regulated through age (age grade), second, warrior institutions in which access is regulated through personal qualities (independent companionship), and third, warrior institutions in which access is regulated through distinctions of rank (dependent companionship). The Homeric warrior institution belongs among the last two categories, mostly the last category (cf. van Wees 1997: 670). It is a company of warriors, an elite team of aristocratic warriors who accompany a war leader of high descent. Such a band of warriors is mentioned in several places but most comprehensively in the catalogue of ships (*Iliad* II: 494ff). Thus the warriors of Myrmidon accompany Achilles, with his foster brother Patroclus as second in command. Nestor's following of warriors is probably the same aristocrats who, at home in the palace in Pylos, sit in the place of honour at the high table during Telemachos' visit (*Odyssey* III: 469-574). The following of warriors surrounding Odysseus also accompanies him on his way home from Troy to Ithaca. The Cretan warriors are led by their king, Idomeneus, etc.

Not surprisingly, here too, the ideal differs from reality. Personal qualities seem definitive for the recruiting of the company of warriors, for internal status differentiation, and for such an important issue as leadership; in particular, martial ability must constantly be proven on the battlefield. Nevertheless, wealth, birth and political position are the most important criteria for the composition of the group of warriors. Men of the people are apparently excluded from the company. Each king is the born leader of a permanent, dependent following of aristocratic warriors, and in that context, honour, respect, achievements and fame are less important. Furthermore, the size of the company of warriors depends on the position of its leader in the superior political hierarchy; thus Agamemnon's following is by far the largest. In addition there is a superior company: Agamemnon has surrounding him – besides a company of noble warriors from his own kingdom – a sworn guard of princely warriors, namely the other Achaean kings.

Actually, this structure calls to mind the Germanic retinue (*Gefolgschaft*, *hird*) in the 5th century AD. In the first centuries AD, the military alliance of warriors is gradually developed by the Germanic tribes from a loosely knit, independent band of warriors, built around mutual respect, prestige, and martial abilities to a dependent, hierarchic system of aristocratic warriors who are bound by oath to a powerful royal family, i.e. a system of retainers (cf. Steuer 1982: 55ff; Hedeager 1990: 184ff; Jørgensen 1999: 156ff). First and foremost, however, the Homeric military institutions of aristocrats make us think of the Late Bronze Age military system in the Aegean area as it can be roughly reconstructed from the Linear B texts. We are now a few centuries earlier than the focal point of Homer's epics, but the connection seems evident despite interpretative difficulties.

The clay tablets from the Pylos palace from around 1200 BC contain information about a series of offices in the Pylian state administration (Page 1959: 183ff; Chadwick 1976: 71ff; Bennet 1997): *Wanax* is thus the highest and most lucrative post, probably the king himself. Besides the *wanax*, there is another important post, the *lawagetas*, which could be intended for the heir to the throne, or maybe the king himself in another role. The actual word means 'leader of the people' but as 'the people' in Homer often refers to the warriors, the best translation is perhaps army leader or 'leader of the warriors'. In addition, there is a third category, the so-called *hequetai* which, directly translated, means companions. Here we are probably being introduced to the company of warriors who in later interpretations are the king's drinking companions and close attendants in war and peace. The *hequetai* of the tablets are often named with their father's name as well as their own, and this careful use of the patronymic indicates an especially high social position. The clay tablets from Knossos from around 1400 BC (Page 1959; Chadwick 1976; Bennet 1997, op. cit.) tell us that *hequetai* had the right to keep slaves, wore distinctive garments and drove special chariots, which, of course, supports their interpretation as warrior aristocrats with a close connection to the king or his deputy commander.

FIG. 2: *Restored wall paintings with warrior processions and fighting scenes in the second megaron (Hall 64) of the palace of Pylos in Messenia. It was destroyed around 1200 BC. The megaron may well have been the seat of* lawagetas, *the leader of the warriors and the* hequetai, *the following of warriors – as discernible in the Linear B texts (after Borchhardt 1972).*

The walls in the secondary megaron in the southwest wing of the Pylos palace were decorated with warrior scenes, both battles between infantrymen and processions with warriors and chariots (Fig. 2). While the primary megaron was most likely the king's reception hall, with the warrior scenes in mind, it is natural to assume that the secondary megaron was the seat of the commander, *lawagetas* (Davis and Bennett 1999: 117f), and his attendants, the *hequetai*.

Male institutions of warriors exist in many contemporary and historical societies where they can be more or less bounded and more or less exclusive in their recruitment. An interesting aspect is the built-in contrast on two levels in these warrior institutions: first of all, it is an exclusive club, which almost by definition excludes women and children, but in many cases also men of non-aristocratic descent. Although the military club thus comprises a potentially oppressive extension of power, there are often well-defined obligations based on the well being of society; including religious, political and defensive functions. Second, war is a more or less important part of the activities of the association. A martial appearance, in which a masculine expression is intensified through spectacular weaponry, often misrepresents, to some degree, a life which also includes peace. The Homeric military institution of warriors also has such contrasting features. In Homer, the military institution occupies above all, a male domain, which is in sharp contrast to a female domain based on a stereotyped feminine ideal.

Odysseus and Penelope: gender in Homer

In Homer's society, women have no access to the company of warriors and regardless of their position in the social hierarchy, men and women live in separate worlds. Men's actions and feelings are more closely connected to those of other men than to those of women, whether their wives or other women in or outside the household (Finley 1972: 148).

Within the aristocracy there are two gender ideals, a masculine and a feminine. These gender ideals are represented in Homer by King Odysseus and his wife Queen Penelope, who both live up to the ideal in an exemplary manner in everything that they do. Whereas Odysseus is the cosmopolitan warrior who fights his own and others' battles, Penelope stays at home and guards the family and its properties. Significantly, the Homeric word *kredemnon* means both city wall and the marriage veil of a woman (Morris 1987: 192). In a way, Penelope is a heroine, defined on the basis of specific female values and certain moral codes. The division of labour between the two main characters is explained in the *Odyssey*, where Penelope relates a conversation with her husband shortly before his departure to Troy. Odysseus says:

Therefore I do not know whether the god will bring me back, or whether I shall be cut off there in the land of Troy: so let all here be your care. Be mindful of my father and my mother in the halls even as you are now, or yet more, while I am far away. (*Odyssey* XVIII: 265-68)

The Homeric epics are more or less one long idolisation of the heroic man. Not many verses are dedicated to the lives and world of women, which, however, are clearly present as a necessary contrast. While the warrior hero is the ideal for male aristocrats, the female aristocrat finds her ideal in a narrow gender identity as a peaceful, caring person who looks after the home during the frequent absences of her husband. Early European written sources thus document the presence of what you could call symmetric gender ideals, in which a violent, extroverted, masculine cosmopolitan is contrasted with a peaceful feminine counterpart in the domestic sphere. This ideal is surprisingly similar to certain gender stereotypes in our own world, and it is kept alive in the same way by separate gender domains, where women and men to a great extent do what culture expects.

In Homeric society, the war hero cannot exist without his female counterpart, who gets her identity from *oikos*, the Homeric word for the domestic domain. The war hero even rejects any connection to *oikos*, which stands for peace, and therefore life. In a way, the war hero denies life, achieving his identity through violence and death. However, he exists *qua* his opposite – the peaceful, home-fixated woman. Thus men and women ideally each rule their world:[6]

Masculine	Feminine
International domain	Private domain – *oikos*
Adventure	Stability
War	Peace
Change	Reproduction
Death	Life
Lion[7]	Human being

In Homeric society, too, gender ideals are kept alive by social practice in that area, at the same time as they help to maintain a certain image of society. And as a rule, concepts of gender are very close to the reality of the actors. The gender ideal – Odysseus-Penelope – has its roots deep in the real world, in which women typically are from Venus and men from Mars, to use a modern expression. The Trojan women stay inside the palace or stoically watch their men on the battlefield from towers and walls of the citadel[8] (e.g., *Iliad* III). Neither do the wives, women servants and female slaves of the Achaean heroes move outside *oikos*, by which is meant the palace at home and the camp by the beach. Women servants and female slaves are typically booty of war, captured during raiding expeditions during the period leading up to the siege of Troy. The ideals and roles of the aristocracy spread downwards in society, in that ordinary women typically work as servants in the palaces, while men's jobs are out in the countryside, even though not primarily in battle. There is, however, a clear division of labour according to gender, originating in and interacting with the contrasting male and female ideals. The image of two gender domains is supported by material culture, which in the epics gives a 'true' picture, with magnificent weapons for men and more peaceful things for women.

The female role is the most constant, but there are interesting deviations which far from fit the ideal: dangerous women with special abilities and a disturbing power over men, such as Helena, Calypso, Circe and Cassandra. There are also goddesses who can join in the turmoil of the battle, apparently without surprising anyone; especially the virgin warriors Pallas Athena and Artemis, but also the more feminine of the species, like Aphrodite, Iris, and Hera. The world of the gods is definitely not a simple reflection of the world of humans.

Amongst the men, there are several examples of warriors who cannot quite live up to the ideal of a war hero, and they are exposed to public ridicule. Magnificent, terrifying weapons cannot change the fact that the beautiful Prince Paris is not made for the horrors of war and the role of hero. Helena mocks him and incites him to fight (*Iliad* III: 42), since her honour, too, is at stake. The woman in the role as the one who is ultimately responsible for the honour of the family and therefore must incite to revenge and war (without participating herself) is found in many tribal societies, such as Polynesian chiefdoms, Arabic Bedouin tribes, Scottish Highland clans and village communities in the Balkans and in Greece. The province of Mani at the southernmost point of Peloponnese is, like Crete, known for its warrior culture, vendettas, and endemic war on a family- and clan level. Women's attitude towards war and violence in Mani can be expressed thus: 'If I were a man, you would see bloody revenge!' (Ambatsis and Ambatsis 2000: 21, my translation from Swedish). Women in Homer and in these tribal societies therefore not only aid in creating and recreating war and violence as legitimate means to achieve certain goals, but also certain standards concerning what women and men can do.

Separate gender domains – as they exist in many cultures – in themselves tell us nothing about relations of dominance, namely because their exact content and the relationship between them are culturally specific. In Homer, however, there is not only a gender asymmetry in the sense that the male domain takes up a disproportionately large amount of space in the story, but also in the sense that men have distinct authority over women. Women in the Achaean camp have no say at all; they are objects who can be moved around and given away in the same way as material things (e.g., *Iliad* I: 183f, 345ff; IX: 128f). After marriage the aristocratic women have acquired a certain authority – if nowhere else, then within their own household. In the royal household of the Phaiakians (*Odyssey* VIII) Queen Arete clearly has a certain influence, but it is King Alcinous who has the final word (see Finley 1972: 103).

Unmarried women of high descent are, however, completely reified, as they fill the role of services and gifts for the sealing of alliances. With no more ado, Agamemnon offers the difficult, sulky Achilles important members of his household, who are obviously his private property:

Three daughters I have in my well-built hall Chysóthemis, Laodíke, and Ifiánassa. Of these let him lead to the house of Peleus which one he will, without gifts of wooing, and I will furthermore give a dower full rich, such as no man ever yet gave with his daughter. (*Iliad* IX: 144-48)

But another facet of the Homeric dominance pattern in the area of gender is of course that women themselves contribute to a great degree to the system by doing exactly what culture expects.

Homeric society

In the Homeric epics, the *Iliad* and the *Odyssey*, the narrator gives the audience an insight into a martial culture, which is described from the social élite's point of view. After that, the story is primarily about men and war. Access to the resources of society is hierarchical, and family, inheritance, power, and wealth structure the social hierarchy.

The ideology of society introduces certain unrealistic elements, especially in how an ideology of equality reigns amongst the aristocracy. We are given the impression that personal qualities, especially martial ability, give access to social position and power, but for much of the time, the hierarchy amongst the aristocracy is a foregone conclusion. The ideology of equality is probably one of denial, which is cleverly manipulated by the high king, but at the same time it is a part of the aristocratic lifestyle and self-image amongst 'equals', because it gives a certain freedom of action and can be used to question power. It is with this background in mind that the constant status rivalry between the heroes should be understood.

In principle, material culture supports the hierarchy by directly reflecting it, but is also used in social strategies which attempt to shift the balance of power. Certain things have special stories and memories of the past attached, and these symbolic qualities reflect on the user and become a part of his personality. Things and personal belongings create and recreate different types of social identity, although within fixed limits.

The Homeric society can best be summarised as an aristocratic warrior culture in which war is an important part of social reproduction, both on an ideological and a socio-economic level. War is described with strong ideological undertones, and reality is misrepresented, in that status-charged duels are given excessive space and attention. A stereotypical warrior ideal, which again confirms a certain type of male identity, is thus reproduced, while the less flattering side – namely looting, piracy and pillaging of foreign settlements – is an ideologically suppressed, but economically necessary, part of society.

The classic male institution of warriors is an important part of society's social organisation, acting on two political levels: partly as a company of followers for individual princes, partly as a superior following of princes for the paramount king. The whole system of followers is strongly asymmetric in its structure. It is implicit in the nature of the system that it excludes women, children, and certain groups of men, and that it misrepresents a reality which is also peaceful. This is because there are also peaceful moments at home and away, and because the warrior groups also take part in peaceful activities, such as those of a religious or sporting nature. However, war takes up a lot of space in Homeric society, and doubtless contributes to society's being held fast in monotonous and very limited ideas of, and standards for, what men and women can do.

Thus two contrasting gender ideals face each other, based on separate, but existentially dependant gender domains. The male aristocrat is portrayed as a martial adventurer and cosmopolitan, while the female aristocrat is described as peaceful and domestic in everything she does. There is, however, a clear asymmetry, as men exercise authority over women; actually, women are reified. There is a great degree of accordance between ideals and reality in the gender area, indicating that power in this case is not contested. To a very large extent, Homer's men and women act – like ourselves – in accordance with the stereotypes of culture, also in fields where, strictly speaking, they need not. Thus society and its culturally determined ideas of gender roles are apparently reproduced with no opposition, and these repetitive patterns are legitimated through different types of cultural consumption, including personal appearance and material culture.

From Homer to archaeology

Homer's epics demonstrate a significant point for archaeology, namely that 'things', especially equipment for war, can have extraordinary value because of the tales they have attracted during their life cycle. By utilising such a memorable object, the actor is enriched by the attached history, simultaneously ensuring a continuation of the tales. It is reassuring to see that in this particular elitist setting material culture really makes culture material, not only by reproducing the hierarchical social order, but also – though to a more limited extent – by being utilised in strategies that question the same order. Through their visibility material things are in themselves powerful by influencing in turn the actions and thoughts of human actors. However, archaeology can gain further assistance from Homer's epics.

Considered as a narrative – note the parallel to the Homeric epics – the burial finds of the illiterate Bronze Age societies in temperate Europe, almost regardless of time and place, mainly tell of the social élite. It is difficult to separate ideal from reality in the archaeological burial material, and often it is not possible to do so from the graves alone, which must be seen as *presentations* of gender, war and society. Certain aspects can possibly be misrepresented consciously or unconsciously, actually in the same way and for the same reasons as they were by Homer. War can be more than status rivalry, and

peace can also be a part of the image. Likewise, gender ideals can seem clearer than everyday gender identities, and monumental graves can be an attempt to legitimise hierarchies which are not generally accepted. It is, of course, best to include other archaeological sources into a 'complete' context, wherever this is possible. To examine structures of duration and change over a long period of time is also a way of evaluating ideals compared to realities in the illiterate past.

Warfare, male companionship, overt material and immaterial rivalry, and opposing gender ideals and identities are perhaps universal components of elitist culture, whereas the constitution of power between coercion and persuasion and between cooperation and resistance varies from case to case. The archaeologically most interesting thing about Homeric society is the way its various constituents are wholly integrated: how it is virtually impossible to understand one element of social practice without considering other elements, and how social organisation and ideology exhibit various levels of agreement and disagreement depending on the extent to which power is questioned, i.e. whether or not power is in need of being legitimised. It is, for instance, easy from a superficial reading of the text to assume that society is based on the outcome of overt status rivalry rather than on other less manipulative mechanisms, or that warfare is predominantly enacted as honourable sporting duels between aristocratic champions. Only a more in-depth analysis reveals that violent and cruel acts of warfare are equally important constituents of society.

My main point is here that superficial interpretations of for instance Bronze Age burials would lead to similar mistakes.

Concluding reflections on ideology and action

Ideology has played a key role in much recent archaeology and furthermore intersects inevitably with the issue of warfare and warriors. The concept of ideology is therefore discussed briefly as a concluding remark to the issue of Homer and archaeology and particularly in relation to the interplay of social action and structure highlighted by modern sociology. A central problem is most clearly revealed when looking at the constitution of Homeric society. Roughly analogous to the situation in the earliest

Scandinavian Bronze Age Homer's epics portray a society in which significant social fields related to kingship, politics and alliance, subsistence and exchange, and indeed warfare, are affected by a discord between what might be termed an egalitarian ideology and an institutionalised hierarchy. The egalitarian ideology could, alternatively, be regarded as a structural residual from a previous social order, still being active in certain situations of power enactment and therefore part of the social system as such.

The disharmony in Homer's epics between, on the one hand, actions and discourses of an egalitarian nature and, on the other hand, actions and discourses in tight accordance with a fairly rigid social hierarchy bring forth the difficult discussion of the position of ideology in society. Homer, in other words, highlights the impossible choice between ideology and reality. The general relationship between burial and society has evoked much discussion in more recent archaeology due to the high proportion in this discipline of sources to ritual practices, even resulting in a major research field labelled 'archaeology of death' (see Tarlow 1999: 1ff for a recent survey). The positions have tended to follow paradigmatic shifts.

Much post-processual archaeology has notably had a strong idealist tone in that ideology is believed to direct the social world, as opposed to a generally earlier materialist attitude that ideology and religion merely reflect the current mode of production. The question has been whether the burial record reflects social structure or ideology, but it is not unusual to find expressions that mix these two terms such as 'ideological reproduction of social structure' (e.g., Sørensen 2000: 85). The implies that we are supposedly not dealing with real social structure, but with ideology, or an ideologically distorted version of social structure – even cases of 'false consciousness'. It also illustrates the ambiguity of the whole matter. In Bourdieu's work a similar division is re-found in what he calls a double reality between what people say and what they actually do – a recurring theme in his studies of the Algerian Kabyle (e.g. Bourdieu 1998: 92ff). The entire discussion about the 'double truth' as one between saying and doing – between discourse and action – is definitely difficult to transfer directly to archaeology, not least because of the material and fragmentary nature of the sources. Bourdieu, in contrast to Giddens, emphasises the ideological component of social structure probably because of his schooling in ethnographic fieldwork.

The structurationist approach possibly offers a key to the solution of this problem, namely its basic assumption of a double existence of structure and agency. Ideology, says Giddens, cannot be conceptualised independently of action, structure and discourse, and refers only to circumstances of asymmetries in the systems of domination, notably when sectional interests are being signified and legitimised (Giddens 1984: 33). Structuration theory is weakly developed at this point, but it may follow that action can be more or less loaded with ideological thinking. It may prove useful, and this makes sense also in the Homeric context presented above, to distinguish roughly between two types of action: a habitual, culture-bound, and almost objective kind of action, which is accepted and carried out as 'the way things should be done' without questioning as opposed to a strategic, political and ideological kind of action, which is constantly negotiated and which continuously questions the social conditions of action[9]. Both kinds of action are of course signified, for example through material culture, but their meaning is different.

Some funerary activities in prehistory obviously possess an ideological tone rooted in strategic thinking, whereas other funerary activities appear much more habitual and culture-bound, and this is also valid for activities in other domains; ritual, domestic or otherwise. Strategic, political and ideologically-influenced action and interaction seem especially frequent during shifts in systems of domination – thus agreeing with Giddens' second dictum about the nature of ideology. Both types of action are, however, socially embedded in that that they are simultaneously preconditioned by, and produce, social structure. Homer highlights this point, which in the study of the longue durées and conjunctures of prehistory gains a long-term perspective. What we can see in European prehistory is, in fact, time-space patterns of human action – more often continuous and reproductive than changeable and strategic, but no matter how much or little motivated all these practices involve past, present and future social structures.

Acknowledgement

An early draft of this paper was translated into English by Patricia Lunddal, Faculty of Arts at the University of Aarhus.

NOTES

1 For quotes I have used A.T. Murray's translations of the *Odyssey* (revised edition 1995) and the *Iliad* (1946). I have, however, consulted Otto Steen Due's new translation of the *Iliad* into Danish (1999a; 1999b), and I have assessed the Greek text myself. Murray's translation of the Homeric *laoì* has been changed from 'people' to 'warriors', since Homer uses laós-laoì with this meaning, synonymous with the aristocratic upper class of society. Likewise in the passage with the boar's tusk helmet, I have made small corrections to Murray's original translation. This is in keeping with Due's new translations as well as Wilster's (1979a; 1979b) old translations of the poems into Danish.

2 The historical position of Homeric society is much debated. My view is close to Moses Finley's, namely that Homeric society has its basis in 10th-9th century Greece after the fall of the complex Mycenaean state societies but prior to the formation of the early Greek polis in the 8th century BC. This is an intermediate phase, which seems to have features in common with the chiefdom type of society described in ethnography and social anthropology. The poems were probably written down late in the 8th century BC, in connection with or after the invention of the alphabet (Morris 1987: 23f). However, 6th century BC Athens of the age of Peisistratos is also a possibility (see Finley 1972: 44ff).

3 Except the Linear A and B tablets, which are administrative lists of a redistributive political economy. The tablets completely lack the overriding character of the epics.

4 Some elements can, however, only be explained by assuming that the epics must preserve a memory of much earlier occurrences. The tower-shaped shields for instance go out of date after the Shaft Grave Period. Perhaps a continuous oral epic tradition existed, even dating back to early Mycenaean times of the 17th and 16th centuries BC.

5 Kurt Raaflaub (1997) has recently argued for the prevalence of a more egalitarian structure in the society of the poems. This is in contrast to Moses Finley's more elitist view, which is more in line with my own readings of the epics, especially the Iliad.

6 Michael Shanks (1993: 97) finds similar binary oppositions on a proto-Corinthian perfume bottle.

7 The lion is often used as a synonym of the war hero; e.g., Hector is a lion (Bloedow 1999).

8 Later Classical vase paintings show an advanced situation in which the citadel of Troy is being attacked: here the women no longer act passively, but defend themselves with kitchen implements (Lise Hannestad, pers. comm.). It also shows that function is dependent on context, and is a very mobile quality in material things.

9 I have discussed this classification of action with Torsten Kolind, who suggested it in the first place.

BIBLIOGRAPHY

- Ambatsis, I. and Ambatsis, J. 2000. 'Vore jag karl skulle du få se på hämnd! Om blodshämnd och fostbrödralag i Grekland'. *Medusa*, 3, pp. 19-28.

- Bennet, J. 1997. 'Homer and the Bronze Age'. In: I. Morris and B. Powell (eds.): *A New Companion to Homer*. Leiden, New York and Köln: Brill. pp. 511-33.

- Bloedow, E.F. 1999. '"Hector is a lion": New light on warfare in the Aegean Bronze Age from the Homeric Simele'. In: R. Laffineur (ed.): *Polemos. Le ontexte guerrier en Égée à l'age du Bronze. Actes de la 7e Rencontres egéenne internationale Université de Liège, 14-17 avril 1998. Université de Liège*. Université de Liège, Liège and University of Texas, Austin, pp. 285-293.

- Borchhardt, J. 1972. *Homerische Helme. Helmformen der Ägäis in ihren Beziehungen zu orientalischen und europäischen Helmen in der Bronze- und frühen Eisenzeit*. Mainz am Rhein: Verlag Philipp von Zabern.

- Bourdieu, P. 1998 (1994). *Practical Reason. On the Theory of Action*. Cambridge: Polity Press.

- Chadwick, J. 1976. *The Mycenaean World*. Cambridge: Cambridge University Press.

- Davis, J.L. and Bennet, J. 1999. 'Making Mycenaeans: warfare, territorial expansion, and representations of the Other in the Pylian Kingdom'. In: R. Laffineur (ed.): *Polemos. Le contexte guerrier en Égée à l'age du bronze. Actes de la 7e Rencontres egéenne internationale Université de Liège, 14-17 avril 1998. Université de Liège*. Université de Liège, Liège and University of Texas, Austin, pp. 105-118.

- Dickinson, O. 1994. *The Aegean Bronze Age*. Cambridge: Cambridge University Press.

- Due, O.S. 1999a. *Homers Iliade på dansk*. Copenhagen: Gyldendal.

- Due, O.S. 1999b. 'Iliaden - krig og digt'. *Sfinx*, 22(3), pp. 129-33.

- Ehrenreich, B. 1997. *Blood Rites. Origins and History of the Passions of War*. New York: Metropolitan Books.

- Finley, M.I. 1972. *The World of Odysseus*. London: Pelican.

- Finley, M.I. 1973. *The Ancient Economy*. London: Chatto and Windus.

- Foucault, M. 1985. *The Archaeology of Knowledge*. London: Tavistock Publications.

- Frankenstein, S. and Rowlands, M. 1978. 'The internal structure and regional context of Early Iron Age Society in South-West Germany'. *Bulletin of the Institute of Archaeology*, London, 15, pp 73-112.

- Giddens, A. 1984. *The Constitution of Society. Outline of the Theory of Structuration*. Cambridge: Polity Press.

- Hedeager, L. 1990. *Danmarks Jernalder. Mellem Stamme og Stat*. Aarhus: Aarhus Universitetsforlag.

- Jørgensen, A.N. 1999. *Waffen und Gräber. Typologische und chronologische Studien zu skandinavischen Waffengräbern 520/530 bis 900 n.Chr*. Copenhagen: Det Kongelige Nordiske Oldskriftselskab.

- Lorimer, H.L. 1950. *Homer and the Monuments*. London: Macmillan and Co. LTD.
- Mallory, J. 1989. *In Search of the Indo-Europeans. Language, Archaeology and Myth*. London: Thames and Hodson.
- Morris, I. 1987. *Burial and Ancient Society. The Rise of the Greek City-state*. Cambridge: Cambridge University Press.
- Morris, I. and Powell, B. (eds.) 1997. *A New Companion to Homer*. Leiden, New York and Köln: Brill.
- Murray, A.T. 1946. *The Iliad*. Cambridge (Mass.): Harvard University Press.
- Murray, A.T. 1995. *The Odyssey*. Revised edition. Cambridge (Mass.): Harvard University Press.
- Page, D.L. 1959. *History and the Homeric Iliad*. Berkeley: University of California Press.
- Raaflaub, K.A. 1997. 'Homeric society'. In: I. Morris and B. Powell (eds.): *A New Companion to Homer*. Leiden, New York and Köln: Brill. pp. 624-48.
- Ravn, M. 1993. 'Analogy in Danish Prehistoric studies'. *Norwegian Archaeological Review*, 26(2), pp. 59-90.
- Rowlands, M. 1980. 'Kinship, alliance and exchange in the European Bronze Age'. In: J. Barrett and R.J. Bradley (eds.): *Settlement and Society in the British Later Bronze Age*. (British Archaeological Reports, 83). Oxford: British Archaeological Reports, pp. 15-55.
- Runciman, W.G. 1999. 'Greek Hoplites, warrior culture, and indirect bias'. *Journal of the Royal Anthropological Institute* (N.S.), 4, pp. 731-51.
- Shanks, M. 1993. 'Style and the design of a perfume jar from an Archaic Greek city state'. *Journal of European Archaeology*, 1(1), pp. 77-106.
- Snodgrass, A.M. 1964. *Early Greek Armour and Weapons*. Edinburgh: Edinburgh University Press.
- Steuer, H. 1982. *Frühgeschichtlichen Sozialstrukturen in Mitteleuropa*. Göttingen: Vandenhoeck und Ruprecht.
- Sørensen, M.L.S. 2000. *Gender Archaeology*. Cambridge: Polity Press.
- Tarlow, S. 1999. *Bereavement and Commemoration. An Archaeology of Mortality*. Oxford and Malden: Blackwell Publishers Ltd.
- Vandkilde, H. 2000. 'Material culture and Scandinavian archaeology: A review of the concepts of form, function, and context'. In: D. Olausson and H. Vandkilde (eds.): *Form, Function and Context. Material culture studies in Scandinavian archaeology*. (Acta Archaeologica Lundensia. Series in 8.o. 31). Lund: Almqvist and Wiksell International, pp. 3-49.
- Vandkilde, H. 2003. 'Commemorative tales: archaeological responses to modern myth, politics, and war'. *World Archaeology*, 35(1). pp. 126-44.
- Vandkilde, H. and Bertelsen K.B. 2000. 'Krig og samfund i et arkæologisk og socialantropologisk perspektiv'. In: O. Høiris, H.J. Madsen, T. Madsen and J. Vellev (eds.): *Menneskelivets Mangfoldighed. Arkæologisk og Antropologisk Forskning på Moesgård. Moesgårds Jubilæumsskrift*. Aarhus: Jysk Arkæologisk Selskabs Skrifter. pp.115-26.
- Wees, H. van 1992. *Statuswarriors. Warfare in Homer and History*. Amsterdam: J.C. Greben Publishers.
- Wees, H. van 1997. 'Homeric Warfare'. In: I. Morris and B. Powel (eds.): *A New Companion to Homer*. Leiden, New York and Köln: Brill, pp.668-93.
- Whitley, J. 1995. 'Tomb cult and hero cult. The uses of the past in Archaic Greece'. I: N. Spencer (ed.): *Time, Tradition and Society in Greek Archaeology. Bridging the 'Great Divide'*. London and New York: Routledge.
- Wilster, C. 1979a (1836). *Homers Iliade* (oversat til dansk). Copenhagen: Museum Tusculanums Forlag.
- Wilster, C. 1979b (1837). *Homers Odyssee* (oversat til dansk). Copenhagen: Museum Tusculanums Forlag.
- Wylie, A. 1985. 'The reaction against analogy'. *Advances of Archaeological Method and Theory*, 8, pp. 63-111.

Index

Ardennes 287-288, 303
Arete, Queen 524
Argentoratum (Strasbourg) 231
Argos 519
Ariovistus 230-231, 235
aristocracy 40-41, 64, 154, 400-
402, 503, 518-519, 522-524
Aristotle 38, 461
Arklöf, Jackie 429
Armageddon 426
armament, renewal process in
495
Arminius 231, 235
armour 60, 72, 295, 298-299,
314-315, 318, 483-484, 490,
504, 508, 511, 528
arms race 107, 123, 133, 159
arms situation, variation 501
arms technology, changes in
267
arms, imported and adopted
495
army, acted as state police force
220
army, cohesion of 92
army, standing 40, 249
army, used to protect trade
220, 225
Aron, R. 53
Arrian 38
arrowhead 58, 144-145, 147,
150, 152-153, 338, 359, 365,
367, 404, 494, 506
arrowhead, armour-piercing 41,
53
arrowhead, leaf-shaped 144,
149-150
arrowheads, antler 361
arrowheads, barbed-and-tanged
152-153
arrowheads, bifacial 404
arrowheads, bone 150
arrowheads, bronze 361
arrowheads, flint 404, 494
arrowheads, made of lithic
material 361
arrowheads, oblong triangular
351, 361
arrowheads, obsidian 404
arrowheads, pressure-flaked 65,
404, 518
arrowheads, short heart-shaped
351, 361-362
arrowheads, stone 150, 358,
361
arrowheads, transverse 404
Artayctus 296

Artemis 523
Ascott-under-Wychwood,
England 144-145
Ashanti 264
Ashgrove, Fife 412
Ashton Keynes 337, 340
Asia 43, 47, 51, 142, 202, 221,
226, 283
Asia Minor 283
Asia, Central 51
Assur 44-45
Assyria 43, 54
Aswan 284
Athene Nikephoros 298
Athens 41, 71, 304, 527
Attalid 298
Augustine 38-41, 55
Augustus, Roman Emperor 285
Aulnat, Auvergne 300
Aunjetitz culture, late 279
Auschwitz 426
Australia 88, 132, 137, 139,
202, 242, 458
Australian Museum 205
Australopithecus 305
Austria 143, 278-279, 302, 332,
346, 360, 371, 373, 417, 509-
510
Austria, Lower 279, 371, 373
authority 26, 40, 46, 95, 98,
110, 133, 149, 192, 197, 203,
222-223, 239-241, 247, 256,
264, 266, 268-269, 356, 372,
394, 427, 435, 440-442, 444,
477, 485, 517-519, 524-525
authority, central 239
Auxilia 229, 231, 236
Avatip, Sepik 69, 132, 385-386
awl, copper 359, 418
axe, bone 357, 406-407
axe, bronze cult 493
axe, copper 407
axe, flanged 493, 506
axe, polished 353, 362, 404
axe, shaft hole 404-405, 407,
415, 493
axe, socketed 323, 493, 494
axe, stone 145-146, 169, 325,
353, 358, 362, 366, 404, 410,
484, 493
axe, t-shaped antler 409
axehead 415
Aztecs 223, 226

Baba Mountain 346
Babylon 43, 421

Bakassi Boys 269-270
Bakel, M. van 219
balance, establish/restore/
recreate 174-175, 179, 182
Balkan metaphor 460
Balkans 13, 16, 86, 105, 228,
389, 398, 404, 428-429, 448,
459-461, 464, 466-468, 472,
510, 523
Ballomarius 231
Ballymacaldrack, County
Antrim 153
Ballynagilly, County Tyrone
149
Baltic 494-495, 498
Baluan Island 187, 189-190,
193-196, 198
BaMbuti 115, 129
bands, nomadic 121
Banská Bystrica 346
Barash, D. 116
barbarian 215-216, 236, 276-
277, 282, 290, 300
barbarism 75-76, 294
Bard, K.A. 52
Barker, F. 86
Barnack barrow, England 151
Barnes, J. 83
Barraud, C. 33
Barrow Hills 145, 152, 159
barrow, chambered long 146
barrow, long 144-146
barrow, round 144-145, 152-
154
Barth, F. 125, 395
Barton, Cambridgeshire 292
Baruya 118, 120, 123, 135, 184
Basa 284
Basatanya 405-406, 408-409
Basel 205
Bašić, N. 429
Basque 476
Bates, M.J. 423
Bátora, J. 341, 344, 348-349,
362, 364, 366, 368, 370, 381
battle 127, 168, 173, 179-181,
196, 198, 206, 219, 230-231,
233, 244, 246-247, 249-252,
254-255, 279, 282-283, 285,
289, 296, 298, 304-309, 315-
318, 321-322, 326, 362, 364,
372, 387, 396-397, 423-426,
428, 430, 523
battle of Good Friday 307, 318
battle of Towton 306, 318
battle of Visby 306, 316-318
battle techniques 364

Cheyenne 115-116, 137, 320, 327, 329, 399

chief, embodiment of society 252

chief, Hawaiian, religiously sanctioned 253

chief, paramount 244, 251-254, 256, 401

chief, ritual 256, 326, 501

chief, warrior 326, 501

chiefdom 63, 89, 99, 101, 225-226, 238, 241-242, 246-249, 255, 257, 326, 527

chiefdom, complex 248, 257

chiefdoms, Polynesian 95, 523

chiefs 6, 64, 71, 81, 92, 100-101, 137, 142, 158, 162, 192, 218-222, 224, 226, 232, 237, 239, 241-249, 251-259, 264, 294-295, 320-321, 326, 397, 401, 500

chiefs, Fijian 92, 255

chiefs, Hawaiian 248, 254-255

chiefs, power of 245, 257

chieftain 64, 289, 292, 295, 297, 348, 356, 358

Chilbolton, Wessex 151

Childe, V.G. 27, 59-63, 66, 491

childhood 63, 146, 312, 377, 414-415, 417-418

Chimaltenango 471

Chimbu 182

China 51, 234, 236, 242, 248, 303, 461

chivalry, Christian 424

Chlopice-Veselé 343-344, 362, 382

Chnodomar, king 231

Christian II, King of Denmark 307

Christianity 39, 45, 184-185, 233, 255, 328

christianity, converting to 233, 255

chronology 160, 343, 353, 361, 363

Chute 144, 163

Cicero 38, 41

Cimbri 230, 294

Cimbrian 297

Circe 523

circumscription 52-54, 98-99, 219-223, 225, 240

cire perdu technique 494

citizen-armies 425

city state 42-43, 92, 214, 105, 214, 297, 398, 516-518, 528

city states, Greek 214, 297, 516-518

civil society 31, 39-40, 47, 428

civil War, English 155

Civilis, freedom-fighter 296

civilisation 10-11, 26, 28-29, 32, 34, 50, 64, 70, 81, 106, 233, 424-425, 441-442, 461, 464, 467, 470

civilisation, American 81

clades variana 231, 279

Claessen, H. 89

clans, agnatic 83

Clark, G. 63

class 60, 101, 147, 150, 220, 231, 237-239, 248-249, 312, 321, 348, 368, 371, 380, 396-397, 401, 427, 451, 471, 517-519, 527

Clastres, P. 106, 109, 133, 396, 403, 419, 448

Claudius, Roman Emperor 290

Clausewitz, C. 28, 55, 86, 391, 478

Cnip, Scotland 153

coexistence 390, 448-449, 451-452, 454-457, 459, 464-465, 467

Cohen, R. 217, 219-220, 223

Cohodas, M. 478

Cojtí Cuxil, D. 478

Cojtí Ren, A. 478

Colby, B. 473-474, 476

cold War 24-25, 47-49, 107, 261, 468

Coldrum, England 146

collective representations 25

collective tombs 278

collectivism 398

colonial change 204-205, 208

colonial histories 204

colonial histories, studying 204

colonial oppression 262, 472

colonial pacification 6, 187

colonial power 14, 17, 204, 208

colonialism 6, 66, 139, 191, 201, 203-205, 207-208, 257, 262-263, 266, 470, 472, 475, 479

colonialism, understanding war 201

colonialists 390, 469

colonisation 110, 188, 213, 228, 232, 402, 444

Čolović, I. 431

Colson, E. 121

Colt Hoare, R. 146, 156

combat 6-7, 80, 90, 114, 142, 151, 153-156, 167, 176, 192, 234, 253, 305-307, 309, 311, 313-315, 317, 319, 321, 331, 336, 338, 360-362, 364, 368, 372, 423, 425-426, 428-430, 486, 488, 498, 506-507, 511

combat victim 338

combat, close/face-to-face 114, 253, 314, 319, 322, 360-362, 364, 372

Commodus 231

communism 47, 82, 451, 459

companionship 68, 399-401, 415, 419, 488, 521, 525

comparative procedure 417

compensation 93, 115, 123, 127, 132, 174, 177, 179, 264, 281, 285

compensation payment 132

competition 17, 26, 43, 60, 67, 77, 116-118, 132, 141-143, 153, 170, 181, 184, 195-198, 222, 241-242, 262, 267-268, 371, 400, 428, 519

composite bow 357, 412, 418

concepts of culture and history, defined by outside powers 390, 476

conceptual framework 13, 28

conflict 6, 12, 24, 29, 31, 35, 39-40, 42-44, 46-54, 60, 63, 67, 71, 73, 77, 83, 90, 93, 101, 113-115, 118-121, 127-130, 134-139, 141-144, 148-156, 158, 163, 168, 173-174, 176-178, 182, 184, 187, 195, 197, 212-213, 217-219, 222-223, 225, 228-229, 238, 241, 257, 261-271, 301-302, 320, 322, 326-327, 329, 360, 370, 386, 421, 426, 430-431, 434-435, 438, 441, 443, 445-446, 464, 466-467, 472, 476

conflict management 93, 119, 136

conflict resolution 115, 121, 262-263, 266-267

conflict, characterising social interaction 115

conflict, civilisational 441

conflict, factors behind 212

conflict, global 47-48

conflict, ideological 24, 47-48

conflict, internal 114, 173, 182, 370

conflict, national 441

dagger, Výčapy-Opatovce type 352, 363

dagger, willow-leaf 352, 359, 362-364

daggers, Arreton series 506

daggers, associated w. males 150, 153

daggers, inappropriate as weapons 153

daggers, use in hunting 394, 507

Dahomey 223, 225

Daloz, J.-P. 264

Damm, C. 60

Danebury, Hampshire 282, 286

Danegeld 232

Dani 114-115, 118, 133, 136

Danube River 231, 342, 404, 417, 510

Dark Age 495

Dartmoor 157, 161, 336, 339

Darvill, T.C. 145

Darwin, C. 31

darwinist 32, 133, 425

Das, V. 465

Dawson, D. 142

De Coppet, D. 33

de Waal, F. 32-33

death house 346, 348-349, 355-356

death, attitude towards 516

death, cause of 144, 152-154, 156, 313, 325, 376, 405-406, 488

decapitation 153, 285, 295, 298-299, 308, 322

Decebalus, king 300

decentrality 105

defeat 39, 41, 45-48, 50-51, 54, 92, 99, 113, 120, 123-125, 131, 230, 243, 247, 262, 282-284, 290, 294-295, 298, 301

defenders 150, 244-245, 278-279

defensive equipment 339

defensive means, absence of 502

dehumanisation 281-282

Demarest, A. 470-471

Denmark 4, 6, 11, 63, 65, 71-73, 226, 297, 300, 305-307, 310-312, 317-318, 321, 338, 420, 464, 490, 492, 496, 498, 501, 503-504, 508-509, 513

dental attrition 310-311, 325

Dentan, R. 76, 130

dentition 312

dependency and fellowship 401

depopulation, different reasons 229

deposition 7, 151, 266, 287-288, 298-299, 304, 332, 336-337, 339, 485, 493, 498, 500, 505-511, 513

deposition in wet places 511

deposition of bodies 288, 337

deposition rules 498

deposition, context of 505-506, 509

deposits, ritual 148, 485-486, 489

deposits, secondary 328

Derventa 441

desecration 328, 360

desecration, of war heroes 328

Deskford 299

Didmarton 337

difference, marker of 391

differences, familiarisation of 452, 459, 464

differences, narcissism of minor 386, 447

differential reception 412

dimorphism, sexual 414

Dinka 126, 263, 265

Dinorben, Wales 158, 161

Diomedes 519-520

Dionysius, tyrant of Syracuse 230

Dionysus 296

Dirks 155, 160, 510

discourse 7, 10-11, 15, 18, 57, 62, 77-79, 85-86, 134, 214, 261-263, 265-267, 287, 328, 383, 385-391, 393-398, 400, 402-404, 406, 408, 410, 412, 414, 416, 418, 420, 422, 424, 426, 428, 430, 434-446, 448, 450, 452, 454, 456, 458, 460-462, 464, 466, 468, 470, 472, 474, 476, 478, 484, 515-516, 526

discourse and action 526

discourse and agency 388

discourse and everyday life 389, 439

discourse as signification 396, 403

discourse, dominant 390

discourses, authoritative 389, 437-441, 443-444

disease 136, 163, 203, 207-208, 305, 312, 318, 328-329, 339, 402, 405, 469, 503

disorder 29, 68, 79, 83, 94, 264, 270

displacement 183, 346, 433, 437, 442-443, 473, 476

dispossessed, voice of 477

diversity 134-135, 165, 212, 262, 264, 267, 299, 457

Dizi 266-268, 270

DNA analysis 279

Dobu 78, 81-82, 86

Domesday Book 337

domestic production, control of 371

domination and coercion 402

Donegore Hill, County Antrim 148

Dorchester 151, 164-165, 338, 340, 492

Dorchester-on-Thames, Oxfordshire 156

Dorset 147, 159, 161, 163-165, 337, 339-340

doves 27, 66-67, 115, 117, 127

DR Congo 262

Drekete 246

dress accessories 414, 417-418, 485

dress finery 409-410, 413-414

Drewett, P. 149

drilling 92

Drumman More Lake, County Armagh 156

Dubrave 462

Due, O.S. 516, 527

Duggleby Howe, Yorkshire 145, 150

Duke of Yorks 203

Dumas, Alexandre 425

Dumont, L. 25, 30, 33

Durham, W. 132

Durkheim, E. 31, 35, 465

durkheimians 30-31

Earle, T. 220, 223

earthen mound 410

East Anglia 157

East Germany 498

ecological crises 63, 213

economic structures, different 228

economy, domination of the 224

economy, shift in 370-371

egalitarian 6, 16-17, 25-26, 59, 63, 98, 108, 111, 122, 133, 167-171, 173, 175, 177, 179-

183, 185, 221, 240, 243, 321, 402, 518-519, 526-527

Egypt 41-43, 51-52, 54-55, 284, 295-296, 301-303

Egypt, New Kingdom 284, 295

Egypt, Pharaonic 302

Egyptians 41, 43-44, 424

Eibl-Eibesfeld, I. 31

Ekagi-me 120

Elbe River 229-232, 234, 509

Eleutherna, Crete 285

elite 13, 15, 18, 61, 64, 237, 242, 254, 256, 262, 269, 306, 326-327, 346, 355-358, 360, 368, 370-371, 394, 399-400, 411, 414-415, 417-419, 428, 473, 485, 507, 511, 517-519, 521, 524-525

elite's point of view 524

elitism 396

elitist culture, universal components of 525

Elysium 423

Ember, C. 76, 120

Ember, M. 120

emigration 228-229, 231

empire 38, 42-44, 95, 105, 135, 192, 212, 215, 223, 228-229, 231-235, 264, 295, 402, 470

enclosure, causewayed 143-151, 507

enclosure, ceremonial 149

enclosure, fortified 148

enclosure, hilltop 148

enclosure, palisade(d) 149, 151, 409

enclosure, ringwork 157

enclosure, ritual 149

enclosure, unfortified 148

enclosures, main role of 157

endogamy 415

enemy heads, taking of/ head-taking 298, 300

eneolithic 341, 344, 348, 361, 364, 507

Enga Province 184

Engels, F. 31, 402

England 47, 135, 143-146, 148-161, 163-164, 184-185, 229, 232, 234, 279-280, 282, 286-289, 291, 299, 331, 339, 362, 423

England, central 145, 151, 157

England, central-southern 289, 299

England, eastern 144, 148-149, 151, 153-154

England, northern 144-145, 151-153, 157-158

England, northwest 153

England, southeast 146, 148, 151, 155-158

England, southwest 143, 155

England, western 144, 146, 150, 153, 158

enlightenment 29, 31, 263, 424

Entremont, southern France 297, 300

environment 17, 26, 28, 32, 55, 83, 85, 91, 94, 107, 116-118, 120-122, 125, 129-130, 133, 135-137, 141-142, 161, 164, 180, 185, 187, 239, 279, 335, 382, 439, 451, 512

environment, anarchic 120

environment, natural 32

environment, warlike 28, 83, 116-118, 120, 122, 125, 129, 133, 187

equality 33, 62, 129, 169-172, 179-180, 182-185, 265, 397, 401, 415, 424, 430, 452, 477, 518-519, 524

equality, ideology of 415, 519, 524

equals, first among 401, 518-519

eques 231

Erskine, J.E. 257

Ertebølle 63, 143

essentialisation 469, 477

Essex 148, 159-162, 303

Ethiopia 213, 264, 266-268, 270-271

ethnic cleansing 282, 302, 389, 443, 446, 456, 465, 467

ethnic identification 389

ethnic nationalism 388-389

ethnic resurgence 475, 479

ethnicity 18, 139, 236, 270, 395, 415, 420, 445, 449, 451, 457-458, 464, 466-467, 479, 485

ethnogenesis 227-229, 231-232, 234, 236

ethnohistorical perspective 6, 167

ethnohistory 72, 111, 135, 162, 329, 403, 479

Etton 149, 163

Eudusii 230

Euphrantides 296

Eurasian steppe 62

Europe 6-7, 13, 16, 18, 24, 27, 38, 44-47, 59, 62-65, 71-73, 77-78, 82, 93-94, 105, 108, 139, 142-143, 145, 150, 153, 155, 160-164, 211-212, 214-216, 227, 230, 234-237, 242, 257, 259, 262-263, 268-269, 276, 278-280, 285-286, 289, 293, 295, 298, 300-301, 303-304, 306, 329, 331, 337-340, 344, 346, 361, 363, 369, 372, 381-382, 386-387, 389, 391, 393-395, 397-399, 401-405, 407-413, 415, 417-421, 427-428, 448, 453, 459-461, 464-465, 467-468, 485-487, 489-490, 492-493, 495, 502-504, 507-513, 515-516, 525

Europe, central 15-16, 59, 105, 150, 227, 230, 233, 278, 337, 344, 346, 361, 363, 372, 381, 387, 393, 404, 408-410, 412, 417-418, 485, 487, 492, 494-495, 508, 510, 516

Europe, northern 105, 143, 296, 298, 339, 387, 401-402, 410-411, 460, 487, 489

Europe, protohistoric 215-216

Evens, T. 126

everyday life 17, 68, 72, 243, 320, 389-390, 411, 431, 437, 439, 443-444, 446-448, 451-452, 457, 459, 467, 478

everyday life, reconquering 443-444

evolutionary advantage 26

evolutionary biology 35

evolutionary economics 133

evolutionary psychology 31, 35-36, 141, 158, 160

evolutionary theory 27, 135, 137, 141, 160

evolutionism 84, 101, 226, 257

excavation 147-149, 151, 154, 159-164, 258, 303, 307, 313, 323, 328-329, 331, 334-336, 339, 373, 502

exchange 6, 14, 17, 25-27, 29-31, 33-36, 64, 68, 73, 77-81, 83-84, 86-87, 93, 95, 98-99, 108, 110, 114-115, 118-120, 123-125, 130, 132, 137, 139, 149-150, 168-174, 176, 178-182, 184-185, 187-193, 195-199, 201-202, 204-208, 215, 221-222, 241, 255-256, 263, 329, 386, 410, 502, 520, 526, 528

Gefolgschaft 15, 64, 214, 216, 223, 228, 229, 397, 400, 403, 521

Gelduba/Krefeld 279

gender 7, 14-15, 18, 62, 69, 72, 82, 87, 111, 124, 162-163, 199, 207, 225, 268, 296, 298, 303, 310, 321, 346, 387, 395-397, 399, 404-406, 409-411, 413-415, 417-421, 423-424, 430-431, 451, 467-468, 484-486, 489, 515, 517, 519, 521-525, 527-528

gender and rank 387, 413, 415, 417, 419

gender differentiation 18, 404-405, 409-410, 414, 418, 420

gender distinction 418

gender equality 424, 430

gender stereotype 410, 419, 522

gender, biological sex 414

gender, social 484

gendered life course 405, 421

gendered objects 410

gender-specific position 410

genocide 64, 67, 282, 302, 402, 445, 466, 479

Gent, G. 334

Georgieva, C. 452

Gerloff, S. 506

German New Guinea 202, 208

German Wars of Liberation 425

Germany 46-47, 71, 75, 88, 143, 162, 202, 208, 230, 232, 234-235, 263, 270, 289, 299, 303, 306, 338, 346, 362, 368, 421-422, 435, 461, 496, 498, 506-508, 510, 527

Germany, eastern 506-507

Germany, north 289, 508-509

Germany, southern 496

Geronimo 98, 240, 403, 419

Gerumsberg 493

Gerzon, M. 429

Gettysburg 426, 428

Gibson, A. 152

Gibson, J.W. 428

Giddens, A. 90, 94, 96, 397, 478, 484, 526

Giddings, R. 173

gift 5, 25, 29-31, 33-34, 36, 64, 68, 77-81, 83-85, 87, 114, 118-120, 124-125, 132, 190, 196, 199, 291, 448, 466

gift exchange 30, 33, 64, 68, 79-81, 83-84, 114, 118-120, 124-125, 132

gift, means of social control 81

gift, pacifying 80

gifts to the gods 394, 408, 486

Gilgamesh 7, 387, 423-425, 427, 429, 431

Gimbutas, M. 62-63, 66

gladiator 428

Glenquickan, Scotland 145

Gleser, G.C. 311

Glob, P.V. 60, 62

Gloucestershire 12, 147, 159, 164, 331, 337, 340

goddesses 38, 519, 523

Godeffroy und Sohn 202

Godelier, M. 120

gods 24, 37, 41, 44, 46, 54-55, 71, 78, 160, 244, 246, 251-254, 258, 285, 289, 296, 298-299, 303, 394, 408, 466, 474, 479, 486, 488, 519, 523

Gog Magog Hills, England 153

Gombe Reserve 33

Gommern, Saxony-Anhalt 229

Goodale, J. 207

Gordon, R. 119

Goths 215

Gotland 306, 310-311, 317, 494

Gournay-sur-Aronde (Gournay) 234, 276, 299-300

grave robbing/robberies 360, 370

Gradić, Bosnia 441

Graeco-Roman 286, 298, 300

gramscian 436

Grand Chaco 16, 110, 401-402, 415, 487

Grandtully, Scotland 152

Granhammar, Sweden 492

grave finds, rank interpretation of 501

grave, female 310

grave-goods 153, 159, 229, 287, 289

grave-goods, absence of 287

graves w. weapons, interpretation of 216, 502

graves, best documented contexts 500

graves, chiefs' 500

graves, full-length w. swords 498

graves, megalith 306

graves, plundered 358

graves, spatial arrangement of 348

graves, warriors' 358

Gray, C.H. 429

great men 120, 135

Great Plains tribes 399

great-ape societies 26

Greece 24, 44-46, 51, 214, 235, 401, 490, 505, 511, 516-518, 523, 527

Greece/Greek, ancient 24, 45-46, 281-282, 284-285, 289, 296-297, 401, 505, 516-518, 527

Green, H.S. 152

Gregor, T. 129-130

Gregory of Tours 227

Greuel, P. 126

Greyhound Yard, Dorchester 151, 165

Grinsell, L.V. 156

group for itself 396, 404

group in itself 396

group size 123, 125, 127, 133

grove deity 290

Guantanamo Bay 281

Guatemala 390, 468-469, 471-473, 475-476, 478-479

Guatemala City 475

Guatemalan War 472

guerrillas 427, 473-475, 479

Guichard, V. 292

guilt 427, 443, 473-474

Guinea 5-6, 12, 14, 16-17, 27, 68-69, 73, 75-88, 93, 101, 107-108, 110, 113-114, 118-120, 124-125, 128, 135-139, 167-169, 182-185, 187, 191, 198-199, 201-204, 206, 208, 261, 385-386, 397, 404, 466

Guiscard, Robert 233

Guiscard, Roger 233

Gunawardana, L. 220

Gundomad 231

guns see firearms

Guttmann, E. 157

habitus 263, 266, 269, 415, 457

Habsburg empire 429

Haedui 230

Hagen 79, 87

Hagenbach 232

Hagesteijn, R. 221

Hahl, Governor 202-203

Haile Sellassie, Emperor 266

hair, liminal substance, treatment of 298

Hajduk tradition 429
Hajer, M. 388, 439
Halberd 153, 363, 506-508
Halberds, occurence in hoards 506-508
Hallpike, C. 121, 130, 223
Hallstatt Culture 502
Hambledon Hill, Wessex 144
Hamburg 136, 198, 232, 236, 271, 503
Hamilton, S. 158
hammer axe 353, 357, 408
hammer axe, small heart-shaped 362
hammer axe, w. facetted butt 362
hammer picks 362
hammer, grooved 362
hammers 361-362, 493
Hammurabi 43
Hamn, Sweden 287, 300
happy hunting grounds 423
Harald Klak 232
Harbison, P. 507
Harborough Rocks, Derbyshire 144
Harding, A. 154-156, 326-327
Harris, M. 118, 217
Harrison, L. 265
Harrison, S. 68-70, 77, 80-81, 85, 91-93, 122, 132-133, 183, 385-386, 391, 448
Hartlepool Bay 145, 164
Harudes 230
Hasanlu 279-280
Hassleben-Leuna 232, 236
Hastrup, K. 326
Hawaii 6, 13, 16, 95, 214-215, 237-238, 242, 248-249, 251-259, 401
Hawks 27, 66-67, 115, 117, 127
Haywood Cave, Somerset 145
Hazleton, England 144-145
head-hunting 75, 79, 82, 114
headman 114, 203, 320, 407
heads, severed 297-298, 300-301
healed fractures 145, 151, 360, 366, 377, 414
Hebrew 24, 40, 44-45
Hebrew Bible 24, 40, 45
Hector 527
Hedeager, L. 60-61, 66, 400
Hegel, G.W.F. 24, 39-41, 45-46, 50, 53-54
Hegemonic 436
Heidenreich, C.F. 218

Helbling, J. 69
Helena 523
Helman Tor, Cornwall 163
helmet 316, 486, 517, 520, 527
helmet's cultural identity 520
Hembury Enclosure 147
Henry VIII, King of England 318
Hequetai 401, 521-522
Hera 523
Herder, J.G. 263
Hernádkak, Hungary 338
hero worship, individualising 516
hero/heroine 60, 66, 70, 173, 175, 179, 285, 287, 321, 327-328, 423, 425, 427-429, 488, 515-524, 527
Heuneburg 494
hidden transcripts 391, 436, 442, 446
Hierakonopolis 284
hierarchies, weak 108, 182
hierarchisation 214, 220
hierarchy 16, 94, 97-98, 106, 108-109, 114, 167, 183-184, 198, 214, 217, 222, 224, 240-241, 246, 251-252, 254-256, 264, 284, 395, 400, 402-403, 427, 477, 498, 515-516, 518-522, 524, 526
hierarchy, predetermined 518
Hill, E. 283, 295
Hill, R. 472
Himmerland 494
Hindwell, Wales 151
hird 521
Hiroshima 426
historical development, individual factors in 214
history of war studies 9
history, end of 40
Hittites 43
Hjortspring boat 338
hoard deposit 498, 506, 509-510
hoard, sacrificial 408
hoards 70, 155, 319, 326-327, 394, 408, 486, 489, 495, 498-500, 506-512
hoards, interpretation of 500
Hobbes, T. 121, 218, 263, 337, 419
Hochschild, A. 293
Hodder, I. 68
Hoffmann, S. 46
Hog Cliff Hill, Wessex 157

Hohmichele 494
Holešov 342-344, 351, 353, 361-364, 370-372
Holland 492
holocaust 465-466
Homer 7, 14, 38, 41, 44-45, 48, 51, 214, 281, 285, 296, 303, 397, 400-401, 403, 424, 431, 483, 488-489, 502, 505, 515-517, 519, 521-528
Homer as analogous social context 516
Homer, gender in 522
Homer, material culture and society 519
Homeric poem/epic 43, 283, 397, 401, 403, 424, 488-489, 515-517, 521, 523-526
Homeric society 515-516, 518, 520, 523-525, 527-528
Homeric society, ideology and social organisation in 518
Homeric society, war and aristocratic warriors 520
Homer's epics, social and material world of 489
homicide 23, 90-91, 115, 176-178, 328, 487
Homo duplex 31, 33-35
Homo oeconomicus 31
Honduras 471
honour 15, 24, 38, 45, 49, 64, 158, 161, 173, 235, 243-244, 253, 281, 285, 355, 385, 387, 421, 424, 429-430, 504, 519-521, 523
hoplite 398
Horik I 232
horses 15, 110, 158, 228, 233, 299-300, 402, 410, 487
Horton Camp 337
hostage 288, 290, 474
household 170-172, 174, 180, 229, 247, 320, 451, 477, 522, 524
Howard, M. 77
Huldremose, Denmark 296
Huli 86, 133, 182, 184
human body 34, 282, 287, 295, 395, 484
human history 25, 42, 50, 89, 134, 237, 423-424
human nature 23-25, 29, 31-32, 34, 36, 39-40, 50, 52, 54, 141, 165, 184, 465
human nature, duality of 25
humanitarian aid 434

Hume, D. 45
humiliation 172, 174, 181, 247, 250, 265, 270, 276, 281-283, 285, 288, 290, 292-295, 298-299, 302, 462
humiliation, iconographies of 293-294
humiliation, ritualised, public 292, 302
Hung bodies 300
Hungarian Plain 370, 405
Hungary 338, 406-408, 508
Huns 215
hunters and gatherers 26, 76, 107, 115, 119, 121, 129-130, 132-133, 142, 218, 266, 424
hunting 30, 41-42, 57, 63, 68, 118-119, 130, 134, 137, 156, 168-169, 176, 185, 205-206, 320-321, 326, 344, 361, 394, 415, 423, 484-485, 493, 506-508
hunting and gathering 42, 63, 130, 137, 168-169
hunting, prestige activity 485, 507
Huntington, S. 265
Huron 218-219, 226
Hus, Manus 190
Hussein, Saddam 48
Hutu 228, 391, 447, 467
Hyksos 43
Hynes, S. 423, 426

Iban 114-115, 119, 124, 128, 136, 139
Iceland 326
iconographic presentations, art and rituals 488
iconography 57, 284-285, 287, 289, 292-296, 298, 301, 488
iconography, war- 57, 284, 292
ideals of aggressiveness 398
identification, localistic 451, 454
identities, ascriptions of social position 385
identities, creation of 386
identities, war-specific 385
identity 7, 11, 13-15, 17-18, 30, 34, 50-51, 59, 64-66, 69, 81, 85, 139, 142, 164, 196, 211-212, 215, 236, 247, 263, 266-271, 281-284, 287, 289, 295-296, 298, 301, 320, 362, 378, 383, 385-391, 393-398, 400, 402, 404, 406-408, 410, 412,

414-420, 422, 424-426, 428, 430, 434, 436, 438, 440, 442, 444-454, 456-458, 460-462, 464, 466-472, 474-478, 484-485, 507, 510-511, 515, 517-518, 520, 523-524
identity construction 472
identity politics 270, 467, 469-470
identity, American 81
identity, Bosnian Muslim in everyday practice 450
identity, collective 396
identity, cultural 415, 485
identity, denial of 281, 283, 287, 298
identity, ethnic 263, 320, 395, 448-450, 452, 458, 464
identity, Muslim national in Bosnia 449
identity, social 64, 395
ideologies, martial 515
ideology 28, 32-33, 47, 49-50, 61-62, 72-73, 80, 90, 94, 98, 101, 120, 122-123, 138, 164, 185, 221-222, 224, 226, 320, 390, 393-394, 398, 403, 415, 420, 425, 451, 457-458, 467, 471, 477, 488, 510, 515-516, 518-520, 524-526
ideology of the firstcomer's primacy 222
Idomeneus 521
Ilaga Dani 114, 118, 136
Iliad 14, 214, 281, 285, 303, 387, 397, 400, 424, 431, 502, 515-521, 523-524, 527-528
Illerup, Jutland 229, 234
Ilongot 119, 132, 138
imaging the captured 284
immigration 123, 127, 176, 230
imperialism 30, 47
imprisonment 282, 288, 298, 301
Incas 223
indebtedness 397, 399
Inderøy 280, 319, 322, 327, 329
indigenous people/populations 65, 67, 109, 119, 128, 130-131, 197, 402, 470
individual actor 67
individual graves 305
individual responsibility 437, 441
individual, rights and obligations of 24, 40, 48, 265
individualism 36, 396

individualist 33-34, 85
Indo-European 43, 64, 163
Indonesia 32, 36
Indus 43
inequality 33, 97, 99, 111, 171, 184-185, 262, 266, 403, 472
inference 493, 501
Ingelmark, B. 306
inhumation 152, 156, 279, 306, 322, 346, 489, 492-494, 499, 501-502, 511, 516
institutionalisation, signs of 395
institutions, durability or durée of 395, 397
institutions, egalitarian 16-17, 108, 167, 170-171, 180, 182-183
institutions, egalitarian, impact on war and exchange 167-168, 171, 180, 182
Insubres 230
insult 82, 173, 179, 196-197, 244, 284-285, 287, 293-294, 298-299, 301, 360
interactionist view 213, 263
international law 78, 86
international relations, theories of 113, 133
intruders 279, 487
Inuit 115, 129
Iran 42-43, 280
Iraq 43, 48-49, 51, 431, 454
Iregren, E. 414
Ireland 6, 107, 134, 141, 144-146, 148-150, 153-164, 198, 202, 232, 339, 478, 504, 507-509, 512
Iris 523
Iron Age 43-44, 65, 71, 147, 158, 161, 164, 212, 229, 234-235, 275-276, 282-283, 285-293, 295, 298-304, 306-307, 318, 322, 328, 336-338, 340, 395, 486, 491, 493, 500-501, 503-504, 507, 516, 527
Iron Age, early 275, 328, 395, 493, 516
Iron Age, late 283, 287, 291-293, 298, 300-302
Iron Age, middle 288
Iroquois 106, 123-124, 126, 218-219, 226, 419
irrigation 219-220, 226, 238, 240, 254
Isandra 224

Le Petit Chauvort, Verdun-sur-le-Doubs, Saône-et-Loire 292
leaders make war 214, 238
leaders, successful 233
leadership 26, 86, 97-98, 108-109, 114, 137, 168, 180, 182, 189, 192-193, 196, 222-224, 230-231, 233, 237-241, 248, 256, 261, 317, 397, 399-400, 402, 404, 477, 517, 521
leadership and prowess in battle 196
leadership, ideology of sacred 224
leadership, stories about 193-194, 196
Leki Male, Poland 507
Lengyel 409
lesions 275, 305-307, 312-318, 322, 325, 327, 406
lesions, bone 305, 307, 313-317, 322, 325, 406
lesions, skull 314, 316-317
Leubingen 507
Levant 41-42, 55, 142, 162
Leviathan 31, 36, 66, 71, 77, 86-87, 101, 111, 128, 135-136, 138, 339, 420
Lévi-Strauss, C. 25. 30, 120
Lewis, A.B. 205-206
liberation struggle 47
Liberia 261-262
Liddell, D.M. 147
Liffs Low, England 151
limes 215, 232
Lindow Moss, Cheshire 288
linear B text/tablets 283, 401, 516, 521-522
linear ditch segments 336
Linear Pottery culture 404
Linkardstown, County Carlow 145, 150
Linzgau 231
Lion temple, Musawwarat es Sufra 285
Lion temple, Naqa 284
lion, synonym of war hero 527
Lipan, Manus 193-194, 198
Little Poland 341, 349, 370
Littleton Drew 146
Livy 38, 300, 303
Lizot, J. 32, 116, 118
Llyn Cerrig Bach, Anglesey 283, 291-292
Locke, J. 263
Lockerby, W. 257
Lockwood, W. 451

Logue, P. 149
Long Island, PNG 175
Long Wittenham, Thames Valley 155
Longue durée 397, 424
Lord's Resistance Army 266
Lorengau, Manus 203
Lorenz, K. 31
Los Angeles 36, 185, 270, 421, 427-428, 431, 490
loss, sense of 453
Lou, Manus 190, 193, 196
Louis I, the Pious 232
Lowie, R. 402
loyalties, conflicting 130, 168
loyalty, oath-bound 401
Lugii 231
Luluai 203-204
Lusehøj 502, 504
Lusitani, Iberia 284
Lüneburg 494

Maasai 265
Macedonia 510
Macehead 387, 394, 406-407, 410, 413, 415, 417, 484
Maček, I. 451, 465
MacGaffey, W. 262
Machiavelli, N. 263
Machiavellian 33, 172, 179
Madagascar 51, 224, 226
Madang 203
Mad'arovce cemetery 353
Mad'arovce Culture 358
Mae Enga 69, 114, 118-120, 122-124, 126, 128, 132-133, 137, 184
Maiden Castle 148, 151, 164-165
Main River 230
Majawil Q'ij 475-477
Maji area 267-268
Malaita 119
Malay 192
Malaysia 32
male bonding 398, 410
male clubs, purposes of 398
male clubs/brotherhoods 387, 397-398, 403
male cult 68, 183
male domain 522, 524
male dominance 410-411, 414, 417, 420
male identification, crisis in 427
male point of view 68

male, boys 83, 91, 169, 172, 193, 221, 244, 269, 320, 348, 355, 387, 399, 405-406, 414, 424, 428-429
male, elderly 144, 146, 406
male, high-status, armed 414
male, low-status, unarmed 414
male/men, young 108, 123-124, 142, 144-145, 147, 151-153, 156, 158, 172, 180, 184, 192, 202, 285, 287-289, 296, 300-301, 310, 316-317, 331, 335, 366, 368, 388, 399, 408, 414, 423-426, 428-430, 485, 499, 506
malehood, adult 405
maleness 62, 208, 404, 415
maleness, warlike 62, 415
Malinowski, B. 27, 33, 75-84, 86
Malkki, L. 447
Malkovich, J. 438
Mallaha 41
Malmer, M. 63
Malmö 72, 428
Malnutrition 312, 325, 327, 377
Malo, D. 252
Malory, Sir Thomas 424
Malthus, T. 53
Mam Tor, England 157
Manacles 290-291
Manambu 122, 132
Mangan, J. 423
Mangyan 129-131
manhood, crisis in 427
Mani, Peloponnese 523
Manley, J. 158
Mann, M. 28, 52, 90, 93-99, 108, 239
Mansell Collection 293
Mantaro Valley, Peru 220
Mantel, E. 289
Manuai, Manus 193-195, 198
Manus Culture 190
Manus Province 110, 187, 198
Maori 73, 151, 401, 424
Marcomanni 227, 230-231
Marcomannic Wars 228, 231
Marcus Aurelius 231
Marine Corps (US) 423, 430
Maring 118-119, 125, 127, 136-137
Marne River 230
Maroboduus 228, 231
Mars 298, 523
Marshfield 337
Marsi 231

Marsk Stig Andersen Hvide   317
martial strain   398, 515
Marx, K.   31, 66, 221, 396, 477
Maschio, T.   207
masculine/masculinity   66, 82, 108, 388, 404, 410, 423, 425-430, 517, 522-523
masculinities, violent   428, 517
masculinity, ideal   426, 428
mass deportation   302
mass grave, Sandbjerg   6, 305, 316
mass grave   15, 18, 58, 143, 211, 215-216, 277-279, 305-310, 312, 316-318, 322, 337-338, 368-369, 372-373, 487, 502
massacre   6, 143, 161, 300, 304-307, 309, 311, 313-315, 317-318
Massilia   297
material culture   7, 11, 14-15, 18, 61, 69, 72-73, 204-206, 208, 232, 282-283, 285, 290-291, 295, 298, 301, 340-341, 343, 385-386, 394-395, 399, 412, 415, 417, 419, 481, 483-488, 490, 492, 494, 496, 498, 500, 502, 504, 506, 508, 510, 512, 515-516, 518-520, 522-526, 528
material culture as agent   483
material culture as source of data   483
material culture, ambiguity of 485
material culture, creates social relationships   484
material culture, creator of distinctions   484
material culture, durable   484
material culture, mediator   484
material culture, multifunctionality   485, 487
material culture, silent form of discourse   484
material destruction   433-434
material discourse   386-387, 403
material similarity   412
material transition   57
material, exotic   404, 409
materialist approach/interpretation   141-142
materialist attitude   526
materialist-functionalist approach   67
materiality   484, 489, 515

matrilineal   124, 356, 398
Matupi   202
Maui   248-251, 253
Mauss, M.   25, 29-31, 33-35, 78, 81-83, 119, 191, 196, 448
Maya code   470, 478
Maya culture related to invented tradition   390, 477
Maya Indians   390
Maya past, ownership of   471
Mbuke, Manus   190, 194
McKenna, T.M.   443
McNeill, T.   90
McVeigh, Timothy   428
Mead, M.   32, 77, 81-82, 85, 190-191, 196
meaning   28-29, 35, 40-41, 45, 69-70, 76, 82, 87, 91-92, 95, 100, 164, 194, 208, 213, 229, 265, 269, 292, 298, 301, 319-321, 328, 361, 381, 385, 394, 413, 439, 446, 453, 473, 475-477, 479, 484, 486, 507, 517, 520, 526-527
Mecklenburg   234, 494, 498, 503-504
Medes   43
media manipulation   388, 435-436, 442, 444
media messages, incorporated into personal narratives   443
media, authority of   442
mediation rejected   268
Mediterranean   41, 53, 105-106, 228, 230, 232-233, 235, 421, 455, 467, 479
Mediterranean, eastern   105-106
Meggitt, M.   118-120, 122, 168, 173, 175-176, 179, 183
Melanesia   33, 71, 87-88, 93, 101, 106, 108, 128, 136, 138, 184-185, 187, 191, 198-199, 205, 258, 386, 391, 448, 467
Meldon Bridge, Scotland   151
Melian dialogue   37
Melpa   79
men, aggressive   116, 133
Mengen   202
mercenaries   228-236, 399, 425, 429
Mercer, R.J.   147-149
Mercia   337
Meroe   284, 304
Merovingian Empire   228-229
Merovingian period   236

Mesolithic   63, 73, 116, 132, 141, 143, 146, 159, 161-164, 306, 318, 421
Mesolithic, late   63
Mesopotamia   42-43, 55, 280
Messenia   401, 522
Messina   233
metallurgy   344, 410-411, 494
metaphorical statement   394
metonymic statement   394, 407, 485-486
Mexico   102, 184, 259, 403, 427, 471, 478-479
Michaels, W.   86
Middle Ages   38, 236, 307-308, 310, 312, 317, 322, 430, 483, 517
Middle East   10, 18, 472, 479
Middle Farm, Dorset   337, 339
Middle Farm, Wessex   157
middle-range societies   26
Midlands   159, 288
Mierzanowice Culture   355
Migration period, protohistoric 215
migration, forced   219, 390, 472, 476
migrations   59, 71-72, 168, 175-178, 183, 205, 215, 228, 258, 319
migrations as military campaigns   215, 228
Mikloucho-Maclay, N.   191-192
military   13, 15-17, 24, 28, 40-41, 43, 46-50, 52, 72, 76-77, 85, 89-90, 92-102, 108, 110, 114, 116-117, 120, 122-126, 128, 131-132, 135, 141, 157, 182, 188, 197, 201, 212, 214-215, 218, 220, 223, 225, 228-229, 231-234, 236-242, 249-250, 253-259, 266, 281, 291, 293, 296, 303, 317, 320-321, 326-327, 387-388, 391, 398-400, 402, 423, 425-427, 429-431, 462, 472, 503, 515-516, 518, 521-522
military apparatus, limited   225
military coalitions   233
military institutions, ideological basis for   387
military leaders   98, 240, 242
military organisation   13, 28, 89-90, 92, 94-101, 108, 188, 214, 218, 237-242, 254-257, 320, 398-400, 516

military service 40, 231, 233, 426
military societies 100, 320, 399, 515
military strength 120, 123-125, 128, 182
military superiority 17, 123, 126, 131
military technology 188, 197, 427
military, feminisation of 430
militia 389
militärische Demokratie 402
Milošević, S. 434, 463
Minoan 494
mission stations 202, 293
Mitanni 43
Mobutu 262, 266
Moche art 295
Moche material, Peruvian 283
Moche, Peru 283, 287, 295, 300, 302
mode of production, Asiatic 221
Moka 79, 83, 87
Molokai 248
Momostenango 472, 476
Monaghan, J. 472
Monarchy 39-40
Mongols 51
Montagu Harbour 202
Montejo, V. 470-471
Morava River 342
Moravia 7, 13, 341-344, 346, 360-361, 364, 368, 371-373, 378, 382, 421, 511
Moravian Gate 341
Moravská Nová 356
Morobe 203
Morris, I. 516
mortality 73, 114, 116, 121, 124, 197, 322, 414, 528
Moseley, H.N. 192
Moses 39, 208, 516, 520, 527
Mosse, G.L. 425
moulds, clay 157, 494
moulds, solid 494
Mount Pleasant, Wessex 151
movement of objects and ritual 206
movement, Communist 47
movement, liberation 47
movement, nationalist 469
movements, rebel/guerrilla 96, 262, 265-266, 475
movements, violent youth 269
Mozambique 262, 271

Mrozek, D. 423
Mucking 157, 159-160
Mujahedini 441, 443
Mundugumor 82
Mundurucu 116, 124
Mungiki 269, 271
Murray, A.T. 527
museum collections 204, 208
museum collections, source of evidence 204
Museum für Völkerkunde, Basel 205
muslim 13, 233, 236, 389, 443, 446, 449-454, 456-459, 461-468
muslim, Bosnian 466
muslimness, definition of 463-465
mutilation 262, 285, 364, 366, 372
Mycenae 43, 51, 283, 485, 494, 510, 516-519, 527
Mycenaean period 516-517
Münster, K. 202
Myrmidons 519
Myth of the War Experience 72, 425-426, 431
Mýtna Nova Vés 342, 344, 348-349, 351, 353, 355-362, 364-368, 370-372
Mälardal type 328
Männerbunde 396-398, 420-421, 515
Møller-Christensen, V. 317

Nages 300
Nagyrev Culture 341
Namatanai 203
Napoleon 32, 35, 45, 425
Napoleonic Wars 77
narratives 85, 178, 180, 196, 203, 386-387, 389, 391, 423, 433, 436-437, 439-441, 443-444, 446-447, 465, 467, 488
narratives, of nationalism 440
nation 40, 46, 48, 76, 102, 218, 228, 242, 257, 263, 290, 388, 425, 437, 439-441, 445, 450, 452, 462, 466, 468, 471
national disambiguation 440, 442
nationalism 7, 50, 71, 75, 101, 388-389, 419, 429, 433, 435, 437, 439-445, 449-450, 452, 464-468, 471-473, 478-479
nationalism, Croatian 435

nationalism, dominant discourse 388-389, 437, 442-443
nationalism, exclusive 388, 434, 449
nationalisms, post-Yugoslav 435, 437-439, 441-443
nation-state 40, 90, 93-94, 96, 101, 228, 237, 258, 446, 452, 464, 471, 478
native American 81, 479
Natufian 55, 142
natural state 25, 29-31, 39, 46, 93, 520
nature of Man 84-85
Nazi Germany 75
Neanderthals 305
Near East 24, 40-44, 46, 53, 55, 138
negotiation 38, 125, 180, 244, 287, 327, 396
Nemeti 230
neo-evolutionist approach 63, 67
Neolithic 6, 41-42, 53, 55, 59-60, 63-65, 69-71, 73, 107, 116, 141, 143-154, 158, 216, 219, 275, 306, 323, 336, 391, 402, 404, 410, 484-487, 492-494, 507
Neolithic, early 41, 63, 143-151, 154, 158, 161, 306, 382, 391, 512
Neolithic, late 63, 65, 150-154, 306, 323, 391, 410, 492, 507
Neolithic, middle 60, 70, 149, 154, 306, 404
Neretva 462-463
Nervii, Gallo-Belgic polity 282
Nestor (of Pylos) 401, 519, 521
Netekamani, king 284
Netherlands 58, 72, 218, 226, 280, 296, 338, 340, 417, 490, 504
Nettleship, M. 217
network 34, 94, 96, 110, 130, 139, 153, 169, 178, 187-188, 190-191, 199, 204, 228, 457, 472, 502
Neupotz 232, 280
Nevermann, H. 192
New Britain 202-208
New Guinea 5-6, 12, 14, 16-17, 27, 68-69, 73, 75-88, 93, 101, 107-108, 110, 113-114, 118-120, 124-125, 128, 135-139, 167-169, 182-185, 187, 191,

198-199, 201-204, 206, 208,
385-386, 397, 404, 466
New Guinea, Highlands 68-69,
79, 82-84, 107-108, 113-114,
118-120, 124-125, 128, 168,
175, 82, 208, 386, 397, 404
New Guinea, lowland and
island 68-69, 107, 191, 202-
203
New Ireland 202
New South Wales 202
New Testament 41
New War culture 428, 430
new wars 223, 261-262, 270
Nielsen, P.O. 63
Nigeria 269-270
Nile Valley 42-43, 284
Nile Valley, Upper 42
Nimrod 507
Nitra Basin 342-344, 349, 353,
371-372
Nitra Culture 7, 341-344, 346-
349, 353-356, 358-364, 366,
368, 370-372, 381
Nitra-Čermáň 342, 349, 352,
363
Nižná Myšľa, Slovakia 369,
373, 378
Noco 246
Nomads 15, 60, 121, 129-130,
132, 224, 234
Nordbladh, J. 60-61, 66, 488
Nordic area 498, 508, 510
Nordland 322, 328
Norfolk 161, 291, 302
Normandy 233-234, 236, 426
Normans 227, 232-234
norms and rules of behaviour
397
norms and values, warlike 117-
118, 122
Norse Christians 275
North America 77, 98, 162,
240, 248, 320, 400
North American tribes 399
North Ferriby, England 159
North Sea 229
Northampton, Midlands 288
Northover, P. 332, 339
Northwest Coast Indians 486
Northwest Coast, America 132-
133, 486
Norton Bavant 146
Norby, Schleswig 500
Norway 7, 154, 319-322, 326-
329, 338, 467, 493
Notgrove 144

notions of the past 390, 477
Nubia 42, 284, 303-304
nuclear/atomic bomb 24, 47
Nuer 92, 101, 115, 117, 126,
132, 135-136, 263, 267, 270
Numancia 301
Numidian cavalry 294
Nydam boat 298
Næstved 6, 279, 305, 307-309,
311-312, 316-318

Oahu 248-249, 251, 255, 257
Obii 231
objects, ownership of 206
objects, ritually powerful 206
obsidian 190, 192-193, 195-
196, 206, 404, 406-407
occupation 18, 47-49, 51, 150,
161, 163, 215, 227-228, 230-
232, 234, 253, 318, 326, 400,
428, 465
Odua People's Congress 269-
270
Odysseus 486, 519-522, 527
Odyssey 285, 424, 515-517,
519, 521-522, 524, 527-528
offering in hoards 394
offering, ritual 281, 410
offering, sacrificial 296, 402
offering, votive 485
offering, weapons 306, 394,
402, 486, 512
Ofnet, Bavaria 143
Ogof-yr-Ychen cave 143
Oikos 523
Oklahoma City 428
Olausson, D. 410
Old Testament 41, 44, 46
Olšava valley 342
Olujic, M. 447, 462
Oppenheimer, F. 97
opposition, Europe/Balkans 389
oppositional behaviour 436
oracle 44
oral history 183-184, 197, 253
Orange, southern France 298
Orchomenos 516
order, moral 31-32, 35, 77
order, religious 31
order, social 29-31, 34, 61-62,
77, 83, 86, 93-94, 133, 477,
518, 525-526
Oregon 428
Orgetorix, Helvetian chieftain
289
Orkney 145, 162
ornaments, bone 348

ornaments, copper 348
Orosius 298-299
Orsett enclosure, Essex 148
Ortner, S. 90
Osgood, R. 154-156, 488
Oslonki 409, 420
ossuaries 299
osteological 162, 275, 300, 322,
327, 341, 364, 368-369, 376,
381
osteological analysis 322, 376
Other, the 7, 24, 35, 37, 40, 46-47,
49, 51, 53, 62-63, 87, 89-90, 93,
98-99, 101, 106, 109, 117, 124-
125, 130, 153, 168-169, 172,
181, 188-190, 195, 197, 204,
208, 213-214, 216, 221, 225,
229, 231-232, 234, 239-241,
246, 249-250, 252-253, 262-
264, 284, 293, 308, 313, 320,
322, 346, 362, 370, 376, 385-
386, 389-390, 398, 402, 410,
414, 420, 430-431, 435-437,
448, 459, 462, 467, 471, 474,
476, 485-486, 488, 491, 493,
495, 497, 499-501, 503, 507,
517, 519-520, 526-527
Otherworld 287, 291, 299
Ottenjahn, H. 494
Otterbein, K. 23, 26-27, 66, 76-
77, 85, 120, 129, 398-399
Ottoman Empire 429
Ottonian 227
Over Vindinge, Præstø 61, 338
Oxford University Parks Science
Area 152

Pacific, the 92, 202, 205, 208,
211, 214, 248
pacification 6, 14, 17, 87, 106,
110, 113, 117, 120, 127-130,
136, 138, 187, 201
Painsthorpe 145
Palace of Nestor 401
Palaeolithic 41, 54-55, 141
Palermo 233
Palestine 41-43
palisade 145, 147-149, 151,
165, 192, 195, 229, 243-245,
492
Pallas 285, 523
Pallas Athena 523
Palstave 493, 501
Pan-Mayan Movement 390
Papua 202, 208
Papua New Guinea 6, 12, 14,
17, 68-69, 73, 80, 86-87, 93,

107-108, 110, 135, 167-169, 175, 184-185, 187, 191, 199, 201, 208, 385 *see also New Guinea*

paramilitary groups 428-429, 434

Parijs, P. van 218

Parioi, Manus 193-196, 198

Paris (France) 232

Paris (prince of Troy) 523

Parker Pearson, M. 336

Parkinson, R. 191-192

participation in decision-making 172

partisan 433, 445

past reconstructed 110, 187, 202, 277, 307, 364, 413, 483, 521

pastoral tribe/societies 57, 121, 129, 132, 265-266, 399

paternalism, North American 471

pathology, pathological method 143, 275, 277, 280

patrilineal 120, 123-124, 130, 189-190, 398

patriotism 13, 39, 45, 50, 389, 431, 453-455, 464

Patroclus 281, 285, 519-521

pattern of culture, Apollonian 81-82

pattern of culture, paranoid 81-82

Patton, General 423-424

Pavlides, C. 204-206

Pax Romana 110

payback 175, 177, 196

payback as modality of reciprocity 196

peace 6, 12, 30-31, 36, 39-40, 45-46, 48, 51, 54, 64-65, 67, 69, 75-76, 78-80, 82, 84-85, 87, 91, 96-99, 107, 109-110, 113, 115, 117, 119, 121-125, 127-129, 131, 133-139, 150, 169, 172-174, 179-180, 183, 185, 195-196, 216, 240-241, 249, 253, 256-257, 264-265, 267, 270, 282, 317, 319, 326, 393, 397, 402, 421, 426, 441, 452, 461-462, 465, 469-470, 472, 475-476, 487, 489, 491, 521-523, 525

peace, cheaper to buy 216

peace, limited 51

peaceful 14-15, 17, 25-27, 32, 36, 40, 54, 58-60, 62-67, 69-

70, 72, 76, 80-81, 87, 93, 106-111, 114-119, 121-122, 127-131, 133-136, 151, 162, 174, 176, 184-185, 192, 215-216, 221-222, 305, 319, 386, 390, 414, 425, 428, 431, 470, 478, 487, 491, 501, 511, 523-525

Pearson, K. 78

peasant culture 63

peasant, peaceful 14, 58-59, 63

peat-bog 296, 298-299

Peisistratos 527

Pelopides, Dynasty of 518

Peloponnese 523

Penelope 522

Pentagon 475

Penywyrlod, Wales 144

Perez, G. 473

performance 53, 117, 181, 184, 265, 268, 292, 296, 424, 518, 520

Pergamum 298

peri-mortem treatment 379

Persia 43-44, 47, 276, 296-297

Persia, Achaemenid 276

Persian Gulf 47

Persian Wars 296

personal qualities 192, 387, 397, 399, 409, 518, 521, 524

personhood, corporate 283

personhood, individual 283

personhood, surrender of 292

Peru 220, 295-296, 301

Pešić, V. 429

Peterson, D. 32-33

Petit Chasseur, Sion 506-507

Pezinok-Cajla 346

Phaiakians 524

Philippines 132, 443, 446

Phillip, of Macedonia 230

phrygian caps 294

Pick, D. 465

Piggott, S. 63

pigs 119, 124-125, 136-137, 169, 173-177, 179-182, 184-185, 195, 198, 208, 243, 252, 344

pig's-tusk ornaments 206

Pililo 206

Pilling, England 153

Pilum, Roman 156

Pitt Rivers Museum 12, 203, 205

Plague 305-306, 313, 326

Plains Indian, North American 300, 399

Plato 40, 424, 461

Plavo, Croatia 433-434, 441-442, 445

Plutarch 295-298, 304

Poland 341, 346, 349, 362, 370, 409, 419-420, 492, 498, 503, 507

pole arms 153, 361, 364

polis 518, 527

political elite 15, 262, 443, 451

political unit, large 43, 50, 255

political unit, small 14, 43, 114, 217

politicisation, processes of 450, 469

politics 11, 14, 24, 28, 35, 46-47, 51, 55, 59, 61, 66, 73, 80-81, 86, 88, 90, 93-94, 102, 105, 111, 119, 127, 134, 136-137, 139, 168, 175, 178, 183, 198, 201, 225, 242, 254, 258, 261, 263, 266-267, 269-270, 385, 389-391, 402, 421, 431, 437-441, 444-446, 450, 457, 461-462, 464, 466-467, 469-470, 473, 475-476, 478-479, 490, 526, 528

politics by other means 24, 46, 201, 385

politics of identity 390, 445-446, 478

politics, danger of 438

Polybius 38

Polynesia 249, 258

Popol Vhu, Maya Quiché book of creation 473

Population density 52, 84, 114, 118, 130, 176, 212

Population/demographic growth 52-53, 212, 222, 267

Porsmose 494

Porter, B. 89

Pospisil, L. 120

post-mortem treatment 380

Postumus, Roman general 232, 300

Post-Yugoslav wars 435, 442

potlatch 61, 486

Potterne, Wessex 158

pottery 147, 149-151, 153-154, 157, 159-160, 192, 230, 291, 307, 373, 378, 404, 410, 414-415, 420-421

Poulnabrone, County Clare 144-145

power 5-6, 13-17, 24-25, 31, 33, 38-48, 50-51, 53-55, 57, 59-

62, 64, 70-73, 79-81, 84, 89-91, 93-102, 105, 108-109, 111, 113, 115-117, 119-123, 125-131, 133, 135, 137-139, 150, 164, 168, 174-176, 178-180, 189, 192-193, 197-198, 201, 204, 207-208, 212, 214-215, 220, 223-224, 226, 230, 237-243, 245-249, 251-258, 262-265, 267-269, 271, 283-284, 290, 293, 296, 320-321, 326-328, 356, 368, 371-372, 385, 387-389, 391, 393-394, 396-398, 400, 402-403, 406, 408, 415, 419-420, 426, 430, 433, 435, 439, 442, 444, 448-449, 469-470, 473-478, 483, 485-486, 493, 501, 515-516, 518-520, 522-526
power enactment 403, 526
power field 402
power games 396
power, adjudicative 120
power, authority converted into 239
power, economic 90, 95, 100, 254-255, 265
power, force 98, 102, 128, 223, 258
power, ideological 95, 100, 108, 255
power, imbalance of 79, 116
power, institutionalisation of 238
power, military 90, 95-96, 98, 108, 242
power, networks of 94-96, 98-101
power, persuasion 16, 110, 525
power, political 38, 41, 71, 90, 100-101, 108, 254-255, 258, 435, 448
power, sources of 16, 90, 100, 108, 150, 255-256, 284, 400, 402, 477, 520
powerlessness 7, 130-131, 389, 433, 436-437, 439, 444
practice, level of 100
Prehistoric society, tales of 27, 59, 66
preservation, post-burial 278
preservation, post-excavation 278
prestige 13-15, 55, 76, 108, 117, 122, 128, 142, 153-154, 157-158, 190-191, 197, 201, 207, 217, 220-222, 319-320,

338, 341-342, 344, 355, 364, 370-372, 398-399, 404, 410, 453, 485, 507, 510, 515, 518, 521
prestige goods 13, 128, 153, 157, 222, 338, 341-342, 344, 355, 364, 370-371, 404, 410
prestige goods, control of 222, 342, 355, 371
prestige goods, copper 344
prestige goods, trade of 341, 344, 370
Priestess, Cimbrian 297
Příkazy, Moravia 365-366, 368
primates 32, 35-36, 134, 138
primordiality 25, 29-31, 33-34, 263
Pringle, R. 119
prisoners' dilemma 122-123, 126-128, 131, 133, 196
prisoners of war/prisoners-of-war/POWs 219, 275, 281-287, 290-298, 300, 518
prisoners, Dacian 294, 300
prisoners, Gaulish 283, 298
prisoners, naked 291, 294-295
prisoners, noble 281, 295-296
prisoners, Nubian 284, 292, 294
prisoners, sacrificial 290
prisoners, Taliban 281-282, 302
process of distinction 396
process, level of 28
production monopoly, defended w. force 190
profession 69, 395-396, 484-485
projectile injuries 143, 338
projectile point 144
property, private 39, 45, 238
Przeczyce, Upper Silesia 511
Przeworsk-culture 230
psychological bond 93
Punic War, first 284
punishment 45, 117, 128, 172, 201, 286, 291, 298, 306, 308, 328, 474
punitive raid 195
Pupu, Nitze 168
Pyecombe barrow, England 151
Pylos 401, 516, 521-522

Qaraniqio 255
Q'esla Kub'al 474
Queensland 184, 202, 208
Quidney Farm, Saham Toney, Norfolk 291, 302

Quiver 409, 418, 494, 520

Raaflaub, K. 527
Rabaul 202-204
Radcliffe-Brown, A.R. 70
radiocarbon analysis 332
radiocarbon dating 154, 323
raid 72, 82, 92, 111, 162, 192, 195, 217, 234, 278-279, 327, 397
raid party, assembling 92, 190
raiders, mobile 158
raids, punitive 95
Rambo 427
Rameses II 284
Randers 501
Randsborg, K. 486, 500
Rank 15, 18, 60, 64, 69, 189, 198, 222-223, 234, 245-246, 251, 253, 257-258, 281, 287, 296, 298, 348, 355, 358, 368, 378, 387, 394-395, 400-403, 413-415, 417-419, 485-486, 501, 521
Rapa 219, 222
Rapiers 153, 155, 160, 483, 505, 510
Rappaport, R. 118
Rauto 202
Read, K. 82-83, 86
Read's Cavern, Churchill, Somerset 291
rebellion 6, 44, 92, 223, 231, 252-253, 296, 305, 307-309, 311, 313, 315, 317
reciprocity, negative 196-197
reciprocity, positive 196
reciprocity/reciprocal 15, 29, 32-35, 37-38, 45-46, 48-52, 54, 81, 98, 122-123, 131, 133, 168, 191, 196-197, 214, 222, 224, 241, 247, 256-257, 264-265, 269, 281, 327, 387, 397, 401, 403
recruiting slogans 423
recruitment 118, 127, 133, 202-203, 207-208, 396, 522
Red Sea coast 266
refugee 388, 435
religion 12, 40-41, 45, 50, 65, 199, 246, 252, 254, 262, 264, 298, 319, 387, 424, 445, 449-454, 459, 464, 466-468, 470, 479, 485, 526
Renamo, Mozambique 262
repatriating research 390, 478

researcher, position of the 390-391, 469-472, 477
Resic, S. 66
resignation 389, 437-438, 442, 444
resistance 92, 94, 110, 119, 130-131, 171, 208, 213, 233, 245, 262, 264, 266, 271, 388, 391, 402, 429, 436, 438, 446, 449, 465, 467, 475-477, 525
resistance studies 262
resources, access to 142, 168, 212
responsibility 10, 14-15, 38, 40, 50, 282, 389, 436-438, 440-441, 444, 472, 501
responsibility, abdication of 438
retaliate, necessity to 14, 196, 281
retinue 64, 228, 253, 400, 515, 521
revenge 24, 38-39, 43, 49, 68, 92, 115, 117-118, 132, 134, 172-175, 178, 180, 182, 184, 195-196, 217, 243, 248, 250, 268, 276, 281, 285, 308, 317, 320, 327, 434, 466, 477, 494, 523
Rewa 243, 246-248, 255, 257
Rhine 230-232, 235
Ribemont-sur-Ancre (Ribemont) 276, 289, 299-300
Richards, P. 96
Riches, D. 91
right to self-defence 39
rights, over labour and products 171
rights, to be protected and defended by group 172
ringforts/ringworks 157-158, 162
risk 24, 70, 116, 123, 142, 158, 172, 252-253, 284-285, 292-293, 436, 447, 497
rites of passage/rite de passage 77, 279, 296, 320, 387, 423, 429, 486
rites to unite the clan 173
Ritter Volcano 203
ritual authority 149
ritual behaviour 287, 289, 300-301, 373
ritual burial 380-381
ritual depositions of weapons 486
ritual killings 372-373, 381

ritual sphere 408, 511
ritual war, ritualisation of 15, 61, 65, 67, 182, 429
ritual, alcohol consumption 397, 410-411
ritual, eating and drinking 397
ritualisation 6, 182, 281-282
rituals and violence 325
rituals, post-battle 298
rituals, secret magic 68
rivalry 61, 132, 262, 264, 266, 372, 415, 419, 485, 515-516, 520, 524-525
River Soar, Leicestershire 156
Rivers, W.H.R. 78, 86
Robarchek, C. 130
Robb, J. 154
Robertson and Hernsheim 202
Robinson, M. 335
rock art 72, 154, 321-322, 326, 488-490, 505, 507-508
rock carving 285, 287, 295, 321-322, 488
rock carvings, narratives of ideal nature 488
rock carvings, Scandinavian 287, 295, 321-322, 488
Rodmarton 144, 146, 337
Roland 413, 424
Rollo 232-233
Roman Dacia 302
Roman Empire 38, 44, 105, 212, 215, 228-229, 231-232, 234-235, 295, 402, 470
Roman wars 283
Romania 508
Romanticism 62, 425
Rome 110, 230-231, 235, 276, 284, 286, 290, 293-296, 303
Roquepertuse 300
Rosaldo, M. 132
Rosaldo, R. 119
Rousseau, J.-J. 24, 39-40, 45-46, 50, 52-54, 66
Rowlands, M. 60-61, 66
Rudstone 338
RUF, Sierra Leone 262
Rusava valley 342
Russia 328, 358
Rygh, K. 322

Saale River 232
sacrifice 13, 50, 52, 72, 152, 171, 223, 251, 270, 283, 285, 287-290, 295-299, 301, 303-304, 320, 340, 342, 344, 390, 425, 430, 445, 470, 490, 504

sacrifice for the nation 425
sacrifice, human/people 13, 15, 50, 223, 287-289, 291, 295-297
sacrifice, weapon 61, 276
sacrificial deposits 485
sacrificial victims to Dionysus 296
Saharan Republic 261
Sahlins, M. 68, 80-81, 119, 121, 252
Saka Valley, Enga 168, 178
Sakkudei, Indonesia 32
Šal'a I 349
Šal'a-Dusikáre 363
Šal'a-Veča 352
Saladin 51
Salamis 296-297
Salerno 233
Sale's Lot 144
Salisbury Plain, Wiltshire 159, 336
Salisbury, R. 84
Samburu 265, 420
Samoa 36, 202
Samurai 93, 303
San Bushmen 488
Sandauce 296
Sandbjerg 6, 277-279, 305, 308-314, 316-317
Sandin, B. 119
Sandwich, Lord 248
Saracens 233
Sarajevo 72, 454-455, 464, 467-468
Sarn-y-Bryn Caled timber circle 152
Sau Valley, Enga Province 181
Saudi Arabia 49, 430, 454
Sava River 510
savage, brutal, bellicose, warlike 66-67
savage, myth of peaceful 26-27, 64, 66, 70
savage, noble and peaceful 26-27, 64, 66, 70, 76, 80-81, 84, 242
savagery 28, 75-77, 80-81, 86, 285, 460
Saville, A. 144, 147-149
Savory, H.N. 158
Saxony 234, 506
Saxony-Anhalt 229
Saxo-Thuringia 417
scabbards 499

social practice 14-16, 28, 57, 67, 69, 73, 86, 90-93, 341, 394, 402, 404, 409, 419, 516-517, 523, 525
social structure 11, 16, 67, 86, 90, 98-99, 110, 204, 240-242, 247-249, 251, 256-257, 295, 309, 325, 353, 361, 371, 381, 396, 405, 409, 413, 415, 417, 419, 483, 487, 525-526
socialisation 117, 120, 122, 133, 264, 452
sociality, peaceful 32, 81
societies, acephalous 106, 122, 167, 170, 265
societies, age-graded 266, 399
societies, barbarian 215
societies, Celtic 227
societies, clan 59, 229, 234
societies, complex 99, 215, 222, 239, 255, 399
societies, Germanic 227
societies, hierarchical 182
societies, male 397-398
societies, matrilocal 130, 402
societies, non-state 18, 25, 30, 39, 93, 107, 111, 211, 239
societies, peaceful 63, 109, 129-130, 133
societies, pre-industrial 398
societies, pre-state 26-27, 105, 237
societies, primitive 66, 77, 106, 403, 419
societies, secret 397
societies, state 10, 90, 105, 114, 131-132, 261, 398, 402, 518, 527
societies, tribal 16-17, 30, 105-107, 113-116, 118-121, 123-124, 126-127, 129-133, 167, 216, 230, 233-234, 236, 261, 397, 399, 419, 523
societies, tribal without war 113, 129-130
societies, warlike 117-118, 124, 127-129, 131, 158
societies, without central power 6, 113, 115, 117, 119, 121, 123, 125, 127, 129-131, 133, 135, 137, 139
society as biology 31
society as culture 31
society, big man 413
society, disruptive 65
society, ideal 483
society, level of 100

society, stratified 217, 220-222, 224, 257, 371
sociobiology 31, 35, 116, 133-134, 184
Sodbury Camp 337
Sofaer-Derevenski, J. 405
soldier 15, 294, 388, 396, 398, 423, 425, 430, 452-453, 461
soldier, mechanical cyborg 388, 430
soldiers, female/women 388, 430
soldiers, professional 218, 279, 315, 317
solidarity 82-83, 95, 120, 123, 126, 171, 180, 182, 265, 397-398, 415, 453, 464-465
Solomon Islands 204
Somalia 261, 265
Sone, Manus 193
Sonna Demesne man 154
Sorabji, C. 451
Sorel, G. 263
source critical analysis/view 276, 489, 500, 517
sources which indicate violence 491
sources, historical 192, 215, 278, 326, 328, 398-399
sources, written 227, 230, 236, 279, 307, 317, 379, 486, 523
South Africa 305
South America 98, 110, 240, 401, 415, 421, 487
South Cadbury, Somerset 160, 287, 339
South Dakota 300
South Hornchurch 157, 161
Southeast Asia 221, 226
Southwark, London 288
Sovereignty 13, 40, 45, 97-98, 133, 167, 216, 238, 240
Soviet Union 47-48, 467
Spain 110, 227, 301, 402
Spania Dolina 346
Sparbu 322, 327
Sparta 37
spear 5, 25, 29, 31, 33, 35, 155-156, 158, 175, 179, 190, 192, 194-195, 250-251, 288, 294, 332, 334-335, 338-339, 368, 493, 502, 508, 511
spear points, obsidian 190
spear tips, bone 493
spearhead 61, 156, 159, 307, 332-333, 338, 340, 368, 430, 492, 494, 496, 500, 506-507

spearhead, Bagterp type 494
spearhead, broad-bladed 156
spearhead, bronze 61, 159, 334
spearhead, hollow-cast 156
spearhead, invention of metal age 494
spearhead, Valsømagle type 338
spearheads, Baltic 494-495
spearheads, Irish 494
spearheads, Minoan 494
spearheads, Mycenaean 494
specialist products 410
spectacle 286, 290, 292-293
Speiser, F. 205-206
Spencer, H. 238-240, 256
Spinoza, B. 263
Spittler, G. 119, 121
SPLM, Sudan 262
Sponsel, L. 32
sport 174, 429, 487
Springfield Lyons 157, 160
Springfield, Oregon 428
spurs, Mušov type 231
Sri Lanka 220, 222, 224, 226
Staines 145, 163
Stamm 187, 228
Stammesbildung 228, 236
state administration 402, 521
state and divinity 40, 53
state and justice 38-41, 44
state dominance 110, 411
state expansion (colonial) 106, 109, 119
state formation 6, 11, 13, 16, 51, 97-99, 101-102, 105, 109, 212, 214-215, 217, 219-221, 223, 225-226, 237-243, 248, 255, 257, 259, 261-263, 402, 471, 511
state formation, factors leading to 212
state intervention 109
state power 39-40, 44, 105, 113, 128, 214, 257, 262
state structures, imposed 261
state, aristocratic 401
state, colonial 119, 201, 264, 402
state, early 223, 238
state, pacifying effect 109, 128
state, power monopoly of 119, 128
state, Samurai-organised 401
state, sovereign 25, 40, 45, 90, 451

states and tribal societies, co-existence of 234
states make war 214, 223
states, birth/emergence of 52, 94, 97, 99, 109, 127, 219, 222-223, 227, 237, 239
states, postcolonial 48, 80, 262-264, 269
states, unstable 221
status 14-18, 26-27, 35, 63, 95, 98-99, 101, 108, 116, 120, 123-124, 137, 141-143, 147-148, 153, 158, 170-172, 189-190, 193-194, 196-198, 220-221, 224, 234, 243, 245-247, 249, 251-253, 255-257, 264, 266, 277, 279, 281-283, 285, 287-289, 293, 295-298, 301, 306, 320-321, 327-328, 348, 355, 360, 368, 373, 376, 380, 387, 394-395, 397, 401, 403, 405, 414, 417, 425, 429, 449, 464, 477, 484-485, 501, 507, 511, 515, 517-521, 524-525
status and wealth, differences in 220
status competition/rivalry 17, 141, 143, 196-197, 515, 520, 524
status, motive for war 196
status, warriors 387, 507, 518
status, ways to enhance 252-253
Staxton, Yorkshire 152
Stead, I. 338
Sten Sture 307
Stepleton Enclosure 147, 149
Steuer, H. 400-401
Štiavnické Vrchy Mountains 346
Stillorgan, Ireland 154
Stocking, G. 76, 84, 86
Stolac, Bosnia Herzegovina 13, 389-391, 447-449, 452, 454-465
Stone Age 14, 36, 57, 59, 61, 63-64, 70-71, 87, 101, 164, 421
Stonehenge 146, 150, 152, 154, 160-161, 506, 512
Stora Vikers, Gotland 494
story lines 388-389, 439-441, 443-444
Strabo 294, 297
strategic essentialising 436
strategic interaction 107, 120-121, 126-127, 131

strategies of disambiguation 441
strategies, repositioning 444
strategy, discursive 17
strategy/strategies 17, 25, 31, 38, 48-49, 68-69, 91, 107, 110, 116, 121-122, 125-131, 133, 168, 179-180, 192-193, 202, 220, 238, 250, 252-253, 362, 364, 368, 371-372, 389, 427, 429, 436, 439, 441-442, 444, 463-464, 483, 485, 511, 517, 519-520, 524-525
Strathern, A. 27, 79-81, 83-84, 86
Strathern, M. 27, 77, 80-81, 93, 191
strenght, moral 425
strenght, physical 425
structural approach 67, 448, 452
structural change 5, 17, 60, 89-90, 96-97, 99, 101, 127, 221, 226, 241-242, 256-257
structural condition 121
Structural Functionalism 113
structural Marxist 60
Structuration theory 484, 526
structure 7, 11, 16, 18, 67, 71, 73, 85-86, 90, 95, 98-99, 106, 110-111, 114, 119-121, 133, 136, 148, 163, 182, 198, 203-205, 214, 221, 224, 226, 240-242, 247-249, 251, 256-257, 295, 306, 309, 319, 321, 323, 325, 327, 329, 341-342, 346, 348, 353, 355, 361, 370-371, 378, 381, 395-397, 401, 405, 409, 413, 415, 417, 419, 437, 442, 460, 473, 483-484, 487, 501, 516-519, 521, 524-527
structure-agency approach 67-68
Subalterns 436
subjugation 6, 96-101, 215, 219-220, 237, 239-242, 256-257, 281, 283-285, 287, 291, 293-297, 299, 301, 303
Sudan 261-263, 266-267, 270, 304
Suebi 227-228, 230, 235, 290
Suebi, confederation of 290
Sulka 202
Sund 277-280, 319, 322-329, 338
Suri 266-268, 270
surprise attack 245, 251

Sussex 148-149, 158, 160-161, 421
Sutton Veny bell barrow warrior 154
Swat Pathan 125
Sweden 62, 73, 287, 293, 300, 306-307, 318, 321, 417, 421, 424, 428-429, 461, 492, 504
sweet potato, introduction of 167-169, 176-177
Switzerland 332, 417, 509-510, 513
sword 14, 72, 93, 146, 154-155, 158, 162, 297, 314, 321, 324, 326-327, 484-485, 487, 493-497, 499-502, 505, 508, 510-511, 520
sword-scabbard relationship 499
sword production 14, 494-495
sword, Ballintober type 156
sword, emergence of 487
sword, symbol of power 327-328
swords, bronze-hilted 494, 496-498, 508
swords, cut and thrust 495
Swords, Ewart Park 508-509
swords, flange-hilted 321-322, 326-327, 495-498, 500-501, 508
swords, fully hilted 495, 497, 500-501
swords, griffzungenschwerter 508
swords, grip-plate 501
swords, Hallstatt 508
swords, hexagonal metal-hilted 496, 499
swords, import from Central Europe 495
swords, metal-hilted 496, 498-499, 507
swords, miniature 495, 499
swords, objects associated w. 500
swords, octagonal hilted 495, 498
swords, organic-hilted 508-509
swords, plate-hilted 496, 499
swords, sharpening 321, 488, 496-498, 508
swords, signs of wear 497
swords, solid-hilted 508
swords, standardisation 495
swords, tanged-hilted 496, 499
swords, treatment 493, 495
swords, vollgriffschwerter 508

Sydney 198, 205
symbolism of submission 301
symbols of destruction and change 474
Syracuse 230
Syria 42-43
Sørup 500

Tacitus 64, 227, 235, 290-291, 296-298, 397, 399-400
Tairora 84, 88, 120, 123, 139, 185
Talheim, Germany 143, 145, 216, 278
Taliban 281-282, 302
Tallington, England 154
Talyaga, K. 168
Tamboran cult 182
Tarrant Launceston 144
Tasman, Abel 242
Taylor, T. 291
Tchambuli 82
technological development 49, 64, 93
technological superiority 49
technology 7, 15, 17, 28, 42, 49, 90-91, 93, 96, 100, 102, 110, 167, 179, 188, 197, 205, 212-213, 258, 267-268, 388, 427, 485, 487, 491
technology of war, innovations in 487
technology, new 17
Tedlock, B. 472
Tedlock, D. 476
Tee, Ceremonial Exchange Cycle 79, 169-170, 176, 178, 181-182
Telemachos 521
Tepe Gawra 279-280
Terminator 7, 387, 423, 425, 427, 429-431
Terray, E. 217
territorial gain 142
terror/terrorism 47-51, 86, 115, 266, 268, 281, 433, 435, 473
Teutoburg Forest 279
Thames Valley 144-145, 148, 152, 155, 157-158, 165
theatre 60, 292, 296
Themistocles 296
theories of (tribal) war 115-120
theories of (tribal) war, biological 115-116
theories of (tribal) war, cultural 115, 117

theories of (tribal) war, ecological and economic 115, 118
theories of (tribal) war, historical 115, 119
theories of (tribal) war, political 115, 119
theories, decision 121, 133
theories, neo-realist 133
theory of role of war in Bronze Age 491
theory of war, reproductive 142
things, bearers of history and memory 520
Thirty Years' War 77, 460
Thomsen, C.J. 491
Thomson, E. 470
Thordemann, B. 306
Thornhill, County Londonderry 149
Thorsted 500
Thrapston, England 157
three-age system, Thomsen's 491
Thucydides 24-25, 37-41, 44-46, 48-51, 53-55
Thuringia 230, 232, 339, 506
Thurnam, J. 146
Thurnwald, R. 78
Thwing, England 157
Tilley, C. 60
Tilly, C. 90, 94, 257
Tilshead Lodge 146
Timpe, D. 231
Tinbergen, N. 31
Tiryns 516
Tiszapolgar 405-406
Tiszapolgar-Basatanya 404-407, 415, 419
Tiszawalk-Kenderföld 407
Titan 188, 195
Tito 449
Točik, A. 341-342, 370
Todd, J.A. 205-206
Todorova, M. 459
Toke, M. 173, 175
Tolai 202-203
Toldnes 319, 321-323, 326-328
tolerance 13, 389-390, 448-449, 457-459, 463-465
tolerance, inter-ethnic 448, 458
Toma, N. 473-474
tomb, chambered 144-146, 151
Tonga 242
tools for violence made exclusively as weapons 493
tool-weapons 404
Topol'čany 342

Tor enclosures 147
Tormarton 7, 156-157, 162, 277-279, 331, 333, 335-340, 492
Towton, North Yorkshire 306
trade, role of long distance 215
tradition, invented 390, 477
traditionalists 475-477
Trajan, Roman Emperor 294
Trajan's column 294-297, 300
Transco 334, 339
trauma, facial 153
trauma, high frequency of 366, 370, 381
trauma/traumata 13, 57, 63, 70, 107, 117, 122, 142-143, 146, 150-151, 153-154, 158, 208, 275, 277, 279, 301, 319, 321-325, 327, 331, 338, 360-362, 364-368, 370, 372-377, 381, 387-388, 395, 404-407, 409, 414, 426, 435-436, 442, 444, 451, 453, 473, 487-488, 492-493, 508
trauma/traumata, skeletal 13, 57, 63, 70, 107, 360, 364, 366-368, 370, 395, 487-488
traumas, World War II 388
traumata, direct evidence of violence 492
traumata, war-related 387, 404
Treherne, Paul 411
Trelleborg 306, 318
Trenčin 342
Tribal Zone theory 67
tribalisation 228, 231-232, 234
Tribeč Mountains 346
tribes 6, 15, 26, 29, 57, 64, 68, 80-81, 105-107, 109-110, 119, 134, 142, 168-170, 176-177, 180-181, 184, 208, 227-231, 233-235, 245, 282, 295-298, 341, 397-400, 402, 404, 419, 500-501, 521, 523
tribes, Arabic Bedouin 523
tribes, birth of 227
tribes, constitution of 234
tribes, Gallic 110
tribes, Germanic 110, 228, 297, 400, 402, 500, 521
tribes, Rhenish 296
tribes, Slavic 229
Triboki 230
tribute 93, 98-99, 132, 219, 231, 240-241, 243, 245, 247, 252-256, 258, 371
trinket, amber 417
trinket, bone 417

trinket, copper 409-410, 414, 417-418
trinket, shell 417
trireme 296
Trivers, R. 35
Trobriand Islanders 79
Trobriands 81
Trojan War 38, 44
trophy skulls 143
trophy-taking 285, 298, 300
Trotter, M. 311, 322
Troy/Trojans 51, 281, 285, 296, 519-523, 527
Trudshøj 501
Trundle 148
Trøndelag 322, 328
Tudjman, Franjo 434, 442, 463
Tulloch of Assery 144-145
Tultul 203-204
Tumu, A. 69, 119, 132, 168, 175, 183
Tumulus Culture 338
Tupi-Guarani 106, 419
Turek, J. 414
Turkana 265
Turkey 51, 440, 449
Turney-High, H. 114
Turnus, Prince of the Latins 285
Turton, D. 133
Tutsi 228, 447
Tuzin, D. 180, 182
Tvrdošovce 349, 359-360, 363-364, 370, 372
Twr Gwyn Mawr, Wales 152
Ty-Isaf 144
Tymowski, M. 220, 222

Uganda 266, 270
Ukraine 341, 358
Umboi 203
Umi 102, 249, 252, 259
Umi-a-Liloa 252
UN missions 429
UN peacekeepers 429
understanding, inflexible 414
UNESCO 269, 455
Únětice Culture 362, 365-367
UNHCR 433
unification 40, 46, 51, 228, 451
Upper echelon 414
Uppsala 12, 72, 270, 307, 315-316, 318, 329, 420-421, 467, 479
Ur 43, 503
Urnfield culture 394, 499, 502
Urnfield migration 491

Urnfield period 346, 489, 498, 510
USA 4, 24, 46-49, 51, 430, 460
Ustaše 433-434, 444-445
Utilitarian 33, 156

Vadomar 231
Váh Basin 341-344, 349, 371
Vah River 343, 372
Valdemar Atterdag, King of Denmark 306, 317
Valdemar Sejr, King of Denmark 317
Valhalla 423
Valsømagle 61, 338, 494
Values of civilisation 441
van den Berghe, P. 473-474
van der Dennen, J. 32-33
Vancouver, Captain George 248, 250, 254
Vandkilde, H. 70, 297
Vangiones 230
Vanua Levu 242
Varna I 404
Varus-Schlacht 231
Velasco, M.A. 471
Velatice 511
Velim, Bohemia 337
Velké Žernoseky 509
Vel'ký Grob 342-343, 349, 362-363, 381
Vencl, S. 57, 66
Vendel period 236
vendetta 447
Venezuela 32, 69
Verata 243
Vercingetorix 286, 295, 298
Versailles Treaty 81
Vestfold 328
Věteřov Culture 373
victims 7, 12, 52, 67-70, 75, 91-92, 150, 153-154, 156, 158, 198, 213, 218, 287, 289-290, 292, 294-298, 305-318, 322, 331, 335-339, 368, 376, 378, 380, 389, 405, 436, 446, 448, 461-465, 467, 476, 479, 488, 502
victory 24, 37-38, 43-44, 46-49, 51, 81, 113, 124, 172, 179, 201, 220, 233, 243-244, 251, 266, 296-300, 397, 431, 486
Victory, battlefield 48-49
Victuales 231
Vietnam 47, 55, 66, 72, 423, 426-428, 430-431

Vietnam War 66, 72, 423, 426-428, 431
Viking Age 215, 306
Vikings 232, 236
Vikletice 414, 416, 419
violence 7, 11, 13-17, 23-33, 36-39, 41-44, 46-47, 49-54, 57-58, 60-73, 76-78, 80, 82-87, 89-92, 95-96, 99-102, 106-107, 110-111, 113-115, 117, 121-123, 127-132, 134, 136-139, 142-146, 153-155, 157-165, 172-173, 179, 184, 191, 197-198, 202, 208, 211, 213-214, 216-220, 222-225, 237-238, 254, 256-258, 261-271, 275-279, 282, 299-300, 303, 305, 318-323, 325, 327-329, 335, 338-339, 341, 364, 366, 368-369, 371-372, 376, 385-387, 389-391, 393, 396-399, 402, 405-407, 417, 419-421, 427-429, 433-434, 436, 439, 442-449, 451-453, 455, 457, 459-461, 463-467, 469-472, 477-479, 487, 489, 491-494, 506, 512, 520, 523
violence and counter-violence 115, 128
violence and identification 7, 13, 72, 447-449, 451, 453, 455, 457, 459, 461, 463, 465, 467
violence and ritual 329
violence as idea 29
violence by colonial officers 208
violence, archaeological evidence 321-322, 492
violence, collective 92
violence, context of 278
violence, gang 214
violence, institutionalised 268
violence, inter-communal 53
violence, legitimacy 91-92, 478
violence, meaningful/meaning of 28, 67, 69, 91, 321
violence, monopoly on 208, 214, 237-238, 261, 267
violence, non-political 469
violence, organised 52-54, 95, 100, 217, 257, 320, 366, 402, 469, 491-492
violence, physical 57, 82, 89, 91, 92, 96, 100-101, 173, 393, 487